ACOUSTIC EMISSION

A Bibliography with Abstracts

IFI DATA BASE LIBRARY

COMPUTER TECHNOLOGY
Logic, Memory, and Microprocessors – A Bibliography

A. H. Agajanian

ACOUSTIC EMISSION
A Bibliography with Abstracts

Thomas F. Drouillard

ACOUSTIC EMISSION

A Bibliography with Abstracts

Thomas F. Drouillard

Rockwell International Corporation
Energy Systems Group
Rocky Flats Plant
Golden, Colorado

Edited by

Frances J. Laner

Consultant, Technical Communications
Denver, Colorado

SPRINGER SCIENCE+BUSINESS MEDIA, LLC

Library of Congress Cataloging in Publication Data

Drouillard, Thomas F

Acoustic emission.

(IFI data base library)
I. Includes indexes.
I. Acoustic emission–Bibliography. I. Laner, Frances J. II. Title.
Z5853.A24D76 [TA418.84] 016.6261'27 79-268

ISBN 978-1-4615-8364-6 ISBN 978-1-4615-8362-2 (eBook)
DOI 10.1007/978-1-4615-8362-2

Originally published by IFI/Plenum Data Company in 1979
Softcover reprint of the hardcover 1st edition 1979

PREFACE

The purpose of this bibliography is to list all of the known literature on acoustic emission in a single volume. The bibliography is intended to serve as a reference to the literature for those interested in the acoustic emission phenomenon, for those engaged in materials research, and for those interested in applying acoustic emission as a nondestructive testing method. It provides an invaluable reference to source material for both undergraduate and graduate students interested in the subjects of acoustic emission, microseismic activity, and nondestructive testing. In addition, it provides a valuable desk-top reference for engineers and scientists in the fields of pressure vessel design and maintenance, stress analysis, mechanical engineering, welding engineering, nuclear engineering, metallurgy, and quality control.

In compiling the bibliography every effort was made to search the literature for all the publications and obtain copies, index them, and publish a complete, comprehensive bibliography of the literature on acoustic emission. This bibliography is the most complete ever published. It includes essentially all of the literature on acoustic emission that has ever been published up through the first quarter of 1977. Publications listed are in the form of technical reports and memoranda, journal articles, technical presentations, proceedings, doctoral, master, and bachelor theses, patents, bound volumes, translations, and newspaper articles. Some references, however, are not technical articles, but consist of editorials or staff-written journal articles which are included because they represent a part of the literature. These are enumerated under *Acoustic emission literature (kinds of)* in the Subject Index. Several references do not report directly on acoustic emission, but discuss material that borders on or directly relates to acoustic emission and, therefore, are considered a valuable part of the bibliography. Most references to works published prior to 1950 are not specifically on the subject of acoustic emission. They are referenced here and have been cited in the literature because they do report the observation of some form of acoustic emission activity.

The bibliography contains 1996 entries, most of which are abstracted or annotated. Wherever possible the original abstract by the author of the article has been reprinted, with minor editing to establish some measure of consistency. The second choice in selecting abstract material has been from publications of abstracting and indexing services, such as and in particular *Metals Abstracts* and *Nuclear Science Abstracts.* The remaining abstracts or annotations were prepared by the author. With the exception of 33 references, copies of all cited material have been obtained by the author to establish that they do exist, are currently available, and to authenticate the accuracy of information in this work.

Numerous references are included for acoustic emission utilized in the technologies of microseismic activity, boiling detection, signature analysis, and leak testing. However, no attempt was made to thoroughly search these areas of technology. For the most comprehensive bibliography on microseismic activity the reader is referred to

Reference 700. The most extensive and current list of references on boiling detection may be found in the following publication: H. A. Scarton, J. F. McDonald, and W. M. Carey. "A Review of the Acoustic Detection of Boiling in Nuclear Reactors." in *Noise and Fluids Engineering.* Edited by Robert Hickling. American Society of Mechanical Engineers, New York. 1977.

In compiling material for the bibliography over the past ten years numerous sources of information have been utilized. The most useful sources have been the lists of references cited by each author of an article. These have been augmented by systematic searches of a number of journals and of the following abstracting and indexing publications: *Chemical Abstracts* (American Chemical Society), *Corrosion Abstracts* (National Association of Corrosion Engineers), *The Engineering Index* (Engineering Index, Inc.), *ERDA Research Abstracts* (U.S. Energy Research and Development Administration), *Government Reports: Announcements & Index* (National Technical Information Service), *INIS Atomindex* (International Atomic Energy Authority), *Metals Abstracts* (American Society for Metals and The Metals Society), *Nuclear Science Abstracts* (U.S. Atomic Energy Commission), *Physics Abstracts* (INSPEC: The Institution of Electrical Engineers), and *Scientific and Technical Aerospace Reports* (National Aeronautics and Space Administration). Several on-line computer searches, which became available during the compilation of material, have proved useful but will be more valuable in future searches. These were CHEMCON, a search of *Chemical Abstracts;* COMPENDEX, a search of *Engineering Index;* and NTIS, a search of the National Technical Information Service data base. Another outstanding list of current literature covering all aspects of nondestructive testing is "NDT Info," compiled by the NDT Centre, Atomic Energy Research Establishment, Harwell, U. K., and published in the journal *NDT International.* It has been invaluable in this task. Other bibliographies have furnished many references previously unknown to the author and provided a means of cross-checking references already indexed. Lastly, a valuable source of information has been colleagues throughout the world who knew this bibliography was in preparation and, either voluntarily or in response to the author's request, supplied copies of their reports and lists of their publications.

This volume is comprised of four parts: Bibliography, List of Journals, Author Index, and Subject Index.

The Bibliography consists of 1902 authored entries followed by 94 unauthored entries. Each entry has been assigned a reference number, which provides a cross-reference to the Author Index and Subject Index. Each entry lists the author and all coauthors; the title of the article; all known sources of publication, such as, formal report, journal article, paper in a proceedings; where presented; dates of publication and presentation; languages of publications and translations; and pertinent information on availability. Entries are arranged alphabetically by author, first coauthor, second coauthor, etc., and finally by title. Unauthored entries are arranged alphabetically by title. Authors' names are spelled precisely as they appeared in publication. No first names have been used, only initials. Titles of referenced materials are reproduced in their entirety, including initial articles (a, an, the) which have been considered in alphabetizing.

All entries were published in the English language unless indicated otherwise. In addition to English, articles and reports were published in twelve different languages: Czech, Dutch, French, German, Italian, Japanese, Polish, Portuguese, Russian, Slovak, Slovenian, and Spanish.

Many reports and translations are available to the public from the National Technical Information Service (NTIS), Springfield, Virginia 22151. These reports are so designated by [NTIS] appearing in the reference and can be ordered by giving the report number or the accession number contained in brackets [ ].

Technical articles were published in approximately 300 different journals which are listed alphabetically in the List of Journals. Here, as in the bibliographic entries, titles of journals are given in full. Titles are in agreement with *Chemical Abstracts* and *Engineering Index*. Following the title is the CODEN, a five-charcter code intended to identify and facilitate machine handling of journal titles. The number in brackets [ ] following the CODEN indicates the total number of articles listed in the bibliography that were published in each journal.

The Author Index is comprised of about 1600 names, which include authors, coauthors, editors, and coordinators.

Abbreviations used throughout the work omit the periods. Diacritical marks in foreign words such as the German umlaut and the French accents have been omitted without a change in spelling.

The Subject Index has been developed with the user in mind. *See* and *see also* entries have been liberally used.

It should be realized that subsequent to the period covered by this bibliography, much acoustic emission activity has been going on and many articles and reports have been and continue to be published. The question, therefore, arises concerning plans to update this work. It is the author's intention to continue to compile current literature and periodically publish a supplement to this bibliography. To facilitate this task the author solicits and welcomes complimentary copies of reports, articles, proceedings, etc., from authors, editors, and publishers. Comments on omissions from the time period covered in this volume and recommendations for future publications would be most welcome.

Coal Creek Canyon, Colorado
November 1978

THOMAS F. DROUILLARD

ACKNOWLEDGMENTS

This publication represents the diligent efforts of countless individuals. I acknowledge all persons listed in the Author Index and express my gratitude for their contributions to the technology and to the literature—their work has made this volume possible. Sincere appreciation is expressed to the many authors and colleagues who so generously responded to my numerous requests for information and copies of their reports and papers. Particular appreciation is expressed to L. Bolin, M. Galli, A. T. Green, A. E. Lord, Jr., G. G. Martin, A. A. Pollock, R. S. Sharpe, J. C. Spanner, and P. G. Williams for allowing me to review their bibliographies to cross-check my list for completeness and omissions, and to H. M. Tensi for the photograph of Josef Kaiser.

Gratitude is extended to the Albuquerque Operations Office of the U.S. Department of Energy and to Rockwell International, Energy Systems Group, Rocky Flats Plant for their support and cooperation in this project. I am deeply grateful to my management for providing me with the time and facilities necessary to complete this project.

The task of organizing the material into a publication was begun some five years ago at the urging of Frances J. Laner, formerly Technical Editor at the Rocky Flats Plant, now a consultant. My very special thanks go to Miss Laner for her editorial review of the manuscript and her constant encouragement and sage technical advice in preparing the manuscript and bringing this volume to fruition.

The efforts and assistance of many co-workers at the Rocky Flats Plant have contributed to the preparation of this volume. I wish to acknowledge the excellent facilities of the Rocky Flats Library and the assistance of its staff, especially Elaine A. Anderson, Zola P. Hill, Janet E. Meddles, and Janet A. Nesheim, for their persistent efforts in obtaining much of the referenced materials and conducting the on-line computer searches. To Meredith L. Murbach of Computer and Information Systems I am grateful for using the computer to generate organized lists of reference material and indices. My thanks go to Douglas G. Heberlein for his assistance in obtaining permissions to use copyright material, and to Gilbert A. Riordan for proofreading the manuscript. Preparation of the manuscript in final format was accomplished by the Technical Reports Group at Rocky Flats. I am especially grateful to Linda M. Morales, Patricia L. Griego, and Ida C. Delaney for the monumental task of typewriter-composition and pasteup of the manuscript in camera-ready copy. My sincere appreciation is extended to Mary E. Montoya for her assistance with coding sheets and preparation of indicies, to Helen M. Roberts for her secretarial assistance, and to Clinton R. Heiple for his critical review of the front matter and Subject Index. Also, I wish to thank A. G. Beattie, S. H. Carpenter, M. A. Hamstad, A. E. Lord, Jr., A. A. Pollock, and C. A. Tatro for their review and suggestions on the Subject Index.

Finally I wish to express appreciation to the authors and publishers listed throughout the bibliography for permission to quote material from their publications.

CONTENTS

Bradford H. Schofield

Warren P. Mason

Clement A. Tatro

Josef Kaiser

INTRODUCTION

History of Acoustic Emission

Acoustic emission and microseismic activity are naturally occurring phenomena which man has observed from early times. Although it is not known exactly when the first acoustic emissions were heard, fracture processes such as the snapping of twigs, the cracking of rocks, and the breaking of bones were no doubt among the earliest. But the first acoustic emissions heard by an artisan or craftsman, by which some message was conveyed about the material with which he was working, may well have been pottery-making. The oldest variety of hardfired pottery dates back to as early as 6500 B.C. Potters observed the sounds of cracking of clay vessels cooling too quickly in the kiln. These were audible acoustic emissions by which the potter knew his creations were defective and structurally failing.

In metals it would be reasonable to assume that the first true acoustic emission heard was the *cry* of tin—the audible emission produced by mechanical twinning of pure tin during plastic deformation. This then would have occurred only after pure tin was smelted. Copper is considered to be the oldest metal used by man, dating back as far as the ninth millennium B.C. The deliberate addition of arsenic and then tin as alloying additions to copper heralded the beginning of the Bronze Age somewhere between the fourth and third millennium B.C. The transition from copper to bronze occurred before tin had been consciously produced as a pure metal by smelting tin ore. Tin differs from copper, gold, and silver in that it is never found in nature in its elementary state; tin is always in chemical combination. Its most common ore, tin oxide, had to be smelted to produce metallic tin. It follows, therefore, that any metallic tin found among ancient ruins must have been smelted. The oldest piece of tin found to date is a bangle that was excavated at Thermi in Lesbos and has been dated between 2650 and 2550 B.C. The bangle is 1-5/8 inches in diameter and consists of two strands of pure tin, one wrapped around the other and hammered flat at the end. During the manufacture of this bangle the craftsman no doubt heard considerable tin cry.

In the *Iliad*, written in the 8th century B.C., Homer tells of the Greek god of fire, Hephaestus, throwing copper, tin, silver, and gold into his furnace to make the shield of Achilles.

Tin is also mentioned in the *Old Testament*, Ezekiel 27:12, "Tarshish was thy merchant by reason of the multitude of all kinds of riches; with silver, iron, tin, and lead, they traded in thy fairs."

Probably the first recorded observations of acoustic emission were made by the 8th century Arabian alchemist Jabir ibn Hayyan. In his book *Summa Perfectionis Majisterii* (*The Sum of Perfection, or of the Perfect Majistery*) he writes that Jupiter (tin) gives off a *harsh sound* or *crashing noise*. He also describes Mars (iron) as *sounding much* during forging, which was most likely produced by the formation of martensite upon cooling. Some scholars argue that *Summa Perfectionis Magisterii* may have been written in the 14th century by the Spanish alchemist Geber (the Latinized form of Jabir) who claimed the work to be a translation of the original Arabic in order to give more authenticity to his book. In either case, this is the oldest recorded observation of acoustic emission and was found in *The Works of Geber*, originally published in London in 1678. This was Richard Russell's English translation of the Latin edition published at Berne, Switzerland in 1545. The work is available today in a book entitled, *The Works of Geber, Englished by Richard Russell, 1678, A New Edition With Introduction by E. J. Holmyard, M.A., D. Litt.* published in 1928 by J. M. Dent & Sons Ltd., London and E. P. Dutton & Company, New York.

More recently (in this century) incidental observations of audible sounds emitted by metals during the course of studying other metallurgical phenomena, primarily twinning and martensitic phase transformation studies, are reported in the literature as early as 1916 when J. Czochralski (Reference 344) reported *Zinn- und Zinkgeschrei* (tin- and zinc-cry) as acoustic phenomena associated with twinning.

In 1923 Albert M. Portevin and Francois Le Chatelier (Reference 1433) reported *petit bruit sec* (small sharp noises) were clearly audible at a distance of several metres during discontinuous yielding and Luders band formation in aluminum-copper-manganese alloys.

In 1924 P. S. Ehrenfest and Abram F. Joffe (see Reference 299) observed that the process of shear deformation in rock-salt crystals and in zinc single crystals is accompanied by very regular *knackenden Lauten* (clicking noises), similar to the ticking of a clock.

Robert J. Anderson published a paper "Some Mechanical Properties of Duralumin Sheet as Affected by Heat Treatment" in 1926 in *The Proceedings of the 29th Annual Meeting of the American Society for Testing Materials, Part-II* in which he stated that in tension testing of Duralumin beyond the yield point and near the failure load, increased loading is accompanied by a series of slips which give rise to a series of serrated lines or Luders lines. These slips are accompanied by a series of *audible clicks or sounds.* For thin sheet the sounds are high pitched, *not unlike the tinkle of Japanese glass chimes*, and for thick sheet the sounds are *low in note, being on the order of a grunt.* It is interesting to note that John M. Lessells, in an accompanying discussion of Anderson's paper, expressed a special interest in Luders lines and slippage and the audible sounds that emanate from test bars during tensile testing.

Miss M. V. Classen-Nekludowa (1929, Reference 299) working with rock salt, monocrystalline and polycrystalline brass, and polycrystalline aluminum achieved 10^4 optical magnification of deformation steps which occurred at the moment of the *Knackens* (cracking noise).

Erich Scheil (1929, Reference 1546) reported that the formation of martensite in steel is accompanied by a *deutlich horbaren Gerausch, dessen Klang dem bekannten Zinngeschrei ahnlich war* (clearly audible noise, whose sound was similar to the well-known tin cry).

P. W. Bridgman (1937, Reference 206) observed during rotation under compressive loading that few metals and many non-metals rotate in a noisy manner—some *chatter,* some *squeak,* some make a *grinding noise,* while others *hiss*. He claimed the most important noise, however, was *snapping* which was in most cases superposed on the ordinary phenomena of plastic flow. The metallic elements that showed *snapping* were titanium, strontium, cerium, arsenic, and osmium.

Charles S. Barrett (1947, Reference 93) reported that the transformation to the face-centered cubic form for lithium is accompanied by a series of audible clicks, as in the twinning of tin or magnesium and the formation of martensite. He concluded that the transformation goes by abrupt shear movement in small isolated regions.

The next period in the history of acoustic emission involved three experiments which were conducted for the purpose of detecting and recording acoustic emissions from the specimen. The first experiment instrumented specifically to detect acoustic emission was conducted in Germany and published in 1936 by Fritz Forster and Erich Scheil (Reference 487). They recorded the *Gerausche* (noises) caused by the formation of martensite in 29% nickel steel. The specimen was supported between 1 mm diameter molybdenum wires, one wire acting as a waveguide. The transverse vibrations caused by martensitic transformation in the specimen were transmitted through the wire waveguide to a receiving transducer which transformed the noise into electric current. The electrical signals were processed through an amplifier, were rectified, and were then further amplified and recorded with a galvanometric light-beam recorder.

In the United States, Warren P. Mason, H. J. McSkimin, and W. Shockley (Reference 1139) performed and published the second instrumented acoustic emission experiment in 1948. At the suggestion of Shockley, experiments were directed towards observing moving dislocations by means of the stress waves they generated. Their instrumentation was capable of measuring displacements of about 10^{-8} cm occurring in times of 10^{-6} sec. The experiment consisted of pressing a pure tin specimen directly against a quartz crystal transducer, applying sufficient stress to deform the specimen and cause twinning dislocations, which, in turn, produced acoustic emission in the ultrasonic frequency range. The quartz crystal had uniform sensitivity from a few kilohertz to 5 MHz. The pure tin specimen coupled to a quartz crystal transducer was compressed between the ends of two glass rods which were acoustically matched to the specimen by tapering. Specimen resonances were eliminated and confirmed by the dissimilar fine structure in traces from different specimens. Since the fine structure, that is, the relatively high frequency vibrations in these stress pulses, was not the same for any two traces, it suggested that this was a characteristic of the twinning process rather than some mechanical resonance of the specimen. Similiar experiments with aluminum crystals failed to produce detectable signals.

The third instrumented experiment was performed in England by D. J. Millard (1950, Reference 1175) during his research work for his Ph.D. thesis at the University of Bristol. He conducted twinning experiments on single crystal wires of cadmium. Twinning was detected using a Rochelle salt transducer. He also evaluated other piezoelectric transducers, namely, quartz and ammonium dihydrogen phosphate. Millard's instrumentation provided a system gain of 4×10^4 at 1 kHz.

The early observations of audible sounds and the three instrumented experiments were not directed at a study of the acoustic emission phenomenon itself. The genesis of today's technology in acoustic emission was the work of Josef Kaiser at the Technische Hochschule Munchen in Germany. In 1950 he published his Ph.D. thesis (Reference 959) which reported the first comprehensive investigation into the phenomena of acoustic emission. The objectives of Kaiser's research were to determine from tensile tests of conventional engineering materials what noises are generated from within the specimen, the acoustic processes involved, the frequency levels found, and the relation between the stress-strain curve and the frequencies noted for the various stresses to which the specimens were subjected. He showed that acoustic processes were set in motion in tensile tests of all nine materials tested; namely, tin, lead,

Duralumin, copper, brass, gray iron, steel (St37 12), tool steel (WSt), and dry spruce wood. He also found the frequency of the acoustic emissions at the yield point for particular metals to be approximately 25 kHz. Kaiser used both quartz crystal and Rochelle salt transducers. In his instrumentation he obtained 10^6 electro-acoustic amplification. His most significant discovery, however, was irreversibility which now bears his name, the *Kaiser effect*. He also proposed a distinction between *burst-type* and *continuous* emission. Kaiser concluded that the occurrence of acoustic emissions arises from frictional rubbing of grains against each other in the polycrystalline materials he tested and also from intergranular fracture.

Kaiser continued his research at the Institut fur Metallurgie und Metallkunde der Technischen Hochschule Munchen until his death in March 1958. His work provided the momentum for continued activities at the Institut by several of his co-workers, Heinz Borchers and Hans-Maria Tensi, and also furnished the impetus for further research elsewhere in the world.

The first extensive research into acoustic emission phenomena following Kaiser's work was performed in the United States by Bradford H. Schofield at Lessells and Associates (now Teledyne Materials Research). His involvement came about as a result of a literature survey he and a co-worker, A. A. Kyrala, had conducted in which they came across Kaiser's article in *Archiv fur das Eisenhuttenwesen* (Reference 956). This precipitated a request from Professor John M. Lessells to his friend and adjudicator of Kaiser's thesis, Professor Ludwig Foppl, for a copy of Kaiser's thesis. Subsequently, Schofield began a correspondence with Kaiser which continued until Kaiser's death. In December 1954 Schofield, under a contract with Wright Air Development Center, initiated a research program directed towards the ultimate application of this phenomenon to the field of materials engineering. His initial research work was to verify the findings of Kaiser. The primary purpose of this early work was to determine the source of emissions. The single crystal work performed by Schofield showed conclusively that grain boundary effects were not the only source of emission as Kaiser had stated. As work progressed it became evident that significant differences existed between his findings and those of Kaiser. The existence of specific characteristic spectra relating to the materials condition on the basis suggested by Kaiser was not evident. However, the existence of the acoustic phenomenon and general characteristics of the individual signals were verified. Schofield performed an extensive investigation into the surface and volume effects as related to acoustic emission behavior. From experimental data obtained from oriented single crystals of aluminum (both with and without an oxide layer) and from oriented single crystals of gold he concluded that the surface condition does have a measurable influence on the acoustic emission spectrum. However, his most important conclusion was that the acoustic emission is mainly a volume effect and not a surface effect.

In the fall of 1956 Professor Lawrence E. Malvern at Michigan State University came across a one-page article entitled, "Horbare Schwingungen beim Verformen (audible vibrations from deformation)" in the German book *Fliessen und Kriechen der Metalle* by Wilhelm Spath (Metall-Verlag GmbH, Berlin-Grunewald, 1955). The article references the observations of *Gerauschen* by Joffe and Ehrenfest (Reference 923), Klassen-Nekludowa (Reference 299), Becker and Orowan (Reference 126), and Kaiser (Reference 956). Interested in studying the asperity theory of friction, Malvern suggested to a new faculty member, Clement A. Tatro, that this acoustic technique would be interesting to look into. Consequently, Tatro initiated laboratory investigations into the acoustic emission phenomenon. Then in 1957 he became aware of the work of Schofield and the two began collaborating. While at Michigan State,

Tatro encouraged a number of his graduate students to do their research in acoustic emission. They were Paul S. Shoemaker (1961, Reference 1620), Robert J. Kroll (1962, Reference 1039), Robert G. Liptai (1963, Reference 1082), and Robert B. Engle (1966, Reference 443). Tatro left Michigan State in 1962 for Tulane University where he continued research activity in acoustic emission, fostering two more graduate students, Benny B. McCullough (1965, Reference 1159) and Davis M. Egle (1965, Reference 429). In 1966 he joined the staff at Lawrence Radiation Laboratory (now Lawrence Livermore Laboratory).

Schofield and Tatro encouraged others to become involved in research activity in the field of acoustic emission. That encouragement, combined with the impact of their published work (the first publications in the English language), helped to establish a number of sites of activity in acoustic emission research and application around the country.

Julian R. Frederick at the University of Michigan had been engaged in research in ultrasonics since he was a graduate student under Dr. Floyd A. Firestone in 1939. His interest in acoustic emission, particularly for studying dislocation mechanisms, was excited in 1948 after reading Mason, McSkimin, and Shockley's article "Ultrasonic Observation of Twinning in Tin" in *Physical Review* (Reference 1139). Frederick visited both Schofield and Tatro in the late 1950's to discuss the state-of-the-art. In 1960 he obtained a National Science Foundation grant for two of his graduate students to start their research in acoustic emission, followed by a third student several years later. They were Jal N. Kerawalla (1965, Reference 988), Larry D. Mitchell (1965, Reference 1196), and Anand B. L. Agarwal (1968, Reference 8).

At Aerojet-General Corporation Allen T. Green, Charles S. Lockman, and Richard K. Steele were trying to verify the structural integrity of Polaris rocket motor cases being fabricated for the U.S. Navy. They noticed that audible sounds were consistently evident during hydrostatic testing. Therefore, they decided to instrument the vessels to detect, record, and then post-test analyze these acoustic signals. They conducted their first acoustic emission test in 1961, using contact microphones, tape recorder, and sound level analysis equipment. Later they improved their detection capability with accelerometers and charge amplifiers. As a part of Aerojet's contract with the National Aeronautics and Space Administration, Green, Lockman, and Steele instrumented the Thiokol 260-inch diameter SL-1 motor case to monitor the vessel during hydrostatic proof testing in 1965. Analysis of the tape recorded acoustic emission data clearly showed crack initiation and propagation prior to catastrophic failure at about 56% of proof pressure. Also, from the recorded test data, and by using triangulation methods, they were able to locate the origin of failure within 12 inches. The fact that acoustic emission did indeed predict and locate failure of the vessel immediately prompted a number of materials people at Aerojet to become involved in acoustic emission. These people were Carl E. Hartbower, Philip P. Crimmins, William W. Gerberich, William G. Reuter, George S. Baker, and Douglas W. Bainbridge.

Harold L. Dunegan at Lawrence Radiation Laboratory became interested in acoustic emission after hearing a paper by Tatro and Liptai (Reference 1740) at the Symposium on Physics and Nondestructive Testing held at Southwest Research Institute in 1962. In 1963 Dunegan wrote an internal report (Reference 404) proposing the application of acoustic emission techniques to pressure vessel research. He was later joined by David O. Harris, Albert E. Brown, Tatro, Engle, and Liptai. Their activities involved a broad spectrum of acoustic emission research and applications.

In 1969 Dunegan left the Laboratory and formed Dunegan Research Corporation (now Dunegan/Endevco) which was the first company devoted exclusively to the manufacture of acoustic emission equipment.

At The Boeing Company two groups became involved in acoustic emission work. In 1965 Harvey L. Balderston, at the Space Division, initiated research on incipient failure detection in bearings and signature analysis from other rotating machinery, and research on hydraulic systems for leak detection as well as cavitation and erosion in valves from fluid flow. At Boeing Scientific Research Laboratories, in the summer of 1967, Robert W. Moss engaged Adrian A. Pollock in order to add acoustic emission to an existing project on the study of crack growth in titanium. Pollock was a third-year graduate student from Imperial College in England. He had been the recipient of the first grant in acoustic emission placed with Dr. R. W. B. Stephens at Imperial College in October 1964 by Roy S. Sharpe from the NDT Centre, Atomic Energy Research Establishment, Harwell. During his six-weeks stay at Boeing, working in several areas of acoustic emission research (Reference 1421), Pollock became so enthused about the practical applications of acoustic emission in detecting crack growth that, upon his return to England, he directed the remaining research for his Ph.D. Thesis (1970, Reference 1410) towards the practical application of acoustic emission for materials testing and nondestructive testing.

In early 1965 Dwight L. Parry at Phillips Petroleum Company (now EG&G, Idaho, Inc.) at the National Reactor Testing Station, was involved in the LOFT (loss of fluid test) program, looking for a faster nondestructive testing method than the conventional ones to detect loss of coolant in a nuclear reactor. Based on information in papers by Schofield and Kaiser, Parry applied acoustic emission techniques. In May 1966 he obtained his first definitive acoustic emission data from failure of the embrittled half-scale PM-2A reactor vessel instrumented with accelerometers, charge amplifiers, and a tape recorder. Later, flaw location was successfully accomplished by triangulation techniques on the full-size PM-2A vessel.

In February 1966 the Division of Reactor Development and Technology of the U.S. Atomic Energy Commission sponsored a dual-phase reasearch program between Phillips Petroleum and Battelle-Northwest Laboratories. Phillips Petroleum was assigned the responsibility to develop an "Incipient Failure Detection System." Parry and Dan L. Robinson developed an acoustic emission instrumentation system which would detect, locate, and characterize incipient failure conditions in the reactor pressure vessel and in the primary pressure system components of a nuclear power plant.

The mission of Battelle-Northwest on this AEC program was primarily directed towards basic research of the acoustic emission phenomenon. Philip H. Hutton, who had initiated acoustic emission activities at Battelle-Northwest in October 1965, served as project leader on this four-goal research program. These goals were: (1) detection of acoustic emission against the noise background of an operating reactor, (2) location of the origin of acoustic emissions, (3) characterization of emissions as to the type of failure or flaw growth mechanism in the metal, and (4) prediction of the time of failure. Later, Hutton was joined by Norman J. Dixon, David O. Hunter, and R. Neal Ord. Their research involved studies of specimen resonance, structural evaluation of pressure vessels and piping systems, and basic instrumentation. The latter involved transducer studies, particularly in reactor environments and at elevated temperatures, plus noise level and performance of preamplifiers, cabling, and signal processing components of the instrumentation system.

At Argonne National Laboratory, T. Theodore Anderson became involved in acoustic emission in January 1967 after conducting a literature survey of acoustic boiling detection. His research activities, which were directed at boiling detection in sodium-cooled reactors, were the first coordinated effort in this area. In 1969 he was joined by A. P. Gavin in developing a high-temperature, submersible microphone for use in the liquid sodium environment of a reactor.

Liptai was also at Argonne in 1967 where he was conducting experiments on martensitic phase transformation in single crystals of a gold-cadmium alloy (Reference 1083). During this period of time he collaborated with Dunegan and Tatro and subsequently joined Tatro's group at Lawrence Radiation Laboratory.

With over 15 years of experience in the field of nondestructive testing, particularly in ultrasonic testing, Thomas F. Drouillard joined the Rocky Flats Plant in July 1967, then operated by Dow Chemical Company. His specific assignment was to investigate acoustic emission as a potential NDT method for production applications. Drouillard's initial involvements were with Dunegan and Tatro at Lawrence Radiation Laboratory and in the activities of the Acoustic Emission Working Group.

Discovery of the Microseismic Method

Seismic and microseismic activity are forms of acoustic emission, differing only in frequency and amplitude, earthquakes being the most energetic. The science of seismology to study earthquake activity and the geomechanical study of microseismic activity in rock are strikingly similar to the acoustic emission study of deformation and fracture in metals. Experimental techniques, instrumentation, and emission-producing events are similar. On a more local scale there has developed great interest in the field of rock mechanics and mining engineering to study seismic or microseismic activity associated with rockburst. The possibility of detecting rockbursts in coal mines was conceived as early as 1923 by E. A. Hodgson in Canada. He proposed the use of sub-audible noises to predict rockbursts and earthquakes.

In 1938 Leonard A. Obert of the U.S. Bureau of Mines was conducting an experiment in a lead-zinc mine in northeastern Oklahoma to determine if the seismic velocity in mine pillars was dependent on pillar loading. Throughout the test spurious signals kept triggering the interval time between two geophones. After eliminating possible equipment defects it was discovered that the triggering was being caused by self-generated signals in the rock. At first the significance of these *microseisms* was not recognized. Shortly after the completion of the Oklahoma tests, Obert and Wilbur I. Duvall conducted similar seismic velocity tests in a copper mine in northern Michigan. The same self-generated signals were encountered. However, because of the high stress condition of the mine, numerous rockbursts were occurring. The source of the microseisms was attributed to these rockbursts. Obert and Duvall recognized the significance of microseismic activity generated in rock under high stress, since this phenomenon provided the basis for a method of detecting and delineating areas of high stress that required no knowledge of the mechanical properties of the rock or the state of stress in the rock.

In 1970 H. Reginald Hardy, Jr., from Pennsylvania State University joined the Acoustic Emission Working Group, bringing to the acoustic emission community a knowledge of geophysics and the beginning of a cross-fertilization with geophysical scientists.

Acoustic Emission Working Groups

In the cocktail lounge of the Desert Inn in Richland, Washington, in the spring of 1967, Jack C. Spanner and Allen T. Green were discussing the status of acoustic emission in this country. They observed that a number of researchers were investigating the phenomenon and publishing reports of their work, but there seemed to be a lack of communication. Spanner also observed that there were differences in the terminology and in experimental techniques used, generally reflecting the researcher's educational background and field of expertise. Jointly they conceived of the idea that there was a need to somehow bring these people together to exchange information. Green explained that the Western Regional Strain Gage Committee had successfully served that technology as a forum for the exchange of information for approximately ten years. Using that organization and its constitution and by-laws as a model, Spanner laid the groundwork for the formation of the Acoustic Emission Working Group (AEWG). He contacted, first by phone and then by letter, over a dozen people currently engaged in or with an interest and experience in the field of acoustic emission, inviting them to participate in the formulation and organization of an informal working group on this subject. Consequently, a *shirt-sleeve* meeting was held in the King's Room of the Esquire Inn Motel in Alcoa, Tennessee on November 2, 1967, following the Conference on Incipient Failure Diagnosis for Assuring Safety and Availability of Nuclear Power Plants held in Gatlinburg, Tennessee. Twelve persons attended this formative meeting of the AEWG. In addition to Spanner and Green there were Harvey L. Balderston, Harold L. Dunegan, Julian R. Frederick, Philip H. Hutton, Robert W. Moss, Charles W. Musser, Dwight L. Parry, Herb N. Pederson, Bradford H. Schofield, and Norman K. Sowards. Subsequently, the first meeting of AEWG was held in Idaho Falls, Idaho on February 8, 1968. At this meeting, in addition to the founding members, eight persons were elected to membership to form the Charter membership. They were T. Theodore Anderson, Thomas F. Drouillard, Robert B. Engle, Robert G. Liptai, D. K. Mitchell, R. Neal Ord, Ronald E. Ringsmith, and Richard K. Steele.

The AEWG set the pace as a forum for personal interchange of current activities in this emerging technology. Similar working groups were organized in Japan and Europe. In Japan Eiji Isono at the Central Research Laboratories of Nippon Steel Corporation, together with Morio Onoe at the Institute of Industrial Science, University of Tokyo, organized the Japanese Committee on Acoustic Emission and held their first meeting in December 1969. Isono subsequently published the first paper, a review article, in Japan in 1970 (Reference 895).

The European Working Group on Acoustic Emission (EWGAE) was formed in 1972 as the culmination of a lot of interest by a number of people throughout Europe who were already engaged in acoustic emission research. Acoustic emission research was being performed: in England by Pollock at Cambridge Consultants, by Don Birchon at the Admiralty Materials Laboratory, by Ian L. Mogford and Ian G. Palmer at Central Electricity Research Laboratories, by Peter G. Bentley at Risley Engineering and Materials Laboratory of the U. K. Atomic Energy Authority, and by Brian Harris at the University of Sussex; in Germany by Jurgen Eisenblatter and Peter Jax at Battelle-Institut in Frankfurt am Main; in France by P. F. Dumousseau at the Centre Technique des Industries Mecaniques and by Madame Nicole Chretien and E. G. Tomachevsky at the Centre d'Etudes Nucleaires de Saclay, Commissariat a l'Energie Atomique; in Italy by M. Mirabile at Centro Sperimentale Metallurgico in Rome; in Denmark by Arved Nielsen at the Research Establishment Riso, Danish Atomic Energy Commission; and in the Netherlands by J. C. F. DeKanter at the Technische Hogeschool Delft. The first meeting in Europe on the subject was the Institute of Physics Conference on Acoustic Emission which was organized by Pollock and held in March 1972 at Imperial College in London. Fifteen papers from throughout Europe and the U.S. were presented

(References 144, 166, 342, 412, 531, 799, 998, 1190, 1202, 1405, 1704, 1854, and three unlisted papers). The success of this conference established both the interest and the need for forming a working group. Consequently, Pollock and Birchon organized the European Stress Wave Emission Working Group which held its first meeting in November 1972, at the Admiralty Materials Laboratory in England. During their second meeting at Battelle-Institut in September 1973, the group formally adopted their present name, European Working Group on Acoustic Emission.

Through the forum of these three working groups there has developed an international exchange of information and cooperation in addressing specific problem areas. Also, since the organization of these working groups, there has been a virtual explosion of activity in all areas of acoustic emission technology—in research into the phenomenon itself, in the application of acoustic emission to materials research, in its application to nondestructive testing and fracture mechanics studies, and in the development of instrumentation specifically designed for application to the detection and processing of acoustic emission signals. This bibliography chronicles the achievements of that activity.

Wilbur I. Duvall and Leonard A. Obert

BIBLIOGRAPHY

A

1 **S. J. Acquaviva and R. Chait. "Monitoring Fatigue Failure of a High Strength Titanium Alloy by Acoustic Emission Techniques." AMMRC-PTR-75-2. Army Materials and Mechanics Research Center, Watertown, Massachusetts. June 1975. [NTIS: AD/A-016-867]; pp 1-7 in *Proceedings of the Tenth Symposium on Nondestructive Evaluation.* Southwest Research Institute, San Antonio, Texas. June 1975. symposium held in San Antonio, Texas. Apr 23-25, 1975.**

The effort describes the application of acoustic emission techniques to cyclically loaded specimens in an attempt to detect crack initiation. Both notched and unnotched fatigue specimens were cycled at 30 cps (hertz) in the low cycle fatigue region, i. e., the stress level in each test was above the endurance limit of the material. Fifteen specimens, both notched ($K_t = 4.0$) and unnotched, were tested. Results of the tests on the unnotched specimens indicate acoustic emission signals being recorded 30 to 150 cycles prior to fracture; and the crack initiation and propagation phases appear to be quasi-simultaneous as expected in a high strength material. The tests on the notched specimens indicate acoustic emission signals corresponding to crack initiation being recorded as long as 2500 cycles prior to fracture. As a result of the study, it appears that acoustic emission can be a useful technique for detecting material flaws prior to catastrophic failure.

2 **K. H. Adams, B. R. Bass, J. E. Borhaug, C. H. Goodman, and H. A. Thompson. "An Experimental Investigation of Delayed Acoustic Emission in Beryllium." UCRL-13402. Lawrence Radiation Laboratory, Livermore, California. Mar 1969. [NTIS].**

Specimens of high purity, annealed beryllium were maintained isothermal and subjected to steady tensile loads for extended time periods. During these periods, the acoustic emission from the specimens was recorded. The time period of these steady conditions extended from the time of application of the load, when the emission was a maximum, to the point where the emission activity had decayed essentially into the background noise of the instrumentation. The load on each specimen was increased incrementally so that the uniform tensile stress at the test section of specimen was varied monotonically through the yield strength up to the fracture point. The investigated temperature range extended from an ambient of about 80 °F up to a maximum of 200 °F. Analyses of these delayed acoustic emission data showed that the rate of emission in these tests did not decay linearly with the time. Instead the rate appeared to approach asymptotically in exponential dependence upon the time. Attempts to correlate these data on a semiempirical basis resulted in several indications that the emission phenomena may be related to the creation of vacancies in the crystal lattice, although additional data are necessary to pursue this possibility further. As a part of this project, techniques and procedures were developed using zinc for the purpose of safely performing post-stress, metallographic examinations of the surface of the beryllium specimens. The funds and time of the project were exhausted before these examinations could be performed.

3 **N. J. I. Adams and H. G. Munro. "Techniques for the Recording and Measurement of Crack Growth during Fatigue and Fracture Toughness Testing."** *Strain,* **STRNB. 5(2):68–73. Apr 1969.**

This note presents details of three techniques of monitoring and recording crack growth during fracture testing. The first technique uses special foil gauges; the second extends potential-flow measurements to asymmetric crack growth; and the third uses an acoustic pick-up to detect intermittent crack growth in plane-strain fracture toughness testing.

4 **R. D. Adams. "Fracture Development in Wood Resulting from Bending Stresses and Detected Using the Acoustic Emission Phenomenon." Master of Forestry Thesis. University of British Columbia, Vancouver, British Columbia, Canada. Apr 1969. [available through interlibrary loan from the Agriculture Forestry Library, University of British Columbia].**

A new approach to the problem of failure prediction of wood components and structures is that of fracture mechanics. In order to implement this approach, knowledge of flaw growth (or crack propagation) in wood subjected to various stress systems and environmental conditions is required. The extension of flaws is accompanied by the release of energy in the form of acoustic or stress wave emissions, which can be detected and have been found to be reliable indicators of crack growth. Using acoustic emissions as a measure, flaw growth was investigated in stressed bending specimens of three species. The test pieces contained a variety of knot configurations, seasoning checks, and resin pockets. Three wood moisture contents were used; specimen size was 2-by-4-inch (nominal) cross section and 50-inch length. A detailed description of the system used for detecting, measuring, and recording acoustic emissions is given. Emission activity was measured using an electronic counter. Simultaneous count, load, and defection measurements were recorded and count-deflection and load-deflection curves plotted. The association between acoustic emissions and crack growth is discussed. Four types of integrated count-deflection patterns were found, which generally exhibited two common features. Firstly, a rate-change point, where the count rate increased rapidly, and secondly a count increase just prior to failure. The rate-change point was approximately coincident with the proportional limit in the load-deflection curve, suggesting an association between elastic behavior and microfailure development. Several qualitative relationships between the count and load curves are presented. A correlation between stable crack growth and the entire load-deflection curve is postulated. The presence of checks and resin pockets affected the microfailure pattern, shown by increased acoustical activity, without appreciably influencing modulus of rupture. Green specimens for the most part produced lower counts than drier ones. Also clear or essentially clear boards exhibited lower emission counts. Species differences were not apparent. The estimation of modulus of rupture employing parameters from the count-deflection curves is discussed and compared to that using modulus of elasticity. Precision of the prediction was not improved using count parameters. A method for determining the direction of crack propagation in a fractured bending member is described.

5 **R. D. Adams and J. E. Flitcroft. "Assessment of Matrix and Interface Damage in High Performance Fibre Reinforced Composites/L'evaluation des dummages de la matrice et des interfaces dans les composites renforcees aux fibres a haute resistance a la rupture." Bristol University, Bristol, England; Paper 4B/3 in** *Proceedings of the Eighth World Conference on Nondestructive Testing.* **volume containing Sections 3F, 3B, 4B. International Committee for Nondestructive Testing, Secretariat du Cofrend, Institut de Soudure, 32 Boulevard de la Chapelle, 75880 Paris Cedex 18, France. 1976. conference held in Cannes, France. Sep 6–10, 1976. (paper in English).**

Two nondestructive testing methods, a vibration technique, and stress-wave emission monitoring are examined for assessing matrix and interface damage in high performance fiber reinforced composites.

6 **R. O. Adams and C. R. Heiple. "Acoustic Emission from Welded Beryllium." Internal Report RFP-2491. Rockwell International, Atomics International Division, Rocky Flats Plant, Golden, Colorado. Mar 1, 1976.**

Acoustic emission was monitored during tensile tests of beryllium braze welds to investigate possible correlations between emission patterns and weld strength on fracture location. The amount of machining damage remaining in the weld groove prior to braze welding and the location of the weld deposit relative to the groove centerline were varied to determine the effect of these parameters on the tensile behavior of the welds. Changes in these variables were expected to cause variations in the strength of the welds and the location of the fracture path.

7 **R. O. Adams and C. R. Heiple. "An Energy Calibration Scheme for Acoustic Emission." Rockwell International, Atomics International Division, Rocky Flats Plant, Golden, Colorado. Sep 13, 1976; paper presented at the 16th Acoustic Emission Working Group Meeting, Williamsburg, Virginia. Oct 18–20, 1976.**

Acoustic emission generated during deformation of materials is generally divided into continuous and burst-type emission. Continuous emission is detected as an increase in the noise level of the system monitoring the emission. This indicates that the waves from individual events are small and are superimposed on each other due to a rapid rate of occurrence. The transient elastic waves corresponding to burst emission are large enough and sufficiently separated in time that individual events can be detected. It is reasonable to expect that the average energy of emission bursts from different types of sources within a material will generally differ. A distinction between different mechanisms may then be made on the basis of the energy of the observed bursts. Thus, for materials which exhibit burst emission, appropriate measures of acoustic emission are the burst rate and energy per burst, as a function of strain or other material (or both) or testing variables. Commercial acoustic emission equipment is available with which the burst rate and a measure of the average energy per burst can be obtained. The energy measurement, however, is only relative and depends on system gain, the transducer used, specimen geometry, transducer location, and other variables. The calibration technique described below is an attempt to determine the actual energy release from the events causing emission bursts in beryllium and to quantitatively evaluate the effects of specimen geometry on the apparent energy per burst.

8 **A. B. L. Agarwal. "An Investigation of the Behavior of the Acoustic Emission from Metals and a Proposed Mechanism for its Generation." Ph.D. Thesis. University of Michigan, Ann Arbor, Michigan. 1968; University Microfilms 68–13270.**

The purpose of the investigation was to detect the acoustic emission from metals during loading and unloading in an attempt to pinpoint the source and understand the fundamental mechanisms responsible for its generation. Specimens of 99.99 percent purity aluminum, 1020 steel, 2024 aluminum, and Alcoa 195 aluminum were loaded in uniaxial tension. A piezoelectric cyrstal transducer, PZT-5, was used as a detector in conjunction with an amplifying system in order to record the amplitude and the number of generated acoustic pulses as a function of the induced stress. The effect of microstructural variations through controlled heat treatment of the specimens on the emission behavior has been studied. Substantial emission during loading has been detected from 99.99 percent purity aluminum and from solution-treated 2024 and Alcoa 195 aluminum alloys, only upon the first application of load. Load emission from 1020 steel appears to be concident with the yield point. No significant load emission could be detected from materials in cold-worked condition. Precipitation hardening of 2024 aluminum causes load emission activity to decrease. Overaging, however, tends to cause it to return. These results indicate that the load emission may be due to the activation of Frank-Read sources of a certain minimum size. For the detection system used, the minimum loop length of the source that can be detected has been estimated to be of the order of 10^{-4} cm. The model offers a reasonable explanation for the plenomenon called the *Kaiser effect,* and is consistent with the existing work hardening theories. Cumulative unload emission count has been found to depend upon the maximum stress reached, the volume of the stressed material, and the hardness. It appears to be independent of the surface area and surface finish. Unload emission behavior seems to be identical in tension and compression loading modes and thus appears to be related to shear stresses.

These results do not lead to a consistent model for the explanation of the unload emission data, hence more work needs to be done to determine the source and mechanism responsible for the unload emission.

9 **A. B. L. Agarwal, J. R. Frederick, and D. K. Felbeck. "Detection of Plastic Microstrain in Aluminum by Acoustic Emission."** *Metallurgical Transactions,* **MTGTB, 1(4):1069–1071. Apr 1970.**

Recent acoustic emission tests for studying the behavior of solids under load show that strain hardening reduces the emission of 99.99% Al in a tensile test from an acoustic burst count of 24,000 for an annealed specimen stressed to 4000 psi to zero for a second application of the same stress. Further, room-temperature aging of 2024 Al alloy following a solution heat treatment causes the number of bursts of acoustic emission produced by a stress of 20,000 psi to decrease continuously from 40,000 counts after 2.5 minutes of aging to 100 counts after aging a fresh solution heat treated specimen for 1140 minutes. A model based on the strain sensitivity of the detection system and on dislocation theory is proposed to account for these results. [Abstract 14-0355. *Metals Abstracts*, Vol. 3, 1970].

10 **H. Ahlborn, J. Becht, and J. Eisenblatter. "Anwendung der Schallemissionsanalyse bei Untersuchungen uber die Risszahigkeit von Faserverbundwerkstoffen (Application of Acoustic Emission Analysis for Studying Fracture Toughness of Composites)." Battelle-Institut eV, Frankfurt am Main, Germany. 1973. (in German); paper presented at the Seminar on Schadigungsgrenzen bei GFK (Damage Limits in GFK), Karlsruhe, Germany. Mar 1973; UCRL-Trans-10785. translated for Lawrence Livermore Laboratory, Livermore, California. Nov 1974. [NTIS].**

Acoustic emission was used as indicator for the beginning of crack propagation in notched specimens of unidirectional reinforced epoxy-resin (UD-GRP) during bending and tension tests. Areas of delamination or debonding smaller than 1 mm^2 were detectable and by evaluation of the amplitude-distribution delamination processes can be distinguished from failure due to fiber fracture.

11 **G. A. Alers and D. A. Tiede. "Inspection Techniques for the Shuttle Thermal Protection System." summary of paper in** *Paper Summaries: ASNT Spring Conference, March 28–30, 1977, Phoenix, Arizona.* **American Society for Nondestructive Testing, Columbus, Ohio. 1977.**

The reusable thermal insulation to protect the Space Shuttle during re-entry is in the form of permanently installed bricks or tiles of a low density, high temperature material covering most of the surface area of the vehicle. It is necessary that the adhesive joint between each brick and the aluminum skin be inspected for quality after installation and that no cracks develop during service that would cause a tile to *flake off* during subsequent missions. Several candidate procedures for performing the inspection within the time constraints of the flight plans for the Shuttle were investigated by making measurements of the appropriate physical properties of the tile material. Based on these results, an acoustic emission test and a low frequency ultrasonic pulse-echo test were chosen as the principal candidate techniques for further development. Both of these methods have demanded the introduction of several innovative schemes for overcoming the problems inherent with the unusual properties of the new material being used for the thermal insulation tiles. In particular, the attenuation of acoustic waves is high, the acoustic impedance is low, the acoustic wave propagation constants are both anisotropic and inhomogeneous, and the acoustic emission signals require special data processing techniques for extracting the useful information. All of these problems and the current status of their solutions will be discussed.

12 **H. J. Alheid and F. Rummel. "Acoustic Emissions during Frictional Sliding along Shear Planes in Rock." pp 149–154 in *Proceedings First Conference on Acoustic Emission/Microseismic Activity in Geologic Structures and Materials.* H. R. Hardy Jr., and F. W. Leighton, Editors. Trans Tech Publications, Clausthal, West Germany. 1977. conference held at Pennsylvania State University, University Park, Pennsylvania. June 9–11, 1975.**

13 **V. P. Alikin, L. D. Gomza, and F. M. Kuznetsov. "Acoustic Emission during Galvanization of Steel." *Soviet Electrochemistry,* SOECA, 11(9):1294–1295. Sep 1975; translated from *Elektrokhimiya,* ELKKA, 11(9):1380–1382. Sep 1975. (in Russian).**

A study was made of the applicability of the method of acoustic emission to determine the degree of hydrogenation of steel during electro-deposition of Zn from three different plating baths. Comparison of the number of impulses arising during the period of electrolysis, the change in plasticity of the specimens, and the internal stresses arising in the process of growth of the deposit, indicated the large number of impulses occurring in the first minutes of electrolysis in a cyanide electrolyte are produced by hydrogenation of the base metal. At the same time, the phenomenon of acoustic vibration during Zn deposition from sulphate and fluoroborate electrolytes over the initial 30 minutes of electrolysis is connected with the growth of internal stresses in the Zn deposit. [Abstract 58-0526. *Metals Abstracts,* Vol. 9, 1976].

14 **C. C. Allen, F. F. Lyle, Jr., and E. R. Reinhart. "Technical Support." Task 4, Project II, pp 65–73 in "In–Service Inspection Program for Nuclear Reactor Vessels." R. D. Wylie. SwRI Project 17–2440, EEI Project RP–79. Biannual Progress Report No. 7. Southwest Research Institute, San Antonio, Texas. Aug 25, 1972. [NTIS: PB–212–144].**

15 **M. Amano and K. Ishikawa. "In-Service Nondestructive Inspection of Nuclear Reactor Vessel."** *Genshiryoku Kogyo* **(Nuclear Engineering, Tokyo), GKOGA, 17(11):44–48, 95. Nov 1971. (in Japanese).**

The code of the American Society of Mechanical Engineers was established for pressure boundary equipment for reactor coolants; reactor pressure vessels being the main part. In the ASME code, the equipment to which in-service inspection is to be applied are all of pressure vessels, piping, valves, pumps, and pipe supports. The inspection has been started in the U.S., Sweden, and West Germany. The No. 2 and No. 3 reactors in the Fukushima nuclear power plant are going to be the first large-scale examples of such inspection in Japan. One of the problems is that inspection will be performed in the presence of high radiation levels. Design of the equipment for purposes of inspection and the development of automatic remotely operated supersonic flaw detection techniques have made this inspection possible. One of the methods recently being actively investigated is acoustic emission. This seems to be the dominant in-service inspection method of the future because of its unique advantages.

16 **E. Amendola, G. Grugni, C. Panzani, F. Tonolini, and P. Vidali. "On-Line Acoustic Emission Monitoring of Welds of the Stainless Steel Steam Lines in a Thermal Power Plant." Paper 6-3, pp 442–451 in** *The Third Acoustic Emission Symposium.* **Japan Industrial Planning Association, Dobashi Building, 3-17-8, Akasaka, Minato-ku, Tokyo, Japan. 1976. proceedings of a symposium held in Tokyo, Japan. Sep 16–18, 1976.**

A periodic inspection of the welds of the stainless steel stream lines of two main units of an Italian thermal power plant revealed the presence of significant cracks in most of the welds. In this concern, it was thought that the AE monitoring of at least some welds could give useful information to the progress of the cracking process and therefore provide an effective surveillance. In this perspective, the first step to be accomplished was the demonstration of the feasibility of AE monitoring of the steam-line welds during power operation. The main problems of this application were the high temperature of the steam line (538 °C) and the intense background noise. The high temperature problem was solved by using acoustic waveguides welded on the steam-line wall. Noise rejection was obtained by using specially designed sensors and by high pass filtering the AE signals before preamplifiers. A satisfactory signal-to-noise ratio was achieved for frequency above 0.8 MHz. The most complete set of measurements was performed on one of the two units where three welds were AE monitored for AE source location. The AE monitoring was proved to be feasible and AE sources could be clearly located even in this hostile environment.

17 **R. E. Anderson and R. G. Poe. "A Study of Space Shuttle Structural Integrity Test and Assessment." NASA CR-133859, SD-72-SH-0001-Pt-1. Part 1, Final Report. North American Rockwell, Space Division, Downey, California. Jan 1972. [NTIS: N73-31778].**

The ultrasonics technique for assessing the structural integrity of the primary surface of the space shuttle vehicles is discussed and evaluated. Analysis was made of transducers, transducer coupling test-structure fabrication, flaws, and ultrasonic testing. Graphs of microphone response curves from the initial noise tests, accelerometer response curves from the final noise tests, and microphone curves from the final noise tests are included along with a glossary, bibliography, and results.

18 **R. T. Anderson and T. J. DeLacy. "Nondestructive Testing of Advanced Composites."** ***Metal Progress,*** **MEPOA, 102(2):88–92. Aug 1972.**

Of the various nondestructive tests applied to resin-matrix and metal-matrix composites, radiographic and ultrasonic tests have proved most beneficial so far. Sensitivity must be relatively high, approaching current state-of-the-art thresholds. Since the complex internal structure of composites causes much background noise at these high sensitivities, methods of noise suppression and image enhancement are highly desirable. It was also found that high-contrast radiography proved to be useful for routine testing of metal-matrix and thin resin-matrix composites. Microradiography has been used during development phases to aid fracture analysis and to observe defects under load. The transmission reflection method of ultrasonic testing detects open bonds between matrix layers satisfactorily. However, the method is only effective for testing fairly thin sections of uniform cross section. Thermal methods work well for finding interlaminar defects and measuring resin ratios of resin-matrix composites. Such methods are not reliable for testing metal-matrix composites, however. Eddy current methods are effective for measuring volume-fractions on metal-matrix material, but have limited application for defect detection. Methods which should be investigated include holography and stress-wave acoustic emission. Vibrational methods, particularly those which can be confined to small areas, may also prove useful. Data analysis and display techniques require immediate attention. Computers and other mechanical electronic devices might be integrated with testing and inspection equipment to facilitate testing.

19 **R. T. Anderson, T. J. DeLacy, and R. C. Stewart. "Detection of Fatigue Cracks by Nondestructive Testing Methods." GDCA-DBG 73-002. General Dynamics, Convair Aerospace Division, San Diego, California. Mar 1973. [NTIS: NASA CR-128946]; paper presented under title "Detection and Characterization of Fatigue Cracks and Crack Growth in 2219 Aluminum Alloy Sheet Materials" at the American Society for Nondestructive Testing National Fall Convention, Chicago, Illinois. Oct 1–5, 1973.**

The effectiveness was assessed of various NDT methods to detect small tight cracks by randomly introducing fatigue cracks into aluminum sheets. The study included optimizing NDT methods, calibrating NDT equipment with fatigue cracked standards, and evaluating a number of cracked specimens by the optimized NDT methods. The evaluations were conducted by highly trained personnel, provided with detailed procedures, in order to minimize the effects of human variability. These personnel performed the NDT on the test specimens without knowledge of the flaw locations and reported on the flaws detected. The performance of these tests was measured by comparing the flaws detected against the flaws present. The principal NDT methods utilized were radiographic, ultrasonic, penetrant, and eddy current. Holographic interferometry, acoustic emission monitoring, and replication methods were also applied on a reduced number of sepcimens. Generally, the best performance was shown by eddy current, ultrasonic, penetrant and holographic tests. Etching provided no measurable improvement, while proof loading improved flaw detectability. Data are shown that quantify the performances of the NDT methods applied.

20 **T. T. Anderson. "Reactor Surveillance and Monitoring Techniques: Acoustic Surveillance." pp 24–38 in "Instrumentation and Control Activities at Argonne, Annual Report for Period Ending June 30, 1971." ANL-7905. Argonne National Laboratory, Argonne, Illinois. Dec 1971. [NTIS].**

Detectors and techniques are being developed for monitoring and interpreting the in-core acoustic frequency spectrum of a liquid metal fast breeder reactor (LMFBR). The current principal objective is to reliably detect coolant boiling in an LMFBR core. Since acoustic signals, including the sounds of boiling, are distorted and masked by the coolant and core components before they reach the reactor exterior, they must be sensed as closely as possible to the source, and then transmitted in as distortion-free a manner as possible to the exterior of the reactor vessel. Two techniques for monitoring the in-core acoustic signals that minimize signal distortion are being investigated: (1) Use of acoustic sensors immersed in the coolant at the core outlet; and (2) Use of waveguides to transmit the core noise to acoustic or vibration sensors outside the core region. The acoustic spectrum information contained in the signals from these techniques will be used to reliably detect off-normal conditions, so that corrective action can be taken. The corresponding research and development activities in this acoustic surveillance program include: (1) Construction and evaluation of sodium-immersible high-temperature radiation-resistant piezoelectric transducers in a prototypical environment, and integration of signal information and processing equipment into a complete operational boiling detection system; and (2) Waveguides capable of transmitting desired in-core acoustic information to distant detection and processing equipment. Also under investigation is the acoustic noise spectrum of typical liquid-metal systems as detected by each technique, and its relation to the noise generated in reactor systems.

21 **T. T. Anderson, J. P. Bobis, and A. P. Gavin. "Acoustic Detection of Sodium Voids in the LMFBR Heat Transfer Simulation Loop." Technical Memorandum ANL-CT-75-32. Argonne National Laboratory, Argonne, Illinois. Feb 1975.**

Acoustic sensors located in the upper plenum of the LMFBR Heat Transfer Simulation Loop (HTSL) were used to measure absolute acoustic pressure levels. These sensors, a sodium-immersed microphone and a waveguide were exposed to 900 °F sodium for 400 hours of experiment operation and were unaffected after a year of immersion in sodium. The spectral distribution of the acoustic signals over the frequency range 200 Hz to 100 kHz was about 1×10^{-3} bars during sodium boiling, and less than 1×10^{-4} bars during noncavitating flow. Most of the flow signal occurred in the frequency range below 10 kHz while most of the boiling signal was in the frequency range 10 to 30 kHz. Maps of bubble position versus time were provided from the HTSL data system. Acoustic impulse signals were observed at times corresponding to: (1) bubble growth, (2) bubble disappearance, and (3) arrival of bubbles at the plenum.

22 **T. T. Anderson, J. P. Bobis, and W. W. Managan. "Acoustic Detection of Boiling Sodium in the Fuel Failure Mockup (FFM) for Test Bundles 5D and 3B." Technical Memorandum ANL-CT-76-5. Argonne National Laboratory, Argonne, Illinois. July 1975.**

A series of out-of-pile tests have been performed on the 19-pin electrically-heated bundles 5D and 3B in the Fuel Failure Mockup (FFM), a sodium loop facility at Oak Ridge National Laboratory (ORNL). Through these tests, heat transfer data are being obtained for the first scheduled in-pile experiment, Pl, in ANL's Sodium Loop Safety Facility (SLSF). Three types of acoustic sensing devices are installed in the FFM facility for the Pl-support test series to detect the onset of boiling sodium in the test section assembly. Waveguide stubs, 3/8-inch diameter by 12 inches long, have been welded to the outer pipe wall of the test section assembly to accommodate two types of sensors: (1) hydrophones inserted in oil-filled cups on the end of waveguide stubs which were located at inlet, middle, and exit of the test section assembly; and (2) a magnetic microphone pickup on the end of a waveguide stub which was located at the center plane of the electrically heated zone. The first types were developed by ORNL I&C Division and Professor R. F. Saxe of North Carolina State University; and the second type was installed by the FFM project staff. The third type, two ANL HT-9 microphones, are located in the sodium, one above and one below a flow baffle plate at the sodium exit from the test section assembly. Magnetic tape recordings were made of acoustic and thermohydraulic data on paralleled ANL and ORNL recorders. The ANL recordings have been analyzed for frequency spectra and signal level before and during flow coastdowns, at steady low flow without gas injection, and at steady low flow with 0.003 and 0.012 pounds per hour argon gas injection.

23 **T. T. Anderson, A. P. Gavin, and J. P. Bobis. "Acoustic Measurements in the EBR-II Primary System: Short Term Test No. 1, March–April 1974." Technical Memorandum ANL-CT-75-35. Argonne National Laboratory, Argonne, Illinois. Apr 1975.**

Two sodium-immersed detectors, a microphone, and an acoustic waveguide, have been developed, installed, and used for monitoring acoustic background of Experimental Breeder Reactor No. II (EBR-II) in Experiment MT-3. The microphone signal is free of structural resonances which appear in the signals from accelerometers mounted on the waveguide and various reactor components. Sound pressure level decreased by an order of magnitude when the primary sodium coolant flow was reduced from 100 to 50%. At full flow and independent of reactor power, the 1-millibar rms sound pressure as sensed by the sodium-immersed microphone had the characteristics of wideband noise over the frequency range 0.6 to 40 kHz.

24 **T. T. Anderson, A. P. Gavin, and J. P. Bobis. "Acoustic Measurements in the EBR-II Reactor Primary Sodium Tank."** *IEEE Transactions on Nuclear Science,* **IETNA, NS-22:671–675. Feb 1975.**

The sodium-immersed detectors, a microphone, and an acoustic waveguide have been developed, installed, and used for monitoring acoustic background of Experimental Breeder Reactor No. II (EBR-II). The microphone signal is free of structural resonances which appear in the signals from accelerometers mounted on the waveguide and various reactor components. Sound pressure level decreased by an order of magnitude when the primary sodium coolant flow was reduced from 100 to 50%. At full flow and independent of reactor power, the 1-millibar rms sound pressure as sensed by the sodium-immersed microphone had the characteristics of wideband noise over the frequency range 0.6 to 40 kHz.

25 **T. T. Anderson, A. P. Gavin, J. R. Karvinen, C. C. Price, and K. J. Reimann. "Detecting Acoustic Emission in Large Liquid Metal Cooled Fast Breeder Reactors." pp 250–269 in *Acoustic Emission, ASTM STP 505.* American Society for Testing and Materials, Philadelphia, Pennsylvania. 1972. proceedings of symposium held in Bal Harbour, Florida. Dec 7–8, 1971.**

Research at Argonne National Laboratory, Argonne, Illinois, on acoustic emission and closely related areas is oriented to goals of the liquid metal cooled fast breeder reactor (LMFBR) program. These goals include acoustic monitoring of reactor core and power plant components, acoustic detection of boiling in liquid sodium, and acoustic emission testing of materials at elevated temperatures. For monitoring of sodium-filled components, several types of acoustic emission sensors are being developed for operation in sodium at temperatures to 650 °C. These sensors contain the high temperature, radiation resistant, and piezoelectric-material lithium niobate which is clamped or bonded to the sensor housing. Reactor components have been monitored acoustically with both high temperature (620 °C rating) accelerometers and with standard accelerometers. The acoustic surveillance tests have shown: (1) flowrate-dependent noise in a sodium-cooled nuclear reactor (EBR-II) and in a sodium component test facility (CCTL); (2) sodium-oxide binding within an EBR-II primary pump; and (3) a loose drain tube in the EBR-II primary sodium to secondary sodium intermediate heat exchanger.

26 **T. T. Anderson, A. P. Gavin B. W. Spencer, and R. E. Holtz. "Acoustically Detected Events during Treat Test R7." Technical Memorandum ANL-CT-76-7. Argonne National Laboratory, Argonne, Illinois. July 1975.**

As part of the LMFBR component technology development program at Argonne National Laboratory, work has been under way on development of a high-temperature sodium-immersible microphone for measurement of acoustic pressures. Acoustic surveillance with high-temperature microphones and vibration transducers has potential as a backup technique to existing design methods to prevent local malfunctions from leading to substantial fuel failure. For example, an acoustic transducer immersed in an LMFBR upper plenum in addition to providing instanteous mechanical-acoustic information for the upper core region might have potential also to detect the sound produced by boiling sodium from a subassembly malfunction. Development of acoustic pressure transducers (microphones) at ANL has successfully proceeded to the stage of application and evaluation in various sodium systems.

The ANL-developed microphone has been used to characterize superheat and saturation boiling sodium in the LMFBR Heat Transfer Simulation Loop, and to monitor the EBR-II primary sodium system as the first application to an LMFBR in the U. S. The TREAT R-series experiments have provided an opportunity to evaluate the microphone's detection capability under simulated LMFBR accident conditions, the subject of this report. This constitutes the first in-pile application of the high-temperature microphone to a nuclear-heated reactor safety experiment.

27 **T. T. Anderson and T. A. Grate. "Acoustic Boiling Detection in Reactor Vessels."** *IEEE Transactions on Nuclear Science,* **IETNA, NS-16 (Part I):260-265. Feb 1969; paper presented at the 15th Nuclear Science Symposium, Montreal, Canada. Oct 23-25, 1968.**

Reactor coolant which boils in a fast reactor or high-flux research reactor, endangering reactor safety, can be detected acoustically. As a contribution to design of acoustic boiling detectors, the tank wall of a reactor is shown theoretically to affect standing waves in that tank. Experimental resonance frequencies of a 3-foot diameter tank demonstrate that for low-pressure reactor tanks, the pressure at standing-wave frequencies is nearly zero at the tank wall, in contrast to the results for a perfectly rigid wall.

28 **T. T. Anderson and F. H. Just. "Acoustic Boiling Detection in Fast Breeder Reactors." Argonne National Laboratory, Argonne, Illinois. 1969; paper presented at the American Society for Metals Engineering Congress, Philadelphia, Pennsylvania. Oct 13-16, 1969.**

If flow of liquid sodium is locally impeded within the nuclear-heated core of a liquid-metal-cooled fast breeder reactor, a fuel element could develop a hot spot to cause boiling of sodium coolant. An acoustic sensor, located remotely from that hot spot, can detect acoustic energy created by growth and collapse of the sodium vapor. Effectiveness of the acoustic boiling detection method depends upon discriminating the signal from boiling in the presence of background noise. The relative importance of background noise up to 100 kilohertz (kHz) is shown from measurements involving accelerometers externally mounted on two large sodium systems. These systems are the primary sodium coolant system of the Experimental Breeder Reactor No. 2 (EBR-II) in Idaho and the Core Components Test Loop (CCTL) at Argonne in Illinois. In both sodium systems, acoustic background is related to pump speed and sodium flow rate. The future use of high-temperature acoustic sensors, immersed in the sodium coolant, appears feasible for acoustic detection systems.

29 **T. T. Anderson, T. P. Mulcahey, and C. Hsu. "Survey and Status Report on Application of Acoustic-Boiling-Detection Techniques to Liquid-Metal-Cooled Reactors." ANL-7469. Argonne National Laboratory, Argonne, Illinois. Apr 1970.**

Boiling in the core of Liquid Metal Fast Breeder Reactors can be detected by either of two principal methods, acoustic or neutron monitoring. This report summarizes literature through June 1967 concerning acoustic methods. In the acoustic monitoring method for boiling detection, either acoustic waveguides or high-temperature acoustic sensors are recommended.

30 **J. L. Andre, N. Chretien, and J. Monier (to Commissariat a l'Energie Atomique, France). "Dispositif de localisation de sources d'emission acoustique (Device for Locating Acoustic Emission Sources)." French Patent 2,253,219. June 27, 1975. (in French).**

The invention relates to a device for locating acoustic emission sources. A two dimensional sensor network, with diamond-shaped meshes, is placed on the surface of a structure in which acoustic wave emitting sources are to be located. The sensors are arranged according to two groups, each of which is connected to a clock and a counter. Every signal fed into a mesh of the network inhibits all the other sensors not belonging to said mesh; the location of the source within the diamond-shaped mesh is achieved by triangulation. This can be applied to the detection of flaws in metal structures, e.g., in nuclear reactor vessels.

31 **D. D. Ang and M. L. Williams. "The Dynamic Stress Field Due to an Extensional Dislocation." pp 36–52 in *Proceedings of the Fourth Midwestern Conference on Solid Mechanics.* University of Texas Press, Austin, Texas. Sep 1959. conference held at the Unviersity of Texas, Austin Texas. 1959.**

The paper deals with the wave patterns due to a plane dislocation moving in an infinite elastic medium. The whole range of velocities from zero to infinity is considered, and there is a fundamental difference in the wave patterns for dislocation velocities below the velocities of sound and those greater than either or both of these. The problem is formulated in terms of Fourier integral equations and solved in closed form. The stresses are given explicitly. In particular they are singular on the wave front and at the dislocation–inverse square root and inverse first power of the distance away, respectively.

32 **M. S. Antsyferov, Editor. *Seismo–Acoustic Methods in Mining.* Consultants Bureau, New York. 1966. translated from Russian by S. E. Hall. original Russian text was published for the Skochinskii Mining Institute of the Academy of Sciences of the USSR and the Gosplan State Committee for the Fuel Industry by Nauka Press, Moscow, Russia. 1964.**

Includes a collection of 23 papers reporting work done at the Skochinskii Mining Institute's Geophysical Research Laboratory and the seismo-acoustic station of the Donets Council of National Economy. Much of the field work was carried out in Donbass coal mines which are active with sudden rock bursts involving coal and gas. Described are various sensors and instruments for seismo-acoustic studies in both laboratory and field, and the results of such studies. Contributing authors include: M. S. Antsyferov, N. G. Antsyferova, G. K. Boiko, V. S. Ivanov, G. M. Ivanova, Ya. Ya. Kagan, A. G. Konstantinova, I. M. Lavrov, V. M. Makarov, S. V. Mirer, Yu. V. Motsur, L. G. Mysina, R. F. Nikitchenko, N. B. Parshikov, L. B. Pereverzev, E. V. Petrosyants, and L. N. Shevehenko.

33 **A. E. Arave. "Stress Wave Analysis Instrumentation." pp 76–81 in "Instrument Development Branch Annual Report. 1968." R. G. Morrison, Editor. IDO–17299. Phillips Petroleum Company, Idaho Falls, Idaho. July 1969. [NTIS].**

Stress-wave emission is being developed in the aerospace and nuclear industries as a tool to determine when fatigue (microcracking) takes place and to locate the fatigue area. At the National Reactor Testing Station, Idaho Nuclear Corporation is doing research to determine the effect of hydrogen embrittlement upon the fatigue of boiling water ractor vessels. Hydrogen-embrittled samples are being placed under constant load, and the incubation period, from the time the load has been applied until fatigue occurs, is being measured. Also, the rate of cracking is being measured. The measurements are being made by stress-wave emission techniques. Described also is the instrumentation put together and developed to obtain data for each sample. Guidance was given by Aerojet-General Corporation to start the program.

34 **M. Arii. "Characteristics of Acoustic Emission under Electro–Slag Weld Cracking Test." Paper No. 6 in *Proceedings of the U.S.–Japan Joint Symposium on Acoustic Emission.* Japanese volume. Japan Industrial Planning Association. Kawamura–Building 6F, 3–21–6 Akasaka, Minato–ku, Tokyo, Japan 107. 1972. symposium held in Tokyo, Japan. July 4–6, 1972. (paper in Japanese, abstract available in English).**

Generally, in the thick plate welding of high carbon steel, many cracks may possibly be occurring. In addition, it is difficult to nondestructively inspect such cracks for the thick plate. In this paper, an attempt is made to detect the welding cracks by using the acoustic emission system. The method of welding is a simplified electroslag welding that is generally used in thick plate welding. And as a specimen, a circular rod was used so as to have a large restraint force under welding of a thick plate. The materials used were high carbon steel (S35C) and low carbon steel (SS41); the former has a trend of weld cracks and the latter has not. Acoustic emission was not measured under welding (which has much mechanical and electrical noises), but after welding was measured for four hours. As a result, acoustic emission which was attributed to cracks under welding was comparatively detected by use of high natural frequency transducer. The frequency of acoustic emission from slag cracks and weld cracks after welding was meaured and the difference was confirmed in both frequency distributions.

35 **M. Arii, H. Kashiwaya, and T. Yanuki. "Crack Opening Displacement Testing by Use of Acoustic Emission." pp 177–181 in *Significance of Defects in Welded Structures.* T. Kanazawa and A. S. Kobayashi, Editors. University of Tokyo Press, Bunkyoku, Tokyo, Japan. 1974. proceedings of the Japan–U.S. Seminar on the Significance of Defects in Welded Structures, held in Tokyo, Japan. Oct 15–19, 1973.**

Methods for plane strain fracture toughness (K_{IC}) testing of high strength steels had been established by the ASTM (E399-70T) and BISRA Fracture Toughness Committee

(DD3: 1971). On the other hand, the crack opening displacement (COD) testing is currently being used in the U.S., Britain, and Japan for estimating the fracture toughness of the high-toughness low strength steels. Brittle material fractures at a critical value of stress intensity factor or crack opening displacement without any prior crack growth. In the more ductile materials, however, unstable fracture initiation may be preceded by a period of slow crack growth from the notch tip. Such slow crack growth may be an essential prelude to unstable fast fracture and thus COD at the start of slow crack growth can also be regarded as a critical value of the fracture characteristics. It is thus desirable to detect the onset of slow crack growth, here a COD test may also be used to assess the resistance to fracture initiation of materials. Unfortunately, such information cannot be obtained from the conventional load-displacement diagram in a COD test since the COD is usually taken as that corresponding to the maximum load in the load-displacement diagram. Where the slow crack growth precedes unstable fracture initiation, the critical COD is found by detecting the initiation of slow crack growth which usually occurs prior to the maximum load. This instant of slow crack growth initiation has been detected by using the electrical potential method. An attempt was made to apply the acoustic emission techniques to the COD test for obtaining the onset of slow crack growth. Findings are presented.

36 **M. Arii, H. Kashiwaya, and T. Yanuki. "Slow Crack Growth and Acoustic Emission Characteristics in COD Test."** *Engineering Fracture Mechanics,* **EFMEA, 7(3):551–556. Sep 1975; pp 23.1–23.15 in** *Combined Non-Linear and Linear (Micro and Macro) Fracture Mechanics Applications to Modern Engineering Structures.* **Volume 1. H. Liebowitz and T. Yokobori, Organizers, 1976. preliminary proceedings of the U.S.–Japan Cooperative Science Program International Seminar, sponsored by the U.S. National Science Foundation and the Japan Society for the Promotion of Science, held at Tohoku University, Sendai, Japan. Aug 12–16, 1974. [for information on availability of proceedings, contact Professor T. Yokobori, Research Institute for Strength and Fracture of Materials, Tohoku University, Sendai, Japan or Professor H. Liebowitz, School of Engineering and Applied Science, George Washington University, Washington, D.C.].**

Brittle material fractures at a critical value of stress intensity factor or crack opening displacement (COD) without any prior crack growth. In the ductile materials, however unstable fracture initiation may be preceded by a period of slow crack growth from the notch tip. Such slow crack growth may be an essential prelude to unstable fast fracture and thus COD at the start of slow crack growth can also be regarded as a critical value of the fracture characteristics. An attempt was made to apply the acoustic emission techniques to the COD test. Acoustic emission techniques can be used to detect the onset of slow crack growth and to obtain a relation between COD and AE characteristics in the COD test. [reprinted by permission of Pergamon Press Ltd].

37 **B. H. Armstrong. "Study of Acoustic Emission Prior to Earthquakes."** *Bulletin of the Seismological Society of America,* **BSSAA, 59(3):1259–1279. June 1969.**

On the basis of popular reports of animal agitation prior to earthquakes, a semiquantitative study has been carried out of the possibility of the emission of acoustic strain and fracture radiation prior to earthquakes. Such emission has been observed in the laboratory and in mines, and the bearing of these reports on the animal behavior reports is discussed. The comparative auditory vibration response of humans and animals is discussed insofar as quantitative data are available. Theoretical estimates are derived of the frequency of the sound emitted by preliminary fracturing in a highly stressed region. The results for rockburst conditions in mines are consistent with the frequency spectra of the observed signals. Application to earthquake conditions indicates some emission should occur over high audible to low ultrasonic frequencies. Estimates of the intensity of such emission are also obtained to ascertain if it might be detectable on a practical basis. These results lead to transmission range estimates which, for rock Q-values at normal pressure, indicate a short range with at best a marginal possibility of detection under certain favorable circumstances. Since, however, rock Q-values are known to increase with pressure, it is pointed out that this may lead to a substantial increase in local transmission ranges within regions of high stress. This effect might permit the practical observation of preliminary emission for shallow earthquakes whose strain fields extend to the surface.

38 **M. Arrington. "Acoustic Emission."** ***British Journal of Non-Destructive Testing*, BJNTA, 17(1):10–14. Jan 1975.**

The growth of acoustic emission methods, from the first experiments to an industrial technique is outlined. The origins of emission, the instrumentation used, and the methods of data processing and presentation are described. Typical testing techniques are described and methods of improving the signal-to-noise ratio for continuous monitoring are detailed. Finally the present and possible future industrial applications of acoustic emission are reviewed.

39 **M. Arrington. "Acoustic Emission: An Introduction for Engineers."** ***Chartered Mechanical Engineer,* CHMGA, 22(4):53–55. Apr 1975; *Interface,* 9(1). Summer 1975.**

Most engineers are familiar with the various techniques of NDT which embrace among others the use of X-rays, gamma rays, eddy current probes, fluorescent dyes, and ultrasonics. Following the development of electronics, it was 25 years ago that a German worker in the field applied the new discipline to the science of NDT. The present article illustrates how far this application has now progressed.

40 **M. Arrington. "Acoustic Emission from Metals."** ***Metallurgist and Materials Technologist,* MMTCA, 8(7):387–389, 391. July 1976; "L'emission Acoustique des Metaux." *Qualite, Revue Pratique de Controle Industriel,* QRPCA, (82):21–24. Dec 1976. (in French).**

Acoustic emission is explained, giving account of the energy storage and release mechanisms. Emission mechanisms can be associated with deformations such as dislocation movement or crack growth. Acoustic emission monitoring has been carried out during proof tests, mainly in the nuclear and aerospace industries.

41 **M. Arrington. "Acoustic Emission from Real and Model Composites." pp 3-1 to 3-5 in "Newsletter on the Non-Destructive Testing of Composite Materials and Adhesive Bonded Joints." Issue Number 2. Royal Aircraft Establishment, Farnborough, Hants, England. July 1973.**

A brief introduction is given to two aspects of a research project undertaken for a PhD degree at Sussex University. Debond Tests in Single Fibre Specimens: One of the most critical properties of any composite is the fibre-matrix interfacial bond strength, as this controls both the transfer of load into the reinforcing phase and the work of fracture of the composite. With glass fibre reinforcing a plastic matrix, direct measurements of this interfacial bond are easily made using a pull-out test, but for carbon fibre reinforced plastics measurements are much more difficult. Initially interlaminar shear strength (ILSS) measurements have been used to show trends in bonding, an increase in the ILSS being correlated with an improved interfacial bond strength. Wadsworth and Spilling have measured the distribution of the lengths of broken fibres in a model system; under conditions of high strain it is expected that the fibres will break into small lengths of between 1c and 21c. This gave a bond strength of 6 MNm^{-2} for a weak interface. It was felt that a more direct measurement of interfacial strength would be preferable. Experiments were undertaken on pull-out tests, but these proved unsuccessful. After this, Broutman's work on single fibre test specimens was considered the most promising approach. In these tests, single fibres are embedded in specially shaped resin specimens, which are tested in compression resulting in debonding. Because of its high sensitivity acoustic emission was chosen as a monitoring system.

42 **M. Arrington and B. M. Evans. "Acoustic Emission Testing of High Alumina Cement Concrete."** ***NDT International,*** **NDITD, 10(2):81-87. Apr 1977.**

This paper summarizes the current problem associated with the use of high alumina cement (HAC) concrete in buildings and the resultant inspection problems. The results of a preliminary feasibility study on the application of acoustic emission testing to assessing structural integrity are presented. This study consisted of two sets of on-site tests plus some supporting laboratory work. The initial results indicate that AE monitoring of load tests may provide a method for estimating the characteristic strength of HAC concrete beams.

43 **M. Arrington and A. A. Pollock. "Some Aspects of the Practical Application of Acoustic Emission." pp 183-186 in** ***Ultrasonics International 1973 Conference Proceedings.*** **IPC Science and Technology Press Ltd, Guildford, Surrey, England. 1973. conference held at Imperial College, London, England. Mar 27-29, 1973.**

The ability of acoustic emission to locate flaws and assess structural integrity is now starting to be exploited. The advantages and limitations of acoustic emission as a nondestructive testing tool are outlined and the following important aspects of practical operation are discussed: (1) Calibration techniques for checking the reliability and reproducibility of measurements made at different times; and (2) The use of carefully planned procedural sequences to quantify spurious emission and increase confidence in the validity of the data. Anticipated future developments include improved methods of failure prediction, the grading of components by acoustic emission, and the use of computerized systems to locate emission sources in real-time and provide rapid interpretation of emission data.

44 **M. Ashkin. "The Scientific Basis of Acoustic Emission." Westinghouse Electric Corporation, Research and Development Center, Pittsburgh, Pennsylvania. 1972; paper presented at the 1972 Institute of Electrical and Electronics Engineers Ultrasonic Symposium. Boston, Massachusetts. Oct 4–7, 1972. [paper not published in proceedings];** ***IEEE Transactions on Sonics and Ultrasonics,*** **IESUA, SU–20:48, Jan 1973. [abstract only].**

The phenomena of acoustic emission in solids will be reviewed from two related points of view. The first emphasizes its use as a tool in material science investigations, while the second covers the basic material science investigations which appear significant for the further development of acoustic emission technology. Acoustic emission experiments will be compared with other measurements to show the unique or complementary role of the emission technique. Some theoretical models useful for describing the emissions will be presented. [reprinted from *IEEE Transactions on Sonics and Ultrasonics*].

45 **J. H. Ashton, J. A. Regalbuto, and Z. R. Wolanski. "Correlation of Diffusion Bonded Ti–6Al–4V Properties with Acoustic Emission, Pulse–Echo and Fractographic Data." pp 665–678 in** ***Titanium Science and Technology.*** **R. I. Jaffee and H. M. Burte, Editors. Volume I. Plenum Press, New York. 1973. proceedings of the 2nd International Conference organized by the Metallurgical Society of the American Institute of Mining, Metallurgical, and Petroleum Engineers, the American Society for Metals, and the Institute of Metals, in association with the Academy of Sciences of the USSR and the Japan Institute of Metals, held in Cambridge, Massachusetts. May 2–5, 1972.**

A series of 6Al-4V titanium pieces was diffusion bonded utilizing hot-retort cold press and vacuum hot-press techniques. The time, temperature, and pressure parameters were varied with each specimen. The bonded specimens were (a) examined metallographically in the bond area before destructive testing, (b) examined ultrasonically before testing, (c) monitored for acoustic emission during tensile pulling, and (d) examined in detail using optical and electron fractography techniques, as well as metallography at the bondline, after failure. Results indicated that pulse-echo techniques can be correlated with tensile test data at high bond strengths. Acoustic emission monitoring of the tensile tests indicated potential for determining diffusion-bond homogeneity. Correlation between microscopic and macroscopic analysis of each specimen was successful. Parent metal tensile specimens which were heat cycled with the diffusion-bonded specimens showed surface cracking after pulling. Preliminary investigation into this phenomenon indicates surface contamination from an alpha stabilizer.

46 **H. H. Askwith. "Preservice and Inservice Inspection Planning Using Acoustic Emissions as an NDE Tool." Bechtel Power Corporation, Houston, Texas. 1974; paper presented at the 13th Acoustic Emission Working Group Meeting, Laguna Hills, California. Aug 19–21, 1974.**

Early attention must be paid to the planning and scheduling aspects of the American Society of Mechanical Engineers Section XI examinations. This report points out some of the pitfalls that occurred at several nuclear generating plants. Preparations for ASME Section XI examinations of nuclear power stations begin with the writing of the Preliminary Safety Analysis Report (PSAR), specifically, its technical specifications section. The technical specifications address the specific code requirements by indicating the various components which will be subjected to the examinations and the type of nondestructive testing that will be performed on each. Generally speaking, a considerable amount of time lapses between the writing of the PSAR and the beginning of baseline examinations in the field. Often these examinations are completely forgotten until late in the construction phase of the plant. The first thing that is impacted by the baseline examination requirements is the construction schedule which often must be modified to reflect the time spans that are associated with them. On reviewing several nuclear sites, the examinations were lengthy, in fact as much as 12 weeks, and required a considerable number of manhours of construction support. This support may have to be drawn from an already established manpower force. Additionally, the baseline examinations will require access to equipment which in many instances, will interfere with other activities that are scheduled to be performed on it. In other words, the baseline examinations can become disruptive unless planned and factored into the construction schedule early in the project.

47 **H. H. Askwith and R. S. Hickman. "Acoustic Emission Examination of ASME Class 2 and 3 Buried or Underground Piping." Bechtel Power Corporation, Houston, Texas. 1975; paper presented at the 14th Acoustic Emission Working Group Meeting, San Antonio, Texas. Apr 28–30, 1975.**

During the last six years the detection of acoustic emissions has proved to be a useful tool to detect flaws in stressed piping systems. The paper will discuss the use of AE on buried or underground piping, ASME Class 2 and 3, in nuclear power generating stations. Current inspection techniques, as required by Section XI of the ASME Boiler and Pressure Vessel Code, are hydrotesting and visual examination for Class 3 lines; and ultrasonic, hydrotest, or visual for Class 2 lines. Adoption of AE in place of presently used inspection techniques could mean elimination of the added costs of valves and Seismic Category I manways or tunnels which are now used, by allowing direct burial of the piping. This could result in multimillion dollar savings for each plant. One type of system currently available for such use consists of a multichannel acoustic emission monitoring system. Current technical problems that must be resolved and field test data that must be acquired before such system can be applied to meet the requirements of the ASME Code, Section XI, will also be discussed.

48 **W. J. Astleford and E. C. Schroeder. "Develop and Deliver a Reactor Noise Simulator." Task 5, Project II, pp 75-89 in "In-Service Inspection Program for Nuclear Reactor Vessels." R. D. Wylie. SwRI Project 17-2440, EEI Project RP-79. Biannual Progress Report No. 7. Southwest Research Institute, San Antonio, Texas. Aug 25, 1972. [NTIS: PB-212-144].**

Final installation and checkout of the Experimental Beryllium Oxide Reactor (EBOR) flow-noise simulator was accomplished during the period of January 9 to 20, 1972. At the end of this period, the system was placed on operational status since duplication of the San Onofre hydraulic noise spectrum had been achieved. The noise simulator is composed of two major subsystems: (a) Pumping system, and (b) Spectrum detection instrumentation. A detailed description of the system and a discussion of the results obtained with this system are contained in the report.

49 **M. Asty, P. Bernard, and J. Monier. "Localisation des Sources d'Emission Acoustique: Possibilities, Limites (Location of Sources of Acoustic Emission: Possibilities, Limitations)." pp 258-278 in** ***Journees d'Etudes sur l'Emission Acoustique.*** **P. F. Gobin, R. Goutte, and C. Guillaud, Organizers. Institut National des Sciences Appliquees de Lyon, Avenue Albert Einstein, 69621 Villeurbanne, Lyon, France. Mar 1975. (paper in French).**

It is necessary to dispose of a system capable of data acquisition and processing in real time. The coordinates of emission sources must be calculated either immediately after the detection of information or after a brief storage time. Emphasis is laid on the various parameters liable to affect the measurement precision: transducers (type, selectivity, form of signal), threshold device (dynamics, influence on the precision), and screening device (influence on the number of data received). Four-transducer patterns aie now in common use: square, centered equilateral triangle, lozenge mesh. . . . Each geometry possesses zones of indetermination. The accuracy on the coordinates of the source varies according to the position of this source with respect to the four-transducer mesh, which leads to a case-by-case study of the arrangement and dimensions of the meshes placed on the structure to be observed. Detection and localization equipment must be designed as a whole system flexible and easy to adapt to any structure.

50 **I. I. Averbukh, V. M. Bobrenko, A. Yu. Detkov, N. A. Glukhov, V. V. Gradinar, V. N. Kolmogorov, Yu. N. Pranitskaya, P. Ph. Shapavalov, and V. E. Vainberg. "Devices for Testing Physical and Mechanical Parameters of Materials and Constructions." All-Union Scientific Research Institute of Nondestructive Testing, Kishinev, USSR. 1973; Paper J-33 presented at the 7th International Conference on Nondestructive Testing, Warsaw, Poland. June 4-8, 1973.**

The paper describes different types of equipment used for determining the degree of anisotropy in samples, as a function of loading. In one case, the velocities of longitudinal waves and the two principal shear waves are measured; in the other, acoustic emission is observed.

51 **I. I. Averbukh and V. V. Gradinar. "Inspecting the Strength Characteristics of Composite Materials."** *Soviet Journal of Nondestructive Testing*, **SJNTA, 9(3):261-264. May-June 1973; translated from** *Defektoskopiya*, **DEFKA, (3):12-15. May-June 1973. (in Russian).**

Article deals with the acoustic emission technique for testing composites. Several NDT methods are used for testing composites, e.g., ultrasonic pulse-echo and through-transmission techniques, impedance, resonance, and vibration methods. Although these detect some types of flaws, they fail to detect small voids and strength characteristics in-service. Use has been made of acoustic emission generated during straining two layers bonded together. The acoustic emission was much larger for samples with porosity in the bonding layer. The same applies to strength. With increasing bond strength the intensity of acoustic emission during straining decreases.

52 **I. I. Averbukh and V. E. Vainberg. "Acoustic Emission as a Function of Strain in Various Materials."** *Soviet Journal of Nondestructive Testing*, **SJNTA, 9(4):401-406. July-Aug 1973; translated from** *Defektoskopiya*, **DEFKA, (4):25-32. July-Aug 1973. (in Russian).**

Article deals with the acoustic emission characteristics related to the deformation of various materials. A simple apparatus for measuring the intensity of acoustic emission is described. The authors found in their experiments that the basic information of acoustic emission is in the intensity. They investigated the character of acoustic emission related to the deformation of steel, aluminum, niobium, copper, and brass, and found dissimilar responses. An investigation of the Kaiser effect lead to finding a reversed Kaiser effect. Also the influence of strain rate and the relation between the Bauschinger effect and acoustic emission are discussed.

53 **I. I. Averbukh and V. E. Vainberg. "Range of Application of Ultrasonic Emission."** *Soviet Journal of Nondestructive Testing,* **SJNTA, 8(4):422-425. July-Aug 1972; translated from** *Defektoskopiya,* **DEFKA, (4):49-54. July-Aug 1972. (in Russian).**

The authors describe their success in eliminating external laboratory noises transmitted through air by bandpassing, the elimination of grip noise in tensile specimens by preloading, a study of the emission from specimens with various size drilled holes, observations of the Kaiser effect, and creep in plastics.

54 **I. I. Averbukh and V. E. Vainberg. "Use of Propagation of Stress Waves for Determining the Thickness of Components."** *Soviet Journal of Nondestructive Testing,* **SJNTA, 7(1):105–106. Jan–Feb 1971; translated from** *Defektoskopiya,* **DEFKA, (1):132–134. Jan–Feb 1971. (in Russian).**

Stress waves occur due to friction of the specimen surface. In the friction process, small particles of material break away and hence surface variations occur. In the case of slight friction of the component, it is possible to detect considerable propagation at frequencies corresponding to the resonance frequency over the thickness of the plane-parallel component and at the harmonics of this frequency. Harmonics appear because the waves of ultrasound undergo multiple reflection at the specimen surface. For each number of the harmonic in the spectrum of the vibrations, there is a maximum. The thickness of the specimen can be determined according to the distance between the maxima. This article describes the technique used and shows the results of the investigation.

55 **I. I. Averbukh, V. E. Vainberg, O. N. Chegorinskaya, and V. V. Gradinar. "Use of Ultrasonic Emission in Nondestructive Inspection."** *Soviet Journal of Nondestructive Testing,* **SJNTA, 8(1):19–23. Jan–Feb 1972; translated from** *Defektoskopiya,* **DEFKA, (1):26–32. Jan–Feb 1972. (in Russian).**

Article presents the experimental results of investigations of the phenomenon of ultrasonic emission. The peculiarities of the emissions of defective and defect-free samples, of the emissions of plastic and bimetal samples, and of the emissions in the heat treatment of various types of steels and in welding are discussed.

56 **P. Azou and C. Francois. "Low Frequencies Acoustic Study in Strain of Metals/ Etude acoustique aux basses frequences de la deformation des metaux." Institut Superieur des Materiaux et de la Construction Mecanique, Paris, France; Paper 3K/4 in** *Proceedings of the Eighth World Conference on Nondestructive Testing.* **volume containing Sections 3K1, 3H1, 3G. International Committee for Nondestructive Testing, Secretariat du Cofrend, Institut de Soudure, 32 Boulevard de la Chapelle, 75880 Paris Cedex 18, France. 1976. conference held in Cannes, France. Sep 6–10, 1976. (paper in French).**

B

57 **C. F. Bacon. "Acoustic Emission along the San Andreas Fault in Southern Central California."** *California Geology,* **CGEOA, 28(7):147–154. July 1975;** *Materials Evaluation,* **MAEVA, 34(5):16A–18A, 20A, 22A, 23A, 30A, and 31A. May 1976.**

The California Division of Mines and Geology (CDMG) has completed the first phase of a pilot study, begun in 1972, to monitor audio-frequency microseismic activity along a major fault zone. Preliminary results indicate that this method may prove helpful as a tool for predicting earthquake activity. The idea that audible noise or even subaudible *sounds* may precede major earthquakes is not new. The literature describes numerous instances of the extreme agitation of animals over periods of many hours before actual major earthquake events. Many stories and folklore, dating back to the employment of *earthquake listeners* by the early Chinese emperors, describe premonitory experiences by humans. To test the credibility of these stories, Armstrong (1969) compared the auditory response of man and animals and discussed in mathematical terms the possibility that they might be capable of detecting sounds that occur prior to rockbursts and perhaps before earthquakes. He concluded that stress buildup in the rocks prior to an earthquake may improve the acoustic transmission characteristics of the rock and may even form *sound channels* so that audio-frequency premonitory noises may be heard or detected at the surface. He suggested that sound with frequencies up to a few kilohertz may travel far enough to be detected and to give useful warnings of shallow focus earthquakes for highly seismic and populated areas. Microseismic acoustic detection methods have been used in mining and slope failure studies since the development, in 1939, of equipment to detect the subaudible popping and cracking of rock which increases in volume and frequency of *pops* just prior to a tunnel wall failure or rockburst. Although some recent workers have listened along faults for possible changes in the amount of acoustic (sound) emission as indicators of impending fault movement, the results were inconclusive. Possible reasons for the inconclusive results obtained in these experiments include poor acoustic transmission characteristics of the rock or soil materials beneath the site chosen for the experiment; lack of acoustic emission due to insufficient stress build up at the time of testing; or, presence of creep zones which relieve stress continuously and prevent its building up to levels associated with acoustic emission. Extraneous noise such as that caused by car doors, aircraft, wind, blasting, internal combustion engines, construction activity, thermal expansion and contraction of rock, and insects and animals can also mask audible frequency microseismic activity. Probably the greatest single reason for inconclusive results has been that insufficient time and effort were spent in monitoring at a given location. The CDMG's present pilot study was designed, not with the idea of spending long periods monitoring particular stations, but with the idea of determining the factors that constitute a good acoustic listening station. Where should the station be located? What should be its physical relationship to the fault being monitored? What kind of geologic materials should be present at and near the site? What kinds of fault activity are most likely to create detectable sounds? What kinds of equipment should be used and in what configuration? What operational problems can be expected and how can they be minimized? The following discussion emphasizes the answers obtained so far and indicates which questions need further consideration.

58 **W. A. Baeslack III, G. Fayer IV, S. Ream, and C. E. Jackson. "Quality Control in Arc Stud Welding."** *Welding Journal,* **WEJUA, 54(11):789–798. Nov 1975.**

Investigations show that an applied magnetic field can counteract arc deflection in stud welding of C steels and that acoustic emission techniques promise to be effective in monitoring weld quality. Four factors affecting weld quality are discussed. Experiments to study required energy input, control of arc deflection, and nondestructive testing methods are described. Weld current and time are important but need not be controlled beyond manufacturer's recommendations. Heat transmission measurements and ultrasonic testing were not effective but measurements of the *arc noise* acoustic signal are closely related to weldment quality. Further study of this NDT method is recommended. [Abstract 22-0494. *Metals Abstracts,* Vol. 9, 1976].

59 **I. M. Bagay, M. M. Kondratenko, and Z. D. Kovalyuk. "Acoustic Wave Emission during Electrical Erosion."** *Applied Electrical Phenomena,* **AECPB, (1):26–28. 1974; translated from** *Elektonnaya Obrabotka Materialov,* **EOBMA, (1):26–28. 1974. (in Russian).**

Acoustic emission occurs during electrical erosion of Bi single-crystal specimens. The generators of individual acoustic pulses are parts of the crystal surface that experience the shock of the electron beam created by spark discharge. The interaction between the Rayleigh type waves that are radiated and defects in the crystal leads to changes in state of all the surfaces bounding the crystal and in the layers near the surface. A metallographic criterion is given for this change in dependence on the crystallographic orientation of the experimental planes. The influence is demonstrated of acoustic radiation of thin films of composition $Pb_{1-x}Sn_xTe$ on their electrical properties.

60 **C. D. Bailey. "Acoustic Emission for In-Flight Monitoring on Aircraft Structures." Lockheed-Georgia Company, Marietta, Georgia. 1974; paper presented at the American Society for Nondestructive Testing National Fall Conference, Detroit, Michigan. Oct 21–24, 1974;** *Materials Evaluation,* **MAEVA, 34(8):165–171. Aug 1976.**

A prototype acoustic system for detecting unstable crack growth in structure has been developed and flight tested on a C-5A aircraft. The system consists of transducers for monitoring 24 locations on the wing structure, a Signal Processing and Display Unit (SPADU) located on the flight deck, and a warning light on the pilot's annunciator panel. Design criteria were established by conducting fracture tests on large scale structural components. Signals generated by various size-crack extensions were measured to determine the detectability limits and signal attenuation in aircraft structure. Structural-borne noise measurements were made during flight to establish signal-to-noise criteria. . . a separate but related program is in progress. The objective is to develop an in-flight system capable of monitoring stable crack growth in inaccessible locations such as under fastener heads and at faying surfaces in laminated joints. Although the programs were conducted on C-5 aircraft structure, the results have direct application to all large aerospace vehicles.

61 **C. D. Bailey, J. M. Hamilton, Jr., and W. M. Pless. "Acoustic Emission of Impact Damaged Graphite-Epoxy Composites." Lockheed-Georgia Company, Marietta, Georgia. 1977; summary of paper in** *Paper Summaries: ASNT Spring Conference, March 28–30, 1977, Phoenix, Arizona.* **American Society for Nondestructive Testing, Columbus, Ohio. 1977.**

Acoustic emission source location techniques were used to monitor an impact damaged area during tensile tests of graphite-fiber reinforced epoxy specimens. The specimens had also been fatigue cycled; some had been impact damaged prior to fatigue cycles and some after the fatigue cycles. The specimens were 16 plies and approximately 0.088 inches (2.44 mm) thick. Eight plies were in the zero direction (applied stress direction) and eight plies were in the ±45° direction. In all zero-direction plies four 1-inch (25.4-mm) wide strips of graphite equally spaced 2 inches (50.8 mm) apart, were removed and replaced with Kevlar. The AE data generated in the impact damaged area are analyzed as a function of the fatigue cycling and the ultimate tensile strength. The results show that the Kaiser effect does not hold true for the impact damaged area and that AE techniques have excellent potential for locating and assessing the damage in large graphite-epoxy composite structural components.

62 **C. D. Bailey, J. M. Hamilton, and W. M. Pless. "AE Monitoring of Rapid Crack Growth in a Production-Size Wing Fatigue Test Article."** *NDT International,* **NDITD, 9(6):298–304. Dec 1976.**

Acoustic emission source location techniques were successfully used on a production-size aircraft wing fatigue test article to monitor crack growths in the range from 0.25 to 1.6 mm per load cycle. The AE data showed good correlation with the crack length data. Analyses of these correlations show that AE monitoring has the potential for being used to determine crack length over this range of crack growth rates. The fatigue test article was constructed of 7075-T6511 aluminum alloy. The test period lasted 14 days and the results demonstrated that AE has application to aircraft structures.

63 **C. D. Bailey and W. H. Lewis. "From Creaking Cracks to Breaking Beams-A Review of Acoustic Emission for Aircraft Structures." pp 118–128 in** *Materials on the Move*, **National SAMPE Technical Conference Series, Volume 6. Society for the Advancement of Material and Process Engineering, Azusa, California. 1974. confernece held in Dayton, Ohio. Oct 8–10, 1974.**

The paper describes three separate areas of activity currently being pursued at the Lockheed-Georgia Company in acoustic emission testing technology. An acoustic emission system known as the Flight Structural Monitoring System was developed and used to detect major cracking in full-size aircraft structural components during dynamic loading. This system is currently being flight tested on a C-5 Galaxy. A flaw locator was used to detect small crack extensions in structural specimens during fatigue testing. High precision measurements of structural-borne noise are currently being made on aircraft structure during flight to determine crack initiation and monitor crack growth.

64 **C. D. Bailey and W. M. Pless. "Acoustic Emission for Detecting Cracks in Aircraft Structure." Lockheed-Georgia Company, Marietta, Georgia. 1974; paper presented at the American Society for Metals Materials Testing and Quality Control Conference, Detroit, Michigan. Oct 21–24, 1974. [This paper is essentially the same as presented at the ASNT National Fall Conference on the same date. See Reference 60 for publication of the paper].**

A long-range program is in progress to develop an on-board aircraft instrumentation system capable of detecting and monitoring the growth of fatigue cracks. These cracks usually originate in locations which are inaccessible to interrogation by the usual ultrasonic and eddy current NDT techniques such as under fastener heads and at the faying surfaces of lap joints. Two significant milestones in this program have been achieved: (1) crack extensions have been detected and monitored during fatigue testing of complex structural specimens assembled with mechanical fasteners, and (2) structure-borne noise measurements have been made on an aircraft during flight which verifies that the constant level of the background noise is low enough to permit detection of the acoustic emission from fatigue crack extensions. These measurements were made at 9 locations which included both high-noise and low-noise areas, and indicate that it is feasible to develop an on-board acoustic emission system.

65 **C. D. Bailey and W. M. Pless. "Acoustic Emission – Monitoring Fatigue Cracks in Aircraft Structures." ISA Reprint 76239. Instrument Society of America, Pittsburgh, Pennsylvania. 1976; paper presented at the 22nd International Instrumentation Symposium, San Diego, California. May 25–27, 1976. [copy of reprint also available as Technical Report DE 76-3 from Dunegan/Endevco, San Juan Capistrano, California].**

Solutions have been found to the problem of detecting AE signals generated by microscopic size crack extensions in the presence of the inherent noise in aircraft structure. Cracks have been located and crack growth monitored in several sizes of aircraft structural test specimens including a full-size wing fatigue test article. This was accomplished in a shop environment without acoustic isolation. Special instrumentation was developed and measurements of the noise level over the frequency range from 0.1 to 1.0 MHz were made at 9 locations on an aircraft during flight. Results show that noise is not a limiting factor in using AE techniques on aircraft during flight. Requirements for an aircraft on-board AE monitoring system are discussed.

66 **C. B. Bailey and W. M. Pless. "Acoustic Emission Structure-Borne Noise Measurements on Aircraft during Flight." Lockheed-Georgia Company, Marietta, Georgia. 1975; paper presented at the American Society for Nondestructive Testing National Fall Conference, Atlanta, Georgia. Oct 13–16, 1975; *Materials Evaluation,* MAEVA, 34(9):189–195, 201. Sep 1976.**

Measurements were made at 9 locations to determine the frequency and level of the structure-borne noise in the airframe of a large jet aircraft during flight. The purpose of the measurements was to determine the feasibility from a noise limiting standpoint of using acoustic emission techniques to detect onset of fatigue damage. Measurements were made on four flights at both 10,000 and 20,000 feet of altitude. Test results show that structure-borne noise will not prevent the use of an acoustic emission system which operates above 500 kHz. The main landing gear area exhibited the highest noise level. The pylon attachment area near the front beams exhibited a moderate noise level. The center wing and the vertical stabilizer areas exhibited below moderate noise levels. The high-noise area may require AE systems which operate at or above 1.0 MHz to achieve a signal-to-noise ratio of 2.

67 **C. D. Bailey and W. M. Pless. "Acoustic Emissions used to Nondestructively Determine Crack Locations in Aircraft Structural Fatigue Specimens." Lockheed-Georgia Company, Marietta, Georgia. 1973; pp 224-232 in** ***Proceedings of the Ninth Symposium on Nondestructive Evaluation.*** **Southwest Research Institute, San Antonio, Texas. July 1973. symposium held in San Antonio, Texas. Apr 25-27, 1973.**

Acoustic emission techniques and sound-travel time determinations were used to locate cracks in a structural fatigue test specimen. The test section was a lap joint 1.5 inches wide by 18 inches long, and assembled with 34 bolts. Two transducers, one at each end of the test section, were used to detect the acoustic emissions from the crack extensions. The differences in the times required for acoustic emission pulses to reach the transducers were computed to determine the locations of cracks. The specimen was subjected to a spectrum of cyclic load conditions which simulated a total of 150,000 aircraft flight hours. The tests were conducted in a shop area and no special precautions were made to acoustically isolate the specimen from the loading fixture or from the noisy shop environment The specimen was monitored for acoustic emissions approximately 10% of the total fatigue test time. At periodic intervals, the fatigue tests were stopped and ultrasonic NDI inspections were conducted to locate any cracks that may have developed. At the end of the fatigue tests the bolts were removed and the holes were inspected for cracks using an eddy current bolt-hole probe. The specimen was then moved to the Metallurgical Laboratory and given a thorough investigation. The specimen was sectioned and the size and location of all cracks were determined with the aid of optical microscopes. A total of 15 crack locations indentified by acoustic emission test results were verified including two cracks on a metal surface inside the lap joint, one crack 0.005 by 0.008 inches in a bolt hole, and several cracks under fastener heads.

68 **C. D. Bailey and W. H. Sproat. "C-5A, Determination of Detectability Limits. Task H of Phase II: Flight Structural Monitoring System." Report No. LG74ER-0044. Lockheed-Georgia Company, Marietta, Georgia. Mar 26, 1974.**

Tests were conducted to establish the detectability limits of the Flight Structural Monitoring System (FSMS) for crack extensions occurring in a large aircraft structural component. Eight crack extensions which had fracture surface areas ranging from 0.11 to 3.0 square inches in a lower surface mid-panel of a C-130 center wing box were monitored. Instrumentation consisted of a spectrum analyzer-peak detector designed to detect and analyze the high frequency pulses generated by 8 FSMS transducers strategically located on the structure. The locations were selected to determine transducer response characteristics at various distances from the crack extensions. Sound attenuation in the structure ranged from 11 db per foot near the crack to 2 db per foot at 8 to 16 feet from the crack. Most of the peak voltage responses generated by the transducers located near the crack were greater than 6.7 volts. Peak voltage generated by transducers 8 feet away ranged from 100 millivolts to greater than 1.0 volt. Values at 16 feet ranged from 10 to 370 millivolts. Generally the voltage output response decreased with increasing frequency over the range from 35 to 140 kHz.

69 **E. C. Bailey and E. R. Reinhart. "The Influence of ASME Section XI and EEI–RP on In-Service Inspection." Paper C121/74, pp 176–183 in** ***Periodic Inspection of Pressurized Components.*** **Conference Publication 8, 1974. published by Mechanical Engineering Publications Limited for the Institution of Mechanical Engineers, London, England. 1975. conference held in London, England. June 4–6, 1974.**

Electric utility, equipment manufacturer, material supplier, and commercial research laboratory engineers have fostered the development of acoustic emission systems to signal the start or growth of cracks within pressure boundary materials. They have also supported improvement of ultrasonic inspection equipment and development of acoustic holography. The United States Atomic Energy Commission provided an unused reactor pressure vessel as a test facility. The commercially available acoustic emission systems, improved ultrasonic inspection equipment, and acoustic holography system are nondestructive examination tools which will reduce radiation exposure for manual inservice inspection of power reactor pressure boundary materials and increase safety assurance during operation.

70 **N. Bailey. "The Prospects of Crack Detection by Stress Wave Emission."** ***The Welding Institute Research Bulletin*****, 13(8):231–235. Aug 1972.**

Article assesses the possibilities of detecting cracks of different types, and other defects, in the light of preliminary tests carried out at The Welding Institute. Although the main area of interest is in weld quality monitoring, stress wave emission has other applications, some of which are briefly mentioned.

71 **D. W. Bainbridge and G. S. Baker. "Influence of Strain Rate on Acoustic Emission from Hardened D6aC during Incremental Crack Growth." Internal Memorandum. Aerojet–General Corporation, Sacramento, California. Nov 8, 1965.**

Incremental crack growth can be produced in hardened D6aC by bending specimens which contain an appropriate localized defect. This type of crack growth is accompanied by the emission of acoustic noise with appreciable amplitude. Bend tests were conducted on D6aC hardened 1550 °F, 15 minutes, air cooled, and tempered 400 °F, 60 minutes to determine the dependence of acoustic emission on the strain rate at 70 °F. The tests were monitored with an oscilloscope, taped, and for each of four strain rates the pulses were discriminated and counted. The discrimination was based upon pulse height and load increment. It was shown that the acoustic pulse rate was essentially proportional to the strain rate for a given pulse size and for a given load over two decades of pulse and strain rates. Further, there were no obvious singularities in pulse distribution in the one second preceding fracture. Subject to the range limitation of these experiments, the acoustic testing of pressure vessels should be conducted at a low strain rate. This procedure will yield information equal to that of the fast test and provides more time for *in situ* analysis.

72 **B. C. Baird. "Incipient Failure Detection in Antifriction Bearings." The Boeing Company, Houston, Texas. 1972; paper presented at the 27th Annual Petroleum Mechanical Engineering and Pressure Vessels and Piping Conference, New Orleans, Louisiana. Sep 17-21, 1972.**

Antifriction bearing failures in a typical production facility are generally detected either shortly before a catastrophic failure is to occur or right after the failure occurs. The normal method used by plant personnel to detect an antifriction bearing failure before it occurs is simply by listening to the bearing's sounds. In order to avoid catastrophic failure and to be able to predict a bearing failure well before it occurs, so that predictive maintenance can be performed, an instrumentation technique more sensitive than the human ear to incipient bearing failures is required. The paper examines several such techniques which have been utilized for performing this task. Among these techniques, the use of high frequency acoustic energy detection has been found to provide the earliest detection of incipient failures in antifriction bearings. The use of high frequency acoustic detection in a typical petrochemical plant is described.

73 **G. S. Baker. "Acoustic Emission and Prefracture Processes in High-Strength Steels." AFML-TR-67-266. Aerojet-General Corporation, Sacramento, California. Mar 1968. [NTIS: AD-830-858].**

A program was conducted to investigate the practical application of stress-wave monitoring techniques to the detection of crack initiation or growth (or both) by determining the correlation between stress-wave emission and prefracture mechanical processes occurring in high-strength steels under stress as affected by material properties and test conditions. The D6aC steel in two heat-treatment conditions (600 and 1075 °F temper) and Grades 200 and 300, 18 percent nickel maraging steel, were investigated using standard accelerometer techniques at a sensitivity level of approximately 0.01 g. It was found that stress-wave activity at this level was a positive indication of nucleation of cracks of greater than 10^{-5} square inches or the growth of existing cracks by a similar area. The effects of varying test temperature from −100 to 250 °F and varying strain rate from $\dot{\epsilon} = 0.01$ to $\dot{\epsilon} = 0.0001$ min^{-1} were investigated. The stress-wave activity was found to be little affected by test temperature for temperatures above the material brittle-ductile transition region and to vary roughly proportionally with strain rate. For test temperatures in the ductile-brittle transition region, the stress-wave activity decreases with temperature. The application of stress-wave monitoring techniques to the monitoring of the hydrotest of solid rocket chambers is discussed. It is concluded that the techniques provide a means of nondestructively detecting and locating propagating flaws in structures under stress, allowing test or service termination prior to catastrophic failure.

74 **G. S. Baker and D. W. Bainbridge. "Metals Physics Program." Final Report on Work Order 8708–46. Period covered: December 1, 1964 through November 28, 1965. Aerojet–General Corporation, Sacramento Plant, Sacramento, California. 1965.**

The sources of stress-wave emissions (> 0.1 g) in 4340 and D6aC steels were investigated to establish the relationships between signals recorded and fracture processes occurring in structures tested using the Aerojet-General Corporation Stress-Wave Detection System. The results indicate that at the 0.1 g level of sensitivity of stress-wave emissions recorded are predominately due to subcritical crack growth leading to rapid failure with section stress significantly below the yield stress. If the section stress exceeds the yield stress, thus causing extensive plastic deformation, cracks apparently can grow to critical size without the emission of a significant number of stress-wave pulses. Cracking of inclusions or inclusion-metal boundary separation and plastic deformation processes do not result in detectable (0.1 g) stress-wave emissions. It was shown in 4340 steel that the stress-wave emission reached a peak after intermediate tempering temperatures. For tempering temperatures below 300 °F and above 800 °F, the stress-wave emission was too sparse or irratic to use as a reliable precursor to failure. In tests on D6aC, it was shown that the stress-wave emission was primarily dependent on strain and mostly independent of time, the total number of pulses to failure being insensitive to an order of magnitude change in strain rate. The feasibility of using stress-wave detection techniques in studies of stress-corrosion cracking was demonstrated for the 4340 steel.

75 **G. S. Baker and D. W. Bainbridge. "Sources of Prefracture Acoustic Emissions." Internal Memorandum. Aerojet–General Corporation, Sacramento, California. Oct 25, 1965.**

Catastrophic failure of a structure will be preceded by crack nucleation and growth to critical size, or, if a crack nucleus is present, by subcritical crack growth alone. Each of these steps will be preceded and accompanied by plastic deformation, fracturing of inclusions, and separations at inclusion-metal interfaces, either wide-spread or on a highly localized level. These processes will all give off characteristic stress-wave emissions signaling the imminence of complete fracture. Whether or not these emissions are detected, of course, depends on the sensitivity of the monitoring devices. Aerojet has chosen a sensitivity level for its Stress-Wave Detection System which is too low to detect plastic deformation and inclusion noises, but high enough to detect the stress-wave emission from subcritical cracks growing in an elastically stressed structure. These of course are the cracks which result in failure of a structure at less than the design stress and thus are of primary importance.

76 **G. S. Baker and A. T. Green. "Stress Wave and Fracture of High-Strength Metals." Aerojet-General Corporation, Sacramento, California. 1967; Paper 1-2, pp 30-64 in "Proceedings of the Symposium on Correlation of Material Characteristics with Systems Performance." P4/US/1/68. Volume 1. 1968. symposium held at the U.S. Air Force Conference Facility, Orlando AFB, Florida. May 10-12, 1967. [proceedings are no longer available; however, the work reported was performed under USAF Contract AF33(615)5027 and reported by G. S. Baker in Report AFML-TR-67-266 (see Reference 73)].**

Crack growth and fracture in a high-strength metal are accompanied by the emission of stress waves (acoustic emission). These stress waves can be detected by appropriate transducers and used to detect and locate growing subcritical flaws before they result in failure of a structure. The results of the application of acoustic monitoring techniques to studies of fracture in laboratory-size samples and to subscale and full-scale rocket motor cases will be presented. The sources of stress-wave emissions from samples undergoing plastic deformation and subcritical crack growth have been identified. The level and character of stress-wave emissions to be expected from such sources will be discussed in terms of material properties. The results of tests on subscale and full-scale cases with both natural and artifically induced flaws are given. These tests include chambers which both passed and failed under hydrotest including 260-inch diameter large rocket chambers fabricated from 18-percent nickel maraging steel.

77 **H. L. Balderston. "Incipient Failure Detection: A System Concept." pp 247-255 in** *Incipient Failure Diagnosis for Assuring Safety and Availability of Nuclear Power Plants.* **E. P. Epler and P. P. Roux, Coordinators. U.S. Atomic Energy Commission, Division of Technical Information, Washington, D.C. Jan 1968. [NTIS: CONF-671011]. proceedings of conference sponsored by Oak Ridge National Laboratory, held in Gatlinburg, Tennessee. Oct 30-Nov 1, 1967.**

The detection of incipient failure of a part, component, subsystem or system is related to the detection of latent defects which already exist in a device, develop during operation of the device, or appear as a result of operational environment, or human error. Defects may also appear as a result of unexpected stress or other unexpected operational hazards. Incipient failure detection is directed to the most basic causes of failure, that is, structural defects or chemical contaminates. Further, these two basic causes of failure apply to structural assemblies, mechanical assemblies, electrical and electronic assemblies. Specifically, the detection techniques may include the use of vibration, acoustical energy, electrical noise, infrared, ultraviolet, X-ray, and gamma radiations. This paper indicates how existing knowledge and new research can be combined to develop an incipient failure detection subsystem with broad applicability to aerospace vehicles and ground-equipment installations. On December 1, 1965, the BOMITE (Boeing Microelectronic Integrated Test Equipment) concept developed by the author had been completed. BOMITE was an integrated test equipment concept designed to be built into, and made a part of, operational equipment and to give that equipment a self-test-for-failure capability. At various times during the development of the concept, it was observed that tests of impending failure would have been even more valuable. In fact, a file was made covering techniques which might be usable and work done by others in nondestructive testing, anticipatory failure detection, physics of failure, and other areas wherein usable techniques had been or were being developed. In electronics, the detection of incipient failure appeared to be limited to discrete parts and was limited to techniques usable in sorting *mavericks* or *deviates* from *lots* on large numbers of similar parts. Information was accumulated on sorting techniques applicable to resistors, capacitors, diodes, transistors, field effect transistors, and selenium rectifiers. Electrical noise appeared to be the tool most widely employed in the above techniques. In structures, vibration, ultrasonic, acoustic emission, X-ray, and infrared radiations appeared to be most often employed; in mechanical equipments sonics and ultrasonics were the most commonly used tools. It appeared that *noise* in its widest definition including vibration, acoustical, electrical, infrared, ultraviolet, X-ray, and gamma radiations would provide the techniques necessary. It was planned to utilize existing techniques wherein possible, modify existing techniques wherein necessary, and develop new techniques to fill in the gaps to enable the development of an incipient failure detection subsystem.

78 **H. L. Balderston. "Incipient Failure Detection: The Detection of Certain Contaminating Processes." Boeing Company, Seattle, Washington. 1967; paper presented at the Sixth Reliability and Maintainability Conference, Cocoa Beach, Florida. July 17–19, 1967.**

Three separate and distinct electrolytic and one galvanic process were identified by visual inspection, metallographic, electron microprobe, and X-ray diffraction analysis in a clocked, flip-flop integrated circuit flat pack or the associated printed circuit test jig (or both) (two on flat pack and two on circuit board). These four processes were all found to be detectable by the use of noise measurements in microvolts per root cycle at 1000 Hz. The direct current applied for noise measurement to the integrated circuit devices was 100 microamperes, as compared to the 6 to 8 milliamperes required for normal operation. After initial experimentation, the devices were caused to fail in a laboratory ambient environment, followed by an acceleration of the rate of electrolytic reaction through the use of essentially 100 percent relative humidity, versus the upper specification limit of 80 to 98 percent relative humidity. In all cases, failures of the devices were preceded by a rising characteristic of high noise readings, 10 to 100 times normal (0.017 to 0.028 microvolts = normal), which allowed predictions of failure to be made. The correlation between measured noise level and extent of contamination was calculated and was found to be high (0.631). It was found that linear and polynominal regression equations could be developed to enable calculation of the degree of contamination indicated by measured noise readings. While this paper reports on the detection of contaminating processes in integrated circuits, the ultimate objective of this research effort was to develop an incipient failure detection subsystem which can be applied to aerospace vehicles and ground equipment. Hopefully, the objective will be attained through monitoring certain characteristics present and associated with incipient failure in operational equipment. A highly significant by-product of the research effort was the identification of at least four of the contaminating processes, the principal pre-existing and process-produced contaminants, and the conditions which appear to accelerate the processes. Further work is being done to develop techniques for the control or elimination (or both) of these contaminants and associated destructive processes. The research clearly indicated the necessity for exercising rigid clean-room controls to prevent contamination during the assembly of integrated circuits into subassemblies. While the processes proved to be extremely slow under ambient conditions, the processes were extremely fast where condensation of water vapor occurs (essentially 100 percent relative humidity). However, even in the first case, it is postulated that failures could occur as soon as six months. With actual mositure present between the leads or with the accumulation of water, as under essentially 100 percent relative humidity, the first failures could occur in eight hours with the bulk of failures falling in a 40 to 50 hour period and some not occurring for 150 hours or more. The use of certain nonhydrogenous conformal coatings appears to eliminate or control at least three of these destructive processes. Other currently employed conformal coatings appear to have little or no effect upon the processes reported herein. This matter is being studied further and will be reported in a later paper. These destructive contaminating processes may well be a greater source of integrated circuit failure than any other currently known source.

79 **H. L. Balderston. "Incipient Failure Detection: The Detection of Incipient Failure in Certain Hydraulic Components." Boeing Company, Seattle, Washington. 1969; paper presented at the Eighth Reliability and Maintainability Conference, Denver, Colorado. July 1969.**

80 **H. L. Balderston. "On-Board Incipient Failure Detection Sub-System." Boeing Company, Seattle, Washington. 1969; paper presented at the Air Transport Association Subcommittee Meeting on Nondestructive Testing, Minneapolis, Minnesota. Sep 1–Oct 2, 1969.**

81 **H. L. Balderston. "Report on Initial Research on Incipient Failure Detection in Certain Bearings for On-Condition Maintenance, Phase II." Report 2-5022-255. Boeing Company, Seattle, Washington. Dec 19, 1967.**

The purpose of this work is to determine if sonic or ultrasonic detection (or both) techniques can be developed to provide information with respect to incipient failure detection in ball bearings.

82 **H. L. Balderston. "Report on Initial Research on Incipient Failure Detection in Certain Hydraulic Components for On-Condition Maintenance, Phase II." Report 2-5022-256. Boeing Company, Seattle, Washington. Jan 2, 1968.**

The purpose of this work is to determine if sonic or ultrasonic detection (or both) techniques can be used to provide quantitative hydraulic fluid-leakage flow information.

83 **H. L. Balderston. "The Broad Range Detection of Incipient Failure Using the Acoustic Emission Phenomena." pp 297–317 in** ***Acoustic Emission, ASTM STP 505.*** **American Society for Testing and Materials, Philadelphia, Pennsylvania. 1972. proceedings of symposium held in Bal Harbour, Florida. Dec 7–8, 1971.**

The broad range detection of incipient failure in structural, mechanical, hydraulic, pneumatic, electrical, and electronic systems is based upon the detection of the two basic causes of failure. These causes are structural degradation and chemical contamination occurring in the structure of the material of which the systems listed above are composed. The detection of incipient failure allows maintenance to be programmed at the convenience of the user, thus preventing downtime of equipment during use periods. The acoustic emission phenomena are employed in the detection of incipient failure in all types of equipment covered in this report. Specifically, the examples chosen are structural, mechanical, pneumatic and hydraulic components, subassemblies, or specimens as applicable. From these examples, the broad range incipient failure detection concept based on the acoustic emission phenomena is developed. The experiments included detection of incipient failure in bearings, hydraulic assemblies, pneumatic subsystems, stress corrosion in aluminum, and ductile tensile failures in carbon steel, 4330M steel, and in boron-epoxy composite material.

84 **H. L. Balderston. "The Detection of Incipient Failure in Bearings."** ***Materials Evaluation,*** **MAEVA, 27(6):121–128. June 1969.**

The detection of incipient failure, in general, is related to the detection of the basic causes of failure. In this paper, defects which lead to failure are created or simulated (or both) and are detected, utilizing resonant frequency techniques falling in the ultrasonic frequency range. Rotational frequencies of the bearing are calculated and identified; resonant frequencies are calculated and identified. Correlation is shown between root mean square amplitude at these frequencies and the severity of the defects incorporated. Frequency translation techniques are employed to bring within the range of hearing, the ultrasonic resonant frequencies used to detect defects. A tape recording has been made of translated frequency data which enables the listener to differentiate between the various defects which cause failures within the bearings.

85 **H. L. Balderston and R. E. Hine. "Incipient Failure Detection: The Detection of Incipient Failure in Structural Material Samples." Boeing Company, Seattle Washington. 1970; paper presented at the Air Transport Association Subcomcittee Meeting on Nondestructive Testing, San Antonio, Texas. Sep 1970.**

Incipient failure detection, in general, is related to the detection of the occurrence or presence of two basic causes of failure; chemical contamination or structural degradation (or both). These basic causes apply not only to static structures but also to dynamic mechanisms, hydraulic components, as well as electrical and electronic systems. Accordingly, the detection of incipient failure is related to the detection of chemical deterioration and structural degradation in any or all of the above mentioned sub-systems. The paper deals with the detection of acoustic emission from structural material samples which can be related to the early detection of stress corrosion and ductile deformation in tensile elements. In the case of stress accelerated processes, the attack upon the structural materials by the corrosive agent results in high level acoustic emission which is easily detectable. Following the initial attack, the structural assembly develops a protective coating of contamination products which slows the reaction considerably. Subsequent increases in acoustic emission are indicative of the exposure of new material to the corrosive agent when cracking, scoring, or flaking occurs. Acoustic emission is also emitted with the extension or enlargement (or both) of surface cracks formed initially. The continuation of surface cracking, the extension, elongation, and deepening continues under stress accelerated corrosion with the release of large bursts of acoustic emission. This emission can be readily detected with proper transducers and equipment. There appears to be a relationship of count rate per unit time with the extent of crack nucleation and propagation occurring in real time. Prior to failure of the sample, an increase in acoustic emission rate appears. The initial attack by chemical contamination can be detected, as well as the progression of damage due to stress corrosion as the process continues. The source of the acoustic emission phenomena is explained in the paper and is accepted by most authorities in the field today. The relationship of acoustic emission in fracture phenomena is dealt with briefly. The relationship of acoustic emission to ductile tensile failures is considered and the results of certain experiments with structural elements under tensile stress are tabulated. Previous work which has been published and referenced in this paper covers the Detection of Incipient Failure in Bearings, the Detection of Incipient Failure in Hydraulic Systems, the Detection of Incipient Failure due to Certain Contaminating Processes- -in electronic circuitry- -, and An On-board Incipient Failure Detection Subsystem.

86 **D. W. Ballard. "Nondestructive Testing's Contributions to Nation's Materials Needs." SC-M-720686. Sandia Laboratories, Albuquerque, New Mexico. Oct 1972.**

Nondestructive test methods have an important role in the conservation of critical materials, equipment, and capital resources. Unfortunately, this role has been largely limited to the high-technology, mission-oriented products and has not been widely recognized in the consumer products and mass-production industries. The paper is a compilation of technical inputs on NDT uses that have demonstrated their contributions to the nation's materials needs throughout the design, fabrication, and application phases of a product's life cycle. It was prepared as background information for the National Commission on Materials Policy (NCMP), a Congressional Committee which is studying materials problems and policies in the U.S. to the year 2000. Similar technical inputs are being supplied to NCMP by representatives of nineteen materials societies through the new Federation of Materials Societies (FMS).

87 **V. M. Baranov, V. A. Luk'yanenko, Yu. V. Miloserdin, and K. I. Molodtsov. "Apparatus for Recording Acoustic Signals from Cracks Initiated in Brittle Materials."** ***Strength of Materials,*** **SMTLB, 4(4):505-508. Apr 1972; translated from** ***Problemy Prochnosti,*** **PPCNB, 4(4):119-122. Apr 1972. (in Russian).**

In the practice of laboratory research, a method is known for testing brittle materials for thermal strength by immersing the heated specimen in a cold liquid (water) and recording the elastic waves initiated during cracking of the specimen with the use of a piezoelectric transducer (pickup). The main drawback of this method is in the development of an intense noise, connected with the boiling of the liquid, confusing the signal produced by cracking. The situation is further complicated by the fact that the sound generated during the boiling of the liquid has a wide frequency range from 20-30 Hz to 40-50 kHz and above. Particularly intense acoustic radiation is observed during the immersion of strongly heated specimens. It is particularly difficult to record the acoustic signals connected with crack initiation, since the cracks appear in the shape of a fine network and generate sound which is weaker than the noise of the boiling liquid. In the apparatus described, the essential improvement in the useful signal-to-noise ratio is achieved by reducing the low-frequency component of the spectrum of signals picked up by the piezoelectric transducer. Moreover, as shown by preliminary studies, the amplitude of the signal pulses produced by cracking of the specimen is reduced less than the amplitude of the signals from boiling.

88 **F. Bardenheuer, H. vom Ende, P. G. Oberhauser, and E. E. Hofmann. "Schallintensitatsmessung zur Uberwachung der Schlackenbildung und die Reaktionsablaufe beim Aufblasen von reinem Sauerstoff auf phosphorarmes Roheisen (Measurement of the Intensity of Noise for Controlling Slag Formation and Reaction Processes when Top-Blowing Pure Oxygen on Pig Iron Poor in Phosphorus)."** ***Stahl und Eisen,*** **STEIA, 88(12):613–621. June 13, 1968. (in German).**

Slag formation and changes in slag composition occurring in oxygen top-blown converter practice were studied in a 40-ton converter and with low phosphorus iron by continuously recording the sound of the blast. Since oxygen supply, final phosphorus content, and the period of forming are related, the latter is shown by a change in the sonic curve. Forming can be advanced by increasing the lance distance. Slag conditions can thus be controlled as well as basicity and rate of oxidation during the blow. Metal-slag reactions are accelerated when 70 to 80 percent of the blow time has elasped, at which point dicalcium silicate becomes liquid releasing lime. [Abstract 60862s reprinted by permission of the American Chemical Society from *Chemical Abstracts,* Vol. 69, 1968].

89 **R. A. Bareiss and B. H. Schofield. "Acoustic Emission under Applied Stress." Contract AF-33(616)-5640. Progress Report No. 1. Lessells and Associates, Inc., Boston, Massachusetts. July 15, 1958.**

The work concerns the study of the sounds generated in materials as a result of stress applied to the material. The purpose of the study is to determine the source and originating mechanism of the acoustic signals emitted. The report covers the period of April 1, 1958 to July 15, 1958. See Reference 1575 for final report on this contract.

90 **S. Barkhoudarian and C. C. Scott. "Preliminary Study of Feasibility of Acoustic Detection of Small Sodium-Water Reactions in LMFBR Steam Generators." APDA-256. Atomic Power Development Association, Inc., Detroit, Michigan. May 1970. [NTIS: N71-10403].**

An evaluation of acoustic techniques for the detection of sound resulting from water leaks into liquid Na is conducted. Acoustic spectra over the range of 100 Hz to 80 kHz are detected. A leak as small as 0.00008 lb/sec is detected by the acoustic instrumentation. Since this is the smallest leak rate used in the tests, it is not to be interpreted as the smallest leak rate that might be detected by acoustic methods. Detection time is found to be essentially instantaneous.

91 **J. T. Barnby and T. Parry. "Acoustic Emission from a Notched Glass-Fibre-Reinforced Polymer in Bending."** ***Journal of Physics D: Applied Physics,*** **JPAPB, 9(13):1919-1926. Sep 11, 1976.**

Acoustic emission from notched three-point bend specimens of a unidirectionally reinforced and a crossply reinforced polymer has been measured. The unidirectional material gave no acoustic emission prior to catastrophic fracture whereas crossply material showed a gradual accumulation of acoustic emission, proportional to the square root of the area of a damage zone at the notch tip, preceding fracture. The acoustic emission correlated with the stress intensity factor of the sharpened notch although the correlation was particular to the angle of the fibres to the notch.

92 **B. Barrachin, N. Chretien, A. Prot, R. Roche, R. Saglio, and E. Tomachevski. "Programme d'Etudes Engage par le Commissariat a l'Energie Atomique dans le Domaine de l'Application des Controles Non Destructifs a la Surete des Reacteurs (Program of Studies Begun by the Commissariat a l'Energie Atomique in the Domain of the Application of some Nondestructive Controls to the Safety of Reactors)." Commissariat a l'Energie Atomique, Saclay, France. Dec 5, 1972. (in French); paper presented at Symposium on the Principles and Standards of Reactor Safety. Julich, Germany. Feb 5-9, 1973. [NTIS: CEA-CONF-2222].**

The use of ultrasonics for improvement of testing procedures was studied with particular emphasis on the development of methods particularly suited to the testing of thick welds. Some examples were given which show the efficiency of this technique in the case of inspection controls in the presence of stainless steel coatings. Developments relative to acoustical emission included the laboratory study of the acoustical emission of various reactor vessel steels; study of the modifications caused by irradiation; development of an acoustical emission detection chain and processing of the data in order to locate the emission sources; and study of transducers associated with this chain. Other studies briefly mentioned were: acoustical holography (a method of examination which should remove many of the uncertainties in the characterization of faults detected by ultrasonics) ringing method which should complement the detection of serious defects, and improvement of the radiographic image qualities to extend the range of radiography and obviate some inhibitory inconveniencies.

93 C. S. Barrett. "A Low Temperature Transformation in Lithium." *Physical Review,* **PHRVA, 72(3):245. Aug 1, 1947.**

In a brief report under *Letters to the Editor,* the author reports that "the transformation (of lithium) to the face-centered cubic form is accompanied by a series of audible clicks, as in the twinning of tin or magnesium and the formation of martensite." He concluded that the transformation occurred by abrupt shear movements in small isolated regions. A more detailed report of this work was published by C. S. Barrett and O. R. Trautz (1948) (see Reference 94).

94 C. S. Barrett and O. R. Trautz. "Low Temperature Transformations in Lithium and Lithium-Magnesium Alloys." *Transactions of the American Institute of Mining and Metallurgical Engineers,* **TAIMA, 175:579–605. 1948.**

The authors report that during transformation on cooling of annealed lithium and lithium-magnesium alloys, the transformation process proceeds by discrete steps which are accompanied by audible clicks. Strain-induced transformation to face-centered cubic is accompanied by audible clicks as well as the reversion to body-centered cubic on heating when the heating is sufficiently rapid. The clicks can be heard easily with the unaided ear, however, the authors used a phonograph pickup attached to the rod or to a wire embedded in the specimen.

95 K. Barron. "Detection of Fracture Initiation in Rock Specimens by the Use of a Simple Ultrasonic Listening Device." *International Journal of Rock Mechanics and Mining Sciences,* **IJRMA, 8:55–59. 1971.**

A commercially available device (Hewlett-Packard Ultrasonic Translator, Type 4905A) for detecting ultrasonic vibrations in the frequency band 36 to 44 kHz has been used in a study of the microseismic activity in laboratory rock specimens under triaxial load conditions. Results indicate that the apparatus provides a useful alternative to the volumetric strain technique for detecting fracture initiation in rock specimens. [reprinted by permission of Pergamon Press Ltd from *International Journal of Rock Mechanics and Mining Sciences*].

96 P. M. Bartle. "Research Study of Stress Wave Emission in Weld Evaluation and Monitoring." Report Misc. 31/1/74 (3329/13/73). The Welding Institute, Abington, Cambridge, England. Feb 1974. [available only to Research Members of The Welding Institute].

Emission assessment has been studied to ensure that techniques are available which give correlation between the assessed value and the magnitude of the event producing the emission; event and ring-counting techniques proved unacceptable but amplitude assessment systems were satisfactory. Useful results were obtained with an amplitude sorting system, quantitative data are required; steps were taken to produce a more acceptable system. Significant emissions cover an amplitude range of 10^6; this affects assessment requirements and means a change is required in the basic approach to weighting emission sources when using location techniques. Application studies resulted in a degree of quantitative correlation between crack area or fatigue behavior of the assessed value of the emission detected.

97 **P. M. Bartle. "Some Studies of Stress Wave Emission Proof Load Monitoring of Aluminum-Steel Diffusion Bonded Joints (Summary)." Report Misc. 39/11/75 (7233/2A/75, 7225/2A/75). The Welding Institute, Abington, Cambridge, England. Nov 1975. [available only to Research Members of The Welding Institute].**

98 **P. M Bartle. "Stress Wave Emission: A Cricital Review of Literature to February 1971." Report C329/1/71. The Welding Institute, Abington Hall, Cambridge, England. Apr 1971.**

Stress wave emission is one of the three forms of energy release which occurs as a result of deformation or rupture. This review specifically relates to metals although the techniques can be applied to nonmetals. It has been known for many years that stress wave emission can occur during deformation and rupture of materials. It is only in the last two decades, however, that this subject has received sufficiently detailed study to allow practical use to be made of the emissions detected. The literature which covers many aspects, both theoretical and practical, shows that much has been deduced regarding the sources of emissions. The potential areas of application are manifold, and include the detection of the formation of defects during welding, the proof testing of structures, the in-service inspection of structures, and the use of stress wave emission as a tool for material assessment, e.g., fatigue, stress corrosion, phase transformation, and residual stress. The literature can be regarded as showing that: (1) Stress wave emission is a broad topic involving many scientific disciplines, (2) Frequently the detailed situation is technically complex and ill understood, and (3) Results have been over generalized and therefore conflict with each other. In view of these, the author in examining the essential features of stress wave emission has tried to apply an element of logic to the results. The review makes direct reference to some 40 odd papers and articles, but others exist and a separate bibliography is to be issued covering all known references to stress wave emission.

99 **P. M. Bartle. "Stress Wave Emission: An Introduction."** ***The Welding Institute Research Bulletin*****, 13(7):203–208. July 1972.**

Stress wave emission or acoustic emission is one of three forms of energy release that can occur when materials are deformed or ruptured. The classic example of stress wave emission is *tin cry*, which is audible to the unaided ear, and which occurs when pure tin is deformed. Techniques are being developed to allow stress wave emissions to be used to detect and locate the formation or extension of defects, cracks or areas of stress concentration during destructive testing, and for weld quality monitoring. The present position is that various equipments are available, either custom built or as standard commercial units, with which emissions can be detected and, possibly, their source located. Interpretation of results obtained tends to be rather speculative, however, except possibly in the most carefully thought-out experiments. This article is intended to introduce readers to the basic concepts of stress wave emission techniques.

100 **P. M. Bartle. "Stress Wave Emission Assessment–Three Generations of Equipment Development."** *The Welding Institute Research Bulletin,* **16(9):255-261. Sep 1975. [available only to Research Members of The Welding Institute].**

The premise that stress-wave emission monitoring will be of little real use until the emission detected can be quantitatively assessed is not universally true, but does usually apply. Consequently, equipment must be developed which allows truly quantitative assessment. The Welding Institute has already shown that the values obtained by conventional emission assessment systems can be meaningless, and has progressed significantly with the development of techniques which can give quantitative correlation between the assessed values of the emission detected and the significance of the events producing the emission (irrespective of the emission amplitude ranges and distributions encountered). These advances are traced through three generations of emission assessment equipment developed over the last few years.

101 **P. M. Bartle. "Stress Wave Emission: Defining Its Capabilities."** *Metals and Materials,* **MEMTA, pp 35-37. Mar 1976.**

Article reviews the progress made to date in acoustic emission as an NDT technique and highlights the importance of the technique as an aid in locating defect areas in large structures.

102 **P. M. Bartle. "Stress Wave Emission Monitoring: Assessment, Correlation and Calibration." Paper C79/74, pp 37-45 in** *Periodic Inspection of Pressurized Components.* **Conference Publication 8, 1974. published by Mechanical Engineering Publications Limited for the Institution of Mechanical Engineers, London, England. 1975. conference held in London, England. June 4-6, 1974.**

The potential of stress wave emission monitoring for allowing the determination or prediction of behavior or properties of materials, joints, or structures depends upon the proving of equipment and techniques. Work over the past threee years has allowed requirements in this respect to be more clearly defined. The factors found to define system requirements are outlined, and early equipment is reviewed in the light of these needs. An essential requirement is that a high degree of correlation between the assessed value of the emission detected and the significance of the events producing the emission be established. Developments towards meeting system requirements are described and results indicating that the necessary correlation can be obtained are discussed. The basic needs for standardization of assessment techniques and calibration procedures are emphasized and other outstanding problem areas are highlighted.

103 **P. M. Bartle and P. V. C. Watkins. "Stress Wave Emission: A Review of Developments at The Welding Institute to January 1973." Report P/64/73 (7081/3/73). The Welding Institute, Abington, Cambridge, England. Dec 1973. [available only to Research Members of the Welding Institute].**

Sections include: General review of developments, equipments and procedures; emission assessment; amplitude monitoring; and acceptance and standardisation. Areas of application of emission monitoring are tabulated with details of the form of test monitored, results obtained, difficulties encountered, and future potential.

104 **J. Baruch, S. Yaron, and S. Golan. "Detection of Cracking during Welding by Acoustic Emission/La detection des fissures lors du soudage par emission acoustique." Israel Institute of Metals, Haifa, Israel; Paper 3K/13 in** ***Proceedings of the Eighth World Conference on Nondestructive Testing.*** **volume containing Sections 3K2, 3H2, 5A. International Committee for Nondestructive Testing, Secretariat du Cofrend, Institut de Soudure, 32 Boulevard de la Chapelle, 75880 Paris Cedex 18, France. 1976. conference held in Cannes, France. Sep 6–10, 1976. (paper in English).**

An extremely sensitive acoustic emission technique is described which can detect hot cracks occurring during welding.

105 **D. Basile, G. Schweitzer, and G. Volta. "Experiences pilotes pour l'evaluation en cours de fonctionnement de la fiabilite de composants et systemes mecaniques et electromecaniques (Pilot Experiments for Assessing the Reliability of Mechanical and Electromechanical Components and Systems during Operation)." Paper IAEA-SM-127/44, pp 215–233 in "Proceedings of the Symposium on the Performance of Nuclear Power Reactor Components." International Atomic Energy Agency, Vienna, Austria. Apr 1970. symposium held in Prague, Czechoslovak. Nov 9–13, 1969. (paper in French).**

The paper gives an overall description of the techniques and procedures implemented at the EURATOM Joint Research Centre at Ispra to develop an effective system for quantitative evaluation of the reliability of a reactor primary circuit. This work particularly concerns the primary circuits of pressure-tube reactors moderated with heavy water. The problems discussed are: identification of the components so as to achieve a uniform statistical basis; a description of the conditions of use; the collection and working up of data during operation; and preventive detection of faults.

106 **B. R. Bass. "Absorption of Low Ultrasonic Sound in Three Liquids by a Pulse Technique." M.S. Thesis. Tulane University, New Orleans, Louisiana. 1966; published in "Final Report on Acoustic Emission Research at Tulane University under UCRL P.O. 2412203. Part 2." C. A. Tatro and B. R. Bass. UCRL-13231, Part 2. Lawrence Radiation Laboratory, Livermore, California. June 1966. [see Reference 1738 for abstract].**

107 **M. N. Bassim and D. R. Hay. "Acoustic Emission during Fracture of High Voltage Cables." summary of paper in** *Paper Summaries: ASNT Spring Conference, March 28–30, 1977, Phoenix, Arizona.* **American Society for Nondestructive Testing, Columbus, Ohio. 1977.**

Acoustic emission was used to study the plastic deformation and fracture of high-voltage electrical cables. These cables consist of a set of individual aluminum wires surrounding a steel core. Their deformation behavior is characterized by a small plastic deformation followed by breakage of the individual aluminum wires well before significant deformation of the steel takes place. Single channel acoustic emission equipment, as well as a three-channel system for flaw location was used simultaneously for monitoring several cables which were either new or used. Despite the high attenuation of acoustic waves and the complex structure of these cables, flaw location was possible and acoustic emission was correlated to whether the cables were new or used and to the presence or absence of artificially induced flaws.

108 **M. N. Bassim and D. R. Hay. "Acoustic Emission Monitoring of a Linepipe Segment during Burst Testing." Failure Control International, Lachute, Quebec, Canada. 1976.**

Acoustic emission was used to monitor a pipe segment submitted to a hydrostatic burst test. A multichannel system with three active channels and a minicomputer for data analysis were used in this investigation. The results obtained show two areas of acoustic activity, one of them is believed to be from extraneous mechanical noise while the other, which develops near the yield point, is due to plastic deformation. Subsequent examination of the burst pipe has shown that failure occurred at the location determined by acoustic emission.

109 **M. N. Bassim and D. R. Hay. " 'Post-Mortem' Detection Avoided by New Acoustic Emission System."** *Canadian Consulting Engineer*, **CCEGB, 15(9):46. Sep 1973.**

110 **M. N. Bassim, D. R. Hay, and J. Lanteigne. "Effect of Crack Size on Acoustic Emission during Mechanical Testing." pp 1636–1640 in** *Proceedings of the Second International Conference on Mechanical Behavior of Materials.* **American**

Society for Metals, Metals Park, Ohio. 1976. conference held in Boston, Massachusetts. Aug 16-20, 1976.

Acoustic emission is a technique currently used to determine incipient failure of structures and components. The source of such emissions in the elastic energy released in the material during crack initiation and propagation and during plastic deformation in the crack tip zone in more ductile materials. In flaw-bearing materials, acoustic emission increases rapidly at the point of instability before fracture occurs. The cumulative acoustic emission count has been related to the fracture toughness parameter K_I by an equation of the form $N = AK_I^m$, where A is a proportionality constant and the exponent m having a theoretical value of 4 and varying experimentally from 4 to 8 depending on the material. This relation is applicable to materials exhibiting a certain ductility provided the stress intensity factor is corrected to account for the large plastic zone at the crack tip. In ductile materials containing a sharp crack, acoustic emission increases markedly with the development of a lateral contraction zone at the side of the specimen. In this investigation, acoustic emission parameters, namely, the cumulative count, the energy of the acoustic signal expressed in terms of the root mean square of the amplitude, and the waveform characteristics, such as the amplitude and pulse-width distribution are presented as a function of the crack size and the extent of loading to fracture. The objective is to determine the acoustic emission characteristics of a material prior to fracture.

111 **M. N. Bassim, D. R. Hay, and J. Lanteigne. "Waveform Analysis of Acoustic Signals from a Mild Steel."** *Materials Evaluation,* **MAEVA, 34(5):109-113. May 1976.**

Tensile specimens of AISI 1015 steel, with and without a sharp flaw, were pulled in tension to fracture while the acoustic signals emitted were detected using a wideband transducer and recorded for later analysis. It was found that the acoustic signals emitted from flawed specimens had a slightly higher amplitude, longer time duration and lower frequency content than those from unflawed specimens. The results show that various characteristics of the acoustic emission signal can be correlated to the state of strain and nature of the emission source.

112 **C. Bathias, B. Brintet, and G. Sertour. "Can be Detected the Initiation of Fatigue Cracks by Acoustical Emission? Application to High Strength Light Alloys for Aeronautics/L'amorcage des fissures de fatigue peut-il etre mis en evidence grace a l'emission acoustique? Application aux alliages legers a haute resistance utilises en aeronautique." Aerospatiale, Laboratoire Central, Suresnes, France; Paper 3K/9 in** *Proceedings of the Eighth World Conference on Nondestructive Testing.* **volume containing Sections 3K2, 3H2, 5A. International Committee for Nondestructive Testing, Secretariat du Cofrend, Institut de Soudure, 32 Boulevard de la Chapelle, 75880 Paris Cedex 18, France. 1976. conference held in Cannes, France. Sep 6-10, 1976. (paper in French).**

The study carried out on different types of test pieces (shot blasted, anodized, and notched) has shown that acoustic emission is a precise means for monitoring fatigue damage up to crack initiation.

113 **J. L. R. Bauza. "Los ensayos no destructivos dentro de la tecnologia nuclear (Nondestructive Tests within Nuclear Techology)."** *Energia Nuclear* **(Madrid), ENNCA, 15(72):298–304. 1971. (in Spanish).**

114 **J. B. Beal. "Ultrasonic Emission Detector Evaluation of Strength of Bonded Materials." pp 61–78 in** *Nondestructive Testing: Trends and Techniques.* **NASA SP-5082. National Aeronautics and Space Administration, Office of Technology Utilization, Washington, D.C. 1967. [NTIS].**

Conference papers are presented on nondestructive testing techniques using infrared microscopy, ultrasonic emission detection, and X-ray television. Development of nondestructive techniques for determining bond strengths is reviewed. Following a literature search of state-of-the art techniques, the nature of bonding was investigated to determine what properties affect adhesion bond strength and what nondestructive technique best measures these strength properties. A photomicroflaw technique, developed to simulate poor bond strength through controlled surface contamination, involves the coating of specimen surfaces with a photo-sensitive emulsion and exposing the surface to an intense light source through a guide. The importance of proper process control during bonding operations and the effects of bonding contamination on bond strength are stressed. Equipment evaluation testing is reported, and the advantages of a developed ultrasonic emission detector are noted. This detector is applicable to metal-to-metal and metal-to-nonmetal bonds in laminate and sandwich core structures. Operation of the detector, a heterodyne ultrasonic receiver operating in the 26 to 36 kHz range, is detailed.

115 **C. Beatson. "When Trouble Cries Out Acoustics Get the Message."** *Engineer,* **ENGIA, 234:30–32. Feb 3, 1972.**

Survey article on the acoustic emission activities being conducted in Great Britain by the United Kingdom Atomic Energy Authority (UKAEA) facilities at the Dounreay Experimental Research Establishment, at the Risley Engineering and Materials Laboratory, at the NDT Centre at Harwell, and also by Cambridge Consultants Ltd. These activities include boiling detection in nuclear reactors and in chemical and process engineering facilities, wear and failure prevention between the top of fuel assemblies and their standpipe containments, vibration detection and signature analysis of plant facilities, metals research, failure modes and fracture of metals, adhesive bond strength, detection and location of flaws in reactors and large pressure vessels, and periodic proof testing.

116 **A. G. Beattie. "Acoustic Emission in InTl Alloy."** *Bulletin of the American Physical Society*, **BAPSA, 16:385. 1971. [abstract only]; paper presented at the March 1971 Meeting of the American Physical Society, Cleveland, Ohio. Mar 29–Apr 1, 1971.**

The spontaneous emission of acoustic energy as a function of temperature has been studied in single crystals of the alloy In-23 at % Tl. As previously reported, acoustic energy is emitted at the martensitic fcc → fct phase transition, but only in well-annealed samples. Much larger amounts of energy were found to be emitted at temperatures below 210 K, with both rising and falling temperatures. Possible correlations of this emission with the temperature and frequency dependences of the acoustic attenuation and velocity will be discussed.

117 **A. G. Beattie. "An Analysis of the Frequency and Energy Characteristics of Acoustic Emission Signals from Tensile and Structural Tests." SAND74-5489. Sandia Laboratories, Albuquerque, New Mexico. 1974. [copy available from author]; paper presented at the 1974 Society for Experimental Stress Analysis Spring Meeting, Chicago, Illinois. May 11–16, 1975.**

Results are reported of studies of the spectral characteristics of acoustic emission from tensile specimens of several metals and a hydrostatic test of a commercial refrigeration bottle. The emission signals were electronically digitized and a large computer used to calculate their spectra. The spectral content of the signals as a function of their relative energy was also investigated. Finally, an attempt was made to remove the frequency characteristics of the transducer from the spectra of the tensile tests.

118 **A. G. Beattie. "Characteristics of Acoustic Emission Signals Generated by a Phase Transition." SC-DC-72-1650. Sandia Laboratories, Albuquerque, New Mexico. 1972. [NTIS]; pp 13–17 in** ***1972 Ultrasonics Symposium Proceedings.*** **Catalog No. 72 CHO 708-8SU. Institute of Electrical and Electronics Engineers, Inc., New York. 1972. symposium sponsored by U.S. Atomic Energy Commission and Institute of Electrical and Electronics Engineers Group on Sonics and Ultrasonics, held in Boston, Massachusetts. Oct 4–7, 1972.**

Studies were made of the acoustic emission signals generated by the fcc to fct phase transformation in the alloy, indium-23 at % thallium. The total detected energy was measured as a function of both polarization and wave propagation direction with respect to the plane of transformation of the sample. The frequency content of the emission signals was also measured as a function of direction and polarization. The results put definite limitations on the types of mechanisms which could generate these signals.

119 **A. G. Beattie. "Deep Rock Disposal Experiment No. 1: Acoustic Studies." SAND75-0162. Sandia Laboratories, Albuquerque, New Mexico. Mar 1975. [NTIS].**

A series of acoustic studies were conducted as a part of the Deep Rock Disposal Experiment No. 1 (DRD-1). These studies included ultrasonic and acoustic emission experiments. The acoustic emission experiments included both continuous monitoring and source location. The thermal design of DRD-1 subjected the acoustic transducers to temperatures in excess of 600 °C near the end of the experiment. This temperature required the use of $LiNbO_3$ transducers and the development of new transducer bonding techniques. After an initial shakedown period, the acoustic experiments proved, in general, to be successful. While this shakedown period prevented the acquisition of a complete set of acoustic data, the data which were obtained proved the value of the acoustic experiments and supports their inclusion in future DRD-1 type experiments. The experiments did show the need for some changes in the experimental apparatus, but these modifications are understood and can be implemented without great difficulty.

120 **A. G. Beattie. "Energy Analysis in Acoustic Emission." Sandia Laboratories, Albuquerque, New Mexico. 1975;** ***Materials Evaluation*****, MAEVA, 34(4):73–78. Apr 1976.**

Three possible methods of acoustic emission energy analysis are examined. These are digital processing, signal-length processing, and electronic processing. Each method is discussed and the three are compared. The comparison indicates that electronic processing is the method of choice. Experimental studies of energy analysis versus ringdown count analysis are used to illustrate where energy analysis should be superior to ringdown count analysis and where both methods will give approximately the same results.

121 **A. G. Beattie and R. A. Jaramillo. "The Measurement of Energy in Acoustic Emission." Sandia Laboratories, Albuquerque, New Mexico 1972;** ***Review of Scientific Instruments,*** **RSINA, 45(3):352–357. Mar 1974.**

A circuit for measuring the energy in randomly occurring bursts of RF energy is described. The circuit is linear to within 10% (over a 15 db dynamic range) for frequencies up to 1.5 MHz. Examples are given of its use with acoustic emission data from a martensitic phase transformation in the indium-thallium system and from a nondestructive test on a filament-wound vessel.

122 **A. G. Beattie, R. T. Johnson, Jr., and R. K. Quinn. "Spontaneous Generation of Acoustic Signals during Switching of Amorphous Semiconductors." SLA-73-5262. Sandia Laboratories, Albuquerque, New Mexico. 1973. [NTIS]; pp 881–887 in** ***Proceedings of the 5th International Conference on Amorphous and Liquid Semiconductors.*** **J. Stuke and W. Brenig, Editors. Taylor and Francis Ltd., London, England. 1974.**

Studies of electrical switching in bulk semiconducting chalcogenide glasses have shown that acoustic emission (spontaneously emitted acoustic waves) is produced during the switching process. The experiments were performed on samples with compositions $Ge_{15}Te_{80}As_5$ and $Ge_5Te_{45}As_{50}$. Switching was both across the surface of the samples and through the thickness (bulk) of the samples. During switches where the current rise is faster than 30 μsec, a burst of acoustic emission is produced 2 to 25 (±2.5) μsec preceding the rise in current. This emission is associated with the initiation of the switching process. Another characteristic type of emission is seen which occurs primarily after the rise in current, and is shown to have its origin in propagating cracks in the sample. These studies give information both about the physics of the swtiching process and the integrity of the material after the switch.

123 **P. W. R. Beaumont. "A Fracture Mechanics Approach to Failure in Fibrous Composites."** ***Journal of Adhesion,*** **JADNA, 6:107-137. 1974; paper presented at the Symposium on Interfacial Bonding and Fracture in Polymeric, Metallic and Ceramic Composites, held at the University of California, Los Angeles, California. Nov 13-15, 1972.**

The paper presents a review of the fracture processes which occur in brittle thermosetting resins containing long strong fibers of carbon, boron, and glass. Important material variables affecting the fracture behavior of the fibrous composite include fiber strength and strength of the fiber-matrix inerface. Other parameters which must be considered include component geometry, environment, loading mode, and nondestructive inspection capability. The effects of microstructural features including fiber volume fraction, together with a consideration of the fiber-resin interface are discussed in relation to the fracture stress and toughness of the composite material. The influence of the fiber-resin interface and environment on slow crack growth in carbon fiber reinforced epoxy resin is also reviewed.

124 **J. Becht and J. Eisenblatter. "Acoustic Emission Analysis: A New Non-Destructive Testing Method for Highly Reliable Structures." Battelle-Institut eV, Frankfurt am Main, Germany. 1973; paper presented at the European Space Research Organization 3rd Symposium on Large Structures for Manned Spacecraft. Frascati, Italy. Oct 22-26, 1973.**

After discussion of the physical causes of acoustic emission and the experimental procedures of acoustic emission analysis (AEA), numerous fields of application of AEA are described. These are: plastic deformation of ferrous and nonferrous metals, crack formation during stress-relief annealing, crack formation during cyclic loading, control of welding processes, stress-corrosion cracking and, finally, microcrack formation in various fibre-reinforced composite materials. The advantages and disadvantages of AEA are compared with those of other nondestructive testing methods.

125 **J. Becht, J. Eisenblatter, and P. Jax. "Werkstoffprufung mit der Schallemissionsanalyse (SEA) (Nondestructive Testing by Acoustic Emission)."** *Zeitschrift fur Werkstofftechnik/Journal of Materials Technology*, **ZWKTA, 4(6):306–314. 1973. (in German).**

After the discussion of the physical principles and the experimental procedures of acoustic emission techniques, examples of their applicability and capability are presented. These concern plastic deformation of metals (ferrous and nonferrous), crack formation during heat treatment for stress relief, crack formation during cyclic loading, control of weld processes (pressure and periodic testing and surveillance of reactor pressure vessels), stress-corrosion cracks, and finally microcrack formation in different fiber-reinforced composite materials. Advantages and disadvantages of the method are compared with those of other NDT methods. [reprinted by permission of Verlag Chemie GmbH].

126 **R. Becker and E. Orowan. "Uber sprunghafte Dehnung von Zinkkristallen (Discontinuous Expansion of Zinc Crystals)."** *Zeitschrift fur Physik*, **ZEPYA, 79:566–572. 1932. (in German).**

The plastic deformation of a zinc crystal occurs in jumps. Their amplitude remains practically independent of the force and the variation of velocity is given by the change in their number. The dependence of the effect on temperature is studied. [abstract from *Chemical Abstracts*, 27:1251. 1933. Copyright by American Chemical Society].

127 **D. Bedere and D. Miannay. "Analyse de l'Emission Acoustique Produite au Cours de la Fissuration de Corrosion sous Contrainte de l'Alliage U–Mo a 10% Mo en Poids (Analysis of the Acoustic Emission Produced during Stress Corrosion Cracking of U–Mo Alloy with 10 wt % Mo)."** *Journal of Nuclear Materials*, **JNUMA, 48(2):183–192. Sep 1973. (in French).**

The acoustic emission accompanying stress-corrosion cracking in air, at ambient temperature, of a U-10 wt % Mo alloy has been investigated by a counting technique. The acoustic output is due to the plasticity and microcracking which accompany macroscopic cracking. The output increases slightly in the course of a test and is not a unique function of the stress intensity factor. The practical interest of this method for nondestructive testing is discussed.

128 **W. B. Bedesem and J. S. Clarke. "Prevention of Catastrophic Brittle Fracture of Heavy-Wall Pressure Vessels." ASME Publication 68–PVP–4. American Society of Mechanical Engineers, New York. 1968.**

The petroleum and petrochemical industry recently have experienced several brittle failures of heavy-wall pressure vessels during hydrostatic testing. The failures were rather unexpected since, in most cases, the materials involved had more than adequate toughness, according to currently used criteria, to be safe against brittle fracture. The paper discusses current criteria for brittle fracture prevention and their limitations, as well as proposed requirements for minimizing the risk of further brittle fractures in heavy-wall vessels. The requirements involve a fracture mechanics leak-before-break toughness criterion that possibly explains the recent failures. When the criterion cannot be satisfied, as is likely beyond certain thickness limits, protection against brittle fracture can be attained only by assuring that there are no defects present greater than a critical size. More extensive use of standard inspection methods and of more sophisticated techniques may be required to ensure that there is no defect to cause brittle fracture during hydrotest or to grow to critical size in service.

129 **J. Beges, F. Erdmann-Jesnitzer, and D. Rehfeldt. "Acoustic Emission Analysis for Hydrogen Induced Delayed Cracking in Weld Metal." pp 322-337 in** *Schallemission: Anwendung bei der Untersuchung, Prufung and Uberwachung metallischer Werkstoffe.* **Deutsche Gesellschaft fur Metallkunde eV, 637 Oberursel/Taunus bei Frankfurt/Main, Adenauerallee 21, Germany. 1974. proceedings of symposium held in Munich, Germany. Apr 18-19, 1974.**

The influence of the hydrogen on weld metal cracking has been widely accepted. The whole last decade is filled up with different research works of hydrogen and its influence on mechanical behaviour of steels, other nonferrous metals, and different alloys. Presented are new procedures together with a broader approach of the hydrogen cracking mechanism. The hydrogen-induced delayed cracking was evaluated with the observations of arised acoustic emission in the root layer of the high tensile weld metal subjected to variable degrees of the hydrogen concentration, compared with the crack opening and the change of the electrical resistance.

130 **G. Belisario, S. Capuano, and A. Di Pietro. "Su Alcune Caratteristiche Della Emissione Sonora Durante Prove di Fatica su Vibroforo (Some Characteristics of Sound Emission during Fatigue Tests Performed with a Vibrophone)."** *La Metallurgia Italiana,* **MITLA, 62(1):11-16. Jan 1970. (in Italian); RFP-Trans-127. translated for Dow Chemical U.S.A., Rocky Flats Division, Golden, Colorado 1973. [NTIS].**

Vibrophone fatigue tests were carried out on 0.4% C steel tensile-test pieces containing transverse holes. The sound emission spectrum was recorded and analyzed. Results show that the acoustic spectrum of sound emitted changes appreciably during tests and this is most marked immediately before failure. [Abstract 22-0903. *Metals Abstracts,* Vol. 3, 1970].

131 **R. L. Bell. "A Progress Report on the Use of Acoustic Emission to Detect Incipient Failure in Nuclear Pressure Vessels." Technical Report DE-74-1. Dunegan/Endevco, San Juan Capistrano, California. Jan 1974;** *Nuclear Safety,* **NUSAA, 15(5):554-571. Sep-Oct 1974.**

Significant advances in the technology and application of the acoustic emission phenomenon to detection, location, and characterization of incipient failure mechanisms in pressure vessels have taken place over the last several years. In addition to programs carried out by development laboratories and private research facilities, the government and the utility industry have also sponsored programs that have greatly increased the scope and applicability of the technique to nuclear pressure vessel inspection. Both periodic hydrotest monitoring with AE and continuous AE surveillance have been shown to be practical for application to the nuclear pressure vessels of the BWR and PWR type. Each of these two approaches have their limitations and advantages, but in general appear to provide more information about the integrity of the nuclear pressure vessel than any other NDT technique presently used. While R&D work is continuing with regard to better determining the severity of the AE sources, considerable field application is already providing useful baseline and operating information.

132 **R. L. Bell. "Acoustic Emission Transducer Calibration: Transient Pulse Method." Endevco, Pasadena, California 1972; paper presented at the American Society for Nondestructive Testing National Fall Conference, Cleveland, Ohio. Oct 16–19, 1972; Technical Report DE 73-3. Dunegan/Endevco, Livermore, California. Feb 1973.**

Paper discusses the history of attempts to calibrate acoustic emission transducers, and explains similarities and differences between accelerometers, ultrasonic transducers, and acoustic emission transducers. The logical progression of techniques is culminated in a present technique currently only little known, but which allows unbiased comparisons between transducers of all types. Relationships between calculated sensitivities, the calibrated sensitivities, and actual acoustic emission tests will be presented and discussed with practical application in mind. The application of acoustic emission techniques to NDT applications requires that standards be developed for the customer to be sure of what he is buying and what he is analyzing.

133 **R. L. Bell. "Investigation of Acoustic Emission Techniques for Continuous Monitoring." pp 49–87 in "A Joint Industry–Utility Program to Develop NDT Techniques for Nuclear Reactor Inspection: An Addendum Report to Steering Committee EEI Project RP-79." Compiled by E. R. Reinhart. Southwest Research Institute, San Antonio, Texas. May 21, 1973. [NTIS: PB-224-698/1].**

Dunegan/Endevco participation in the EEI Project RP79 began in January 1972 and continued to the final hydrotest on October 31, 1972. During those nine months, studies were completed on mechanical wave-propagation characteristics of the EBOR vessel, methods of transducer attachment to the vessel, transducer design to allow acoustic emission data acquisition within the simulated noise environment of the San Onofre power plant, development and evaluation of a real time, computerized monitoring system, and analysis of acoustic emission information generated during flaw growth in the reactor vessel wall and in stainless steel piping. These studies culminated in a final system which was evaluated at the EBOR facility in August, September, and October 1972. The results of those tests indicate that detection of in-wall cracks as well as stress corrosion cracking in the EBOR stainless steel piping is feasible in an environment such as was found at San Onofre. Significant advances in transducer design have made this possible. But even more important to the operational plant were the results of quantifying the acoustic emission information received. By analyzing the acoustic energy content from the crack growth in a unique way, it appears possible to predict not only qualitative crack growth, but also to quantitatively measure it.

134 **R. L. Bell. "The Use of Acoustic Emission to Detect Incipient Failure in Pressure Vessels (Der Nachweis entstehender Fehler in Druckgefassen durch Schallemission)."** ***Kerntechnik*****, KTAPA, 16(5):216–222. May 1974. (in English).**

The application of acoustic emission technology to field evaluation of nuclear reactor pressure vessels has been successfully accomplished. At present, both the acoustic emission monitoring of hydrotests and continuous acoustic emission surveillance are being used for this purpose. A number of technological advances have been made in the past several years that have provided new hardware and basic material information. These include new transducers, full-scale computerized location systems, various location algorithms, material studies of reactor steels, fatigue crack growth studies in reactor pressure vessels, etc.

135 **R. L. Bell, M. P. Kelly, and H. L. Dunegan. "Acoustic Emission Monitoring of Nuclear Pressure Vessels."** ***Transactions of the American Nuclear Society*****, TANSA, 19(Supplement 2):14–15. 1974; paper presented at the Conference on Nondestructive Testing in the Nuclear Power Industry, held at Ohio State University, Columbus Ohio. Sep 23–25, 1974.**

The monitoring of nuclear reactor pressure vessels with acoustic emission techniques has been under intensive study for more than five years. At present, both continuous surveillance and preoperational work are being accomplished in the USA. There are also significant programs in Europe and Japan. The purpose of this paper is to review these field applications and their impact on current and future nondestructive testing of nuclear pressure containment systems.

136 **R. L. Bell, M. P. Kelly, and H. L. Dunegan. "Acoustic Emission Monitoring of Proof Tests and Simulated In-Service Nuclear Pressure Vessels."** ***Transactions of the American Nuclear Society*****, TANSA, 16:85–86. June 1973; paper presented at the 19th Annual Meeting of the American Nuclear Society, Chicago, Illinois. June 10–14, 1973.**

During the past year, Dunegan/Endevco has participated in the EEI-TVA RP79 project and in the heavy-section steel test program. The information obtained from this participation has led to new technology which can provide increased efficiency and safety for operating nuclear power plants. The objectives of the program were to evaluate the use of acoustic emission techniques to assure structural integrity of nuclear pressure vessels. Two alternative approaches were considered: acoustic emission monitoring of hydrotests, and continuous acoustic emission monitoring of in-service vessels.

137 **S. B. Bellosillo. "Acoustic Emission: Another Tool for the Materials Analyst."** *Canadian Research and Development,* **CRDVA, 5:23–28, 49. May–June 1972.**

Paper is primarily intended to interest materials investigators in the use and possibilities of the technology in basic materials evaluation and to show the importance of furthering the knowledge in this field of activity; and also to show application-oriented investigators the potential of the techniques for utilization in various applications. An historical review is included in this paper, followed by discussion of a basic acoustic emission detection system that can be set up in conjunction with a standard mechanical test. Techniques and applications are briefly discussed at the end of the paper, primarily to show coverage of what has been done, and presented also are some ideas of what may be expected from future developments.

138 **V. M. Belov, Yu. B. Drobot, and A. P. Drozdov. "Detection of Cracks in Welded Joints between Tubes and Tube Plates by Measuring the Parameters of Stress Wave Emission."** *Welding Production,* **WEPRA, 21(8):41–44. Aug 1974; translated from** *Svarochnoe Proizvodstvo*, **SVAPA, (8):27–29. Aug 1974. (in Russian).**

A method of detecting cracks in welded joints between tubes and tube plates during their static and cyclic loading is described. The method is based on recording the stress waves, generated at the instant of crack formation by piezogenerators attached to the workpiece. From the intensity and the number of pulses, the parameters and the number of defects formed can be assessed. The method was tested on welded joints made from high Ni-alloy Kh20N45 tubes, with tube plates made from steel 0Kh18N10T. The results of these tests are presented and discussed. [Abstract 22-0525. *Metal Abstracts,* Vol. 8, 1975].

139 **V. M. Belov, Yu. B. Drobot, A. P. Drozdov, and Yu. D. Kon'kov. "Detection of Crack Formation in a Weld by Acoustic Emission."** *Soviet Journal of Nondestructive Testing*, **SJNTA, 10(4):395–399. July–Aug 1974; translated from** *Defektoskopiya*, **DEFKA, (4):29–34. July–Aug 1974. (in Russian).**

The possibilities of detecting crack formation during the welding of stainless steel are investigated. The experimental apparatus and procedure are described. The noises that are characteristic of the welding process are classified. Results are given that show the possibility of detecting cracks in a weld by recording the average level of the acoustic emission.

140 H. Benioff. "Earthquake Source Mechanisms." *Science,* SCIEA, 143(3613): 1399–1406. Mar 27, 1964.

141 M. N. Bentley and G. Green. "A Method of Calibration for Acoustic Emission Measurements." in *Proceedings of the Institute of Acoustics.* Institute of Acoustics, 47 Belgrave Square, London, SW1X 8QX, England. 1976. acoustic emission meeting held at Chelsea College, London, England. Dec 20–21, 1976.

Studies of the intrinsic noise of an acoustic emission system, and studies of its response to a particular impulsive stimulus, have been shown to enable comparison to be made with similar data obtained with a system having different response, so that each system can be calibrated. In this way, data obtained from differing systems can be compared, and it is suggested that it may be possible to establish whether any acoustic emission system is capable of determining particular components in a signal.

142 P. G. Bentley, T. E. Burnup, E. J. Burton, A. Cowan, and N. Kirby. "Acoustic Emission as an Aid to Pressure Vessel Inspection." Paper C30/72, pp 54–66 in *Periodic Inspection of Pressure Vessels.* Institution of Mechanical Engineers, London, England. 1972. proceedings of conference arranged by the Applied Mechanics Group of the Institution of Mechanical Engineers, held in London, England. May 9–11, 1972.

The development by the United Kingdom Atomic Energy Authority of acoustic emission monitoring as a nondestructive tool to aid pressure vessel inspection has progressed in three main ways. Firstly, an experimental facility, SWEL, has been built and tested which is capable of detecting and locating small flaws from the acoustic signals received by up to 10 sensors mounted on the pressure vessel during a pressure test. Secondly, measurements have been made using early components of this facility to gain experience on large pressure vessels at manufacturers' works. Thirdly, a large number of measurements have been made in the laboratory of emission rates from artificial defects in test pieces and pressure vessels of low- and medium-strength steels. Thus, the capability is now available to make a 100% inspection of pressure vessels over a wide range of engineering steels; and there is the prospect, with further development, of monitoring during operation the severity of any defects relative to vessel failure.

143 P. G. Bentley, T. E. Burnup, and D. G. Dawson. "Acoustic Emission Measurements on the EBOR Vessel." pp 137–152 in "A Joint Industry-Utility Program to Develop NDT Techniques for Nuclear Reactor Inspection: An Addendum Report to Steering Committee EEI Project RP-79." Compiled by E. R. Reinhart. Southwest Research Institute, San Antonio, Texas. May 21, 1973. [NTIS: PB-224-698/1].

The Edison Electric Institute (EEI) and Tennessee Valley Authority have sponsored Research Program RP79 to upgrade acoustic emission for on-line monitoring of pressurized water reactors. A part of this program is a series of evaluation tests using the Experimental Beryllium Oxide Reactor (EBOR) vessel provided by the U.S. Atomic Energy Commission at the National Reactor Test Site, Idaho. Personnel from REML were invited to attend the tests, first to observe the proof hydro in November 1971 and then to participate by taking measurements during February–March 1972. This report covers the measurements made during the latter period and is intended for dissemination to participants and to members of the EEI management committees and other interested parties. The EBOR test facility, now devoted primarily to acoustic emission, represents a major step forward in the general development of these techniques although it is oriented toward U.S. water reactors. Briefly, the vessel is a 10-foot diameter, 23-foot high, 6-inch thick structure with a flanged head, all in unclad 1% Cr - 0.5% Mo. Internals have been removed together with sections of insulation. The system is fitted with many facilities for acoustic emission testing, including: (a) Bend bars, about 1-inch square, welded to the vessel with a notch adjacent to the surface which is propagated by an oscillatory mechanism; (b) In-vessel flaws, introduced artificially, are grown by an external hydraulic pressure cycling device; (c) Provision to hydro-pressure the vessel with both natural and artificial in-vessel flaws present; and (d) Noise generation by a hydraulic-cavitation source mounted through the top dome. The remaining major facility, of being able to operate the system at elevated temperatures, was not used during REML participation.

144 **P. G. Bentley, T. E. Burnup, D. G. Dawson, and J. Parker. "Acoustic Emission Measurements in Large Industrial Pressure Vessels." United Kingdom Atomic Energy Authority, Risley Engineering and Materials Laboratory, Risley, Warrington, Lancashire, England. 1972; paper presented at the Institute of Physics Conference on Acoustic Emission, Imperial College, London, England. Mar 14, 1972. [proceedings were never published].**

Acoustic emission measuring systems, including source location, have been under development at REML for several years. This experience is incorporated in the SWEL mobile for work on industrial systems, typically reactor pressure vessels. System tests on a 3-inch thick vessel are described, particularly emphasizing source location and the advantages this gives in eliminating spurious signals. Preliminary work on an application to reactor pressure vessels has concentrated on determining background noise levels and attenuations. The measurements on the 50-foot diameter pressure vessel at Berkeley power station were made at various operating conditions up to full power and the frequency content determined. The measurements also included a source location analysis which gave an unexpected source of signal.

145 **P. G. Bentley, E. J. Burton, A. Cowan, D. G. Dawson, and T. Ingham. "Acoustic Emission and Pressure Vessel Failure." pp 643–654 in** ***Proceedings of the 2nd International Conference on Pressure Vessel Technology. Part 2: Materials, Fabrication and Inspection.*** **American Society of Mechanical Engineers, New York. 1973. conference held in San Antonio, Texas. Oct 1–4, 1973.**

The development of the UK Atomic Energy Authority acoustic emission monitoring system has continued, directed towards its use to prevent pressure vessel failure in materials of general engineering practice. The acoustic facility, SWEL, can locate sources on large vessels to ±3 inches and has a two-dimensional display, on line, of the emission sources. Measurements have been made of the location of artificial and natural defects in the EBOR vessel, USA. The emission, after the first cycle, from cracks grown by pressure cycling of artificial defects in experimental vessels, 3 inches and 1 inch thick, has been shown to be very low. A systematic investigation has been made of the acoustic emission from a range of steels and a ranking derived in terms of fracture toughness and tensile strength. The use of acoustic emission as a general inspection tool has been surveyed. Its present complexity and cost limit its use to specific large vessels, especially those in which failure could lead to a high cost from loss of production. In-service monitoring is not considered yet to be practicable and this view has been confirmed from background noise measurements on British Gas-Cooled Reactors. The main present use for the technique is on the validation of pressure vessels with defects in brittle regions during pressurization tests, before operation and during shutdown periods. Doubts have been raised of the signal strengths from flaws in the parent plate of ductile materials. Further tests are required on experimental vessels taken to failure to establish confidence.

146 **P. G. Bentley and D. G. Dawson. "Acoustic Emission Measurements on Thick Walled Ductile Steel Pressure Vessels." Paper No. 10 in *Reactor Inspection Technology*. British Nuclear Energy Society, London, England. 1975. proceedings of the Symposium on Reactor Inspection Technology, sponsored by the British Nuclear Energy Society, Western Branch, held in Bristol, England. Feb 26, 1975.**

The development of the U.K. Atomic Energy Authority acoustic emission monitoring system has continued, directed towards its use to prevent pressure vessel failure in materials of general engineering practice. The acoustic facility, SWEL, can locate sources on large vessels to ±3 inches and has a two-dimensional display, on line, of the emission sources. Measurements have been made of the location of artificial and natural defects in two experimental ductile steel vessels and on the U.S. EBOR test vessel. Defects of a brittle nature are shown to emit strongly and can be readily detected. Defects in parent plate are shown to be quiet when significant crack growth is occurring, in accord with results from metallurgical specimens. No signals are detected during periods of yielding around defects in vessels in spite of indications from specimens that low level emission occurs.

147 **P. G. Bentley, D. G. Dawson, and J. A. Parker. "Instrumentation for Acoustic Emission." TRG Report 2482(R). The Reactor Group, United Kingdom Atomic Energy Authority, Risley Engineering and Materials Laboratory, Risley, Warrington, Lancashire, England. Dec 1973. [this report is identical to Reference 148].**

Instrumentation has been developed to detect and analyse acoustic emission from steel structures under stress, particularly pressure vessels, which gives an indication of defects that would otherwise lead to failure. Sensitive transducers and low noise amplifiers are used to measure the weak signals from structural steels and to condition them to a form suitable for on-line computer analysis. An array of 10 to 20 transducers is used to locate emitting sources under computer control and two simultaneous displays show this information after spurious signals have been removed. The instrumentation is built into a self-contained mobile laboratory for studies at manufacturer works and industrial sites.

148 **P. G. Bentley, D. G. Dawson, and J. A. Parker. "Instrumentation for Acoustic Emission Testing of Steel Pressure Vessels." Session 3, pp 1-29 in *The Second Acoustic Emission Symposium.* Japan Industrial Planning Association, Dobashi Building 5F, 3-17-8 Akasaka, Minato-ku, Tokyo, 107 Japan. 1974. proceedings of a symposium held in Tokyo, Japan. Sep 2-4, 1974. [for single copy of paper see Reference 147].**

Instrumentation has been developed for application to steel pressure vessels and is built into a self-contained mobile laboratory SWEL. The system includes sensitive transducers developed specially for this application and analysis facilities for source location with an array of 20 transducers. The location is performed with flexible control from a software programed digital computer and data are presented to show sources on-line for initial evaluation and off-line for greater accuracy. Count rates from the analysis are presented simultaneously to show the total validated signals and the activity of any source selected for detailed monitoring during the course of a test. The system has been evaluated on thick-walled pressure vessels.

149 **F. L. Bentzen. "Operation and Disposal." Task 3, Project II, pp 53-64 in "In-Service Inspection Program for Nuclear Reactor Vessels." R. D. Wylie. SwRI Project 17-2440, EEI Project RP-79. Biannual Progress Report No. 7. Southwest Research Institute, San Antonio, Texas. Aug 25, 1972. [NTIS: PB-212-144].**

The operation of the Nondestructive Test Method Evaluation Facility (NDTF) began in November 1971 with a hydrostatic test to 1875 psi. This was followed by tests to various pressures and flow conditions to evaluate the capability of acoustic emission systems. A noise simulator was installed to provide noise typical of that found in operating nuclear reactor systems to further evaluate the capability of the systems for on-line use. Two high temperature runs were made, one with wall temperatures to 450 °F and one to 510 °F, to provide a temperature environment typical of operating nuclear systems. A slot was cut in the vessel wall and a crack later grown from this slot to provide a source of acoustic emission for testing the Prototype Reactor Surveillance System and other acoustic emission equipment.

150 **F. L. Bentzen and C. V. Mullen. "Development of a Nondestructive Test Method Evaluation Facility." Task 1, Project II. pp 10–17 in "In-Service Inspection Program for Nuclear Reactor Vessels." R. D. Wylie. SwRI Project 17-2440, EEI Project RP-79. Biannual Progress Report No. 4, Southwest Research Institute, San Antonio, Texas. Jan 7, 1971. [NTIS: PB-198-274].**

A contract was signed with Idaho Nuclear Corporation to design, construct, and operate a facility for evaluation of acoustic emission and other NDT equipment. The mothballed Experimental Beryllium Oxide Reactor (EBOR) facility will be converted from a gas-cooled reactor facility to a pressurized water system capable of providing temperatures, pressures, and system noise approaching those found in large pressurized water reactors. The tentative maximum operating conditions for the completed facility are 1460 psig and 590 °F (saturation conditions) with or without flow. The basic design is essentially complete and a flow sheet has been developed. Mechanical and electrical components are being obtained from surplus stocks to reduce the overall project costs. Specifications are being written for those components that are to be purchased.

151 **F. L. Bentzen and C. V. Mullen. "Development of a Nondestructive Test Method Evaluation Facility: Design." Task 1, Project II. pp 22–23 in "In-Service Inspection Program for Nuclear Reactor Vessels." R. D. Wylie. SwRI Project 17-2440, EEI Project RP-79. Biannual Progress Report No. 5. Southwest Research Institute, San Antonio, Texas. May 28, 1971. [NTIS: PB-200-518].**

Work continued on the contract for the design, construction, operation, and disposal of the Nondestructive Test Method Evaluation Facility. The design was completed and request-for-bid packages have been prepared. A schedule has been developed for the remainder of the construction task. Construction will be complete, testing devices will be installed, and the facility will be ready for testing acoustic emission systems by August 15, 1971.

152 **F. H. Berg, L. R. Boyd, K. A. Smith, and J. Wallach. "State-of-the-Art Report on Acoustical Monitoring of Nuclear Reactors." GEAP-13979. General Electric Company, Breeder Reactor Department, Sunnyvale, California. July 1973. [TIC].**

A state-of-the-art review is presented on several LMFBR related areas within the field of acoustic monitoring. These areas are: (1) vibration monitoring; (2) boiling detection; (3) stress emission monitoring; and (4) position indication and identification. A discussion on signal processing is included. The references include both domestic and foreign literature sources. Each area is presented in light of the present thoughts and techniques in use as indicated by the literature. The state-of-the-art and limitations of these five areas are indicated.

153 **P. Bernard and N. Chretien (to Commissariat a l'Energie Atomique). "Dispositif de localisation des sources d'emission d'ondes de contrainte (Acoustic Emission**

Source Location Equipment)." French Patent 2,164,031. June 2, 1973. (in French).

Equipment is described for localizing small area stress wave emission sources. It consists of at least three detectors designed to press against the surface of the material studied at specially defined places. Each of these detectors is provided with a transducer furnishing an electrical signal whose amplitude depends on the stress waves; an amplifying and filtering unit is supplied for each detector; unit provides a single pulse each time the threshold is exceeded; and means are incorporated for determining the time interval between the instant when a pulse appears on one of the units and the occurrence of corresponding pulses on the other units from information about this instant. An important application of the invention is in the surveillance of reactor vessels.

154 **S. Bhattacharya and K. Schroder. "A New Method of Detecting Fatigue Crack Propagation in Ferromagnetic Specimens."** *Journal of Testing and Evaluation,* **JTEVA, 3(4):289–291. July 1975.**

The paper describes a new method for detecting fatigue crack growth in ferromagnetic specimens. This method is similar to the Barkhausen effect type of measurement where discontinuous voltage pulses are induced in a pickup coil surrounding a specimen when the specimen is magnetized slowly. As a fatigue crack grows, similar voltage pulses are generated. Simultaneous display of the Barkhausen-type magnetic signals and the acoustic emission signals during fatigue crack propagation show that they are correlated. [reprinted by permission of the American Society for Testing and Materials, Copyright].

155 **R. Bianchetti, M. A. Hamstad, and A. K. Mukherjee. "Correlation between Acoustic Emission and Microstructure in a Dispersion Hardened Aluminum Alloy." pp 1625–1630 in** *Proceedings of the Second International Conference on Mechanical Behavior of Materials.* **American Society for Metals, Metals Park, Ohio. 1976. conference held in Boston, Massachusetts. Aug 16–20, 1976.**

Tensile tests were used to study the acoustic emission from several plates of 7075-T6 aluminum. Specimens from one plate produced a large burst-type emission in addition to the continuous emission. Metallographic and fracture surface studies indicated that the burst-type emission resulted from the brittle fracture of large (20 to 60 micrometres) primary inclusions in the microstructure.

156 **R. Bianchetti, M. A. Hamstad, and A. K. Mukherjee. "Origin of Burst-Type Acoustic Emission in Unflawed 7075-T6 Aluminum." UCRL-13661 and UCRL-77289. Lawrence Livermore Laboratory, Livermore, California. Sep**

10, 1975. [NTIS: UCRL-13661]; *Journal of Testing and Evaluation,* **JTEVA, 4(5):313–318. Sep 1976.**

Tensile tests were used to study the acoustic emission from several 7075-T6 aluminum plates. Specimens from one plate produced a large burst-type emission in addition to the continuous emission. Metallographic and fracture surface studies indicated that the burst-type emission resulted from the brittle fracture of large (20 to 60 μm) primary inclusions in the microstructure. [reprinted by permission of the American Society for Testing and Materials, Copyright].

157 **H. Bibring and F. Sebilleau. "Etude Acoustique de la Transformation Allotropique du Cobalt (Acoustical Study of the Allotropic Transformation of Colbalt)."** *Memoires Scientifiques de la Revue de Metallurgie,* **MRMTA, 56:609–613. Nov 1959. (in French).**

The acoustical investigation by means of piezoelectric excitation shows the allotropic transformation accompanied by acoustic impulses. Each impulse is due to the sudden formation of martensitic elementary domains. Recordings made during thermal cycling show clearly the various characters of the allotropic transformation and their difference between heating and cooling cycles. [Abstract 538d. *Chemical Abstracts,* Vol. 54, 1960. Copyright by American Chemical Society].

158 **L. W. Bickle and A. J. Smiel. "Applicability of Acoustic-Emission Techniques to Civil Engineering Research." AFWL-TR-74-299. University of New Mexico, Eric H. Wang Civil Engineering Research Facility, Albuquerque, New Mexico. June 1975. [NTIS: AD/A-012-210].**

When a solid undergoes a plastic deformation or fracture, part of the stored strain energy is released as elastic stress waves which can be sensed by sophisticated instrumentaiton. These stress waves are termed acoustic emission even though the amplitudes and frequencies involved are usually outside the audible range. This study demonstrated that acoustic emission technology is becoming increasingly more suitable for field applications in civil engineering. Further research relating to reinforced concrete, soils, paving materials, and passive transducers is clearly justified.

159 **K. Bierney. "Listening to Stress in Materials."** *Engineering,* **ENGNA, 215(12): 966–969. Dec 1975.**

The article presents a brief explanation of what acoustic emission is, the ways in which it can be used, and its potential applications for the continuous monitoring for structural integrity of offshore platforms and nuclear reactor pressure vessels.

160 **R. C. Bill. "An Acoustic Emission Study of the Deformation Mechanisms of Polycrystalline Aluminum and Copper." Ph.D. Thesis. University of Michigan, Ann Arbor, Michigan. 1970; University Microfilms 71-15, 095.**

In this paper, critical evaluation is given to possible microscopic deformation mechanisms that may account for the observed acoustic emission behavior of 99.99% Al and 99.9% Cu specimens. Observed microstructural features, in particular grain size and the presence of subgrains, are considered in conjunction with acoustic emission behavior in evaluating these mechanisms. During microstrain, i.e., strains less than 5×10^{-4}, both aluminum and copper are seen to give little emission, and the emission behavior is not seen to be sensitive to the grain size. From this observation, and from stress-strain measurements, it was concluded that grain interior obstacles such as forest dislocations provide the most significant barrier to slip. The conclusion was also reached that dislocation source activation is not typical of events that occur during the microstrain regime. During the transition from microstrain to macrostrain, and in the macrostrain regime up to strains of 2% the grain size is found to have a strong effect on the acoustic emission behavior of 99.99% aluminum specimens. This effect is much stronger than the effect grain size has on the stress-strain behavior. Examination of various behavioral models leads to the conclusion that unstable dislocation sources located near the grain boundaries are activated when slip progresses from one grain to an adjacent grain. The key observation leading to this conclusion is that the acoustic emission activity declines at a rate inversely proportional to the grain size as the grain size increases beyond a value defined by the equipment sensitivity. Hence, since the grain boundary surface area in a specimen is inversely proportional to the grain size, the acoustic emission activity is proportional to the grain boundary surface area. A similar, though less well defined, acoustic emission behavior is observed in polycrystalline specimens of 99.9% copper. A relation is observed between the probable time of duration of slip events in 99.99% aluminum and the frequency of the acoustic emissions detected from a specimen. This constitutes an important connection between the data obtained from acoustic emission studies and the dislocation velocity, which control the time duration of a slip event in a specimen with a fixed glide distance.

161 **R. C. Bill and J. R. Frederick. "Use of Acoustic Emission in Nondestructive Testing: Fourth Semiannual Report 01971-4-P, March 1, 1970 - August 31, 1970." Advanced Research Projects Agency Contract F33615-68-C-1703. University of Michigan, Department of Mechanical Engineering, Ann Arbor, Michigan. Dec 1970. [NTIS: AD-879-708].**

Acoustic emission may be defined as the noise given off spontaneously by solid materials as a result of a sudden relaxation of stresses within the material. Stress relaxation can occur as a result of the nucleation or propagation of cracks, or as a consequence of various elastic or plastic deformation processes. The principal elastic or plastic deformation mechanisms that are sources of acoustic emission in solids are (1) the slip of existing dislocations in a metal, (2) the activation of dislocation sources, (3) twinning, and (4) grain boundary slip. This report describes the results of an investigation into the effects of one microstructure parameter, namely grain size, on acoustic emission.

162 J. R. Birchak and C. W. Chagnon. "Acoustic Emission Kaiser Effect in Zircaloy 4 Fuel Rods." Babcock and Wilcox Company, Lynchburg Research Center, Lynchburg, Virginia. 1977.

Zirconium nuclear fuel rod cladding is produced by welding end caps to tubing. The Kaiser effect has been investigated as a technique for evaluating these fuel rods. Rods having an end cap welded to one end were monitored for acoustic emissions during pressurization from the other end. Acoustic emissions were generated at the welded end during the first pressurization. Upon repressurization, the rod was initially quiet, as expected from the Kaiser effect. Three anomalous effects, however, were observed as the pressure was further increased: (a) although a few of the rods generated the anticipated acoustic emission bursts immediately after the previous peak pressure was exceeded, most of the rods were quiet until the pressure had exceeded the previous peak by 40 to 80 psi; (b) if on the other hand the rod was thermally aged after the first pressurization, then acoustic emission bursts resumed at pressures significantly below the previous peak pressure; and (c) similarly bursts were generated at pressures below the previous peak if the rod were impacted on the end cap between the first and second pressurization, which produced a compressive stress in the weld regions subject to tensile stress during the first pressurization. These three effects demonstrate that the resumption of bursts can begin at pressures either above or below the previous peak, depending on thermal and mechanical treatment between successive pressurizations.

163 D. Birchon. "The Potential of Acoustic Emission in NDT." ***British Journal of Non-Destructive Testing,*** **BJNTA, 18(3):66–71. May 1976.**

Acoustic emission and its characteristics as an NDT technique are defined. Generation, propagation, and attenuation of the waves are discussed, together with theoretical and practical requirements for their detection and recording. Energy partition between the emission and the phenomena initiated by the release is explained in the light of deformation and fracture processes within a strained mild-steel ferrite-pearlite matrix. A prototype defect-location system for pressure-vessel, monitoring etc., is illustrated, and possible extension of acoustic emission to fields other than NDT are outlined. [Abstract 22-1332. *Metals Abstracts,* Vol. 9, 1976].

164 D. Birchon, R. Dukes, and J. Taylor. "Acoustic Emission Monitoring of a Pressure Vessel (Correlation of Results with Those of Ultrasonic Inspection)." Paper C86/74, pp 88–94 in ***Periodic Inspection of Pressurized Components.*** **Conference Publication 8, 1974. published by Mechanical Engineering Publications Limited for the Institution of Mechanical Engineers, London, England. 1975. conference held in London, England. June 4–6, 1974.**

Results of some defect location studies on a pressure vessel are reported and correlated with those of ultrasonic inspection. Good agreement was observed, with a probability greater than 90 percent that a defect location detected would be confirmed by ultrasonics. This good agreement is considered to result from the use of peak sensing rather than the more commonly used leading-edge triggering technique. Attention is drawn to the influence of the defect extension process upon the ease of detection, contrasting the difficulty of detecting slow crack growth with the ease of detection of pulses originating from the fracture of hard particles or their separation from the matrix, and to the influence of the Kaiser effect, which can mean that a flaw may not be detectable unless previously applied stress levels are exceeded, or flaw growth has occurred since the previous inspection, or some metallurgical recovery process has operated.

165 D. Birchon, R. Dukes, and J. Taylor. "Some Aspects of Defect Location and Assessment in Pressure Vessels Using Acoustic Emission Techniques." pp 669–684 in *Proceedings of the 2nd International Conference on Pressure Vessel Technology. Part 2: Materials, Fabrication and Inspection*. **American Society of Mechanical Engineers, New York. 1973. conference held in San Antonio, Texas. Oct 1–4, 1973.**

Some of the factors which have been studied in support of the design and construction of the UK Admiralty Materials Laboratory automatic defect-location system are discussed. Particular attention has been paid to producing a rugged piece of equipment for field use, to reduce the costs and complexity of hard and software, and to reduce the number and specialized competence required of the operators, in an attempt to produce a tool for engineers; the capabilities and limitations of the technique as applied to pressure vessel inspection therefore figure largely in the work. Some aspects of relating signals to their sources are discussed with special relevance to the shortcomings of *ring* counting—the technique most widely used at present. The paper concludes with a note on a simple design of controllable emitter for proving defect-location ability.

166 D. Birchon, R. H. Warren, R. Dukes, and J. England. "Some Aspects of the Detection, Processing and Interpretation of Stress Waves in Structural Materials." Admiralty Materials Laboratory, Holton Heath, Poole, Dorset, England. 1972; paper presented at the Institute of Physics Conference on Acoustic Emission, Imperial College, London, England. Mar 14, 1972. [proceedings were never published].

Outlined are MOD(N) interests in the structural validation of pressure vessels and critical structures by stress wave analysis and the development of an on-line defect detection and location plotting system. Some of the problems of validation by this technique and the performance requirements of the equipment are discussed. The development of stress wave sensors and signal amplification equipment are used as practical examples of how such requirements may be met.

167 D. Birchon, R. H. Warren, and P. M. Wingfield. "Structural Validation: A Positive Approach to NDT." *Chartered Mechanical Engineer*, **CHMGA, 19(6):54–60. June 1972.**

Structural validation is the term given to the assessment of the health of a structure, in the light of the duties to be performed. The method is an overall approach, intended to observe only those aspects of flaw behaviour relevant to a structure's service duties and to assessing their significance, either to establish the minimum extent of repair, or to establish confidence in operating the structure with a known remanent strength. Stress-wave emission analysis is the most promising technique at present available for development in structural validation of pressure vessels, hulls, etc., in metals and composites. [reproduced in part by permission of the Council of the Institution of Mechanical Engineers from *The Chartered Mechanical Engineer*].

168 **W. Bismuth. "Bruit de Frottement d'une Roue de Metro Roulant sur un Rail Circulaire (Sound of Rolling Friction from a Subway Wheel on a Circular Rail)." pp 166–172 in *Journees d'Etudes sur l'Emission Acoustique.* R. F. Gobin, R. Goutte, and C. Guillaud, Organizers. Institut National des Sciences Appliquees de Lyon, Avenue Albert Einstein, 69621 Villeurbanne, Lyon, France. Mar 1975. proceedings of a conference held in Lyon, France. Mar 17, 1975. (paper in French).**

169 **H. W. Blake and C. M. Davenport. "Acoustic Flaw Triangulation on a Thin Spherical Shell: An Analytical Solution." Y-1806. Union Carbide Corporation, Oak Ridge Y-12 Plant, Oak Ridge, Tennessee. Dec 30, 1971. [NTIS].**

An analytical solution for acoustic flaw triangulation on thin spherical shells has been prepared. A FORTRAN program is provided which takes as input the differences in the arrival times of acoustic energy at the transducers and evaluates the spherical coordinates of the flaw. Only three experimental time differences need be measured. For this measurement, four transducers are required. One transducer is placed at (R, 0, 0) in spherical coordinates; the other three are spaced 90 degrees apart around the equator.

170 **W. Blake. "Design, Installation and Operation of Computer Controlled Rock Burst Monitoring Systems." pp 157–167 in *Proceedings First Conference on Acoustic Emission/Microseismic Activity in Geologic Structures and Materials.* H. R. Hardy, Jr. and F. W. Leighton, Editors. Trans Tech Publications, Clausthal, West Germany. 1977. conference held at Pennsylvania State University, University Park, Pennsylvania. June 9–11, 1975.**

Microseismic techniques for monitoring rock-burst prone mine structures are being accepted by the mining industry. The installation and operation of automatic computer controlled rock burst monitoring (RBM) systems are now realities, but the versatility of such systems must be increased and expanded to allow more of the data analysis to be performed on-line by the computer. Basic guidelines for selecting the components of a computer controlled RBM are presented in detail. Installation procedures are described to emphasize the importance of the initial installation. The operation of the RBM, including on-site data analysis, is discussed at length to show the importance and need for having the capability to instantaneously update and evaluate new RBM data.

171 **P. J. Blalock. "Give Your Rotating Equipment a Physical." Dow Chemical U.S.A., Louisiana Division, Plaquemine, Louisiana. 1972; paper presented at American Society for Nondestructive Testing National Fall Conference, Cleveland, Ohio. Oct 16–19, 1972.**

A practical program for on-stream nondestructive examination of rotating machinery is one that works. It accomplishes objectives in the most efficient manner. It is an educational process that includes management, process, maintenance, and operational personnel. It has specialists working in the field with all these persons, examining the cause and effect, and determining a solution to a particular problem. It saves downtime and money. This paper presents Dow Chemical U.S.A. Louisiana Division's approach to nondestructive examination of rotating machinery. The approach here includes only one plan and, as any plan should be, is flexible and can be changed to suit all needs. The important thing is that the job was organized so that it was correctly completed in the least amount of time.

172 **T. H. Blewitt, R. R. Coltman, and J. K. Redman. "Low-Temperature Deformation of Copper Single Crystals."** *Journal of Applied Physics,* **JAPIA, 28(6): 651–660. June 1957.**

Single crystals of copper were deformed at 4.2 and 77.3 K. At 4.2 K, after a large strain produced by normal slip, jerky flow (discontinuous slip) occurs. From the study of reactor irradiated crystals, it was deduced that a packet of 30 slip lines, each containing 10^4 dislocations, was released to form each jerk of the discontinuous flow. For samples of certain orientation, the region of discontinuous flow was followed by a region of deformation twinning. The fact that deformation twinning was taking place was verified by X-ray methods. The twinning elements were determined to be the (111) plane and the [112] direction. At 77.3 K, discontinuous slip was not observed. In a limited range of orientations, twinning has been observed at 77.3 K. The effects of reactor irradiations on discontinuous slip and on twinning were also studied and it was determined that the occurrence of twinning was unaffected by the irradiations. The authors make specific reference to the observation of acoustic emission on p 653: "Beyond the region of normal slip, it is apparent from the discontinuities in the load-elongation curve that a high velocity localized extension occurs. This region has been designated as the discontinuous slip region. It should be noted that the sample apparently continues to work-harden with increasing deformation. A clicking noise was heard, which coincided with each discontinuity in the load-elongation curve. The way in which these discontinuities appear on the strip chart is shown."

173 **D. B. Board. "Incipient Failure Detection in High-Speed Rotating Machinery." pp 8–18 in** *Proceedings of the Tenth Symposium on Nondestructive Evaluation.* **Southwest Research Institute, San Antonio, Texas. June 1975. symposium held in San Antonio, Texas. Apr 23–25, 1975.**

During U.S. Army funded research of state-of-the-art diagnostic and prognostic techniques for potential application to the CH-47C Chinook Medium Lift Helicopter, a new technique for high frequency vibro-acoustic emission analysis was evaluated on three CH-47 drive system transmission in a regenerative test stand. Test results indicate that this new technique for high frequency vibro-acoustic analysis shows excellent potential for early stage *in situ* detection and identification of faults in complex, high-speed, rotating machinery. Of the five faults indicated and verified by teardown inspection during this test, only one was known at the outset of the test. All five were detectable through analysis of test data without any need for baseline information. This capability for *in situ* fault detection and isolation without the need for baseline data and with decisions based on engineering rationale rather than rationalized pattern recognition, indicates an important superiority over the more classical forms of low frequency vibration analysis that have been employed to date for helicopter transmission diagnosis and prognosis.

174 **F. Bodson, P. Flahaut, M. Fremoit, and J. Bonte. "Detection des Tapures de Trempe par Emission Acoustique (Detection of Quench Cracks by Acoustic Emission)."** *Qualite, Revue Pratique de Controle Industriel,* **QRPCA, 14(73):33–36. June 1975. (in French).**

175 **F. Bodson, P. Flahaut, M. Fremoit, and J. Bonte. "Influence des fluides de trempe sur la formation des tapures dans deux aciers 55S7 et mise en evidence par emission acoustique de leur delai d'apparition (Influence of Quenching Fluids on the Formation of Shrinkage Cracks in Two 55S7 Steels and Acoustic Emission to Reveal Their Delayed Appearance)."** *Traitement Thermique,* **TRTHA, (104):83–88. Apr 1976. (in French).**

A study was made of the conditions for the appearance of shrinkage cracks during quenching in various fluids of 10 to 24-mm diameter specimens in two 0.55% C low-alloy steels. Cracks appear notably after quenching specimens of sufficiently small diameter in water at a sufficiently high temperature. Acoustic emission tests have enabled the moment at which cracks appear to be specified. This occurs particularly at the end of the martensitic transformation in the centre of specimens. [Abstract 56-0571. *Metals Abstracts,* Vol. 9, 1976].

176 **V. V. Bogorodskii, V. P. Gavrilo, V. S. Grigor'ev, and A. V. Gusev. "Sound Generation Mechanisms in the Breaking of Ice Samples in a Liquid."** *Soviet Physics–Acoustics,* **SOPAA, 15(2):161–164. Oct–Dec 1969; translated from** *Akusticheskii Zhurnal,* **AKZHA, 15(2):184–188. Apr–June 1969. (in Russian).**

The spectral characteristics of noise generated in water during the breaking of ice are analyzed, with the ice regarded as a solid having widely varying physicomechanical properties, which are determined chiefly by the temperature, structure, and salinity. Hypotheses are advanced regarding the mechanism of generation of the sound field in water during the breaking of ice.

177 **H. J. Bohnstedt. "Betriebsuberwachung von Kernkraftwerkskomponenten durch Gerausch und Schwingungsanalyse (On-Load Surveillance of Nuclear Power Plant Components by Noise and Vibration Analysis)."** *Maschinenschaden,* **MSCNA, 48(3):95–96. 1975. (in German).**

The report published by the European Atomic Energy Commission reviews the experience gathered on on-load surveillance techniques, based on noise and vibration analysis, in respect of nuclear power plant components, light-water reactors, and steam turbines. The theoretical principles are discussed, the measuring devices described, and the applied surveillance techniques elucidated. A summary is then given of the test results obtained by the various organizations involved in the preparation of this report. Finally, a comparative analysis of these tests results is presented.

178 **V. S. Boiko, R. I. Garber, L. F. Krivenko, and S. S. Krivulya. "Acoustic Radiation by Twinning Dislocations."** *Soviet Physics–Solid State,* **SPSAA, 12(6):1387–1388. Dec 1970; translated from** *Fizika Tverdogo Tela,* **FTVTA, 12(6):1753–1755. June 1970. (in Russian).**

Acoustic radiation by twinning dislocations has been observed during the twinning of calcite. A comparison of signals obtained during the different stages of elastic twinning shows different time dependences for the acoustic radiation in different stages. The differences are presumably due to features of the velocity distribution of the dislocations and their motion near the crystal surface. The signals obtained during the expansion of the residual twinning layer are of lower amplitude and are shorter in duration, so their structure is revealed. The correlation between the nature of the signals and the dislocation velocities near the surface implies that the radiative part of the field of moving dislocations is most probably being observed. [reprinted by permission of the American Institute of Physics].

179 **V. S. Boiko, R. I. Garber, L. F. Krivenko, and S. S. Krvulya. "Sound Emission of Twinning Dislocations during their Emergence from the Crystal."** *Soviet Physics–Solid State,* **SPSSA, 11(12):3041–3042. June 1970; translated from** *Fizika Tverdogo Tela,* **FTVTA, 11(12):3624–3626. Dec 1969. (in Russian).**

An elastic twin represents an aggregation of a large number of dislocations of one type which on removing an external load leave the crystal under the action of the forces of surface tension, and their velocities can be sufficiently large. The experimental setup allows the establishment of a relation between the exodus of twinning dislocations and the appearance of the sound signal. The signal amplitude increases with the length of the twin. Thus, an increase in amplitude can be considered to be a result of increased velocity of motion of the dislocation. [Abstract 83644y. *Chemical Abstracts,* Vol. 72, 1970. Copyright by American Chemical Society].

180 **V. S. Boiko, R. I. Garber, L. F. Krivenko, and S. S. Krivulya. "Transition Emission of Sound by Dislocations."** *Soviet Physics–Solid State,* **SPSSA, 15(1):238–239. July 1973; translated from** *Fizika Tverdogo Tela,* **FTVTA, 15(1):321–323. Jan 1973. (in Russian).**

181 **L. Bolin. "A Bibliography on Acoustic Emission." Report LiTH-IKP-R-44. Linkoping University, Department of Mechanical Engineering, Linkoping, Sweden. 1974. [available from author].**

Bibliography contains 510 references, many with abstracts, and covers the time period from 1920 into 1974. References are listed under one of 22 subjects which best describe the major subject covered in the report or article. The subjects and number of references for each section are: Historical-17; Feature articles-20; General-132; Tensile tests on virgin metallic materials-24; Fracture or crack growth tests in metallic materials-43; Theoretical-11; Amplitude and frequency distribution-10; Test equipment-32; Fatigue-17; Creep-2; Phase transformation-10; Stress corrosion and hydrogen embrittlement-12; Bauschinger and Portevin-Le Chatelier effects-8; Composite materials-24; Ceramic materials-13; Weld monitoring-25; Structural integrity analysis: general-64, piping-2, filament wound constructions-4, mechanical cables-5; Miscellaneous applications-31; and Bibliographies-4.

182 **M. M. Bolles. "Investigation and Design of a Fluid Coupled Acoustic Transducer System for the Detection of Acoustic Emission." Ph.D. Thesis. Tulane University, New Orleans, Louisiana. 1967; University Microfilms 67–17902; published in "Final Report on Acoustic Emission Research at Tulane University under UCRL P. O. 2412203, Part 3." M. M. Bolles, B. R. Bass, H. A. Thompson, and K. A. Adams. UCRL–13231, Part 3. Lawrence Radiation Laboratory, Livermore, California. Nov 1966. [see Reference 183].**

Acoustic emission is a term used to describe the phenomenon of the spontaneous production of elastic stress waves in some crystalline materials when subjected to monotonically increasing state of stress. These stress waves are of extremely small amplitudes, and can be separated from background noise only with great difficulty. The most generally used method for the detection of acoustic emission is a tensile specimen with a piezoelectric crystal placed on one end of the specimen. The resulting signal is then electronically processed for display and analysis. In this investigation, a fluid-coupled transducer system has been designed and evaluated in an attempt to increase the signal-to-noise ratio for a typical acoustic emission pulse. The system employed consisted of a fluid-filled, elliptic cylindrical container with the longitudinal axis of the specimen arranged collinearly with one of the focal axes of the container. A cylindrical piezoelectric crystal was located in a similar manner along the other focal axis. The container served as a reflector and was designed to focus the acoustic energy emitted by the specimen onto the crystal. For the purpose of comparison, an additional piezoelectric crystal was affixed to the end of the specimen. An analysis of the results showed that the signal-to-noise ratio was generally higher for the fluid-coupled transducer system. The analysis also indicated a correlation between the direct-coupled and fluid-coupled systems of approximately 60 to 75 percent relative to the least active of the systems.

183 **M. M. Bolles, B. R. Bass, H. A. Thompson, and K. H. Adams. "Final Report on Acoustic Emission Research at Tulane University under UCRL P.O. 4212203, Part 3." UCRL-13231, Part 3. Lawrence Radiation Laboratory, Livermore, California. Nov 1966. [see also Reference 182].**

The first section contains a discussion of the attempts made to employ the high pressure system. The system was used to internally pressurize test specimens in an effort to obtain acoustic emission under a biaxial state of stress (Goal No. 2). Section II of the report contains the results of additional work performed on the selection of optimum transmission fluids which was performed after Part II of the complete report had been submitted. Both Part II and Section II of Part III contain work done toward the PhD dissertation of Milton M. Bolles. In this work, the subject of the optimum design for mounting a fluid-coupled transducer is treated and a comparison of the experimental results obtained with both direct and fluid-coupled transducers is presented. Section III of the report is in dissertation form and sets forth the work which was done toward the accomplishment of Goal Nos. 3 and 5 of the contract. Section IV contains the shop drawings and a list of materials which were used in the construction of the reflector assembly surrounding the remote-coupled transducer.

184 **Yu. I. Bolotin and V. M. Belov. "Weld Quality Control by Acoustic Emission during Electron-Beam Welding."** *Welding Production,* **WEPRA, 23(4):28–30. Apr 1976; translated from** *Svarochnoe Proizvodstvo,* **SVAPA, (4):29–31. 1976. (in Russian).**

Detection of defects directly during electron-beam welding is of considerable importance, as this would make it possible to effect a repair of a defect without subsequent depressurisation of the chamber. A method of recording acoustic oscillations from developing defects for assessment of the quality of electron-beam welding has been described by Kharchenko *et al.* But inspection by this method can be effected only during cooling of the weld and in the audible frequency range. The presence of noise of an industrial nature restricts the application of the method and the associated apparatus. Equipment operating in the frequency range 70 to 1500 kHz may be applicable for this purpose. The employment of acoustic emission for the inspection of welded joints during welding is currently at the initial stage of development. The necessary prerequisites exist for applying this method to the detection of defects during the welding and cooling of welds. The mechanical load required for the generation of AE is created as a result of the structural inhomogeneities of the weld material and the nonuniformity of the thermal conditions. Substantial stresses develop in the weld material and in the heat-affected zone, and they give rise to AE in defective regions. Welding was carried out at a turbine plant on type LUCH-1 electron-beam welding equipment on specimens made of steel Kh12.

185 **Yu. I. Bolotin, Yu. B. Drobot, V. V. Nechaev, V. M. Belov, and B. K. Gavrilenko. "The Use of Acoustic Emission for Examination of Deformation Processes in Welded Joints."** *Welding Production,* **WEPRA, 22(2):12–14. Feb 1975; translated from** *Svarochnoe Proizvodstvo,* **SVAPA, (2):9–11. Feb 1975. (in Russian).**

Acoustic emission was used to study deformation and fracture during the automatic submerged-arc welding of 30KhGSNA steel.

186 **Yu. I. Bolotin, V. A. Greshnikov, A. A. Gusakov, Yu. B. Drobot, and V. P. Chentsov. "Use of Stress Wave Emission for Nondestructive Testing of Materials and Articles."** *Soviet Journal of Nondestructive Testing,* **SJNTA, 7(6):617–632. Nov–Dec 1971; translated from** *Defektoskopiya,* **DEFKA, (6):5–25. Nov–Dec 1971. (in Russian).**

Survey of the literature on methods and means of nondestructive testing of the quality of materials and articles using stress wave emission. It considers the physical principles of the method, the principles for the design of measuring apparatus, some results of applied experimental investigations, and examples of the use of nondestructive testing in practice. An evaluation is given of the methods and the areas for their practical application.

187 **Yu. I. Bolotin, L. A. Maslov, and V. I. Polunin. "Establishing a Correlation between the Size of a Crack and the Amplitude of Acoustic Emission Pulses."** *Soviet Journal of Nondestructive Testing,* **SJNTA, 11(4):509–512. July–Aug 1975; translated from** *Defektoskopiya,* **DEFKA, (4):119–122. July–Aug 1975. (in Russian).**

A method and apparatus for determining the dimensions of cracks in metal samples by reference to the amplitude of acoustic emission pulses are described, and corresponding tests carried out with Cr steel 40Kh and duralumin D16-AT are reported. A theoretical expression linearly relating the length of a spasmodically propagating crack to a simple function of the amplitude of acoustic emission is derived. The difference between the crack lengths estimated in this way and those found by direct measurement is not greater than 30% for the steel samples and 7% for the Duralumin. [Abstract 22-0434. *Metals Abstracts,* Vol. 9, 1976].

188 **H. Borchers and J. Kaiser. "Akustische Effekte bei Phasenubergangen im System Blei–Zinn (Acoustic Effects during Phase Transformations in the Lead–Tin System)."** *Zeitschrift fur Metallkunde,* **ZEMTA, 49(2):95–101. Feb 1958. (in German).**

In the lead-tin system the applicability of the electroacoustical method is investigated for the determination of the melting points of pure metals, of liquidus and solidus, of eutectic line, and of the solid solubility line. The method gives additional information in the case of solidification processes which can also be observed otherwise. It clearly indicates, however, boundaries in the solid state which are more difficult to determine by standard methods.

189 **H. Borchers and H. M. Tensi. "Eine Verbesserte Piezoelektrische Methode zur Untersuchung von Vorgangen in Metallen bei Mechanischer Beanspruchung und bei Phasenanderung (Improved Piezoelectric Method for Investigation of Processes in Metals during Mechanical Stressing and Phase Changes)."** *Zeitschrift fur Metallkunde,* **ZEMTA, 51(4):212-218. Apr 1960. (in German).**

Piezoelectric crystals (quartz, $BaTiO_3$) were used to translate mechanical impulses emanating from metals under stress into proportional electric impulses which could be amplified and recorded. The application used for this electroacoustical test method is described, and schematic illustrations are given. The curve of mechanical changes occurring during tensile stress tests in Zn showed that with increasing stress, a rapid increase in impulses occurred up to a maximum. In the region of plastic deformation, the impulses dropped rather sharply. The piezoelectric method was used to study the melting and crystallization of metals and to detect phase changes. During cooling of a contracting molten metal system no impulses were recorded. Soon after solidification begins, the first impulses were recorded and quickly reached a maximum, coinciding with the end of the solidification process. If the whole metal mass is in the solid state, all impulse sending stops abruptly. Metals such as Bi and Sb send out impulses only during melting. During melting, these two metals undergo phase changes and contraction which may be responsible for the recorded impulses. In the system Cd-Zn, two impulse maxima are recorded in the region of phase separation. One maxima is correlated to a one-phase region, but not all alloys display the same behavior. Seed formation and crystal growth in metals do not give recordable impulses. [Abstract 14035f reprinted by permission of the American Chemical Society from *Chemical Abstracts,* Vol. 54, 1960].

190 **H. Borchers and H. M. Tensi. "Piezoelektrische Impulsmessungen bei der Phasenanderung von Unlegierten Proben der binaren Legierungen Blei-Zinn, Blei-Antomon und Kadmium-Zink (Piezoelectric Impulse Measurements during Phase Reactions in Unalloyed Specimens of the Binary Alloys Lead-Tin, Lead-Antimony and Cadmium-Zinc)."** *Metall,* **MTLLA, 17(8):784-788. Aug 1963. (in German).**

Piezoelectric impulses were measured during phase transformations of unalloyed Bi and Zn and of binary Pb-Sn, Pb-Sb, and Cd-Zn alloys to determine usefulness of these *acoustic* recordings; it was found that transformations accompanied by dilatation are recorded feebly at best; transformations accompanied by volume contraction during cooling cause impulses only toward end of reaction, those during melting are registered from beginning to end; decomposition in solid state could not be observed. [abstract reprinted from *Engineering Index,* p 1197, 1963].

191 **H. Borchers and H. M. Tensi. "Piezoelektrische Impulsmessungen wahrend der Mechanischen Beanspruchung von Al Mg 3 und Al 99 (Piezoelectric Impulse Measurements during Mechanical Stressing of Al Mg 3 and Al 99)."** *Zeitschrift fur Metallkunde,* **ZEMTA, 53(10):692–695. Oct 1962. (in German).**

The piezoelectric measurement of mechanical impulses that appear during plastic deformation of samples consisting of AlMg3 and Al99 is reported. By means of different heat treatment, the formation of the yield strength and the delivery of impulses can be varied. The Kaiser effect, or, the determination of the preload by means of piezoelectric measurements and the connection between mechanical impulses and the energy of deformation are discussed using macroscopic atomic theories.

192 **J. F. Borst. "Acoustic Emission: Is It a Promise or Mirage?"** *Nuclear Engineering International,* **NEINB, 22(254):54–56. Mar 1977.**

The promise of acoustic emission is that is offers . . . a technique which does not require people to go and look for defects, but rather lets the defects themselves tell their story . . . answers which could potentially quantify a defect not by size but directly in terms of its potential severity to the integrity of the structure . . . a tool which would permit listening to changes in the indications to allow reactor shutdowns and corrective actions to be made as conditions require.

193 **R. J. Bossert and P. M. Biggs. "Acoustic Emissions Indicate Weld Quality of Pressurized Vessels." Bendix Corporation, Kansas City, Missouri. Sep 10, 1974; Appendix B. pp 45–53 in** *IMOG (Interagency Mechanical Operations Group) Subgroup on Environmental Testing: Reports and Meeting Minutes.* **Meeting No. 26 held at Bendix Corporation, Kansas City, Missouri. Sep 10–11, 1974. proceedings published Apr 1975. [OUO, not available to the public].**

194 **S. F. Botten. "Acoustic Emission and Its Practical Applications."** *FWP Journal,* **FWPJA, 15(3):31, 32, 34, 36, 38, and 40. Mar 1975.**

Acoustic emission provides early warning of the presence of subcritical cracks, enabling the prevention of impending failure which could be catastrophic. The mechanism of noise formation by materials under stress is described. Methods of recording this noise with transducers and reducing background electrical noise to detect crack growth are given with schematic diagrams. The use of geometric arrays of transducers to locate the position of a crack is outlined. [Abstract 22-0861. *Metals Abstracts,* Vol. 8, 1975].

195 **K. D. Boultinghouse. "A Stress Corrosion–Environmental Load Frame and Instrumentation." SC–DR–710845. Sandia Laboratories, Albuquerque, New Mexico. Jan 1972.**

A load frame with features to apply uniaxial loads, to control the environment of the test, and to monitor crack propagation has been designed to allow the study of stress corrosion cracking.

196 **Yu. S. Boyarskaya, I. I. Averbukh, M. S. Kats, E. I. Purich, and Yu. N. Dubrovin. "Use of the Acoustic Emission Method to Evaluate Material Brittleness."** *Industrial Laboratory,* **INDLA, 40(5):681–683. Nov 1974; translated from** *Zavodskaya Laboratoriya,* **ZYDLA, 40(5):550–553. May 1974. (in Russian).**

The average number of emission pulses (N) per crack (T) on a crystal remains constant over a wide crack concentration range. The N/T increases with increasing stress (P) applied to a crystal. This is explained by an increase in the number of cracks and their length with increasing load. The P dependence of N/T is linear. For LiF crystals, an anisotropy of N/T was observed. The N/T was 2-3 fold larger when the crystals were scratched along [100] than along the [110] direction on the (001) face. This is attributed to a different mechanism of brittle fracture of the crystal. Based on the N/T anisotropy of LiF crystals, the N/T value can be used not only for determining the brittleness of crystals but also for determing the degree and type of brittle fracture of crystals. [Abstract 160431s reprinted by permission of the American Chemical Society from *Chemical Abstracts,* Vol. 81, 1974].

197 **R. W. Boyle, A. M. Sullivan, and J. M. Krafft. "Determination of Plane Strain Fracture Toughness with Sharply Notched Sheets."** *Welding Journal,* **WEJUA, 41(9):428s–432s. Sep 1962; ASME Publication 62–MET–13. American Society of Mechanical Engineers, New York. 1962.**

In sharply notched sheet fracture specimens under load, crack extension is often signaled by an abrupt readily discernible burst of growth from the notch tip. Fracture toughness, calculated at this point, can be closely identified with that developed under known conditions of plane strain. Present results on 7075-T6 Al, Zn, Mg alloy indicate that this new technique permits plane-strain fracture toughness to be validly determined with specimens only one-fourth the size of notched rounds conventionally employed for this purpose. With compliance gage instrumentation, specimens of only one-eighth this size yield useful results. Of acoustic emission, the authors state that the observance of a distinct break in the load-deflection curve of a sharply notched sheet specimen corresponding to a *pop-in* of a crack from the notch can be used to evaluate the plane-strain fracture toughness of the material. It may often be observed by an audible click.

198 **W. F. Brace. "Current Laboratory Studies Pertaining to Earthquake Prediction." Massachusetts Institute of Technology, Department of Geology and Geophysics, Cambridge, Massachusetts. 1967.**

Laboratory studies in three areas may prove useful in earthquake prediction. These studies deal with the source mechanism, with microfracturing, and with changes in rock properties under stress. Of the various processes which produce sudden stress drops in rocks, brittle fracture and frictional stick-slip are being currently studied. Both are accompanied by numerous small elastic shocks which begin at stresses well below the fracture or sliding stress. Although the shocks in both cases are due to microfracturing, the pattern of microfracturing prior to brittle fracture differs in a characteristic way from that before frictional sliding. Just before fracture, the number and intensity of shocks increase markedly, and the locations from which the larger shocks emanate cluster along the eventual fault plane. Neither feature has been observed prior to frictional stick-slip. Large changes in sound velocity have been observed at certain phase changes. Electrical resistivity and magnetic susceptibility are strongly affected by stress: in rocks partially saturated with water, electrical resistivity may decrease rapidly at low stress; in fully saturated rocks, it may decrease rapidly at high stress, just before fracture. Little change takes place just before frictional sliding of saturated rocks.

199 **P. H. Braddock. "New Method for NDT Uses Ultrasonic Waves from Loads."** ***Metalworking News,*** **pp 1 and 4. Oct 5, 1964. [available from Fairchild Publications, Inc., New York].**

A short news report on ultrasonic sound generated by loading. The test is designed to be a quality control device for predicting structural integrity of metals and other materials by proof loading at a fraction of design load. Ralph M. Parker, research engineer associated with George M. Swett Company and Western Inspection Service, is the inventor and holds patent rights.

200 **B. T. Brady. "Seismic Precursors before Rock Failures in Mines."** ***Nature,*** **NATUA, 252(5483):549-552. Dec 13, 1974.**

Seismic precursors, such as anomalous V_p values or seismic activity (or both), whose behavior is qualitatively similar to those reported to precede earthquakes are observed before rock failures in dry underground mines. Similar processes may be involved during failure of rock in the mine and the earthquake, and diffusion of fluids into the focal region of a potential earthquake may not be a necessary condition to produce most of the precursors reported.

201 **W. J. Bratina. "Ultrasonic Attenuation Compared to Acoustic Emission in the Study of Deformation Mechanisms." Paper 11.5, pp 233-237 in** ***Ultrasonics***

International 1975: Conference Proceedings. **IPC Science and Technology Press Ltd, Guildford, Surrey, England. 1975. conference held at Imperial College, London, England. Mar 24–26, 1975.**

The present state of acoustic emission is largely empirical. The theory in terms of mechanisms is still speculative, whereas the physical picture and the mathematical theory of ultrasonic attenuation are well-developed. In order to assess more completely the possible deformation mechanisms for iron and various iron-base alloys, an attempt was made to correlate ultrasonic attenuation data with published acoustic emission results. The phenomena investigated included tensile deformation, yielding, work hardening, dislocation depinning and generation over the temperature range 77 to 500 K. Further, the effects of strain aging and strain rate as studied by both techniques are discussed. Finally, restressing of a previously deformed material is analysed.

202 **F. R. Breckenridge, C. E. Tschiegg, and M. Greenspan. "Acoustic Emission: Some Applications of Lamb's Problem." NBSIR 74-565. Institute for Basic Standards, National Bureau of Standards, Washington, D.C. Sep 1974. [NTIS: AD/A-009-541];** *Journal of the Acoustical Society of America,* **JASMA, 57(3):626–631. Mar 1975.**

A method for obtaining the signatures (waveforms) of certain acoustic emission events has been developed. The waveform is that at the source, free of contamination by ringing of the specimen, apparatus, and transducer. The technique is based on the comparison of two signals at the transducer, one from the event in question and one from an artificial event of known waveform. The apparatus is also adapted to the calibration of transducers in a certain sense. The configurations of source (real or simulated acoustic-emission event) and receiving transducer correspond to those of some special cases of Lamb's problem. As a byproduct, the results may be of some interest to seismologists.

203 **A. Bret. "Detection Acoustique dans le Reacteur Rapsodie (Acoustic Detection in the Rapsodie Reactor)." pp 413–432 in** *Journees d'Etudes sur l'Emission Acoustique.* **P. F. Gobin, R. Goutte, and C. Guillaud, Organizers. Institut National des Sciences Appliquees de Lyon, Avenue Albert Einstein, 69621 Villeurbanne, Lyon, France. Mar 1975. proceedings of a conference held in Lyon, France. Mar 17, 1975. (paper in French).**

Acoustic emission studies in the Rapsodie reactor showed that it is possible to reveal phenomena such as: argon carryover in the sodium coolant circuit, variation of the operating distance between core ceiling and fuel elements, or variations in the reactor operating conditions: power, sodium temperature, and primary pump flow rate.

204 A. Bret, M. Gaillard, and V. Lechel. "Acoustic Detection of Bubble Emission in the Rapsodie Reactor." ERDA-TR-60. CEA Centre d'Etudes Nucleaires de Cadarache, Saint-Paul-les-Durance, France. 1975; translation of French paper presented at the European Nuclear Conference, Paris, France. Apr 21-25, 1975. [NTIS: ERDA-TR-60, CONF-750411-39].

The acoustic detection instrumentation of the Rapsodie reactor is described and the results of acoustic measurements obtained during the study of the operational parameters are presented. Finally, it is shown how phenomena such as argon entrainments can be detected with a great sensitivity.

205 P. W. Bridgman. "Effects of High Shearing Stress Combined with High Hydrostatic Pressure." *Physical Review,* PHRVA, 48:825-847. Nov 15, 1935.

The author observed during rotation under compressive loading of 57 elements that nonmetals usually emitted grinding and snapping noises while metals almost always were quiet. No instrumentation was used to observe the acoustic emissions.

206 P. W. Bridgman. "Flow Phenomena in Heavily Stressed Metals." *Journal of Applied Physics,* JAPIA, 8:328-336. May 1937.

The author reports that few metals and many of the nonmetals, instead of rotating smoothly under torsion, rotate with protest, usually in a noisy manner. Some materials chatter, some squeak, some make a grinding noise, while others hiss. The most important of these acoustic emissions is snapping which occurs at more or less regular intervals in many materials during rotation. Snapping usually accompanies plastic flow with no way of predicting when the next snap will occur. The rate of snapping was observed to be strain-rate dependent in some materials. In brittle materials, rotation produced a succession of jumps and snaps. The author explains this acoustic emission phenomenon as internal rupture. Snapping was found to be common among salts, but much rarer in metals. Snapping was observed in titanium, strontium, cerium, arsenic, and osmium.

207 P. W. Bridgman. *Studies in Large Plastic Flow and Fracture.* Harvard University Press, Cambridge, Massachusetts. 1964. pp 283-284 and 288. [book originally published in 1952 by McGraw-Hill].

In Chapter 16, under "Shearing Combined with Approximately Hydrostatic Pressure," the author discussed the acoustic emission phenomenon of snapping caused by internal rupture. This is a brief discussion of material presented in more detail in References 205 and 206.

208 G. H. Brigman. "Theoretical Study of Acoustic Wave Generation by Friction." pp 317-318 in "1970 Applied Research Program." Research Summary APR-15. General Dynamics, Convair Aerospace Division, Fort Worth, Texas.

209 **J. S. Brihmadesam and N. J. McCormick. "Acoustic Emission Analysis of Stress Corrosion Cracking in Type 304 Stainless Steels."** *Nuclear Technology,* **NUTYB, 29(1):127–135. Apr 1976.**

Past experiments to demonstrate the feasibility of using acoustic emissions analysis to detect stress-corrosion crack growth during proof testing have been performed for Type-304 stainless steel. There was little or no acoustic emission at operating stress levels, where the operating stress intensity factor was 58.0 ksi$\sqrt{\text{inch}}$ but emission increased to a maximum at the proof stress intensity factor of 73.0 ksi$\sqrt{\text{inch}}$ for specimens that were given stress relief anneal and at a proof stress intensity factor of 102.0 ksi$\sqrt{\text{inch}}$ for specimens heat treated at higher temperatures.

210 **B. J. Brindley and R. P. Harrison. "Acoustic Emission Characteristics of a Mild Steel." Laboratory Note No. RD/L/N 137/72. Central Electricity Research Laboratories, Leatherhead, Surrey, England. June 15, 1972.**

The acoustic emission characteristics of a low carbon strongly aging mild steel have been studied at room temperature. The onset of failure can be predicted in this material. It is suggested that acoustic emission recovery depends upon metallurgical changes. It has been shown that the cracking of oxide films can give rise to emissions. The results are discussed in terms of their relation to proof testing of pressure vessels.

211 **B. J. Brindley, R. P. Harrison, and I. G. Palmer. "A Study of Acoustic Emission from Mild Steel." Laboratory Memorandum No. RD/L/M 381. Central Electricity Research Laboratories, Leatherhead, Surrey, England. July 27, 1972; paper presented at the Annual Meeting of the International Institute of Welding, Toronto, Canada. July 13, 1972.**

The memorandum summarizes some results of a study of acoustic emission during deformation and fracture of a carbon-manganese pressure vessel steel and a strongly aging mild steel. Detailed results are reported by I. G. Palmer in Reference 1344 and by B. J. Brindley and R. P. Harrison in Reference 210. Preliminary results on a nuclear reactor pressure vessel weld metal are also reported.

212 **B. J. Brindley, J. Holt, and I. G. Palmer. "Acoustic Emission–3: The Use of Ring-Down Counting."** *Non-Destructive Testing.* **NDETA, 6(6):299–306. Dec 1973. [This is the third in a series of articles on acoustic emission. See References 1716, 1404, 336, and 712 for Parts 1, 2, 4, and 5, respectively].**

The authors describe the principle of ringdown counting of acoustic emission signals and discuss its advantages and limitations. They consider the experimental factors which influence the ringdown count and describe methods for their investigation. They determined experimentally the dependence of ringdown count on signal-peak height and used it with the peak-height distribution to obtain the dependence of total ringdown count on system gain for a particular test situation. They also describe the application of ringdown counting to the study of the metallurgical variables affecting the acoustic-emission activity of steels. They show how, in the case of plastic deformation, it can be used to determine the plastic zone size and hence the size of defects.

213 **C. R. Brinkman and J. M. Beeston. "The Effect of Hydrogen on the Ductile Properties of Irradiated Pressure Vessel Steels." IN-1359. Idaho Nuclear Corporation, National Reactor Testing Station, Idaho Falls, Idaho. Feb 1970. [NTIS].**

The influence of hydrogen on the mechanical properties (i.e., ductility, fracture strength, and tendency towards delayed failure) was investigated for several irradiated pressure vessel steels. Included in this investigation were ASTM A302B, A542, and HY-80 steel irradiated at or to fluences of from 8×10^{18} to 4×10^{20} n/cm^2 $E > 1$ MeV. Specimens from plate sections of these steels, which had been quenched and tempered and some from A542 which were given prestrain and heat-treatment modifications, were prepared and tested. The effects of strength level from irradiation, heat treatment, and microstructure were thus determined. Reductions in ductility and true fracture strength occurred with increasing hydrogen content, but were not extensive at strength levels less than 180 ksi in specimens containing 1 to 2 ppm hydrogen. This concentration, however, produced a marked effect on the ductile properties when the strength level was increased by irradiation hardening or heat treatment beyond this threshold range. Irradiation hardening increased the magnitude of the decrease in notched strength resulting from a given hydrogen content in all of the steels and conditions tested. Hydrogen-induced delayed failure, however, did not occur to any large extent in HY-80, A302B, and A542 steel in the normal quenched and tempered condition even after irradiation to fluences in excess of 10^{20} n/cm^2 $E > 1$ MeV and hydrogen concentrations of up to 4 ppm.

214 **B. Brintet and Guebin. "La Maintenance des Structures en Aeronautique au Moyen de l'Emission Acoustique (Maintenance of Aeronautical Structures by Means of Acoustic Emission)." pp 92–109 in** ***Journees d'Etudes sur l'Emission Acoustique.*** **P. F. Gobin, R. Goutte, and C. Guillaud, Organizers. Institut National des Sciences Appliquees de Lyon, Avènue Albert Einstein, 69621 Villeurbanne, Lyon, France. Mar 1975. proceedings of a conference held in Lyon, France. Mar 17, 1975. (paper in French).**

215 **W. Brockmann and T. Fischer. "Acoustic Emission Analysis as a Method of Testing Adhesive Metal Joints." pp 277–300 in** ***Proceedings of the 22nd National SAMPE Symposium, Volume 22: Diversity – Technology Explosion.*** **Society for the Advancement of Material and Process Engineering, Azusa, California. 1977. symposium held in San Diego, California. Apr 26–28, 1977.**

The acoustic emission analysis of stressed and aged metal bonds leads to a better knowledge of the deformability of adhesives and their influence on the crack propagation than technological test methods. In their acoustic emission behavior in the unaged and aged state, different adhesives, for example, epoxy-nitrile and epoxy-nylon adhesives show different properties. The aging process in cleavage specimens in a humid environment needs only a few days.

216 **J. J. Broderick, R. F. Burchill, and H. L. Clark. "Design and Fabrication of Prototype System for Early Warning of Impending Bearing Failure." NASA-CR-123717, MTI-71TR1. Mechanical Technology Incorporated, Latham, New York. Jan 1972. [NTIS: N72-28501].**

Ball-bearing performance tests run on several identical ball bearings under a variety of load, speed, temperature, and lubrication conditions are reported. Bearing temperature, torque, vibration, noise, strain, cage speed, etc., were monitored to establish those measurements most suitable as indicators of ball-bearing health. Tape records were made under steady-state conditions of a variety of speeds and loads. Sample sections were selected for narrow-band spectral analysis with a real-time analyzer. An artificial flaw was created across the inner race surface of one bearing using an acid-etch technique to produce the scratch. Tape records obtained before and after established a characteristic frequency response that identifies the presence of the flaw. The signals found most useful as indicators of performance degradation were ultrasonic outputs.

217 **A. E. Brown and R. G. Liptai. "Round-Robin Testing of Acoustic Emission Source." UCRL-73272. Lawrence Radiation Laboratory, Livermore, California. June 29, 1970. [NTIS]; pp 318-331 in *Acoustic Emission, ASTM STP 505.* American Society for Testing and Materials, Philadelphia, Pennsylvania. 1972. proceedings of symposium held in Bal Harbour, Florida. Dec 7-8, 1971.**

The use of a martensitic phase transformation in gold-47.5 at % cadmium alloy as a source of acoustic emissions is discussed. The use of this source as a standard provided a means of comparing acoustic emission data from several laboratories. The results of round-robin testing of this alloy at various laboratories are also presented.

218 **C. R. Brown. "Acoustic Emission from GRP." pp 7-1 to 7-5 in "Newsletter on the Non-Destructive Testing of Composite Materials and Adhesive Bonded Joints." Issue Number 4. Royal Aircraft Establishment, Farnborough, Hants, England. July 1975.**

An investigation of acoustic emission from specimens of GRP was begun in 1974. The emphasis has been on the use of acoustic emission for materials assessment rather than for defect location. Preliminary experiments were made with GRP beams of approximate dimensions: length 300 mm; width 10 mm; thickness 6 mm. These had been cut from 10-mm sheets to give batches of nominally identical specimens. The original sheets were classified in terms of matrix type: epoxy or polyester, resin-hardener ratio, and density of square woven-roving reinforcement. Most of these experiments were performed with an apparatus comprising a Dunegan type D140-30B transducer, a Dunegan type 802P 60db preamplifier, and a Dunegan type 301 signal conditioner. This system gave a ringdown count with a threshold of 140 V (referred to the transducer) for signals in a 100 to 300 kHz pass band. The limitations of this system (for the study purposes) were clear at the outset. The ideal technique would have been to measure the energy of the acoustic emission signals in the hope of relating this to the energy released in the (thermodynamically) irreversible processes responsible for failure. As the ringdown count for a given event is roughly proportional to the logarithm of the energy associated with that event, the integrated (or summed) ringdown count for many events (e.g., for a tensile test) is at best a measure of the logarithm of the product of the energies (a quantity which does not lend itself to analysis or prediction). Further, statistical analysis of the results of large numbers of identical experiments has shown that the ringdown count may be a much more dispersive quantity than are the macroscopic physical properties of the specimens. But the physical properties of nominally identical GRP specimens already show appreciable scatter; so the results obtainable with the combination of specimens and equipment could only take the form of empirical relationships based on the analysis of adequate numbers of *identical* experiments.

219 **J. W. Brown. "An Investigation of Microseismic Activity in Rock under Tension." M.S. Thesis. Pennsylvania State University, Mining Department, University Park, Pennsylvania. 1965.**

220 **J. W. Brown and M. M. Singh. "An Investigation of Microseismic Activity in Rock under Tension."** ***Transactions of the Society of Mining Engineers of AIME,*** **TAMMA, 233:255–265. 1966.**

221 **W. F. Brown, Jr. and J. E. Srawley. "Current Status of Plane Crack Toughness Testing." NASA TM X-52209. National Aeronautics and Space Administration, Lewis Research Center, Cleveland, Ohio. 1966. [NTIS: N66-28019]; paper presented at the American Society for Testing and Materials Annual Meeting, Atlantic City, New Jersey. June 29, 1966.**

The report deals with the design and testing of crack-notched specimens for determination of the resistance of materials to unstable opening-mode crack extension under plane strain conditions. The entire text in the report concerning acoustic emission is as follows: "Present techniques for detection of crack sounds provide no way to relate the acoustic emission to the amount of crack propagation that has occurred. Also it may be difficult to eliminate sources of extraneous noise. At present the acoustic method best serves as a supplement to the previously described techniques. For example, the presence or absence of sound indicates whether a deviation in linearity of the load-displacement record is due to crack propagation on a fine scale or to plastic flow at the crack tip."

222 **W. F. Brown, Jr. and J. E. Srawley. p 40 in** ***Plane Strain Crack Toughness Testing of High Strength Metallic Materials, ASTM STP 410.*** **American Society for Testing and Materials, Philadelphia, Pennsylvania. 1967.**

The objective of this publication is to present a state-of-the-art survey of the analytical and experimental basis for determination of the plane-strain crack toughness of metallic materials. It is anticipated that the information presented will serve as a basis for formulating recommended practices for K_{Ic} testing. At present acoustic emission testing best serves as a supplement to other techniques. For example the presence or absence of sound indicates whether a deviation in linearity of the load-displacement record is due to crack propagation on a fine scale or to plastic flow at the crack tip. The acoustic method has the greatest inherent sensitivity to crack extension and responds only to actual crack movement.

223 **R. H. Bryan. "Acoustic Emission Monitoring and Post-Test Ultrasonic Examination of Intermediate Vessel V-9." Section 5.1, pp 59-68 in "Quarterly Progress Report on Reactor Safety Programs Sponsored by the NRC Division of Reactor Safety Research for October–December 1974, II. Heavy-Section Steel Technology Program'" G. D. Whitman. ORNL-TM-480. Volume II. Oak Ridge National Laboratory, Oak Ridge, Tennessee. Mar 1975. [see Reference 981].**

Intermediate vessel V-9, which was tested August 28, 1974, was monitored for acoustic emission by three companies, each working independently of the others. During the pressurization of V-9, the acoustic emission observations were reported orally to the ORNL testing staff with one report subsequently written and included in this section. In the last stages of pressurization, the incidence of acoustic emissions from the region of the cylinder-to-hemisphere weld, in comparison with emissions from the neighborhood of the prepared flaw, raised questions as to whether a rupture along the weld might precede rupture at the flaw. The vessel ruptured at the prepared flaw at a pressure of 26,900 psig, about 2.8 times design pressure. After the test of vessel V-9, the cylinder-to-hemisphere weld was examined ultrasonically.

224 **P. Buchman. "On Acoustic Emission from Ferroelectric Crystals."** *Solid-State Electronics,* **SSELA, 15(1):142-144. Jan 1972.**

Acoustic emission was observed from ferroelectric crystals during polarization reversal and phase transition. These signals originate from deformation and strain as a result of motion of domains or changes in the dimensions of the unit cell. Single crystals of $BaTiO_3$ and TGS were employed in the experiments. Contact to the $BaTiO_3$ crystals was made via a thin film of gold to which fine wires were attached by means of a conductive epoxy. The acoustic emission signals were detected by a lead zirconate titanate

(PZT) transducer operated at a resonance of 100 kHz. The amplified acoustic emission signal was fed into a count rate meter and displayed on an X–Y recorder. The coupling between the ferroelectric crystals and the transducer was made via silicone oil. [reprinted by permission of Pergamon Press Ltd from *Solid-State Electronics*].

225 **J. S. Buck and L. J. Graham. "Acoustic Emission Testing Experience at the EBOR Facility." pp 15–38 in "A Joint Industry-Utility Program to Develop NDT Techniques for Nuclear Reactor Inspection: An Addendum Report to Steering Committee EEI Project RP-79." Compiled by E. R. Reinhart. Southwest Research Institute, San Antonio, Texas. May 21, 1973. [NTIS: PB-224-698/1].**

Atomics International (AI) participated in the EBOR test program March 12-25 during the first vessel wall crack propagation test, and September 11-15 during a special test involving crack propagation in the lower vessel stainless steel safe-end pipe. The first test was part of the scheduled RP79 test program, and was open for monitoring on a noninterference basis to interested industrial organizations. The second test series was part of a utility-industry follow-on program; as such the test effort was funded jointly by utilities and AI. Acoustic emission testing methods and instrumentation developed at the AI, Science Center, and Space Divisions of Rockwell International were used during this experimentation. The test plans were directed towards obtaining information which would best complement the experience gained through laboratory testing and testing of other vessel structures, and which would most directly apply to in-service inspection of nuclear reactor systems. To give an understanding of the overall approach to the inspection application, some of the applicable laboratory experience is also presented, along with the descriptions of testing methods, instrumentation, data, and conclusions of the EBOR tests. The main test objectives in this program were: (1) To monitor the acoustic emission signals from crack propagation in the carbon steel vessel wall and the stainless steel safe-end pipe, and to compare the signal characteristics to those of the simulated San Onofre operational noise background; and (2) To perform location of a vessel wall crack by acoustic triangulation, using only portable recording instrumentation on-site, and then performing data processing at Rockwell facilities in Los Angeles, California.

226 **J. S. Buck, L. J. Graham, and P. Morgan. "Application of Acoustic Emission Technology for Detection of Active Cracks in Power Turbine–Generators." AI–75–24. Rockwell International, Atomics International Division, Canoga Park, California. 1975.**

The test program, performed as a joint effort by the Los Angeles Department of Water and Power (LADWP) and Rockwell International Corporation, has shown a basic feasibility for use of acoustic emission technology for the detection of active cracks in steam-turbine generators. Measurements were made to show that simulated stress-wave acoustic emission signals from sources located within the rotor assembly would transmit to transducers attached at the bearings, and also to show that a steam-turbine operational acoustic noise background would not mask acoustic emission signals present. The conclusion of feasibility resulting from these measurements was further supported by the monitoring, during a special turbine operational test, of acoustic signals which are believed to have been caused by the growth of active cracks (cracks insignificant in size relative to turbine integrity).

227 **O. Buck. "Characterization of a Propagating Crack by Ultrasonic Techniques."** *International Journal of Fracture Mechanics,* **IJFMA, 8(1):121-124. Mar 1972.**

A system is briefly described in which ultrasonic flaw detection as well as acoustic emission are used simultaneously in an attempt to characterize the properties of a propagating crack during fatigue cycling. Thus information is obtained on: (a) crack depth and acoustic emission as a function of the number of fatigue cycles; and (b) crack opening and closure as well as acoustic emission during any load cycle as a function of the momentary load at fatigue cycle rates up to about 10 Hz. Some of the results obtained thus far on part-through crack (PTC) specimens are presented.

228 **O. Buck, C. L. Ho, and H. L. Marcus. "Plasticity Effects in Crack Propagation."** *Engineering Fracture Mechanics,* **EFMEA, 5(1):23-34. Feb 1973; paper presented at the Symposium on Fracture and Fatigue, George Washington University, Washington, D.C. May 3-5, 1972.**

Paper describes experimental results of crack propagation studies in part-through crack specimens. Crack extension under cyclic loading is characterized during the loading cycle by the interaction of the crack with a surface wave. The surface wave signal contains information about the absolute crack length, changes in crack length, and the way the crack front opens and closes during cyclic loading. Simultaneously, compliance measurements have been carried out. The results of the experiments are as follows: (1) Direct measurements of the crack depth as a function of the number of load cycles and of different holding times at various loads were obtained; (2) Going through the load cycles, the surface wave signal displays a change in the *apparent* crack depth versus load. This is interpreted as crack closure at low load levels; and (3) Changes in the loading spectrum cause time-dependent relaxation and retardation effects which are not only reflected in the crack growth rate but also in the crack-closure behavior. The results are discussed in terms of Elber's model which suggests that internal residual strains are responsible for this crack closure. [reprinted by permission of Pergamon Press Ltd].

229 **J. J. Buggy, D. L. Dill, and E. Paxson. "Surveillance and In-Service Inspection Recommendations for the FTR: Techniques, Program, Development." WARD-3977-1. Westinghouse Electric Corporation, Advanced Reactor Division, Madison, Pennsylvania. Feb 1969. [NTIS: N70-27575].**

Surveillance and in-service inspection of the FTR reactor internals and reactor system components are recommended to assist in preventing failures that could affect the preservation of public health and safety, and to guard against failures that could have significant effects on plant availability, repair cost, and time. For the reactor internals, the recommended inspections include: Continuous capability for diagnostic monitoring using noise analysis; Frequent in-vessel general visual inspection under sodium using an acoustic viewing system; A materials program consisting of continuous irradiation of specimens in the FTR with periodic ex-reactor testing: Infrequent in-vessel detailed visual inspection under sodium using an acoustic viewing system; and Infrequent, and only as required depending upon reactor operation and other inspection results, in-vessel dimensional measurements, visual inspection, and nondestructive testing.

230 **L. R. Bunnell, J. C. Crowe, and P. E. Hart. "Acoustic Emission Monitoring of Surface-Damaged Ceramic Materials." pp 341-342 in *The Science of Ceramic Machining and Surface Finishing.* S. J. Schneider and R. W. Rice, Editors. NBS-SP-348. May 1972. proceedings of a symposium sponsored by American Ceramic Society, Office of Naval Research, and National Bureau of Standards, held at NBS, Gaithersurg, Maryland, Nov 2-4, 1970. [NTIS: N72-29492 20-15].**

The acoustic emission response of Lucalox, fused silica, single crystal alumina, Pyroceram, and soda-lime-silica glass to surface damage is reported. Low temperature annealing at 200 °C reduces the response levels; annealing at 400 °C removes all of the induced acoustic activity.

231 **A. R. Bunsell. "Acoustic Emission for Proof Testing of Carbon Fibre-Reinforced Plastics." *NDT International,* NDITD, 10(1):21-25. Feb 1977.**

Recent studies of the behavior of flat unidirectional carbon fibre-reinforced plastics (CFRP) specimens have given insight into the failure mechanisms of this type of material when subjected to steady or varying loads. Progressive damage, as monitored by the acoustic emission activity, under both types of loading conditions, is found to reduce with time or numbers of cycles until the material becomes stabilized. Before this state is reached damage arises from the relaxation and inelastic behavior of the matrix and if sufficient numbers of fibres fracture the specimen will fail. After stabilization however the material will not fail under those loading conditions. These results have helped in the interpretation of the behavior of CFRP hoops and pressure vessels and have indicated how the monitoring of the acoustic emission activity may be used as a proof testing technique for this type of structure.

232 **A. R. Bunsell and M. Fuwa. "An Acoustic Emission Proof Test for a Carbon Fibre Reinforced Epoxy Pressure Vessel/Un essai de caracterisation du comportement des capacites sous pression en carbone resine par les techniques**

d'emission acoustique." Ecole Nationale Superieure des Mines de Paris, Centre des Materiaux, Paris, France; Paper 4B/6 in ***Proceedings of the Eighth World Conference on Nondestructive Testing.*** **volume containing Sections 3F, 3B, 4B. International Committee for Nondestructive Testing, Secretariat du Cofrend, Institut de Soudure, 32 Boulevard de la Chapelle, 75880 Paris Cedex 18, France. 1976. conference held in Cannes, France. Sep 6–10, 1976. (paper in English).**

The absence of acoustic emission during load cycling of a CFRP pressure vessel can be taken as a sign that the structure is stable and can be used with confidence.

233 **A. R. Bunsell, M. Fuwa, and B. Harris. "Acoustic Emission Studies of Filament Wound Carbon Fibre Reinforced Rings and Pressure Vessels." pp 9–1 to 9–15 in "Newsletter on the Non-Destructive Testing of Composite Materials and Adhesive Bonded Joints." Issue Number 4. Royal Aircraft Establishment, Farnborough, Hants, England. July 1975.**

Following earlier work on the use of acoustic emission techniques for studying fatigue and fracture in unidirectional plate samples of CFRP, experiments have been carried out on filament wound rings and pressure vessels. The hydraulic expansion to failure of NOL type rings represents the first step away from model systems to more realistic structures and it appears that the rings behave in exactly the same way as tensile samples. More significant is that polar wound pressure vessels in the form of oblate spheroids also show the same kind of acoustic emission patterns. The study reported here concerns the development of a proof testing technique for the pressure vessel based on the experience gained on the flat plates, details of which have been given in earlier Newsletters (Nos. 2 and 3).

234 **A. R. Bunsell, M. Fuwa, and B. Harris. "Evaluation Studies of Acoustic Emission." pp 7–1 to 7–7 in "Newsletter on the Non-Destructive Testing of Composite Materials and Adhesive Bonded Joints." Issue Number 3. Royal Aircraft Establishment, Farnborough, Hants, England. June 1974.**

An important part of the work is the evaluation of acoustic emission as a means of monitoring damage within composite materials. While the approach taken has not led a detailed frequency analysis of emission, it has led to closer consideration of the nature of these emissions in terms of the energy liberated by events within the specimen. Any deformation process within a material will result in stress wave emissions but this will only be a fraction of the total energy involved. Plastic strain, heating, the creation of new surfaces, and frictional forces will all share in contributing to the total energy involved, while reflection at interfaces and at the boundaries of the specimens will further complicate the interpretation of the acoustic emission recorded at the surface. These stress wave emissions are monitored by a piezoelectric ceramic transducer coupled to the surface of the specimen by silicone grease to obtain good acoustic energy transfer. The transducer is usually designed so as to be unaffected by low frequency vibrations and so overcomes problems of general laboratory noise. The transducer used in this work is a resonant transducer with a fundamental frequency of about 120 kHz.

235 **A. R. Bunsell, M. Fuwa, and B. Harris. "Non-Destructive Testing of Fibre Reinforced Material at Sussex University." pp 9-1 to 9-6 in "Newsletter on the Non-Destructive Testing of Composite Materials and Adhesive Bonded Joints." Issue Number 2. Royal Aircraft Establishment, Farnborough, Hants, England. July 1973.**

Since the previous newsletter, a serious study was made of the failure of CFRP using the Dunegan stress wave emission equipment. Ideally the aim of this research is to develop a method of nondestructive testing using acoustic emission as the principal diagnostic tool and to obtain, if possible, a better understanding of failure mechanisms at the microscopic level. Resin preimpregnated sheets consisting of Type 1, surface-treated carbon fibre and epoxy resin, supplied by Ciba-Geigy Ltd., were used in this initial series of experiments. These sheets were cut, stacked unidirectionally, and formed into plates with volume fractions between 40 and 50% by hot pressing. The test specimens cut from these plates were 2-mm thick and 20-mm wide, with a gauge length of 50 mm excluding the end tabs. The width was reduced to 10 mm in later experiments since there was apparently no advantage in using the wider samples. Preliminary tests were made to determine the optimum specimen shape and three designs are used. Two designs use opposing notches cut across the width of the specimen to reduce the load-bearing cross section. The sharper notch type is V-shaped with sides at 45° to a plane normal to the surface. The second type of notch is rounded with a radius of 9.5 mm. A third type of specimen without notches was also examined. All of the notched specimens in these first experiments have shown genuine tensile failures in the notched regions. The fracture surface, as examined in a scanning electron microscope, shows regions of brittle failure as well as areas of fibre pull out. There is evidence that continuous crack propagation is not occurring, as the points of fracture initiation in the fibres themselves are randomly distributed.

236 **E. J. Burton. "Acoustics Diagnostics and Nuclear Power Plants."** ***Journal of the British Nuclear Energy Society,*** **JBNSA, 13(2):183-192. Apr 1974.**

The status of acoustic diagnostics is outlined and potential applications are considered in relation to reactor systems ranging from fast to thermal. The latter include the use of stress-wave emission in detecting faults in pressure vessels. The technique is capable of detecting small flaws in pressure vessels of high strength materials or in a brittle region of a low-strength material during hydrostatic testing without interfering with works production schedules. In addition the techniques can play a role in ensuring the integrity of the secondary circuits, e.g., leaks between the intermediate sodium circuit and secondary water circuit. Other potential applications include detection of leaky fuel elements in gas-cooled reactors and monitoring of turbines, particularly in relation to broken blades. [Abstract 22-0858. *Metals Abstracts,* Vol. 7, 1974].

237 **E. J. Burton. "In-Service Inspection and Monitoring of LMFBRs: The Development of Acoustic Monitoring Techniques for LMFBRs."** *Journal of the British Nuclear Energy Society,* **JBNSA, 15(3):265–270. July 1976.**

Acoustic monitoring techniques are being developed for nuclear reactor systems, one specific example of considerable international importance being the use of acoustic emission to prevent pressure vessel failure in PWRs. The techniques have an added interest in sodium-cooled reactors because of the need to provide information that might have been obtained by optical methods but for the opacity of sodium. This paper is concerned not to give a comprehensive review of techniques but to give a selected presentation of topics which are either of special importance or ones in which significant advances have been made recently. The topics are discussed in terms of the instrumentation technique to achieve the anticipation or the monitoring of incipient events of various modes of failure. Manufacturing defects may be identified by ultrasonic inspection or by stress wave emission. Mechanical failures arising from maloperation may be detected by visualization under sodium, using ultrasonics. Mechanical failures arising from material loss by corrosion or erosion may be anticipated by the detection of acoustic events. Overheating leading to core damage may be detected from the noise of sodium boiling. Mechanical failure from fatigue may be anticipated from an understanding of the vibrational modes of the sodium and its coupling with the structure.

238 **W. M. Bustin (to ESSO Research Engineering Company). "Pipeline Leak Detector." U.S. Patent 3,503,433. Apr 28, 1970.**

A pipeline leak detector using the ultrasonic characteristics of typical leaks eliminates background noise by filtering out the audio-frequency noise. The ultrasonic spectrum of the leak is electrically demodulated to provide a DC signal proportional to the noise level rather than the noise itself. Means are provided for detecting predetermined ranges of the ultrasonic frequency spectrum and beating a variable frequency oscillator frequency thereagainst to produce a difference frequency in the audio range for subsequent demodulation. The leak detector is mounted on a plurality of roller supports and includes sound insulation means isolating the microphone from the remainder of the detector.

239 **W. R. Butler and F. J. Munno. "Prompt Detection of Boiling through Acoustics."** *Transactions of the American Nuclear Society,* **TANSA, 14(2): 776–777. Oct 1971.**

Several recent articles have reported on efforts to measure the state of boiling in large reactor systems by detecting the sound waves generated by boiling. These articles report difficulties that stem from background noise tending to obscure the low intensity sound generated by the boiling phenomena. In this investigation, discrete frequency sound waves are injected into a fueled medium with sufficient intensity to create sound signals that are well above the background noise level. The attenuation of the injected sound signals was found to be a strong function of the boiling parameters in the fueled medium. The change in sound wave intensity at the reactor point can be correlated with the state of boiling within the fueled core and can serve as a prompt and direct measure of the state of boiling throughout the core. Experimental studies have been conducted to measure the attenuation of sound waves through water with varying conditions of boiling at atmospheric pressure. The results indicate that acoustic instruments can be effectively used to measure the state of boiling. The onset of boiling, for example, occurs at a ΔT of about 17 °F on the heat flux curve and 12 °F on the sound intensity curve. The use of degassed water should bring these two breakpoints closer to coincidence. A second set of experiments was conducted in a subcritical reactor assembly with boiling conditions simulated by air bubbles in the water moderator. The results indicate that the attenuation of sound waves through local regions of simulated boiling is identifiable even in a complex heterogeneous medium of fuel-water mixtures. The curves identified as S_0, S_1, S_2, and S_3 are for the sound source located within the boiling simulator and at three radial positions out to 6 inches from the surface of the boiling simulator. For these source locations, the sound intensity falls off by about a decade when the boiling simulator is turned on. Curves S_4 and S_5 are for the sound source located at 8 and 10 inches from the boiling simulator; they indicate a loss of sensitivity for boiling detection as the sound source is moved away from the boiling region. The concept remains promising, however, because dozens of sound sources and receivers can be distributed throughout the core in a fashion similar to the current use of miniature in-core neutron detectors.

240 **J. D. Byerlee. "Acoustic Emission in Rock during Fluid Injection." pp 633-637 in *Advances in Rock Mechanics.* Volume II, Part A: Reports of Current Research. National Academy of Sciences, Washington, D.C. 1974. proceedings of the 3rd Congress of the International Society for Rock Mechanics, held in Denver, Colorado. Sep 1-7, 1974.**

A large-scale field experiment at the Rangely oil field has been under way since 1969 to determine whether earthquakes can be controlled by controlling the fluid pressure in active fault zones. Concurrently, laboratory emperiments have been carried out to study the acoustic emission generated during fluid injection into a fault zone in a sample of Weber Sanstone, the reservoir rock at the oil field. As in the field experiments, the laboratory experiments show that when the fluid injection pressure is increased, the elastic shock activity increases, and it decreases when the injection pressure is decreased. Also, the results from a number of cycles of fluid pressure increase and decrease have shown that the elastic shock activity decreases as the stress is relieved by movement on the fault. Thus, it is clear that earthquakes can be generated by fluid injection into active faults, and the earthquake hazard may be reduced as the tectonic stresses are relieved.

241 **J. D. Byerlee and D. Lockner. "Acoustic Emission during Fluid Injection into Rock." pp 87-98 in *Proceedings First Conference on Acoustic Emission/ Microseismic Activity in Geologic Structures and Materials.* H. R. Hardy, Jr. and F. W. Leighton, Editors. Trans Tech Publications, Clausthal, West, Germany. 1977. conference held at Pennsylvania State University, University Park, Pennsylvania. June 9-11, 1975.**

The source of acoustic emission was located during water injection into a sandstone sample while it was subjected to a constant confining pressure and differential stress. For the first 10,000 minutes, the shocks were randomly distributed throughout the sample. During the next 460 minutes, they had their origin at the end of the sample where the water was being injected. Failure then occurred rapidly and the fracture front advanced through the sample as the water flowed through the fractured material of reduced permeability and maintained a low effective stress at the fracture front. A notable result of the experiment was that the most intense activity occurred at the fracture front and once it passed a region, the acoustic emission decreased even though the differential stress was maintained at a constant value and the pore pressure in that region was increasing as the water flowed into the sample.

242 **J. D. Byerlee and L. Peselnick. "Elastic Shocks and Earthquakes."** *Naturwissenschaften,* **NATWA, 57:82–85. 1970.**

C

243 **J. D. Cadman and R. E. Goodman. "Landslide Noise."** ***Science,*** **SCIEA, 158(3805):1182–1184. Dec 1, 1967.**

Acoustical monitoring of real landslides has revealed the existence of subaudible noise activity prior to failure and has enabled prediction of the depth of the seat of sliding when conducted in boreholes beneath the surface. Recordings of noise generated in small slopes of moist sand, tilted to failure in laboratory tests, have been analyzed to determine the foci of discrete subaudible noise events. The noises emitted shortly before failure were plotted close to the true sliding surface observed after failure. The foci of earlier events lay either within the central portion of the sliding mass or in a region behind the failure surface. The head and toe zones were devoid of strong seismic activity. [Copyright 1967 by the American Association for the Advancement of Science].

244 **J. D. Cadman, R. E. Goodman, and C. Van Alstine. "Research on Subaudible Noise in Landslides." Geotechnical Engineering Report of Investigation, National Science Foundation Grant GK 109. University of California, Department of Civil Engineering, Berkeley, California. June 27, 1967.**

The report presents the results of an investigation of subaudible rock noise phenomena, conducted by the geological engineering group of the University of California, Berkeley from 1964 to 1967. This work was sponsored by a grant (GK 109) from the National Science Foundation. The main effort in this project has been in the area of rock noises in landslides. Further experinces were gained with noise occurrence in landslides in the field, and basic instrumentation was improved. In addition, a provocative series of experiments were conducted with noises in small scale landslides in the laboratory. Other areas of investigation not reported here included rock noise monitoring in active faults, study of rock noises in field loading tests, in large laboratory unconfined compression tests, and in smaller laboratory triaxial compression tests. Subaudible rock noise consists of acoustic energy released from a rock being strained. These natural sonic oscillations in rock under stress have been of considerable interest since the pioneering work of Obert and Duvall at the U.S. Bureau of Mines in 1939, which suggested important applications of mine safety. Interest in the application of subaudible rock noise to landslides is more recent. Section 2 reviews previous work in these areas and related investigations in seismology and rock mechanics.

245 **W. M. Carey. "Acoustic Detection of the Collapse of a Sodium Vapor Bubble in an Infinite Sea of Sodium." Technical Memorandum ANL-CT-76-21. Argonne National Laboratory, Argonne, Illinois. Dec 1975.**

A discussion of the problem of sodium-vapor bubble collapse is presented. The physics of vapor collapse is given in light of the work by Peppler *et al.* Theoretical estimates of the sound source level based on the work by Rayleigh and Judd are compared to an approximate pressure-volume work approach and recent experimental observations. Reactor ambient noise and transmission loss considerations are presented in regard to their impact on this detection problem. A methodology is proposed which considers the importance of the sound source level, ambient noise, transmission loss, and a detection threshold providing a means by which the feasibility of sodium-vapor bubble collapse detection in an operating LMFBR may be assessed. The interrelationships between the detection threshold and the probability of detection and false alarm are discussed and applied to a standard acoustic square law detection system. This analysis clearly illustrates that the feasibility of such a detection system is strongly dependent on the knowledge of expected sound levels, ambient noise levels, and the transmission loss between the source and receiver. Furthermore, requirements of a high degree of probability of detection and a low probability of false alarm necessitated a high signal-to-noise ratio for a single sensor system, but that the probability of false alarm requirement could be relaxed for systems multiple independent sensors. Finally, the need for additional experimental and theoretical information is presented in terms of sound source levels, ambient noise, and a means for determining transmission loss.

246 **W. M. Carey and R. W. Albrecht, Coordinators. "Proceedings of a Seminar on the Potential for LMFBR Boiling Detection by Acoustic–Neutronic Monitoring." ANL–CT–76–34. Argonne National Laboratory, Argonne, Illinois. June 1976. seminar held at Argonne National Laboratory. Apr 8–9, 1976.**

A seminar involving ten technical presentations by principal investigators was held to assess the current scope of ERDA-sponsored programs, to determine the feasibility of sodium-boiling detection in LMFBRs, and to establish areas in need of additional research and development. The consensus was that (1) feasibility of boiling detection by acoustic, neutronic, and acoustic-neutronic monitors has been demonstrated in U.S. and European programs; (2) additional research and development is needed in areas of reactor noise, cavitation, and the effects of noncondensable gases on sound source levels and transmission; (3) the role of acoustic-neutronic monitors from the standpoint of reactor surveillance rather than reactor safety is a viable approach to be adapted; and, in particular (4) a need exists for an operational LMFBR demonstration system. As a result of this seminar, an integrated national program will be developed. Authors and coauthors of the work presented include: R. W. Albrecht, T. T. Anderson, R. A. Bennett, J. P. Bobis, W. M. Carey, Y. S. Cha, C. E. Cohn, C. K. Day, S. W. Eisenhawer, H. K. Fauske, A. P. Gavin, R. E. Holtz, W. H. Leavell, F. E. LeVert, W. W. Managan, R. F. Saxe, W. H. Sides, Jr., and B. W. Spencer.

247 **W. M. Carey, T. T. Anderson, J. P. Bobis, A. P. Gavin, and D. R. Pedersen. "Acoustic Monitoring of the SLSF P1 Test."** ***Transactions of the American Nuclear Society,*** **TANSA, 23:320–321. 1976.**

Two ANL high-temperature microphones were monitored to observe acoustic backgound and transient pressure pulses during the Sodium Loop Safety Facility (SLSF)

P1 in-reactor experiment. The experiment simulated an LMFBR loss-of-piping-integrity (LOPI) transient on a 19-element, end-of-life, UO_2 fuel assembly. This test required that all sensors be capable of operating at an elevated temperature in a sodium environment with internal pressures of 1.4 MPa. The lithium niobate transducer normally utilizes an oxygen vent line. However, in this experiment thermal conditioning and oxygen impurities in the helium were found to be sufficient to enable the microphone to function for 90 days without the vent line. A similar microphone was also tested under gradually increasing internal pressure in an acoustically excited furnace. After a two-week internal pressurization at 650 °C, the test microphone was found to operate satisfactorily to 1.75 MPa, recover from pressurization to 3 MPa, and rupture at 13.2 MPa (1900 psi).

248 J. M. Carlyle. "Acoustic Emission in Fiber Reinforced Composites." Master of Technology Thesis. Brunel University, Uxbridge, England. Sep 1974; NADC-75082-30. Naval Air Development Center, Air Vehicle Technology Department, Warminster, Pennsylvania. May 15, 1975. [NTIS: AD/A-012-382].

Composite materials present complex testing problems to materials engineers because of the subtle nature of their failure initiators. However, acoustic emission monitoring offers a simple means of studying the effects that microscopic defects have on composites, since the emissions are a direct result of material deformations. This dissertation examines the research and development applications that acoustic emission monitoring has for fiber-reinforced composites. Specific emphasis is placed upon obtaining information from fundamental failure processes, but practical engineering applications are reviewed as well, including fatigue-crack detection and structural-integrity evaluation. A comprehensive history of acoustic emission is presented to enable studies in composites to be related to previous work in metals and geological materials. Basic principles of the phenomenon are covered, including the generation of emissions, the propagation of stress waves through materials, the production of electrical signals from mechanical stress waves, and signal processing to obtain data on microscopic sample deformations. An experiment is discussed which indicates that six-ply, 0°, ±45° carbon fiber-reinforced plastic can be successfully monitored for incipient failure. The failure precursor takes the form of a drastic reduction in the acoustic emission count rate at approximately 99% of the sample failure load, and several explanations are offered as to the origin of this dip.

249 J. M. Carlyle. "Imminent Fracture Detection in Graphite-Epoxy Using Acoustic Emission." Naval Air Development Center, Warminster, Pennsvlvania. 1976; *Experimental Mechanics,* EXMCA, 18(5):191–195. May 1978.

An experiment designed to detect incipient failure in graphite-epoxy tensile specimens is described. Tests using eighteen samples of six different graphite-epoxy compositions in six-ply balanced $[0/+45/-45]_S$ laminates indicate that a failure precursor does exist. This precursor takes the form of a sudden reduction in the acoustic emission output at 99% of the ultimate tensile load and evidence indicates that the reduction is the result of a change in the fundamental failure mechanism. The shape of the acoustic emission count rate curve is analyzed and found to correlate well with micromechanical fracture activity.

250 **J. M. Carlyle and W. R. Scott. "Acoustic-Emission Fatigue Analyzer."** *Experimental Mechanics,* **EXMCA, 16(10):369–372. Oct 1976.**

A new technique has been developed for acoustic-emission fatigue monitoring which facilitates the interpretation of dynamic micromechanical failure processes in materials, through the use of a new method for data presentation. In contrast to conventional acoustic-emission fatigue records in which the emission rate is plotted as a function of the number of fatigue cycles, the new technique produces a graph in which each acoustic event is plotted within a coordinate system of load versus number of cycles. This approach results in the simultaneous recording of the relationship between sample load, acoustic-emission rate, and number of fatigue cycles. Monitoring of 300M steel has established that it is possible to distinguish between fatigue-crack propagation, fatigue-crack closure, and testing-machine noise using a single transducer.

251 **J. L. Carpenter, Jr. and W. F. Stuhrke. "NDE–An Effective Approach to Improved Reliability and Safety–A Technology Survey." NASA CR-134963. Martin Marietta Corporation, Orlando, Florida. June 1976. [NTIS: N76-25577].**

This technology survey report is comprised of technical abstracts for about 100 significant documents relating to the nondestructive testing of aircraft structures of related structural testing and the reliability of the more commonly used evaluation methods. Particular attention is directed toward six NDE techniques: acoustic emission; liquid penetrant; magnetic particle; ultrasonics; eddy current; and radiography. The Introduction of the report includes an overview of the state-of-the-art represented in the documents that have been abstracted.

252 **S. H. Carpenter. "Acoustic Emission from Plastic Deformation." pp 643–665 in "Proceedings of the ARPA/AFML Review of Quantitative NDE." D. O. Thompson, Program Director. AFML-TR-75-212, SC595.10AR. Science Center, Rockwell International, Thousand Oaks, California. Jan 1976. [NTIS]. conference held at the Science Center, Rockwell International, Thousand Oaks, California. July 15–17, 1975.**

The fact that low level bursts of noise or acoustic emission are generated during the plastic deformation of crystalline solids has been known for many years. The most familiar example is the so-called *tin cry* observed when crystals of tin are deformed. To date, a multitude of mechanisms, generally relating to dislocation processes, have been proposed to explain the acoustic emission phenomena related to plastic deformation. Most often these mechanisms are not verified by correlation with data on other physical properties. Results will be presented which indicate that by proper selection of test procedures it might be possible to identify particular emissions with particular deformation processes. Results will also be presented on simultaneous measurements of acoustic emission and dislocation damping which indicate that in some materials the majority of the acoustic emission is not related to dislocation processes, but to unique microstructural features. A long range goal is to develop the capability to determine and evaluate polycrystalline microstructures from acoustic emission measurements.

253 **S. H. Carpenter and A. Hadjicostis. "Sources of Acoustic Emission Generated during Plastic Deformation." summary of paper in** *Paper Summaries: ASNT Spring Conference, March 28–30, 1977, Phoenix, Arizona.* **American Society for Nondestructive Testing, Columbus, Ohio. 1977.**

Acoustic emission generated during plastic deformation of a wide variety of sample materials has been investigated. Measurements have been conducted to investigate the effects of purity, crystal structure, texture, and other microstructural variations. Sample materials have ranged from pure single crystals to commercial alloys. The combination of proper and complete test procedures, the simultaneous measurement of the dislocation damping, and the digitization of the data for accurate analysis have allowed for correlation of the acoustic emission data with various tests or metallurgical parameters (or both). Such correlations provide useful information in trying to identify the sources of the acoustic emission.

254 **S. H. Carpenter and F. P. Higgins. "The Identification of Sources of Acoustic Emission Operative during Deformation." Paper 4-3, pp 288–304 in** *The Third Acoustic Emission Symposium.* **Japan Industrial Planning Association, Dobashi Building, 3-17-8, Akasaka, Minato-ku, Tokyo, Japan. 1976. proceedings of a symposium held in Tokyo, Japan. Sep 16–18, 1976.**

The measurement and subsequent analysis of the acoustic emission generated during plastic deformation of crystalline solids have been used as the primary tool in numerous studies of basic deformation processes. The success of these studies is somewhat open to question since from the data generated a multitude of different mechanisms have been proposed as being the generating source of the measured acoustic emission. If acoustic emission techniques are to be developed into useful investigative tools it is necessary to establish methods and procedures which can be used to accurately identify and correlate sources with their particular emission. This is a difficult task since the measured emission may be comprised of emission from several sources which may be operating either independently or cooperatively. Two procedures are readily available to the investigator which can help in the analysis and identification of actual sources of emission. First, a better and more complete selection of testing procedures. is needed. In many cases it may be possible to affect a separation of emissions from different sources by properly selecting a series of testing procedures. An example of this technique will be discussed in detail. A second procedure is to simultaneously measure an additional experimental parameter sensitive in a known way to the operation of the proposed mechanisms. In the majority of investigations reported to date, the acoustic emission data are identified and correlated only with the load deflection curve. While this type of information is useful, the load deflection curve is not sensitive enough to different microscopic deformation mechanisms to be of much use in identifying sources of emission. The simultaneous measurement of the mechanical damping and modulus defect have been carried out while measuring the acoustic emission generated during the deformation of pure iron. These parameters are sensitive in a known way to both magnetic and dislocation mechanisms operative under an applied stress. These data can then be used to separate and identify emissions and then to correlate them with possible sources. Examples of this procedure will be discussed in detail.

255 **J. J. Carter. "Everything You Always Wanted to Know About–Acoustic Emission."** ***Metals Australia*****, MAUSB, 5(10):353-356. 1973.**

Informal article which presents a brief, nontechnical description of the theory of sound; sound transmission, reflection and absorption; ultrasonics; and finally acoustic emission.

256 **J. J. Carter. "Non-Destructive Detection and Evaluation of Stress Corrosion Cracking." pp 61-77 in** ***Effects of Chemical Environment on Fracture Processes*****, Proceedings of the 3rd Tewksbury Symposium on Fracture. C. J. Osborn and R. C. Gifkins, Editors. Department of Metallurgy, University of Melbourne, Parkville, Victoria, Australia. 1974. symposium held at the University of Melbourne. June 4–6, 1974. [proceedings available from D. S. Mansell, Department of Mining and Metallurgy, University of Melbourne, Parkville, Victoria 3052, Australia].**

Stress wave emission is compared with conventional methods of nondestructive inpection as means of detecting and evaluating stress corrosion cracking. The methods are described briefly and their advantages and limitations outlined. In the case of stress wave emission, some experimental results are included, as well as a brief review of appropriate literature. It is concluded that, for the present, stress wave analysis can usefully complement but not replace the older inspection methods. Complete evaluation of stress corrosion cracking requires additional knowledge from fields such as fracture mechanics and materials engineering.

257 **J. J. Carter. "Stress Wave Metallography."** ***Journal of the Australian Institute of Metals*****, JAMTA, 19(1):74-87. Mar 1974.**

Stress wave emission may be observed when metals undergo phase transformation, deformation, or fracture. It is therefore associated with structural changes and may be used to study those changes. Publications relevant to the sources of stress wave emission during phase transformation or deformation are reviewed critically, and courses of investigation which might help to discriminate these sources are suggested.

258 **J. W. Carter. "Acoustic Emission: A New Measurement Concept." MHSMP-75-11. Mason and Hanger-Silas Mason Company, Amarillo, Texas. Mar 1975. [NTIS].**

259 **E. Catier. "L'emission Acoustique: un stethoscope pour le controle de qualite (Acoustic Emission: A Stethoscope for Quality Control)."** ***Traitements de Surface*****, TRSUA, (129):7-15. Apr 1974. (in French).**

The principles of acoustic emission testing are presented. The materials likely to release an emission of acoustic waves when subjected to thermal, mechanical, or chemical action are numerous. The occurrences giving rise to an acoustic emission fall into two categories: the weak energy phenomena associated with plastic deformation (twinning, grain boundary slip, etc); and the much more energetic phenomena associated with fracture mechanisms (initiation and propagation of cracks, inter and transgranular fracture). A description is given of the equipment used for analyzing and detecting acoustic emission. A few examples of applications are: surveillance of the integrity of structures (pressure vessels, reactor vessels, etc.); and testing of bonded joints and welds.

260 **J. Cekada. "Raziskave akusticne emisije pri utrujanju strojnih delov (Research of Acoustic Emission due to Fatigue of a Mechanical Part)." Diplomsko delo st. 2313. Fakulteta za strojnistvo, Univerza v Ljubljani, Ljubljana, Yugoslavia. 1975. (in Slovenian).**

261 **A. Cete. "Seismic Source Location in the Ruhr District." pp 231-241 in** ***Proceedings First Conference on Acoustic Emission/Microseismic Activity in Geologic Structures and Materials.*** **H. R. Hardy, Jr. and F. W. Leighton, Editors. Trans Tech Publications, Clausthal, West Germany. 1977. conference held at Pennsylvania State University, University Park, Pennsylvania. June 9-11, 1975.**

Seismic source location in the Ruhr District by minimum number of seismic stations is of great importance for investigation of rock bursts. As known from field experiments, commonly the calculated sources deviate from the actual sources. The deviations not only depend on station coordinates, velocities, and arrival times of seismic waves but also on the configuration of the seismic network and the position of the source in relation to the seismic network. The deviations are given as isolines for different errors in location parameters typical in field work. The accuracy required for seismic source location can be obtained if these isolines are taken into consideration in planning and setting up the seismic network. General comments concerning the instrumentation being used to detect rock bursts in the Ruhr District are presented.

262 **A. J. Chabowski, W. H. Bowyer, J. M. Cook, and C. T. Peters. "Complementary Methods for Assessing the Strength of High Alumina Cement Concrete Structures/Les methodes complementaires pour evaluer la solidite des structures on beton fondu." Building Research Establishment, Fulmer Research Institute. London, England; Paper 4B/12 in** ***Proceedings of the Eighth World Conference on Nondestructive Testing.*** **volume containing Sections 3F, 3B, 4B. International Committee for Nondestructive Testing, Secretariat du Cofrend, Institut de Soudure, 32 Boulevard de la Chapelle, 75880 Paris Cedex 18, France. 1976. conference held in Cannes, France. Sep 6–10, 1976. (paper in English).**

The pull-out and acoustic emission tests to assess the *in situ* strength of high alumina cement concrete structures are described and discussed. Where acoustic emission is detected in structures at lower loads than the design load, assessment of strength using the pull-out test may be required.

263 **D. G. Chakrapani and E. N. Pugh. "The Transgranular SCC of a Mg–Al Alloy: Crystallographic, Fractographic and Acoustic–Emission Studies."** ***Metallurgical Transactions A,*** **MTTAB, 6A(6):1155–1163. June 1975.**

A study has been made of the propagation of transgranular stress-corrosion cracks in a Mg- 7.5-wt % Al alloy tested at room temperature in an aqueous NaCl-K_2CrO_4 solution; the studies were carried out on single cracks which initiated and propagated within single grains in coarse-grained bend specimens. The resulting fracture surfaces were flat, and two-surface analysis established their orientation to be $\{31\bar{4}0\}$. Specimens fractured at liquid-nitrogen temperature cleaved on $\{10\bar{1}0\}$, indicating that the $\{31\bar{4}0\}$ are not the normal cleavage planes in this material. The stress-corrosion fracture surfaces were cleavage-like in appearance, containing shallow steps which were matching and interlocking on opposite faces. Discrete acoustic signals were emitted during crack propagation, and these are considered to result from discontinuous crack advance. It is concluded that stress-corrosion cracking in this system occurs by discontinuous cleavage on $\{31\bar{4}0\}$ planes. [Copyright 1975 by the American Society for Metals and the Metallurgical Society of AIME].

264 **R. H. Chambers and S. A. Hoenig. "New Techniques in Nondestructive Testing by Acoustical and Exo-Electron Emission: Semiannual Progress Report, 1 July 1968–28 February 1969." Advanced Research Projects Agency Contract F33615–68–C–1707. University of Arizona, Engineering Experiment Station, Tucson, Arizona. Feb 28, 1969. [NTIS: AD–691–230].**

The purpose of the acoustic emission investigation was to determine two points: (1) the upper range of the frequency spectrum of acoustical emissions and whether it is limited by the rapidly rising attenuation usually associated with rising frequency, and (2) the existence of any recognizible peaks or valleys in the acoustical power spectrum identifiable with a particular mechanism of acoustical emission, for example, with initial microcrack size. A sample calculation shows that high-frequency stress waves should be expected from microcrack formation and propagation. Depending upon the nature of the stress concentration, frequencies as high as 100 MHz should be expected. Furthermore, as cracks extend, there should be a corresponding decrease in the frequencies of the maximum in the power spectrum. The acoustical output of quartz crystal transducers attached to deforming specimens has been analyzed by a spectrum analyzer. Although this method does not measure the frequency content of individual transient pulses, it does measure the average frequency spectrum of several hundred pulses.

265 **R. H. Chambers and S. A. Hoenig. "New Techniques in Nondestructive Testing by Acoustical and Exo-Electron Emission: Semiannual Progress Report, 1 March 1969–31 August 1969." Advanced Research Projects Agency Contract F33615-68-C-1707. University of Arizona, Engineering Experiment Station, Tucson, Arizona. Aug 31, 1969. [NTIS: AD-720-357].**

During the course of investigating the frequency response of various piezoelectric transducers for use in acoustical emission studies, the need arose for a wide frequency range acoustical generator. The report describes the new generator and applies it to the calibration of several quartz and PZT-5 transducers. A nearly white noise is generated by the fracture of small (10×10^{-3} inches) diameter silicon carbide particles between a steel rotating plunger and a quartz or aluminum anvil or backing plate.

266 **R. H. Chambers and S. A. Hoenig. "New Techniques in Nondestructive Testing by Acoustical and Exo-Electron Emission: Semiannual Progress Report, 1 September 1969–30 April 1970." Advanced Research Projects Agency Contract F33615-68-C-1707. University of Arizona, Engineering Experiment Station, Tucson, Arizona. Apr 30, 1970. [NTIS: AD-720-358].**

Report describes investigations of the frequency distribution of acoustical emission from growing fatigue cracks in 24-ST aluminum sheet. The fatigue cracks were introduced by a high-amplitude, low-cycle bending mode of deformation. Efforts were also devoted to investigating the post fatigue exo-electron emission from nickel wires.

267 **R. Chance. "Acoustic Emission In-Process Electron Beam (EB) Weld Inspection Technique." Note No. ADN 09-01-70.4. Grumman Aerospace Corporation, Advanced Development Program, Bethpage, New York. Nov 1970. [NTIS: AD-884-725 (USGO)].**

Note describes a preliminary evaluation of the acoustic emission technique for nondestructive, in-process monitoring of electron beam weld quality. Acoustic emission analysis was chosen for this study because it could become the optimum technique for real-time, nondestructive inspection of EB welds. It is the only nondestructive method that is capable of detecting weld defects as they are formed in the vacuum chamber, and is insensitive to defect orientation. When perfected, the technique would allow inspection of EB welds on-line; it may make immediate weld repairs practical by pinpointing the location of defects while the work piece is still in the vacuum chamber. Results on 28 runs made thus far indicate that an effective application of the acoustic emission technique would require electrical grounding and acoustic isolation of the weld back-up plate. The restrictions this requirement places on tooling need further study. The limited data obtained make flaw detection from acoustic emission records possible, but classification and sizing methods must still be developed.

268 **T. Chance. "Production AE Testing of Oil Sealed Pressure Containment Systems." p 46 in** *Paper Summaries: ASNT 36th National Fall Conference, September 27–30, 1976, Houston, Texas.* **American Society for Nondestructive Testing, Columbus, Ohio. 1976. [abstract only].**

Production AE testing of petroleum equipment on a daily production basis is discussed, with particular regard to RAM and annular type blow-out preventors. Problems of plant environment, equipment location, placement of sensors, ultrasonic coupling problems, and data interpretation problems are presented with solutions used in these areas. Examples of flaws detected by acoustic emission testing, the resulting changes in manufacturing procedures, and the improvement in product quality due to these changes are shown.

269 **R. M. Chaney. "AE Application for Proof Testing Pressure Vessels." p 46 in** *Paper Summaries: ASNT 36th National Fall Conference, September 27–30, 1976, Houston, Texas.* **American Society for Nondestructive Testing, Columbus, Ohio. 1976. [abstract only].**

In an attempt to keep abreast of the latest techniques and equipment, three newly constructed Ultrafiner reactors (pressure vessels), were monitored with acoustic emission during hydro-testing in the fabricator's shop. An advanced acoustic emission system that consisted of a flexible array of transducers, preamplifiers, signal conditioner, a computer, an output CRT display, and a thermal printer were used. AE parameters and other test parameters were monitored, processed, stored, and displayed on command to assess the condition of the vessels before they were placed in service. All emission sources found in the three vessels were identified and determined to be noncritical conditions. With minimal time and effort the vessels were shown to be sound structures.

270 **F. H. Chang, D. E. Gordon, R. H. McDaniel, and A. H. Gardner. "The Characterization of Damage Growth in Advanced Composites by Acoustic**

Emission Monitoring." pp 180–181 in *Paper Summaries: ASNT 36th National Fall Conference, September 27–30, 1976, Houston, Texas.* American Society for Nondestructive Testing, Columbus, Ohio. 1976.

Acoustic emission monitoring (AEM) in nondestructive evaluation (NDE) uses piezoelectric sensors for the detection of stress waves generated during material deformation. Before it can be applied to heterogenous materials such as advanced composites, parameters measurable by AEM must be related to failure mechanisms in the advanced composites. This paper presents the methodology and experimental results of acoustic emission monitoring of graphite epoxy composite fracture specimens undergoing tensile testing. Special features of the AEM system used in the study include spatial noise filtering and emission signal amplitude distribution capabilities. Signals detected by the sensors are recorded by a magnetic tape recorded for frequency spectrum analysis. Fracture specimens were fabricated from graphite epoxy composites with different ply orientations chosen to exhibit fiber failure, linear, and nonlinear matrix failures. A slit was machined by ultrasonic method in the center of each specimen to serve as a damage initiating site. Results obtained by periodic X-ray inspection, using an opaque additive, provided a basis for the interpretation of the AEM results. Three parameters measurable by the AEM system, accumulated counts, amplitude distribution, and frequency spectrum, provided indications of the occurrence of critical damage growth at a load level as early as 55 percent of the ultimate. Examples of the correlation of these parameters with matrix failures, ply delaminations, and massive failures will be presented. Each case can be verified by the actual observation of damage growth from results of X-ray inspection. The physical relationship between the stress wave generation and the failure mechanisms in the composites will be discussed.

271 **H. H. Chaskelis. "Acoustic Emission as an Index of Submarine Hull Integrity." NRL Memorandum Report 2709. Naval Research Laboratory, Washington, D.C. Jan 1974. [NTIS: AD-774-490].**

Two large flat steel plates (one 6 X 20 feet, 2-inch thick of HY-80; the other 8 X 18 feet, 2-½ inch thick of HY-140) are used as test specimens to realistically evaluate with respect to size the wave propagation characteristics of acoustic emissions. Wide-band transducers, amplifiers, and video-recorders are used to capture and store the acoustic signals generated by various means. The video-tapes are then analyzed to replay to an oscilloscope and a spectrum analyzer. Preliminary data reduction indicates that sound propagates faithfully over appreciable distances. However a distinction is not apparent, in the frequency domain, between emissions from crack extension and signals from other sources not related to this phenomenon.

272 **H. H. Chaskelis, W. H. Cullen, and J. M. Krafft. "Acoustic Emission from Aqueous Stress Corrosion Cracking in Various Tempers of 4340 Steel." NRL Memorandum Report 2608. Naval Research Laboratory, Washington,**
277 **D.C. June 1973. [NTIS: AD-764-048]. [This is same article as Reference 273].**

A commercially available system is employed to detect and count acoustic emissions emanating from the aqueous stress-corrosion crack propagation in 4340 steel. Standard ASTM E399 compact tension specimens are prepared in four tempering temperatures: 204, 316, 427, and 538 °C (400, 600, 800, 1000 °F). Crack length is monitored with a notch-opening clip gage. Comparisons show the time rate of emission events to increase with stress intensity in rough correspondence to, but much more rapidly than, the crack velocity. Tempering back tends to suppress the count rate as a warning of fast fracture instability. Some similarity is observed between the areal rate of acoustic emission and factors which influence degree and intensity of plastic flow instability in the fracture process zone.

273 **H. H. Chaskelis, W. H. Cullen, and J. M. Krafft. "Acoustic Emission from 4340 Steel during Stress Corrosion Cracking." pp 31–44 in** *Fracture Toughness and Slow-Stable Cracking, ASTM STP 559.* **American Society for Testing and Materials, Philadelphia, Pennsylvania. 1974. proceedings of the 1973 National Symposium on Fracture Mechanics, Part–1, sponsored by ASTM Committee E-24 on Fracture Testing of Metals, held at University of Maryland, College Park, Maryland. Aug 27-29, 1973. [This is same article as Reference 272].**

A commercially available system is employed to detect and count acoustic emissions emanating from the aqueous stress-corrosion crack propagation in 4340 steel. Standard ASTM E 399 compact tension specimens with short notches, $a/W = 0.25$ versus 0.50 standard, are prepared in four tempering temperatures: 204, 316, 427, and 538 °C (400, 600, 800, and 1000 °F). Crack length is monitored with a notch-opening clip gage. Comparisons show the time rate of emission events to increase with stress intensity in rough correspondence to, but much more rapidly than, the crack velocity. Tempering back tends to suppress the count rate as a warning of fast fracture instability. Some similarity is observed between the areal rate of acoustic emission and factors which influence degree and intensity of plastic flow instability in the fracture process zone. [reprinted by permission of the American Society for Testing and Materials, Copyright].

274 **V. P. Chentsov. "Transducer for Studying Stress–Wave Emission Signals."** *Soviet Journal of Nondestructive Testing,* **SJNTA, 10(6):669–672. Nov–Dec 1974; translated from** *Defektoskopiya,* **DEFKA, (6):56–59. Nov–Dec 1974. (in Russian).**

Data are presented on the power of SWE signals in the creep stage of materials. Requirements are given for SWE signal transducers. The choice of a piezoelectric ceramic sensing element is justified. A design is described for a high-sensitivity SWE transducer and the amplitude-frequency responses are shown for three models of this design. Recommendations are made regarding the application of thin piezoelectric ceramic disks for the detection of SWE signals in the frequency range from 80 to 150 kHz and the use of the piezoelectric transducers' resonant characteristics.

275 **T. T. Chiao, M. A. Hamstad, and E. S. Jessop. "Tensile Properties of an Ultrahigh-Strength Graphite Fiber in an Epoxy Matrix." UCRL-75209, Rev. 1. Lawrence Livermore Laboratory, Livermore, California. Oct 1, 1974. [NTIS] ; paper presented at the American Society for Testing and Materials Conference on Composite Reliability, Las Vegas, Nevada. Apr 15-16, 1974.**

The fiber performance and potential as a reinforcement for fiber composites of a special PAN-based graphite fiber were evaluated by testing the fiber's tensile properties in an epoxy matrix. Representative strand samples were taken from 30 spools (~4 kg) of single-end, 1500-filament fiber to make over 5000 fiber/epoxy strand specimens using the filament-winding process. Characteristics studied were fiber uniformity, strength and modulus distributions at room and liquid nitrogen temperatures, stress-strain behavior, the effect of strain rate on fiber strength, and acoustic emission during tensile loading to failure. The fiber was found to have a 3560-MPa (517 ksi) failure stress, a 1.7% failure strain, a 206-GPa (30×10^6 psi) modulus, and a density of 1.77 Mg/m^3 at 23 °C. Liquid nitrogen temperature and various strain rates had no significant effect on fiber tensile properties.

276 **T. T. Chiao, M. A. Hamstad, M. A. Marcon, and J. E. Hanafee. "Filament-Wound Kevlar 49/Epoxy Pressure Vessels." NASA CR-134506, UCRL-51466. Lawrence Livermore Laboratory, Livermore, California. Nov 9, 1973. [NTIS].**

The performance of an organic fiber (Kevlar-49) has been investigated. First, a detailed study of the strength of fiber-epoxy strands was completed. Variables studied included: fiber uniformity, strength distribution at room and LN_2 temperatures, stress-strain characteristics, strain rate effect on the fiber strength, stress-rupture behavior, and aging under no load. Second, the composite performance in filament-wound pressure vessels was studied. Variables studied included: spherical versus cylindrical vessels, cylindrical vessel design and size, winding patterns, reproducibility, pressurization rate, fatigue, and effect of liquid hydrogen temperatures. Third, a study was made of the acoustic emission generated during tensile tests of the fiber-epoxy strands, tensile tests of fiber-epoxy elongated NOL rings, burst tests at room temperature of the cylindrical and spherical pressure vessels, and vessel fatigue tests. Fourth, a microstructural characterization of the failure mode of the fiber was carried out using transmission and scanning electron microscopy and optical microscopy. Fracture surfaces of bare fibers were compared with epoxy impregnated strands from tensile tests, pressure vessels, and NOL rings.

N. P. Chironis. "Sounds of Stress Warn of Failure about to Happen." ***Product Engineering,*** **PRENA, 37(12):44-47. June 6, 1966.**

Staff-written article discusses acoustic emission monitoring of the 260-inch Thiokol solid-fuel rocket case during hydrostatic pressure test at Newport News Shipyard.

278 **K. A. Chishko. "Acoustic Emission during Annihilation of Prismatic Dislocation Loops and Kinks in Line Dislocations."** *Ukrainskii Fizicheskii Zhurnal,* **UFIZA, 19(8):1264–1271. Aug 1974. (in Russian).**

The sound emission accompanying the annihilation of prismatic dislocation loops (or bends in straight dislocations) in metal crystals is analyzed theoretically. Each annihilation, or act of straightening, is accompanied by an elastic pulse which propagates through the crystal at a specific velocity. It is accompanied by sound emission, the frequency and amplitude of which depends on the elastic and other physical constants of the material, the positioning of the dislocation relative to other dislocations, the surface of the material, and possible impurities. Under certain critical conditions the sound intensity may be high. [Abstract 13-0090. *Metal Abstracts,* Vol. 8, 1975].

279 **J. F. Chretien, P. Bae, F. Ferrieux, and P. Vigier. "Etude Experimentale de l'Emission Acoustique de Certains Alliages d'Aluminium (Experimental Studies of the Acoustic Emission from Certain Aluminum Alloys)." pp 119–143 in** *Journees d'Etudes sur l'Emission Acoustique.* **P. F. Gobin, R. Goutte, and C. Guillaud, Organizers. Institut National des Sciences Appliquees de Lyon, Avenue Albert Einstein, 69621 Villeurbanne, Lyon, France. Mar 1975. proceedings of a conference held in Lyon, France. Mar 17, 1975. (paper in French).**

Preliminary experiments were performed to study the possibility of applying acoustic emission to the characterization of physical properties of aluminum alloys. The effects of some parameters such as impurity concentration, thermal treatments, and precipitation state are revealed. These experiments showed that acoustic emission is a sensitive means for studying or discovering microscopic deformation phenomena.

280 **J. F. Chretien and N. Chretien. "A Bibliographical Survey of Acoustic Emission."** *Non-Destructive Testing,* **NDETA, 5(4):220–224. Aug 1972.**

Paper analyzes the research carried out to investigate acoustic emission as applied to inspection and nondestructive testing of structures in recent years in leading laboratories in the U.S. The study is intended to provide technicians interested in applying this method to their problems with precise information on the techniques used at present. Cited are 52 references.

281 **N. Chretien. "Avenir de l'emission acoustique (Future of Acoustic Emission)."** *Centre de Documentation Siderurgique, Circulaire d'Informations Techniques,* **CDSCA, 32(4):877–884. 1975. (in French).**

The characteristics of acoustic emissions from stressed solids, and the methods of detection and signal processing are presented. Sources of emission discussed include plastic deformation, fracture, and phase transformation in metals; applications of acoustic emission monitoring include welding process monitoring and control, and surveillance of highly stressed structures.

282 **N. Chretien. "Emission Acoustique (Acoustic Emission)." CEA-CONF-2339. CEA Centre d'Etudes Nucleaires de Saclay, 91-Gif-sur-Yvette, France. Apr 1973. (in French). [NTIS].**

The mechanisms involved in the generation of stress waves are reviewed. Theoretical models are given for the events resulting in the generation of waves. The apparatus necessary to detect and exploit the models of constraint is described. Some applications of the technique to nuclear structures and materials are given.

283 **N. Chretien. "Results of Acoustic Emission Monitoring of the EBOR Vessel." pp 39-48 in "A Joint Industry-Utility Program to Develop NDT Techniques for Nuclear Reactor Inspection: An Addendum Report to Steering Committee EEI Project RP-79." Compiled by E. R. Reinhart. Southwest Research Institute, San Antonio, Texas. May 21, 1973. [NTIS: PB-224-698/1].**

The check of a nuclear pressure vessel's integrity, before its installation and during its active life, is one section of the research and development program on the acoustic emission technique. The aim is the correlation of acoustic emission with other parameters (stress, strain, defects development, etc.) in performing classic mechanical and metallurgical tests on samples so as to apply the knowledge to the study of a loaded pressure vessel's behavior. This is the reason for the interest in the EEI program's EBOR vessel tests.

284 **N. Chretien, B. Audenard, P. Bernard, G. Clerc, and P. Gallice (to Commissariat a l'Energie Atomique). "Dispositif de surveillance de l'evolution des contraintes dans un materiau solide (Device for Monitoring Behavior of Stresses in Solid Materials)." French Patent 2,155,143. Apr 24, 1973. (in French).**

A device for monitoring stress relaxation in solid materials (metals, alloys, concretes, ceramics, and metal-ceramic composites) subjected to thermal or mechanical influences and to radiation is described. The device consists of a detector which is applied to the surface of the material. This detector is provided with a transducer furnishing an electrical signal, the amplitude of which is a function of the stress waves, and whose passband corresponds to the wave frequency spectrum under consideration. An amplifying and filtering unit processes the electrical signal. The system incorporates a low threshold circuit discriminator which provides a pulse each time the threshold is exceeded, and also a counter. The main industrial uses of this device are in the nondestructive testing of materials, the tests to the rupture point carried out on parts subjected to irradiation embrittlement, and in the permanent or periodic monitoring of parts operating under permanent stresses such as pressurized nuclear reactor vessels.

285 **N. Chretien and P. Bernard. "L'Emission Acoustique: Technique Nouvelle de Controle de Qualite (Acoustic Emission: Technique for Quality Control)."** *Qualite, Revue Pratique de Controle Industriel,* **QRPCA, 11(57):33–37. Nov 1972. (in French).**

286 **N. Chretien, P. Bernard, and B. Barrachin. "Inspection of Steel Pressure Vessels by Acoustic Emission." pp 655–667 in** *Proceedings of the 2nd International Conference on Pressure Vessel Technology. Part 2: Materials, Fabrication and Inspection.* **American Society of Mechanical Engineers, New York. 1973. conference held in San Antonio, Texas. Oct 1–4, 1973. [NTIS: CEA-CONF-2519].**

The inspection of steel pressure vessels by acoustic emission is discussed in terms of: the importance of verifying the integrity of pressure vessels; equipment and methods for acoustic emission detection and the processing of data obtained as acoustic emission signals; the relation between fracture mechanisms and acoustic emission in steels; rupture mechanisms in test pressure vessels; hydraulic tests on industrial pressure vessels to evaluate the reliability of acoustic emission testing for detecting defects before leaks or ruptures occur; and current laboratory research on acoustic emission testing. It is concluded that the inspection of steel pressure vessels by detection of acoustic emission presents attractive features. First, it is a global control method and it is easy to apply to the pressurized equipment to be monitored. Secondly, the method ensures, if not continuous monitoring in the present status of its development, at least an efficient monitoring easy to adapt to normal operating conditions of the plant (pressure rises after shutdown, normal pressure transients, regulatory tests). Lastly, it is a selective method since it deals only with defects that are evolving, and consequently are hazardous.

287 **N. Chretien, P. Bernard, and J. Fayolle. "Mesure de l'Energie des Signaux d'Emission Acoustique: Etalonnage et Essais (Measurement of Energy in Acoustic Emission Signals: Calibration and Testing)." pp 188–214 in** *Journees d'Etudes sur l'Emission Acoustique.* **P. F. Gobin, R. Goutte, and C. Guillaud, Organizers. Institut National des Sciences Appliquees des Lyon, Avenue Albert Einstein, 69621 Villeurbanne, Lyon, France. Mar 1975. proceedings of a conference held in Lyon, France. Mar 17, 1975. (paper in French).**

The possibility of using an Audimat-W device for analyzing the electric energy of signals delivered by a piezoelectric sensor for acoustic emission was investigated. The tests performed revealed that the 7075-T651 aluminum alloy can be used as a reference material.

288 **N. Chretien, P. Bernard, P. A. Ruault, and J. P. Pascaud. "Comportement a la pression d'un defaut reel de soudure prelevee sur chantier: Essai d'utilisation de la technique d'emission acoustique (The Effect of Variations**

in Pressure on a Weld Defect in Pipe Taken from the Operating Environment: Use of the Acoustic Emission Technique)." *Soudage et Techniques Connexes,* **SOTCA, 27(7/8):309-315. July-Aug 1973. (in French).**

A 609 mm dia by 11.2 mm thick X52 HRR steel tube specimen containing a tube-to-tube weld with a ~6 cm long defect was subjected to cyclic variations of pressure ranging from a minimum of 45 bar to a maximum of 67 bar at a frequency of 48 cycles/h. After 32,000 cycles the pressure was increased in 10-bar increments to 160 bar, then reduced to atmospheric pressure. The specimen was then pressurized until it ruptured (at 206 bar). Crack extension of the defect was monitored throughout the test by acoustic emission and ultrasonic testing.

289 **N. Chretien and D. Miannay. "Emission Acoustique (Acoustic Emission)." Commissariat a l'Energie Atomique, Centre d'Etudes Nucleaires de Saclay, Gif-sur-Yvette, France. Dec 1972. (in French). [NTIS: CEA-CONF-2188].**

Acoustic emission is the creation and spontaneous propagation of stress waves in a solid when it is deformed, fissured, or structurally transformed. The instrumentation in the detection and analysis chain is described, and the treatment of the data is discussed. The application of the method to plastic deformation, rupture, and phase transformations is considered. The technological applications to the detection of fissuration caused by welding, nondestructive control, and surveillance of strongly stressed structures are discussed.

290 **N. Chretien and J. Monier (to Commissariat a l'Energie Atomique, France). "Dispositif multicapteurs de localisation simultanee de source d'emission d'ondes de contrainte (A Multisensor Device for Simultaneously Locating Several Acoustic Emission Sources)." French Patent 2,262,303. Sep 19, 1975. (in French).**

The invention relates to a multisensor device for simultaneously locating several acoustic wave sources, wherein are associated the time intervals (t, t′) between two signals corresponding to the same source situated within a pseudo-lozenge comprising four sensors, and by means of which events corresponding to each (t, t′) are memorized in a computer. That device can be applied to the detection of wave source emission in a nuclear reactor vessel during a pressure-test.

291 **N. Chretien and C. Perennes. "La Localisation des Sources d'Emission Acoustique (Location of Acoustic Emission Sources)." pp 279-322 in** *Journees d'Etudes sur l'Emission Acoustique.* **P. F. Gobin, R. Goutte, and C. Guillaud, Organizers. Institut National des Sciences Appliquees de Lyon, Avenue Albert Einstein, 69621 Villeurbanne, Lyon, France. Mar 1975. proceedings of a**

conference held in Lyon, France. Mar 17, 1975. (paper in French); ***Bulletin d'Informations Scientifiques et Techniques, Commissariat a l'Energie Atomique,*** **BUIAA, (211):85-95. Feb 1976. (in French).**

This study is concerned with the localization of acoustic emission sources. After a definition of the acoustic event, the electronic features are reviewed. Several solutions of the localization problem are examined: graphical, analytical, or numerical methods. It appears that the square pattern formed by five sensors with a central sensor leads to the best precision on the largest area.

292 **N. Chretien and E. Tomachevsky. "Application de l'Emission Acoustique au Centre d'Etudes Nucleaires de Saclay." Annexe III, pp 455-474 in "Non-Destructive Tesing for Reactor Core Components and Pressure Vessels." IAEA-145. International Atomic Energy Agency, Vienna, Austria. 1972. report of a panel meeting held in Vienna, Austria. Nov 29-Dec 3, 1971. (in French). [NTIS: CONF-711114].**

Studies at Saclay for evaluating acoustic emission testing as a nondestructive quality control technique are reviewed briefly. The acoustic characteristics of a number of materials were determined. Among these were austenitic steel, carbon steel, and the light alloy AG 5. The application of the method to the quality control of welds was studied. Laboratory and site tests were made to determine its use in the detection of structural faults. The use of the method for the location of emission sources was also studied. The studies showed that this technique can be used for the analysis of the behavior of materials under stress, fabrication quality control, total quality control at the end of fabrication, inspection at the end of site installation, periodic inspection, and monitoring during operation.

293 **R. H. Christensen. "Cracking and Fracture in Metals and Structures." pp 326-374 in** ***Proceedings of the Crack Propagation Symposium.*** **Volume 2. College of Aeronautics, Cranfield, England. 1962. symposium held in Cranfield, England. Sep 1961. [reproduced copy of paper from proceedings available from Linda Hall Library, Kansas City, Missouri].**

The history and significance of cracking and fracture in engineering structures is first reviewed. The complete elimination of fatigue cracks and minute flaws in aerospace structure is not practical in the present state of technology. The *failsafe* design concept provides multiple load paths and therefore insures adequate structural safety in the presence of cracks. Experimental and analytical investigations of crack propagation and residual strength were conducted. In the experimental phase, testing was done on a variety of steels, a titanium alloy, a nickel alloy, and a coated refractory metal at elevated and cryogenic temperatures. Cracks were propagated in plate specimens containing centrally located (crack starter) notches. The rate of cracking was observed. Most residual strength measurements were made on fatigue-cracked specimens, rather than on specimens containing only machined slots. Natural cracks were found to reduce residual strengths more severely than manufactured slots. Test results are included on crack strengths under biaxial stressing of nickel-alloy sheet materials and on one of the steels. Studies of crack growth based on the appearance of fracture surfaces were undertaken. Findings from this study support the general conclusion that the crack propagation phase of fatigue damage is more predictable and less subject to scatter than the crack nucleation phase. Most scatter in fatigue data results from scatter in the microcrack nucleation period. At elevated temperatures, cracking is more uniform than at low temperatures, where cracking of less ductile materials tends to be sporadic. Visible crack-growth periods at elevated temperatures were found to be greater than at low temperatures under the same stress conditions, even though total fatigue life at the higher temperatures is less. An expression for residual strength is given and is based on the effective width of the plastic zone existing at the end of a crack. Good agreement with test results is demonstrated. The formula contains a parameter called the plastic-zone notch resistance factor, R_p, which is a measure of the notch resistance of the material. The expression for the rate of crack propagation also is based on considerations of plasticity. The expression accounts for the observed fact that crack-growth rates approach infinity as the crack length approaches critical. Values predicted by the formula correlate well with test results for crack propagation under constant amplitude, as well as programed loads. It is recommended the additional experimental and analytical methods and techniques be developed to understand the dynamics of fracture. Such studies will be necessary in order to correlate (1) modes of fracture in varying structural environments, and (2) fracture mechanism as a function of stress fields. A crystal accelerometer was used to detect microsounds that accompany micro-cracking in stressed metal. The micrometal quakes were monitored on headphones. [The major portion of research work reported in this paper is given in Report ASD-TR-61-207 by R. H. Christensen and P. H. Denke, (see Reference 294)].

294 R. H. Christensen and P. H. Denke. "Crack Strength and Crack Propagation Characteristics of High Strength Metals." ASD-TR-61-207. Douglas Aircraft Company, Long Beach, California. Jan 1962. [NTIS: AD-275-395].

Results of a fracture testing program for high strength sheet metals are presented. More than 500 sheet-stock panels ranging in width from 1 to 18 inches and thickness from 0.020 to 0.100 inches were tested. These panels contained centrally located cracks which were generated principally by fatigue loading at various exposure times under a variety of environments. Fracture strength of the cracked panels were determined for rupturing temperatures ranging from −340 to 2500 °F. The experimental results were studied analytically. A semiempirical expression for residual strength is derived. This expression is based on a modification of the formula given by Crichlow for the effective width of the plastic zone. An expression for the rate of crack propagation also is presented. These formulas are shown to agree well with test results and are suitable for design applications. A digital analysis of the elastic and plastic stress and strain distribution in the cracked plate was performed. The analytical method is discussed and results are presented. Brief mention is made of a technique and setup used for measuring micro-sounds or micrometal quakes that accompany microcracking in stressed metal. The setup used is similar to that used by H. E. Romine at the Naval Research Laboratory.

295 **H. G. Christmas. "Acoustic Emissions Evaluation of Vessels at ARCO's Cherry Point Refinery." Atlantic Richfield Company, Ferndale, Washington. 1975;** *Materials Evaluation,* **MAEVA, 34(6):22A–25A. June 1976; paper presented at the American Society for Nondestructive Testing National Fall Conference, Atlanta, Georgia. Oct 13–16, 1975.**

Four catalytic reactor vessels, seven heat exchangers and two high pressure separators were tested at the Atlantic Richfield Company Cherry Point Refinery utilizing acoustic emission during a pressure test to 105 percent of normal operating pressure. The test was conducted in two groups, using process hydrogen and with the skin temperature of the reactor vessels slightly in excess of 275 °F. The first group of vessels was tested in a total elapse time of approximately ten hours. The second group of vessels was tested in a total elapse time of slightly less than six hours. All reactors, heat exchangers and high pressure separators were analyzed as being of high structural integrity under the conditions of the imposed test.

296 **Y. P. Chugh, H. R. Hardy, Jr., and R. Stefanko. "An Investigation of the Frequency Spectra of Microseismic Activity in Rock under Tension." Pennsylvania State University, College of Earth and Mineral Sciences, University Park, Pennsylvania. 1968; paper presented at the Tenth American Institute of Mining, Metallurgical, and Petroleum Engineers Symposium on Rock Mechanics, held at the University of Texas, Austin, Texas. May 20–22, 1968.**

Unfortunately, a basic understanding of the phenomenon of microseismic activity is extremely limited. Although a considerable number of field studies have been carried out, these have for the most part been mainly empirical. Furthermore, only a limited number of laboratory studies have been conducted, and many of these are incomplete or suspect due to lack of adequate instrumentation. Realizing the tremendous potential of this technique as a tool for studying the mechanical behavior of rock, the Department of Mining at the Pennsylvania State University in 1964, with the assistance of National Science Foundation Grant No. GK-151, initiated an extensive laboratory program in an effort to gain further understanding of both the mechanisms and the basic laws governing microseismic activity. The initial phase of this program involved the study of noise rate and energy rate in Indiana limestone, Crab Orchard sandstone, and Barre granite under incremental uniaxial tensile stress. The results of this phase have been presented recently by Brown and Singh (see Reference 220). Phase two of the program involved the study of the frequency spectra of microseismims observed in the same three rock types under incremental uniaxial tensile stress. Studies were carried out over a wide range of applied stress and moisture conditions. This latter research forms the subject matter of the present paper.

297 **S. Claes. "La Forme des Signaux d'Emission Acoustique et Leur Role dams les Essais de Localisation (The Shape of Acoustic Emission Signals and their Role in Localization Testing)." pp 215–257 in** *Journees d'Etudes sur l'Emission Acoustique.* **P. F. Gobin, R. Goutte, and C. Guillaud, Organizers. Institut**

National des Sciences Appliquees de Lyon, Avenue Albert Einstein, 69621 Villeurbanne, Lyon, France. Mar 1975. proceedings of a conference held in Lyon, France. Mar 17, 1975. (paper in French).

The analysis of elastic energy transport in matter is presented. Such a study is necessary in localization tests and to establish relations between the shape of acoustic emission signals and the characteristics of the emission source.

298 **R. N. Clark and T. A. Mathieson. "Use of Microprocessors in Analysis of Acoustic Emission Weld Monitoring Data." pp 265–274 in** *Proceedings of the Eleventh Symposium on Nondestructive Evaluation.* **Southwest Research Institute, San Antonio, Texas. 1977. symposium held in San Antonio, Texas. Apr 20–22, 1977.**

The use of acoustic emission as a nondestructive technique for real-time weld evaluation creates a need for supporting analysis equipment. This equipment is needed to extract the information required for setting parameters on acoustic emission weld monitors, as well as exploring the application of acoustic emission to various welding materials and processes. The equipment requirements are fast analysis turn around time, simplicity of operation, and maximum information extracted from the weld. The authors are developing a system for this analysis using wideband tape recordings from a modified video tape recorder and a microprocessor-based computer. Analysis is made of amplitude frequency content and pulse-related information of the acoustic emission bursts, and presentation of the analysis is designed for simplicity and efficiency in the operator-computer interaction. The paper presents current results of this procedure, a description of the analyzer system and associated equipment, and possibilities for future work using microprocessors in this and in other areas.

299 **M. Classen-Nekludowa. "Uber die sprungartige Deformation (Sudden Deformation of Crystals under Strain)."** *Zeitschrift fur Physik,* **ZEPYA, 55:555–568. 1929. (in German); "On the Stepwise Deformation." UCRL–TRANS–11291. translated for Lawrence Livermore Laboratory, Livermore, California. Oct 1977. [NTIS].**

The mechanism of the plastic deformation of crystals is studied. The deformation occurs in sudden jumps of a few thousandths of a millimetre and is accompanied by a uniform noise like the ticking of a watch. The interval between jumps may be a few seconds and the rate of deformation in this interval is zero for the whole crystal. [abstract from *Chemical Abstracts,* 23:4606. 1929. Copyright by American Chemical Society].

300 **F. J. Climent, R. Cribbs, and T. Eckert. "Nondestructive Testing and Performance Monitoring Techniques (Updated Instrumentation Techniques for Monitoring Embankmens and Concrete Structures for Civil Works)." Final**

Report on Contract SA No. 157161 with the State of California, Department of Water Resources. Aerojet–General Corporation, Sacramento, California. Nov 1968.

Three tasks are covered in this report, namely: Task I, Strong Motion Accelerometers Installed in Embankments Internally; Task II, Detached Remotely Interrogated Piezometers and Stress Cells; and Task III, Measurement of Density, Thickness, and Detection of Voids in Concrete Walls, Linings, and Slabs. Discussions of pertinent instrumentation used and description of prototype systems recommended for testing and included.

301 **T. G. Cochran. "Influence of Oxidation upon the Kaiser Effect of Acoustic Emission." USAMC–ITC–02–08–75–112. Army Materiel Command, Department of Maintenance Effectiveness Engineering, USAMC Intern Training Center, Red River Army Depot, Texarkana, Texas. Apr 1975. [NTIS: AD/A-009-283].**

The objective of this research was to examine acoustic emission during reload of steel specimens. In particular, the role of oxidation upon acoustic emission during reload was investigated. Sixteen samples of AISI 1020 steel were loaded to 3100 pounds. Acoustic emission signals were recorded during loading. Then samples were divided into two groups and one of these groups immersed in Sears SAE 10-40 motor oil to retard oxidation. After two weeks, all specimens were reloaded. The tests were again monitored for acoustic emission. Data reduction involved comparing the signal to a threshold voltage and counting events when the signal equalled or exceeded the threshold. Summation counts of acoustic emission were tabulated for each test. The sign test was used to compare summation counts during load against summation counts during reload. To compare the difference between the two groups during reload, the Wilcoxon test for independent samples was used. Conclusions indicate that a significant decrease occurs in acoustic emission during reload. No significant difference was evident in acoustic emission from the two groups during reload indicating no effect of oxidation on acoustic emission.

302 **C. Coddet, J. F. Chretien, and G. Beranger. "Adaptation de la methode d'emission acoustique a l'etude de mecanismes de fissuration au cours de l'oxydation des metaux et alliages a haute temperature. Application a l'oxydation du titane (Adaption of the Acoustic Emission Method to the Study of Fissuration Mechanisms during the Oxydation of Metals and Alloys at High Temperature. Application to Titanium Oxydation)."** *Comptes Rendus Hebdomadaires des Seances de l'Academie des Sciences, Serie C: Sciences Chimiques,* **CHDCA, 282(17):815–818. May 3, 1976. (in French).**

A method of recording the acoustic emission from a specimen during the growth of an oxide film is described schematically and the results of its application to the oxidation of Ti at 860 °C in air are reported. The acoustic signals correspond to stages of stress relief in the growing oxide film and can be related to changes observed by thermogravimetry and scanning electron microscopy. [Abstract 22-0199. *Metals Abstracts,* Vol. 10, 1977].

303 **R. A. Collacott. "Monitoring to Determine the Dynamics of Fatigue Testing."** *Journal of Testing and Evaluation,* **JTEVA, 4(3):181–187. May 1976.**

These investigations were begun to monitor the sonic response during the fatigue life testing of cantilever specimens; they led to the discovery that the power spectral density from the second overtone (3 X fundamental) changed with time, originally exceeding that from the fundamental. Overtones may have a pronounced influence on fatigue testing, and the experimental results should lead to consideration of the possible dynamic effects introduced by such overtones. Tests show that changes can be detected as a consequence of work hardening and crack growth. Possibly this method may be developed as a nondestructive testing technique. [reprinted by permission of the American Society for Testing and Materials, Copyright].

304 **J. Colomer. "Generalites sur les methodes de controle non destructif des soudures (Principles of Nondestructive Testing Methods of Welds)."** *Metaux* **(Corrosion, Industrie), MTUXA, 49(587–588):283–294. 1974. (in French).**

305 **J. Colomer and G. Lucas. "Le controle non destructif au C.E.A. (Nondestructive Testing at the C.E.A.)."** *Bulletin d'Informations Scientifiques et Techniques, Commissariat a l'Energie Atomique,* **BUIAA, (211):5–15. Feb 1976. (in French).**

The different nondestructive testing methods used at the CEA are presented: X-ray or gamma radiography, X-ray stress analysis, neutron radiography, ultrasonic testing, eddy currents, electrical testing, microwaves, thermal testing, acoustic emission, optical holography, and tracer techniques.

306 **M. R. Compton. "A Real-Time Software Operating System for a Computer Controlled Acoustic Emission Flaw Detector." BNWL-SA-4501. Battelle-Northwest, Richland, Washington. Sep 1972; paper presented at the American Society for Nondestructive Testing National Spring Conference, Los Angeles, California. Mar 12–15, 1973;** *Materials Evaluation,* **MAEVA, 31(7):121–127. July 1973.**

A computer-controlled acoustic emission flaw monitor has been developed to do in-service inspection monitoring of reactor pressure vessels. To efficiently accommodate the infrequency, the randomness and the burst characteristics of acoustic emission flaw signals, a real-time software operating system for the computer were developed to do the on-line data acquisition, the on-line determination of flaw locations, and the on-line accumulation of flaw activity status. The flaw source location is determined by an advanced and simplified technique. The flaw activity status contains the total number, the total energy, and the energy release rate of flaw signals in a defined region. The flaw acitivity status is reported periodically in printouts for operating personnel. The monitor has been field tested for a year and has demonstrated the feasibility of in-service inspection of reactor pressure vessels.

307 **N. G. W. Cook. "The Seismic Location of Rockburst." pp 493–516 in** ***Rock Mechanics.*** **C. Fairhurst, Editor. Pergamon Press, New York. 1963. proceedings of the Fifth Symposium on Rock Mechanics, held at the University of Minnesota, Minneapolis, Minnesota. 1962.**

308 **R. R. Corle. "The Use of Acoustic Emission to Improve Motor Case Structural Reliability." Lockheed Propulsion Company, Redlands, California. 1973; AIAA Paper No. 73–1258 presented at the American Institute of Aeronautics and Astronautics/Society of Automotive Engineers 9th Propulsion Conference, Las Vegas, Nevada. Nov 5–7, 1973. [photocopy of paper available from AIAA Library, 750 3rd Avenue, New York, NY 10017, also as Technical Report AIAA-73-1258 from Dunegan/Endevco, San Juan Capistrano, California].**

A program was conducted to determine the flaw detection sensitivity of acoustic emission relative to the critical flaw size that will cause rocket motor case failures. Data from seven different metals tested revealed that acoustic emission can detect flaws ranging from 10% of the critical size for electron-beam-welded 4335V steel to 100% for 6061-T6 aluminum. The program also demonstrated that acoustic emission can detect flaw instability in sufficient time to discontinue a hydrotest prior to catastrophic failure. In addition, a cursory evaluation of applying acoustic-emission technology to verify rocket-motor nozzle reliability was successful.

309 **R. R. Corle and J. A. Schliessmann. "Flaw Detection and Characterization Using Acoustic Emission." Lockheed Propulsion Company, Redlands, California. 1972; paper presented at American Society for Nondestructive Testing National Fall Conference, Cleveland, Ohio. Oct 16–19, 1972;** ***Materials Evaluation,*** **MAEVA, 31(6):115–120. June 1973.**

A test program was conducted to evaluate acoustic emission techniques for use in detecting flaws during proof testing of rocket motor cases. Steel sheet specimens which contained tight fatigue cracks of various sizes were tested. Selected specimens were inspected by standard radiographic and magnetic particle techniques to determine the sensitivity level of these methods. Each specimen was then loaded to simulate proof testing. An acoustic emission signature was recorded for each specimen during proof testing and was found to be a function of the flaw size. By evaluating the signature, unflawed specimens could be distinguished from flawed specimens, and the flawed specimens could be ranked on the basis of flaw size, with flaws as small as 0.035 inches (0.9 mm) deep being detected. By comparison, flaws as large as 0.16 inches (4.1 mm) deep and 2.6 inches (66 mm) long were not detected by radiographic techniques, and surface flaws as large as 0.1 inches (2.5 mm) deep and 0.1 inches (2.5 mm) long were not detected by magnetic particle inspection.

310 **H. Corte, O. Kallmes, and D. Jarrot. "Mechanical Failure of Ideal 2-D Fibre Networks."** *Paper-Maker and British Paper Trade Journal,* **PMBPA, 142(1):61-68, 72. July 1961.**

Article deals with evaluating the mechanical strength and failure of 2-D sheet paper, i.e., thin heterogeneous structure whose thickness is comprised of no more than two fibers thick. Acoustic emission was used to investigate the details of fracture of fiber networks. The principal mode of failure was the bond between two fibers. A microphone was used as the sensor to detect airborne acoustic signals.

311 **G. Cortellazzi, S. Boffi, G. Caglioti, and F. Rossitto. "A Lattice-Dynamics Approach to the Acoustic Signal by a Uniformly Moving Dislocation."** *Journal of Applied Physics,* **JAPIA, 44(4):1518-1523. Apr 1973.**

The acoustic signal generated by a uniformly moving edge dislocation is interpreted as the linear response of the crystal to the excitation produced by this motion. The proposed approach is based on the atomistic model of Frenkel and Kontorova, but the rigidity of the substrate is here removed. The expectation value of the atomic displacement induced in the crystal is thus directly evaluated in terms of lattice dynamics via a normal mode expansion. Numerical predictions are presented for several selected dislocation velocities in a specific case. These model calculations could help in interpreting experimental results.

312 **B. Cox. "A Correlation between Acoustic Emission during SCC and Fractography of Cracking of the Zircaloys."** *Corrosion,* **CORRA, 30(6):191-202. June 1974.**

The fractorgraphy of all transgranular stress-corrosion cracking (SCC) failures of the Zircaloys is generally similar and consists of areas of cleavage separated by regions of ductile tearing. Minor differences arise due to processes which occur after the crack front has passed (i.e., due to etching in some environments). Major differences are apparent between room temperature and high temperature [200 to 400 °C (392 to 752 °F)] fractures in the nature of the ductile process which interconnects the regio is of cleavage. Two types of acoustic emission have been recorded during SCC tests: (1) A continuous *swishing* noise, which control experiments have shown to arise either from electrochemical noise or from dislocation generation during plastic deformation, and (2) a discontinuous, *cracking* noise, which may result from either brittle crack propagation, oxide film cracking, or twinning. Since neither type of acoustic emission can be related uniquely to a single process which is occurring, its diagnostic value in the SCC of hexagonal metals is apparently limited.

313 **B. Cox. "A Correlation between Acoustic Emission during SCC and the Fractography of Cracking of the Zircaloys in Various Media."** ***Corrosion,*** **CORRA, 28(12):473. Dec 1972. [abstract only]; paper presented at the Corrosion Research Conference, Anaheim, California. Mar 19–21, 1973. [see Reference 312 for published article].**

Two types of acoustic emission have been observed during SCC of the zircaloys in a number of media. These parallel two types of emission observed during mechanical testing. The first is a continuous *swishing* type of emission, which has been related to dislocation motion in mechanical tests, but which may also result from electrochemical processes. The second type of emission is a discontinuous cracking noise, which may be due to brittle cracking in a transgranular SCC process, but which could also be caused by twinning (e.g., during a purely mechanical test). The discontinuous acoustic emission which is typical of SCC failures at 300 °C can be distinguished from that due to twinning, however, by a careful metallographic search for twins after failure has occurred. The acoustic emission from zircaloy specimens during SCC in a number of environments has been correlated with the fractography of the failures and with control experiments, and the present situation in explaining the observed features will be discussed.

314 **B. Cox. "Acoustic Emission during SCC of Zirconium Alloys." paragraph 4.2.2, pp 78–79 in "Chemistry and Materials Division Progress Report, January 1 to March 31, 1973." AECL-4509, PR-CMa-24. Atomic Energy of Canada Ltd., Chalk River Nuclear Laboratories, Chalk River, Ontario, Canada. Apr 1973. [NTIS].**

Research is reported in the fields of solid-state science, general chemistry, physical chemistry, and materials science. New previously unreported results include anisotropy of damage in ion-implanted GaAs and GaP, cross section, and excitation curve for $^{127}I(n, 2n)^{126}I$; sources of He production in thermal neutron-irradiated Ni-based alloys; calculated bulk and perpendicular surface Debye temperatures of metals from Morse potentials; solubility of $NiTeO_3$; ^{60}Co activities and Fe deposition rates on CANDU fuel bundles, electrochemical H/D separation factors on Fe, Mo, and W electrodes; H yields and isotope effects from Febetron radiolysis of water vapor; comparison of calorimetric and chemical dosimeters for reactor dosimetry; spectra and solvated electron yield in pulse-irradiated biphenyl–ethylamine and pyrene–methylamine solutions; ^{39}K NMR in methylamine; stress-corrosion cracking of Zircaloy and Zr alloys in aqueous chloride, organic, and iodine–organic solutions; oxidations and polarization of Zr alloys in fused salt; electron diffraction and microscopy of ZrO_2 films grown in NaCl and iodine–containing environments; thermoluminescence of CO-annealed ThO_2; dislocations produced by hydride precipitation in Zircaloy-2; ^{64}Cu diffusion in α-Zr and α-Zr-O alloys; thermal desorption studies of oxygen-platinum interaction; effects of urea and guanidine on H_2 and D_2 evolution at Fe; and isotope effect on current–potential relations at Fe.

315 B. Cox. "Evnironmentally Induced Cracking of Zirconium Alloys." Report AECL-4821. Atomic Energy of Canada Ltd., Chalk River Nuclear Laboratories, Chalk River, Ontario, Canada. 1973; *Reviews on Coatings and Corrosion,* RVCCB, 1(4):366–422. 1973.

Although a relatively smaller amount of work has been reported on the SCC (stress corrosion cracking) and LME (liquid metal embrittlement) of zirconium alloys than on titanium alloys, it is evident that both alloy systems are susceptible to cracking in similar environments. Mechanistic studies on cracking of zirconium alloys have been few, and show some significant differences from similar studies on titanium alloys. Thus, it is not possible to find any convincing evidence for hydrogen participating in the transgranular SCC processes in zirconium alloys and electrochemistry seems to be unimportant in many environments (e.g., I_2 vapour). Since, for LME, the reduction in bond strength resulting from adsorption of the liquid metal atoms at the crack tip appears to be an acceptable mechanism, the conclusion is that a similar adsorption of halogen atoms in a suitable chemical form is the most probable explanation for transgranular cracking processes. Intergranular cracking appears to be readily explicable on the basis of preferential chemical or electrochemical attack at grain boundaries.

316 J. I. Craig, W. C. Strahle, and J. Palfery. "Audible and Ultrasonic Acoustic Emissions from Composite Solid Propellants." AFOSR Interim Scientific Report on Grant No. AFOSR-75-2805. Georgia Institute of Technology, School of Aerospace Engineering, Atlanta, Georgia. July 1975. [NTIS: AD/A-017-583].

An apparatus to test the audible and ultrasonic acoustic emissions from composite solid propellants was designed, constructed, and checked out. Tests were conducted at 300 psia for eight aluminized HTPB-AP propellants supplied by the Air Force Rocket Propulsion Laboratory. The tests were conducted under both nitrogen and air pressurization. For the audible emissions in the range 100 to 2500 Hz, there is a marked effect of the pressurization fluid. In air, the noise output is 10^4 higher than in nitrogen. A definite possibility exists that the noise output is responsible for observed chamber pressure roughness level. For the ultrasonic signal in the range 50 to 300 kHz, there are strong spectral peaks observed but their origin is not clear and they do not correlate with observed deflagration behavior. For audible signals, a striking effect of AFCAM aluminum coating is the reduction of signal strength.

317 J. I. Craig, W. C. Strahle, and J. Palfrey [*sic*]. "Audible and Ultrasonic Acoustic Emissions from Composite Solid Propellants." AFOSR-TR-75-0172. Georgia Institute of Technology, School of Aerospace Engineering, Atlanta, Georgia. Jan 1975. [NTIS: AD/A-005-714].

The primary task of this grant was to design, build, and check out an acoustics apparatus to measure both audible and ultrasonic acoustic emissions from burning composite solid propellants at pressures to 1000 psia. The check out was achieved with propellants furnished by the Air Force Rocket Propulsion Laboratory. Preliminary results are reported on the acoustic emissions from these propellants. The apparatus description is also given.

318 J. Crha, J. Adamek, and K. Mazanec. "Pouziti emise napetovych vln k sledovani rozvoje defektu v ocelovych konstrukcich (Investigating Crack Propagation in Steel Structures by Stress-Induced Acoustic Emissions)." *Strojirenstvi,* STRJA, 25(1):54-58. 1975. (in Czech); BISI-13495. translated by the British Industrial and Scientific International Translation Service, The Metals Society, London, England. Sep 1975. [available from the American Society for Metals, Metals Park, Ohio].

The recording and analysis of acoustic emissions, over a wide range of frequencies, are shown to be a valuable method for the detection of cracking in stressed bodies or in welds, especially in bodies stressed as a whole, such as pressure vessels. The apparatus will operate automatically and needs no means for excitation (as ultrasonic testing does). It can be built up from readily obtainable components.

319 P. P. Crimmins. "Correlation of Stress-Wave-Emission Characteristics with Fracture in Aluminum Alloys." Report 1246-Q-3. Aerojet-General Corporation, Sacramento, California. June 1969.

The report describes the progress during the third quarter of a program to investigate the correlation of stress-wave-emission characteristics with fracture in selected aluminum alloys. During this reporting period, single-edge-notch tensile and part-through-crack tensile specimen tests were performed. Described are the complete instrumentation systems employed for these tests and the stress-wave-emission fracture characterization data obtained. The titles of articles reviewed as part of the continuing literature survey are also indicated in this report.

320 P. P. Crimmins. "Development of a Nondestructive Testing Technique to Determine Flaw Criticality." Advanced Research Project Agency Contract F33615-68-C-1705. Aerojet Solid Propulsion Company, Sacramento, California. Sep 1970. [NTIS: AD-875-390].

A brief description is presented on the program scope for additional studies to develop a nondestructive testing technique to determine flaw criticality. The results of limited instrumentation evaluations for obtaining real-time stress wave emission data are discussed. The fabrication of specimens to be tested at Aerojet and during concurrent programs under ARPA sponsorship at the University of Michigan and North American Rockwell Corporation is also discussed.

321 **P. P. Crimmins and A. T. Green. "Correlation of Stress–Wave–Emission Characteristics with Fracture in Aluminum Alloys." Report 1246–Q–1. Aerojet–General Corporation, Sacramento, California. Dec. 1968.**

The report describes the progress during the first six months of a program to investigate the correlation of stress-wave-emission characteristics with fracture in selected aluminum alloys. During this period, the required test materials were obtained. Specimen fabrication was initiated and completed for the parent metal specimens; weldments for the specimens to be preflawed in the weld deposit were also completed. The report briefly describes the overall program and the initial results of a continuing literature survey to be performed throughout the program. Work planned for the next program quarter is also indicated.

322 **P. P. Crimmins and A. T. Green. "Development of a Nondestructive Testing Technique to Determine Flaw Criticality." Advanced Research Projects Agency Contract F33615–68–C–1705. Aerojet–General Corporation, Sacramento, California. Feb 1969. [NTIS: AD–850–604].**

A brief description is presented of the overall program to develop a nondestructive testing technique to determine flaw criticality. The results of initial evaluations of stress-wave emission monitoring instrumentation components are also presented. On the basis of the results obtained, a stress-wave emission counter-printer system has been obtained for use in subsequent crack-growth stress-wave emission evaluations. Sensors, filters, and frequency spectrum analyzers have also been obtained for these evaluations. This report outlines the first 8 months of a 20-month program.

323 **P. P. Crimmins and A. T. Green. "Stress Wave Analysis Technique." pp 31–41 in "Nondestructive Test Techniques Conference." Aerojet–General Corporation, El Monte, California. July 1969. proceedings of an Aerojet–General in–house conference held at the Corporate Information Center, El Monte, California. May 1, 1969.**

The Stress Wave Analysis Technique (SWAT) has been extensively employed as a laboratory technique for detecting and monitoring the occurrence of crack growth in structural metals and composites, and the mechanisms by which fracture occurs. Stress-wave-emission analysis has also been successfully applied to the detection and location of propagating flaws in structures both during testing and under service conditions. In such applications, the stress waves which accompany the energy release associated with crack extension are detected and analyzed to qualitatively determine the onset of crack instability while at the same time triangulate, using seismic techniques, to the source of the emission. Prior to the development of SWAT, the detection and location of defects in structures were possible only by conventional nondestructive testing methods such as magnetic particle, dye penetrant, ultrasonic, X-ray, and proof-testing techniques. Considerable judgment is necessary to establish whether the defects found are of a deleterious, propagating type. Repairs are made in accordance with accepted standards, and flaws that are not of the deleterious propagating type are repaired. Consequently, repair itself may be more detrimental to service life than the original defect; thus, a more serious condition may exist after the repair of innocuous flaws than if the defects were not removed.

324 **R. D. Crooks. "Detection of Incipient Cracking in Large Tubular Heat Exchanger by Acoustic Emission Techniques." pp 45–46 in** ***Paper Summaries: ASNT 36th National Fall Conference, September 27–30, 1976, Houston, Texas.*** **American Society for Nondestructive Testing, Columbus, Ohio. 1976. [abstract only].**

Acoustic emission techniques employing the Trodyne Corporation multisensor comprehensive data MSCD System have been used to evaluate the presence, location, and magnitude of thermal shock cracks in nickel alloy tubes used in a large multipass process water heater. The method proved successful in pinpointing several defective areas in the tubes which would otherwise have been located only after heat cracking progressed to the point of actual leakage. The discussion includes details of the heat exchanger construction for complex tube arrangement which precluded use of more common nondestructive testing methods. The tube failure mode and causes of prior tube failure are reviewed together with a preliminary acoustic emission testing program intended to establish the feasibility of this technique.

325 **N. O. Cross. "A Users Experience and Views on Acoustic Emission Testing of Pressure Vessels." ESSO Research and Engineering Company, Florham Park, New Jersey. 1972; paper presented at the 27th American Society of Mechanical Engineers Petroleum Mechanical Engineering Conference, New Orleans, Louisiana. Sep 17–21, 1972.**

Acoustic emission testing is gaining recognition as a supplementary nondestructive method for assuring the safety of heavy-wall pressure vessels during their initial acceptance hydrostatic test. Experience with this technique has demonstrated an ability to pinpoint not only large rejectable defects but normally acceptable small weld discontinuities. Common acceptance of AET for pressure vessel testing will, however, require additional development and standardization to firmly establish its ability to reliably identify and locate growing flaws before they become critical and lead to failure of the pressure vessel under test. More widespread use of this method is anticipated as the cost of instrumentation and testing is reduced, as standard application techniques evolve, and as consistent and meaningful standards of interpretation are developed. Experience with early AET system designs applied during the hydrotesting of several vessels, pitfalls encountered and eventually overcome in the design, construction, laboratory and full-scale field application of a grass roots AET system serve as the basis for general recommendations regarding minimum requirements for effective application of AET within the petrochemical industry.

326 **N. O. Cross. "Acoustic Emission Technique for Insuring Safe Hydrostatic Tests of Pressure Vessels." ASME Publication 70–PET–31. American Society of Mechanical Engineers, New York. 1970.**

The petroleum and petrochemical industries have recently experienced several brittle failures of heavy-wall pressure vessels during hydrostatic testing. The failures were unexpected since, in most cases, the materials involved had more than adequate toughness according to currently used standards. The paper discusses the application of acoustic emission testing as a means for locating defects that could cause brittle fracture during hydrotest or grow to critical size in service.

327 **N. O. Cross. "Display and Analysis of Real Time Data from Acoustic Emission Tests of Pressure Vessels." Session 1, pp 1–65 in *The Second Acoustic Emission Symposium.* Japan Industrial Planning Association, Dobashi Building 5F, 3–17–8 Akasaka, Minato-ku, Tokyo, 107 Japan. 1974. proceedings of a symposium held in Tokyo, Japan. Sep 2–4, 1974.**

Within thc past several years, a variety of pressure vessels and containment systems has been acoustic-emission tested (AET) on a commercial basis. The AE tests have been more frequently applied to vessels fabricated to stringent material and construction requirements. Such structures have correspondingly extensive radiographic (RT), ultrasonic (UT), and surface inspection techniques specified to assure the detection and removal of unacceptable flaws. Despite these high inspection standards and manufacturing quality control, vessel owners are finding AET a valuable and needed supplementary inspection method which serves the following purposes: (1) To locate fabrication flaws not discovered by the usually specified nondestructive testing method employing UT and RT; (2) To find service induced flaws (i.e., stress cracks, fatigue cracks, areas of embrittlement) in the structure; and (3) To requalify (and upgrade) existing vessels for more severe operating conditions. The primary purpose of this paper is to compare the various types of AE response related to material properties and flaw growth in pressure vessels with other AE test parameters. The analysis presented is based on actual and hypothetical cases and aims at showing several potential limitations in using currently applied emission summation, rate, and graphic displays of emission sites as the means for estimating the relative importance or nature and size of structural discontinuities. The major limitation in AE data evaluation at present is due to the missing link between the well-documented emission response from small, artificially flawed fracture mechanics specimens and the generally unknown AE signature from real, complex, and unique weld discontinuities in large fabricated structures. Data are presented which attempt to show that the variable and large total-detected emissions from the entire volume of a vessel can mask emissions from discrete locations on a vessel. Several new data display concepts are presented which may improve real time appraisal of AE locations. Other means of regulating the rate of emission production such as automatically programing vessel pressurization are also discussed. Reviewed also are some of the factors that make AET such a potentially powerful and essential inspection tool in assuring safe and reliable operation of increasingly critical, internally pressured structures. These factors involve design principles, materials properties, fabrication procedures, and limitations inherent in conventional inspection techniques.

328 **N. O. Cross, L. L. Loushin, and J. L. Thompson. "Acoustic Emission Testing of Pressure Vessels for Petroleum Refineries and Chemical Plants." pp 270–296 in** *Acoustic Emission, ASTM STP 505.* **American Society for Testing and Materials, Philadelphia, Pennsylvania. 1972. proceedings of symposium held in Bal Harbour, Florida. Dec 7–8, 1971.**

Acoustic emission testing is being used by fracture mechanics and nondestructive test engineers to establish the design adequacy, structural integrity, and operating safety of petroleum and petrochemical equipment. Used as a tool in the application of fracture mechanics concepts, AET of heavy-wall pressure vessels during hydrostatic testing also provides the means of assuring against catastrophic brittle fracture by locating growing defects. The state-of-the-art in developing effective electronic instrumentation and flaw-site triangulation methods has reached the point where AET should be considered competitive with other, more commonly specified test methods. For onstream surveillance, AET has the potential for the detection of service induced flaws and for detecting potentially detrimental changes in material properties. The amplitude and frequency content of environmental noise can, however, limit applications of this type. Although techniques are available for minimizing the reception of unwanted electrical and mechanical sources of noise, additional effort is needed in this area to facilitate widespread inservice use. Other areas requiring further research and development effort and the development of standards include flaw signature analysis, signature analysis characteristics for various materials, simplified instrumentation, acoustic parameter selection, the effect on signals of environmentally induced materials property changes, and data display.

329 **H. A. Crostack. "Untersuchungen zum Festigkeitsverhalten von Punktschweissverbindungen mittels Schallemissionsanalysen (Investigations of the Mechanical Behavior of Spot Welded Joints by Means of Acoustic Emission Analysis)." Doctor–Engineer Thesis. Universitat Dortmund, Dortmund, West Germany. Jan 1976. (in German).**

330 **H. A. Crostack. "Weld Quality Control Measurements of EB–Weldments by Means of Acoustic Emission." pp 27–47 in "EWGAE–ISPRA: Proceedings of the Third Meeting of the European Working Group on Acoustic Emission."**

A. C. Lucia, Editor. Report EUR-5513e. Commission of the European Communities, Directorate General Scientific and Technical Information and Information Management, Luxembourg, Belgium. 1976. meeting held at the Joint Nuclear Research Center, Ispra, Italy. Sep 25-26, 1974.

The mechanical properties of EB-weldments of high strength steels are influenced by cracks which arise during and after the actual welding process, and as a result of phase transformations. These cracks oftentimes are so small that they cannot be detected by normal NDT-methods such as X-ray, ultrasonics, and magnetic particles. However, in these high strength steels even the smallest cracks can lead to failure as has been revealed during linear-elastic fracture toughness tests on these steels. The position of defects in a weldment has extreme importance because the crack propagation is influenced by the structure as well as the state of stress. Information about the causes of crack initiation can be obtained from the time elapse before and the corresponding temperature history curves. Thus, for example, martensitic transformation can attribute to higher crack sensitivity. Acoustic emission analysis can be applied to investigate the origins of variations in welded joint properties. Research aims to characterise weld quality as a function of emitted sound from cracks. The AE measurement during the initiation of small cold cracks in high strength steels is difficult since sound emission during martensitic phase transformations is simultaneously detected. Additional disturbing noise may come from the welding process or from the welding machine. For these reasons, the application of AEA is to be evaluated to determine whether all types of responses can be analysed and whether different signals can be differentiated from one another. The crack sensitivity of welded high strength steels is related to the welding parameters. Since the EB-welding process is usually automated, the weld quality control by maintaining a specific AE value after welding would be a useful mode of application of this control method. Using the high strength low alloyed steels 38 CrMoV 51 and 45 CrMoV 67, AE signals were recorded during and after the EB-welding and then were analysed. Both steels are prone to cracking as a martensitic structure forms during welding. Initially the noises of the welding process and from the machine itself, as well as the acoustic emission from the base materials during simulated heating and cooling, were investigated before the full welding process was researched. In this way, the influence on the AE of the welding parameter and of circular beam deflection with different frequencies and radii was found. The various noises could be differentiated, some by AEA, others by frequency analyses and pulse-slope. Characteristic values for the quality control as a function of the different welding parameters could also be found by using a summation of counts and registering each pulse and corresponding time elapse. The AE can therefore yield information about cracks, about the geometry of the weldment, and about the volume of martensite.

331 **H. A. Crostack and E. Roeder. "Anwendung der Schallemissionsanalyse zur Bestimmung von Vorgangen der Rissbildung und Rissausbreitung (Application of Acoustic Emission Analysis for the Determination of Crack Formation and Propagation)."** ***Maschinenschaden,*** **MSCNA, 48(1):17-22. 1975. (in German).**

Acoustic emission testing is useful in the testing of mechanical structures as well as in the study of materials. It allows the early detection and localization of critical cracks and thus ensures timely remedial measures. The various possibilities of application of this method are indicated and explained.

332 J. C. Crowe. "Techniques for Nondestructive Evaluation of Radioactive Waste Capsule End Cap Welds." BNWL-SA-3758. Battelle-Northwest, Richland, Washington. Feb 1971. [NTIS]; *Materials Evaluation,* **MAEVA, 30(5):93–98. May 1972; paper presented at American Society for Nondestructive Testing Spring Conference, Los Angeles, California. Mar 8–11, 1971.**

Solid radioactive waste storage requires a high degree of container integrity. The paper describes the techniques developed at Battelle-Northwest to nondestructively inspect end cap welds for cesium and strontium waste containers. Ultrasonic methods have been defined for locating weld anomalies and acoustic emission techniques are currently being evaluated as a possible additional method to monitor weld quality. Initial ultrasonic and acoustic emission data taken with prototype equipment exhibit good correlation with destructive analysis of sample capsule welds. The ultrasonic inspection system is designed to incorporate a novel data presentation technique. The ultrasonic scanning and data readout system produce a readable, easy-to-interpret record of the weld inspection. The scanning technique can be adapted to X–Y recorder printout or high speed CRT mapping methods. Future development areas are suggested which could extend the recording method to present depth-of-flaw or signal amplitude information in other ultrasonic inspection applications.

333 C. Crussard. "Application de la Theorie des Ondes Explosives a la Croissance de la Martensite (Application of the Theory of Explosive Waves to the Growth of Martensite)." *Comptes Rendus Hebdomadaires des Seances de l'Academie des Sciences,* **COREA, 240:2313–2315. June 13, 1955. (in French).**

Presented are calculations showing that a shock wave and reaction wave may propagate in a solid at about the velocity of sound. This indicates the probability that martensite is formed by this mechanism.

334 G. Curtis. "A Broadband Polymeric Foil Transducer." *Ultrasonics*, **ULTRA, 12(4):148–154. July 1974.**

If a 0.001-inch (0.025-mm) thick foil of polyethylene teraphthalate is metallized upon one side and a DC voltage is applied to the metallizing, then the foil will adhere firmly by electrostatic attraction to any conducting surface. In acoustic emission experiments with such a foil as a capacitance detector, it was observed that by cycling the polarizing voltage it was possible to increase the sensitivity of the device until it was only 30 db below that of a damped PZT transducer. Such an enhanced response probe has been found to have a frequency response which is flat from 10 kHz to 5 MHz. It also has the advantages of being lightweight; needs no coupling agent when applied to conducting substrates; is stable over long peiods; can be made in any shape or array of shapes; and will conform to curved surfaces.

335 **G. J. Curtis. "A Review of Current Ultrasonic Nondestructive Testing Developments in the United Kingdom." AERE-R 7946. United Kingdom Atomic Energy Authority, Research Group, Atomic Energy Research Establishment, Harwell, Oxfordshire, England. Feb 1975.**

336 **G. Curtis. "Acoustic Emission–4: Spectral Analysis of Acoustic Emission."** *Non-Destructive Testing*, **NDETA, 7(2):82-91. Apr 1974. [This is the fourth in a series of articles on acoustic emission. See References 1716, 1404, 212, and 712 for Parts 1, 2, 3, and 5, respectively].**

The author discusses the significance and measurement of the frequency and shape of acoustic emissions. He considers theoretical descriptions of acoustic emission and compares them with experimental evidence. He considers the implications of this understanding of the nature of acoustic emission and illustrates these implications by a description of a program of investigation at AERE Harwell. This program added to knowledge of acoustic emission and the author considers the results in the light of other experiments. He concludes that spectral analysis could complement the use of amplitude distribution and ringdown counting but not enough is yet known of the general meaning of the spectrum of an acoustic emission.

337 **G. J. Curtis. "Acoustic Emission Energy Relates to Bond Strength."** *Non-Destructive Testing,* **NDETA, 8(5):249-257. Oct 1975.**

The author presents the results of experiments on adhesively bonded aluminum-alloy lap joints where stress, far less than needed to break the joint, causes acoustic emission. The lap shear strength of those joints bonded with Redux 775 proved to be related to the cumulative acoustic emission energy per unit volume of the adhesive layer for different thicknesses of adhesive. What evidence there is for those joints bonded with the more ductile BSL 308 points to a similar relationship. The results demonstrate that acoustic-emission energy monitoring has potential as a nondestructive test of bond strength and merits further examination to establish the true relationship between bond strength and the total acoustic-emission energy released in the fracture of various structural adhesive systems.

338 **G. J. Curtis. "Acoustic Emission in Stressed Epoxy Bonded Structures." AERE-R 7684. Final Report. United Kingdom Atomic Energy Authority, Research Group, Atomic Energy Research Establishment, Harwell, Oxfordshire, England. Mar 1974.**

An acoustic emission spectral and energy analysis sytem has been derived which makes use of a polymeric foil tranducer with a flat bandwidth from 10 kHz to 10 MHz. To guard against the interference of radiated electrical transients and radio stations, the Instron static test machine used to fracture specimens is housed in an electrically screened room. A low resistance earth mat and main transient filter have been installed in order to reduce interference from 50 Hz, 100 Hz, and switching transients. Once detected, acoustic emissions are amplified and pass into two circuits, one for spectral analysis and the other for energy analysis. A PDP8/I computer on line to the experiment collects digitised emissions and records them on magnetic tape. Upon completion of the experiment, the experimenter can examine each emission together with its spectra upon a storage oscilloscope. Consideration has been made of the energy radiated by various geometries of acoustic emission source. Equipment has been derived which uses the approximation that the energy in the emission is proportional to the square of the area under the derived voltage pulse. Either the cumulative area or the area in each millisecond time interval is displayed upon a chart recorder. Use has been made of this sytem to examine the spectra and energy of acoustic emissions derived from shear failure of Redux 775 and BSL 308 bonded aluminum lap joints. The major conclusions are as follows: (1) In Redux 775 emission shapes are varied. Spectral activity ranges from 500 Hz to 1 MHz. Varying the strength of the joint by changing its glue-line thickness shows that the stronger joints exhibit greater high frequency activity for longer into fracture than do weaker ones; (2) Graphing the cumulative area under the acoustic emission derived voltage pulse at fracture against the strength of the joint, produces a curve suggesting an inverse sixth power law between cumulative energy at break and the strength of the joint. Experiments were made mainly with Redux 775, but some made upon two extremes of strength of BSL 308 adhesive tended to verify this result. [reprinted by permission of the United Kingdom Atomic Energy Authority, copyright by U.K.A.E.A.].

339 **G. J. Curtis. "Current Ultrasonic Nondestructive Testing Developments in the United Kingdom." pp 691–698 in** ***1974 Ultrasonics Symposium Proceedings.*** **Catalog No. 74 CHO 896-1SU. Institute of Electrical and Electronics Engineers, Inc., New York. 1974. symposium held in Milwaukee, Wisconsin. Nov 11–14, 1974.**

The current direction being taken by ultrasonics research in nondestructive testing is towards determination of defect significance. In any given situation, the significance of a defect is a question for fracture mechanics to solve. Inevitably, however, the solution will contain some reference to defect position, size, shape, and orientation. The purpose of the paper is to review some of the techniques currently under investigation which, it is hoped, will aid in the nondestructive assessment of some or all of these parameters. The techniques are: the direct visualisation methods of Schlieren, photoelasticity, holography, and automated *B* and *C* scanning; and the indirect methods of spectroscopy, pulse attenuation, and acoustic emission. As attempts at quantisation stand or fall on knowledge of transducer performance, some space is given to transducer development and behaviour assessment.

340 **G. J. Curtis. "Evaluation of Adhesion by Ultrasonic Techniques." pp 59–62 in** ***Ultrasonics for Industry 1970: Conference Papers.*** **Iliffe Science and Technology Publications Ltd., Surrey, England. conference held in London, England, Oct 20–21, 1970.**

To date no general way exists of predicting the strength of any material, composite, or otherwise. Not surprisingly, since the mechanisms of adhesion are still ill-defined, neither is there a general way of predicting the strength of an adhesive joint. Four acoustic techniques which have been marginally successful are reviewed and some of the features worth further exploration are noted.

341 **G. J. Curtis. "Offshore NDT Forum."** *British Journal of Non-Destructive Testing,* **BJNTA, 19(2):91-92. Mar 1977.**

Letter from author correcting his contribution of the Offshare NDT Forum.

342 **G. J. Curtis. "The Characterisation of Failure Mechanisms by Means of Acoustic Emission." Atomic Energy Research Establishment, Harwell, Berkshire, England; paper presented at the Institute of Physics Conference on Acoustic Emission, Imperial College, London, England. Mar 14, 1972. [proceedings were never published].**

The absence of a broad bandwidth transducer has made the use of acoustic emission monitoring as a guide to failure mechanism an arbitrary and qualitative technique. No knowledge of true emission shape exists. It is not clear how much the shape is modified by the geometry of the structure or the materials of which it is made. Presented with an unknown bandwidth of emissions modified or otherwise by the modes of vibration of the specimen, counting procedures introduce unknown bias. The purpose of this paper is to describe a transient spectrum analysis system which was developed and which makes use of a broad bandwidth transducer, (DC to 10 MHz). From its application to isotropic solids some conclusions are drawn of true emission shape and modification due to specimen material, geometry, and distance of the transducer from the source of emissions. With these results as a basis, some insight is gained into the failure mechanisms in single and multiinterfacial composite materials.

343 **J. P. Cusick. "Audiofrequency Acoustic Emission."** *Bulletin of the American Physical Society,* **BAPSA, 8(Ser.II):237. 1963. [abstract only]; paper presented at the March 1963 Meeting of the American Physical Society, San Francisco, California.**

The acoustic emission of NaCl single crystals during mechanical testing has been measured in the frequency range, 300 to 10,000 cycles per second (cps). Preliminary data indicate the observed emission to occur at about 3000, 6000, and 9000 cps. The higher frequency components correspond to displacement amplitudes considerably smaller than that at 3000 cps. The observed emission has been measured during 4-point bending experiments. Simultaneous measurement of the load-deflection curve for the mechanical test suggests that the emission is associated with the slip process.

The presentation was based on preliminary research which was never completed and no report was ever written.

344 **J. Czochralski. "Die Metallographie des Zinns und die Theorie der Formanderung bildsamer Metalle."** *Metall und Erz,* **MERZA, XIII (N.F.IV), (18):381-393. Sep 22, 1916. (in German).**

Author reports on the acoustic phenomenon of *tin cry* from both tin and zinc produced by *umklappen* of the lattice structure during twinning.

D

345 **M. Dadian. "Emission d'ondes sonores au cours de chargement du fer en hydrogene cathodique (The Emission of Sound Waves during the Cathodic Hydrogen Generation of Iron)."** *Mecanique Electricite,* **MQELA, 52(229):26. Jan 1969. (in French).**

346 **M. Dadian. "Mise en evidence d'ondes sonores se produisant dans le fer au cours du chargement en hydrogene cathodique (Sound Waves Produced in Iron when Charged with Cathodic Hydrogen)."** *Comptes Rendus Hebdomadaires des Seances de l'Academie des Sciences, Series C: Sciences Chimiques,* **CHDCA, 265(21):1155–1157. Nov 20, 1967. (in French).**

This article reports that the emission of sound waves by iron during electrolytic charging with hydrogen are attributed to the formation of internal cracks.

347 **L. W. Dahlke, W. E. Jorgenson, H. J. Saxton, and A. R. Willis. "The Performance and Inspection of Flawed Pressure Vessels." SAND 76-8654. Sandia Laboratories, Livermore, California. June 1976. [NTIS]; pp 19–20 in** *Paper Summaries: ASNT 36th National Fall Conference, September 27–30, 1976, Houston, Texas.* **American Society for Nondestructive Testing, Columbus, Ohio. 1976.**

The effects of flaws on pressure vessel performance and the ability of NDT techniques to detect and actively monitor these flaws prior to and during burst testing were studied. Twenty flawed and unflawed spherical pressure vessels fabricated from several steels and a titanium alloy were evaluated. The flaws ranged from circumferential machined slots on the inner surface to weld voids and lack of penetration. This interdisciplinary study employed (1) radiography to characterize the defects, (2) acoustic emission and holography during proof-testing to detect the flaws and unsatisfactory structural performance, (3) radiography and acoustic emission to monitor flaw and vessel growth during hydrostatic burst testing, and (4) extensive materials and structural evaluations. The study showed that radiography is an excellent technique for monitoring defect and vessel growth during burst testing. Holography was found to be capable of detecting inner wall defects spanning 20 percent or more of the wall thickness. Except for the titanium vessels, acoustic emission failed to detect flaws in these vessels. In addition, the insensitivity of tough materials to small flaws was verified.

348 **A. T. D'Annessa and J. S. Owens. "Acoustic Emission Characteristics of Heat Treat Cracking in Rene 41 Weldments."** *Welding Journal,* **WEJUA, 51(3):156s–168s. Mar 1972.**

The paper is concerned with the characterization of post-weld heat treat cracking in Rene 41 weldments- -a well-known problem often encountered with complex or highly restrained superalloy weldments. The purpose is to describe *when* and *how* heat-treat cracking occurs in Rene 41. To accomplish this, an acoustic emission technique was developed for *in situ* monitoring of crack-susceptibility test specimens during heat treatment. Accordingly, several test parameters were established which were found to be capable of discriminating characteristics of the cracking process, thus making it possible to study subcritical and gross-cracking events. Acoustic emission data were used to screen the effects of welding process and pre-weld base metal processing variable on cracking behavior, and to provide a means for correlating the relative crack susceptibility of differing microstructures.

349 A. T. D'Annessa and J. S. Owens. "Effects of Furnace Atmosphere on Heat Treat Cracking of Rene 41 Weldments." *Welding Journal*, **WEJUA, 52(12):568s–575s. Dec 1973.**

The influence of furnace atmosphere on the susceptibility to postweld heat treat cracking of the precipitation hardenable nickel base super-alloys has been the subject of previous studies. Although the results of these studies varied somewhat, an increased resistance to heat treat cracking was reported using inert atmospheres and vacuum. Due to experimental shortcomings, these prior results were qualitative in nature and, therefore, subject to some speculation. These shortcomings were overcome in recently completed Air Force sponsored research programs with the development of a dissimilar alloy crack-susceptibility specimen and an acoustic emission technique for monitoring cracking events during heat treatment. The subject specimen was found to be reliably reproducible from both a fabrication and test result standpoint, and the acoustic emission technique provided a method for obtaining quantitative information from crack-susceptibility tests. Consequently, data were obtained which more precisely characterized the influence of furnace environments on post-weld heat treat cracking. The primary objective of these studies was to determine the effects of oxygen on the susceptibility of Rene 41 to heat treat cracking. Possible benefits of excluding oxygen from the furnace environment have been reported by other investigators, including the suggestion that the absence of oxygen from the furnace atmosphere during heat treatment would eliminate postweld cracking. The results reported here are from tests conducted with hydrogen, helium, and vacuum furnace environments. These data are compared with results of air atmosphere control tests and tests conducted using oxygen atmospheres. The scope of work described includes crack-susceptibility evaluations and metallographic (replica and scanning electron microscopy) and acoustic emission analyses.

350 A. T. D'Annessa and J. S. Owens. "Procedures for Avoiding Heat-Treat Cracking in Nickel-Base Superalloy Weldments." AFML–TR–70–224. Lockheed–Georgia Company, Marietta, Georgia. 1970.

Reported is a study concerned with the development of procedures for avoiding post-weld heat-treat cracking in nickel-base superalloy weldments. Circular-path type crack-susceptibility test specimens, utilizing a dissimilar alloy frame, were developed for *on-heating* and *isothermal* cracking studies. An acoustic emission system was assembled and a technique developed for monitoring subcritical and gross-cracking events during post-weld heat treatment. Heat-treat cracking of Rene 41 weldments was found to be a two-stage process with subcritical microcracking occurring early in the aging temperature range followed later by gross cracking which occurred by slow (stable) crack growth or rapid (unstable) crack extension. Acoustically defined crack-susceptability test parameters were found to be responsive to variations in base metals, heat-treating procedures; and welding process parameter manipulations thus provided a method for qualifying crack-susceptability test results. The post-weld heat-treat cracking of Rene 41 weldments was found to be influenced by base-metal rolling direction, welding process variables, pre-weld metallurgical condition of the base metal, and heat-treat atmosphere. A *1975 °F for 25 minutes, water-quench* solutioning treatment was found effective for avoiding *on-heating* gross cracking in a specification grade and a low-carbon heat of Rene 41. Microcracking obtained with specimens given pre-weld *solution-air cool* treatment indicated that cracking behavior was influenced by post-solution treatment cooling rates. On-heating and isothermal crack data indicated that Rene 41 sheet material converted from Rotating Electrode Process (REP) powder bar, a powder metallurgy product, has a greater resistance to gross cracking than conventional wrought product sheet. Tensile and stress-rupture properties obtained for REP powder sheet compare faborably with those of conventional Rene 41 sheet. Crack-susceptibility results indicated that the solutioning treatment used for Rene 41 improved the resistance of Astroloy to *on-heating* cracking; gross-cracking temperatures were elevated by 100 to 125 °F for pre-weld solutioned Astroloy sheet.

351 **M. S. Darlow and R. H. Badgley. "Early Detection of Defects in Rolling-Element Bearings." SAE No. 750209. Society of Automotive Engineers, Warrendale, Pennsylvania. 1975.**

Results of studies which demonstrate the feasibility of the high-frequency resonance technique (HFRT) for defect analysis of rolling-element bearings in virtually any bearing system are presented. Specific emphasis is placed on helicopter engine and transmission applications. The HFRT is a method for separating ball-pass or roller-pass frequency peaks from background noise by isolating and demodulating high-frequency spectrums of accelerometer signals. The process is also called envelope detection. Tests were conducted with bearings from a UH-1 helicopter transmission to show the effectiveness of the HFRT explicitly for that application and implicitly for other bearing applications. The resonance peaks for these tests were selected by inspection of raw data signals and were found to correlate reasonably well with predictable bearing-race natural frequencies. [Abstract 22-0661. *Metals Abstracts,* Vol. 9, 1976].

352 **M. S. Darlow, O. A. Shinaishin, and S. J. Acquaviva. "Acoustic Emission Detection of Fatigue Crack Initiation and Propagation in Notched and Un-notched Titanium Specimens." AMMRC-CTR-75-1, MTI-75TR18. Mechanical Technology Incorporated, Latham, New York. Jan 1975. [NTIS: AD/A-007-224].**

A brief explanation of the acoustic emission process and its theoretical background are presented, and its applicability to this investigation. Both static and dynamic (fatigue) tests were run. The fatigue specimens were run unnotched and notched in reverse bending and tensile fatigue and the total number of acoustic emissions were plotted versus number of fatigue cycles. Cracks formed in unbroken notched specimens from reverse bending and tensile fatigue were metallurgically examined in order to develop a generic relationship between acoustic emission count and fatigue crack size. The point of crack initiation was clearly and consistently defined by the commencement of acoustic emission throughout the entire spectrum of tests conducted. The titanium specimens used in these tests were brittle, causing the process of crack propagation, from crack initiation to specimen failure, to be rapid. In general, the cracks formed in notched specimens propagated to failure much slower than those formed in unnotched specimens. For example, the unnotched specimens run in tensile fatigue lasted generally less than 100 cycles from crack initiation to specimen failure, while the notched specimens run in tensile fatigue took, on the average, more than 2000 cyles to fail after crack initiation. It is possible, based on the results of this work, to develop a simple mechanism to stop a test at the earliest indication of crack presence. The experimental results show a faster rise in acoustic emission counts where there is a higher level of localized stress. This conclusion may also prove valid for higher uniform stress, but no experimental verification was attempted in this work. Several areas for future investigation have been identified. These areas, which are important to practical utilization of acoustic emission in assessing titanium structural integrity under dynamic loads, include: the separation of acoustic emission signals due to crack initiation (nucleation) from those due to crack growth; the evaluation of the effect of notches on the acoustic emission data and the comparison of the behavior of notched specimens to unnotched specimens; and the evaluation of the relationship between the acoustic emission rate and the initiation of a crack.

353 **G. J. Dau. "Recent Developments in Nondestructive Testing and their Implications to the Industry." BNWL-SA-2217. Battelle-Northwest, Richland, Washington. Nov 19, 1968; paper presented at the Student Chapter of the American Nuclear Society, University of Arizona, Tuscon, Arizona.**

This is the text of a talk and covers a general discussion on acoustic emission, holography, and eddy current testing.

354 **G. J. Dau. "The Sounds of Failure."** ***Industrial Research,*** **IDRSA, 13(4):40–44. Apr 1971.**

355 **G. J. Dau. "Thoughts on Emerging NDE Technology for the Nuclear Power Industry."** ***Metal Progress,*** **MEPOA, 109(6):35–38. June 1976.**

Nondestructive evaluation methods which offer significant advances, particularly as applied to the nuclear power industry, are discussed with illustrations of their specific capabilities. Topics covered include use of computers, especially with adaptive learning networks; eddy current inspection with multiparameter technique; computer and optical processing of radiographic images for image enhancement; reconstruction of three-dimensional details by tomographic analysis of penetrating radiation; stress profile measurement by ultrasonics; and correlation of acoustic emissions to structural integrity. [Abstract 22-1556. *Metals Abstracts,* Vol. 9, 1976].

356 **G. J. Dau and K. E. Stahlkopf. "Nondestructive Evaluation Program for Primary Pressure Boundary." Electric Power Research Institute, Palo Alto, California. 1975; paper presented at the American Society for Nondestructive Testing 35th National Fall Conference, Atlanta, Georgia. Oct 13–16, 1975. [copy available from EPRI, Records and Report Center, P.O. Box 10412, Palo Alto, California 94303].**

A valuable contribution to assurance of safe and reliable operation of nuclear power plants is provided by the use of nondestructive testing methods. In addition to the safety aspects, these methods can also make a significant economic impact. To illustrate this, assume the modest goal for an NDE-R&D program is to save one day's outage time for each nuclear power plant per year. Assigning a value of $300,000 per day for replacement power costs for the 53 plants in operation today, achievement of the above goal produces an annual savings of about 16 million dollars. The purpose of this report is to summarize the projects the Electric Power Research Institute (EPRI) has under way for the development of improved or new nondestructive evaluation (NDE) techniques for the primary pressure boundary. Specific project summaries are given under the following four headings: Steam Generator, Pressure Vessel, Piping and Components, and Property Measurements. Each summary contains a brief description of the project, the name of the prime contractor, duration and cost, and EPRI technical project monitor.

357 **T. J. Davis. "Electronic Interference Reduction in Acoustic Emission Testing." BNWL-1596. Battelle-Northwest, Richland, Washington. Nov 1971. [NTIS].**

The report describes techniques which may be used to significantly enhance signal-to-noise ratio for acoustic emission monitoring in severe electrical environments. Methods for applying common mode rejection and optimizing signal shielding and grounding are presented. A new and useful approach for isolating instruments from unregulated 60 Hz power lines is also discussed.

358 **M. I. Dawes. "Ultrasound Detectors: Leaks/Machine Noise/Electrical Discharge."** ***Ultrasonics,*** **ULTRA, 5(1):1–7. Jan 1967.**

Article describes the instrumentation and applications of ultrasonic detectors used for detection of leaks, machine noise, and electrical discharge. Included are photographs and descriptions of four commercially available leak detectors.

359 **C. K. Day. "An Investigation of Acoustic Emission for Defect Formation in Stainless Steel Weld Coupons." BNWL-902. Battelle-Northwest, Richland, Washington. Jan 1969. [NTIS].**

Results of an exploratory investigation to determine capabilities and limitations of acoustic emission techniques for detecting weld defects are presented. Acoustic emissions reveal cracks while they form during the welding operation; data indicate that gross porosity can also be detected with this technique. Interference signals, similar to acoustic emissions from bonafide weld defects, can be produced by scale flaking away from the weld surface, by electromagnetic interference, and by impact on the weld material by tools, welding rods, and other metal objects. Suggestions for reducing the effects of interference signals are included.

360 **C. K. Day. "An Investigation of the Plastic Bursts of Microstrain in Zinc as Sources for Acoustic Emission." M.S. Thesis. Washington State University, Pullman, Washington. 1969.**

An investigation of the plastic bursts of microstrain from zinc single crystals has been conducted to determine if acoustic emission is associated with these plastic bursts. A sensitive acoustical monitoring system was constructed and used in conjunction with the single crystal tests. The results indicate a need for additional sensitivity when testing basal-oriented zinc single crystals under the conditions described within this thesis. Signals which had the same form as acoustic emission were detected during tests of a nonbasally oriented specimen, but additional tests would be needed to determine if these signals were acoustic emission.

361 **C. K. Day. "In-Flight Aircraft Fatigue Detection." Y-58034. Battelle-Northwest, Richland, Washington. Mar 1968.**

A limited research program was conducted to investigate the technical feasibility of detecting aircraft structural fatigue under in-flight conditions. Continuous monitoring of airframes for incipient failure now appears possible with the application of a technique known as acoustic emission. To demonstrate the method, photographs of in-flight acoustic emission signals along with their spectral analysis are shown. In particular, it is demonstrated that in-flight emission produced by fatigue may be detected nondestructively. The purpose of this feasibility study was to determine if acoustic emission signals resulting from the onset of fatigue of aircraft structures could be detected under the various environmental conditions existing during aircraft flight. Problems related to the environment included structural vibration, electrical interference, and airflow noise.

362 **D. S. Dean. "Rocket Propulsion Establishment." p 11-1 in "Newsletter on the Non-Destructive Testing of Composite Materials and Adhesive Bonded Joints." Issue Number 1. Royal Aircraft Establishment, Farnborough, Hants, England. Jan 1973.**

The programs of work here have the objectives of (1) getting a better understanding of the mechanism of stress wave emission in composites and (2) developing if practical, SWE as an NDT technique for the inspection of composite rocket motor cases. To meet (1) above, RPE has recently placed an extramural contract with the U.K. Atomic Energy Authority Research Group, Materials Physics Division. The objectives are to correlate the nature of an emission, i.e., amplitude, frequency content, decay rate, etc., with defined changes in carbon fibre and other composite specimens. With regard to (2), RPE's first six-months work has been an experimental phase in which emissions were generated in a number of standard test pieces including Durestos, carbon fibre, glass fibre, etc., by flexural strength tests. Different monitoring techniques including real-time digital analysis and video recording of emissions for subsequent analysis have been tried out. Phase 2 of the work will be to develop a master monitoring channel from which it is hoped a better understanding of any important parameters will be obtained, e.g., by using a data transient recorder to obtain a Fourier analysis of emission pulses. Also the range of samples will be extended to include some more nearly approaching real motor cases. Phase 3 will be concerned with the development of a fault location channel.

363 **D. S. Dean and L. A. Kerridge. "An Immersion-Technique for the Detection of Acoustic Emission in Carbon-Fibre Reinforced Plastics Pressure Vessels."** ***NDT International,*** **NDITD, 9(5):233-238. Oct 1976.**

Conventional techniques of detecting stress wave emission by wall-mounted transducers give rise to gross inaccuracy in the recording of amplitude, number, spectral content, and location of emissions. The paper describes an alternative immersion system which overcomes most of these limitations and is more convenient for production testing.

364 **D. S. Dean and L. A. Kerridge. "An Immersion Technique for the Detection of Stress Wave Emission in Carbon Fibre Reinforced Plastic Pressure Vessels." RPE Memorandum No. 672. Rocket Propulsion Establishment, Westcott, Aylesbury, Bucks, England. Jan 1976. [copy of report, designated RPE Memo 672, BR-52031, can be obtained from the Defence Research Information Centre, Station Square House, St. Mary Cray, Orpington, Kent BR5 3RE, England]. [see Reference 363 for publication].**

Conventional techniques of detecting stress wave emission by wall-mounted transducers give rise to gross inaccuracy in the recording of amplitude, number, spectral content, and location of emissions. An alternative immersion system is described which overcomes most of these limitations and is more convenient for production testing.

365 **D. S. Dean and L. A. Kerridge. "Stress Wave Emission." pp 8-1 to 8-4 in "Newsletter on the Non-Destructive Testing of Composite Materials and**

Adhesive Bonded Joints." Issue Number 4. Royal Aircraft Establishment, Farnborough, Hants, England. July 1975.

Work has continued on SWE, primarily applying the technique to the NDT of CFRP rocket motor cases and, secondly, to controlled test vessels or specimens in support of that work. The facility has been developed considerably but the philosophy behind it remains one of versatility. Since it is not yet known which parameters are most important, all options are kept open, priority going to functional need as it becomes apparent. For example, the rate of pressurization was soon found to be of significance, therefore, the manual hydro pump has been replaced by a fully automatic hydro-electric pumping system. This greatly facilitates extended cyclic fatigue testing. Also, instrumentation has been extended to five monitoring channels with fibre optic recording of up to sixteen parameters, such as totals, rates, rms, mean squares, pressure, audible emissions, etc. A real-time IMH2 spectrum analyser with limited storage is also available. A number of CFRP rocket motor cases, scheduled for programs other than SWE measurements, has been monitored during proof testing and data are being built up. The problem is complex, with so many possible sources of emissions which are not likely to affect ultimate integrity. It is considered for instance, that noise arises from hoop cracks in the outer resin coating. However, from continued analyses of the data obtained so far, it appears from motor cases taken to proof pressure and a few subsequently taken to burst, that there are significant emissions which can possibly be used as a basis for prediction of ultimate failure. Acoustic emission, in a slightly different context, has also proved effective as an early warning of motor case porosity. Although no quantitative measurements as to the degree of porosity are made, the technique has permitted cases being withdrawn from proof test and relined before undue damage was done. In support of the work, and as a means possibly of understanding better the problem encountered, SWE measurements are also being made on hydraulically tested CFRP oblate spheroids. The program is a controlled one inasmuch as some spheroids contain irregularities such as broken fibres, voids, resin rich-depleted areas, etc. It is mainly from this work that it is hoped to substantiate the validity or otherwise of any failure prediction technique for CFRP motor cases.

366 E. de Atley. "'Listening' to Metal for Telltale Signs of Wear." ***Electronic Design,*** **ELODA, 18(18):30. Sep 1, 1970.**

367 G. R. DeBaise, A. W. Porter, and R. E. Pentoney. "Morphology and Mechanics of Wood Fracture." ***Materials Research and Standards,*** **MTRSA, 6(10):493–499. Oct 1966.**

Wood fracture begins with stable extension of intrinsic flaws at low stress values. The growth of flaws is dependent upon the kind and direction of the loading stress. Flaws extend with time for creep and relaxation conditions in a characteristic manner for each type of test. Crack propagation in the radial- and tangential-longitudinal planes for the opening cleavage mode of fracture has similar strain energy release rates, which are functions of temperature and moisture content. In the radial-longitudinal plane, slow fracture forms a smooth surface and occurs predominantly between the middle lamella and the primary cell wall and between the latter and the S1 cell wall. The surface of failure for rapid fracture is through all the cell wall layers which exposes the cell lumens. The transition from slow to rapid fracture usually occurs at a ray crossing or at tracheid ends. [reprinted by permission of the American Society for Testing and Materials, Copyright].

368 **J. C. F. DeKanter. "Sprekend Materiaal: Stand van Zaken van een Veelbelovende Teckniek de Akoestische Emissie Analyse (Emitting Materials: State of Affairs of a Very Promising Technique–the Acoustic Emission Analysis.)" Laboratorium voor Metaalkunde, Technische Hogeschool Delft, The Netherlands. Fall 1971. (in Dutch); RFP-Trans-101. translated for Dow Chemical U.S.A., Rocky Flats Division, Golden, Colorado. Aug 1972. [NTIS].**

The paper reviews the development and presents a survey of the state-of-the-art in the field of acoustic emission. The introduction contains an explanation of what acoustic emission means and why it augurs so well for the future. Some highlights from its history are quoted. Two exploratory fields have now been opened up to acoustic emission analysis; one is nondestructive testing, and the other is the fundamental work on solid-state deformation mechanisms. Chapter I deals with the source of acoustic emission. There are four deformation modes; elastic, microplastic, plastic, and cracking, which are discussed in terms of acoustic emission. Dislocation kinetics are mentioned. In Chapter II, the detection and discrimination of acoustic emission signals are discussed. A brief description is given of several electronic components. Different signal analysis methods such as *primary pulse analysis* and *total signal analysis* are discussed. Discrimination possibilities such as localization coincidence, frequency, rise-time analysis, and scope- and audio-monitoring are discussed. Chapter III deals with the interpretation of acoustic emission signals. Acoustic emission analysis helps in detecting and discriminating different deformations. The question of continuous versus burst emissions is illustrated with some experimental examples. Chapter IV deals with acoustic emission generated by crack propagation and initiation. A brief description is given of fracture mechanics. Detection of crack growth due to stress corrosion and fatigue is discussed in brief, and the importance of the periodical overload principle as proposed is emphasized. In Chapter V, the possibilities of acoustic emission in the field of nondestructive testing and safeguarding of pressure vessels are discussed. Interpretation of acoustic emission signals during proof-testing is discussed in general. Chapter VI gives a brief survey of experience and information gathered during the subsequent visits of the author abroad.

369 **I. De Lotto and S. Golinelli. "Content Addressable Memories in Scientific Instruments."** ***Nuclear Instruments and Methods,*** **NUIMA, 130(2):487–490. Dec 15, 1975.**

Presented is the content-addressable-memory feature of a new system designed in the laboratories for nondestructive testing of nuclear reactor pressure vessels based on acoustic emission analysis. The content addressable memory is divided into two parts: the first selects the most frequent events among incoming ones (FES: Frequent Event Selection memory), and the second stores the frequent events singled out (FEM: Frequent Event Memory). The statistical behavior of FES is analyzed, and experimental results are compared with theoretical ones; the model presented proved to be a useful tool in dimensioning the instrument store capacity.

370 **J. D. Desai and W. W. Gerberich. "Analysis of Incremental Cracking by the Stress-Wave Emission Technique."** *Engineering Fracture Mechanics*, **EFMEA, 7(1):153–165. Mar 1975.**

One of the most fruitful areas for application of the acoustic emission technique lies in characterization of yielding and fracture processes. For example, what are the microscopic details of the macroscopic slow-crack growth process that precedes crack instability? Fundamental to answering this is the ability to detect and quantify the microcracking phenomena. It was found that the emission of a single elastic stress wave could be correlated to a load drop, ΔP, occurring during crack growth. Furthermore, this load drop could be interpreted via a theoretical compliance analysis in terms of area swept out by the advancing crack. It is proposed that any discrete stress wave emission associated with a fracture process can be interpreted in terms of an incremental fracture area, ΔA. Crack-line loaded or compact tension specimens of 7075-T6 aluminum were used for the experimental investigation. A theoretical relationship between the load drop and the crack growth step was derived for this test specimen configuration. [reprinted by permission of Pergamon Press Ltd].

371 **P. de Santis, T. Papa, and D. Sette. "Acoustic Emission in Isothermal Martensitic Transformations."** *Acoustica*, **ACUSA, 29(5):317–318. 1973.**

Martensitic transformations of iron alloys are of great importance in metallurgy. In this letter, considered was the application of an acoustic emission technique to an isothermal martensitic transformation. Particular consideration was given the case of an Fe-Ni-Cr alloy of the type commonly used as commercial stainless steel.

372 **M. T. Destribats, A. M. Touffait, M. Asty, R. Saglio, A. Samoel, and M. Roule. "In-Service Inspection of P.W.R. Nuclear Vessels/Inspection en service des cuves des reacteurs P.W.R." Centre d'Etudes Nucleaires de Saclay, Saclay, France; Paper 5C/12 in** *Proceedings of the Eighth World Conference on Nondestructive Testing.* **volume containing Sections: 5C, 3J, 1D. International Committee for Nondestructive Testing, Secretariat du Cofrend, Institut de Soudure, 32 Boulevard de la Chapelle, 75880 Paris Cedex 18, France. 1976. conference held in Cannes, France. Sep 6–10, 1976. (paper in French).**

Described are the methods and equipments developed at the CEA for in-service inspection of PWR nuclear vessels; these include focussed ultrasonics and acoustic emission.

373 **A. Yu. Detkov. "Using the Acoustic–Emission Method to Check the Strength of Fiberglass Rings."** *Soviet Journal of Nondestructive Testing,* **SJNTA, 12(5):488–493. Sep–Oct 1976; translated from** *Defektoskopiya,* **DEFKA, (5):34–40. Sep–Oct 1976. (in Russian).**

A method is proposed for determining the ultimate loads on rings made of fiberglass by registering and analyzing the acoustic-emission pulses in the high-frequency range. The influence of the loading rate on the parameters of the acoustic-emission pulses is studied. It is shown to be possible to detect cleavage in fiberglass by means of the acoustic emission. A comparison is made between the ultrasonic and the acoustic-emission methods for the prediction of the ultimate loads.

374 **H. R. Deveraux and D. L. Robinson. "The Application of Acoustic Techniques for the Integrity Analysis of Piping." pp 204–214 in** *Prevention of Structural Failure: The Role of Quantitative Nondestructive Evaluation.* **T. D. Cooper, P. F. Packman, and B. G. W. Yee, Editors. Materials/Metalworking Technology Series, Number 5. American Society for Metals, Metals Park, Ohio. 1975.**

The application of acoustic emission in the nondestructive testing of piping systems is described. The apparatus used, leak detection, and discontinuity detection techniques are given. The major variance in systems is the placement of the detectors which can vary from 200 to 1000 feet apart for pipe depending on its condition. In complex systems, detectors must be placed to avoid blind spots caused by *T's*, sampling, and feeder lines. Tranducers are available for remote sensing in buried locations. Stressing of the pipe for flaw detection can be produced by internal pressure or by passing a heavy vehicle over buried pipes. Some typical results are given for defective welds in a 16-inch buried gas transmission line. [Abstract 22-1156. *Metals Abstracts,* Vol. 8, 1975].

375 **E. Dickhaut and J. Eisenblatter. "Acoustic Emission Measurements during Electron Beam Welding of Nickel–Base Alloys." ASME Paper No. 74–GT–45. American Society of Mechanical Engineers, New York. 1974; paper presented at the American Society of Mechanical Engineers Gas Turbine Conference, Zurich, Switzerland. Mar 31–Apr 4, 1974;** *Journal of Engineering for Power, Transactions of the ASME,* **JEPOA, Series A, 97(1):47–52. Jan 1975.**

During EB welding of nickel-base superalloys, flaws may occur under unfavorable conditions. In welds of Waspaloy *Cameron* flaws appeared in the heat-affected zone, while flaws in the weld seam were observed in the case of Inconel 713 *Nimocast.* The flaws in Waspaloy could not be detected by conventional NDT methods because of their position at the root of the weld. Therefore, for direct observation of flaw formation in these alloys, acoustic emission measurements were made during EB welding of these alloys. Two different types of acoustic emission were found. In nickel-base superalloys and in austenitic steel, background noise occurs as so-called burst signals which follow one another at short time intervals; these signals correlate exactly with markings (grooves) on the surface of the weld seam. This type of acoustic emission depends on the local heat transfer (beam current per unit time and area). Apparently this background noise is caused by weld solidification. Flaw formation, on the other hand, is indicated by single burst signals having a much higher amplitude than those due to solidification. In the case of Waspaloy, this second type of acoustic emission was observed only at fairly high speeds of welding (i.e., higher than 25 mm/sec). This result is consistent with evaluations of micrographs.

376 **P. F. Dingwall. "Stress Wave Emission." pp 13–2 to 13–3 in "Newsletter on the Non-Destructive Testing of Composite Materials and Adhesive Bonded Joints." Issue Number 1. Royal Aircraft Establishment, Farnborough, Hants, England. Jan 1973.**

In-house work in the Structures Department has been concerned with carbon fibre reinforced plastics. The aim of the program is to collect data which will enable assessment of material quality to be carried out, with a subsequent extension to the testing of simple structures. Importance is attached to the assembly of a library of experimental results upon which, hopefully, predictions of quantity may be subsequently based. Extension of the work to structures will also include assessment of the quality of bonded joints and some work on this aspect will be necessary. At present, laboratory work is confined to the testing of unidirectional tensile specimens in Type 1 and Type 2 fibres.

377 **P. F. Dingwall. "Transducers for the Detection of Stress Wave Emission." Technical Memorandum Structures 829. Royal Aircraft Establishment, Farnborough, Hants, England. Feb 1973.**

Memorandum discusses types of transducers used in acoustic emission, their characteristics and design, and their response to induced stress wave stimulus.

378 **P. F. Dingwall and D. L. Mead. "Stress Wave Emission." pp 9–3 to 9–12 in "Newsletter on the Non-Destructive Tesing of Composite Materials and Adhesive Bonded Joints." Issue Number 3. Royal Aircraft Establishment, Farnborough, Hants, England. June 1974.**

The new wide bandwidth amplitude sorter, which was developed under contract by the University of Surrey, has now been delivered and is in the process of evaluation. One signal and up to 8 guard transducer inputs are provided, although, due to noise problems only 3 guard transducers may be used at present, if the maximum sensitivity is required. The instrument consists of two almost identical channels which independently process the signal and guard pulses from the tranducers, which are connected to the instrument *via* low noise pre-amplifiers. The pre-amplifiers feed an attenuator (which also includes a summing circuit in the case of the guard channel) and after amplification and rectification, the pulses are level detected and used to drive logic outputs. Both the signal and guard channels are interlocked in such a way that whichever channel is stimulated first, the logic output of the other channel is inhibited for a fixed period of time; this may be pre-set to a value depending on transducer spacing and stress wave velocity.

379 **P. F. Dingwall and D. L. Mead. "Stress Wave Emission." pp 10-1 to 10-6 in "Newsletter on the Non-Destructive Testing of Composite Materials and Adhesive Bonded Joints." Issue Number 4. Royal Aircraft Establishment, Farnborough, Hants, England. July 1975.**

The wide-band amplitude sorter mentioned in Newsletter No. 3 has now been evaluated and is in use. Apart from this, the instrumentation remains essentially as described in Newsletter No. 1. The electrical resistance of a unidirectional CFRP specimen in the fibre direction is dependent, among other things, on the number of continuous fibres. In order to determine at what stage during a tensile test fibres break and to characterise the stress wave emissions in terms of amplitude, the changes in specimen resistance was monitored throughout a number of stress wave emission tests using a conventional Wheatstone bridge. Work has started on the characterisation by amplitude and frequency of emissions occurring from fibre failure or shear failure (or both) of CFRP. The three-point bend test was chosen in order that the failure mode could be controlled by varying the rig span to specimen thickness (S/d) ratio.

380 **P. F. Dingwall and D. E. W. Stone. "Some Problems in the Calibration of the Acoustic Emission Sensors." pp 87-113 in "EWGAE-ISPRA: Proceedings of the Third Meeting of the European Working Group on Acoustic Emission." A. C. Lucia. Editor. Report EUR-5513e. Commission of the European Communities, Directorate General Scientific and Technical Information and Information Management, Luxembourg, Belgium. 1976. meeting held at the Joint Nuclear Research Center, Ispra, Italy. Sep 25-26, 1974.**

Because of the nature of the emissions arising from carbon fibre reinforced plastics (CFRP), it has been a practice to use a relatively wide band system 20 kHz to 1 MHz. Thus for transducers, needed are the frequency response and the sensitivity, i.e., the magnitude of the response to a standard stimulus. In this paper, attention will be concentrated on the former, although some approaches to the latter problem will be discussed.

381 **M. Dobovsek. "Raziskava akusticne emisije mehansko obremenjene snovi v korozijsko aktivnem okolju (Research of Acoustic Emission of Mechanically Stressed Material in Corrosion Active Surrounding)." Diplomsko delo st. 2059. Fakulteta za strojnistvo, Univerza v Ljubljani, Ljubljana, Yugoslavia. 1974. (in Slovenian).**

The work describes some characteristic properties of acoustic emission caused by corrosion of steel and aluminum in salt water and HCl solution.

382 **W. A. Dowden. "Acoustic Emission for Nondestructive Testing: A Reading List." The Nondestructive Testing Centre, Atomic Energy Research Establishment, Harwell, Berkshire, England. July 1970.**

383 **T. J. Downing. "The Acoustic Analysis of Materials in Dynamic Applications." Memorandum Report M67-6-1. Frankford Arsenal, Philadelphia, Pennsylvania. Sep 1966. [NTIS: AD-641-720].**

The sound spectrograph, which plots a pattern showing signal energy-level variation in both frequency and time, was utilized as the primary analysis equipment during this period. To produce a meaningful pattern, magnetic tape slowdown methods and special signal-handling electronics were employed for spectrogram axis-scale control and time-displacement referencing. Spectrographic patterns were made for comparative analyses of proper and defective (tooth missing) gear-train operation. Significantly different patterns were obtained demonstrating that effects hidden in the waveform become evident in the pattern. Sound spectrography is exploited fully in the analysis of system characteristics where either fixed excitation is operated on by a time-varying transfer function, or controlled excitation having time-varying spectral content is applied to a fixed transfer function. The former class includes all rotational and reciprocating equipment and equipment powered by periodic excitation. The latter class includes all passive (no moving parts) devices excited by programed forcing functions.

384 **N. I. Dreiman. "Acoustic Emissions from Iron-Manganese Alloys."** ***Metal Science and Heat Treatment,*** **MSHTA, (6):410-412. June 1968; translated from** ***Metallovedenie i Termicheskaya Obrabotka Metallov,*** **MTOMA, (5):78-80. May 1968. (in Russian).**

Discussion of the measurement of frequency spectrum and acoustic damping rate of mechanically generated sounds radiating from plates of iron-manganese alloys.

385 **N. I. Dreiman. "Determining Internal Friction in a Metal from the Characteristics of Its Sound Emission during Its Vibration."** ***Soviet Journal of Nondestructive Testing,*** **SJNTA, 11(2):236-239. Mar-Apr 1975; translated from** ***Defektoskopiya,*** **DEFKA, (2):129-132. Mar-Apr 1975. (in Russian).**

Some acoustic methods of inspection are based on recording the sound emission of metals and alloys during elastic (normally transverse) vibrations. The parameters of the sound emitted by different metal parts with excitation of their oscillations are a criterion for measuring the length, width, diameter, thickness, modulus of elasticity, Poisson's ratio, density, modulus of rupture, development of a crack, or intergranular corrosion. In this article, the possibility of determining the value of internal friction from the characteristics of the sound emitted by a metal during excitation of its vibration by a mechanical impulse is given a theoretical basis and experimentally confirmed.

386 **N. I. Dreiman. "Sound Emission by Metals and Alloys."** *Industrial Laboratory,* **INDLA, 38(1):52–54. Jan 1972; translated from** *Zavodskaya Laboratoriya,* **ZYDLA, 38(1):46–47. Jan 1972. (in Russian).**

Methods and apparatus for studying the sounds emitted by metals and alloys under various conditions of stresses and strains are described. Sound emission may be simulated by creating forced oscillations in the sample by impacting the specimen with a ball using a device to control the impact force and the point of impingement. Sound pressure level recordings of such oscillations are presented and their possible causes are discussed. The device could be modified for potential use as an acoustic emission calibration technique. [Abstract 32-0461. *Metals Abstracts,* Vol. 5, 1972].

387 **N. I. Dreiman and G. A. Kherodinashvili. "Dependence of the Sound Emission of Structural Steels 35 and 45 on Heat Treatment."** *Metal Science and Heat Treatment,* **MSHTA, 15(11–12):1074–1076. Nov–Dec 1973; translated from** *Metallovedenie i Termicheskaya Obrabotka Metallov,* **MTOMA, (12):55–56. 1973. (in Russian).**

The theory of sound emission from C steels subjected to repeated sharp impacts is presented, the intensity and spectrum of the sound emission being related to the general physical, chemical, and mechanical properties of these materials. The theory is tested against experimental measurements carried out for two varieties of C steel: St 35 and St 45. Theory and experiment are in qualitative agreement. The rate of attenuation of the sound vibrations in the steel depends, in particular on the internal friction, yield strength, Young's modulus, Poisson's ratio, the momentum of the impacts, and the geometrical dimensions of the sample and striker. [Abstract 32-0592. *Metals Abstracts,* Vol. 7, 1974].

388 **Yu. B. Drobot. "Second All-Union Seminar on Acoustic Emission."** *Soviet Journal of Nondestructive Testing,* **SJNTA, 12(2):230. Mar–Apr 1976; translated from** *Defektoskopiya,* **DEFKA, (2):143–144. Mar–Apr 1976. (in Russian).**

This is a brief report on the Second All-Union Seminar on Nondestructive Monitoring of States of Stress and Strain in Materials and Components by Acoustic Emission held October 7-9, 1975, in Khabarovsk, Russia. The seminar was organized by the State Committee on Standards, the Council of Ministers of the USSR, the Khabarovsk Regional Board of the Engineering Industry Society, and the Khabarovsk Interdisciplinary Center for Scientific Information. Over 250 persons participated in the seminar in which 129 papers were presented.

389 **T. F. Drouillard. "Acoustic Emission: A Bibliography for 1970–1972." RFP-2059. Dow Chemical U.S.A., Rocky Flats Division, Golden, Colorado. Mar 15, 1974. [NTIS]; pp 241–284 in** ***Monitoring Structural Integrity by Acoustic Emission, ASTM STP 571.*** **Edited by J. C. Spanner and J. W. McElroy. American Society for Testing and Materials, Philadelphia, Pennsylvania. 1975.**

The bibliography includes nearly all references in the literature on acoustic emission published during the three-year period, 1970 through 1972. Included in the 412 references are several for each of the associated technologies including: signature analysis, boiling detection, cavitation, leak detection, seismology, and rock mechanics. Information has also been obtained from eight abstracting and indexing services searched in compiling the bibliography. The bibliography has been arranged alphabetically by author and is cross-referenced with a list of approximately 400 authors. Also included is a subject index. Technical articles listed were published in some 90 different journals and in 8 different languages.

390 **T. F. Drouillard. "An On-Line Bibliographic Search of Acoustic Emission Literature." Dow Chemical U.S.A., Rocky Flats Division, Golden, Colorado. Apr 1975; paper presented at the 14th Acoustic Emission Working Group Meeting, San Antonio, Texas. Apr 28–30, 1975.**

Currently available to the public is an on-line search of bibliographic material on acoustic emission. In this presentation, the discussion included the service that is available, and the organizations providing data bases on acoustic emission. A description of the specific searches made were presented, giving details on the search strategy used, the information obtained in each search, and the approximate costs involved.

391 **T. F. Drouillard. "Crack Detection in PIGMA Welding of Beryllium." Dow Chemical U.S.A., Rocky Flats Division, Golden, Colorado. Apr 1971; paper presented at the 7th Acoustic Emission Working Group Meeting, Atlanta, Georgia. May 20–21, 1971.**

A series of pressurized inert gas metal arc (PIGMA) welds using aluminum filler metal were made in a beryllium pipe butt weld and monitored for acoustic emission to determine the feasibility of detecting crack formation and propagation during and immediately after the welding process. By inserting short segments of copper wire into the root of the weld groove, gross cracking was produced in the Be. Cracking during welding and post-weld cool down produced burst-type emissions which were detected.

392 **T. F. Drouillard. "Determination of Crack Susceptibility of Beryllium by Acoustic Emission." ETDL-100-73-1. Dow Chemical U.S.A., Rocky Flats Division, Golden, Colorado. June 25, 1973; paper presented at the 11th Acoustic Emission Working Group Meeting, Richland, Washington. July 18-20, 1973.**

It is difficult to establish uniform weld parameters for joining pressed powder beryllium because of its inconsistent material properties. Each pressing, or log, is inherently nonhomogeneous, exhibiting throughout the log variations in density, chemistry, and mechanical properties. Also, there is a variation in properties of material made to the same specification by different manufacturing processes. To date, there exists no meaningful way of assessing the weldability of pressed powder beryllium. Reported is a technique to determine by acoustic emission the susceptibility to cracking of a beryllium specimen subjected to a thermal shock. The specimen is clamped between two flat plates, each containing a circular relief hole, providing both restraint and a heat sink to the specimen. Access of a GTA welding torch to the specimen is through the relief hole in the top plate. The specimen is subjected to a welding environment for a short period of time while monitoring the process for acoustic emission. The acoustic emission produced during the process is caused by gross crack propagation and is a measure of the weldability and fracture toughness of the beryllium material.

393 **T. F. Drouillard. "Presentation of Acoustic Emission Test Data from Welded Beryllium Fracture Specimens." RFP-1617. Dow Chemical Company, Rocky Flats Division, Golden, Colorado. Nov 2, 1970; paper presented at the 6th Acoustic Emission Working Group Meeting, Argonne, Illinois. Nov 12-13, 1970.**

Acoustic emission test data for real-time interpretation are presented in two ways: audible signals of emission events and plotting emission data against stress or time. The audible presentation is valuable because different emission events have characteristic sounds that can be identified through experience in the same manner as one can distinguish different persons by the sound of their voices. In tension testing to determine the onset of plastic deformation, crack propagation, and incipient failure, total emission and the rms voltage of emission events are the most meaningful parameters to plot. The rms voltage is sensitive to both continuous emission, indicative of gross plasticity in a specimen, and to burst-type emission, indicative of single-emission producing events such as fracture, twin formation, and martensitic phase transformation. In defect-free beryllium weldments, continuous emission from gross plasticity is indicative that failure will occur in base metal while burst emissions from crack initiation and propagation are indicative that failure will occur in the fusion line or heat-affected zone. From the summation plot, it becomes apparent that failure is imminent as the curve approaches a vertical direction. Failure can be predicted within about 10% of the fracture stress.

394 **T. F. Drouillard. "The Sound of Imminent Failure." Dow Chemical U.S.A., Rocky Flats Division, Golden, Colorado. Dec 1972; paper presented at the 10th Acoustic Emission Working Group Meeting, Pasadena, California. Dec 6–8, 1972.**

The only warning that premature failure of a weld joint would soon occur was the audible sound of emission. The paper discusses the importance of the audible sound of emission events and suggests that valuable spectral content is available for use in analyzing complex structures during fracture tests and proof loading.

395 **T. F. Drouillard, R. G. Liptai, and C. A. Tatro. "Industrial Use of Acoustic Emission for Nondestructive Testing." pp 122–149 in** ***Monitoring Structural Integrity by Acoustic Emission, ASTM STP 571.*** **Edited by J. C. Spanner and J. W. McElroy. American Society for Testing and Materials, Philadelphia, Pennsylvania. 1975. [see Reference 396 for presentation].**

The industrial use of acoustic emission is presented. Emphasis is given to the characteristics of acoustic emission as a unique nondestructive testing method for use in the evaluation of the structural integrity of materials, components, and engineering structures. Several applications of acoustic emission testing are discussed.

396 **T. F. Drouillard, R. G. Liptai, and C. A. Tatro. "Nondestructive Evaluation Concepts Applied to Industrial Use of Acoustic Emission." RFP-2060. Dow Chemical U.S.A., Rocky Flats Division, Golden, Colorado. Jan 28, 1974. [NTIS]; paper presented at the American Society for Testing and Materials Symposium on Monitoring Structural Integrity by Acoustic Emission, Ft. Lauderdale, Florida. Jan 17-18, 1974. [see Reference 395 for publication].**

Various nondestructive testing and evaluation concepts as applied to the industrial use of acoustic emission are presented. Emphasis is given to the characteristics of acoustic emission as a unique NDT method for use in the evaluation of the structural integrity of materials, components, and engineering structures. Also included is a brief review of some of the problems pertaining to the introduction of emission techniques to the industrial atmosphere. Several applications of acoustic emission testing are discussed.

397 **L. B. Dufour and M. J. G. Broekhoven. "Verslag van Twee Akoestische Emissie Lokalisatie–Experimenten aan een Drukvat (Two Acoustic Emission Localization Experiments on a Pressure Vessel)."** *Lastechniek,* **LASTA, 40(5):122–135. May 1974. (in Dutch, summary in English).**

In a cooperative venture by members of the Dutch Working Group on Acoustic Emission Analysis to gain experience in source location, two tests were performed on the KEMA pressure vessel using the Dunegan/Endevco 1010 multi-channel source location system. The vessel (1220 mm O.D., 2790 mm long, 13.3 mm wall thickness, with two nozzles) was constructed from components normally used in high-pressure gas transmission pipelines. During fabrication all components were cold prestrained without subsequent heat treatment. No significant emissions were detected from artificial surface defects in the prestrained areas. However, distinct detection and location was obtained from a number of brittle weld beads deposited on the vessel, as well as from several natural weld defects in non-prestrained vessel welds.

398 **J. C. Duke, Jr., and R. E. Green, Jr. "Simultaneous Acoustic Emission and Ultrasonic Attenuation Monitoring of the Mechanical Deformation of Aluminum." pp 1646–1650 in** *Proceedings of the Second International Conference on Mechanical Behavior of Materials.* **American Society for Metals, Metals Park, Ohio. 1976. conference held in Boston, Massachusetts. Aug 16–20, 1976.**

It has been well-documented in the scientific literature that the two nondestructive testing techniques, acoustic emission and ultrasonic attenuation, are extremely sensitive to the microstructural alterations associated with the mechanical deformation of materials. Both techniques have been extensively exploited in basic research experiments concerned with obtaining information about the fundamental mechanisms causing plastic deformation, microcrack formation, crack growth, and fracture and in applied technical investigations concerned with detection of defects causing failure in structural materials. Theoretical treatments have also appeared in which the observed experimental results for both acoustic emission and ultrasonic attenuation are attributed to similar deformation mechanisms. Yet, despite the extensive investigation and similarity in description of the fundamental mechanisms acting, only one brief study has previously employed the two techniques simultaneously. Effort has been devoted here to comparing the results obtained utilizing these two methods in light of earlier findings. Such a comparison yields more information about the mechanisms responsible for the mechanical deformation of metals than is obtained by either technique alone. In addition, previously unreported details of the behavior, evident from examining the information from the two techniques individually, are discussed. The work has been limited to polycrystalline aggregates of aluminum alloys because of the technological importance of such face-centered cubic materials and because these materials have played an important role in previous investigations where either ultrasonic attenuation or acoustic emission monitoring was used.

399 J. C. Duke and R. A. Kline. "The Influence of the Surface Layer on Acoustic Emission." ***Scripta Metallurgica,*** **SCRMB, 9(8):855–857. Aug 1975.**

Acoustic emission during plastic deformation of 70-30 yellow brass and 1100 Al was investigated. Tensile tests were performed. Specimens were loaded, reloaded, loaded after surface removal, and reloaded without additional surface removal. Acoustic emission during the first loading was much higher than upon reloading. After polishing, emissions were restored to approximately their original level. These observations were in accordance with the Kaiser effect and resulted from the presence of a thin material surface layer. The initial layer was formed upon machining. [Abstract 32-0118. *Metals Abstracts,* Vol. 9, 1976].

400 R. Dukes. "Admiralty Materials Laboratory Defect Location System." pp 1-1 to 1-4 in "Newsletter on the Non-Destructive Testing of Composite Materials and Adhesive Bonded Joints." Issue Number 2. Royal Aircraft Establishment, Farnborough, Hants, England. July 1973.

A defect location system developed by the Admiralty Materials Laboratory was recently demonstrated to the national press. When a defect in a piece of material moves or extends, it produces a pulse of energy some of which appears as a wave moving through the material. These emissions are known as stress wave emissions (or acoustic emissions in U.S. terminology). Since the velocity of propagation of the group wave is essentially constant, measurement of its time of arrival at a series of transducers at known locations enables the position of the original source of the emission to be calculated. The AML Defect Location System (DLS) was designed (in support of the Navy Department Nuclear Power Plant Program) to carry out this type of location on large welded structures and is now able to locate defects in flat plates, cylinders, and spheres, and combinations of these. The stress wave emissions are picked up by up to 20 piezoelectric transducers placed on the structure and each transducer signal is transmitted via 100 metres of cable to a low noise preamplifier, main amplifier, and timing channel. By choice of the piezoelectric chip size, the resonant frequency of the transducer can be chosen to make use of quiet areas in the overall noise spectrum and the main amplifier is switchable over a series of discrete centre frequencies in the range 0.1 to 1.0 MHz to accommodate this. The effect of electrical interference is minimised by using specially screened transducers and carefully designed preamplifiers connected by a super screened cable (developed by AEA Winfrith), which enables the system to operate with long cables between the transducer and the preamplifiers. A deliberate decision was taken to locate the preamplifier at the equipment end of the main cable to minimise the effect of any thermal or radiation induced noise which might be encountered at the structure; this is at variance with normal practice which is to mount the preamplifier as close as possible to the transducer. In the event, signal levels $<3\ \mu V$ peak-to-peak at the transducer can be handled adequately.

401 P. Dumousseau. "Essais non Destructifs par Emission Acoustique: Revue des Applications et Exemples Pratiques (Nondestructive Testing using Acoustic

Emission: Review of Applications and Practical Examples)." pp 56-80 in *Journees d'Etudes sur l'Emission Acoustique.* P. F. Gobin, R. Goutte, and C. Guillaud, Organizers. Institut National des Sciences Appliquees de Lyon, Avenue Albert Einstein, 69621 Villeurbanne, Lyon, France. Mar 1975. proceedings of a conference held in Lyon, France. Mar 17, 1975. (paper in French).

A review of industrial applications of acoustic emission is presented: control during material fabrication, material working, and welding, control and welded joints, quality control, defect localization, failure prediction, and control under operating conditions.

402 **P. F. Dumousseau. "Some Physical Remarks on Acoustic-Emission Application/Quelques remarques physiques sur l'application de l'emission acoustique." Centre Technique des Industries Mecaniques, Senlis, France; Paper 3K/12 in *Proceedings of the Eighth World Conference on Nondestructive Testing.* volume containing Secions 3K2, 3H2, 5A. International Committee for Nondestructive Testing, Secretariat du Cofrend, Institut de Soudure, 32 Boulevard de la Chapelle, 75880 Paris Cedex 18, France. 1976. conference held in Cannes, France. Sep 6-10, 1976. (paper in French).**

Description follows of various examples showing that it is possible to improve, by a convenient signal-processing, the methods based on the emission character of defective parts. Discussion of some deficiencies of the ASME proposed standard concerning, particularly, the material behavior, the defect severity, and the accuracy of location determination.

403 **P. F. Dumousseau. "Statistical Study of Ring-Down Counting Techniques of Characterization of Materials. Influence of Peak Amplitude Function." pp 141-143 in "EWGAE-ISPRA: Proceedings of the Third Meeting of the European Working Group on Acoustic Emission." A. C. Lucia, Editor. Report EUR-5513e. Commission of the European Communities, Directorate General Scientific and Technical Information and Information Management, Luxembourg, Belgium. 1976. meeting held at the Joint Nuclear Research Center, Ispra, Italy. Sep 25-26, 1974.**

Dispersions on results of ringdown counting are a considerable problem for applicability of AE. A statistical study of plastic deformation has been performed on 5 series of tensile tests, concerning various polycrystalline materials of industrial use (steels with or without yield point instability, aluminum alloys with or without Portevin-Le Chatelier effect). Evident limitations in the practical use of AE are due to dispersion of ringdown counts. But if AE is related to strain behaviour of materials (length of yield point instability for example), it is shown that dispersion is in the same range. The root mean square, is shown to be a less dispersive method and useful for normalization of results. The main interest of this method is the ability to locate, *with a small dispersion,* the position of maxima or minima significant of the particular transitions in the behaviour of metals. It means that AE is *not* a dispersive method of determining some materials characteristics. The influence of the amplitude distribution function on definition of counting and energy is discussed. Various models of peak amplitude laws are considered (power, Gaussian, Rayleigh,) (η^2 laws) and results of computations are presented. They show a dependence of counting and energy measurements to the peak amplitude law. It is also shown that, for a given law, counting and energy techniques lead to different results. Practical applications of the mention to quality control of weld and crack growth study are presented also.

404 H. L. Dunegan. "Acoustic Emission: A Promising Technique." UCID-4643. Lawrence Radiation Laboratory, Livermore, California. Dec 9, 1963.

Strained metals say: *ouch!* Consequently, methods have been developed for hearing their cry when strained to the *elastic limit.* This could have far reaching significance in the area of pressure vessel research. Recording the sound emitted as a vessel is pressurized shows promise for a unique method of determining the fracture point of the vessel without taking it to failure. The purpose of this report is twofold: First, to acquaint the reader with facts concerning the acoustic emission that occurs when metals are strained; and second, to describe an approach being taken on practical applications of acoustic emission theory to pressure vessel research. Hopefully a technique will be developed to predict the failure of a pressure vessel while under proof test, without taking the vessel to failure. If practical ways can be developed to achieve this goal, another weapon will have been added to the arsenal of the materials engineer. It has been known for many years that tin under stress emits sound (tin crying). Early experiments indicated that the crying was related to twinning in the crystalline structure. More recent evidence has shown the existence of acoustic emission in metals that do not exhibit twinning. Some of this experimental evidence is presented, along with Lawrence Radiation Laboratory approach for practical applications of this phenomena.

405 H. L. Dunegan. "Acoustic Emission Monitoring of Vessel V-2." Technical Report DE 74-8. Dunegan/Endevco, San Juan Capistrano, California. Aug 12, 1974.

Dunegan/Endevco participated during pressurization tests of Vessel V-2, September 1972. A 10-channel computer interfaced acoustic emission system was used to monitor acoustic emission from the vessel as it was taken to failure. The results presented were reconstructed from the permanent punched-tape records made during the test and the voice audio record recorded by the operator on tape during the course of the test.

406 H. L. Dunegan. "Acoustic Emission Testing of 12-Nickel Maraging Steel Pressure Vessels." June 7, 1970. Appendix B, pp B-1 to B-18 in "Development of Optimum TIG Welding Procedure for Production of Weldments in 12% Nickel

Maraging Steel." P. M. Schmidt and R. S. Snow. NASA CR-72981. General Dynamics/Electric Boat Division, Groton, Connecticut. July 1971; Technical Report DE-73-1. Dunegan/Endevco, San Juan Capistrano, California. 1973. [see also Reference 1552]; pp 635-642 in *Proceedings of the Second International Conference on Pressure Vessel Technology, Part 2: Materials, Fabrication and Inspection.* **American Society of Mechanical Engineers, New York. 1973. conference held in San Antonio, Texas. Oct 1-4, 1973.**

Acoustic emission data are obtained from three-point bend fracture toughness specimens of 12-nickel maraging steel, and two pressure vessels of the same material. One of the pressure vessels contained a programmed flaw which was extended by fatigue cycling. It is shown that the flawed vessel had similar characteristics of the fracture specimen, thereby allowing estimates to be made of its nearness to failure during a proof test. Both the flawed and unflawed pressure vessels survived the proof pressure and 5 cycles to working pressure. The acoustic emission response, obtained from the flawed vessel during the proof cycle and 5 cycles to the working pressure, indicated that it was near failure. It did not survive the second proof pressure cycle as the flaw was extended through the vessel wall causing a leak to occur.

407 **H. L. Dunegan. "Incipient Failure Diagnosis by Use of Acoustic Emission." p 30 in** *Proceedings of Mechanical Failures Prevention Group, Eleventh Meeting.* **W. T. Sawyer. Chesapeake College, Wye Mills, Maryland. Apr 13, 1971. conference held in Williamsburg, Virginia. Apr 7-8, 1970. [NTIS: AD-724-475].**

A popular approach to predicting lifetimes of structures undergoing cyclic loading or otherwise subjected to environments conducive to crack growth has been the use of fracture mechanics. A pre-existing flaw is assumed, and crack growth rate and fracture toughness of the material are obtained, thereby allowing failure predictions to be made by knowing the stress history, cyclic rate, and flaw size. Problems in real structures are: (1) Stresses, in complex structures difficult to know, geometry, and residual stresses; (2) Flaw size and shape may be impossible to measure; and (3) The environment future difficult to anticipate. One experimental approach used to establish an upper limit on the flaw size is the proof test. If the structure survives proof test, it is know that the largest flaw present is at least smaller than one required to cause failure. With this knowledge, the crack growth rate information is used to calculate a lifetime based on the flaw being near the critical size at the proof load. The best cyclic life one can expect on this basis is 10,000 cycles at 2/3 the proof load. In many instances this is not enough assurance. A new technique exists which is based on recording the acoustic emission from a structure that has been periodically stressed to a value higher than the normal working load. The three key factors to successful predictions are: (1) The irreversible nature of acoustic emission; (2) The ability to relate the acoustic emission directly to K without knowledge of either stress or flaw size; and (3) The ability to periodically load the structure to a value higher than the working load without generating background noise in the bandpass of the acoustic emission system.

408 H. L. Dunegan. "Listening to. . .the Liberty Bell." Acoustic Emission Applications Report. Dunegan/Endevco, San Juan Capistrano, California. Aug 1976.

Brief report on the monitoring of the Liberty Bell for propagation of its famous crack during moving of the Bell from its location in Independence Hall in Philadelphia to a specially built pavilion just outside the Hall for the bicentennial year 1976. Results of acoustic emission monitoring showed that no extension of the main crack occurred during the moving operation.

409 H. L. Dunegan. "Practical Applications of Acoustic Emission." Technical Report DE-74-5. Dunegan/Endevco, San Juan Capistrano, California. Apr 1974; pp 270–291 in *Schallemission: Anwendung bei der Untersuchung, Prufung and Uberwachung metallischer Werkstoffe.* Deutsche Gesellschaft fur Metallkunde eV, 637 Oberursel/Taunus bei Frankfurt/Main, Adenaueralle 21, Germany. 1974. proceedings of symposium held in Munich, Germany. Apr 18–19, 1974.

Report discusses the application of acoustic emission to (1) proof-test monitoring of pressure vessels by use of multiple channel acoustic emission instrumentation, (2) monitoring of aircraft for fatigue crack growth and structural failure, and (3) monitoring of the burn rate of solid propellants by acoustic emission techniques.

410 H. L. Dunegan. "Quantitative Capabilities of Acoustic Emission for Predicting Structural Failure." pp 86–113 in *Prevention of Structural Failure: The Role of Quantitative Nondestructive Evaluation.* T. D. Cooper, P. F. Packman, and B. G. W. Yee, Editors. Materials/Metalworking Technology Series, Number 5. American Society for Metals, Metals Park, Ohio. 1975.

The paper reviews the capability of acoustic emission to estimate the stress intensity factor of a growing crack, and therefore provide predictive ability for determining failure. Crack growth due to rising load, fatigue, and hydrogen embrittlement is discussed. Also discussed are the techniques for flaw location measurements.

411 H. L. Dunegan. "Technical Report No. 1001 on Model P-100 Passive Pressure Transducer." Dunegan Research Corporation, Livermore, California. 1969.

412 H. L. Dunegan. "The Impact of Acoustic Emission on the Technical Community." Dunegan Research Corporation, Livermore, California; paper presented at the Institute of Physics Conference on Acoustic Emission, Imperial College, London, England. Mar 14, 1972. [proceedings were never published].

Acoustic emission, although in its infancy compared to other test methods, is fast becoming recognized as a tool of many uses. It is one of the few tools available that crosses over so easily into the fields of fracture mechanics, materials testing, and quality assurance. Examples are given of present acoustic emission investigations being conducted in each of these fields. Some forecasts of its potential applications to such problems as mine safety and inspection of large structures, such as bridges and dams, will also be discussed.

413 **H. L. Dunegan. "Ultrasonic Acoustic Emission from Materials."** ***IEEE Transactions on Sonics and Ultrasonics,*** **IESUA, SU-16:32. Jan 1969. [abstract only]; paper presented at the 1966 Institute of Electrical and Electronics Engineers Ultrasonic Symposium, Cleveland, Ohio. Oct 12-14, 1966.**

Techniques for measuring and recording acoustic emission from materials are presented. Results of acoustic emission from metals under stress are shown to be related to microdynamical processes due to the movement of dislocations. Results are also shown of acoustic emission from specimens containing sharp cracks. The acoustic emission from these specimens is then correlated to the stress intensity factor at the crack tip through the use of linear fracture mechanics. Applications of this information for the nondestructive proof testing of structures containing flaws are demonstrated. Potential applications of acoustic emission are also shown for the inspection of structures undergoing subcritical flaw growth due to fatigue or stress corrosion.

414 **H. L. Dunegan. "Using Acoustic Emission Technology to Predict Structural Failure."** ***Metals Engineering Quarterly,*** **MENQA, 15(1):8-16. Feb 1975; paper presented at the 2nd Annual American Society for Metals Materials/Design Forum: "Prevention of Structural Failure: The Role of Quantitative NDE." Port St. Lucie, Florida. Apr 9-11, 1974.**

The problems associated with assuring structural integrity of large complicated structures, such as bridges and power plant pressure vessels, have been the cause of much concern to the prime contractors manufacturing these structures, and to the licensing agencies and code writers whose responsibility it is to assure public safety. This concern is rightly justified since striving for greater efficiencies continues in the use of materials which call for higher strength-to-weight ratios, higher operating temperatures and pressures, and economics in manufacturing. Each of these factors requires that better inspection techniques and more quantitative information be available for making a judgment of operating efficiency, as well as safety. One approach taken to help solve some of the problems is sponsored research in linear fracture mechanics and nondestructive testing technology. Much progress has been made in both areas over the last decade, such that the combination of design based on linear fracture mechanics, coupled with a thorough NDT inspection, is sufficient precaution in most cases to prevent structural failures. Problems that can arise in applying this technique come about by the inability, in some instances, to accurately determine the stresses under operating environments and the size and shape of the flaw. Both kinds of information are needed to determine the stress intensity factor at a flaw site, and therefore predict structural failure. The purpose of this report is to show how acoustic emission techniques can be used to estimate the stress intensity factor of a growing crack and therefore provide predictive ability for determining failure. [Copyright 1975 by the American Society for Metals].

415 **H. L. Dunegan, A. E. Brown, and P. L. Knauss. "Differential Transducer." U.S. Patent 3,593,048. July 13, 1971.**

A transducer such as a piezoelectric crystal is halved and, with one half inverted, the parts are joined together with insulation therebetween. An electrical connection is made between faces of the two halves on one side of the unit and electrical conductors extend separately from the two halves on the other side of the unit to a differential amplifier. Ambient electrical fields do not influence the output of the transducer.

416 **H. L. Dunegan, A. E. Brown, and P. L. Knauss. "Piezoelectric Transducers for Acoustic Emission Measurements." UCRL-50553. Lawrence Radiation Laboratory, Livermore, California. Dec 13, 1968. [NTIS].**

An analytical expression for estimating the frequency response of a transducer, which is based on the size of the piezoelectric element and the piezoelectric constants of the material, is presented. Also, transducer design and fabrication techniques for acoustic emission transducers that can be used in the 100 kHz region are described including a transducer which utilizes a split piezoelectric element to achieve a differential response.

417 **H. L. Dunegan and A. T. Green. "Factors Affecting Acoustic Emission Response from Materials."** *Materials Research and Standards,* **MTRSA, 11(3):21–24. Mar 1971; pp 100–113 in** *Acoustic Emission, ASTM STP 505.* **American Society for Testing and Materials, Philadelphia, Pennsylvania. 1972. proceedings of symposium held in Bal Harbour, Florida. Dec 7–8, 1971.**

The paper presents selected examples of the acoustic emission response from widely differing materials with varied mechanical history, heat treatment, and state of stress. The evidence points to a need for characterization of the acoustic emission response of materials, prior to application of the techniques for determining structural integrity so that all parameters are represented under similar conditions.

418 **H. L. Dunegan and D. O. Harris. "Acoustic Emission: A New Nondestructive Testing Tool." UCRL-70750. Lawrence Radiation Laboratory, Livermore,**

California. Jan 16, 1968. [NTIS]; pp 203–238 in *Proceedings: 1968 Symposium on the NDT of Welds and Materials Joining.* American Society for Nondestructive Testing, Evanston, Illinois. 1968. proceedings of the Third Annual Symposium on Nondestructive Testing of Welds and Materials Joining, Los Angeles, California, Mar 12, 1968; *Ultrasonics,* ULTRA, 7(3):160–166. July 1969.

A new approach to nondestructive testing is proposed which is centered around the concept of utilizing the transducer action of a flaw in a stress field. This situation results in local plastic deformation occurring in the vicinity of the flaw which in turn gives up energy in several different ways. Thermal emission, exo-electron emission, and acoustic emission are the three primary forms that this energy takes upon release. The first two of these are mentioned only briefly as challenging new areas of research, while the remainder of this report is devoted to the subject of acoustic emission. It is shown that acoustic emission from unflawed tensile specimens can be treated from a dislocation dynamics approach. A good correlation between the acoustic emission and a proposed theoretical model for mobile dislocation density as a function of plastic strain is shown for aluminum. A previously developed theoretical model for the plastic volume around a sharp crack in an elastic stress field predicted that the summation of acoustic emission should vary as the fourth power of the stress intensity factor at the crack tip. Experimental data from 7075-T6 aluminum are presented and agrees well with this model. It is further shown that acoustic emission data from flawed specimens, interpreted from a fracture mechanics viewpoint, can find direct application for nondestructive quality assurance tests on engineering structures.

419 **H. L. Dunegan and D. O. Harris. "Acoustic Emission Techniques." Chapter 3, pp 38–75. *Experimental Techniques in Fracture Mechanics.* A. S. Kobayashi, Editor. SESA Monograph No. 1, published jointly by Iowa State University Press, Ames, Iowa and Society for Experimental Stress Analysis, Westport, Connecticut. 1973.**

A brief historical review of acoustic-emission studies on both rocks and metals is provided, and the suitable transducers and electronics for acoustic-emission studies are described. The onset of plastic deformation in materials is controlled by the presence of stress raisers, such as cracks and other types of flaws, and hence acoustic emission provides a useful tool for flaw detection. The uses of acoustic emission for detection of subcritical flaw growth, such as that which occurs during pop-in, stress-corrosion cracking, fatigue, and hydrogen embrittlement are described. Numerous examples of the use of acoustic emission to study fracture mechanics in Al alloys, AISI 4340 and other high-strength steels, and in Be and U alloys are explained and illustrated. [Abstract 31-0619. *Metals Abstracts,* Vol. 7, 1974].

420 **H. L. Dunegan, D. O. Harris, and C. A. Tatro. "Fracture Analysis by Use of Acoustic Emission." UCRL-70323. Lawrence Radiation Laboratory, Livermore, California. Feb 6, 1967. [NTIS]; paper presented at the First National**

Symposium on Fracture Mechanics, Lehigh University, Bethlehem, Pennsylvania. June 19–21, 1967; ***Engineering Fracture Mechanics,*** **EFMEA, 1(1):105–122. June 1968.**

The report contains results of acoustic emission studies on flawed and unflawed specimens of aluminum and beryllium. Acoustic emission from the flawed specimens is found to begin at stress levels far below the general yield stress. A theoretical model given here indicates that the total number of acoustic emission signals from a specimen containing a crack should be proportional to the fourth power of the stress intensity factor obtained from a sharp-crack fracture mechanics analysis. This is in disagreement with experimental data from single-edge-notched fracture toughness specimens of the two materials, which indicate that acoustic emission varies more like the sixth to eighth power of the stress intensity factor. An example is given to show how the acoustic emission data obtained on fracture toughness specimens can be used to nondestructively test the fracture strength of an engineering structure. [reprinted by permission of Pergamon Press Ltd].

421 **H. L. Dunegan, D. O. Harris, and A. S. Tetelman. "Detection of Fatigue Crack Growth by Acoustic Emission Techniques." UCRL-71457. Lawrence Radiation Laboratory, Livermore, California. Mar 10, 1969; pp 20–31 in** ***Proceedings of the Seventh Symposium on Nondestructive Evaluation of Components and Materials in Aerospace, Weapons Systems and Nuclear Applications.*** **Western Periodicals Company, North Hollywood, California. July 1969. symposium held in San Antonio, Texas. Apr 23–25, 1969;** ***Materials Evaluation,*** **MAEVA, 28(10):221–227. Oct 1970.**

It has been well-established in tensile tests that acoustic emission is an irreversible process associated with plastic deformation. Widely recognized also is that plastic deformation occurs at the root of a sharp crack in a structure that is stressed. Acoustic emission tests performed on cracked fracture toughness specimens have confirmed that acoustic emission associated with the plastic zone at the crack tip is also nearly irreversible for stress intensities at the crack tip of less than one half the critical stress intensity value required to cause unstable fracture. The report shows how this irreversible feature can be utilized in a practical NDT test on a structure undergoing crack growth due to cyclic loading or other environmental effects. The technique is based on periodic acoustic emission monitoring of a structure being loaded back to its proof stress.

422 **H. L. Dunegan and C. A. Tatro. "Acoustic Emission Effects during Mechanical Deformation." Chapter 12, pp 273–312 in** ***Techniques of Metals Research.*** **R. F. Bunshah, Editor. Volume 5, Part 2. Interscience Publishers, New York. 1971.**

Discussed are the history of acoustic emission, the techniques of detection, the special mechanical techniques, the special precautions concerning specimen design, and the correlation with existing microplasticity theories and AE potential for application in both basic and applied research. [reproduced in part by permission of publisher].

423 **H. L. Dunegan and C. A. Tatro. "Passive Pressure Transducer Utilizing Acoustic Emission." UCRL–14981, Rev. 1. Lawrence Radiation Laboratory, Livermore, California. Apr 3, 1967. [NTIS]; *Review of Scientific Instruments,* RSINA, 38(8):1145–1147. Aug 1967.**

A passive pressure transducer has been developed which utilizes the acoustic emission effect. This effect is noted when a suitable solid material is placed under increasing stress, causing permanent deformation processes to occur which are accompanied by emission of acoustic energy. These processes are irreversible, so if the stress is released and then reapplied, no acoustic energy will be emitted on the second cycle until the stress reaches its previous maximum level. The instrument described here contains a diaphragm of anodized aluminum which is first exposed to the pressure that is to be measured and then withdrawn and read by subjecting it to steadily increasing known pressure while monitoring it for acoustic emission. The pressure at which acoustic emission begins is taken as the pressure originally measured by the instrument. Pressure transducers of this type are inexpensive and require no calibration. Accuracy is within a few percent and can probably be improved. Possible applications include deep-ocean depth measurements, measurements of peak acceleration, and *telltale* uses for monitoring peak pressures.

424 **H. L. Dunegan and C. A. Tatro (to H. L. Dunegan). "Peak Pressure Measurement by Acoustic Emission." U.S. Patent 3,447,378. June 3, 1969.**

The invention disclosed and claimed herein relates to the determination of maximum pressures by a passive-pressure transducer utilizing acoustic emission of materials. The invention provides for the placement of a solid material such as a diaphragm in a position to experience applied pressure establishing plastic deformation and the subsequent determination of the maximum applied pressure by determination of reapplied pressure upon such material at which acoustic emission commences.

425 **H. L. Dunegan, C. A. Tatro, and D. O. Harris. "Acoustic Emission Research: Status Report, December 1963–August 1964." UCID–4868, Rev. 1. Lawrence Radiation Laboratory, Livermore, California. Nov 24, 1964. [NTIS].**

The results of a new experimental approach to detecting and recording the acoustic emission of materials when subjected to strain is discussed. Specimens in the form of diaphragms are loaded by means of hydraulic pressure. The acoustic energy emitted from the specimens is detected with transducers having fundamental frequencies to 1 megacycle. The detection is accomplished by immersing the specimen in water and using the water for coupling the acoustic energy to the transducer. Results are shown of acoustic emission of anodized 6061 aluminum and beryllium diaphragms. It is demonstrated that the yield point of the aluminum and the development of microcracks leading to failure in beryllium can be recorded. Three possible applications of the acoustic emission phenomenon are given.

426 **H. L. Dunegan and A. S. Tetelman. "Acoustic Emission."** ***Research/Development,*** **REDEA, 22(5):20–24. May 1971. [reprint available as Technical Report DRC 71–4. Dunegan/Endevco, San Juan Capistrano, California].**

Acoustic emission was used for nondestructive testing of ordnance and pressure vessels, for determining the efficiency of welding and adhesive-bonding processes and for studying the microscopic processes of fatigue, stress-corrosion cracking, and composite failure. Steels, Ti, Al, concrete, wood and fibre-reinforced resins were investigated. Applications of the method to detect flaws, predict susceptibility to stress-corrosion cracking and hydrogen embrittlement, measuring coating thickness on 6Al-4V Ti specimens, and to detect high-temperature failure in Rene 41 tensile specimens are described. [Abstract 22-0999. *Metals Abstracts,* Vol. 4, 1971].

427 **H. L. Dunegan and A. S. Tetelman. "Acoustic Emission: Noise Detection on Stressed Materials as Non–Destructive Testing Tool."** ***Materialprufung,*** **MTPRA, 14(5):153–155. May 1972. (in German). [German translation of Reference 426].**

The construction principles of acoustic emission curves are outlined, and the application of the technique is discussed. Results have shown that the acoustic emission signal (ΣN) is directly related to the stress intensity factor (k) of stressed materials, so that the technique can be used, e.g., in the detection of cracks in turbines. Suitable materials for testing are: Steel, Ti, Al, concrete, wood, and fibre-reinforced resins. [Abstract 22-0113. *Metals Abstracts,* Vol. 6, 1973].

428 **H. L. Dunegan and A. S. Tetelman. "Non–Destructive Characterization of Hydrogen–Embrittlement Cracking by Acoustic Emission Techniques." UCRL–71654. Lawrence Radiation Laboratory, Livermore, California. Nov 10, 1969. [NTIS];** ***Engineering Fracture Mechanics,*** **EFMEA, 2(4):387–402. June 1971; Technical Bulletin DRC–106. Dunegan Research Corporation, Livermore, California; paper presented at American Society for Metals Materials Engineering Congress, Philadelphia, Pennsylvania. Oct 13–16, 1969.**

Acoustic emission techniques are used to determine the rate of cracking of high-strength steel due to hydrogen embrittlement, making it possible to predict failure in certain structural components undergoing hydrogen attack. The acoustic emission rate from a hydrogen-charged, linear-compliance, fracture-toughness specimen of high-strength 4340 steel was recorded as a function of the stress-intensity factor at the crack tip in the specimen. The structural component (a hydrogen-charged bolt of the same material) was then torqued down in a simple fixture and the acoustic emission was recorded as a function of time. The acoustic emission rate obtained from a near-critical value of the stress-intensity factor of the fracture-toughness specimen was used to predict within ±15% the time for the onset of unstable failure of the bolt. The most significant result of this investigation is that regardless of the initial loading, nucleation time for cracking to begin, or any other factors connected with geometrical difference in specimens, the acoustic emission rate can be used to determine the time for the stress intensity factor at the onset of rapid fracture.

E

429 D. M. Egle. "A Comprehensive Analysis of an Acoustic Emission Detection System." Ph.D. Thesis. Tulane University, New Orleans, Louisiana. 1965; University Microfilms 66–1550.

The term *acoustic emission* is used to describe a low-level emission of sound waves which occurs in some materials when subjected to a slowly increasing stress. The emissions are produced in a bar with a section of reduced diameter and with enlarged ends or attachments bonded to the bar to transmit the tensile load. The sound waves are detected by a piezoelectric crystal placed on one end of the bar. The mechanical configuration of the test specimen and the nature of the piezoelectric transducer modify the crystal response to such an extent that it bears little resemblance to the acoustic emission source pulse. The purpose of this investigation is to analyze wave propagation and the transducer response in this detection system. The response of the system to longitudinal waves is found using the wave method combined with Fourier integral methods. The distorting effects of the several parts of the system are determined experimentally using pulse testing techniques. The system's response to bending wave excitation is also evaluated experimentally. Photographs of oscilloscope traces of the crystal response of acoustic emission pulses are presented and the Fourier transforms of these pulses are calculated by numerical means. A comparison of these transforms to the responses of the detection system to longitudinal and bending waves indicates that the acoustic emission source excites both types of waves. It is concluded that the bending wave source pulse is as large or possibly larger than the longitudinal source strain pulse.

430 D. M. Egle. "Detecting High Cycle Fatigue Crack Growth Using Acoustic Emission." University of Oklahoma, Department of Aerospace, Mechanical, and Nuclear Engineering, Norman, Oklahoma. 1973; pp 51–54 in *Fracture and Flaws, Proceedings of the 13th Annual Symposium.* New Mexico Section, American Society of Mechanical Engineers. 1973. symposium jointly sponsored by New Mexico Section ASME, American Society for Metals, and the University of New Mexico College of Engineering, held Mar 1-2, 1973. [proceedings available from G. M. Ferguson, 819 Cagua, N.E., Albuquerque, NM 87108].

A series of experiments with controlled displacement cyclic flexural deformation of notched bars of 7075-T6 aluminum show crack growth rates of 8μin/cycle can be detected using acoustic emission. A simple modification of a conventional acoustic emission detection system allows this to be done in the presence of a moderate amount of mechanical noise.

431 D. M. Egle and A. E. Brown. "A Note on Pseudo-Acoustic Emission Sources." UCRL-75685. Lawrence Livermore Laboratory, Livermore, California. June 14, 1974; *Journal of Testing and Evaluation*, JTEVA, 4(3):196–199. May 1976.

Paper describes an experimental comparison of the stress waves generated by three sources of simulated acoustic emission events: a piezoelectric generator, an electric spark, and a pulsed laser. The stress waves produced by the electric spark indicate that the stress generating mechanism is the thermally excited pressure pulse in the air surrounding the spark. Each of the sources is shown to be capable of producing stress waves of less than 1-μs duration, but the piezoelectric generator appears to be the most advantageous technique. [reprinted by permission of the American Society for Testing and Materials, Copyright].

432 **D. M. Egle and A. E. Brown. "Considerations for the Detection of Acoustic Emission Waves in Thin Plates." UCRL-75684. Lawrence Livermore Laboratory, Livermore, California. June 14, 1974. [NTIS]**; *Journal of the Acoustical Society of America,* **JASMA, 57(3):591–597. Mar 1975.**

A series of simulated acoustic emission experiments in thin plates and bars conducted to assess the effects of structural geometry on the detection of emission is described. It is shown that the structure introduces peaks in the received signal spectra corresponding to thickness resonances that may be estimated from Lamb wave theory. Spectra of acoustic emission signals from a thin-walled pressure vessel show a peak at the frequency corresponding to the lowest symmetric plate mode. It is concluded that a significant portion of the energy detected by a surface contact transducer on a thin-walled structure concentrates at these resonances and should be accounted for in choosing a detection system.

433 **D. M. Egle, J. R. Mitchell, K. H. Bergey, and F. J. Appl. "Acoustic Emission for Monitoring Fatigue Crack Growth." pp 638–1 to 638–7 in** *Advances in Instrumentation.* **Volume 27, Part 2. Instrument Society of America, Pittsburgh, Pennsylvania. 1972. proceedings of the 27th Annual Instrument Society of America Conference, New York. Oct 9–12, 1972;** *ISA Transactions,* **ISATA, 12(4):368–374. 1973.**

Recent studies have shown that the stress waves (or acoustic emission) generated by propagating fatigue cracks in a structure are a sensitive indicator of impending failure in the structure. The low intensity of these stress waves poses a major problem in using acoustic emission techniques in practical situations because of the relatively high level of background noise generated by other mechanisms. The paper reports the modifications to the conventional acoustic emission detection system which were helpful in distinguishing crack emission from the noise caused by rubbing in mechanical joints and from noise caused by low frequency vibration associated with roller bearings.

434 **D. M. Egle and C. A. Tatro. "Analysis of Acoustic Emission Strain Waves."** *Journal of the Acoustical Society of America,* **JASMA, 41(2):321–327. Feb 1967.**

An acoustic-emission loading and detection system is described and the detection system's harmonic response to both longitudinal and flexural waves is evaluated. Photographs of acoustic-emission response pulses are presented, and it is shown that the flexural waves are of considerable importance in describing the detection system's response to the acoustic-emission source pulse. An estimate indicates that the energy of the flexural part of one acoustic-emission pulse is two orders of magnitude greater than the energy of the londitudinal part.

435 **J. Eisenblatter. "Acoustic Emission as a Non-Destructive Testing Method for Incipient Failures in Nuclear Reactor Pressure Vessels (Schallemissionsanalyse als zerstorungsfreies Prufverfahren zur Feststellung von Fehlern in Kernreaktordruckgefassen)." Battelle-Institut eV, Frankfurt am Main, Germany. 1971; pp 529-530 in** ***Proceedings of the First International Conference on Structural Mechanics in Reactor Technology.*** **EUR 4820, Volume 4. Commission of the European Communities, Brussels, Belgium. 1972. conference held in Berlin, German. Sep 20-24. 1971.**

The method applied is based on the measurement of elastic waves emitted by propagating cracks by means of piezoelectric transducers. As the acoustic emission signals have to be detected in the presence of sonic and ultrasonic noises generated by a working reactor, measurements were carried out at frequencies up to 1 MHz to find out a frequency band of low noise level. Measurements with notched specimens showed that there were clear relationships between the size of the plastic zone at the crack tip and the rate or intensity of acoustic emission. Experiments with pressurized notched tubes proved that acoustic emissions can be noticed in case of ductile crack growth before brittle fracture some time before the failure occurs. [Abstract 15065. *Physics Abstracts,* Vol. 75, No. 936, 1972].

436 **J. Eisenblatter. "Bibliographie uber Schallemission und Mikroseismik (Bibliography on Acoustic Emission, Stress Wave Emission, Microseismicity)." Battelle-Institut eV, Frankfurt am Main, Germany. Mar 1972.**

The bibliography contains 270 references to the literature on acoustic emission and microseismicity for the period from 1917 to 1972. Many of the references are taken from bibliographies by A. T. Green and J. Ross (see Reference 636) and by J. L. Knill, J. A. Franklin, and A. W. Malone (see Reference 1016). References are in English or German, according to their origin.

437 **J. Eisenblatter. "Industrielle Anwendung der Schallemissionsanalyse (Industrial Use of Acoustic Emission Analysis)." pp 222-253 in** ***Schallemission: Anwendung bei der Unterschung, Prufung und Uberwachung metallischer Werkstoffe.*** **Deutsche Gesellschaft fur Metallkunde eV, 637 Oberursel/Taunus**

bei Frankfurt/Main, Adenauerallee 21, Germany. 1974. proceedings of symposium held in Munich, Germany. Apr 18–19, 1974. (paper in German).

The paper deals with applications of the acoustic emission analysis which may become interesting for industrial utilization. Especially described are the failure detections in large pressure vessels and the weld control during the welding process.

438 **J. Eisenblatter. "Schallemissionsanalyse: Ein neues zerstorungsfreies Prufverfahren (Acoustic Emission Analysis: A New Non-Destructive Method of Evaluation)."** *Ingenieur Digest.* **11(10):62–67. 1972. [reprint of article, in German, available from author at Battelle-Institut eV, Frankfurt am Main, Germany]; UCRL-Trans-10728. translated for Lawrence Livermore Laboratory, Livermore, California. [NTIS].**

The physical causes of the emission of acoustic waves in metals, especially in the course of cracking, are discussed. The development of acoustic emission analysis as a nondestructive method of evaluation has now reached the stage where a number of interesting applications are possible. Studies of plastic deformation in metals, fault detection in the pressure examination of large boilers, evaluation of a special type of crack formed during stress-relief annealing (under-cladding cracks), and differentiation between various types of fracture processes in glass-fiber-reinforced plastics are described.

439 **J. Eisenblatter, W. Heide, P. Jax, H. Jost, and R. von Klot. "Acoustic Emission as an Inspection Method for Nuclear Reactor Vessels." Paper C120/74, pp 168–175 in** *Periodic Inspection of Pressurized Components.* **Conference Publication 8, 1974. published by Mechanical Engineering Publications Ltd., for the Institution of Mechanical Engineers, London, England. 1975. conference held in London, England. June 4–6, 1974.**

Different reasons for acoustic emission (e.g., plastic deformation, crack propagation during cyclic and normal tensile loading, and friction at the crack tip during fatigue loading) are presented and the differences between the various types of acoustic emission are discussed in detail. A quantitative correlation between the rate of crack propagation and the rate of acoustic emission has been established. Acoustic emission was evaluated during hydrostatic proof tests of large vessels, and during the different operating phases of light-water nuclear reactors. Particle loss in a stream generator was indicated by increasing levels of high-frequency noise.

440 **J. Eisenblatter, W. Heide, H. Jost, and R. von Klot. "Forschungsarbeiten zur Kontinuierlichen Uberwachung der Rissausbreitung in Druckbehaltern. Tiel II: Weitere Untersuchungen uber die bei der Rissausbreitung Emittierten Schallwellen (Research Work on the Continuous Monitoring of Crack Extension in Pressure**

Vessels. Part 2: Further Investigations in Respect to the Acoustic Waves Emitted During Crack Extension)." Research Report 31/1. Battelle-Institut eV, Frankfurt am Main, Germany. Mar 1971. (in German); Risley-TR-2161. translated for Risley Engineering and Materials Laboratory, UKAEA, Risley, Warrington, Lancashire, England. July 28, 1971. [Available from National Lending Library for Science and Technology, Boston Spa, Yorkshire, England as Translation NLL-Risley-TR-2161-909.1.9F].

Part 2 reports the development and application of broadband probes of lithium sulphate and of resonance probes with high thermal resistance for measuring acoustic emission. Signals were recorded during measurements on notched fracture mechanics specimens for amplitude load cycles and during tensile tests following the production of fatigue cracks. The steels investigated included 22 Ni-Mo-Cr 37 and Ck 45 steels, which are the steels most frequently employed for reactor pressure vessels in Germany. High acoustic pulse rates were observed only during slow ductile tearing; the magnitude was dependent on the speed at which the crack propagated. It was also found that during internal pressure tests on notched Type St 35.8 steel tubes, ductile crack growth prior to brittle fracture was indicated by an increase in acoustic pulse rate.

441 **J. Eisenblatter, P. Jax, and H. J. Schwalbe. "A Survey of Acoustic Emission Research in Europe and Investigations Conducted at Battelle-Frankfurt." Battelle-Institut eV, Frankfurt am Main, Germany; paper presented at Session 7 of the Second Acoustic Emission Symposium held in Tokyo, Japan. Sep 2-4, 1974. [paper included as sparate volume with proceedings, available from Japan Industrial Planning Association, Dobashi Building 5F, 3-17-8 Akasaka, Minato-ku, Tokyo, 107 Japan].**

Paper presents a brief survey of study groups in Europe engaged in the field of acoustic emission. Special consideration is given to the research work on acoustic emission conducted at the Battelle-Frankfurt Laboratories. The results of several completed investigations are reported in detail.

442 **R. K. Elsley. "Exploiting the Gold Mine of Information in Ultrasonic Signals." summary of paper in** ***Paper Summaries: ASNT Spring Conference, March 28-30, 1977, Phoenix, Arizona.*** **American Society for Nondestructive Testing, Columbus, Ohio. 1977.**

Traditional ultrasonic nondestructive testing converts the radio-frequency (rf) waveform from a receiving transducer into a video signal, thereby sacrificing much of the information contained in the RF waveform. By retaining all of the information, and doing special types of signal processing, it is possible to determine much about defects encountered during an inspection process. In situations of fairly simple geometry such as occur in the parallel layers used for adhesive bonds, the interaction of ultrasound with the medium is well understood, and test data can be compared to theoretical expectations to detect the occurrence of subtle defects.

In more complicated situations such as are found in scattering from *defects* in solids or in acoustic emissions, data processing techniques such as pattern recognition can discover ways of estimating the physical properties of the defect or of recognizing flaw growth from acoustic emission signal characteristics.

443 **R. B. Engle. "Acoustic Emission and Related Displacements in Lithium Fluoride Single Crystals." Ph.D. Thesis. Michigan State University, East Lansing Michigan. 1966; University Microfilms 67-7535.**

Acoustic emission is the result of lattice vibrations caused by mechanisms that govern deformation of crystalline materials when they are stressed. Observations were made using single crystals of lithium fluoride oriented for easy-glide deformation when loaded in direct shear. Acoustic-emission pulses and displacement pulses derived from rapid step displacements were observed. The largest displacements (10×10^{-6} inches) were found to occur in coincidence with large acoustic-emission pulses. Smaller displacements (20×10^{-8} inches) and acoustic-emission pulses were observed to occur together, though not in coincidence, throughout the tests. Some large acoustic-emission pulses had no displacements associated with them. On the basis of these results, a mechanism for the acoustic-emission process is proposed and is shown to be consistent with the observations of others who have studied acoustic emission in crystalline materials. Estimates of dislocation group velocities are made which agree with previous dislocation-velocity measurements in lithium fluoride. The proposed mechanism involves the interaction between piled-up groups of dislocations and the obstacles that cause the pile-up. The pinning interaction causes an increase in local strain energy stored in the region of the obstacle. When the driving stress on the leading dislocation, composed of the applied stress and additional stress due to the pile-up itself, is large enough to cause breakaway and acceleration of part of the group, the local strain energy is available to excite lattice vibrations that appear as acoustic emission. A secondary emission process arises during collisions between coherent groups of moving dislocations and obstacles in their slip planes. Finally, an alternative-type emission occurs at high stress levels after large plastic deformation when the stress concentrations at the leading edge of dislocation groups causes crack nucleation and propagation.

444 **R. B. Engle and H. L. Dunegan. "Acoustic Emission: Stress Wave Detection as a Tool for Nondestructive Testing and Material Evaluation." UCRL-71267. Lawrence Radiation Laboratory, Livermore, California. Sep 13, 1968. [NTIS];** ***International Journal of Nondestructive Testing,*** **IJNTA, 1:109-125. June 1969; pp 165-181 in** ***Physics and Nondestructive Testing.*** **W. J. McGonnagle, Editor. Volume 2. Gordon and Breach Science Publishers, New York. 1971.**

Methods for nondestructive testing and material evaluation using acoustic-emission signals are reported. Experimental results show that emission from flawed specimens is characteristically different from emission from flaw-free specimens. Experiment also shows that acoustic emission from flawed specimens is normalized by the stress intensity factor that is basic to the study of brittle fracture. These results suggest applications to nondestructive testing and material evaluation that are being investigated at the Lawrence Radiation Laboratory, Livermore. Work with some of these applications is described; notably a proposed acceptance test and fatigue crack extension monitoring, in the area of NDT; and determination of a critical stress intensity factor for stress corrosion cracking, in the area of material evaluation. Data from tensile tests of 7075-T6 aluminum suggest a correlation between the acoustic-emission rate and an expression describing mobile dislocation density as a function of plastic strain. Further search is intended for correlations between acoustic-emission behavior and dynamic dislocation parameters. Some of these planned efforts are described. A theoretical base is expected to lead to further applications and clearer interpretation of present NDT results.

445 **D. Ensminger.** *Ultrasonics.* **Marcel Dekker, Inc., New York. 1973. pp. 240–243.**

In Chapter 7, Nondestructive Testing–Basic Methods and General Considerations, a brief description of acoustic emission is presented based on some of the early reports of J. R. Frederick, B. H. Schofield, P. H. Hutton, and D. M. Romrell and L. R. Bunnell, most of which were published in *Materials Evaluation.*

446 **F. Erdmann-Jesnitzer, J. Beges, and D. Rehfeldt. "Akustische Analyse in Schweissforschung und Praxis als neue Methode der Gutesicherung (Acoustic Analysis in Welding Research and Practice as a New Method of Quality Control)." Part-1.** *Schweisstechnik* **(Vienna), SWTEA, 27(2):25-28. Feb 1973. (in German). [for Part-2 see Reference 447].**

447 **F. Erdmann-Jesnitzer, J. Beges, and D. Rehfeldt. "Akustische Analyse in Schweissforschung und Praxis als neue Methode der Gutersicherung (Acoustic Analysis in Welding Research and Practice as a New Method of Quality Control)." Part-2.** *Schweisstechnik* **(Vienna), SWTEA, 27(3):45–48. Mar 1973. (in German). [for Part-1 see Reference 446].**

The design of equipment suitable for monitoring acoustic emissions from cracks in welded assemblies as they form and the application of this technique to hydrogen cracking in high-restraint mild steel CO_2 welds are discussed. From the results of this work, a quantitative relationship between weld hydrogen content and the onset of hydrogen (cold) cracking is proposed. [Abstract 55-0079. *Metals Abstracts,* Vol. 7, 1974].

448 **M. D. Erickson. "Containment Vessel Flaw Detection Test." BNWC-94. Battelle-Northwest, Richland, Washington. May 14, 1965. [Official Use Only–not released to public].**

Data needed for the first development phase of a nondestructive testing technique were obtained during the pressure-proof testing of a large containment vessel. A total of six ultrasonic bursts were recorded and analyzed. Analysis indicated that the origin of the sound was different in each case. No definite conclusions can be drawn about the validity of the testing technique at the present time. Further transducer and instrumentation system development will be required to determine the value of this technique.

449 **R. Ettemeyer and D. Hauptfleisch. "Schallemissionsanalyse am Primarsystem des Kernkraftwerks Gundremmingen (Acoustic Emission Analysis on the Primary System of the Gundremmingen Nuclear Power Station)."** *Atomwirtschaft-Atomtechnik,* **AWAKA, 20(4):182–186. Apr 1975. (in German).**

Article summarized the inservice acoustic emission proof pressure evaluation of the primary system of the Gundremmingen nuclear power station which has been in operation for more than seven years. The work was performed by D. L. Parry of Exxon Nuclear Company. For detailed test report see References 1365 and 1933.

450 **A. G. Evans. "Acoustic Emission: A Summary of Current Understanding." pp 463–481 in "Proceedings of the Interdisciplinary Workshop for Quantitative Flaw Detection." D. O. Thompson, Program Manager. AFML-TR-74-238, SC595.2ITR. Science Center, Rockwell International, Thousand Oaks, California. Nov 1974. [NTIS: AD/A-003-672]. Workshop held at the Science Center, Rockwell International, Thousand Oaks, California. June 17–20, 1974.**

451 **A. G. Evans. "High Temperature Acoustic Emission in Structural Ceramics."** *American Ceramic Society Bulletin*, **ACSBA, 53:395. Apr 1974. [abstract only]; Paper 18-S2-74 presented at Symposium 2 of the American Ceramic Society 76th Annual Meeting, Chicago, Illinois. Apr 28–May 2, 1974.**

Acoustic emission measurements have been obtained accompanying high temperature fracture mechanics and strength measurements. The acoustic emission rate has been related to the crack growth rate, exhibiting approximate proportionality for all systems studied. A detailed study of the amplitude dependence of the acoustic events has enabled features of the macro- and micro-crack propagation processes to be elucidated.

452 **A. G. Evans. "Residual Stress Measurement Using Acoustic Emission."** *Journal of the American Ceramic Society*, **JACTA, 58(5–6):239–243. May–June 1975.**

Analysis is presented which shows how acoustic emission caused by microcracking expects to vary during an elastically reversible load cycle. The acoustic emission rate during unloading can be predicted directly from the acoustic emission rates observed during the loading and holding periods of a test cycle. Any discrepancy between the measured and predicted rates during unloading indicates irreversible behavior. A measure of the severity of the residual stress that results from irreversibility can be determined from the measured acoustic emission.

453 **A. G. Evans, J. R. Clifton, and E. Anderson. "The Fracture Mechanics of Mortars." Science Center, Rockwell International, Thousand Oaks, California, and Center for Building Technology, Institute for Applied Technology, National Bureau of Standards, Washington, D.C. 1975; paper presented at the American Society for Metals Conference on Nondestructive Evaluation in the Nuclear Industry, Denver, Colorado. Dec 1-3, 1975.**

A fracture mechanics study of plain and polymer impregnated mortars has shown that the (slow and rapid) macrocrack propagation resistance of these materials is not significantly affected by mortar processing variables, such as water-cement ratio and curing time, but is strongly enhanced by polymer impregnation. Acoustic emission measurements have indicated the important role of microcracking in the fracture of both plain and impregnated mortars; with the susceptibility to microcracking being substantially retarded by polymer inpregnation.

454 **A. G. Evans and L. J. Graham. "A Model for Crack Propagation in Polycrystalline Ceramics."** *Acta Metallurgica,* **AMETA, 23(11):1303-1312. Nov 1975.**

A basic model for macrocrack propagation in polycrystalline ceramics is developed, based on experimental acoustic emission amplitude studies. The model has implications concerning the role of the microstructure on the crack propagation resistance of ceramic materials. [reprinted by permission of Pergamon Press Ltd].

455 **A. G. Evans and H. Johnson. "A Fracture Mechanics Study of Zinc Selenide for Laser Window Applications." Institute for Materials Research, National Bureau of Standards, Washington, D.C. 1974.**

Fracture mechanics studies of chemically vapour-deposited zinc selenide have shown that this material has a low fracture toughness and is susceptible to water (moisture) dependent slow crack growth. Biaxial flexure strength studies have indicated that fracture (in this configuration) is controlled primarily by surface located flaws, and that macrofracture mechanics adequately describes the propagation of typical pre-existing flaws in this material. Acoustic emission studies have shown that crack growth can be consistently detected by this technique, and that fracture in air is preceeded by precursor events, up to 2s prior to catastrophic propagation. Failure diagrams generated from the slow crack growth parameters have shown that the lifetime requirements of the laser window can be achieved by overload proof testing to a proof ratio of 1.35, when the proper proof test precautions are implemented.

456 A. G. Evans and M. Linzer. "Failure Prediction in Structural Ceramics Using Acoustic Emission." Institute for Materials Research, National Bureau of Standards, Washington, D.C. 1973; *Journal of the American Ceramic Society*, **JACTA, 56(11):575-581. Nov 1973;** *Journal of Research of the National Bureau of Standards, Section B: Mathematics and Mathematical Physics,* **JNBBA, 78B(3):175. 1974. [abstract only].**

Crack propagation in a typical structural ceramic (porcelain) is found to be accompanied by acoustic emission. Two types of emission are detected. The first is due to the slow growth of the fracture initiating flaw, with the emission rate depending primarily on the crack velocity. Failure prediction using this source of emission can only be effective, however, if low level emission—which can be related exclusively to the crack growth—can be detected. The second source of emission, which occurs during bulk stressing, is the cracking associated with second phase particles (quartz particles in porcelain) due to the combined action of the applied stress and local thermal and mechanical stress. An analysis for predicting emission rates is developed which forms the basis for using this source of acoustic emission in failure prediction.

457 A. G. Evans, M. Linzer, H. Johnson, D. P. H. Hasselman, and M. E. Kipp. "Thermal Fracture Studies in Ceramic Systems Using an Acoustic Emission Technique." *Journal of Materials Science,* **JMTSA, 10(9):1608-1615. Sep 1975. [NTIS: AD/A-020-247].**

An acoustic emission technique for measuring the failure time in thermal shock experiments is described. The technique offers a unique opportunity to measure the heat-transfer coefficient of the test system and hence, to obtain a fully quantitative measure of the peak surface stress generated during the test. Measurements on soda-lime glass have demonstrated that rapid thermal fracture in the material occurs when the surface stress attains a critical value equal to the propagation stress for the most deleterious surface flaw. The effects of slow crack growth on the failure time are also investigated and correlated with recently developed theory. [reprinted by permission of Chapman & Hall Ltd].

458 A. G. Evans, M. Linzer, and L. R. Russell. "Acoustic Emission and Crack Propagation in Polycrystalline Alumina." *Materials Science and Engineering,* **MSCEA, 15(2-3):253-261. Aug-Sep 1974.**

Acoustic emission is measured during the fracture of polycrystalline alumina. It is shown that acoustic emission is obtained during macrocrack growth, which can be used for failure indication. Acoustic emission is also obtained owing to the formation of nonpropagating grain-size microcracks at the surface. This emission can mask that due to macrocrack growth, but only for short failure times or at high stresses. Models are presented which relate the acoustic emission to the various crack propagation processes.

459 **A. G. Evans, H. Nadler, and K. Ono. "An Acoustic Emission Study of the Fracture of Zinc Selenide."** *Materials Science and Engineering,* **MSCEA, 22(1):7–14. Jan 1976.**

Acoustic emission data obtained on coated and uncoated discs of zinc selenide during progression to fracture suggest that final fracture should easily be detected (without significant interference from external sources) by transducers tuned to receive signals at 500 to 700 kHz. Studies of precursor activity have implications for failure prediction using fracture mechanics, but have not yielded sufficiently definitive spectral results to suggest approaches for consistently detecting precursors by acoustic emission.

460 **A. G. Evans, S. M. Wiederhorn, M. Linzer, and E. R. Fuller, Jr. "Proof Testing of Porcelain Insulators and Application of Acoustic Emission."** *American Ceramic Society Bulletin,* **ACSBA, 54(6):576–581. June 1975.**

An investigation of proof testing has been conducted on large cone and post porcelain insulators. The cone configuration was unsuitable for overload proof testing to the loads needed for effective lifetime predictions. The only merit of subjecting this configuration to a proof test is the assurance of no immediate macrocracking during installation. This is achieved by testing to a load marginally larger than the service load. Conversely, the post configuration exhibited the basic prerequisites for effective proof testing but detailed stress analysis of this system is needed before specific proof testing procedures can be recommended. Finally, acoustic emission measurements were a major asset for monitoring both macrocracking events and stress development during the proof tests.

461 **R. C. Ewing. "Acoustic-Emission Tests Reveal Critical Vessel Flaws."** *Oil and Gas Journal,* **OIGJA, 69:90–93. Sep 13, 1971.**

Acoustic-emission tests of heavy-wall pressure vessels can be used to locate weld defects as well as other defects in the structure. During hydrostatic testing, sensors on the outside of the vessel pick up any vibrational energy emitted. Random burst-type emissions signal deformation of the material under stress, for example a growing flaw in a weld. Thus not only the size and location of the flaw are significant, but also the associated stress intensity. Flaws are located by triangulation calculations, for which the data may be fed into a small computer.

F

462 C. C. Feng. "A Technique for Checking Calibration Change with Temperature for Acoustic Emission Transducers." Engineering Report No. 76-1C. Endevco, San Juan Capistrano, California. Jan 16, 1976.

Described is a test technique developed to check the sensitivity and frequency response variations of acoustic emission transducers due to ambient temperature changes from −300 to +1000 °F. Previously, there has been no established procedure for characterizing an acoustic emission transducer at temperature extremes. The more familiar room temperature calibration methods such as the spark impulse method, the ultrasonic method, the grinding powder method, or the sandblast method are difficult to implement at a temperature extreme. This testing technique provides stress waves which can be applied independent of temperature. They are low amplitude, transient, and of short duration. This is achieved by impacting small *microspheres* back and forth in a metallic container. Both the test transducer and this small container are mounted to a vibration exciter and are placed within a temperature chamber. It is then vibrated at a low frequency causing the beads to impact the container to which the transducer is attached. The vibration controls the velocity of impact, the number of impacts, and thus the transducer output. The vibration is monitored by a reference accelerometer mounted on that portion of the vibration exciter outside the temperature chamber. This impact process is theoretically repeatable within 1 db over the temperature range of 0 to 1400 °F. The signal output of an AE transducer is taken at various temperatures and is either analyzed with a spectrum analyzer or by detecting an rms voltage through a bandpass filter. Any variation from the room temperature test data is an indication of the effect of temperature on the transducer output. It should be noted, however that this procedure does not relate to other specific calibration methods but is a relative sensitivity test. In the next section, the test set-up, the test procedure, and some test data will be presented. A review on impact theory will also be given in the Appendix.

463 C. C. Feng. "Acoustic Emission Transducer Calibration—Spark Impulse Calibration Method." Engineering Report No. 74-7-C. Endevco, San Juan Capistrano, California. 1974.

Report discusses the progress which has been made in the spark-impulse calibration technique of acoustic emission transducers since the work reported by R. L. Bell in Reference 132. The more recently developed test methods are described, as well as the physics and some of the characteristics of the spark-impulse technique.

464 P. Fenici, N. Kiesewetter, and P. Schiller. "Continuous Acoustic Emission from Aluminum." pp 114–125 in "EWGAE-ISPRA: Proceedings of the Third Meeting of the European Working Group on Acoustic Emission." A. C. Lucia,

Editor. Report EUR-5513e. Commission of the European Communities, Directorate General Scientific and Technical Information and Information Management, Luxembourg, Belgium. 1976. meeting held at the Joint Nuclear Research Center, Ispra, Italy. Sep 25-26, 1974.

Continuous acoustic emission of aluminum single crystals and polycrystals during tensile tests at constant cross-head speed and at room temperature is measured with a root mean square level recorder. By means of the Kaiser effect, it is shown that the continous emission is related to the plastic deformation. The plot of continuous emission against strain takes different shapes for pure single crystals, pure polycrystals, and impure polycrystals. The measured voltages have about the same value for pure single and polycrystals and are considerably greater than that for impure polycrystals. A method is developed to distinguish between continuous emission and burst.

465 **A. T. Fikkers. "Influence of External Restraint on Cold Cracking."** *Metal Construction and British Welding Journal*, **MCBWA, 6(6):188-193. June 1974.**

A simple method of testing for susceptibility to hydrogen-induced cracking is described. It is based upon the registration of the transient thermal stresses resulting from external restraint. The test results can be compared with those of Japanese investigations of the intensities of restraint in real structures. The cooling rate after welding, the hydrogen content of weld-metal, and the external restraint have been varied in an investigation into the behavior of eight types of normalized steel.

466 **V. I. Filatov, V. V. Gradinar, and I. I. Averbukh. "Inspection of Embrittlement in Electrodeposites by the Acoustic Emission Method."** *Industrial Laboratory,* **INDLA, 39(10):1638-1639. Oct 1973; translated from** *Zavodskaya Laboratoriya,* **ZYDLA, 39(10):1259-1261. Oct 1973. (in Russian).**

An acoustic method of testing the brittleness of electrodeposited iron, chromium, and nickel protective coatings approximately 15 and 50-μm thick on one side of flat structural steel specimens is proposed. The method is based on recording the number of pulses of the elastic waves which arise during the formation and development of microcracks, using standard acoustic apparatus. Experiments using this technique are described and the results are compared with those of other methods. The acoustic method gives a sensitivity and accuracy which are perfectly adequate from the point of view of practical engineering, and it is much simpler than conventional methods.

467 **K. Fink. "Stand und Entwicklungstendenzen der zerstorungsfreien Prufung von Schweissverbindungen (The Current State and Development Trends in the Nondestructive Testing of Welded Joints)."** *Schweissen und Schneiden,*

SCSCA, 24(9):375–380. Sep 1972. (in German); HB No. 8919. "Present State of, and Trends in, the Nondestructive Testing of Welds." Henry Brutcher Technical Translations, Altadena, California. 1973.

The NDT of defects in welds can be divided into three sections: (a) detection of material inhomogeneities; (b) identification of the length, shape, and size of such inhomogeneities; and (c) evaluation of the identified inhomogeneities. Methods used for NDT are magnetic particle, dye-penetrant, and direct current potential techniques for surface defects and radiographic, acoustic emission, and ultrasonic pulse-echo techniques for internal as well as external defects. The current developments in NDT are reviewed and a table comparing the different methods, indicating the types of defect, e.g. cracks, pores, inclusions, etc., that can be detected by each technique is presented. The mechanization and automation of NDT of welded joints are discussed and a brief account of the relevant national and international standards and recommendations concerning the testing of welded joints is given. [Abstract 55-0543. *Metals Abstracts,* Vol. 6, 1973].

468 V. M. Finkel, I. S. Guz, I. A. Kutkin, and V. I. Bazaikin. "Recording of Rayleigh Waves Generated by a Moving Crack." ***Industrial Laboratory,*** **INDLA, 36(2):288–289. Feb 1970; translated from** ***Zavodskaya Laboratoriya,*** **ZVDLA, 36(2):216–217. Feb 1970. (in Russian).**

A method is described of recording the Rayleigh waves associated with moving cracks in metal plates and other materials and thus establishing the characteristics of crack propagation. The method enables the frequency and energy spectra of the waves generated in the rupture of materials to be directly recorded. The characteristics of crack propagation bear a one-to-one relation to the spectra produced. Some typical practical examples are presented. [Abstract 22-0640. *Metals Abstracts,* Vol. 3, 1970].

469 V. M. Finkel, I. S. Guz, I. A. Kutkin, and A. Ya.Volodarskii. "Generation of Rayleigh Waves by a Moving Crack." ***Soviet Physics–Solid State,*** **SPSSA, 12(8):1837–1840. Feb 1971; translated from** ***Fizika Tverdogo Tela,*** **FTVTA, 12(8):2300–2305. Aug 1970. (in Russian).**

The fracture of Plexiglas samples has experimentally disclosed the existence of Rayleigh waves radiated from the vertex along the edges of a growing crack. A spectral analysis of the experimental oscillograms on an M-20 electronic computer has shown that the spectrum of the Rayleigh pulses falls in the band from 0 to 200 kHz in the material investigated. An energy evaluation of the relative energy-frequency distribution indicates that the maximum energy transport capability occurs at frequencies up to 100 kHz. The authors demonstrate the influence of the fracture conditions on the parameters of the Rayleigh waves. [reprinted by permission of the American Institute of Physics].

470 **V. M. Finkel and S. V. Serebryakov. "Radiation of Sonic and Ultrasonic Waves during Growth of Cracks in Steel."** *Physics of Metals and Metallography,* **PHMMA, 25(3):169–175. 1968; translated from** *Fizika Metallov i Metallovedenie,* **FMMTA, 25(3):543–548, Mar 1968. (in Russian).**

The sonic impulses are in the frequency range of 3 to 24 kHz, but the ultrasonic components in the 60 to 100 kHz range. A correlation was found between the emitted and the elastic energies (which is stored up by the solid before rupture). There is a tendency for the frequencies to increase with decrease of the annealing temperature. One of the main reasons for these sonic emissions is the irregular and intermittent growth of the cracks. [Abst. 21413 b. *Chemical Abstracts,* Vol. 69, 1968. Copyright by American Chemical Society].

471 **R. M. Fisher and J. S. Lally. "Appendix A: Discussion." pp 617–620, appended to chapter: "Agenda Discussion: High-Speed Dislocations." J. Weertman and J. Krafft. pp 609–617 in** *Dislocation Dynamics.* **A. R. Rosenfield, G. T. Hahn, A. L. Bement, Jr., and R. I. Jaffee, Editors. McGraw-Hill Book Company, New York. 1968; proceedings of the Second Battelle Institute Materials Science Colloquia, held in Seattle, Washington and Harrison, British Columbia. May 1–6, 1967.**

The appendix is a brief discussion of the work of R. M. Fisher and J. S. Lally which is reported in Reference 472.

472 **R. M. Fisher and J. S. Lally. "Microplasticity Detected by an Acoustic Technique."** *Canadian Journal of Physics,* **CJPHA, 45(2):1147–1159. Feb 1967; paper presented at the International Conference on Deformation of Crystalline Solids, Ottawa, Canada. Aug 22–26, 1966.**

Evidence for the formation of slip lines and slip bands has usually been obtained from slip-line replica and etch-pit techniques. An apparatus has been built to detect the formation of small slip events within a crystal by monitoring small longitudinal oscillations of the test specimen which occur when the load is momentarily relaxed. Each load drop results in a burst of acoustic energy, some of which are detected and recorded electronically. Results are presented of measurements of the frequency and magnitude of acoustic pulses from single crystal specimens during plastic flow. They suggest that slip events occur cooperatively and that dislocation velocitites during yielding may be as high as 10^3 centimetres per second. Data are given for Mg single crystals, Cu, Fe, low carbon steel, 2.8% Si steel, Cu-7 Al alloy, and brass. [reprinted by permission of the National Research Council of Canada].

473 **C. E. Fitch, Jr. "Acoustic Emission Signal Analysis of Flat Plates." BNWL-1008. Battelle-Northwest, Richland, Washington. Apr 1969. [NTIS].**

An acoustic emission signal from a deformation or fracture mechanism is assumed to initiate as essentially a point source for detection frequencies in the range of a few megahertz or below. Approximate mathematical expressions for predicting the number of signal reflections and their time separations, and thus total expected signal durations, were derived and studied for flat specimen geometries. A 0.5-inch thick A-212-B steel plate was used to determine whether experimental observations could be predicted from approximate relationships derived. Both artificial and real acoustic emission sources were tested under a variety of conditions. Observations of good experimental agreement may provide a basis for optimizing signal detection in flat geometries. A preliminary study of pulse shapes and pulse durations was also conducted in experimental attempts to determined source depth as well as range.

474 **C. E. Fitch, Jr., P. H. Hutton, and H. N. Pedersen. "Crack Detection in Pressure Piping by Acoustic Emission." pp 3.1–3.9 in "Nuclear Safety Quarterly Report–November, December 1968, January, 1969 for Nuclear Safety Branch of USAEC Division of Reactor Development and Technology." BNWL-1009. Battelle-Northwest, Richland, Washington. Mar 1969. [NTIS].**

Currently planned basic study of acoustic emission wave propagation in flat-plate geometry has been completed with a preliminary demonstration that the pulse shape produced by a given transducer design when excited by an acoustic emission signal can be predicted. This latter information is valuable in establishing nominal pulse length and gating periods for data analysis. It also has potential use for later more sophisticated techniques for describing a flaw in terms of acoustic emission. The overall work has provided significant guidance to optimized detection techniques. Wave propagation in curved plate remains to be investigated for unique features. A solid-state tuned preamplifier has been designed and successfully used. This has the advantages of reducing system electronic noise and of alleviating the problem of preamplifier saturation by high level, low-frequency noise signals. Data have been collected from acoustic emission generated during burst testing a 24-inch outside diameter by 1.5-inch wall, Type 316 stainless steel pressure vessel at the Battelle-Columbus test site. As the part-through-wall flaw began to open, the emission rate profile correlated well with the curve profile from a crack-opening displacement gauge. Acoustic emission is being applied locally to investigate potential slow crack growth by fatigue in known high stress locations or process piping. A few positive indications have been obtained to date. An acoustic emission monitor system installed on the Containment System Experiment simulator vessel was used to monitor a blowdown test sequence. No emission was detected from known flaws in the vessel heads.

475 **E. R. Fitzgerald. "Mechanical Resonance Dispersion and Plastic Flow in Crystalline Solids."** *Journal of the Acoustical Society of America,* **JASMA, 32(10):1270–1289. Oct 1960.**

An explanation is given for the mechanical resonance dispersions observed at audiofrequencies in many crystalline solids. The resonances are considered to result from alternating simultaneous slip occurring on transient slip planes with mean lifetimes of

10^{-3} to 10^{-4} seconds. The slip along such planes is opposed by displacement-dependent, velocity-dependent, and acceleration-dependent stresses. These last two stresses are found to be of magnitudes which can result from the interchange of atoms across a slip plane with a consequent decrease in momentum of the fast moving (slipped) part of the sample undergoing deformation. If the rate of transfer is independent of the slip process, a velocity-dependent (viscous) stress results, but if some of the atoms are induced to transfer as a direct result of the slip itself, then an acceleration-dependent stress is present. These slip planes, because of their sudden formation, short lifetimes, and inertial coefficients, do not contribute to the static or ultrasonic compliance, but do affect the dynamic compliance at audio-frequencies. The transient slip planes are supposed to be generated by the movement of dislocations across the sample. Thus, a concept of combined consecutive and simultaneous slip is suggested in which dislocations remain an essential part of the mechanism, but are not the sole source of deformation. An interesting consequence of the theory is that a critical stress for plastic flow or, alternatively, for brittle fracture can be predicted from values of the dynamic compliance (or modulus). Predicted values of critical stress (for both single crystals and polycrystalline metals) made from the dynamic modulus at 100 cycles per second are much closer values than those made on the basis of either the static or ultrasonic moduli. In particular, it is found that the predicted values for brittle fracture stress agree almost exactly with observed breaking stresses for single crystals of sodium chloride, germanium, glycine sulfate, and, at two temperatures and orientations, for natural quartz crystals. The occurrence of negative absorption or induced emission is discussed, and a means of generating sustained, self-excited, audio-frequency oscillations in a crystalline solid is outlined. The possibility of mechanical failure at an intrinsic material resonance is considered briefly and some other aspects of the mechanical behavior of crystalline solids are also examined. Concerning the observation of acoustic emission the following excerpts are quoted from the article:

> In heated sodium chloride (~400 °C), each jump was accompanied by a noise like the ticking of a clock. The ticks were spaced at regular time intervals and the frequency of the ticks increased as the stress increased. The course of plastic slip should thus be accompanied by the generation of audio-frequency vibrations which could be detected by a high-sensitivity microphone and analyzed for the relative intensities of various components. It has already been noted that ticking noises are heard during the plastic deformation of sodium chloride at high temperatures. Similar clicks are heard during the bending of tin or indium specimens, and in the case of tin a high audio-frequency *cry* is also heard. This has been attributed to a rapid succession of clicks because of twinning, but according to the present concepts similar *cries* might be expected also during translational slip, and a suggestion is made that discrete audio frequencies may be generated during slip and could be observed with a sensitive microphone or other detector and analyzed. Sharp clicks and other noises during slip at high temperatures have been noted by many previous investigators but the observed sounds never have been systematically analyzed into component frequencies.

476 **E. R. Fitzgerald.** ***Particle Waves and Deformation in Crystalline Solids.*** **Interscience Publishers, a Division of John Wiley and Sons, New York. 1966. pp 54–63.**

477 J. M. Fitz-Randolph. "Acoustic Emission Characterization of the Fracture Process in a Boron–Epoxy Composite." M. S. Thesis. University of California at Los Angeles, Los Angeles, California. 1971. [Thesis No. 3395, UCLA Library].

The effect of fiber orientation and temperature on the failure strength and notch sensitivity of a boron-epoxy composite has been investigated by means of slow-bend testing. The behavior of the composite in both longitudinal (parallel to fiber axis) and transverse (perpendicular to the fiber axis) directions has been evaluated in the temperature range –196 to +150°C. A temperature rise from –196 to +150°C causes the composite strength for each fiber orientation to decrease by at least 50%. The composite was found to be insensitive to notches perpendicular to the fiber axis. The acoustic emission per broken fiber during the fracture process was obtained by a compliance calibration technique and by monitoring the electrical resistance of the sample. It was found that acoustic emission can be directly correlated to the strain energy release rate. Emission bursts were found to be characteristic of crack extension. An examination of theoretical fracture toughness, G_{IC}, and experimental work of fracture, γ_F, values suggested that the energy released by fiber during crack propagation is approximately equal to the work done in pulling broken fibers out of the resin matrix.

478 J. M. Fitz-Randolph, D. C. Phillips, P. W. R. Beaumont, and A. S. Tetelman. "Acoustic Emission Studies of a Boron–Epoxy Composite." ***Journal of Composite Materials,*** **JCOMB, 5:542–548. Oct 1971.**

Acoustic emission and stress wave emission are terms used to describe the release of stored acoustic energy from a material when it undergoes plastic deformation, phase transformation, or fracture. This energy is released in discrete pulses characteristic of the process. If a transducer is attached to a specimen of the material, emission above the threshold of the transducer can be amplified and read as acoustic emission counts. The response of the count-rate measuring device depends upon the amplifier gain setting, and the number and amplitude of the pulses monitored by the transducers. Using a totalizer, the count rate can be integrated to give the sum of counts. Count-rate and sum of counts (ΣAE) printouts produce a pattern of energy release characteristic of a given process. The authors describe how acoustic emission has been correlated with the breaking of fibres during the fracture of a boron-epoxy composite. [Abstract 19211. *Physics Abstracts,* Vol. 75, No. 937, 1972].

479 J. M. Fitz-Randolph, D. C. Phillips, P. W. R. Beaumont, and A. S. Tetelman. "The Fracture Energy and Acoustic Emission of a Boron–Epoxy Composite." ***Journal of Materials Science,*** **JMTSA, 7(3):289–294. Mar 1972.**

The fracture surface energy (γ) of a boron fibre-epoxy resin composite has been measured by three different techniques: work of fracture, linear elastic fracture mechanics, and compliance variation. Significant differences were obtained by the different methods. The compliance data were analyzed to give γ at different stages of crack propagation. It was observed that γ decreased as the crack entered the material and that this variation of γ could be correlated with the pull-out length of fibres and acoustic emission generated during fracture. The fracture surface energy is explained in terms of a debonding model.

480 **P. Fleischmann, R. Goutte, and C. Guillaud. "Emission Acoustique dans le Domaine du Megahertz: Description de la Chaine d'Acquisition: Premiers Resultats (Acoustic Emission in the Megahertz Region: Description of the Data Acquisition System: Preliminary Results)." pp 323–335 in *Journees d'Etudes sur l'Emission Acoustique.* P. F. Gobin, R. Goutte, and C. Guillaud, Organizers. Institut National des Sciences Appliquees de Lyon, Avenue Albert Einstein, 69621 Villeurbanne, Lyon, France. Mar 1975. proceedings of a conference held in Lyon, France. Mar 17, 1975. (paper in French).**

The role played by transfer functions of both the sensor and the sample in the processing of acoustic emission signals is underlined.

481 **P. Fleischmann, D. Rouby, F. Lakestani, and J. C. Baboux. "A Spectrum Analysis of Acoustic Emission." *Non-Destructive Testing,* NDETA, 8(5): 241–244. Oct 1975.**

The authors describe a data acquisition system to determine the frequency spectrum between 500 kHz and 3 MHz of the initial acoustic emission. They measured the transfer function of the piezoelectric transducer and of the specimen for their results using a computer. They present an application of this device to the study of the emission of a low-carbon steel.

482 **E. Fontana, G. Grugni, C. Panzani, B. Pirovano, G. Possa, and F. Tonolini. "Acoustic Emission Monitoring during Hydrotests of a Thin Wall Pressure Vessel." pp 126–140 in "EWGAE–ISPRA: Proceedings of the Third Meeting of the European Working Group on Acoustic Emission." A. C. Lucia, Editor. Report EUR-5513e. Commission of the European Communities, Directorate General Scientific and Technical Information and Information Management, Luxembourg, Belgium. 1976. meeting held at the Joint Nuclear Research Center, Ispra, Italy. Sep 25–26, 1974; *Energia Nucleare,* ENNLA, 22(5): 253–259. May 1975.**

Results are presented of an acoustic emission monitoring during hydrotests of a thin-wall steel pressure vessel. Location of acoustic sources was based on longitudinal wave-front detection. The careful calibration of the three sensors instrumentation system used for acoustic source location was found to be useful, and allowed an accurate location error analysis. Acoustic emission in the hydrotests was found to be due mainly to stress release in weld seams.

483 **E. Fontana, G. Grugni, B. Pirovano, G. Possa, and F. Tonolini. "Controllo Non Distruttivo di Recipienti a Pressione Mediante Analisi dell'Emissione Acustica nel Corso della Prova a pressione Idrostatica (NDT Technique for Pressure Vessels, Based on Stress Wave Emission Analysis during Hydrotest)."** ***Energia Nucleare*** **(Milan), ENNLA, 21(10):550–558. Oct 1974. (in Italian).**

Paper presents a relatively new NDT technique for pressure vessels, based on stress wave emission detection and analysis during hydrotest. The instrumentation commonly used for detection and signal conditioning is briefly outlined. Methods for acoustic source location are reported. The state-of-the-art of this fast progressing inspection technique is surveyed.

484 **M. G. Fontana and K. F. Graff. "Corrosion Cracking of Metallic Materials, Part II. Acoustic Emission: Experiment and Theory." AFML-TR-72-102, Part II. Ohio State University, Department of Metallurgical Engineering, Columbus, Ohio. Aug 1972. [NTIS: AD-751-529 or N73-16570].**

The aspects of acoustic emission associated with the theory of the stress waves emitted from cracking and the experimental measurements of the acoustic emission pulses are reported. In the theoretical phase of the work, a survey of the applicable literature in the area of stress waves from moving cracks is reviewed. Certain simple theoretical arguments pertaining to the energy release from an increment of crack growth are reported. In the experimental phase, an evaluation of a number of ultrasonic nondestructive test transducers for acoustic emission work was carried out and an optimum transducer selected. A complete acoustic emission test system was then constructed and several tests performed by the conclusion of the project.

485 **K. M. Foreman and R. A. Oman (to Grumman Aerospace Corporation). "Acoustic Emission Flow Measurement System." U.S. Patent 3,958,458. May 25, 1976.**

Pseudo-sound created by the interaction of fluid flowing over a discontinuity in a conduit is sensed by a transducer. The transducer signal is reduced to a frequency-amplitude distribution and the amplitude at the characteristic frequency is used to calculate fluid flow rate using a dipole model. The apparatus is non-intrusive and passive. Several transducers may be serviced by a single central processor remote from the transducer site. Alternative apparatus provides for the determination of change in fluid flow direction and fluid density.

486 **R. Forestier. "Etude dans le domaine plastique, la Metallo-Dyno-Phonie: procede d'investigations et de controle application aux tractions uniaxiale et triaxee (A Study in the Plastic Domain, Metallo-Dyno-Phonic: A Process of Investigation and Control with Application to Uniaxial and Triaxial Tractions)."**

C795–CPS Paris 6–72. OTUA (Office Technique pour l'Utilisation de l'Acier), Paris, France. Apr 1972. (in French); RFP–Trans–124. translated for Dow Chemical U.S.A., Rocky Flats Division, Golden, Colorado. 1973. [NTIS].

487 **F. Forster and E. Scheil. "Akustische Untersuchung der Bildung von Martensitnadeln (Acoustic Study of the Formation of Martensite Needles)."** *Zeitschrift fur Metallkunde,* **ZEMTA, 28(9):245–247. Sep 1936. (in German); HB No. 375. Henry Brutcher Technical Translations, Altadena, California. Aug 12, 1936. (in English).**

An application for obtaining oscillographs of the noises produced in the formation of martensite needles is described. The curves obtained indicate a time of formation of 0.002 seconds. The number of needles formed depends on the rate of cooling, first increasing then decreasing as the rate is increased, and increases with decreasing grain size. [Abstract 1739-8. *Chemical Abstracts,* Vol. 31, 1937. Copyright by American Chemical Society].

488 **F. Forster and E. Scheil. "Untersuchung des zeitlichen Ablaufes von Umklappvorgangen in Metallen (Investigation of the Time History of** ***Umklapp*** **Processes in Metals)."** *Zeitschrift fur Metallkunde,* **ZEMTA, 32(6):165–173. June 1940. (in German); UCRL–Trans–750(L). translated for Lawrence Radiation Laboratory, Livermore, California. Oct 1961.**

The phrase *umklapp or flop-over process* is used for rapid changes or transformations which cause a noticeable or measurable noise. The changes investigated were: martensite formation in Fe-Ni alloys, β-α transformation of Zr, and twinning upon deformation of Bi. A cathode-ray oscillograph with special amplifier was used for the measurements; photographic records were made. These processes progress in jumps; with each jump the resistance changes rapidly, followed by oscillations. In the martensite formation (decrease in resistivity), the resistance at first passes through a maximum for undeformed wires and a minimum for cold-worked wires. During twinning, the speed of the flop-over and frequency of oscillation depend upon the orientation of the monocrystals. During the martensite formation, the structural condition and the wire thickness exert an influence on the oscillations. Transformations of the same character but progressing slowly are called shift-over processes; in this case the progress is steady. [Abstract 3934-8. *Chemical Abstracts,* Vol. 35, 1941. Copyright by American Chemical Society].

489 **K. A. Fowler. "Acoustic Emission Simulation Test Set."** *Materials Research and Standards,* **MTRSA, 11(3):35–36. Mar 1971.**

The design and construction test set, which employs a thin-wire magnetrostrictive transducer, is described. The device enables one to check important system parameters in acoustic emission testing. [Abstract 46681. *Physics Abstracts,* Vol. 74, No. 920, 1971].

490 **K. A. Fowler. "Acoustic Emission Transducers and an Experimental Method of Verifying Their Performance." Technical Memorandum No. 5. Panametrics, Waltham, Massachusetts. Sep 23, 1970; paper presented at a joint meeting of the Boston Chapters of American Society for Nondestructive Testing, Institute of Electrical and Electronics Engineers, American Welding Society, American Society for Metals, American Society for Testing and Materials, and American Society of Mechanical Engineers. Boston, Massachusetts. Sep 23, 1970.**

The generic types of transducers useful in acoustic emission monitoring are reviewed. The basic modes of ultrasonic wave propagation and the interrelationships between structure geometry, propagation mode, material, and transducer on the apparent form of acoustic emission signals are considered. An experimental method of comparing transducers and of determining the parameters that must be considered in the design of practical acoustic emission monitoring systems is proposed.

491 **K. A. Fowler and E. P. Papadakis. "Observation and Analysis of Simulated Ultrasonic Acoustic Emission Waves in Plates and Complex Structures." pp 222–237 in *Acoustic Emission, ASTM STP 505.* American Society for Testing and Materials, Philadelphia, Pennsylvania. 1972. proceedings of symposium held in Bal Harbour, Florida. Dec 7–8, 1971.**

Experiments have been performed with the Panametrics acoustic emission simulation test set which show that acoustic emission signals in plates and shells travel in the plate modes supported by these shapes. The first longitudinal plate mode has the highest group velocity in its low-frequency limit, and signals in that mode and frequency range arrive first, followed by strong signal at the velocity of the first flexural mode in its nondispersive region. In some cases, the flexural signal is much stronger than the longitudinal signal. In general, the received signals are dispersed. The time-frequency-energy output matrix tends to be complicated, as energy of different frequencies is spread over time by the various dispersive modes. Special transducers can be used to discriminate against certain particle motions, suppressing the reception of certain modes, and may be useful in triangulation studies.

492 **C. Francois, A. Nouvet, and J. Pertuis. "Detection Acoustique de la Fissuration de Toles Portant une Attache Soudee d'Angle au Cours d'Essais de Fatigue (Acoustic Detection of Cracks in Fillet Welds Joining Steel Plates during Fatigue Tests)." No. 2854LA. Ministere des Armees, Delegation Ministerielle pour l'Armement, Service Technique des Constructions et Armes Navales, Paris, France. May 11, 1971. (in French); *Sciences et Techniques de l'Armement,* STQAB, 46(4):991–1014. 1972. (in French).**

The acoustic transmission, proper to elastic waves, which results from materials plastic distortion connected with the spreading of cracks, is intercepted by a piezo-electrical detector and is recoded with the help of an appropriate system. During strain tests on angle-weld connected sheet irons, it has been possible to display the cracks' appearance and spreading. The acoustic action of the test piece, where the welds are tested in compression, is ten times as high as the recorded tensile testing action. Finally, attention was given to the incidence of the proffered chances due to acoustic transmission on the fracture mechanism.

493 **J. A. Franklin. "Some Effects of Loading Technique on the Acoustic Emission from Stressed Rock." M. Sc. Thesis. Imperial College, Department of Geology, London, England. Oct 1965. [available in positive 35-mm microfilm from University of London Library, Photographic Department, Senate House, Malet Street London WC1E 7HU, England].**

The nature of acoustic emission from stressed materials is discussed with the aid of an extensive literature review and bibliography. The suitability of various loading systems and techniques for investigation of acoustic emission is assessed, and details are given of a hydraulic machine and automatic load-deformation measuring system designed and built for this purpose. Tests are described in which specimens of seven rock types were loaded in compression in order to obtain acoustic data that would supplement knowledge gained from the literature. These tests were also intended to assist in the development of apparatus and instrumentation and to facilitate the design of future emission experiments. Methods of analyzing and interpreting the acoustic data are described. Various parameters of the experimental emission data are examined to determine those most suitable for the study of deformation processes. Possible applications of emission techniques are suggested, with particular reference to the field of rock mechanics.

494 **G. E. Frantti. "Seismissions and Surface Waves Related to Geologic Structures." pp 271-289 in *Proceedings First Conference on Acoustic Emission/Microseismic Activity in Geologic Structures and Materials.* H. R. Hardy, Jr. and F. W. Leighton, Editors. Trans Tech Publications, Clausthal, West Germany. 1977. conference held at Pennsylvania State University, University Park, Pennsylvania. June 9-11, 1975.**

The paper describes field studies related to stability and ground subsidence associated with undergound mines. Also included are studies into the effect on surface waves of near surface structural inhomogeneities.

495 **J. L. Frarey. "Practical Application of the Sonic Analyzer." pp 340-348 in *Incipient Failure Diagnosis for Assuring Safety and Availability of Nuclear Power Plants.* E. P. Epler and D. P. Roux, Coordinators. U.S. Atomic Energy Commission, Division of Technical Information, Washington, D.C. Jan 1968. [NTIS: CONF-671011]. proceedings of conference sponsored by Oak Ridge National Laboratory, held in Gatlinburg, Tennessee. Oct 30-Nov 1, 1967.**

Curtiss-Wright began the development of a sonic analyzer in 1960. The objective of the program was to produce an analyzer capable of detecting defects in the rotating elements of machinery such as jet engines and transmission systems. Requirements for such an analyzer were that it should be light-weight, portable, and simple to use. In addition, to insure ease of operation, it should require neither mechanical nor electrical connections to the machine under test. Basically this objective has been achieved. The authors believed they have arrived at the panacea for all maintenance problems. It should come as no surprise to a group such as this that this is not the case. Nevertheless, the authors believe that it is a valid claim for the sonic analyzer; and that it is capable of making a significant contribution in the diagnosis of engine condition. With the present state of the art in electronics, it is relatively easy to build black boxes that will give an indication when certain conditions have been exceeded. Computer capability encourages the amassing of vast amounts of data and allows the use of sophisticated tests to these data. Computer systems can also explain if the problem can be fixed at the next scheduled maintenance, if parts are available, and what the down time will be. The most important factor in any of these systems is what characteristic of the parameter or group of parameters can be used to give an indication that some maintenance action is required. In other words, what is measured, and what limits are established for these measurements. The paper discusses the what and how questions for the sonic analyzer and addresses the question of setting limits.

496 M. C. Fraser. "Acoustic Emission Characteristics of Selected Alloys." Battelle-Northwest, Richland, Washington. Apr 20, 1967.

Subaudible sound and ultrasound generated within a metal when it is stressed both below and within the plastic range are phenomena attracting considerable interest as a tool for studying material behavior and as a way of monitoring for overload. The report describes the application of Nortec equipment to such studies. The purpose of the work was to determine deformation dependent sonic-emission characteristics of steel, stainless steel, aluminum and titanium alloys, and to evaluate the Nortec NDT 101 amplifier-stretcher unit's capabilities with respect to sonic emission studies. The ultrasound was detected with a piezoelectric crystal, the output from which was pre-amplified, fed into a Nortec NDT 101 amplifier-stretcher unit, and then recorded on a strip chart.

497 C. L. Frederick. "Acoustic Emission Monitoring of Welds." Nortec Corporation, Richland, Washington. 1972; paper presented at U.S.-Japan Joint Symposium on Acoustic Emission, sponsored by Japan High Pressure Institute and Japanese Society for Non-Destructive Inspection, Tokyo, Japan. July 4–6, 1972; paper presented at American Society for Nondestructive Testing National Fall Conference, Cleveland, Ohio. Oct 16–19, 1972.

Acoustic emission monitoring has shown an increased use by research laboratories, quality assurance, testing, and inspection groups. This monitoring, which complements rather than replaces traditional test methods, is a highly sensitive real-time process. Monitoring is performed as the material under evaluation is changing, defects are forming or growing, or material is being deformed or transformed. A source of stress is required to cause the emissions. The above requirements, which can be provided by the process and the ability to examine large areas in remote locations without mechanical scanning, make acoustic emission a useful tool in weld monitoring during the weld operation.

498 C. L. Frederick. "Acoustic Information Analysis." Nortec Corporation, Richland, Washington. 1972; paper presented at U.S.-Japan Joint Symposium on Acoustic Emission, sponsored by Japan High Pressure Institute and Japanese Society for Non-Destructive Inspection, Tokyo, Jap. July 4–6, 1972.

Paper describes acoustic emission processing techniques and includes definitions for a list of acoustic emission terms.

499 J. R. Frederick. "Acoustic Emission as a Technique for Nondestructive Testing." *Materials Evaluation,* **MAEVA, 28(2):43–47. Feb 1970.**

A promising method for nondestructive testing and inspection is to observe the acoustic emission from materials when they are stressed. Observations can be made on tensile specimens or on manufactured parts. Except for the noise produced by crack nucleation or propagation, the exact mechanisms which cause acoustic emission are not clear at the present time. The amount of noise depends upon microstructure and the level of the applied stress. The technique appears to offer considerable promise for the *in-service* monitoring of critical structures. The cumulative load emission is plotted for hot-rolled 1020 and 1045 steels, cold-rolled 1018 steel, 2024, 7075, and Alcoa 195 aluminum alloys.

500 J. R. Frederick. "Some Basic Mechanisms of Acoustic Emission from Metals." University of Michigan, Department of Mechanical Engineering, Ann Arbor, Michigan. 1969; paper presented at the American Society for Metals Materials Engineering Congress, Philadelphia, Pennsylvania. Oct 13-16, 1969.

Acoustic emission from metals can result from several mechanisms. Emission which is produced by crack nucleation and propagation is finding use as a nondestructive testing technique. Other sources of emission are: (1) the melting or solidification of solids in which a reduction in volume occurs, (2) phase transformations in the solid state which involve a volume change, (3) sudden slip which occurs either during the application or the removal of a stress, or (4) twinning. Emission due to slip has been observed at stress levels that are only a small fraction of the gross yield stress. The amount of emission, observed as a result of slip, is dependent on the stress history and the microstructure of the metal. Various models have been proposed to explain observed variations of emission with strain and with microstructure.

501 **J. R. Frederick. "Use of Acoustic Emission in Nondestructive Testing." AFML-TR-72-114. Final Report. University of Michigan, Department of Mechanical Engineering, Ann Arbor, Michigan. Oct 1972. [NTIS: AD-758-216].**

A model based on the activation of dislocation sources has been developed for predicting the effect of microstructure on acoustic emission. The model predicts a minimum dislocation source length and slip distance below which no emission will be detected. Experimental results obtained on various metals and alloys having different microstructures substantiate the model. The effect of grain size on the acoustic emission from 99.99% aluminum and 99.9% copper suggests that macroscopic yielding in these metals occurs with the help of dislocation pile-ups that have been held up by grain boundaries. Experimental results obtained under conditions of low background noise level show that some metals produce acoustic emission when the applied load is removed. The amount of emission that is observed correlates with the magnitude of the Bauschinger effect in the metal. In those metals that show an *unload* acoustic emission effect, there is the possibility of using it to investigate the magnitude of the residual stress in a specimen. Creep and fatigue phenomena can also be studied by the use of acoustic emission. However, for high strength materials the low signal level for any events other than those associated with the final propagation of a crack makes it difficult to obtain significant data that are useful for the purpose of predicting fatigue or creep life.

502 **J. R. Frederick. "Use of Acoustic Emission in Nondestructive Testing: First Semiannual Report 01971-1-P, July 1, 1968–February 28, 1969." Advanced Research Projects Agency Contract F33615-68-C-1703. University of Michigan, Department of Mechanical Engineering, Ann Arbor, Michigan. May 1969.**

The objective of the program is to investigate the feasibility of the use of acoustic emission as a nondestructive testing tool. A model is proposed in order to explain the acoustic emission which results from dislocation motion. The model is used to predict the minimum dislocation source length and the minimum slip-region radius that is capable of producing detectable acoustic emission. Agreement is obtained between the model and experimental results on 99.99% aluminum and 2024 aluminum alloy. Data have been obtained on the acoustic emission from various alloys during the removal of the stress from a test specimen. This effect appears to be related to the magnitude of the Buschinger effect in these materials.

503 **J. R. Frederick. "Use of Acoustic Emission in Nondestructive Testing: Second Semiannual Report 01971-2-P, March 1, 1969–August 31, 1969." Advanced Research Projects Agency Contract F33615-68-C-1703. University of Michigan, Department of Mechanical Engineering, Ann Arbor, Michigan. Nov 1969. [NTIS: AD-863-195].**

Acoustic emission may be defined as the noise given off spontaneously by solid materials as a result of a sudden relaxation of stresses within the material. Stress relaxation can occur as a result of the nucleation or propagation of cracks, or as a consequence of various elastic or plastic deformation processes. The principal elastic or plastic deformation mechanisms that are sources of acoustic emission in solids are: (1) the slip of existing dislocation in a metal, (2) the activation of dislocation sources, (3) twinning, and (4) grain boundary slip. The report describes the results of an investigation into the effects of the microstructure on acoustic emission. The characteristics of the emission, observed in some materials when the load is removed from a specimen, are described, and an explanation of this phenomenon on the basis of dislocation motion is given.

504 J. R. Frederick. "Use of Acoustic Emission in Nondestructive Testing: Third Semiannual Report 01971-3-P, September 1, 1969-February 28, 1970." Advanced Research Projects Agency Contract F33615-68-C-1703. University of Michigan, Department of Mechanical Engineering, Ann Arbor, Michigan. June 1970. [NTIS: AD-872-256].

Acoustic emission may be defined as the noise given off spontaneously by solid materials as a result of sudden relaxation of stresses within the material. Stress relaxation can occur as a result of the nucleation or propagation of cracks, or as a consequence of various elastic or plastic deformation processes. The principal elastic or plastic deformation mechanisms that are sources of acoustic emission in solids are (1) the slip of existing dislocations in a metal, (2) the activation of dislocation sources, (3) twinning, and (4) grain boundary slip. The report describes the results of an investigation into the effects of one microstructure parameter, namely grain size, on acoustic emission. It also describes a method by which acoustic emission phenomena may possible be used to determine the amount of residual stress in metals.

505 J. R. Frederick and D. K. Felbeck. "Dislocation Motion as a Source of Acoustic Emission." pp 129-139 in *Acoustic Emission, ASTM STP 505.* American Society for Testing and Materials, Philadelphia, Pennsylvania. 1972. proceedings of symposium held in Bal Harbour, Florida. Dec 7-8, 1971.

The sudden strain increment that results from the unpinning of dislocations or from the activation of dislocation sources can result in observable acoustic emission. A model has been developed that relates the amount of acoustic emission to the size of a dislocation source, the distance the dislocations slip, and to the grain size. The model is able to explain the effect of the microstructure of the metal on the amount of observed acoustic emission. Data obtained on the acoustic emission from various metals support the model. In general, the greatest amount of emission occurs during the application of stress to a test specimen and particularly during yielding. However it is also possible to obtain acoustic emission during removal of a stress in those materials that show a Bauschinger effect.

506 **S. W. Freiman, K. R. McKinney, and H. L. Smith. "Slow Crack Growth in Polycrystalline Ceramics." pp 695–676 in** *Fracture Mechanics of Ceramics.* **R. C. Bradt, D.P.H. Hasselman, and F. F. Lange, Editors. Volume 2: Microstructure, Materials, and Applications. Plenum Press, New York. 1974. proceedings of the Symposium on the Fracture Mechanics of Ceramics, held at Pennsylvania State University, University Park, Pennsylvania. July 11–13, 1973.**

It is shown in this paper that slow crack growth is a common phenomenon in ceramics. Examples of this behavior in PZT, barium titanate, and several aluminas of various purities and grain sizes are presented and also a relationship between slow crack growth behavior and microstructure. The use of these fracture mechanics data for prediction of strengths is discussed. Finally, discussed is the use of acoustic emission measurements for analyzing crack propagation in ceramics.

507 **L. B. Freund. "The Initial Wave Front Emitted by a Suddenly Extending Crack in an Elastic Solid."** *Journal of Applied Mechanics, Transactions of the ASME,* **JAMCA, Series E, 39(2):601–602. June 1972.**

This note should be read in conjunction with another paper *(Journal of the Mechanics and Physics of Solids,* 20:129-140. 1972) which considers the problem of a semi-infinite crack which begins at time t = 0 to propagate uniformly with speed V. This solution generalizes that given by Baker (*Applied Mechanics Review,* 16:Review 1488. 1963) in that the crack is already sitting in a static stress field when it suddenly begins to move; in Baker's problem the load is switched on at t = 0 and at that instant the crack begins to move. The present note complements the above-mentioned paper by working out in detail the stress release waves radiated out from the crack tip as it begins to move. The motivation is that interest in these waves has recently arisen for possible application of new acoustic emission techniques of material testing and failure prediction. [Abstract 4557. *Applied Mechanics Reviews,* 26(6):723. June 1973].

508 **G. R. Fritsche and B. M. Strauss. "Use of the SEM in Correlating Acoustic Emissions with Prefracture Processes." pp 129–136 in** *Scanning Electron Microscopy/1971.* **O. Johari and I. Corvin, Editors. Part 1. Illinois Institute of Technology Research Institute, Chicago, Illinois. 1971. proceedings of the 4th Annual Scanning Electron Microscope Symposium, Chicago, Illinois, Apr 27–29, 1971.**

The prefracture behavior of 2¼ Cr-1 Mo steel was investigated using acoustic emission signals in conjunction with scanning electron microscopy of fracture surfaces. The objective was to study the fracture characteristics of this material in order to appraise the brittle fracture tendencies of refinery pressure vessels. Acoustic emission signals were monitored during tensile loading of 2-inch tension test specimens using a PZT-5 piezoelectric sensor and signal amplifier. The fracture mechanism was studied subsequent to the mechanical test by examination of portions of the fracture surfaces in the SEM. The fracture behavior was characterized by an initial linear-load displacement curve which was followed by a gradual deviation from linearity prior to failure. The deviation from linearity was accompanied by a threefold increase in the rate of acoustic emissions. The scanning electron micrographs of the fractures indicated that the increase in acoustic emissions coincided with the beginning of slow crack propagation. The abrupt change in acoustic emissions at the initiation of slow crack propagation was thought to be the result of microvoid coalescence as evidenced by an array of small voids which were found surrounding the dimples on the fracture surfaces. This work has added a measure of credibility to acoustic emissions as a fracture testing tool.

509 D. N. Fry, R. C. Kryter, and J. C. Robinson. "Investigation of Boiling Detection in the HFIR Using Out-of-Core Instrumentation." *IEEE Transactions on Nuclear Science,* **IETNA, NS-18 (Part-1):345–350. Feb 1971; paper presented at Institute of Electrical and Electronics Engineers 2nd Nuclear Power Systems Symposium. New York. Nov 4–6, 1970.**

Analysis of neutron fluctuations produced by coolant pressure oscillations and of neutron noise was investigated as two possible boiling detection methods during a boiling experiment in the High Flux Isotope Reactor (HFIR). Neither method detected boiling because the void volume due to boiling was 0.007 cubic inches, which is less than the detection threshold for either method.

510 R. Frydman, R. Pascual, and R. M. Volpi. "Acoustic Emission due to Dislocation and Grain Boundaries." *Scripta Metallurgica,* **SCRMB, 9(11):1267–1270. Nov 1975.**

In the early acoustic-emission studies of metals, Kaiser pointed out that the shear displacement occurring at grain boundaries was the major source of emission in plastically deformed polycrystals. Later work on single crystals showed however that dislocation glide is a source of acoustic emission as well. At present there is little evidence about the relative contributions of both grain boundaries and lattice dislocations, to the acoustic emission observed during plastic deformation of polycrystals. In order to estimate these contributions, a superplastic alloy (Zn-O, 4% Al) was used in the present investigation. Superplastic materials are known to deform in a superplastic or normal mode depending on grain size d, deformation temperature T, and strain rate ϵ. The experimental evidence indicates that superplastic deformation involves mainly grain-boundary sliding, in association with some accomodation mechanism, the contribution of lattice dislocations to the total deformation being small. This is borne out by the rounded appearance of the grains and the almost complete absence of dislocations within the grains after superplastic deformation. On the other hand, the normal mode involves dislocation motion, as seen for instance by the formation of dislocation networks and slip lines. The superplastic alloy was tested in the superplastic and normal regime by changing d, T, and $\dot{\epsilon}$ and the acoustic emission was simultaneously recorded. It is found that superplastic deformation results in little acoustic emission, whereas intense acoustic activity is observed during normal deformation. This is taken as an indication that the main contribution of acoustic emission in polycrystals arises from dislocation motion, and not from grain boundaries. [reprinted by permission of Pergamon Press Ltd].

511 T. Fuji. "Acoustic Emission." *Keikinzoku Yosetsu,* **KEYOB, 10(6):251-258. 1972. (in Japanese).**

Presented is a review of wave forms, measurements, properties, and applications of acoustic emission as a nondestructive testing method.

512 T. Fuji. "Report of Acoustic Emission Study Group Visisted to U.S.A." *Journal of N.D.I.* **(***Hi-Hakai Kensa,* **Japan), HIHKA, 21(4):242-247. Apr 1972. (in Japanese).**

513 T. Fuji and T. Mori. "Analysis and Application of Delayed Cracking by Use of Acoustic Emission." Paper No. 11 in *Proceedings of the U.S.-Japan Joint Symposium on Acoustic Emission.* **Japanese volume. Japan Industrial Planning Association, Kawamura-Building 6F, 3-21-6 Akasaka, Minato-ku, Tokyo, Japan 107. 1972. symposium held in Tokyo, Japan. July 4-6, 1972. (paper in Japanese, abstract available in English).**

The paper discusses the application of acoustic emission for research of two kinds of delayed cracking: (1) Heel cracking of fillet welds in high strength steel (50 kg/mm^2 minimum tensile strength), and (2) transverse cracking of submerged arc-welded butt joints in high strength steel (80 kg/mm^2 minimum tensile strength). Heel cracking is easily reproduced by using an unrestrained T-section test piece. The vertical and horizontal plates are tack-welded together at both ends and then a fillet weld bead is made on one side. Total count and count rate are recorded up to 110 minutes after welding starts with a welding electrode, such as LBM 52 (Kobe Steel Co., for 50 kg/mm^2 tensile strength steel). The cracking ratio was 60%, and count rate reached a peak value at about 20 minutes after welding started and then reduced. Using NC 37 (Kobe Steel Co., for 18 Cr-8 Ni steel), no heel cracking occurred, producing no indication of count rate. In order to verify these results, an experiment was conducted involving 18 T-section specimens of the same material. The vertical plate of each specimen was removed at various times after the welding start, and the degree of weld cracking was investigated. Most cracking occurred between 10 and 60 minutes after welding starts. These results are in good agreement with the initial test evaluation. Transverse cracking sometimes occurs in weld metal of submerged arc welding with high tensile strength steel with 80 kg/mm^2 minimum tensile strength. In order to reproduce transverse cracking, welding flux which has a tendency to cause cracking was used. Using a double-V groove weld-joint test piece, temperature, total count, and count rate were recorded for up to 17 hours after welding was completed. Count rate reached a peak value at about three hours and almost all emission occurred within seven hours after welding was completed.

After radiographing, the test piece was gradually cut off on the weld line. In order to recognize transverse cracking, fluorescent magnetic particle testing was conducted on the cutting surface of weld metal. Test data were also recorded using a V-groove weld-joint test piece with a backing strip. In this case, count rate reached a peak value one hour after welding was completed and then reduced gradually. Almost all emission occurred within nine hours after welding was completed. Transverse cracking was detected by radiography and fluorescent magnetic particle testing of the sectioned weld.

514 **T. Fuji, Y. Otsuka, and K. Ishihara. "Acoustic Emission at the Fatigue Test of Model Spherical Tank." Session 9, pp 44–61 in *The Second Acoustic Emission Symposium.* Japan Industrial Planning Association, Dobashi Building 5F, 3-17-8 Akasaka, Minato-ku, Tokyo, 107 Japan. 1974. proceedings of a symposium held in Tokyo, Japan. Sep 2–4, 1974.**

There are now more than 2000 spherical tanks of high tensile strength steel in Japan and most of them are in Kanagawa Prefecture. When the materials to be contained are propane, ammonia, etc., stress corrosion cracking may occur, but even for the case of oxygen, natural gas, etc., which do not cause cracking, sometimes cracks may be found during in-service inspection of tanks. For the reason mentioned, Industry Safety Section of Kanagawa Prefecture planned an experimental model of a spherical tank to find out the causes of initiation and propagation of cracks, and also to conduct hydraulic fatigue tests. By means of acoustic emission and ultrasonic testing, crack initiation and propagation were observed from artificial notches and welds in test plates which are patched up on spherical tank by welding. Reported will be the example of circumstances of AE at fatigue tests of a model spherical tank.

515 **T. Fuji, Y. Otsuka, and K. Ishihara. "Acoustic Emission at the Fatigue Test of Spherical Tank." *Pressure Engineering, Journal of Japan High Pressure Institute (Kotsu Gasu),* KOGAA, 13(2):65–71. 1975. (in Japanese). [see Reference 514 for English translation].**

516 **T. Fuji and E. Yamamoto. "Simplified Multichannel Acoustic Emission Equipment for the Testing of the Spherical High Pressure Tank." Session 2, pp 67–81 in *The Second Acoustic Emission Symposium.* Japan Industrial Planning Association, Dobashi Building 5F, 3-17-8 Akasaka, Minato-ku, Tokyo, 107 Japan. 1974. proceedings of a symposium held in Tokyo, Japan. Sep 2–4, 1974.**

Rapid progress has been made in Japan in the application of acoustic emission for tesing spherical high-pressure vessels after the first AE symposium which was held in Tokyo in 1972. Regrettably, however, this is still lagging behind other advanced countries in the application of a multichannel acoustic emission equipment designed for testing high-pressure vessels. The multichannel AE equipment employed to test high pressure vessels requires more than ten channels. At present, few companies or research laboratories in Japan possess the multichannel AE equipment. In the application of this equipment, the peripheral equipments are used to indicate the location of AE waves being generated, expanding the AE system and greatly increasing the number of control knobs. Moreover, in actual measurement, qualified engineers for computer system operation are required. To solve this problem, designed and manufactured was a simplified 16-channel acoustic emission system. Such flexibility as is obtainable from the AE equipment incorporating a computer is not available from this system, but it is designed exclusively for AE wave measurement, not requiring complex procedures and thus being rather practicable in application. The equipment is slated to be finally completed in November 1974, and is planned to be applied for measuring the AE waves being generated in spherical high-pressure vessels during the test.

517 **T. Fujimura. "Applicability of the Acoustic Emission Method of the Reactor Pressure Vessel." Paper No. 7 in** ***Proceedings of the U.S.-Japan Joint Symposium on Acoustic Emission.*** **Japanese volume. Japan Industrial Planning Association. Kawamura-Building 6F, 3-21-6 Akasaka, Minato-Ku, Tokyo, Japan 107. symposium held in Tokyo, Japan. July 4-6, 1972. (paper in Japanese, abstract available in English).**

The goal was to apply the acoustic emission method to reactor pressure vessels in the near future. Tensile tests were performed on mild steel to establish basic acoustic emission data. Based on experimental results and material presented in various reports, the following matter is discussed: (1) Experience and experimental results about observation of acoustic emission phenomena by tensile tests, in particular, electric noise problems and the detection system; (2) The reason for applying acoustic emission testing for integrity surveillance of reactor pressure vessels; (3) The Kaiser effect did not hold true in all of the experimental work, which is a requisite to applying acoustic emission testing to reactor pressure vessels; (4) Acoustic emission test results offer no means of estimating crack size and shape; (5) The acoustic emission record indicates the severity of a flaw and incipient failure; and (6) Flaw location in tensile specimens by arrival-time analysis.

518 **G. Funke and Z. Pawlowski. "Acoustic Emission and Fractographic Analysis Applied to the Estimation of Crack Growth during Low-Cycle Fatigue/ Emission acoustique et fractographie applique pour l'estimation de developement des fissures pendant la fatigue oligocyclique." IPPT-PAN, Warszawa, Poland; Paper 3K/10 in** ***Proceedings of the Eighth World Conference on Nondestructive Testing.*** **volume containing Sections 3K2, 3H2, 5A. International Committee for Nondestructive Testing, Secretariat du Cofrend, Institut de Soudure, 32 Boulevard de la Chapelle, 75880 Paris Cedex 18, France. 1976. conference held in Cannes, France. Sep 6-10, 1976. (paper in English).**

Attempts are described to differentiate between incubation and crack propagation stages, as well as to determine the damage cumulation from acoustic emission data.

519 **M. Fuwa, A. R. Bunsell, and B. Harris, "An Evaluation of Acoustic Emssion Techniques Applied to Carbon-Fibre Composites."** *Journal of Physics D: Applied Physics,* **JPAPB, 9(2):353-364. Feb 1, 1976.**

A study has been made of the use of acoustic emission techniques to monitor failure processes within carbon-fibre reinforced plastics and to establish some of the limitations of the technique. The type of signal to be expected from a resonant transducer of the type most widely used is discussed, and simulation studies confirm this analysis. Bundles of different types of fibres have been broken in order to investigate the type of emission obtained when fibres fracture. Sources of extraneous emission arising from the grips and from shear damage at notch tips have also been studied. Acoustic emission detection is shown to be a fruitful source of information about deformation processes in this kind of material provided due consideration is given to its limitations.

520 **M. Fuwa, B. Harris, and A. R. Bunsell. "Acoustic Emission during Cyclic Loading of Carbon-Fibre-Reinforced Plastics."** *Journal of Physics D: Applied Physics,* **JPAPB, 8(13):1460-1471. Sep 11, 1975.**

Acoustic emission detection has been used to study the damage occurring in unidirectional carbon-fibre reinforced plastics (CFRP) during cyclic loading and stress relaxation at high stress levels. The results suggest that true fatigue process does not occur in CFRP and that the damage sustained during cycling is of the same kind as that which occurs in ordinary tensile loading. As a result either of many reversals of load or of holding a sample at constant load, the rate of emission of stress waves associated with fibre fractures diminishes with time until the sample becomes silent. A sample in this state is relatively safe from further damage, and a procedure of proof testing based upon this result is suggested.

G

521 **C. F. Galambos and C. H. McGogney. "Opportunities for NDT of Highway Structures."** ***Materials Evaluation,*** **MAEVA, 33(7):168–175, 180. July 1975; paper presented at the Conference on Emerging Techniques in Nondestructive Field Testing, sponsored by the U.S. Department of Transportation and the American Society for Nondestructive Testing, held at the Transportation Systems Center, Cambridge, Massachusetts. Aug 13–14, 1974.**

Statistical information on the number and kinds of highway bridges existing on the several highway systems in the United States is presented, along with the laws, regulations, and practice of inspection of these bridges. Present applications of nondestructive testing methods are reviewed as they apply to bridge inspection, and the needs and opportunities for other NDT techniques are pointed out. Some of these opportunities are: (1) application of acoustic emission to fatigue crack detection, (2) high frequency vibration analysis for determining the structural integrity of individual members, (3) development of instrumentation for the detection of corrosion-related flaws inside prestressed concrete members and (4) adaption of techniques for the determination of the existing stress in members, joints, and fasteners.

522 **M. Galli. "Stress Wave Emission: A Bibliographical Survey, N.1." Report EUR-5616e. Euratom J.R.C., Technology Division, Ispra, Italy. 1976.**

The report gives an inventory of papers and publications which deal with the theory and practice of stress wave emission (acoustic emission). The bibliography contains 477 references, listed alphabetically by author and covers a period up to the end of 1972.

523 **M. Galli. "Stress Wave Emission: A Bibliographical Survey, N.2." Report EUR-5617e. Euratom J.R.C., Technology Division, Ispra, Italy. 1976.**

This report gives an inventory of 274 papers and publications, listed alphabetically by author, which deal with the theory and practice of acoustic emission. The bibliography covers the period 1973 and 1974. It also contains titles of papers published before 1973 but not contained in the first bibliographical survey, Report EUR-5616e (see Reference 522).

524 **J. H. Ganton and A. R. Milne. "Temperature- and Wind-Dependent Ambient Noise under Midwinter Pack Ice."** ***Journal of the Acoustical Society of America,*** **JASMA, 38(3):406–411. Sep 1965.**

Measurements of ambient noise under shorefast pack ice during winter revealed two distinct noise-producing mechanisms. One mechanism was the production of tensile cracks at the surface of the sea ice by thermal stresses caused by decreasing air temperatures. The resulting ice-cracking sounds produced impulsive noise. The second mechanism was the interaction of the granular ice surface with the wind, which produced Gaussian noise. Separate pressure spectra are presented for underwater noise generated by each mechanism. The spectra and the statistical properties of the underwater noise appeared to be related to the local meteorology and the temperature changes within the ice cover.

525 **C. G. Gardner. "Developmental Methods." Chapter 11, pp 263-269 in "Nondestructive Testing: A Survey." NASA SP-5113. Technology Utilization Office, National Aeronautics and Space Administration, Washington, D.C. 1973. [NTIS or U.S. Government Printing Office, Washington, D.C. 20402 (Stock No. 033-000-00471-1)].**

The principles and applications of nondestructive tests are reviewed. The methods discussed include the use of liquid penetrants, magnetic particles, X-ray radiography, ultrasonic vibrations, eddy currents, and strain sensing. Brief accounts are included of three nondestructive evaluation methods that have been introduced within the past few years: (1) acoustic emission, (2) the use of coherent (laser) light, and (3) ultrasonic holography.

526 **C. G. Gardner. "Nondestructive Evaluation: A Survey of NASA Contributions, Chapter 1 (Introduction), Chapter 11 (Developmental Methods), Cover Page, Acknowledgements, and Contents." NASA CR-129245. Southwest Research Institute, San Antonio, Texas. Oct 1972. [NTIS: N73-12491].**

A survey of nondestructive evaluation (NDE) techology, which is discussed in terms of popular demands for a greater degree of quality, reliability, and safety in industrial products is presented as an overview of the NDE field to serve the needs of middle management. Three NDE methods are presented: acoustic emission, the use of coherent (laser) light, and ultrasonic holography. This is the same text included in Reference 525.

527 **J. E. Garnier, A. S. Rupaal, and J. L. Bates. "Application of Acoustic Emission to Evaluation of Dielectric Breakdown in Insulator Material."** ***American Ceramic Society Bulletin,*** **ACSBA, 55(5):533-534. May 1976.**

Acoustic emissions (AE) are generated by energy released from dynamic processes occurring inside a material. When an electrical field is applied to a material, localized electrical discharges can occur at regions of enhanced electrical stress even though the average electrical field is below that required for complete breakdown. The occurrence and energy dissipated in these events can be difficult to detect by the usual methods of current and voltage monitoring, especially when high voltages are present. However, if a portion of energy released during these discharges is converted into mechanical energy, this would appear as pressure waves (AE) propagating through the sample medium. These pressure waves can be detected by a transducer that converts mechanical waves into voltage pulses which can be detected by standard AE techniques. An analysis of these waves could then provide information on processes involved which destroy the insulating capacity of the material.

528 **B. Gasc and E. G. Tomachevsky. "Surveillance des Reacteurs par les Methodes de Detection d'Emission d'Ondes de Contrainte et de Detection Acoustique (Reactor Surveillance by Stress-Wave Emission Detection and Acoustic Detection)." IAEA-SM/168/D-6. CEA Centre d'Etudes Nucleaires de Saclay, France. 1973. (in French). [NTIS: CEA-CONF-2256]; paper presented at IAEA Symposium on Nuclear Power Plant Control and Instrumentation, Prague, Czechoslovakia. Jan 22-26, 1973; pp 479-494 in** *Nuclear Power Plant Control and Instrumentation 1973.* **International Atomic Energy Agency, Vienna, Austria. 1973.**

529 **A. Gatti, R. L. Mehan, and M. J. Noone. "Development of a Process for Producing Transparent Spinel Bodies." Naval Air Systems Command Contract N00019-71-C-0126, Final Report. General Electric Company, Space Division, Philadelphia, Pennsylvania. Dec 1971. [NTIS: AD-736-980].**

Transparent spinel material has been made and processing variables documented. Present studies involved continued improvement of reproducibility and the characterization of the process and the material it produces and included the substitution of commercially available spinel or alumina raw materials, the kinetics of volatile additive removal, and the documentation of acoustic emissions during mechanical testing from spinel, alumina, and Lucalox alumina.

530 **J. Gaulin (to the Commissariat a l'Energie Atomique, France). "A Pick-up of Acoustic Emission from Test Pieces Subjected to Tensile Stress." British Patent 1.414,537. Nov 19, 1975.**

531 **B. Gauss, R. Matera, G. Piatti, and G. Volta. "Acoustic Emission, Strain Rate and Density of Cracks in Steels." Euratom CCR, Ispra, Italy; paper presented at the Institute of Physics Conference on Acoustic Emission, Imperial College, London, England. Mar 14, 1972. [proceedings were never published].**

Stress-wave emission has been measured on specimens of low carbon, AISI-304L, and AISI-310 steels during deformation under tensile load at various strain rates

(0.02 to 0.6%/min). On AISI-310 steel specimens, the corresponding fractional volume of microcracks was measured. A relationship between N, the total number of bursts, the strain ϵ, the strain rate $\dot{\epsilon}$, and the fractional volume of cracks was searched. The slope of the curve N(ϵ) for low values of strain is steeper in the case of low strain rate, but it reaches sooner an asymptot which is lower than in the case of high strain rate. The same kind of trend is found for N as a function of the fractional volume of microcracks and of the strain rate. The relationships for SWE suggest a rate-process phenomenological model of which a preliminary outline is given.

532 **A. P. Gavin and T. T. Anderson. "Design and Application of Immersible Microphones for LMFBR Acoustic Monitoring."** *Transactions of the American Nuclear Society,* **TANSA, 19:335. 1974.**

Argonne National Laboratory in Illinois has developed a rugged microphone for use as an immersed acoustic monitor in liquid-metal fast breeder reactors (LMFBR). These microphones have been tested extensively in water, in air up to 1200 °F, and in sodium at 1200 °F. They have been successfully installed and employed as acoustic monitors in several operating liquid-metal systems.

533 **A. P. Gavin and T. T. Anderson. "High-Temperature Acoustic Sensors for Boiling Detection."** *IEEE Transactions on Nuclear Science,* **IETNA, NS-18 (Part-1): 340-344. Feb 1971; paper presented at Institute of Electrical and Electronics Engineers 2nd Nuclear Power Systems Symposium. New York. Nov 4-6, 1970.**

A system utilizing immersed acoustic noise detectors to detect boiling is one possible solution to the problem of detecting loss of cooling in a fuel channel of a liquid metal cooled fast-breeder reactor. Two models of a symmetric drive sensor incorporating lithium niobate as the piezoelectric element have been fabricated and tested for acoustic response in water at room temperature and in an air furnace up to 600 °C. Sensitivities of both models were 500 pC/bar, with acoustic response to 100 kHz when immersed in water. The sensors were not adversely affected by exposure in the furnace. Decreasing internal resistance with increasing temperature limits the useful range of present models to about 540 °C.

534 **A. P. Gavin, T. T. Anderson, and J. J. Janicek." Sodium Immersible High Temperature Microphone Design Description." Technical Memorandum ANL-CT-75-30. Argonne National Laboratory, Argonne, Illinois. Feb 1975.**

Argonne National Laboratory has developed a rugged high-temperature microphone for use as a sodium-immersed acoustic monitor in Liquid Metal Fast Breeder Reactors. Microphones of this design have been extensively tested in room temperature water, in air up to 1200 °F, and in sodium up to 1200 °F. They have been successfully installed and employed as acoustic monitors in several operating liquid metal systems. This report describes the design, construction sequence, calibration, and testing of these microphones in a manner such that any or all of the technology can be used by industrial instrument suppliers in the production of similar high-temperature microphones or related devices. The body of the report describes the principle of operation, purpose of various parts, reasons for pertinent design and material requirements, and the construction sequence, with emphasis on steps which are critical to successful operation of the completed microphones. Also described in detail is a conditioning procedure requiring exposure at high temperature with controlled internal-gas composition, which is vital to successful operation at design conditions. Material and fabrication quality assurance requirements essential for production of consistently reliable microphones and for safety of the systems into which the units are to be installed are discussed, emphasizing their importance in production of components whose major use is expected to be in nuclear reactor safety systems.

535 **A. P. Gavin, T. T. Anderson, and W. W. Managan. "Effects of Fast Reactor Irradiation on Lithium Niobate Microphones."** *Transactions of the American Nuclear Society,* **TANSA, 23:432. 1976.**

Evaluation of data obtained during exposure and by postirradiation examination and calibration of six Argonne National Laboratory (Illinois) high-temperature (HT) microphones exposed in the Experimental Breeder Reactor II (EBR-II) core for six months confirms that integrated neutron exposure up to 3×10^{22} nvt at temperatures from 700 to 1130 °F had no major effect on microphone performance. The radiation resistance is attributed to annealing of radiation damage at the high operating temperature.

536 **R. L. Geisler. "Acoustic Emission System for Solid Propellant Burn Rate Measurements." U.S. Patent Application 427,159. Department of the Air Force, Washington, D.C. Dec 21, 1973. [NTIS: AD-164-700].**

Covers a solid-strand burn rate analyzer for measuring solid propellant burn rates, sensing acoustic signals created by thermal fracture or deflagration of the solid oxidizer utilizing a piezoelectric pick-up transducer mounted on the propellant case, a system of signal amplification, a wave-form distribution converter, and an X–Y strip chart for recording test results.

537 **W. W. Gerberich. "A Discussion of Slow Crack Growth Associated with Plane-Strain Instability."** *ASM Transactions Quarterly,* **ASMQA, 59(4):899-908. Dec 1966.**

A review of the critical aspects of fracture pertaining to thick sections is made. Detailed discussion of subcritical crack growth prior to plane-strain instability and slow crack growth occurring after *pop-in* is presented. Several examples of the different modes of cracking in titanium and steel alloys are discussed in terms of the applicability of current fracture mechanics concepts. Specifically, it is shown that stresses less than half that required by plane-strain instability may promote microfracture which may be associated with brittle second phases or time-dependent embrittlement mechanisms, such as hydrogen embrittlement or stress corrosion. It is also shown that if the thickness and toughness conditions are appropriate, a plane-strain *pop-in* crack may be arrested once it has traversed the material thickness and that further growth occurs in a plane stress mode. [reprinted by permission from *ASM Transactions Quarterly*, copyright by American Society for Metals, 1966].

538 **W. W. Gerberich. "On Continuum Models of Ductile Fracture." UCRL-19086. Lawrence Radiation Laboratory, Berkeley, California. Oct 1969. [NTIS]; *Journal of Materials Science*, JMTSA, 5:283-294. Apr 1970.**

Several fracture criteria are reviewed with respect to ductile fracture. It is suggested that both critical crack-tip displacement and critical fracture strain criteria may describe the fracture of a ductile second-phase rod in a ductile matrix. As a first approximation, this is experimentally verified by observations of ductile stainless steel fibers fracturing in an age-hardened aluminum matrix. For 0.05, 0.10, and 0.20 volume fraction composites, the average fracture strains are calculated to be 1.15 as compared to a measured average of 0.93, while the average critical crack-tip displacement is calculated to be 0.0204 inches as compared to an *observed* average of 0.0157 inches. The statistical variation in the fracture strain was not sufficiently small to allow any choice between those proposed criteria. In fact, both the experimental and theoretical evidence point to the equivalency of these criteria.

539 **W. W. Gerberich and D. G. Atteridge. "Analysis of the Ductile-Brittle Transition Temperature in Fe-Binary Alloys." COO-2212-1. Annual Technical Progress Report I. University of Minnesota, Department of Chemical Engineering and Materials Science, Minneapolis, Minnesota. Mar 1973. [NTIS].**

The overall thrust of the program is to understand the fracture and flow mechanisms in Fe-X systems from a fundamental viewpoint. Within this scope, experimental techniques involving the recording of elastic waves emanating from a group of dislocations (acoustic emission) or from a moving crack front (stress waves) are being developed. The first year's program was addressed largely to (1) setting up test facilities, (2) preparing material and test specimens, and (3) exploring various approaches to the analysis of discontinuous plasticity and fracture events. In addition, several adjunct studies of a combined theoretical and experimental nature, such as a hydrogen embrittlement analysis, were initiated. In the report, six sections are covered. The first three sections, I, II, III, are devoted entirely to the present program with the aim of using the acoustic emission technique as an experimental tool in understanding: (1) dynamical dislocation processes; and (2) discontinuous crack growth processes. Sections IV and V are more theoretically oriented in that a basic understanding of embrittlement processes either due to hydrogen phenomena (IV) or cohesive bond energy considerations (V) is desired. Finally, section VI presents the current experimental status of the program within the context of the original proposal.

Within the first year, the findings showed: (1) studies of the Portevin Le-Chatelier effect with the acoustic emission technique may lead to establishment of quantitative parameters associated with dislocation dynamics. This could prove invaluable when applying the technique to the study of such processes as low temperature solute softening in bcc alloys; (2) the quantitative jump distance of a microcrack process, such as nonpropagating cleavage microcracks, may be theoretically analyzed in terms of a compliance analysis; (3) an experimental comparison involving relationship (2), elastic stress waves and the load drops occurring during crack growth processes with much greater accuracy than has been previously possible; (4) the ductile fracture process can be described in terms of dislocation dynamics; and (5) the total cohesive bond energy is not the major consideration in establishing the ductile-brittle transition of transition-metal binary alloy systems.

540 W. W. Gerberich and D. G. Atteridge. "The Study of Dislocation Dynamics Associated with Ductile–Brittle Transition and Hydrogen Embrittlement Phenomena Observed in BCC Metals." COO–2212–2. Annual Technical Progress Report II. University of Minnesota, Department of Chemical Engineering and Materials Science, Minneapolis, Minnesota. Mar 1974. [NTIS].

Laboratory setups for acoustic emission and stress-wave emission capabilities allow the recording of dislocation and cracking events. Stress relaxation, dynamic loading, load drop, crack-opening displacement, and other detailed flow and fracture tests can be run. Environmental setups include low-temperature and gaseous environment capabilities. Theoretical and experimental analyses were carried out of both stress-intensity threshold and crack growth rate which are conditions for hydrogen-assisted cracking. Several test programs have evolved which should resolve the role of both the stress field and the long-range diffusion of hydrogen in the embrittlement process. The ability to associate a single stress-wave emission with a single discontinuous fracture event of finite area was developed, enabling *in situ* studies of such growth processes. For studying the ductile-brittle transition, all test specimens were prepared and readied for testing except for pre-fatigue cracking of some fracture toughness specimens. Results are given on the ductile-brittle fracture transition in Fe-base alloys.

541 W. W. Gerberich and D. G. Atteridge. "The Study of Dislocation Dynamics Associated with Ductile–Brittle Transition and Hydrogen Embrittlement Phenomena Observed in BCC Metals." COO–2212–3. Annual Technical Progress Report III. University of Minnesota, Department of Chemical

Engineering and Materials Science, Minneapolis, Minnesota. Mar 1975. [NTIS].

Program consisted of three main objectives associated with the different types of microscopic fracture modes: (1) Ductile Fracture–in which the need was to understand the microscopic hold coalescence process leading to crack initiation and growth; (2) Plastically-Induced Cleavage–in which the aim was to tie together fracture mechanics, dislocation dynamics, solute softening, and cohesive energy concepts into a comprehensive ductile-brittle transition theory; and (3) Hydrogen-Assisted Intergranular Fracture–in order to verify what constitutes a viable embrittlement model, investigations were made of both stress field effects, as well as concentration and microstructural effects. Theoretical concepts of both crack growth kinetics and threshold conditions were developed.

Three significant developments in the study of flow and fracture pertain to dislocation dynamics, brittle fracture, and hydrogen-assisted crack growth: (1) Dislocation Dynamics–Stress relaxation at small strains promises to connect up change of strain rate data at large strains to acoustic emission phenomena at low strains. It was found that logarithmic differentiation of the acoustic emission representation of mobile dislocation density gives an estimate of strain rate sensitivity. This compares favorably with strain-rate sensitivity exponents taken from change of strain rate tests. The ramifications of this to the study of dislocation dynamics could be considerable; (2) Brittle Fracture–Ductile-brittle transition analysis points towards a relatively simple criterion for predicting behavior using strain-rate sensitivity, solute-softening, cohesive-energy, and fracture mechanics concepts. Initial evaluation of impact and yield data from Fe, Fe-Ni, and Fe-Si alloys indicates that a single parameter controls cleavage. Thus far, this parameter separates impact data into the brittle and ductile fracture regimes; (3) Hydrogen Embrittlement–Tests in both cathodically-charged and gaseous-hydrogen environments are providing evidence for a unified concept of hydrogen induced slow crack growth. It is shown that increased grain size slows kinetics with growth rate. This tentatively supports theories based upon relatively long-range diffusion of hydrogen to prior austenite grain boundaries. Additional evaluation of cathodically-charged specimens has demonstrated that growth rate can decrease with increasing stress intensity during Stage II growth. This is explained in terms of plastic deformation at the crack tip. Crack-tip blunting can alter both the severity of the plastic constraint and the position of maximum triaxiality. This effect must be incorporated into all embrittlement theories which have stress-induced diffusion as a major component.

542 **W. W. Gerberich, D. G. Atteridge, and J. F. Lessar. "An Acoustic Emission Investigation of Microscopic Ductile Fracture."** *Metallurgical Transactions A*, **MTTAB, 6A(4):797–801. Apr 1975.**

A high-strength 4340 steel fracture-toughness specimen was heat treated to give a ductile-rupture type of slow crack growth under rising load. For evaluation of the step-wise growth process, the specimen was instrumented with acoustic stress wave emission (SWE) detection equipment. The resulting crack area swept out by the advancing crack was correlated to the magnitude and number of the acoustic emission pulses. A crack growth model was developed which accounts for the direct relationship between crack area swept out and the sum of the individual SWE amplitudes, and for the experimentally observed bimodal distribution of the SWE amplitudes. The model postulates that slow crack growth takes place in a stepwise mechanism. This involves a repeated two-step process where the first step is the formation of a multitude of individual thumbnail cracks and the second step is the simultaneous interconnection of these thumbnail cracks to form a new continuous crack front. [Copyright 1975 by the American Society for Metals and the Metallurgical Society of AIME].

543 W. W. Gerberich and G. S. Baker. "Toughness of Two-Phase 6 Al-4 V Titanium Microstructures." pp 80-99 in *Applications Related Phenomena in Titanium Alloys, ASTM STP 432.* **American Society for Testing and Materials, Philadelphia, Pennsylvania. 1968.**

A duplex heat treatment which provides different shapes, sizes, and volume fractions of alpha platelets in 6 Al-4 V titanium is presented. It is shown that a 40 percent increase in fracture toughness may be attained without any significant decrease in yield or ultimate strength. With respect to toughness, critical aspects of both the interstitial oxygen content and the shape, size, and spacing of the alpha phase are discussed. Metallographic and fractographic evidence supports the contention that the second-phase alpha in platelet form acts as a crack arrestor and causes a deviation in the crack path, resulting in additional energy absorption. A tentative hypothesis for fracture in two-phase microstructures is presented based upon the theoretical crack-tip displacement. Preliminary evidence indicated that the critical toughness, size, and spacing of the second phase may all be related to the crack-tip displacement concept. [reprinted by permission of the American Society for Testing and Materials, Copyright].

544 W. W. Gerberich, G. S. Baker, W. G. Reuter, P. P. Crimmins, and C. E. Hartbower. "Evaluation of Selected Titanium Alloys as a Function of Environment and Heat Treatment." Final Report on U.S. Navy Contract N600(61533)-65927. Aerojet-General Corporation, Sacramento, California. Aug 1967.

Fracture toughness and stress-corrosion-cracking studies were performed using a 1-inch thick plate of low interstitial (0.07 wt % O_2) 6 Al-4 V titanium. Evaluations were conducted in both the annealed condition and a duplex heat-treated condition which produces an elongated-platelet alpha in a beta matrix. Stress-wave emission analysis of the discontinuous crack-growth process in a rising load and environmentally induced slow-crack growth was also performed for both the 6 Al-4 V and 7 Al-2 Cb-1 Ta titanium alloys. These studies showed that the duplex heat treatment slightly decreased the strength properties in comparison to the annealed. However, the duplex treatment increased the apparent plane-strain fracture toughness (K_{Ic}) by about 45% and increased the threshold stress intensity for stress-corrosion cracking (K_{Iscc}) by about 70%. Stress-wave emission was detected in both nonenvironmental, rising-load tests, and in sustained-load tests in a simulated sea-water environment. These data demonstrated that stress-corrosion cracking was at least partially by a jump process and not entirely a dissolution mechanism.

545 **W. W. Gerberich and C. E. Hartbower. "Characterization of Fatigue-Crack Growth by Stress-Wave Emission." Aerojet-General Corporation, Sacramento, California. Nov 30, 1965.**

Precracked, notched-bend specimens of grade-200 18% nickel maraging steel have been subjected to pure-bend cyclic loading and the resulting fatigue-crack growth monitored by means of accelerometer-type transducers. The recorded stress-wave emission confirmed the discontinuous nature of fatigue-crack growth and permitted observations on the effect of stress-intensity factor, cycling rate, and overload.

546 **W. W. Gerberich and C. E. Hartbower. "Feasibility Study for Measuring Fatigue-Crack Growth Rate in Welded HY-80 Steel Using Stress-Wave Emission." Final Report on Contract N600(167)-64934(X) (FBM). Aerojet-General Corporation, Sacramento, California. July 15, 1966. [DDC: AD-885-455L].**

The results of a study conducted to determine the feasibility of monitoring fatigue crack growth in HY-80 weld and heat-affected-zone metal utilizing a stress-wave detection system are presented. A semi-elliptical PTC flaw in a beam subjected to pure-bending fatigue was grown in three increments to a depth of approximately one-half the thickness of the 0.70-inch thick by 2.0-inch wide samples. Incremental fatigue crack growth was detected and monitored by means of the stress waves emitted from the crack, proving the feasibility of the stress wave detection system. A qualitative relationship between SWE and fatigue crack growth indicated that increasing crack growth increments gave increasing numbers of SWE. Expressions for the stress-intensity factors at the surface and bottom of the flaw were derived, allowing quantitative comparison of the crack growth rates of the surface and depth dimensions of the flaw. The data further suggested that there may be a quantitative relationship between the summation of SWE and amplitudes and an increment of crack extension for a given number of cycles. Recommendations are also contained herein for follow-on studies to establish quantitative relationships between SWE and cracking in submarine-hull materials under variable test and service conditions so that a technique may be developed for detecting flaw initiation and growth in submarine hull components during fabrication, hydrotest, and service.

547 **W. W. Gerberich and C. E. Hartbower. "Monitoring Crack Growth of Hydrogen Embrittlement and Stress Corrosion Cracking by Acoustic Emission." pp 420-438 in** ***Proceedings of Conference: Fundamental Aspects of Stress Corrosion Cracking.*** **published by National Association of Corrosion Engineers, Houston,**

Texas. 1969; paper presented at Conference on Fundamental Aspects of Stress Corrosion Cracking, Ohio State University, Columbus, Ohio. Sep 11-15, 1967.

A detailed investigation was made of the slow crack growth process for D6aC high strength steel tested in the hydrogenated condition, in distilled water and in combination. The experimental technique involved a combination of stress wave monitoring to determine the number of and time intervals between crack steps; a crack opening displacement gauge to incrementally measure the crack growth rate; and electron fractography to interpret the morphological nature of the growth process. It was found that crack growth occurred as discontinuous steps for all test conditions. Also, the fractographic evidence showed all test conditions to result in similar failure modes with both intergranular and cleavage areas prominent. From the stress wave analysis, it was determined that the crack growth process could be approximated by an average jump being related to the theoretical crack tip displacement. The time intervals between these steps could be approximately related to hydrogen diffusion as the controlling mechanism. Phenomenological analyses of the diffusion-controlling mechanisms for both hydrogen embrittlement and delayed failure in distilled water were made. These indicated that a difference in the boundry conditions for hydrogen diffusion could account for the observed differences in the time-dependent cracking. The experimental data exhibited a synergistic effect when the steel was both hydrogenated and subjected to the distilled water environment. The fact that superposition of the individual theoretical treatments also indicated a synergistic effect gives added significance to the hypothesis that hydrogen diffusion is the controlling mechanism in both types of embrittlement.

548 W. W. Gerberich and C. E. Hartbower. "Some Observations on Stress Wave Emission as a Measure of Crack Growth." *International Journal of Fracture Mechanics,* **IJFMA, 3(3):185-192. Sep 1967.**

A technique using a tape recorder and oscilloscope for measuring stress waves emitted during slow crack growth is briefly described. The number and size of the stress waves generated bear a unique relationship to the amount of crack growth. From elasticity theory, a semi-empirical relationship is developed. This relationship is valid for about 100 data points obtained from crack growth in high strength Al, Ti, and steel alloys.

549 W. W. Gerberich, C. E. Hartbower, and P. P. Crimmins. "Spontaneous Strain Aging as a Mechanism of Slow Crack Growth." *Welding Journal,* **WEJUA, 47(10):433s-443s. Oct 1968; paper presented at the American Welding Society 49th Annual Conference, Chicago, Illinois. Apr 1-5, 1968.**

The paper is a sequel to an earlier paper describing a new technique for monitoring crack growth, step-by-step, as induced by hydrogen embrittlement, stress corrosion, and fatigue. Here, spontaneous strain aging is discussed as a less-known mechanism of slow crack growth in steel. Single-edge-notch tensile specimens were exposed to moderately elevated temperature (400 to 500 °F) under sustained load. Compared to hydrogen embrittlement and stress-corrosion cracking, the embrittlement at elevated temperature was attained by relatively few movements of the crack front. Light and electron fractography confirmed that the large stress waves occurring during the periods of sustained load were associated with gross movement of the crack front. At the start of the holding period the time between stress waves (Δt_S) was short, with Δt_S increasing with time at load until finally a steady-state condition was reached where the secondary incubation time was relatively constant. After the D6aC heat treatment which effectively solutionized the carbides (1750 °F for 0.5 hours), the steady-state secondary incubation time was approximately 80 seconds as compared to 1650 seconds in the heat-treat condition with massive carbides (1550 °F for 0.25 hours). The crack growth rate also varied by a factor of 20 for the two heat-treat conditions. It thus appears that the embrittlement is a function of the amount of carbon in solid solution; with an effective solutionizing heat treatment, the kinetics of both secondary-incubation time and crack-growth rate suggested the embrittling mechanism to be carbide precipitation. Consistent with other phases of the research, a relationship was found between crack growth rate as measured by a crack opening displacement gage and rate of stress-wave emission in the elevated-temperature testing.

550 **W. W. Gerberich and Y. Katz. "Measurement of Sonic Emissions from Environmentally Induced Cracks."** *Journal of the Electrochemical Society,* **JESOA, 116(8):286C (Abstract No. 61). Aug 1969. [abstract only]; Abstract No. 61, pp 175–177 in "The Electrochemical Society, Inc. Fall Meeting, Detroit, Michigan, October 5–9, 1969: Extended Abstracts of Battery Division." The Electrochemical Society, New York. 1969; paper presented at the Fall 136th Meeting of the Electrochemical Society, Inc., Detroit, Michigan. Oct 5–9, 1969.**

Sonic emissions associated with environmentally induced discontinuous crack motion are discussed. For the environment-metal combinations studied, it is shown that stress-corrosion-cracking is a discontinuous process and that the number of stress waves or counts per unit time is related to the crack growth rate. From these observations, it is hypothesized that many stress-corrosion-cracking processes result in an apparent activation energy which may be made up of a chemically related term and a mechanically related term.

551 **W. W. Gerberich and W. G. Reuter. "Theoretical Model of Ductile Fracture Instability Based on Stress-Wave Emission." Office of Naval Research Contract N00014–66–C–0340. Final Report. Aerojet–General Corporation, Sacramento, California. Feb 1969. [NTIS: AD–684–137].**

The report discusses a postulated, general interpretation of fracture in terms of how dislocation mobility affects the local fracture stress. However, it is difficult to correlate this concept directly to the slow crack growth process which occurs in many complex materials. Also, there has been little experimental evidence of dislocation dynamics and discontinuous crack motion in complex materials which would allow such a model to be established. This program was initiated to study these characteristics in an effort to establish such a fracture criterion.

552 **J. H. Gieske. "An NDT Investigation of the Behavior of Acoustic Emission from a Brazed Metal to Ceramic Bond." SC-DC-713903 and SC-R-722793. Sandia Laboratories, Albuquerque, New Mexico. 1971. [NTIS: SC-DC-713903]; Paper-643 presented at the 26th Annual Instrument Society of America Conference, Chicago, Illinois. Oct 4-7, 1971; pp 643-1 to 643-6 in** ***Advances in Instrumentation,*** **Volume 26, Part 2. Instrument Society of America, Pittsburgh, Pennsylvania. 1971;** ***ISA Transactions,*** **ISATA, 11(1):24-30. 1972.**

The bond under investigation was made by brazing the metal electrodes of a voltage insulator tube onto the ends of a ceramic cylinder. A nondestructive testing method for evaluating the bond between metal and ceramic was sought in order to detect poor bonds. The behavior of acoustic emission from the bond as it is put under stress was used as the nondestructive technique for this purpose. A significant difference in the acoustic emission behavior of different bonds was found. Both large-amplitude bursts and small-amplitude bursts were exhibited with the greatest total count of 160. The larger bursts were characteristic of a large area of failure, while the small bursts represented minor failures which caused no significant overall bond weakening. Some bonds exhibited small acoustic emission bursts until catastrophic bond failure. In most cases, such bonds also had the highest ultimate strengths.

553 **K. E. Gilchrist and D. Wells. "Acoustic Emission from Graphite under Stress." TRG Report 1505(C). United Kingdom Atomic Energy Authority, Reactor Group, Risley, Warrington, Lancashire, England. Aug 1967;** ***Carbon,*** **CRBNA, 7(5):627-631. 1969.**

Graphite samples have been tested for acoustic emission under stress. In tests on isotropic graphite, no acoustic emission was detected until the compressive stress had almost reached the fracture stress. Two tests on Pile Grade A anisotropic graphite revealed acoustic emission long before the final fracture of the specimen. Stress cycling of a Pile Grade-A graphite specimen to increasing stress levels showed that on each cycle, emission only occurred after the maximum stress on the previous cycle had been exceeded. It was observed that the Poisson's ratio of anisotropic Pile Grade-A graphite increased with compressive stress, whereas in the isotropic graphite it was essentially independent of stress.

554 **P. P. Gillis. "Dislocation Mechanisms as Possible Sources of Acoustic Emission."** ***Materials Research and Standards,*** **MTRSA, 11(3):11-13. Mar 1971.**

The generation of packets of sound waves in solid bodies is discussed, and some possible sources of acoustic emissions in crystalline solids are enumerated. Sources involving dislocation motion are considered to be generally important and are discussed qualitatively.

555 **P. P. Gillis. "Dislocation Motions and Acoustic Emissions." pp 20-29 in** ***Acoustic Emission, ASTM STP 505.*** **American Society for Testing and Materials,**

Philadelphia, Pennsylvania. 1972. proceedings of symposium held in Bal Harbour, Florida. Dec 7-8, 1971.

The possible role of dislocations in producing acoustic emissions is discussed from a kinetic point of view. A general theory of the acoustic power of dislocation motions is developed. For deformations that tend towards homogeneity and are dependent on dislocation motions, the implications of this theory are qualitatively described and shown to be corroborated by selected experimental results.

556 **P. P. Gillis and M. A. Hamstad. "Some Fundamental Aspects of the Theory of Acoustic Emission."** *Materials Science and Engineering,* **MSCEA, 14(2):103-108. May 1974.**

A dislocation process that seems likely to be of importance in the generation of acoustic emissions is described. A simple mathematical theory, which describes this process, is developed so that order of magnitude estimates can be made. These estimates indicate that acoustic emission events may be associated with only a small fraction of the total plastic strain. Hence, it may be difficult to correlate acoustic emission with parameters related to total strain.

557 **J. M. Giraud. "Tentative d'Application de l'Emission Acoustique au Controle des Collages (Attempt to Apply Acoustic Emission for Testing Bonds)." pp 173-187 in** *Journees d'Etudes sur l'Emission Acoustique.* **P. F. Gobin, R. Goutte, and C. Guillaud, Organizers. Institut National des Sciences Appliquees de Lyon, Avenue Albert Einstein, 69621 Villeurbanne, Lyon, France. Mar 1975. proceedings of a conference held in Lyon, France. Mar 17, 1975. (paper in French).**

558 **A. G. Glover, J. Holt, and J. A. Williams. "The Examination of a Mild Steel Pressure Vessel under Creep Loading Using Acoustic Emission." pp 118-132 in** *Schallemission: Anwendung bei der Untersuchung, Prufung und Uberwachung metallischer Werkstoffe.* **Deutsche Gesellschaft fur Metallkunde eV, 637 Oberursel/Taunus bei Frankfurt/Main, Adenauerallee 21, Germany. 1974. proceedings of symposium held in Munich, German. Apr 18-19, 1974.**

The use of acoustic emission techniques for high temperature applications is outlined. The monitoring of a mild steel header operating at 565 °C and 2.52 hbar using stress wave emission is described in detail. The possible sources of emission are identified, and it is suggested that the more important ones arise from oxide cracking and crack growth.

559 **V. N. Glygalo, V. V. Kirsanov, L. S. Kravtsova, and A. G. Leshchinskii. "Acoustic Emission in the Deformation of Zirconium."** *Soviet Journal of Nondestructive Testing,* **SJNTA, 11(3):392–393. May–June 1975; translated from** *Defektoskopiya,* **DEFKA, (3):140–142. May–June 1975. (in Russian).**

The relationship between the degree of deformation of Zr and the acoustic emission from the surface and interior is considered. A close relationship is established between the structural state and the acoustic emission. In ductile Zr samples, the intensity of the acoustic emission varies smoothly with degree of deformation; in material made brittle by heat treatment or nuclear irradiation, there is a peaked form of emission-deformation characteristics may be used as a criterion, for instance, in revealing irradiation damage suffered by Zr reactor parts. [Abstract 22-0162. *Metals Abstracts,* Vol. 9, 1976].

560 **P. Gobin and D. Rouby. "Bases Physiques de l'Emission Acoustique (Physical Basis of Acoustic Emission)." pp 12–35 in** *Journees d'Etudes sur l'Emission Acoustique.* **P. F. Gobin, P. Goutte, and C. Guillaud, Organizers. Institut National des Sciences Appliquees de Lyon, Avenue Albert Einstein, 69621 Villeurbanne, Lyon, France. Mar 1975. proceedings of a conference held in Lyon, France. Mar 17, 1975. (paper in French).**

It is shown that the acoustic emission associated with plastic deformation in crystalline materials relates to the existence of fast local deformation phenomena. A simple tensile test does not account for these phenomena. Various mechanisms of discontinuous deformations involving irreversible deformation motions are envisaged. The analysis of the emission spectrum can be used to measure the instantaneous velocity of the dislocation. The acoustic emission was also studied in cracked materials, composite materials, and during martensitic transformation.

561 **D. J. Goddard, J. Holt, and I. G. Palmer. "Sample Ringdown in Acoustic Emission Tests." Laboratory Note No. RD/L/N 156/73. Central Electricity Research Laboratories, Leatherhead, Surrey, England. June 25, 1973.**

The problem of sample ringdown in acoustic emission tests and its effect on the data obtained are discussed. The use of electrical measurements in the investigation of the effect is analyzed and illustrated. Finally the results of an actual test on a pressure vessel steel are used to show that the combination of signal amplitude distribution and sample-controlled ringdown waveform are consistent with the observed dependence of the emission count on system gain.

562 **D. G. Goddard and J. A. Williams. "The Application of Acoustic Emission Techniques to Liquid Metal Embrittlement." Report RD/L/N50/71. Central Electricity Research Laboratories, Leatherhead, Surrey England. 1971.**

Report was published under the title "The Embrittlement of Annealed and Cold Worked Al-20% Zn in Liquid Gallium under Zero Extended Stress." See Reference 563.

563 **D. J. Goddard and J. A. Williams. "The Embrittlement of Annealed and Cold-Worked Al–20% Zn in Liquid Gallium under Zero External Stress."** *Journal of the Institute of Metals,* **JIMEA, 99:323–324. Oct 1971.**

It is well-known and documented that the presence of a liquid metal on a solid polycrystalline material can cause embrittlement. Several mechanisms have been suggested involving the penetration of liquid metal along a grain boundary and the effect on this of applied stress. One experimental characteristic of such embrittlement is the variation in crack-growth rates between material with and without residual stress. One of the most efficient embrittlement couples is the gallium/aluminum (or aluminum alloy) system. In this instance, failure occurs at high velocities and the details of the cracking process in the system have not been extensively studied for that reason. In such cases, acoustic emission techniques offer advantages because of the rapid response achievable and this note reports an investigation of the emissions from strained and annealed Al-Zn alloy after application of gallium under zero external stress.

564 **W. E. Goff. "Aircraft Production: Non-Destructive Testing–1."** *Flight International,* **FLIND, 105(3382):14. Jan 3, 1974.**

565 **L. W. Gold. "The Cracking Activity in Ice during Creep."** *Canadian Journal of Physics,* **CJPHA, 38(9):1137–1148. Sep 1960.**

Observations were made at one temperature on the cracks that form in ice during creep under constant compressive load. The ice had a hexagonal symmetry with respect to the grain boundaries. The load was applied perpendicular to the long axis of the grains. A piezoelectric crystal was used to detect the cracking activity in the ice. Two stages of cracking were observed. The first occurred during the transient period of the creep, and the plane of these cracks tends to be parallel to the grain boundaries and to the direction of the stress. The rate at which these cracks formed decreased markedly as the creep rate approached a constant value. Above a certain stress, the creep rate continuously increased with time. Under this condition the second stage of cracking was observed. These cracks tend to be more irregular in direction and to occur in planes that are at 45° to the applied stress. [reprinted by permission of the National Research Council of Canada].

566 **S. Golinelli. "A Minicomputer in an Acoustic Emission Source Location System."** *Energia Nucleare,* **ENNLA, 22(12):642–646. Dec 1975.**

The article presents the computer system of the acoustic emission instrumentation developed at CISE and described in a previous article (see E.N., May 1975, pp. 253-259). An on-line minicomputer is used to process the acoustic emission data and to provide the actual source locations. The data relevant to the most active sources are automatically selected by content addressable memory features, and used by the location algorithm. In particular, an analysis is made of the content addressable memory lay-out and its statistical behavior is presented; finally, the error propagation effect of the hyperbolic triangulation algorithm is described.

567 M. J. Golis, R. P. Meister, J. L. Crowe, and G. J. Posakony. "Investigation of Techniques to Nondestructively Test Reinforced Plastic Composite Pipe." U.S. Army Mobility Equipment Research and Development Center Contract No. DAAK02-74-C-0008. Interim Report. Battelle Columbus Laboratories, Columbus, Ohio. Dec 31, 1973. [NTIS: AD-774-294].

The report covers the work done during Phase I of a comprehensive investigation of methods to inspect reinforced plastic composite pipe (RPCP). It discusses the fundamental principles of ten potential techniques of pipe inspection, fabrication installation, and the results obtained during selected experiments designed to verify the performance characteristics of each respective method. The methods considered likely candidates for the inspection task included optical viewing, ultrasonics, radiometry, and microwave monitoring. Other methods considered during the study included holography, acoustic emission, thermal inspection, and radiography. The flaws considered most detrimental to RPC pipe were undercure, resin starved and/or resin rich areas, delaminations, unbonded inclusions, and folds in the prepreg tape. Each of the test techniques was evaluated, based upon its ability to detect each of these conditions as well as its ease of implementation into a mobile van from which the pipe would be fabricated in the field. Using a weighted averaging technique, it was found that the ultrasonic method and the optical monitoring are the best for finding RPC pipe flaws. Both liquid-coupled and air-coupled ultrasonics were found sensitive to the critical condition of undercure. When transparent liner material is used in the pipe, optical monitors are also capable of detecting this condition. Additional work is recommended for the comprehensive use of these techniques on typical pipe samples. Work plans include the recording of C scans using the ultrasonic techniques and further verification of the capabilities of the optical method.

568 R. E. Goodman. "Subaudible Noise during Compression of Rock." ***Bulletin of the Geological Society of America,*** **BUGMA, 74(4):487-490. Apr 1963.**

Acoustical monitoring of rock specimens during cyclic compression tests has allowed recognition of similar patterns of behavior for dissimilar types of rock. During first loading of sandstone and quartz diorite cylinders, oscillations in the audio-frequency range emanate from the specimens throughout the test. Upon repetition of loading, the number of sonic disturbances decreases. However up to 60 percent of the initial high rate of sonic activity during compression is recovered by the rock after a period of rest. After several cycles of loading, the attainment of particular levels of stress is consistently accompanied by sonic disturbances. [reprinted from the *Bulletin of the Geological Society of America*].

569 **R. E. Goodman and W. Blake. "Rock Noise in Landslides and Slope Failures."** *Highway Research Record,* **HIRRA, 119:50–60. 1966.**

When a rock specimen is placed under load in the laboratory, acoustic disturbances, known as subaudible rock noise (SARN) are emitted. With a sensitive pickup on the specimen and a high grain amplifier, these noises can be detected and, in fact, a number of people have studied rock noises in the laboratory. An application of this phenomenon is the detection of noises from stressed rock around tunnels and mines as a safety measure in underground work. With the support of the California Department of Highways, the authors have investigated the possible application of rock noise monitoring to landslide and slope stability problems, and discuss findings. After constructing and field testing a suitable rock noise detector, a number of recently active landslides and highway cut slope failures in Northern California were monitored to determine whether or not slides emit detectable rock noises. A variety of types and conditions of material was represented in the study, including sheared shale, soft sandstone and claystone, serpentine, peridotite, gabbro, metamorphic rocks, disintegrated granite, volcanic flows, interflow zones of volcanic ash and stream sediments, and ancient slide debris. At each rock noise-monitoring location a preliminary subjective estimate of the state of activity of the slide was made. *Stable* denotes cuts thought to have been in a safe condition at all times; *active* denotes cuts in which there were recent slide scarps or other diagnostic forms of microrelief, groundwater flowing at springs or drains, or recent tension cracks at the surface; *inactive* denotes cuts in which there had obviously once been movements and in which movements might recur, but which appeared to be in a safe state at the time of observation. A definite correlation existed between the estimated state of activity and the rock noise rate in these slide areas. Most of the areas of high noise rate experienced later movements, whereas none of the quiet areas did.

570 **R. Gopal. "Acoustic Emission Monitoring System Evaluation of Prairie Island." Paper C119/74, pp 162–167 in** *Periodic Inspection of Pressurized Components.* **Conference Publication 8, 1974. published by Mechanical Engineering Publications Limited for the Institution of Mechanical Engineers, London, England. 1975. conference held in London, England. June 4–6, 1974.**

The paper describes recent advances in instrument systems for detection, location, and display of acoustic emission sources in nuclear power plants and discusses results of measurements and system performance during hydrostatic testing, hot functional tests, and precritical operations at Prairie Island nuclear power plant.

571 **R. Gopal. "Acoustic Monitoring Systems Evaluation in Nuclear Power Plants."** *Transactions of the American Nuclear Society,* **TANSA, 19**

(Supplement 2):16–17. 1974; paper presented at the Conference on Nondestructive Testing in the Nuclear Power Industry, held at Ohio State University, Columbus, Ohio. Sep 23–25. 1974.

572 **R. Gopal. "Acoustic Monitoring Systems to Assure Integrity of Nuclear Plants." pp200–220 in** *Monitoring Structural Integrity by Acoustic Emission, ASTM STP 571.* **Edited by J. C. Spanner and J. W. McElroy. American Society for Testing and Materials, Philadelphia, Pennsylvania. 1975. proceedings of a symposium held in Ft. Lauderdale, Florida. Jan 17–18, 1974.**

A comprehensive program is under way at the Westinghouse Electric Corporation to develop and evaluate acoustic monitoring systems as a nondestructive testing tool for nuclear power plants that will assure structural integrity with minimum interference to normal power plant activities. A prime requirement for instrumentation for nuclear reactors is that it should function reliably in severe environments for several years. Sensors and instrumentation for source location and display developed for this application are described. Results of tests performed in major test facilities for vessel flaw growth, vessel rupture, and pipe rupture under various operating conditions are summarized. Evaluation performed in nuclear power plants both during hydrotests and operation of plant is presented. Laboratory tests for studying relationships between acoustic emission and fracture mechanics technologies are presented along with a summary of wave propagation results. Measurements, to date, indicate that an acoustic monitoring system has practical applications in nuclear power plants for improving safety and availability of plants. [reprinted by permission of the American Society for Testing and Materials, Copyright].

573 **R. Gopal. "Final Report of Westinghouse Participation in Acoustic Emission Tests at the EBOR Site." pp 153–187 in "A Joint Industry/Utility Program to Develop NDT Techniques for Nuclear Reactor Inspection: An Addendum Report to Steering Committee EEI Project RP-79." Compiled by E. R. Reinhart. Southwest Research Institute, San Antonio, Texas. May 21, 1973. [NTIS: PB-224-698/1].**

The report summarizes the Westinghouse Nuclear Energy Systems participation in the EEI and TVA EBOR Program designed to develop a real time system for monitoring reactor pressure vessels. Several real time systems were developed and demonstrated at ambient and elevated temperatures. In addition, basic information was obtained on stress wave propagation and acoustic failure signatures generated by pop-in of a large flaw in a reactor pressure vessel wall.

574 **R. Gopal. "On-Line Acoustic Emission Monitoring System for Nuclear Power Plants." pp 273–274 in** *Proceedings of the U.S.-Japan Joint Symposium on Acoustic Emission.* **English volume. Japan Industrial Planning Association,**

Kawamura-Building 6F, 3-21-6 Akasaka, Minato-ku, Tokyo, Japan 107. 1972. symposium held in Tokyo, Japan. July 4-6, 1972.

A program is under way at Westinghouse to develop and evaluate an Acoustic Emission Surveillance System for applications in nuclear power plants. The objective is to come up with an on-line system for assuring the integrity of the pressure boundary of a plant. Major phases of the program are (1) Sensor and Electronics Development, (2) Analysis System, and (3) Measurements at Nuclear Plants and Test Facilities. This presents a brief discussion of the three phases of the program.

575 **R. Gopal. "Progress in the Development and Application of Acoustic Surveillance Instrument Systems."** *Transactions of the American Nuclear Society,* **TANSA, 16:233-234. 1973. [summary only].**

A program is under way at Westinghouse to develop and evaluate an acoustic surveillance system for applications in nuclear power plants. The overall objective is to develop a system for assuring the integrity of the pressure boundary and preventing any rupture in the primary system. Major phases of the program include (a) transducer and electronics development, (b) feasibility demonstration with in-vessel flaw propagation, pipe rupture, and vessel rupture tests, and (c) application of the system in an operating plant.

576 **R. Gopal. "Surveillance Systems for Nuclear Power Plants."** *Transactions of the American Nuclear Society,* **TANSA, 17:378. Nov 1973.**

Paper describes the progress at Westinghouse in the development and applications of systems for detection of abnormal primary system conditions via vibration and acoustic surveillance techniques. The objective is to develop instrument systems that will improve availability and prevent catastropic failures through detection of malfunctions at incipient stages.

577 **R. Gopal and C. J. Ross. "Acoustic Emission Monitoring Systems for Primary System Surveillance." pp 480-500 in** *Topical Meeting on Water-Reactor Safety, Salt Lake City, Utah, March 22-28, 1973.* **Compiled by G. A. Freund. Aerojet Nuclear Company, Idaho Falls, Idaho. 1973. [NTIS: CONF-730304 (entire proceedings)].**

Some background in acoustic emissions, recent advances in instrumentation for detecting severe environmental acoustic emission, results of acoustic emission measurements, the measurements taken on the Prairie Island Unit No. 1 Vessel, and areas of future endeavor are presented.

578 **I. Grabec. "Akusticna Emisija (Acoustic Emission)."** *Strojniski Vestnik/ Mechanical Journal,* **STJVA, 21(3-4):41-48. Mar-Apr 1975. (in Slovenian with English translation).**

The article presents a description of the basic properties of the acoustic emission phenomena and the methods for nondestructive testing of materials that made use of this phenomenon. Experiments with acoustic emission show that its detection permits, in an idealized situation, to determine the occurrence of inelastic changes. In testing techniques, this property can often be applied where every other method has failed. A typical example of such application is detecting a crack in a sample of loaded material. [Abstract 000670. *Engineering Index*, Vol. 74, 1975].

579 **I. Grabec. "Analogy between Triboluminescence of Rubber and Acoustic Emission in Metals."** *Non-Destructive Testing,* **NDETA, 8(5):258-260. Oct 1975.**

The author describes the triboluminescence of rubber and demonstrates it by records of the intensity of emitted light and the variation of strain with stress of a rubber band. The marked increase of the emission of light with onset of plastic deformation and the irreversibility of triboluminescence on repeated deformation suggest an analogy between triboluminescence and acoustic emission. These properties of triboliuminescence, it is suggested, may be applied like sound emission in non-destructive testing.

580 **I. Grabec, J. Cekada, and J. Kolar. "Akusticna emisija pri utrujanju snovi (Acoustic Emission due to the Fatigue of a Material)." Fakulteta za strojnistvo, Unverza v Ljubljani, Ljubljana, Yugoslavia. 1976. (in Slovenian); paper presented at the 13th Yugoslav Congress of Rational and Applied Mechanics, held at Sarajevo, Yugoslavia. June 1976.**

The article describes characteristic properties of acoustic activity due to developing fatigue cracks. The common feature observed on several materials is that there exist three distinctive regions of the fatigue process with respect to the activity. In the initial region, corresponding to low number of load cycles, the acoustic activity grows due to developing cracks. In the next region, a practically constant activity is observed, corresponding to the uniformly growing cracks. In the last, breakdown region, the activity rises again due to the failure of the sample.

581 **I. Grabec and P. Leskovar. "Acoustic Emission of a Cutting Process."** *Ultrasonics,* **ULTRA, 15(1):17-20. Jan 1977.**

The cutting process of aluminum alloy, on a turner's lathe, was characterized by spectral analysis of the sound emitted by the process. In the audible frequency range, the spectrum was found to be discrete while in the ultrasonic range it appeared to be continuous. The influence of the relevant cutting parameters on the spectral distribution are presented and a qualitative explanation of the corresponding effects is given. The possibility of using acoustic emission analysis to evaluate the sharpness of a cutting tool is shown to be hardly realizable.

582 **I. Grabec, M. Perosa, and M. Stupica. "Akusticna emisija mehansko deformiranega fiber glassa (Acoustic Emission of Mechanically Deformed Fiberglas)." Fakulteta za strojnistvo, Univerza v Ljubljani, Ljubljana, Yugoslavia. 1974. (in Slovenian); paper presented at the 12th Yugoslav Congress of Rational and Applied Mechanics, held at Ohrid, Yugoslavia. June 1974.**

The article describes characteristic properties of acoustic emission of mechanically deformed epoxy matrix fiberglas band. The rate of emission bursts grows as a certain power (approximately 3) of deformation. Supposing that each burst represents a microdamage, it is suggested that the damage of the band could be described by a fraction of the total burst numbers recorded at a given and critical deformation. Experimentally proved Kaiser effect then shows that a single cycle of unloading and reloading the load below the maximum previously applied value causes practically no damage.

583 **I. Grabec and A. Peterlin. "Acoustic Emission of a Crazing Polymer."** *Journal of Polymer Science, Part A-2: Polymer Physics,* **JPLPA, 14:651–661. 1976.**

The acoustic emission from a crazing polyvinyltoluene in a tensile and bending experiment is described. Acoustic emission appears as a series of bursts which most likely correspond to the initiation and growth of crazes. The emission intensity is characterized by acoustic activity (pulse rate) measured by the ring-down technique. The average activity increases with strain. During repeated loading the acoustic activity shows a measurable intensity and significant rise only beyond the maximum of the former runs. This is equivalent to the Kaiser effect in metals. Acoustic emission during the creep experiment occurs in three characteristic periods. They are characterized as the relaxation, fatigue, and breakdown periods. Visual observations indicate that the relaxation period corresponds to the initiation, and the fatigue period to the growth of crazes. In the breakdown period, a macroscopic crack develops and the sample fails. [reprinted by permission of John Wiley & Sons, Inc.].

584 **L. J. Graham. "Acoustic Emission Transducer Characterization." SCTR-71-19. North American Rockwell Corporation, Science Center, Thousand Oaks, California. Dec 1971.**

One possible extension of the usefulness of acoustic emission monitoring is in distinguishing different sources of emission from differences in their frequency characteristics.

As part of the effort at the Science Center to determine the broadband frequency responses of several types of commercial and laboratory-designed transducers were determined by exposing them to the output of a *white noise* acoustic noise generator. The principal results of this study were the following: (1) Using the capacitor microphone as a standard, the amplitude of the surface displacement produced by the noise generator was determined to vary smoothly as $1/f^2$; (2) The response of several acoustic emission transducers to this noise generator fell off with frequency more slowly than $1/f^2$. That is, these transducers were somewhat more sensitive to surface displacements at the higher frequencies. In effect, these broadbanded piezoelectric devices have characteristics similar to velocity or acceleration detectors rather than displacement detectors above their fundamental resonant frequency; (3) The resonant characteristics of the common transducers are somewhat smoothed out by broadbanding, but many resonances still appear and the frequency response is irregular; (4) The output of a piezoelectric plate below its fundamental resonant frequency tends to fall off with decreasing frequency as f^2. Some modification in this response is possible with broadbanding; (5) Piezoelectric transducers for acoustic emission detection have better characteristics when made from PZT than from quartz because of the lower Q and higher coupling coefficient of the PZT; and (6) No one transducer was found for acoustic emission detection which is best for all conditions, and the possibility remains of selecting an optimum transducer for each situation.

585 **L. J. Graham. "Broadband Detection and Analysis of Acoustic Emission Signals." SCTR-75-4. Science Center, Rockwell International, Thousand Oaks, California. May 1975.**

586 **L. J. Graham. "Frequency Analysis of Acoustic Noises during Fatigue Testing of the EBOR Nuclear Reactor Pressure Vessel." SCTR-72-6. North American Rockwell, Science Center, Thousand Oaks, California. Apr 1972. [company proprietary].**

587 **L. J. Graham. "Frequency Response Characteristics of Acoustic Emission System Components, (Part 1)." SCTR-73-6. Science Center, Rockwell International, Thousand Oaks, California. Mar 1973.**

The frequency spectrum analysis system of the Science Center was utilized to determine the frequency response characteristics of various components to be used in an acoustic emission triangulation system at the Space Division. Emphasis was placed on the frequency band extending from about 20 kHz up to 2 MHz which covers the range in which most acoustic emission signals are observed. The frequency response of the Dunegan S/D 60 preamplifier was determined along with the response of nine different transducers. Effects due to various acoustic couplants and various interconnecting cables were investigated. Acoustic bursts generated during fatigue tests on 1/16-inch thick sheet aluminum samples were analyzed to determine the characteristics of both the acoustic emissions generated by crack growth in the samples and of extraneous noises in the test environment. Since the acoustic wave transmission characteristics of the material in which the emission signals are generated play a vital role in the frequencies which finally reach the transducers, an extensive investigation of the attenuation versus frequency curves for several different kinds of aluminum panels was conducted. It was found that an insulation layer attached to the aluminum plate by an adhesive produced extremely high attenuation of the emission signals while sprayed-on insulation caused much less attenuation of the acoustic energy.

588 **L. J. Graham. "Frequency Response Characteristics of Acoustic Emission System Components, Part II." SCTR-73-14. Science Center, Rockwell International, Thousand Oaks, California. Sep 1973.**

In order to better discriminate against background noise in an acoustic emission triangulation system, one must know the frequency content and pulse shapes of both the flaw generated noises and the background noises. Also one must know how the acoustic path between the source and the sensor affects the signal. The report describes (1) the broadband characterization of noises during a fatigue test at Space Division; (2) the change in signal shapes due to dispersion in plates; and (3) the acoustic attenuation in a variety of structural elements.

589 **L. J. Graham. "Sources of Acoustic Emission in Aluminum Alloys." pp 631-642 in "Proceedings of the ARPA/AFML Review of Quantitative NDE." D. O. Thompson, Program Director. AFML-TR-75-212, SC595.10AR. Science Center, Rockwell International, Thousand Oaks, California. Jan 1976. [NTIS]. conference held at the Science Center, Rockwell International, Thousand Oaks, California. July 15-17, 1975.**

A combined acoustic emission-fractography study indicates that for several aluminum alloys the only important source of detectable AE is the brittle cleavage fracture of small zones in the materials. In most of the alloys (2024, 2048, 2219, and 7075), these zones are primarily at intermetallic particles in the matrix. A small secondary source of AE is the brittle cleavage of grain or subgrain sized zones of the matrix itself. For alloys in which the size of the intermetallic particles is sufficiently small (6061 and 1100), the latter is the only source, and relatively few AE are generated in these materials. Ductile fracture, characterized by a dimpled, cusp-like appearance of the fracture surface, was found not to be a source of detectable AE. The heat treatment, surface condition, specimen orientation, and specimen origin in the rolled plate materials were found to have a large effect on the number and type of AE generated and on the fracture behavior of the tensile specimens. The frequency spectra of most of the AE bursts were like that of acoustic white noise. This spectrum is predicted by a theoretical model in which the source event is described mathematically as a short duration release in stress occurring over a highly localized region in the material. In addition, three other distinctive spectral types were observed. An interpretation of the sources of the AE having these spectral types has been made based on the fractographic observations and on other mathematical models for the AE generation process.

590 L. J. Graham and G. A. Alers. "Acoustic Emission from Polycrystalline Ceramics." Naval Air Systems Command Contract N00019-72-C-0382, Final Report SC513. 7FR. North American Rockwell, Science Center, Thousand Oaks, California. Dec 22, 1972. [NTIS: AD-754-839].

The principal objective was to determine if some characteristic feature in the frequency spectrum of an individual burst or in the amplitude distribution of a series of bursts could be found and related to the microstructure or type of the ceramic. Also, it was hoped that a relationship between the rate of emission generation could be correlated with the rate of fracture of individual grains during the advance of the crack front. The report describes the results obtained when slow crack growth was induced in samples of polycrystalline magnesia, alumina, Lucalox, and spinel, as well as when attempts were made to produce slow crack growth in silicon nitride.

591 L. J. Graham and G. A. Alers. "Acoustic Emission in the Frequency Domain." Preprint No. SC-PP-73-108. Science Center, Rockwell International, Thousand Oaks, California. Jan 31, 1974; pp 11-39 in *Monitoring Structural Integrity by Acoustic Emission, ASTM STP 571.* Edited by J. C. Spanner and J. W. McElroy. American Society for Testing and Materials, Philadelphia, Pennsylvania. 1975. proceedings of a symposium held in Ft. Lauderdale, Florida. Jan 17-18, 1974.

A means for quickly and easily determining the broadband frequency content of acoustic bursts as short as 20μs in duration has been developed using a video-tape recorder and a standard spectrum analyzer. It is shown by examples from several tests on laboratory specimens and on large structures that the frequency content of an acoustic burst is related to the mechanism which produced it and is not substantially affected by the sample size or by mode conversion due to multiple reflections in the structure. The frequency content of the burst can be changed in two ways, however: (1) by the frequency-dependent attenuation of the propagation medium, and (2) in the cases where the medium is dispersive. Results of measurements on the effect of these factors in a variety of structures are given. Although acoustic emissions from many materials tend to be *white noise,* several examples of acoustic emissions and extraneous background noise bursts having distinctive frequency spectra are given which suggest possibilities for discriminating true acoustic emission signals from background noise on the basis of frequency content alone. [reprinted by permission of the American Society for Testing and Materials, Copyright].

592 L. J. Graham and G. A. Alers. "Broadband Ultrasonic Attenuation Measurements in Unusual Materials." pp 703-706 in *1974 Ultrasonics Symposium*

Proceedings, Catalog No. 74 CHO 896–1SU. Institute of Electrical and Electronics Engineers, Inc., New York. 1974. symposium held in Milwaukee, Wisconsin. Nov 11–14, 1974.

In applying ultrasonic nondestructive testing methods to cases where the ultrasonic path length is long (acoustic emission) or where the material under test is highly attenuating (composites, honeycomb, insulating foam), the attenuation characteristics of the structure determines the ultrasonic frequency range to be used. In order to establish attenuation characteristics quickly over a broad frequency range, an acoustic instrument consisting of a white-noise generator and a damped piezoelectric receiver has been assembled for the frequency range of 0 to 3 MHz. By varying the path length, the frequency dependence of the attenuation can be obtained. The system used to make these measurements will be described and results for a variety of structures and materials will be presented.

593 **L. J. Graham and G. A. Alers. "Characteristics of Acoustic Emission Signals from Ceramics." SC544.14FR. Science Center, Rockwell International, Thousand Oaks, California. Feb 22, 1974. [NTIS: AD–778–015].**

Acoustic emission bursts generated during slow and rapid crack growth in a variety of polycrystalline ceramics, as well as twinning and rapid crack growth in sapphire were characterized by their number, amplitude, pulse shape, and frequency content. The microstructures and fracture surface features of the specimens were examined and correlations were made between these features and the characteristics of the acoustic emission bursts. Nine materials were studied covering a range of preparation techniques, compositions, and grain sizes. These included alumina from three sources, magnesia from three sources, plus spinel, procelain, and silicon nitride from one source each. An empirical correlation was found in which the number of grains fractured to produce a single acoustic emission event increased as the grain size decreased. It was also found that the load required to initiate crack growth was proportional to either the 2/3, 1, or 3/2 power of the ligament length depending upon the amount of transgranular fracture in each specimen. A variety of types of frequency spectra of the individual acoustic emission bursts were observed which in some cases could be identified with a specific generation process such as crack-twin interaction or step-wise transgranular cleavage.

594 **L. J. Graham and G. A. Alers. "Frequency Spectra of Acoustic Emissions Generated by Deforming Metals and Ceramics." North American Rockwell, Science Center, Thousand Oaks, California. Sep 27, 1972; pp 18–21 in *1972 Ultrasonics Symposium Proceedings.* Catalog No. 72 CHO 708–8SU. Institute of Electrical and Electronics Engineers, Inc., New York. 1972; symposium sponsored by Institute of Electrical and Electronics Engineers Group on Sonics and Ultrasonics, held in Boston, Massachusetts. Oct 4–7, 1972.**

In order to distinguish acoustic emissions from background noises and to better understand the mechanism of emission generation, it is important to obtain the frequency spectrum of an individual burst over as wide a frequency range as possible. A Sony video-tape recorder has been modified in such a way that an individual acoustic emission burst can be recorded and subsequently played back into a Hewlett-Packard spectrum analyzer. This system makes possible the rapid determination of the frequency content of the burst from 0 to 3 MHz. It also permits the measurement of the amplitude distribution of any series of bursts emitted during a particular time interval of a mechanical deformation test. Since most of the irregularities observed in the spectra could be traced to internal resonances in the ceramic transducers used, a special capacitor microphone transducer was developed which had smooth frequency response. Results on aluminum alloys, titanium alloys, some steels, and single crystal MgO all showed the same general spectrum. A reactor pressure vessel steel (A533-B) deforming in tension exhibited a maximum in the emitted energy near 1 MHz and the cracking of glass under a spherical indentor exhibited a similar maximum which moved between 0.5 and 1.0 MHz depending upon the applied load.

595 **L. J. Graham and G. A. Alers. "Investigation of Acoustic Emission from Ceramic Materials." Naval Air Systems Command Contract N00019-71-C-0344, Final Report SC513.6FR. North American Rockwell Corporation, Science Center, Thousand Oaks, California. May 19, 1972. [NTIS: AD-745-000].**

The frequency content, amplitude distribution, and rate of occurrence of acoustic emissions generated in model ceramic materials were determined. Single crystal MgO and glass, used as models of a ductile and a brittle ceramic, were mechanically loaded by three different means to produce rapid crack growth, controlled slow crack growth, and plastic deformation in both four-point bending and by cyclic bending at 100 cycles per second. A method was developed for easily determining the frequency spectrum of a single acoustic emission over the frequency range of 1 kHz to 3 MHz. A video-tape recorder was modified and used in conjunction with broadband transducers and amplifiers to record the acoustic emissions on magnetic tape and then play them back in *stop action* into a standard frequency spectrum analyzer. The main results of the tests with the model materials were that no acoustic emissions were detected before rapid crack growth commenced or during the slow crack growth in glass. Rapid crack growth produced acoustic emissions which had characteristic features in their frequency spectra which seemed to depend on the size of the crack produced. Acoustic emissions were generated and were easily detected during plastic deformation of single crystal MgO in four-point bending and were also observed to occur at the stress maxima during cyclic fatigue loading of this material. The frequency spectra of these emissions was much like those produced by a ductile metal.

596 **L. J. Graham and G. A. Alers. "Microstructural Aspects of Acoustic Emission Generation in Ceramics." pp 175-188 in *Fracture Mechanics of Ceramics.* R. C. Bradt, D. P. H. Hasselman, and F. F. Lange, Editors. Volume I: Concepts, Flaws, and Fractography. Plenum Press, New York. 1974. proceedings of a symposium held at Pennsylvania State University, University Park, Pennsylvania. July 11-13, 1973.**

The main objective of the studies described in this paper is to determine the conditions under which acoustic emission signals are generated in ceramics prior to failure and to see if these signals have any characteristic features which can be used to determine the microscopic process taking place within the material. In the first section, the techniques used and the materials studied are outlined. Section II then describes the characteristic features of the acoustic emission signals generated during plastic flow of MgO single crystals and accompanying the fracture of polycrystalline ceramics and glass. Following the work of Gatti *et al.*, the third section describes the establishment of a relationship between the occurrence of an individual acoustic burst and the motion of a slowly growing crack front. The fourth section describes the application of acoustic emission monitoring techniques to the determination of the material parameters appearing in fracture mechanics theories.

597 **L. J. Graham and G. A. Alers. "Spectrum Analysis of Acoustic Emission in A533-B Steel." Science Center, Rockwell International, Thousand Oaks, California. 1973; paper presented at the American Society for Nondestructive Testing National Spring Conference, Los Angeles, California. Mar 12-15, 1973; *Materials Evaluation,* MAEVA, 32(2):31-37. Feb 1974.**

Some striking characteristics have been observed in the frequency content of acoustic emissions produced during the deformation and crack propagation in A533-B pressure vessel steel. Spectrum analysis of individual acoustic emission bursts over the 20 kHz to 2 MHz frequency range show two distinct types of spectra one of which can be associated with plastic flow and the other with crack extension. It was found that these two spectral types were independent of the test procedures and sample geometries used.

598 **L. J. Graham and R. K. Elsley. "Microstructure Effects on Acoustic Emission Signal Characteristics." Project II, Unit II, Task 2. pp 109-114 in "Interdisciplinary Program for Quantitative Flaw Definition." D. O. Thompson, Program Manager. Semi-Annual Report SC595.21SA, Rockwell International, Science Center, Thousand Oaks, California. Feb 1977.**

Certain characteristics of acoustic emission signals from graphite-epoxy composite specimens were correlated last year with the mechanical behavior of the materials. Moisture degradation, which reduced the ultimate strength, resulted in a change in the AE amplitude distribution early in the loading history. Also, AE having distinct frequency spectral types tended to occur at or near singularities in the load curve. In order to transform those observations into a viable NDE tool, two things need to be done: (1) develop a quicker and more objective method for extracting the pertinent correlations from the data, and (2) develop the statistical relationship between the AE characteristics and some mechanical strength parameter such as ultimate strength or remaining lifetime. In these preliminary experiments, it has been demonstrated that (1) A minicomputer based data acquisition system can, in real time, capture a significant number of acoustic emssion signals; (2) The minicomputer can rapidly classify them according to type, taking into account a variety of signal features;

and (3) The computer can also discover relationships between the features of the emissions and what is known about the process going on in the sample.

599 **T. W. Graham. "The Consideration of Initial Flaw Size Distribution and Crack Detection Probabilities in Fracture Mechanics Analysis for Failure Prevention." M.S. Thesis. University of California, Los Angeles, California. 1973.**

The purpose of the thesis is to propose an analytical technique for failure prevention based on fracture mechanics analysis and nondestructive inspection methods. This technique combines distributions of the initial flaw size and crack detection probabilities. The basic fracture mechanics principles are reviewed and the major inspection techniques described. Conventional S-N fatigue test data are first evaluated in terms of a fracture mechanics, crack-growth equation to show the relationship between the probability of failure, stress level, and assumed inherent flaw size. Various aspects of predicting crack growth as a function of the stress intensity parameter are then discussed. These principles are applied to structural designs where periodic inspections are to be employed. It is demonstrated that the incorporation of the crack detection probability as a function of crack size can be used to calculate a net probability of inspections detecting a failure before a flaw can grow from an initial to critical size. The analytical method is then extended to include a distribution of initial flaw sizes, as well as a crack detection probability distribution. Finally, it is shown how to tailor the inspection schedule to optimize the probability of detecting both the large flaws missed during manufacturing inspections, as well as small flaws not detectable by the manufacturing inspection technique employed or that developed in service.

600 **J. M. Grandemange. "Procedes de Detection de l'Emission Acoustique Appliques a la Caracterisation Mecanique des Materiaux Composites Bore–Aluminium (Procedure for the Detection of Acoustic Emission Applied to the Mechanical Properties of Boron–Aluminum Composite Material)." pp 144–162 in *Journees d'Etudes sur l'Emission Acoustique.* P. F. Gobin, R. Goutte, and C. Guillaud, Organizers. Institut National des Sciences Appliquees de Lyon, Avenue Albert Einstein, 69621 Villeurbanne, Lyon, France. Mar 1975. proceedings of a conference held in Lyon, France. Mar 17, 1975. (paper in French).**

The acoustic emission of an aluminum-boron composite material submitted to a characterization test was studied and the results were compared with those of visual and radiographic methods. It was shown that it is possible to separate various types of events by comparing the responses of the different test devices used.

601 **J. M. Grandemange and K. N. Street. "Stable Crack Growth Detection by Acoustic Emission Methods in Notched Boron–Aluminum Composites." Centre des Materiaux, Ecole des Mines de Paris, 91102 Corbeil Essonnes,**

France. 1975; Paper 4.1 presented at the 4th Meeting of the European Working Group on Acoustic Emission, Saclay, Paris, France. Sep 24-26, 1975.

Boron-aluminum fiber-reinforced composite material is characterized by a quasi-elastic low strain to failure fracture behavior. As a consequence, its fracture toughness or sensitivity to notches is of primary importance in its application in aeronautical or other structures. The composite structure itself is unorthodox when compared to classical engineering materials and as a consequence presents unique problems in both experimental and analytical fields. The fibers are essentially completely elastic to fracture and exhibit a statistical variation in breaking strength, yielding a strength-length effect. The matrix on the other hand generally undergoes large scale yielding before final rupture of the composite although its load bearing capacity is only a small fraction of that of the fibers. Acoustic emission stress waves coinciding with the high elastic energy fiber fracture within the composite can be easily monitored. In unnotched material, an effective composite yield strength or proportional limit can be defined at the moment of first fiber breakage. However, the notches can lead to various failure modes depending on the constituents and the fabrication conditions. Typical examples are longitudinal matrix shearing, splitting, or fiber fracture, all within the region of the notch root. Unlike homogeneous materials, the fiber fracture mode in B-Al does not appear to be strongly influenced by notch-root radius up to about 200 to 400 μm. However, a phenomenon of slow crack growth has been found to exist wherein a number of fibers at the notch root break progressively, increasing the notch length, under a rising load, until final instability or complete rupture takes place. As the stable cracking generally takes the form of an internal phenomenon, the possibility was studied of utilizing acoustic emission methods to define the crack length changes. A low-frequency, about 20 kHz, Bruel and Kjaer Type-4344 accelerometer was coupled to a tape recorder as one method and a Dunegan/Endevco 3000 series unit working in the 100 to 300 kHz band was used as a second unit. In the preliminary experiments, the latter set at a low sensitivity of 50 db gave consistently higher emission counts than the accelerometer. Calibration tests on unnotched and notched specimens using X-ray radiography and matrix dissolution techniques showed that the lower sensitivity accelerometer was detecting only fiber breaks. Fractographic examinations suggested that the higher frequency transducers were picking up events related to secondary fracture processes such as fiber pullout and/or matrix-matrix interface failures. A preliminary fracture mechanics approach using *R* curve analyses for *isotropic* materials yielded results in which continuing work is under way to verify whether this approach is useful in predicting final failure in B-Al sheet material.

602 **J. M. Grandemange and K. N. Street. "The Effect of Notch Root Radius and Thickness on the Static Fracture Toughness of B-Al Composites." pp 1019-1050 in *Proceedings of the 1975 International Conference on Composite Materials. Volume 2.* Edited by E. Scala, E. Anderson, I. Toth, and B. R. Noton. Metallurgical Society of AIME, New York. 1976. conference cosponsored by the Society for the Advancement of Material and Process Engineering, Society of the Plastics Industry, Society of Plastics**

Engineers, Institute of Electrical and Electronics Engineers, American Institute of Aeronautics and Astronautics, and American Society for Metals, held in Geneva, Switzerland. Apr 7–11, 1975 and in Boston, Massachusetts. Apr 14–18, 1975.

Unidirectional B-Al of low yield strength matrix is characterized by a brittle-type tensile fracture in the unnotched state when pulled parallel to the fibres. The means to correctly measure the fracture toughness of such a material remains to be determined. As a first step, study was made of the effects of notch root radius, notch length, and specimen thickness utilizing DEN specimens subjected to Mode I opening fracture perpendicular to the fibres. The majority of the work involved pure aluminum as the matrix in an attempt to maximize notch blunting by plastic flow. Some experiments were also performed on alloy-matrix composites. For B-Al (pure), the crack initiation stress, the composite rupture strength, and hence the apparent stress intensity factor K_c did not vary with notch root radius over the range 50 to 5000 μm. On the other hand, these same three properties were found to be dependent on the specimen thickness.

603 **H. Granjon and S. Debiez. "Note on Acoustic Emission in Relation to Cold Cracking in the Welding of Steel/Note sur l'emission acoustique associee a la fissuration a froid en soudage d'acier." Institut de Soudure, Paris, France; Paper 3K/14 in** *Proceedings of the Eighth World Conference on Nondestructive Testing.* **volume containing Sections 3K2, 3H2, 5A. International Committee for Nondestructive Testing, Secretariat du Cofrend, Intitut de Soudure, 32 Boulevard de la Chapelle, 75880 Paris Cedex 18, France. 1976. conference held in Cannes, France. Sep 6–10, 1976. (paper in French). [NTIS: FRNC-CONF-171].**

Described are acoustic emission tests on self-restraint specimens and on implants, for detecting cold cracking when welding steel.

604 **A. M. Green. "Acoustic Emission Tests Begin to be Heard."** *Iron Age,* **IRAGA, 214(12):43–45. Sep 16, 1974.**

Article covers an interview with W. F. Hartman of Trodyne Corporation. The nondestructive testing technique is based on the principle that materials under stress emit subsonic sounds that indicate potential fracture and failure. It can be used during construction or after a part is in service to prevent failure. A prime application is for monitoring quality control of spot welds. [Abstract 045918. *Engineering Index,* Vol. 73, 1974].

605 **A. M. Green. "Progress is Reported in Acoustic Emission Tests."** *Iron Age Metalworking International,* **IAMIA, 14(4):34–35. Apr 1975.**

Article is based on an interview with W. F. Hartman of Trodyne Corporation. Although in its infancy, recording flaw-induced modulation of sounds emitted by stressed materials is proving to be a valuable and economical nondestructive test for pressure vessels, pipe lines, and spot welds. Other applications under study, such as LNG tankers, welding, and on-line quality control only scratch the surface of the technique's potential. Acoustic emission monitoring can be used for monitoring processes or checking parts before or during service. Comparison of sound emission with previous records is a simple method of requalifying pressure vessels or detecting changes in structural integrity of critical installations. By isolating flaws to a specific area, other nondestructive tests need to be made only in that area. [Abstract 22-0868. *Metals Abstracts,* Vol. 8, 1975].

606 **A. T. Green. "Acoustic Emission: A Generic Discussion."** *IEEE Transactions on Sonics and Ultrasonics,* **IESUA, SU-19:391. July 1972. [abstract only].**

Acoustic emissions are the impulsively generated small amplitude elastic-stress waves created by deformations in a material. The rapid release of kinetic energy from the deformation mechanism propagates elastic waves from the source, and these are detected as small displacements on the surface of the specimen. The emissions indicate the onset and continuation of deformation and may be used to locate the source of deformation through triangulation techniques. The sensing transducer is normally constructed from a piezoelectric crystal that converts low-level stress waves in the structure to electrical signals that are amplified, electronically filtered (bandpassed) and processed in a variety of ways. The signals are transient in nature and tend to ring the detection transducer at resonance. This results in an electrical signal that is a damped sinusoid with a carrier frequency strongly dependent on the transducer characteristics. Acoustic emission instrumentation techniques are rapidly expanding in the field of industrial nondestructive inspections, materials, and data processing. [reprinted from *IEEE Transactions on Sonics and Ultrasonics*].

607 **A. T. Green. "Applications of Acoustic Emission." Dunegan Corporation, Livermore, California. 1972; paper presented at the 1972 IEEE Ultrasonics Symposium. Boston, Massachusetts. Oct 4–7, 1972. [paper not published in proceedings];** *IEEE Transactions on Sonics and Ultrasonics,* **IESUA, SU-20:48. Jan 1973. [abstract only].**

Acoustic emission techniques, now refined through a decade of use in the United States, are finding their way into more applications than many casual observers would imagine. In this presentation, mention is made of the common nondestructive test uses of acoustic emission. Brief descriptions of acoustic emission uses in medicine, geology, soil engineering, rodent infestations, corrosion and fatigue damage, and other less known but equally important areas are presented. [reprinted from *IEEE Transactions on Sonics and Ultrasonics*].

608 **A. T. Green. "Characteristics of Acoustic Emission Response from Materials." pp 233–272 in** *Proceedings of the U.S.-Japan Joint Symposium on Acoustic*

Emission. English volume. Japan Industrial Planning Association, Kawamura-Building 6F, 3-21-6 Akasaka, Minato-ku, Tokyo, Japan 107. 1972. symposium held in Tokyo, Japan. July 4-6, 1972.

Just as most materials vary in their physical properties, so do they vary in their acoustic emission response to deformation. Some of the available information which describes the variation in this response to environmental and other test parameters is presented. Data from acoustic emission experiments on smooth tensile, precracked-fracture toughness, stress corrosion, strain rate, elastic modulus, and ductile-brittle fracture have shown information which indicates how these parameters affect the acoustic emission response of the material. The paper tacitly indicates the importance of determining the acoustic emission response from a given material in order to utilize the acquired data to its fullest extent.

609 **A. T. Green. "Detection of Incipient Failure in Pressure Vessel Materials by the Stress-Wave Analysis Technique." pp 188-191 in "Gas-Cooled Reactor Program Semiannual Progress Report for Period Ending March 31, 1969." Compiled and edited by A. W. Savolainen. ORNL-4424. Oak Ridge National Laboratory, Oak Ridge, Tennessee. Oct 1969. [NTIS].**

Aerojet-General Corporation completed compression and tension testing of cylinders made with the three basic concrete mixes. The stress-wave analysis technique (SWAT) was used to monitor cracking during loading. The data obtained are currently being evaluated. Preparations are near completion for tests of a 1 to 2.75-scale mortar model in which SWAT will be used to monitor cracking, both during prestressing and pressure testing.

610 **A. T. Green. "Detection of Incipient Failures in Pressure Vessles by Stress-Wave Emissions."** _Nuclear Safety,_ **NUSAA, 10(1):4-18. Jan-Feb 1969; Technical Report DRC-108. Dunegan Research Corporation, Livermore, California.**

Advances in stress-wave emission techniques for detecting and monitoring failure processes in engineering materials have created considerable interest in their application in systems that will foretell incipient failure of a pressure vessel. Programs on various aspects of the technology are being carried out in a relatively small, but rapidly growing, number of development laboratories. These programs range from pure research into the basic phenomena, such as in materials evaluations programs, to the development of environmentally hardened sensors and the fabrication and use of application hardware and systems. Source location of the stress wave has been demonstrated as being within the capability of present technology. Resolution of the severity of the located defect has not been completely accomplished, although significant progress has been made toward this objective. The in-place, continuous surveillance nature of the technique enhances its value, since operational requirements of a nuclear pressure vessel greatly impede the use of the usual inspection techniques.

Simultaneous applications of other diagnostic surveillance techniques, with the same sensors and hardware, make possible complete monitoring of pressure vessels for the detection of incipient failures.

611 **A. T. Green. "Development of a Nondestructive Testing Technique to Determine Flaw Criticality." Advanced Research Projects Agency Contract F33615-68-C-1705. Aerojet-General Corporation, Sacramento, California. Aug 1969. [NTIS: AD-859-314].**

The report presents a brief description of the overall program to develop a nondestructive testing technique to determine flaw criticality. Continuous-wave and pulse-shock-wave tests have produced emission-sensor response data for comparative evaluation of seven typical sensors. Data processed by two real-time spectrum analyzers are presented. Recent work relevant to nondestructive inspection is reviewed. Plans for future work are discussed.

612 **A. T. Green. "Idaho Nuclear Corporation 20-Inch Valve Pressure Test: Stress-Wave Analysis Technique Report." Report No. 9633:0047. Aerojet-General Corporation, Sacramento, California. Feb 22, 1967.**

Two objectives were attempted during this test program. One objective was to determine the structural adequacy of the valve by pressurizing until failure occurred. The other objective was to apply the Aerojet-General Stress-Wave Analysis Technique (SWAT) during the pressurization testing in order to sense incremental extensions of flaws and to sense imminent failure of the valve. The valve body was pressurized to 2010 psig when leakage at the blind flange joints could no longer be overcome by the flow of the pressurization system. Visual inspection of the valve body has not detected any gross deformity nor other evidence of structural degradation. The only deformation was an elongation of the threaded studs observed by the smaller amount of torque required to remove the blind flanges, as compared with the originally applied torque before testing, and also some distortion of the blind flanges. Stress-wave data acquired during the testing demonstrated that the signal levels from flaw growth in 304 cast stainless steel are detectable.

613 **A. T. Green. "Stress Wave Analysis Technique and Its Application to Pressure Vessels for Pressurization Testing and In-Service Operation Monitoring." pp 40-48 plus 12 pages of figures in "Summary of Transactions, Pressure Vessel Section, 43rd Annual Conference of the California Certified Boiler and Elevator Inspectors Association." issued by the State of California, Division of Industrial Safety, San Francisco, California. 1970. conference held in San Francisco, California. Jan 20-22, 1970.**

614 **A. T. Green. "Stress-Wave Detection: Saturn S-II." NASA-CR-61161. Final Report. Aerojet-General Corporation, Sacramento, California. Dec 1966. [NTIS: X67-18378].**

The ability to sense and record stress-waves, the elastic-waves transmitted from flaw growth, in a Saturn S-II propellant tank bulkhead was demonstrated during a proof-pressure hydrotest. The stress-wave data were used in triangulation techniques to describe the origin of the wave. Tensile specimen tests confirmed that flaw growth in 2014-T6 aluminum alloy procedes incrementally and that the relationship between stress-wave emission and flaw extension can be developed in order to provide an assessment of flaw criticality and structural reliability.

615 **A. T. Green. "Stress Wave Emission: An Important New Tool for the Technical Community." pp 4-7 to 4-17 in** ***Proceedings of the 6th Space Congress on Space, Technology and Society.*** **Volume 1. published by University of Florida, Space Congress Printing Company-Brevard Printing Company, Cocoa Beach, Florida. 1969. symposium held in Coca Beach, Florida. Mar 17-19, 1969.**

Discussion of the application of stress wave emissions as a means of detecting deformations within a material which are undetectable by other techniques. The basic principles of such detection are considered. One of the first stress wave analysis technique (SWAT) developments was accomplished on a pressure vessel. SWAT was also used to establish the origin of fracture in a variety of pressure vessels which were purposely tested to failure as a part of structural design development programs. In conclusion, it is found that the most rewarding attribute of stress wave emission monitoring is that the material undergoing deformation serves as its own transducer: it both generates and transmits the signal of interest to the sensor.

616 **A. T. Green. "Stress-Wave Emission and Fracture in 6Al-4V Titanium." ASM Technical Paper No. W70-10.2. American Society for Metals, Metals Park, Ohio; paper presented at the 1970 Western Metal and Tool Conference, Los Angeles, California. Mar 9-12, 1970; technical brief of paper in** ***Metals Engineering Quarterly,*** **MENQA, 11(1):61-62. Feb 1971.**

The results of investigations to establish stress-wave analysis technique (SWAT) system parameters for the monitoring of thin-walled Ti-6Al-4V pressure vessels are presented. Stress-wave emission characteristics which can be employed as precursors of failure in air, distilled water inhibited with sodium dichromate (500 ppm), and methyl alcohol environments were determined. A SWAT system possessing sufficient sensitivity to detect and monitor stress-wave emissions associated with subcritical crack growth is discussed. Critical stress intensity data and relationships between applied stress intensity and failure time for the various material thickness (0.030 and 0.060 inches) and environments investigated are presented.

617 **A. T. Green. "Stress-Wave Emission and Fracture of Prestressed Concrete Reactor Vessel Materials." Technical Report DRC-71-3. Dunegan Research Corporation, Livermore, California. May 1971; pp 635–649 in *Materials Technology. Volume 1: An Inter-American Approach for the Seventies.* American Society of Mechanical Engineers, New York. Aug 1970. proceedings of the 2nd Inter-American Conference on Materials Technology, sponsored by Southwest Research Institute, held in Mexico City, Mexico, Aug 24–27, 1971; "Emision de ondas de tension y fractura de materiales de concreto pretensado para recipientes de reactores." pp 655–677 in Spanish translation of proceedings. (in Spanish); AGC Report 4190. Aerojet–General Corporation, Materials Technology Department, Sacramento, California. June 1969. [Original and most complete report of work conducted for Oak Ridge National Laboratory under Contract No. W-7405–eng-26, Subcontract No. 3118].**

Applicability of using stress-wave emissions to detect the failure and progression of failure in cylindrical concrete specimens and a model prestressed concrete pressure vessel is presented in this paper. Stress-wave-emission data obtained during structural tests have provided a description of the time relationship between the onset and continuing failure process. Three concrete aggregates, one mortar model of prestressed pressure vessel material, and a model prestressed pressure vessel were tested in the course of the investigation. Fingings of the study have shown that variations in the stress-wave-emission data can determine impending failure and previous structural loading extremes. A limited correlation between stress-wave-emission data and modulus is shown, as is a potential means of observing long-term structural changes in prestressed concrete structures. Variations in the time of stress-wave arrival and emission data can be used to determine anomalous structural responses and maximum cracking locations. Stress-wave-emission data are also shown as being an extremely fine measurement of deviation from elastic response of the mortar model material.

618 **A. T. Green. "Stress–Wave Emission Generated during the Hydrostatic Compression Testing of Glass Spheres." Office of Naval Research Contract N00014–67–C–0333. Aerojet–General Corporation, Sacramento, California. Apr 1969. [NTIS: AD–853–221].**

The test program consisted of hydrostatically compression pressure testing two 10-inch diameter, hollow, glass spheres to failure while applying the Aerojet-General Stress Wave Analysis Technique as a means of obtaining information regarding the origin and propagation of failure. Prior to pressurization to failure, each specimen was visually examined and observed defects were noted as a means of correlating stress-wave emissions with known defects. A system of underwater accelerometers was used as the stress-wave-emission sensor and all emission data were recorded on magnetic tape. Post-test analysis of the data was used to determine the origin and mode of failure. The following program results were obtained: (1) The pressure causing failure of each specimen was obtained, (2) The stress-wave characteristic of increasing rate of emission was used as the criterion to reduce pressure and prevent specimen failure, (3) The time differences between stress-wave arrival at each sensor location was used in conjunction with seismic triangulation techniques to determine the origin of the emission, (4) The underwater accelerometer system for operation under high pressure environments was evaluated as a stress-wave emission sensor, and (5) The origin of failure and mode of failure of each sphere was identified.

619 **A. T. Green. "Structural Testing with Acoustic Emission." pp 9–17 in** *Acoustic Emission Detection and Magnetic Flyer Strain Gaging.* **Society for Experimental Stress Analysis, Westport, Connecticut. 1972. proceedings of Western Regional Strain Gage Committee meeting, held in Marina Del Ray, California. Mar 20, 1972.**

620 **A. T. Green. "Survey of Acoustic Emission Technology: Past, Present and Future." Aerojet–General Corporation, Sacramento, California. 1969; presentation made to the American Society for Metals Materials Engineering Congress, Philadelphia, Pennsylvania. Oct 13–16, 1969. [paper never written].**

A detailed historical review of acoustic emission's development into a technology including a brief introduction to some of the accepted terminology. Important results of past and present programs using acoustic emission are discussed. Current research and practical applications of the technology are described and future uses are projected.

621 **A. T. Green. Testing of Glass Hemispheres Using Aerojet–General Stress–Wave Analysis Technique." Office of Naval Research Contract N00014–67–C–0333. Final Report. Aerojet–General Corporation, Sacramento, California. Sep 1967. [NTIS: AD–825–542].**

The application of the Stress-Wave Analysis Technique to three, hollow, 10-inch diameter glass hemispheres undergoing external hydrostatic pressure tests accomplished the following program objectives: (1) Stress waves were emitted, detected, and recorded during external pressurization of the test specimens; (b) A characteristic increase in the rate of stress-wave emission was noted as indicative of impending failure; (c) The recognizable nature of the rate of emission characteristic was used as the criteria to reduce pressure and prevent complete failure of one test specimen; (d) Corroboration of stress-wave emission being generated by failure processes was obtained through visual observation of the test specimen before and after testing; and (e) Time differences between stress-wave arrival at each sensor location can be used in triangulation to determine the origin of the wave. These tests confirmed the ability of the Stress-Wave Analysis Technique to detect and locate areas of failure in homogeneous, hollow, glass hemispheres under external pressure; and demonstrated that at least one characteristic of impending failure is the rate of stress-wave emission. These tests were conducted under constantly increasing hydrostatic pressure.

622 **A. T. Green and W. N. Carver. "Instrumentation for Acoustic Emission." pp 201-231 in *Proceedings of the U.S.-Japan Joint Symposium on Acoustic Emission.* English volume. Japan Industrial Planning Association, Kawamura-Building 6F, 3-21-6 Akasaka, Minato-ku, Tokyo, Japan 107. 1972. symposium held in Tokyo, Japan. July 4-6, 1972.**

Of importance in any testing program, instrumentation and technique become increasingly so in acoustic emission technology because of the nature of the available signals - transient, low energy elastic stress waves. This paper reviews the instrumentation which has been used during various research and development efforts. With a brief historical review of the instrumentation used in early investigations, the paper continues through the techniques commonly used today. Basic laboratory tests, under controlled conditions are perhaps the easiest application of current instrumentation. Variations in sensor, amplification, noise rejection techniques, and signal processing methods are discussed. For field use, systems where uncontrolled background conditions, usually much larger specimens and therefore multiple sensor instrumentation systems are required, other techniques may now become important. Spatial filters, frequency filters, timing circuits to enable emission source location, and computational procedures become of greater interest and usefulness. Broad review of many of the conceptual ideas in use today and specific discussion of selected techniques are presented. A brief indication of the techniques which appear promising for future use is included.

623 **A. T. Green and P. P. Crimmins. "Correlation of Stress-Wave-Emission Characteristics with Fracture in Aluminum Alloys." Report 1246-Q-2. Aerojet-General Corporation, Sacramento, California. Mar 1969.**

The report describes the progress during the first nine months of a program to investigate the correlation of stress-wave-emission characteristics with fracture in selected aluminum alloys. During this reporting quarter, specimen testing has been conducted, all specimens have been fabricated, and the materials characterization has been completed. This report displays some of the characteristic stress-wave emission data obtained during specimen testing, and adds titles to the continuing literature survey being performed as part of this program. Work planned for the next quarter is also indicated.

624 **A. T. Green and H. L. Dunegan. "Acoustic Emission Analysis of Crack Propagation and Fracture in Pressure Vessels and Pressure Vessel Materials." Technical Report DC 72-1. Dunegan Corporation, Livermore, California. 1972; pp 507-527 in *Proceedings of the First International Conference on Structural Mechanics in Reactor Technology.* EUR-4820, Volume 4. Commission of the European Communities, Brussels, Belgium. 1972. conference held in Berlin, Germany. Sep 20-24, 1971.**

The paper presents the acoustic emission data generated and recorded during metal tensile specimen tests, both unflawed and flawed, model concrete pressure vessel tests, and compressive tests of cylindrical specimens. Irradiation, temperature, prestressing, and internal pressurization test results are discussed for the various specimens, as are other factors which affect the acoustic emission response of the material. Parametric relationships between acoustic emission data and other indices of structural performance are presented.

625 **A. T. Green, H. L. Dunegan, and A. S. Tetelman. "Nondestructive Inspection of Aircraft Structures and Materials via Acoustic Emission." Technical Report DRC-107. Dunegan Research Corporation, Livermore, California. Sep 1970; paper presented at Air Transport Association of America Meeting, San Antonio, Texas. Sep 15-17, 1970; pp 275-289 in** ***International Advances in Nondestructive Testing.*** **W. J. McGonnagle, Editor. Volume 5. Gordon and Breach Science Publishers, New York. 1977.**

Since the original efforts of Dr. Kaiser in Germany in the late 1940's and early 1950's when he determined that acoustic emission test techniques could "determine for any specimen whether it had already been subjected to any stress, the magnitude of this stress, and that it could also be used to determine the elastic limit, yield point, and rupture limit," a great amount of research and development effort has been expended in the United States elaborating on his findings. Acoustic emission techniques have been shown to be one of the most informative methods of determining materials and structural performance currently available. This presentation will limit itself to acoustic emission results obtained on typical aircraft structural materials such as aluminum, titanium, and steel and a new material, metal-honeycomb. Briefly, the effect of temperature, welds, fatigue cycles, and other parameters on acoustic emission data will be illustrated. A damage concept in which fracture mechanics technology merges with acoustic emission technology is presented. The application of acoustic emission systems to a full scale aircraft during structural testing is also described.

626 **A. T. Green and C. E. Hartbower. "Detection of Incipient Failure in Pressure Vessels by Stress Wave Emission." Aerojet-General Corporation, Sacramento, California. 1970; Paper No. 3 presented at the International Symposium on Non-Destructive Testing of Nuclear Power Reactor Components, Rotterdam, The Netherlands. Feb 26-27, 1970. symposium organized by Nederlands Atoomforum, Rontgen Technische Dienst N.V., Rotterdamsche Droogdok Maatschappij N. V. [master copy of proceedings, CONF-700224, available on inter-library loan from Technical Information Center, P. O. Box 62, Oak Ridge, Tennessee 37830, Attn: Master Files, Document Management Branch].**

Progress in the development of stress-wave emission, or acoustic techniques for in-place, continuous monitoring for incipient failure detection in reactor pressure vessels is surveyed. Operational systems that have been developed and utilized on pressure vessel systems are described and their reliability and availability are discussed.

627 **A. T. Green and C. E. Hartbower. "Development of a Nondestructive Testing Technique to Determine Flaw Criticality." Advanced Research Projects Agency Contract F33615–68–C–1705. Aerojet–General Corporation, Sacramento, California. May 1970. [NTIS: AD–869–999].**

The report presents the results of the program to develop a nondestructive testing technique to determine flaw criticality based upon stress wave emission. Nine sensors typically used for emission studies have had relative response calibrations completed under continuous wave and pulse-shock-wave tests and these results are presented. Tests on unflawed, surface-flawed and edge-flawed specimens of D6aC steel, 6Al-4V titanium and bend tests of borosilicate glass have been conducted, with emission data reduced and analyzed for this report. Transient spectrum analysis of the emission data was accomplished over the 180 to 268 kHz frequency band.

628 **A. T. Green and C. E. Hartbower. "Stress-Wave Analysis Technique for Detection of Incipient Failure." RA/FA–DSR. Aerojet–General Corporation, Sacramento, California. Dec 1967. [NTIS: TID/SNA–710]; pp 127–162 in** *Incipient Failure Diagnosis for Assuring Safety and Availability of Nuclear Power Plants.* **E. P. Epler and D. P. Roux, Coordinators. U.S. Atomic Energy Commission, Division of Technical Information, Washington, D.C. Jan 1968. [NTIS: CONF–671011]. proceedings of conference sponsored by Oak Ridge National Laboratory, held in Gatlinburg, Tennessee. Oct 30–Nov 1, 1967.**

Research has shown that crack growth resulting in a sudden release of elastic energy can be detected by a sensor that converts the elastic wave into an equivalent electrical signal. Thus, this technique is unique as a nondestructive inspection method in that the growing defect in the material transmits its own signal, with the sensor acting as the receiver. The receiving sensors used in our studies and applications have been piezoelectric accelerometers, directly calibrated in engineering units. The Stress-Wave Analysis Technique (SWAT) has been employed as a research tool aimed at learning more about the causes of slow crack growth, particularly as affected by fatigue, hydrogen embrittlement, stress corrosion cracking, and spontaneous strain aging. SWAT is now being employed as a field inspection and monitoring process during the pressure qualification testing of rocket motor cases and other pressure vessels. With SWAT one can not only determine if there is a growing flaw, but can also locate the flaw by triangulation techniques. Moreover, in most structures as the crack increases in length, there is an increase in the rate and amplitude of stress-wave emission that presages failure.

629 **A. T. Green, C. E. Hartbower, and P. P. Grimmins. "Subcritical Crack Growth in Beryllium at –320 °F." RN–S–0523. Aerojet–General Corporation, Scaramento, California. Oct 1969. [NTIS]; paper presented at the American Society for Testing and Materials Symposium on Fracture Toughness Testing at Cryogenic Temperatures, Toronto, Canada. June 21–26, 1970.**

Compact tension tests (Manjoine Wedge Opening Loading design) of vacuum-hot-pressed, S200-grade beryllium were conducted at –320 °F. Instrumentation included piezoelectric accelerometers and electric-resistance strain gages. The objective of the tests was to evaluate the subcritical crack growth characteristics of S200-grade beryllium under plane-strain conditions. The stress-wave-analysis technique revealed discontinuous crack growth at –320 °F prior to the final plane-strain (K_{Ic}) instability. The crack toughness of the beryllium tested was 16.4 ksi·inch½ at –320 °F.

630 **A. T. Green, C. E. Hartbower, and C. S. Lockman. "Feasibility Study of Acoustic Depressurization System." Report NAS 7-310. Aerojet-General Corporation, Sacramento, California. Feb 1965. [see Reference 632 for final report].**

A program was conducted to establish the feasibility of using an acoustic technique to detect the growth of a critical flaw, locate the flaw, and initiate a system that would arrest the growth of the flaw, and thus prevent the destruction of a 260-inch diameter rocket motor case during its proof hydrotest. Tensile specimens of 18% nickel, 200 grade, maraged steel, parent and weld metal, with part-through-cracks (surface flaws) were tested to failure; the energy released and detected as stress waves as the flaw propagated was monitored on an oscilloscope and recorded on magnetic tape. Data recordings were also made of stress waves emanating during the proof hydrotest of a 156-inch diameter rocket motor case. Test-specimen design, flaw preparation, weld procedures, instrumentation techniques, test results, conclusions, and recommendations are detailed. The results of Phase O confirm the feasibility of an acoustic depressurization system to prevent the failure of rocket chambers and tankage during hydrotest due to crack growth. Large laboratory test specimens (a 2.8-square inch corss section) containing surface cracks of various sizes and shapes have been successfully unloaded after amplitude and frequency-of-occurrence of stress-wave emission indicated near critical flaw growth. With small surface cracks, over 100 stress waves were detected in some specimens before failure. Furthermore stress-wave amplitudes of 1.0-g units or greater occurred in most specimens considerably before maximum load. Nevertheless, ten specimens were unloaded when the stress-wave impulses as viewed on an oscilloscope indicated near-critical flaw growth; in no case where this was attempted did specimen cracking cause failure. In other words, the cracking that caused 1.0-g units or higher stress-wave emission prior to maximum load was stable crack propagation.

631 **A. T. Green and C. S. Lockman. "Acoustic Analysis of Filament-Wound Polaris Chambers." Aerojet-General Corporation, Sacramento, California. 1963; paper presented at the Fourth Semi-Annual Polaris Glass Reinforced Plastic Research and Development Conference, sponsored by the Special Projects Office, Department of the Navy, held at Aerojet-General Corporation, Sacramento, California. Jan 22-23, 1963.**

The objectives of the acoustic analysis program are outlined and the planned approach to accomplish these objectives is itemized in the presentation. Data acquisition techniques in use at Aerojet-General Corporation and at vendors facilities prior to the program adoption, and revisions to these techniques are described. Data obtained during the velocity of sound phase of the program are presented, techniques and setups are described. The data from recordings made during Polaris chamber hydrotests are shown. Analysis of the data is limited to comparisons between types of glass and resin, doily placement, and proof or burst test. Three dimensional data analysis techniques are described.

632 **A. T. Green, C. S. Lockman, S. J. Brown, and R. K. Steele. "Feasibility Study of Acoustic Depressurization System." NASA CR-55472. Final Report. Aerojet-General Corporation, Sacramento, California. Mar 1966. [NTIS: N66-37273].**

Feasibility was demonstrated for a conceptual system utilizing the elastic waves transmitted from flaw growth in a rocket motor case to indicate potential failure and initiate remedial action to prevent destruction. Tests confirmed the ability to arrest flaw growth based upon recognition of a *critical stress-wave signature* and to locate the flaw in large rocket motor cases.

633 **A. T. Green, C. S. Lockman, and H. K. Haines. "Acoustical Analysis of Filament-Wound Polaris Chambers." Report 0672-01F. Aerojet-General Corporation, Sacramento, California. Sep 16, 1963. [NTIS: AD-418-330].**

Given is the final report on the program to investigate acoustic analysis of Polaris filament-wound chambers. The structural dynamic energy emanating from chambers during hydrostatic pressure tests was recorded by accelerometers and analyzed. Velocity-of-sound measurements were made to determine parameters affecting wave propagation through composite structures. The analysis of data has provided a method for establishing the structural integrity of a chamber after it has been hydrostatically tested to proof pressures. This verification may be made at lower proof pressures than those currently used.

634 **A. T. Green, C. S. Lockman, and R. K. Steele. "Acoustic Verification of Structural Integrity of Polaris Chambers."** ***Modern Plastics,*** **MOPLA, 41(11): 137-139, 178, 180. July 1964.**

Equipment, procedures, and techniques of data presentation for the use of acoustic methods to verify the structural integrity of solid rocket filament-wound motor cases are described. The method can be extended to other filament-wound structures. The proper evaluation of the release of internal energy provides a sensitive nondestructive means of predicting the burst pressure of the finished product.

635 **A. T. Green and C. F. Morais. "The Use of Acoustic Emission for Quality Assurance of Pressure Containment Structures." Paper G 6/3 in** *Structural Mechanics in Reactor Technology, Preprints of the 2nd International Conference.* **Compiled by T. A. Jaeger. Volume 6A, Supplement Volume. Commission of the European Communities, Brussels, Belgium. 1973. conference held in Berlin, Germany. Sep 10–14, 1973. [NTIS: CONF-730942-P6A-S].**

Acoustic emission techniques have advanced from laboratory use to applications which directly involve establishing the quality assurance of pressure containment and many other structures. The authors, working in acoustic emission since 1961, provide a clear description of how these techniques have been developed to where they now represent an entirely new technology. Acoustic emission data acquired from nuclear reactor pressure vessel tests and experimental programs on other small pressure vessels are presented and discussed. The paper presents an orderly description of the advantages and disadvantages of acoustic emission testing. Particular attention is given to the instrumentation used and its importance as to requirements necessary to be able to compare results obtained by others are discussed. The usefulness of spectral analysis and amplitude distribution is described. Actual data acquired during pressure vessel tests, ranging from thick wall nuclear vessels to thin wall aerospace vessels, metallic, nonmetallic, and glass are shown. From a quality assurance viewpoint, multichannel real-time acoustic emission source-location system requirements are reviewed and evaluated. In particular the interplay between the material and the emission is discussed. Finally the use of acoustic emission techniques during the welding fabrication of pressure vessels and other structures is shown to be an extremely useful tool.

636 **A. T. Green and J. Ross. "Bibliography on Stress Wave/Acoustic Emission: 1932–1967." Aerojet–General Corporation, Sacramento, California. Oct 1967.**

Bibliography lists 81 references, arranged alphabetically by author for each year. This was the basis on which the first Acoustic Emission Working Group bibliography was compiled.

637 **A. T. Green and A. S. Tetelman (to Becton, Dickinson Electronics Company). "Acoustic Emission Damage Measurement." U.S. Patent 3,774,443. Nov 27, 1973.**

A precracked gage element is attached to a structural member to be monitored for damage and is continuously or periodically monitored to determine acoustic emission therefrom as a measure of fatigue damage of a structural member.

638 **R. E. Green, Jr. "Ultrasonic Attenuation Detection of Fatigue Damage." pp 187–193 in** *Ultrasonics International 1973 Conference Proceedings.* **IPC**

Science and Technology Press Ltd, Guildford, Surrey, England. 1973. conference held at Imperial College, London, England. Mar 27-29, 1973.

639 R. E. Green, Jr., and R. B. Pond, Sr. "An Ultrasonic Technique for Detection of the Onset of Fatigue Damage." AFOSR-TR-72-1199. Johns Hopkins University, Department of Mechanics and Materials Science, Baltimore, Maryland. June 1972. [NTIS: AD-745-978].

The purpose of the present research is to develop an ultrasonic technique for the detection of the onset of fatigue damage. Particular emphasis is placed on the use of ultrasonic attenuation measurements as a continuous monitor of fatigue damage during cyclic testing of metal specimens. Efforts to date have indicated that the history of attenuation during fatigue of the specimen provides an index as to the onset of failure. Ultrasonic attenuation measurements made simultaneously with fatigue tests on aluminum bars give an early warning of eminent fracture. Such a warning occurs much earlier than detection of energy reflected from the crack by conventional ultrasonic techniques. This is true for specimens which were initially defect-free and for those which initially had structural defects.

640 R. E. Green, Jr. and R. B. Pond, Sr. "An Ultrasonic Technique for Detection of the Onset of Fatigue Damage." AFOSR-TR-73-0901. Johns Hopkins University, Department of Mechanics and Materials Science, Baltimore, Maryland. May 1973. [NTIS: AD-760-787].

The purpose of the research is to develop an ultrasonic technique for the detection of the onset of fatigue damage. Particular emphasis is placed on the use of ultrasonic attenuation measurements as a continuous monitor of fatigue damage during cyclic testing of metal specimens. Efforts to date have indicated that the history of attenuation during fatigue of the specimen provide an index as to the onset of failure. Ultrasonic attenuation measurements made simultaneously with fatigue tests on aluminum bars give an early warning of eminent fracture. Such a warning occurs much earlier than detection of energy reflected from a crack by conventional ultrasonic techniques. This is true for specimens which were initially defect free and for those which initially had structural defects. A systematic comparison is being made between ultrasonic attenuation and other ultrasonic techniques, principally surface wave techniques and acoustic emission techniques.

641 R. E. Green, Jr. and R. B. Pond, Sr. "An Ultrasonic Technique for Detection of the Onset of Fatigue Damage." AFOSR-TR-76-0811. Johns Hopkins University, Department of Mechanics and Materials Science, Baltimore, Maryland. Apr 1976.

This report is a summary of research activities performed under the Air Force Office of Scientific Research Contract No. F44620-71-0062 for the period March 1, 1971 through February 28, 1976. The primary purpose of the research was to develop an optimum ultrasonic technique for the early detection of fatigue damage. As a result of the research conducted, it is concluded that the optimum ultrasonic technique for early detection of fatigue damage is a technique which permits simultaneous measurements of acoustic emission activity and ultrasonic attenuation changes.

642 **R. J. Greenfield. "Amplitudes and Spectra from Underground Sources." pp 405–425 in** ***Proceedings First Conference on Acoustic Emission/Microseismic Activity in Geologic Structures and Materials.*** **H. R. Hardy, Jr. and F. W. Leighton, Editors. Trans Tech Publications, Clausthal, West Germany. 1977. conference held at Pennsylvania State University, University Park, Pennsylvania. June 9–11, 1975.**

The signals generated by various types of sources in infinite elastic media are described with emphasis on features of the radiation patterns which can differentiate between different types of microfracturing mechanisms. Relationships between the fracture size and signal spectra are discussed. Modification of the signal during propagation by an elastic attenuation are described in both the time and frequency domain. Also it is demonstarted that layering of the material between the source and the receiver can greatly effect the amplitude of the received signal. The existence of a void or opening in the vicinity of a seismic source will modify the radiated signal. To study this effect, expressions were developed for the seismic wave fields from sources on the surface of a cylindrical opening. The expressions were employed to demonstrate the way in which the signal is modified by the presence of the opening. Both frequency and time domain effects on the radiated compressional and shear waves are shown. The presence of the opening is shown to effect wavelengths shorter than approximately 15-cavity radii.

643 **A. F. Grenis and A. P. Levitt. "Acoustic Emission of Graphite and Boron Fiber–Reinforced Aluminum Composites during Deformation and Fracture." AMMRC–TR–75–15. Army Materials and Mechanics Research Center, Watertown, Massachusetts. June 1975. [NTIS: AD/A–014–937].**

Experimental acoustic emission techniques of sound detection and analyses of pulse rate and pulse count as a means of evaluating fiber-reinforced composites are discussed. Since the stress-strain relationship for the composite, before fracture, was also obtained using conventional mechanical tension-test equipment, it was possible to detect that region (strain) where significant fiber fracture began for a particular composite specimen. The composite materials of interest included graphite fiber-reinforced aluminum alloy 7000, boron fiber-reinforced aluminum alloy 6061, and boron fiber-reinforced aluminum alloy 2024.

644 **A. F. Grenis and A. P. Levitt. "Fracture Characteristics of Different Fibers in Fiber Reinforced Aluminums as Detected by Acoustic Emission." pp 1058–1084 in** *Proceedings of the 1975 International Conference on Composite Materials. Volume 2.* **Edited by E. Scala, E. Anderson, I. Toth, and B. R. Noton. Metallurgical Society of AIME, New York. 1976. conference cosponsored by the Society for the Advancement of Material and Process Engineering, Society of the Plastics Industry, Society of Plastics Engineers. Institute of Electrical and Electronics Engineers, American Institute of Aeronautics and Astronautics, and American Society for Metals, held in Geneva, Switerzland. Apr 7–11, 1975 and in Boston, Massachusetts. Apr 14–18, 1975.**

Experimental acoustic emission techniques of sound detection and analyses of pulse rate and pulse count as a means of evaluating fiber-reinforced composites are discussed. Since the stress-strain relationship for the composite, before fracture, was also obtained using conventional mechanical tension-test equipment, it was possible to detect that region (strain) where significant fiber fracture began for a particular composite specimen. The composite materials of interest include graphite fiber reinforced aluminum alloy 7000, boron fiber reinforced aluminum alloy 6061, and boron fiber reinforced aluminum alloy 2024.

645 **A. F. Grenis and F. J. Rudy. "Fracture Characteristics of Boron Fiber-Reinforced Aluminum as Detected by Acoustic Emission."** *Army Research and Development Newsmagazine.* **Jan–Feb 1975.**

A brief summary of the work reported by the same authors in Report No. AMMRC TR 73-57 (AD-774-816). See Reference 646.

646 **A. F. Grenis and F. J. Rudy. "Fracture Characteristics of Boron Fiber-Reinforced Aluminum 2024 as Detected by Acoustic Emission." AMMRC TR 73-57. Army Materials and Mechanics Research Center, Watertown, Massachusetts. Dec 1973. [NTIS: AD-774-816].**

Results of experimental studies utilizing acoustic emission techniques with aluminum 2024 and with boron fiber-reinforced aluminum 2024 are discussed. Since the stress-strain relationship for the composite before fracture was also recorded, it was possilbe to detect that region (strain) where significant fiber fracture began for a particular composite specimen. Emphasis was on the study of the characteristics of fiber fracture rather than on determining the ultimate strength of the composite.

647 **S. C. Grigory and S. P. Ying. "Pressure Vessel Investigations: Testing of 6-in.-Thick Flawed Tensile Specimens." pp 48–67 in "Heavy-Section Steel Technology Program Semiannual Progress Report Period Ending August 31, 1971." F. J. Witt, Program Director. ORNL-4764. Oak Ridge National Laboratory, Oak Ridge, Tennessee. Apr 1972. [NTIS].**

Southwest Research Institute continues to test large 6-inch thick tensile specimens containing sharp flaws. A total of 13 such specimens have now been tested. The last three specimens tested contained flaws in a submerged-arc weld. These specimens exhibited tougher behavior at lower temperatures than the plate specimen. Strains based on a 24-inch gage length spanning the flaw exceed 1%, even for the specimens tested at 0 °F. A series of 1-inch thick models of the larger specimens are also being tested. A total of 7 specimens were tested, with 19 additional specimens scheduled for testing. Acoustic emission measurements were also made on the specimens tested at Southwest Research Institute. Plots of emission versus stress intensity were made, and three modes of behavior were noted. The acoustic emissions increase rapidly before yield, exhibit a small plateau near the yield point, and increase again under gross plastic deformation. Both sizes of specimens behaved in much the same fashion, with the larger specimens exhibiting a higher rate of emission than the smaller ones.

648 S. C. Crigory and S. P. Ying. "Pressure Vessel Investigations: Testing the 6-in. Thick Flawed Tensile Specimens." pp 64–81 in "Heavy Section Steel Technology Program Semiannual Progress Report for Period Ending February 28, 1971." F. J. Witt, Program Director. ORNL–4681. Oak Ridge National Laboratory, Oak Ridge, Tennessee. Dec 1971. [NTIS].

During this reporting period, emphasis was placed on the simulated service test activity, although effort was also continued under the complex stress state studies and the specific safety research studies. Activities reported include the development of procedures for flaw sharpening of the simulated service test specimens, the testing of the eight additional large flawed tensile specimens at Southwest Research Institute, and an evaluation of the acoustic emission data obtained during the testing of these specimens. Acoustic emissions were studied as a function of load or stress intensity factor. The cumulative emissions were shown to be an exponential function of the stress intensity. The exponent appeared to vary with test temperature. Expediting of procurement of six intermediate test vessels from Taylor Forge Company was continued, and a summary of the material properties evaluation of the vessel components led to a retempering of shell courses.

649 P. J. Guest. "The Micromechanics of Fracture of Pearlitic Structural Steels." UCRL–19106. Lawrence Radiation Laboratory, University of California, Berkeley, California. Dec 1969. [NTIS].

Few quantitative investigations studying fracture and the related fracture properties of hot-rolled, ferrite-pearlite structural steels have been made. The intent of this investigation was to provide fracture toughness data that can be made available in structural specifications and applied in structural design, as well as to understand the micro-characteristics of fracture. A mechanical and metallurgical analysis has been carried out on the pearlitic, martensitic, and spheroidized carbide structures of plain carbon hypoeutectoid steels in the temperature range of –320 to +150 °F. The quantitative measurement of the microcharacteristics associated with cleavage fracture was determined by the use of the stress wave emission technique, which essentially monitors the elastic wave produced by discontinuous crack movements. The measured mechanical properties were correlated with information obtained from optical and scanning electron microscopy techniques. The analysis shows that as the volume fraction of pearlite is increased from 0 to 1, the plane strain fracture toughness decreased by 65% at +70 °F and by 61% at −320 °F. The data showed there to be a square root of *d* relationship between the plane strain fracture toughness and ferrite grain size. The micromechanical analysis showed that by increasing the volume fraction of pearlite from 0 to 1, the microcrack density increased by a factor of 8 at −100 °F and by a factor of 9 at −320 °F.

650 **F. J. Guild, D. Walton, R. D. Adams, and D. Short. "The Application of Acoustic Emission to Fibre-Reinforced Composite Materials."** *Composites,* **CPSOA, 7(3):173–179. July 1976.**

Applied to metallic materials, acoustic emission is rapidly gaining status as a practical nondestructive test tool. This has resulted from exhaustive research efforts to detect and identify such features as phase transformation, plastic deformation, and crack growth. This review examines the *state-of-the-art* with regard to composite materials where fracture processes are generally more complex and interpretation is correspondingly more difficult.

651 **L. R. Gulley, Jr. "Inspection of Advanced Composite Aircraft Structure." Air Force Materials Laboratory, Wright-Patterson AFB, Ohio. 1975.**

Advanced composite materials (boron-epoxy, graphite-epoxy, etc.) are being used more widely for new aircraft structures, but ways to inspect these structures have been slow to develop. With advanced composite tail structures on the F-4, F-14, -15, and -16 aircraft, the need has arisen to provide inspection methods. The Air Force Materials Laboratory technical efforts have been directed to this area by the extensive background in composites as well as by recent aircraft incidents involving failures of advanced composite structures. In the paper to be presented, discussion will center on the use of techniques including holographic interferometry and thermochromic paints for the detection of corrosion, cell damage, and disbonding in composite rudders and other structures.

652 **B. Gutenberg and C. R. Richter.** *Seismicity of the Earth.* **Princeton University Press, Princeton, New Jersey. 1949.**

653 **I. S. Guz and V. M. Finkel. "Relationship between the Spectrum of Waves Emitted by a Growing Crack and the Energy Reserve at its Tip."** *Soviet Physics – Solid State,* **SPSSA, 14(7):1619–1621. Jan 1973; translated from** *Fizika Tverdogo Tela,* **FTVTA, 14(7):1865–1869. July 1972. (in Russian).**

The parameters of the elastic waves excited by the rupture of Plexiglas samples are investigated. Waves are emitted only when the rate of crack growth varies. The width of the spectrum and the frequencies of the waves carrying the most energy depend on the reserve of elastic energy at the tip of the crack. [reprinted by permission of the American Institute of Physics].

654 **I. S. Guz and A. D. Zotov. "Emission of Elastic Waves during Crack Growth in Steel that has Passed through Different Heat Treatments."** *Strength of Materials,* **SMTLB, 6(4):461–463. Apr 1974; translated from** *Problemy Prochnosti,* **PPCNB, (4):63–65. Apr 1974. (in Russian).**

It is well-known that the instantaneous rupture of a continuity is accompanied by the emission of elastic waves that carry information concerning the state of the material and that may therefore be used successfully for diagnosing crack development. In this connection, it is important to know the relations between the parameters of the waves that form and the fracture process on the one hand, and the properties of the medium in which fracture takes place on the other. It has been shown previously that during fracture, Rayleigh waves as well as longitudinal and transverse waves are emitted. A relation between the parameters of the Rayleigh waves and the stored energy before fracture has been established. In the present paper, examined were the relations between the parameters and the mechanism of emitting longitudinal waves during crack growth in steel that has been subjected to different types of heat treatment. Similar studies were carried out in which sonic pulses, generated by a growing crack, were recorded by means of piezoelectric transducers through an air layer. The method used in this work does not permit unique differentiation of the type of recorded wave, and the presence of an air layer leads to damping and distortion of the investigated pulses. In the present investigations, the air layer has been eliminated, and the results may therefore differ substantially from those cited by V. M. Finkel and S. V. Serebryakov in Reference 470.

H

655 **R. C. Haborak. "An Acoustic Emission System for Monitoring Structural Integrity of Complex Configurations." Note No. ADN 09-01-71.5. Grumman Aerospace Corporation, Advanced Development Program, Bethpage, New York. Sep 1971. [NTIS: AD-893-449-L (USGO)].**

An acoustic emission system manufactured by Battelle-Northwest Laboratories was evaluated for its ability to assess the structural integrity of a complex titanium part undergoing fatigue. The system was successful in detecting and locating early crack formation. The results of this evaluation indicated that modifications must be incorporated into the present instrumentation. These modifications are designed to enhance the quality and reduce the quantity of the results obtained from testing similar and even larger complex structures in an industrial environment. Battelle is presently incorporating the required modifications into their acoustic emission system.

656 **D. J. Hagemaier. "Acoustic Emission Testing: Feasibility Study." LR-DAC-4446. Douglas Aircraft Company, Materials and Process Engineering, Long Beach, California. Apr 29, 1969.**

A feasibility study was conducted to demonstrate acoustic emission test equipment, developed at Battelle-Northwest, for evaluating failure mechanisms in various materials and structures. Tests were conducted on induction-brazed tube joints; cyclic fatigue of typical aircraft fastener specimens; and tensile tests of quartz-phenolic, aluminum, and stainless steel tensile bars. The acoustic emissions (stress waves) produced as the specimens cracked were detected, amplified, time integrated, and recorded. The recorded results varied for the specimens tested, but were basically identical for each specimen category. Also, the results indicate a load-time signature which shows crack initiation and the possibility to predict ultimate failure.

657 **D. J. Hagemaier. "Applications for Specialized Nondestructive Testing Methods."** ***Metal Progress,*** **MEPOA, 100(2):67-68. Aug 1971.**

Testing methods described include acoustic emission, eddy-sonic, holography, leak testing, microwave, Mossbauer effect, neutron radiography, and thermoelectric methods. Their applications are indicated. [Abstract 12744. *Engineering Index,* 71 (Part 1):1761. 1972].

658 **D. J. Hagemaier. "Nondestructive Methods for Rating Materials."** ***Metal Progress,*** **MEPOA, 100(2):72-75. Aug 1971; updated and expanded in** ***Metal Progress,*** **MEPOA, 106(1):184-193. Mid-June 1974, [see also Reference 659].**

Chart of 24 different nondestructive testing methods, including acoustic emission, listing parameters measured or detected, applications, advantages, and limitations for each.

659 **D. J. Hagemaier. "Nondestructive Testing Methods for Materials Evaluation."** *Materials Evaluation,* **MAEVA, 28(6):25A–28A. June 1970, [see Reference 658 for updated and expanded chart].**

Four page chart of 24 different nondestructive testing methods, including acoustic emission, listing parameters measured or detected, applications, advantages, and limitations of each.

660 **D. J. Hagemaier. "Recent Techniques for Nondestructive Materials Evaluation." Douglas Paper 6045. Douglas Aircraft Company, Long Beach, California. 1970; "Prove non distruttive sui materiali." pp 489–502 in** *Enciclopedia Della Scienza e Della Tecnica Mondadori,* **Arnoldo Mondadori Editore, Milano, Corso, Italy. 1973. (in Italian).**

As new methods of NDT are developed and their applications, limitations, and advantages are determined they are added to the existing list of NDT methods for product evaluation. This paper presents a discussion of the following recently developed methods or techniques for nondestructive materials evaluation: Acoustic-emission, eddy-sonic, holography, microwaves, Mossbauer effect, neutron activation analysis, neutron radiography, and contact thermography.

661 **D. J. Hagemaier, H. J. McFaul, and D. Moon. "Nondestructive Testing of Graphite Fiber Composite Structures." Douglas Paper 5785. Douglas Aircraft Company, Long Beach, California. 1970; paper presented at Society of Automotive Engineers meeting. Los Angeles, California. Oct 7, 1970;** *Materials Evaluation,* **MAEVA, 29(6):133–140. June 1971.**

The paper discusses nondestructive testing methods and techniques for detection of material and manufacturing anomalies in state-of-the-art graphite composites. Introduction of composites into large-scale production depends in large measure on the successful application of NDT for process control and monitoring. Visual, liquid penetrant, radiographic, ultrasonic, sonic, thermal, and acoustic emission NDT methods are used to evaluate graphite fiber composite aircraft specimens and structures. Reference standards are fabricated with built-in discontinuities, commensurate with each structural article fabricated. The test specimens and parts were tested by various methods in order to determine optimum discontinuity detection. Tests were conducted in research and development graphite composite aircraft structures consisting of (1) graphite composite horizontal stabilizer assembly, (2) graphite composite flap assembly, and (3) molded graphite flap ribs.

662 D. J. Hagemaier, H. J. McFaul, and J. T. Parks. "Nondestructive Testing Techniques for Fiberglass, Graphite Fiber and Boron Fiber Composite Aircraft Structures." ***Materials Evaluation,*** **MAEVA, 28(9):194-204. Sep 1970; paper presented at American Society for Nondestructive Testing Spring Conference, Los Angeles, California. Mar 9-12, 1970.**

Various nondestructive test methods were evaluated for inspection and evaluation of boron, graphite, and glass-fiber composites for aircraft structures. Typical specimens were evaluated using microscopic, fluorescent penetrant, radiographic, ultrasonic, and thermochromic test methods. Optical microscopic examination is useful for determination of fiber pattern from the edge of a panel. It was concluded that fluorescent penetrant combined with microscopic examination is a useful tool to determine surface defects; radiography combined with magnified photography is an excellent method to determine fiber pattern, fiber gaps, broken fibers, crushed core, and resin-rich areas; and ultrasonic and thermal methods appear to have merit for determining unbonded areas. The results of a literature survey concerning the NDT of composites are presented in abstracted form and indicate significant applications and limitations of various test methods. The direction of future NDT research and development efforts is indicated.

663 C. P. Haigh and A. B. Ponter. "Sound Emission from Boiling on a Submerged Wire." ***Canadian Journal of Chemical Engineering,*** **CJCEA, 49(3):309-313. June 1971.**

The sound accompanying nucleate and film boiling on a platinum wire immersed in a water pool at atmosphere pressure is related to the rate of heat transfer from the wire surface. In particular the total sound intensity is shown to be a function of the heat transfer resulting from bubble evolution. Analysis of the acoustic spectra for saturated and subcooled boiling shows how the level of selected components varies with changes in the heat flux and liquid temperature. The possibility of using an acoustic detection device to detect boiling caused by channel blockage in a liquid-cooled system is discussed. [reprinted from *The Canadian Journal of Chemical Engineering*].

664 B. C. Haimson and K. Kim. "Acoustic Emission and Fatigue Mechanism in Rock." pp 35-55 in ***Proceedings First Conference on Acoustic Emission/ Microseismic Activity in Geologic Structures and Materials.*** **H. R. Hardy, Jr. and F. W. Leighton, Editors. Trans Tech Publications, Clausthal, West Germany. 1977. conference held at Pennsylvania State University, University Park, Pennsylvania. June 9-11, 1975.**

Acoustic emission records taken during monotonic and cyclic loading of rocks are reported. Four to five different rock types were tested in compression, tension, and tension-compression. Failure mechanism in each of these loading modes is different and AE can be utilized to satisfactorily describe the types of failure. In particular, two classes of tensile failure can be easily identified through AE. In addition, the accelerated fatigue failure observed in tension-compression is clearly recognized through AE. In general, the abundance of AE can be correlated to the strain-time behavior during loading. However, since AE appears to be significantly more sensitive to imminent failure it could possibly be used to advantage in some practical applications.

665 **E. T. Hall and D. I. Crecraft. "Bonded Joints and Nondestructive Testing: NDT of Resistance Spots, Roll-Spot, Stitch and Seam Welds."** *Non-Destructive Testing,* **NDETA, 4(3):181–191. June 1971.**

Several ultrasonic techniques have been developed for measurement of weld quality which are largely applicable to the spot welding process. Some of these techniques have been devised for application after completion of welding as a separate process away from the welding machine. Others have been developed for use on the welding machine. Of these latter, some have potential application to in-process quality control. This article describes the welding process and reviews current techniques for quality assessment including ultrasonic techniques either in use or under development. To enable these developments to be viewed against the background of current resistance welding practice, the definitions and terminology of the welding processes are explained and the present position of alternative quality assurance techniques is reviewed.

666 **S. L. Halverson, T. T. Anderson, A. P. Gavin, and T. Grate. "Radiation Exposure of a Lithium Niobate Crystal at High Temperatures."** *IEEE Transactions on Nuclear Science,* **IETNA, NS-17(6):335–340. Dec 1970.**

Lithium niobate is being considered for use as the active element in piezoelectric transducers in a nuclear reactor. For tests in a simulation of this environment, a Y-cut disk of this material 1 cm in diameter and 0.1 cm thick was mounted in a holder designed to allow measurement of piezoelectric properties of the materials, and was exposed to 4×10^9 of ^{60}Co gamma rays at elevated temperatures in air. Exposure was at a dose rate of 1.75×10^7 R/h with approximately one half of the dose accumulated at 281 °C. The piezoelectric properties showed no changes that would affect the usefulness of the material for the intended purpose.

667 **T. Hamada. "Effects of Threshold Level on Probabilities of Detection and Location of AE Sources." Paper 2-3, pp 101–117 in** *The Third Acoustic Emission Symposium.* **Japan Industrial Planning Association, Dobashi Building, 3-17-8, Akasaka, Minato-ku, Tokyo, Japan. 1976. proceedings of a symposium held in Tokyo, Japan. Sep 16–18, 1976.**

In AE location systems, a threshold circuit is used for each sensor and AE sources are located by detecting signals which exceed the threshold level, so that the detecting and locating probability of AE sources are influenced by the threshold to a great extent. Large threshold value (low sensitivity), of course, results only in poor detectability, but too small threshold (high sensitivity) also reduces the location performance, because spurious locations are increased by random combinations of sensed signals. Although the selection of the threshold depends on the practical situations of individual systems, it is desirable to formulate the general effect of threshold based on an appropriate model. The purpose of this paper is to introduce a simple mathematical model for AE amplitude and attenuation, and to describe quantitatively detection probabilities and location performances as a function of threshold. For the simplicity of analysis, a uniform infinite plane is selected as the AE source and propagation medium, and a three-sensor system in a triangle arrangement is employed as an example of no redundancy location.

668 **M. A. Hamstad. "Acoustic Emission from Filament-Wound Pressure Bottles." UCRL-73763. Lawrence Livermore Laboratory, Livermore, California. May 5, 1972. [NTIS]; pp 321-331 in** ***Non-Metallic Materials: Selection, Processing and Environmental Behavior.*** **National SAMPE Technical Conference Series, Volume 4. Society for the Advancement of Material and Process Engineering, Azusa, California. 1972. conference held in Palo Alto, California. Oct 17-19, 1972.**

Acoustic emission records were obtained during burst tests of filament-wound composite pressure bottles to determine how acoustic emission depends on various winding parameters. Winding parameter changes included: (1) different ratios of longitudinal wraps to hoop wraps; (2) different filament materials; (3) interspersed hoop and longitudinal wraps; and (4) bottles wound with single-end, 6-end, 12-end, and 20-end glass yarn. The acoustic emission record gave a graphic display of the microscopic damage that occurred in the composite bottle as a function of pressure.

669 **M. A. Hamstad. "Acoustic Emission Uses in Research and Development of Composite Materials." UCRL-76257. Lawrence Livermore Laboratory, Livermore, California. Nov 21, 1974.**

A comprehensive guide to the use of acoustic emission testing in the research and development of composite materials is presented. Included are discussions of: (1) the unique information to be gained from acoustic emission data, (2) the sources of acoustic emission in composites, (3) successful composite acoustic emission testing, (4) the measurement and the interpretation of acoustic emission from composites, (5) the specific applications of previous acoustic emission data from composites, and (6) the future directions of acoustic emission research in the field of composites.

670 **M. A. Hamstad. "On Energy Measurement of Continuous Acoustic Emission." UCRL-76286. Lawrence Livermore Laboratory, Livermore, California. Dec 26, 1974.**

The measurement of energy in a continuous acoustic emission signal by a true rms (root-mean-square) voltmeter or a tms (true-mean-square) voltmeter is reviewed. Included in this review are (1) a comparison of typical rms and tms voltmeters, (2) a discussion of how these voltmeters measure energy, (3) a discussion of the subtraction of background noise from the true acoustic emission, (4) a presentation of the advantages of the rms or tms measurement techniques, and (5) a discussion of some practical factors in such energy measurements.

671 **M. A. Hamstad, R. Bianchetti, and A. K. Mukherjee. "A Correlation between Acoustic Emission and the Fracture Toughness of 2124–T851 Aluminum." UCRL–78151. Lawrence Livermore Laboratory, Livermore, California. June 9, 1976. [NTIS] ;** ***Engineering Fracture Mechanics,*** **EFMEA, 9(3):663–674. 1977.**

The paper reports on an experimental study of the possible application of acoustic emission as an inspection technique, to determine whether 2124-T851 aluminum plates meets certain, minimum, fracture toughness specifications. Unflawed specimens, taken in the three principal directions, from three separate plates are tested in compression and tension. The true-mean-square or the root-mean-square voltage of the continuous acoustic emission generated during the tests is recorded as a function of specimen strain. The acoustic emission, chemical analysis, fracture surfaces, and large (1 to 20 μm) second-phase particles are studied in relation to fracture toughness. Results indicate that the generated acoustic emission from a longitudinal compression test, long-transverse tension test, and short-transverse tension test all correlate with changes in plate fracture toughness. Based on the experimental results, the monitoring of acoustic emission during relatively simple tension or compression tests of aluminum may be useful in checking whether the material meets a fracture toughness specification. Also, the results indicate that acoustic emission is useful in studies to develop fracture toughness models. [reprinted by permission of Pergamon Press Ltd].

672 **M. A. Hamstad and T. T. Chiao. "A Physical Mechanism for the Early Acoustic Emission in an Organic–Fiber/Epoxy Pressure Vessel." UCRL–74907. Lawrence Livermore Laboratory, Livermore, California. July 25, 1973;** ***SAMPE Quarterly,*** **SAMQA, 5(2):22–27. Jan 1974.**

A physical mechanism is proposed to explain the appearance of the first peak in the acoustic emission count rate during pressurization of organic-fiber/epoxy composite vessels. The acoustic emission from organic-fiber/epoxy elongated NOL rings and simple strands is also discussed. Micrographs indicate that filament breakage, which occurs when the epoxy cracks, is the source of this early acoustic emission peak. It is concluded that acoustic emission testing may be the best way to determine which composite properties might be changed in order to minimize this early filament damage.

673 **M. A. Hamstad and T. T. Chiao. "Acoustic Emission from Stress Rupture and Fatigue on an Organic Fiber Composite." UCRL-74745. Lawrence Livermore Laboratory, Livermore, California. Jan 15, 1974. [NTIS]; paper presented at the American Society for Testing and Materials Composite Reliability Conference, Las Vegas, Nevada. Apr 15-16, 1974.**

A series of tests were carried out to determine the scatter in acoustic emission data gathered during stress rupture and fatigue testing of a simple fiber-reinforced epoxy composite. Single-end organic fiber/epoxy strands were monitored for acoustic emission during tensile tests to failure, stress rupture tests, and dynamic fatigue tests. During the stress rupture and fatigue tests, the plots of summation of acoustic emission counts versus time were found to resemble metal creep curves in that primary, secondary, and tertiary regions could be distinguished. Because of a significant amount of scatter, only a limited correlation can be made between the slope of the summation of acoustic emission in the steady state region and the specimen life during stress rupture and fatigue testing. In addition, flawed specimens were easily sorted out during the tensile tests. These results indicate that acoustic emission data gathered during composite fatigue and stress rupture testing might be useful for life prediction.

674 **M. A. Hamstad and T. T. Chiao. "Acoustic Emission Produced during Burst Tests of Filament-Wound Bottles." UCRL-74528. Lawrence Livermore Laboratory, Livermore, California. Mar 29, 1973;** *Journal of Composite Materials,* **JCOMB, 7(3):320-332. July 1973.**

Acoustic emission was recorded during burst tests of filament-wound, composite pressure vessels. Organic and graphite fibers were tested, and two different epoxy resin systems were used: one with a low and another with a relatively high cure temperature. Acoustic emission was studied for the effects of different winding patterns, artificial flaws, winding-induced fiber fraying, different resins, and different fibers. Small effects produced in the vessels by changes in these variables were greatly magnified when they appeared as changes in acoustic emission. They would, in fact, be difficult or impossible to detect by other test means. [reprinted by permission of Technomic Publishing Company, Inc., 265 Post Road West, Westport, CT 06880].

675 **M. A. Hamstad and T. T. Chiao. "Structural Integrity of Fiber-Epoxy Vessels by Acoustic Emission: Some Experimental Considerations." UCRL-78198. Lawrence Livermore Laboratory. Livermore, California. May 21, 1976. [NTIS]; "Structural Integrity of Fiber-Epoxy Vessels by Acoustic Emission."** *SAMPE Quarterly,* **SAMQA, 8(1):31-45. Oct 1976.**

The application of acoustic emission to the prediction of the burst pressure of pressure vessels wound with organic fiber-epoxy filament has been studied. Experimental aspects of this work included determination of longitudinal and shear velocities, propagation signal losses, resonant versus nonresonant acoustic emission transducers, pressure vessels of different sizes and thicknesses, different epoxy systems, and frequency spectrum analysis. The central conclusion is that the acoustic emission system should be operated with a relatively low frequency band pass to avoid large propagation signal losses.

676 **M. A. Hamstad and A. K. Mukherjee. "A Comparison of the Acoustic Emission Generated by Tensile and Compression Testing of 7075 Aluminum." UCRL-77502. Lawrence Livermore Laboratory, Livermore, California. Nov 20, 1975.**

An experimental study of the acoustic emission generated during tensile and compression testing of unflawed 7075 aluminum has been completed. Variables studied included: specimen orientation in a 13-cm plate, 1, 3, and 4.8% plastic stretches before the aging heat treatment, and 7075-T6 versus 7075-T651. The level of acoustic emission for strains greater than 1.5% was vastly different between compression and tensile results.

677 **M. A. Hamstad and A. K. Mukherjee. "Acoustic Emission Testing of Unflawed 7075–T6 Aluminum." UCRL–75787. Lawrence Livermore Laboratory, Livermore, California. June 28, 1974; *Journal of Testing and Evaluation,* JTEVA, 3(3):167–172. May 1975.**

Tensile tests were performed to study the differences in the continuous acoustic emission from 7075-T6 aluminum plates made by five manufacturers. The grain structures and the chemical compositions of these plates were determined. Results are given on a sixth plate, which emitted burst-type acoustic emission superimposed on the continuous acoustic emission. The effect of orientation with respect to the rolling direction within a single plate was studied. Results from interrupted tensile tests are also reported; these tests were stopped at about 4.5 percent strain for different time periods before they were resumed. The tests indicate that acoustic emission is sensitive to differences between plates of 7075-T6 aluminum that do not show up on a stress-strain curve. Furthermore, the dislocation mechanisms that cause the continuous acoustic emission in this aluminum alloy are altered by room-temperature diffusion-based ageing. [reprinted by permission of the American Society for Testing and Materials, Copyright].

678 **M. A. Hamstad and A. K. Mukherjee. "Acoustic Emission versus Strain and Strain Rate: Measurement and a Dislocation Theory Model for a Dispersion-Strengthened Aluminum Alloy." UCRL–76077. Lawrence Livermore Laboratory, Livermore, California. Oct 5, 1974. [NTIS].**

The essential character of the continuous acoustic emission generated during a tensile test of 7075-T6 aluminum has been studied. Measurement of this continuous acoustic emission by a true rms meter is discussed. The question of subtracting the background electronic noise from the acoustic emission signal is considered, as well as the question of the strain-rate sensitivity of continuous acoustic emission in the alloy. A theory describing the dependence of rms on strain and strain rate is derived using a dislocation theory model and the measurement characteristics of the rms meter. It is concluded that rms is proportional to the square root of the plastic strain rate, but that this half-power dependence is not a material constant.

679 **M. A. Hamstad and A. K. Mukherjee. "The Dependence of Acoustic Emission and Strain Rate for a Dispersion Strengthened Aluminum Alloy." pp 574–578 in** *4th International Conference on the Strength of Metals and Alloys: Proceedings. Volume 2.* **Laboratoire de Physique du Solide – ENSMIM, INPL–Parc de Saurupt, 54042, Nancy Cedex, France. 1976. conference held in Nancy, France. Aug 30–Sep 3, 1976.**

This is a summary paper of the authors' Report UCRL-76077 (see Reference 678).

680 **M. A. Hamstad and A. K. Mukherjee. "The Dependence of Acoustic Emission on Strain Rate in 7075-T6 Aluminum." UCRL-74260 (Rev. 1). Lawrence Livermore Laboratory, Livermore, California. June 29, 1973. [NTIS]; paper presented at the Society for Experimental Stress Analysis 3rd International Congress on Experimental Mechanics, Los Angeles, California. May 13–18, 1973;** *Experimental Mechanics,* **EXMCA, 14(1):33–41. Jan 1974.**

The dependence of root mean square (rms) acoustic emission output on strain rate was determined during tensile tests on 7075-T6 aluminum over a range of strain rates from 0.015 to 0.230 per minute. A special normalization technique was developed to eliminate the inherent differences in the acoustic emission from supposedly identical specimens tested under identical conditions. It was found that the rms acoustic emission output increases approximately linearly with strain rate, and that the slope of the linear dependence decreased with increased plastic strain. The effect of pinned dislocation breakaway on the acoustic emission output versus plastic strain relationship is the subject of a preliminary theory.

681 **M. A. Hamstad and R. G. Patterson. "Considerations for Acoustic Emission Monitoring of Spherical Kevlar-Epoxy Composite Pressure Vessels." UCRL-79356. Lawrence Livermore Laboratory, Livermore, California. May 16, 1977. [NTIS].**

Continuing research is being conducted in the applications of acoustic emission testing for predicting burst pressure of filament-wound Kevlar-49 epoxy pressure vessels. The study focused on three specific areas. The first area involves development of an experimental technique and the proper instrumentation to measure the energy given off by the acoustic emission transducer per acoustic emission burst. The second area concerns the design of a test fixture in which to mount the composite vessel so that the acoustic emission transducers are held against the outer surface of the composite. Included in this study area is the calibration of the entire test setup including couplant, transducer, electronics, and the instrument measuring the energy per burst. In the third and final area of study, the number, location, and sensitivity of the acoustic emission transducers used for proof testing composite pressure vessels are considered.

682 **S. Hanagud and R. G. Clinton. "Acoustic Emission Techniques in the Development of a Diagnostic Tool for Osteoporosis." pp 41–45 in *1975 Ultrasonics Symposium Proceedings.* Catalog No. 75 CHO 994-4SU. Institute of Electrical and Electronics Engineers, Inc., New York. 1975. symposium held in Los Angeles, California. Sep 22–24, 1975.**

It has been demonstrated in the past that bones do emit acoustic emission signals and that these signals can be detected and analyzed. The paper discusses the results of acoustic emission on bone specimens subjected to tension. The tests are done both on normal bones and on bones that have suffered damage similar to that of osteoporosis. The damage similar to that of osteoporosis has been produced by using a solution containing acetic acid. Bone specimens have been prepared by a careful machining process. A simple design for the diagnostic tool has been presented and discussed. The tool is capable of applying a load and measuring emission at selected locations.

683 **S. Hanagud, R. G. Clinton, and J. P. Lopez. "Acoustic Emission in Bone Substance." pp 79–81 in ASME (AMD–Vol 2), proceedings of Biomechanical Symposium, available from American Society of Mechanical Engineers, New York: paper presented at ASME Joint Meeting, Georgia Institute of Technology, Altanta, Georgia. June 20–22, 1973.**

The paper discusses the application of acoustic emission techniques to research on the mechanical properties of bone material.

684 **S. Hanagud, J. I. Craig, and R. Deadwyler. "Acoustic Emission in Sea Ice." pp 4–7 in *1972 Ultrasonics Symposium Proceedings.* Catalog No. 72 CHO 708-8SU. Institute of Electrical and Electronics Engineers, Inc., New York. 1972. symposium sponsored by Institute of Electrical and Electronics Engineers Group on Sonics and Ultrasonics, held in Boston, Massachusetts. Oct 4–7, 1972.**

Acoustic emission in sea ice subjected to slow rates of compression has been studied. Signals in the range 10 to 40 kHz and 100 to 300 kHz have been investigated for different temperatures of sea ice. Correlation of maximum strength with significant emission levels and maximum emission rates have been investigated. Methods of extrapolation of maximum strength from observations at low loads and the appropriate models for statistical distribution of acoustic emission activity have been discussed.

685 **S. Hanagud, G. T. Hannon, and R. Clinton. "Acoustic Emission and Diagnosis of Osteoporosis." pp 77–80 in** ***1974 Ultrasonics Symposium Proceedings.*** **Catalog No. 74 CHO 896-1SU. Institute of Electrical and Electronics Engineers, Inc., New York. 1974. symposium held in Milwaukee, Wisconsin. Nov 11–14, 1974.**

The paper discusses the use of acoustic emission in bone substance as a tool in early diagnosis of osteoporosis. Acoustic emission tests have been conducted on carefully prepared bone specimens subjected to bending loads. The specimens included femur from cattle and cadavers. A special procedure was devised to produce nearly identical specimens. This procedure also avoids any introduction defects in the bone specimens during the machining operations. A comparison of emission patterns from 60 perfect and defective specimens clearly indicated that the development of an effective early diagnostic tool for osteoporosis is possible by using acoustic emission techniques. Acoustic emission signals were detected by using Dunegan-Endevco transducers, which are moderately peaked at 155 kHz. The amplified and filtered signals were analyzed by using a counter. The paper also proposes a method of forecasting the defective bones from early observations by using the probability distributions for counts and count rates.

686 **M. Hanff, N. Lions, and J. Peronnet (to Commissariat a l'Energie Atomique, France). "Dispositif de Detection Acoustique dans un Reacteur Nucleaire au Sodium et Procede d'Etalonnage dudit Dispositif (Acoustical Detector Device for a Sodium Nuclear Reactor and its Calibration)." French Patent 2,150,572. July 22, 1971. (in French).**

A device for detecting local boiling in a sodium nuclear reactor is presented together with a method for calibrating the device. The applications of this acoustical detector device are not limited to the detection of acoustical phenomena accompanying boiling. Cavitation, abnormal vibrations, and all sound-wave emitting phenomena within the sensitivity range and passband limits of the apparatus can be detected. The detector is provided with an acoustical conductor in the form of a tube open at its lower end and immersed in the sodium, and a second acoustical conductor joined in series to the first and in the form of a solid rod extending to the outside of the reactor vessel. An accelerometer detects the signals from the second conductor, an amplifier enabling the signals detected to be fed into the measurement equipment.

687 **A. Hannavy. "New Detection Devices Help Predict Potential Failure."** ***Product Engineering,*** **PRENA, 37:37–44. Mar 27, 1967.**

688 **H. R. Hardy, Jr. "Acoustic Emission as a Nondestructive Tool for Monitoring the Stability of Underground Geologic Structures/Les emissions acoustiques: un moyen non destructif pour controler la stabilite du rocher des installations souterraines." Pennsylvania State University, University Park, Pennsylvania; Paper 5A/3 in *Proceedings of the Eighth World Conference on Nondestructive Testing,* volume containing Sections 3K2, 3H2, 5A. International Committee for Nondestructive Testing, Secretariat du Cofrend, Institut de Soudure, 32 Boulevard de la Chapelle, 75880 Paris Cedex 18, France. 1976. conference held in Cannes, France. Sep 6–10, 1976. (paper in English).**

The paper discusses the application of acoustic emission to various geotechnical problems. These include the structural stability of underground coal and hardrock mines, tunnels, gas storage reservoirs, and mined cavities for storage of fluids.

689 **H. R. Hardy, Jr. "Acoustic Emission from Geologic Structures." Internal Report RML-IR/73-18. Pennsylvania State University, Department of Mineral Engineering, University Park, Pennsylvania. Dec 1973; paper presented at the 12th Acoustic Emission Working Group Meeting. New York. Dec 5–7, 1973.**

690 **H. R. Hardy, Jr. "Applications of Acoustic Emission in Rock Mechanics." Pennsylvania State University, College of Earth and Mineral Sciences, University Park, Pennsylvania. 1969; paper presented at the 4th Acoustic Emission Working Group Meeting, Philadelphia, Pennsylvania. Oct 14–15, 1969.**

Acoustic emission is a process in which a material or structure emits transient vibrations in the audible and subaudible range when stressed. When geologic materials are being considered, the phenomenon is often referred to as *microseismic activity, rock noise,* or *seismoacoustical activity.* Each individual microseismic disturbance is generally referred to as a *microseismim.* The terms microseismic activity and microseismim will be employed exclusively in this paper. In geologic materials the origin of microseismims is not well understood, but it appears to be related to processes of deformation and failure which are accompanied by a sudded release of strain energy. In such a polycrystalline material, microseismims may originate at the microlevel as a result of dislocations or at the microlevel by twinning, grain boundary movement, or initiation and propagation of fractures through and between the grains. It is assumed that the sudden release of stored elastic-strain energy accompanying these processes generates a stress wave which travels from the point of origin within the material, to a boundary where it is observed as a microseismim. Although an understanding of the generating process is important from a fundamental point of view, it is secondary with respect to application. Here the important considerations are as follows: (i) Microseismims originate at locations where the material is mechanically unstable; (ii) They propagate through the surrounding material undergoing attenuation, usually frequency dependent, as they move away from their source; (iii) With suitable apparatus microseismims may be detected at locations a considerable distance from their source;

(iv) Their rate of occurrence, magnitude, and frequency spectra, provide indirect evidence of the type and degree of instability; and (v) Observations obtained from a number of stations (array) make it possible to determine the actual source location. Research workers in the field of rock mechanics have found microseismic activity to be an extremely useful tool. Studies of both a basic and applied nature have been under way for some years.

691 **H. R. Hardy, Jr. "Application of Acoustic Emission Techniques to Rock Mechanics Research." pp 41–83 in** ***Acoustic Emission, ASTM STP 505.*** **American Society for Testing and Materials, Philadelphia, Pennsylvania. 1972. proceedings of symposium held in Bal Harbour, Florida. Dec 7–8, 1971.**

Presented is a comprehensive review of the applications of acoustic emission in the field of rock mechanics. Acoustic emission experiments in rock mechanics were initiated in North America in the late 1930's as part of a study by the U.S. Bureau of Mines into problems associated with mine design and alleviation of rock bursts. More recently acoustic emission techniques had been employed to an increasing degree for the laboratory study of basic deformation and failure mechanisms in geologic materials. The paper outlines the major rock mechanics areas where acoustic emission techniques are being applied, and includes a brief description of a number of basic and applied studies.

692 **H. R. Hardy, Jr. "Emergence of Acoustic Emission/Microseismic Activity as a Tool in Geomechanics." pp 13–31 in** ***Proceedings First Conference on Acoustic Emission/Microseismic Activity in Geologic Structures and Materials.*** **H. R. Hardy, Jr. and F. W. Leighton, Editors. Trans Tech Publications, Clausthal, West Germany. 1977. conference held at Pennsylvania State University, University Park, Pennsylvania. June 9–11, 1975.**

693 **H. R. Hardy, Jr. "Evaluating the Stability of Geologic Structures Using Acoustic Emission." pp 80–106 in** ***Monitoring Structural Integrity by Acoustic Emission, ASTM STP 571.*** **Edited by J. C. Spanner and J. W. McElroy. American Society for Testing and Materials, Philadelphia, Pennsylvania. 1975. proceedings of a symposium held in Ft. Lauderdale, Florida. Jan 17–18, 1974.**

Paper describes the application of acoustic emission to the study of geologic structures. Evaluation of the overall mechanical stability of large-scale geologic structures such as underground and open-pit mines, highway and waterway cuts, petroleum reservoirs, and underground gas storage reservoirs is extremely complex. Acoustic emission appears to be one of the most suitable techniques available for such purposes. A brief review of the associated literature is presented along with a description of a mobile monitoring facility developed by the writer for field use.

Two current field projects, associated with underground gas storage and longwall coal mining, presently under way by the writer are described. [reprinted by permission of the American Society for Testing and Materials, Copyright].

694 **H. R. Hardy, Jr. "Microseismic Activity and Its Application in the Natural Gas Industry." Paper 69-T-17, pp T-147 to T-156 in** ***Transmission Conference Proceedings.*** **AGA Catalog Number X59969. American Gas Association, New York. Fall 1969.**

During the day-to-day operations of the natural gas industry, engineers are concerned with such questions as: (1) What is the optimum pressure for a specific underground storage reservoir? (2) Is a certain storage reservoir mechanically stable at its present pressure? (3) What are the optimum drilling conditions to insure minimum damage to a certain stratum? (4) Is a section of uncased hole mechanically stable or is it in the process of microscopic mechanical deterioration which will eventually lead to macroscopic failure and possible closure of the hole? (5) Is a section of buried pipeline being subjected to unpredicted local ground stresses? and (6) What minimum pressure is required to hydrofracture a specific rock formation? These questions all have one thing in common; mainly, each of the possible trouble areas is extremely remote from the engineer. Even if it were possible to observe the critical area optically (e.g., the use of a bore-hole television camera to study the wall of an uncased bore hole) such a technique would probably only be sufficiently sensitive to show results after the damage had reached a near-terminal stage. In this paper, the possible application of the so-called *microseismic technique* to the investigation of these and other problems in the natural gas industry will be discussed. This technique, although indirect, is extremely sensitive, making it possible to detect potential trouble spots at an early stage when remedial action may be successfully undertaken.

695 **H. R. Hardy, Jr. "Microseismic Techniques: Basic and Applied Research."** *Rock Mechanics/Fels Mechanik/Mecanique des Roches,* **RMFMA, 11(Suppl. 2):93–114. Spring 1973.**

The phenomenon of microseismic activity appears to provide the basis for one of the most useful tools presently available in the field of rock mechanics. During the last six years, the Rock Mechanics Laboratory at The Pennsylvania State University has been involved in studies related to this phenomenon, and the present paper describes a number of the basic and applied research studies undertaken. Emphasis however is on the applied aspects of these studies, including the use of microseismic activity in the laboratory to define failure in pressurized gas storage reservoir models, and in the field to study underground gas storage reservoir stability and to evaluate coal mine strata control techniques. A mobile laboratory has been under development in conjunction with the two field programs and a brief description of this laboratory is included.

696 **H. R. Hardy, Jr. "Model Studies Associated with the Mechanical Stability of Underground Natural Gas Storage Reservoirs." Paper 4-42 in** *Proceedings of the 2nd Congress, International Society for Rock Mechanics.* **Volume 2. 1970. meeting held in Belgrade, Yugoslavia. Sep 21-26, 1970.**

One of the most pressing problems of the natural gas industry today is that of locating and developing sufficient underground storage reservoir capacity to meet peak seasonal demands. In these reservoirs, gas is stored under pressure in the pore spaces and fractures within the rock that originally contained gas, oil, or water. Since it is economically desirable to store gas at the maximum possible pressure, an accurate value for the optimum pressure, namely the maximum pressure to which the reservoir could be pressurized and still remain mechanically stable, is of great importance. An analytical and experimental program has been under way by the author in an effort to develop a suitable criteria for establishing optium storage pressures. The present paper will describe a laboratory model study presently under way as part of this research program. The object of this laboratory study is to determine the optimum pressure as a function of the *in situ* stress state for a variety of model materials and cavity geometries. Experiments are carried out on cylindrical models, approximetely four inches in diameter and six inches in length, containing a central cavity to simulate the storage reservoir. These models are prepared from rock and rock-like materials whose physical properties were accurately determined from earlier tests. In a typical experiment, the model is mounted within a triaxial vessel and loaded under constant axial and confining stresses equivalent to specific *in situ* stress states. Hydraulic fluids introduced into the central cavity are then pressurized until the model becomes unstable and finally fails. Piezoelectric transducers located on the triaxial vessel monitor microseismic activity within the model and provide an indication of its mechanical stability. Strain gages mounted on the internal and external surfaces of the model provide data for comparison with analytical results obtained from elastic theory and finite element analysis.

697 **H. R. Hardy, Jr. "Monitoring the Stability of Geologic Structures Using Near-Surface Microseismic Transducers." pp 357-379 in** *Proceedings First Conference on Acoustic Emission/Microseismic Activity in Geologic Structures and Materials.* **H. R. Hardy, Jr. and F. W. Leighton, Editors. Trans Tech Publications, Clausthal, West Germany. 1977. conference held at Pennsylvania State University, University Park, Pennsylvania. June 9-11, 1975.**

During the last five years the Rock Mechanics Laboratory at Pennsylvania State University has been actively involved in a number of field-oriented microseismic projects. In particular, these include stability studies of underground gas storage reservoirs and an underground long-wall coal mine. In these studies microseismic activity has been detected by near-surface transducer arrays located over the structure under investigation. Details of the various transducer installations employed are described along with typical experimental results.

698 **H. R. Hardy, Jr. and A. W. Khair. "Applications of Acoustic Emission in the Evaluation of Underground Gas Storage Reservoir Stability." pp 77-111 in**

Proceedings of the 9th Canadian Rock Mechanics Symposium. **Mines Branch, Department of Energy, Mines and Resources, Ottawa, Ontario, Canada. 1974. symposium held in Montreal, Canada. Dec 13–15, 1973.**

In light of the recent energy crisis, the desirability of stockpiling fuels close to primary distribution areas has assumed an even greater improtance than in the past. In this regard, the natural gas industry in Canada as well as in the United States is faced with the problem of locating and developing sufficient underground storage reservoir capacity in high demand areas to meet peak seasonal requirements. Since it is economically desirable to store gas at the maximum possible pressure, studies have been under way by the authors to estimate optimum operating pressures for underground gas storage reservoirs and to evaluate their stability under such operating conditions. These studies have comprised two phases which will be briefly described in this paper. Phase I involved an analytical and laboratory study in which the behaviour of small scale models containing reservoir cavities of simple geometry was investigated under equivalent *in situ* stress conditions and internal pressure. Acoustic emission techniques were utilized during the laboratory studies to establish reservoir failure criteria. Phase II of the program has been concerned with full-scale operating storage reservoirs and the investigation of the feasibility of using acoustic emission techniques to monitor the stability of such structures.

699 H. R. Hardy, Jr., R. Y. Kim, R. Stefanko, and Y. J. Wang. "Creep and Microseismic Activity in Geologic Materials." Chapter 21, pp 377–413 in *Rock Mechanics: Theory and Practice.* **American Institute of Mining, Metallurgical, and Petroleum Engineers, New York. 1970. proceedings of the 11th Symposium on Rock Mechanics, held at Berkeley, California, June 16–19, 1969.**

Since 1964, the Department of Mining at Pennsylvania State University, University Park, Pennsylvania, has been carrying out extensive studies associated with the phenomena of microseismic activity and inelastic behavior in geologic materials. Briefly, the program to date has involved the following stages: (1) Study of the microseismic noise and energy rates for selected geologic materials under incremental, uniaxial tension; (2) Investigation of the frequency spectra of microseismic activity of selected geologic materials under incremental, uniaxial tension; (3) Investigation of the time-dependent inelastic deformation of geologic materials and the analysis of such deformation in terms of mechanical models; and (4) Simultaneous study of microseismic activity and time-dependent inelastic deformation in geologic materials under incremental, uniaxial compression. Details of the research involved in Stage 4 of the program constitute the subject of this chapter. It should be noted that the majority of the experimental data quoted forms part of a thesis to be presented by one of the authors (Kim) in partial fulfillment of the requirements for the Ph.D. degree in mining engineering at Pennsylvania State University. During the last two or three years, a number of important papers have been published which deal primarily with the fundamental structure and mechanical behavior of geologic materials. Recent papers have provided research workers in the rock mechanics field with a meaningful overall picture of rock as a structural material. The concept that many geologic materials may be considered basically brittle elastic solids containing various distributions of cracks and pores appears to provide a sound model for the structure of such materials. Bieniawski has developed a meaningful scheme for considering rock fracture in compression. This scheme covers all significant processes taking place in such materials from initial load application to complete failure. The authors suggest that the general concepts developed by Brace, Bieniawski, Walsh, and others provide an extremely useful framework in which the results of current as well as earlier research should be examined. This chapter describes an experimental study carried out to investigate creep and microseismic activity in a number of geologic materials under uniaxial compression and to ascertain what relationships, if any, exist between these two phenomena. For the most part, the chapter deals with the experimental aspects of the problem under study. A detailed consideration of the results of the current study in terms of fundamental rock structure and fracture processes will be delayed, however, until a comparison study being carried out under triaxial compression is completed.

700 **H. R. Hardy, Jr. and F. W. Leighton, Editors.** *Proceedings First Conference on Acoustic Emission–Microseismic Activity in Geologic Structure and Materials.* **Series on Rock and Soil Mechanics. Volume 2 (1974/77) Number 3. Trans Tech Publications, Clausthal, West Germany. 1977. conference held at Pennsylvania State University, University Park, Pennsylvania. June 9–11, 1975. [available in the U.S. from Trans Tech Publications, 16 Bearskin Neck, Rockport, MA 01966].**

Because of the increasing activity throughout the world in basic and applied research in such areas as rock burst mechanics, underground gas storage reservoir stability, stability of earth filled dams, earthquake mechanics, hydrofracturing research, slope stability monitoring, fundamental behavior of geologic materials, strata control in coal and hardrock mines, and comminution, the co-editors thought it timely to bring together people in these areas of work for a *working conference* on acoustic emission-microseismic activity in the area of geomechanics. This volume is the proceedings of that conference, consisting of 25 papers dealing with a wide range of laboratory, field, and analytical aspects of acoustic emission-microseismic activity in the areas of mining, petroleum, and civil engineering, and in geology and geophysics. Included in the volume is a *Master Bibliography* of 212 references specifically related to the area of microseismic activity in geologic structures and materials. For individual papers see References 12, 170, 241, 261, 494, 642, 664, 692, 697, 990, 1002, 1024, 1037, 1064, 1093, 1098, 1152, 1223, 1240, 1305, 1416, 1437, 1512, 1623, and 1774.

701 **G. G. Harman. "Interconnecting Bonding: Non-Destructive Test for Beam-Lead Bonds." pp 31–32 in "Semiconductor Measurement Techology: Quarterly Report, July 1 to September 30, 1974." W. M. Bullis, Editor. NBS Special Publication 400–12. National Bureau of Standards, Electronic Technology Division, Institute for Applied Technology, Washington, D.C. May 1975. [NTIS: AD/A–011–121 or from U.S. Government Printing Office, Washington, D.C. 20402, SD Catalog No. C13.10:400–12].**

The quarterly progress report describes NBS activities directed toward the development of methods of measurement for semiconductor materials, process control, and devices. The emphasis is on silicon device technologies. Principal accomplishments during the reporting period include (1) completion of Hall effect measurements to determine activation energies of the gold donor and acceptor levels in silicon; (2) successful direct measurement of fast interface state density with the circular CCD test structure; and (3) demonstration of the feasibility of the use of acoustic emission as a nondestructive means for testing individual beam-lead bonds. Results are also reported on a holder for semi-automated sheet resistance measurements, progress on development of mathematical models of dopant profiles, analysis of thermally stimulated current and capacitance measurements on junction diodes, X-ray photoelectron spectroscopy, a comparative study of surface analysis techniques, design and fabrication of a test pattern for resistivity-dopant density evaluation, epitaxial layer thickness measurement; on use of the flying-spot scanner, initial work on the scanning low energy electron probe, mathematical modeling of ultrasonic bonding, an improved method for force adjustment and measurement on beam-lead bonders, helium mass spectrometry for leak testing, thermal resistance measurements on Darlington pairs, and transistor thermal response measurements. Supplementary data concerning staff, publications, workshops and symposia, standards committee activities, and technical services are also included as appendices.

702 **D. O. Harris. "A Preliminary Investigation of the Applicability of Acoustic Emission to the Assessment of Residual Stresses in Brittle Coatings." DC–72–TM4. Dunegan Corporation, Livermore, California. Apr 1972.**

Acoustic emission is the term applied to the low level sounds impulsively produced within the bulk of a material as it becomes deformed. Sophisticated and sensitive electronic instrumentation is required to detect and analyze these sounds, which are capable of supplying valuable information on the deformation processes occurring in the material. Acoustic emission techniques have been applied to a wide range of material in a large variety of applications. Brittle coatings applied on a substrate may have considerable residual stresses in them, which could impair their service performance. The measurement of these residual stresses is difficult by current techniques. If a coating with high residual stresses is scratched, these scratches (cracks) can extend over a period of time. Growing cracks in brittle materials have been found to produce much acoustic emission. Hence, the presence of growing cracks (and therefore the presence of residual stresses) can be detected by acoustic emission techniques. Furthermore, the larger the residual stress, the more the cracks will extend in a given period, and the more acoustic emission will be observed. Hence, it is possible that acoustic emission techniques could be used to measure the magnitude of the residual stresses. The purpose of this investigation was to assess the applicability of emission techniques to the detection of residual stresses in coatings applied to a ceramic substrate.

703 **D. O. Harris. "A Review of Aerospace Applications of Acoustic Emission." ISA Reprint 76237. Instrument Society of America, Pittsburgh, Pennsylvania.**

1976; paper presented at the 22nd International Instrumentation Symposium, San Diego, California. May 25–27, 1976. [copy of reprint also available as Technical Report DE 76-1 from Dunegan/Endevco, San Juan Capistrano, California].

A review of aerospace applications of acoustic emission is presented. A brief historical review is given, with emphasis on the aerospace area. A discussion of typical instrumentation systems is included. Some basic experimental observations are presented, before proceeding to a discussion of aerospace applications, such as monitoring of proof tests on rocket motor cases, and rocket motor nozzles. Burn-rate monitoring of solid rocket propellants and in-process weld monitoring are also briefly mentioned. The paper concludes with a discussion of areas of future research required to bring acoustic emission to its full potential as a tool for the assessment of structural integrity.

704 **D. O. Harris. "Acoustic Emission from AMMO Steel." Dunegan/Endevco, San Juan Capistrano, California. Nov 26, 1973.**

The purpose of the investigaiton is to determine the acoustic emission characteristics of AMMO steel. This is the material that was used for a demonstration of the use of acoustic emission for the location of defects in a spherical pressure vessel. A summary of the results of the pressure vessel test is given in the report "Essai de Functionnement du Localisateur a 10 Voies Dunegan/Endevco 1010." by E. G. Tomachevsky, Commissariat a l'Energie Atomique, Division d'Etude et Developpement des Reacteurs, Note Technique EMT 73-254, FE 4111.4.03. Little emission was observed from a defect in the parent metal that was known to be growing. This result was surprising, and a laboratory investigation of the acoustic emission characteristics of this material was therefore conducted. The report gives the results of the investigation. Tests were conducted on unflawed tensile specimens of both the parent and weld materials, and also on notched tensile bars of the parent material. One of the notched bars had a fatigue crack at the notch root.

705 **D. O. Harris. "Acoustic Emission from Unflawed Tensile Specimens of 2219-T87 Aluminum." Technical Memo DE 75-1. Dunegan/Endevco, San Juan Capistrano, California. Mar 4, 1975.**

The alloy, 2219-T87 aluminum, is used extensively in aerospace applications that have been reported to produce only low-amplitude emissions. Since it is desirable to use acoustic emission for nondestructive evaluation of structures fabricated from this material, Dunegan/Endevco undertook to perform some tests to assess its acoustic emission response. Tensile tests on unflawed specimens machined from a 0.5-inch plate were performed. The length and width of the gage section were 2 and 0.25 inches, respectively. The specimens were tested on an Instron at a crosshead velocity of 0.10 inches per minute, and elongation was measured with an extensometer. The specimens were preloaded and Teflon coated pins were used so as to minimize extraneous noise from the pin areas.

706 **D. O. Harris. "Acoustic Emission Monitoring of Wire Rope and Sling Assembly." Technical Report DC-72-3. Dunegan Corporation, Livermore, California. May 25, 1972.**

The purpose of this investigation is to determine the applicability of acoustic emission to the assessment of the integrity of wire rope and a sling assembly. Rising load to failure, stress corrosion, and fatigue tests were performed on wire rope samples, which indicated the ease with which acoustic emission could be detected from these parts. A one-to-one correlation between acoustic emission events and wire breakage was achieved, which has obvious important implications in the testing of wire rope and points out the high degree of usefulness of acoustic emission in this area. Proof tests on a flawed and unflawed sling assembly were acoustically monitored. The acoustic emission was unable to detect the flaw, which is probably due to the blunt nature of the simulated flaw (hacksaw cut) and the conservative design of the part. These results do not preclude the possibility of detecting a more realistic flaw, such as a fatigue crack, and tests are continuing to ascertain the degree of sensitivity to a realistic flaw. Attenuation measurements were conducted, which indicate that multiple sensors are required for complete coverage of the sling assembly.

707 **D. O. Harris. "Analytical Techniques Relating Acoustic Emission to Fracture." SAI/SR-119-PA. Science Applications Incorporated, Palo Alto, California. Aug 1975.**

This paper was prepared for inclusion in the acoustic emission section of the American Society for Nondestructive Testing Handbook. A review of basic linear elastic fracture mechanics is presented, with emphasis on those aspects most relevant to acoustic emission studies. Various proposed models relating fracture mechanics parameters and acoustic emission are presented and discussed. Selected experimental results related to the proposed models are included. The results reviewed indicate that although much additional work remains to be done to obtain definitive and quantitative information on crack extension by use of acoustic emission, acoustic emission techniques provide a useful tool for fracture mechanics research, as well as a viable means for nondestructive evaluation of structures.

708 **D. O. Harris. "Detection of Fatigue Crack Growth in a Sling Assembly by Use of Acoustic Emission Monitoring of a Periodic Proof Test." Technical Report DC-72-3 (Addition 2). Dunegan Corporation, Livermore, California. Sep 11, 1972.**

The report is the final installment of a series that covers work performed to assess the applicability of acoustic emission techniques to the nondestructive evaluation of the structural integrity of wire rope and a sling assembly. This particular installment concerns the use of a periodic overload to detect fatigue crack growth in a sling assembly occurring during cyclic loading. Three transducers were mounted on the sling assembly, which was monitored both during fatiguing and proof loading. A flaw locator utilizing two of the transducers was also used. The amount of emission observed during the proof loading was found to correlate best with visual observations of crack growth. Attention was therefore focused on this parameter, and it was found that ample warning of impending failure was provided. Hence, acoustic emission techniques in conjunction with a periodic overload appear capable of detecting degradation of structural integrity in this type of hardware.

709 **D. O. Harris. "Results of Fatigue with Periodic Overload Tests on Wire Rope." Technical Report DC-72-3 (Addition). Dunegan Corporation, Livermore, California. June 5, 1972.**

The purpose of this brief note is to provide the results of tests conducted since the interim report dated May 25, 1972, was submitted. Since that time a fatigue with periodic overload test has been completed, and the results will be presented here. The original interim report was Dunegan Corporation Technical Report DC-72-3 (see Reference 706).

710 **D. O. Harris. "The Effect of Gain and Frequency Bandpass on Acoustic Emission Observed from Growing Fatigue Cracks (Die Wirkung von Verstarkung und Frequenzbereich bei der Schallemissionsanalyse, beobachtet an wachsenden Ermudungsrissen)." Technical Report DE-74-4. Dunegan/Endevco, San Juan Capistrano, California. Jan 1974; pp 170–188 in** *Schallemission: Anwendung bei der Untersuchung, Prufung and Uberwachung metallischer Werkstoffe.* **Deutsche Gesellschaft fur Metallkunde eV, 637 Oberursel/Taunus bei Frankfurt/Main, Adenaueralle 21, Germany. 1974. proceedings of symposium held in Munich, Germany. Apr 18–19, 1974.**

The effect of gain and frequency bandpass on the acoustic emission observed from growing fatigue cracks was investigated. Results from 7075-T6 aluminum and 4140 steel showed that the emission counts per cycle varied linearly with gain. This suggested an empirical modification of an earlier derived relationship between the acoustic emission counts per cycle and energy released by crack extension per cycle. The modified relationship allowed all published data on materials that deform only by slip (that was obtained by using a particular model of sensor and bandpass) to be plotted as a single curve, thereby indicating the generality of the suggested relation. The use of a spark-bar calibration allowed the effect of using various sensors and bandpasses to be predicted. The predicted results agreed closely with the experimental data, thereby indicating the usefulness and applicability of the spark-bar calibration procedures. A generalized relation between emission counts per cycle, energy released by crack extension per cycle, and such important instrumentation variables as gain, bandpass, sensor sensitivity, and sensor damping was proposed that agrees well with a wide variety of test results. This relation allows optimum instrumentation settings to be selected for a particular application. The results obtained indicate the utility of acoustic emission techniques for in-service monitoring of cyclically loaded structures.

711 **D. O. Harris and R. L. Bell. "The Measurement and Significance of Energy in Acoustic Emission Testing." Technical Report DE-74-3A. Dunegan/Endevco, San Juan Capistrano, California. Sep 1974;** *Experimental Mechanics,* **EXMCA, 17(9):347-353. Sep 1977.**

A technique for measuring the energy sensed at an acoustic emission transducer is presented that utilizes a squaring circuit and digital integrator. Theoretical relationships between energy and other more conventional acoustic emission parameters, such as counts and rms voltage, are derived for certain idealized cases. Experimental results from the following types of tests are presented: (1) unflawed tensile *(continuous* emission); (2) precracked stress corrosion cracking; (3) precracked fracture toughness; and (4) fatigue crack growth. Energy, counts, rms voltage, energy per event, and counts per event measurements are included. In the case of unflawed tensile specimens, energy techniques appeared somewhat superior to counts. In all other cases, a direct relationship between counts and energy was obtained. Energy measurements tended to give a larger weight to higher amplitude events. Other than this, energy measurements appeared to have no advantage over counts. The theoretical relationship predicted between energy/event and count/event agreed well with experimental observations. Overall, the test results presented indicate that energy techniques provide no significant advantage over counting threshold crossings (ringdown counting) in cases in which crack extension in metals is the primary source of acoustic emission.

712 **D. O. Harris and H. L. Dunegan. "Acoustic Emission–5: Applications of Acoustic Emission to Industrial Problems."** *Non-Destructive Testing,* **NDETA, 7(3):137-144. June 1974.**

This final paper in the series on acoustic emission deals with existing industrial applications of this technology. Such applications have increased greatly in recent years and now include such areas as weld monitoring of proof tests on structures and components, and the detection and location of flaws in pressure vessels. The paper presents representative examples drawn from the literature of such applications, and draws comparison with conventional nondestructive testing techniques in some cases. The authors conclude that acoustic emission can provide information that is difficult or impossible to obtain by other techniques, and can therefore cut costs substantially in the quality assurance of a wide variety of industrial products. (For Parts 1, 2, 3 and 4 see References 1716, 1404, 212, and 336, respectively).

713 **D. O. Harris and H. L. Dunegan. "Acoustic Emission Testing of Wire Rope." Technical Report DE-72-3A. Dunegan/Endevco, Livermore, California. 1972; paper presented at American Society for Nondestructive Testing National Spring Conference, Los Angeles, California. Mar 12-15, 1973;** *Materials Evaluation,* **MAEVA, 32(1):1-6. Jan 1974.**

The results of acoustic emission monitoring of tensile, fatigue, and fatigue with periodic overload tests on wire rope are presented. The tensile tests showed a one-to-one correlation between wire breakage and acoustic emission events, and that damaged cables consistently produced more acoustic emission at a given load than undamaged ones. Continuous monitoring of cyclically loaded rope provided warning of impending fatigue failure as early as half the lifetime. The acoustic emission monitoring of a periodic proof test on cyclically loaded rope is also capable of providing warning of impending fatigue failure. The test results presented show that acoustic emission techniques are well suited for studies of the failure mechanics and nondestructive evaluation of wire rope.

714 **D. O. Harris and H. L. Dunegan. "Continuous Monitoring of Fatigue Crack Growth by Acoustic Emission Techniques." Technical Report DE-73-2. Dunegan/Endevco, Livermore, California. Feb 1973; paper presented at the Society for Experimental Stress Analysis 3rd International Congress on Experimental Mechanics, Los Angeles, California. May 13-18, 1973;** *Experimental Mechanics,* **EXMCA, 14(2):71-81. Feb 1974.**

The application of acoustic emission to the detection of fatigue crack propagation in 7075-T6 aluminum and 4140 steel is investigated. The relationship between crack growth rate, cyclic stress intensity factor, load cycling rate, and observed acoustic emission behavior is presented. Crack growth rates of less than 10^{-6} in/cycle could be detected, and acoustic emission counts per cycle was shown to be closely related to the energy released by crack extension per cycle. A quantitative relationship for the threshold conditions for detection of fatigue crack growth is presented which agrees with experimental test results. The results also showed that fatigue crack growth occurs in an accelerating and decelerating manner, even though the stress intensity range remains uniform, and that the count rate passes through a peak that is believed to be associated with a plane strain-plane stress transition. The effects of instrumentation sensitivity and frequency bandpass are also investigated. The results obtained indicate that acoustic emission techniques should be suitable for in-service monitoring of a variety of cyclically loaded structures, even in the presence of high background noises.

715 **D. O. Harris and H. L. Dunegan. "Fracture Toughness and Acoustic Emission Characteristics of N50A Beryllium." ENS 66-253 Lawrence Radiation Laboratory, Livermore, California. 1966.**

The report consists of three independent sections: an introduction to fracture mechanics; a presentation and discussion of fracture toughness data obtained on N50A beryllium; and a presentation and discussion of the characteristics of the acoustic emission observed from the beryllium fracture specimens. The fracture toughness, K_C, for 0.10-inch thick plate was found to be approximately 23 ksi-inch$^{1/2}$; and for 0.30-inch thick plate, about 11 ksi-inch$^{1/2}$. The acoustic emission characteristics were found to be greatly altered by the presence of cracks, and to depend on the crack length. The stress intensity factor was found to be a controlling parameter in the emission studies. The results obtained indicate that acoustic emission obtained on fracture toughness specimens may be applied to the prediction of failure in stressed bodies.

716 **D. O. Harris and H. L. Dunegan. "Fracture Toughness of Beryllium." UCRL-70255. Lawrence Radiation Laboratory, Livermore, California. Jan 4, 1967;** ***Journal of Materials,*** **JMLSA, 3(1):59–72. Mar 1968.**

Fracture toughness data are presented for varying thicknesses of two grades of beryllium, N50A which contains about 0.8 percent beryllium oxide, and S200 which contains about 2 percent beryllium oxide. Values of K_c (fracture toughness) for equivalent thicknesses of these two grades are about the same, with the lower purity S200 tending to show higher toughness. Thus, it appears that the more expensive, higher purity material offers no advantage when failure occurs by the catastrophic growth of preexisting flaws. The value of K_c was found to vary with plate thickness even though all the fracture surfaces were completely flat, indicating that a flat fracture surface does not necessarily ensure that plane-strain conditions prevail. [reprinted by permission of the American Society for Testing and Materials, Copyright].

717 **D. O. Harris and H. L. Dunegan. "Verification of Structural Intergrity of Pressure Vessels by Acoustic Emission and Periodic Proof Testing." UCRL-72783. Lawrence Radiation Laboratory, Livermore, California. 1970; Technical Report DRC-71-2. Dunegan Research Corporation, Livermore, California. May 1971; pp 158–170 in** ***Testing for Predication of Material Performance in Structures and Components, ASTM STP 515.*** **American Society for Testing and Materials, Philadelphia, Pennsylvania. 1972.**

The application of acoustic emission monitoring of periodic proof tests to the detection of structural degradation of pressure vessels is demonstrated. Test results of welded steel and aluminum vessels are presented which show that structural degradation and impending failure due to fatigue crack growth can be detected by this technique. Subcritical crack growth due to stress corrosion or hydrogen embrittlement could be detected with equal ease. Hence, emission monitoring of a periodic proof test can provide a powerful tool for the verification of the structural integrity of pressure vessels, as well as a wide variety of other structures.

718 **D. O. Harris, H. L. Dunegan, and A. S. Tetelman. "Prediction of Fatigue Lifetime by Combined Fracture Mechanics and Acoustic Emission Techniques." UCRL-71760. Lawrence Radiation Laboratory, Livermore, California. Oct 24, 1969; Technical Bulletin DRC-105. Dunegan Research Corporation, Livermore, California. 1969; pp 459–471 in** ***Proceedings of the Air Force Conference on Fatigue and Fracture of Aircraft Structures and Materials.*** **A. H. Wood, R. M. Bader, W. J. Trapp, R. F. Hoener, and R. C. Donat, Technical Editors. AFFDL-TR-70-144. Air Force Materials Laboratory, Wright-Patterson Air Force Base, Ohio. Dec 1970.**

The use of low-level acoustic emission to detect the presence and growth of fatigue cracks is discussed. Low-level acoustic emission is the characteristic and irreversible sound emitted when a material is plastically deformed. The detection technique involves periodic proofing to loads greater than the working load while simultaneously monitoring for acoustic emission. If crack growth has occurred since the previous proof, acoustic emission will be observed while loading from the working to proofing load; larger amounts of crack growth will result in more emission during proof. Since the acoustic emission from a through crack in a plate or sheet is directly related to the stress intensity factor, as also is the fatigue crack growth rate, it is possible to directly correlate acoustic emission, crack growth, and number of fatigue cycles. In a previous publication, it was shown that periodic proof testing could be used to discontinuously monitor fatigue crack growth in wedge-opening-loading (WOL) fracture specimens of trip steel. A procedure was established for predicting the optimum number of cycles between periodic proof tests. The present work was conducted to extend the analysis and predication technique to other materials, especially 7075-T6 aluminum alloy, and also to crack geometries more likely to occur in actual situations, such as a penny-shaped crack. The results obtained are used to determine optimum conditions for periodic proof testing of high-strength aluminum parts.

719 **D. O. Harris and A. T. Green. "Acoustic Emission from Porcelain Panels." Dunegan Research Corporation, Livermore, California. Apr 30, 1971.**

The purpose of this investigation was to demonstrate the applicability of acoustic emission to the assessment of the quality of porcelain enamel. The porcelain panels supplied by the Porcelain Enamel Institute exhibited various degrees of adherence as determined by destructive testing. Currently the results of these destructive tests are not known to DRC personnel. Hence, the conclusions drawn here are made without prior knowledge of the relative degrees of adherence.

720 **D. O. Harris, B. G. Roth, and B. A. Kuhn. "Acoustic Emission Monitoring of Proof Tests on Canisters and Related Hardware." UCID-15777. Lawrence Radiation Laboratory, Livermore, California. Jan 14, 1971. [NTIS].**

Proof tests on canisters and related hardware for underground nuclear tests are conducted routinely at the Nevada Test Site pull-test facility. The investigation was to determine the feasibility and desirability of acoustic monitoring of the proof tests, for it is believed that acoustic emission can provide valuable information on the integrity of structures. The results of five proof tests are presented and discussed herein. The test results indicate that the pull tester is sufficiently quiet to allow acquisition of meaningful acoustic emission data for assessment of the integrity of canisters and related hardware. In addition, preliminary results indicate that the acoustic emission source can be approximately located by comparing the activity on the various channels. However, all of the parts tested *good*, and it remains to be determined whether or not acoustic emission can provide sufficient warning of impending failure to allow detection of an occasional *bad* part. Past experience indicates that this can be done. To definitely demonstrate failure detection, a part must be deliberately damaged or pulled to failure. Tentative plans for such a test are now under way.

721 **D. O. Harris, A. S. Tetelman, and F. A. I. Darwish. "Detection of Fiber Cracking by Acoustic Emission." Technical Report DRC-71-1. Dunegan Research Corporation, Livermore, California. Feb 1971; pp 238-249 in** *Acoustic Emission, ASTM STP 505.* **American Society for Testing and Materials, Philadelphia, Pennsylvania. 1972. proceedings of symposium held in Bal Harbour, Florida. Dec 7-8, 1971.**

A theoretical model is presented that relates acoustic emission to fiber cracking that occurs during a rising-load tensile test on a fiber-reinforced composite. The percentage of broken fibers in an Al_3Ni fiber-reinforced aluminum was measured as a function of tensile strain by optical inspection of the polished surface of strained specimens. The information was then used in conjunction with the proposed model to predict the acoustic emission response of the composite material. These predictions are compared with experimental observations, and a good agreement was obtained between the two sets of results. These results indicate that it is possible to quantitatively relate acoustic emission to the micromechanics of the deformation processes occurring within fiber reinforced composites, thereby demonstrating the applicability of acoustic emission to materials studies and also to nondestructive evaluation of the integrity of composite materials.

722 **S. D. Hart. "Inspection of Selected Areas of Engineering Structures by Acoustic Emission." pp 65-73 in** *Acoustic Emission.* **Edited by R. W. Nichols. Applied Science Publishers Ltd, Ripple Road, Barking, Essex, England. 1976. proceedings of a colloquium on acoustic emission held during the 1975 Annual Assembly of the International Institute of Welding, Tel Aviv, Israel. July 7-13, 1975.**

A review is presented of the use of acoustic emission as a tool for the surveillance of selected areas of an engineering structure. This would include areas known to contain flaws to determine whether cracks propagate from them during operation of the structure, or areas which are expected to sustain high loads of a cyclic nature where fatigue would be a factor, or areas of particular interest during a proof test. The main direction of discussion in this paper is towards techniques, rather than acoustic emission per se, because the various other aspects of acoustic emission are discussed in other papers.

723 **S. D. Hart. "On the Possibilities for the Use of Acoustic Emission for Non-Destructive Testing of Welds and Welded Structures/Sur les possibilites d'utiliser l'emission acoustique pour l'examen non destructif des soudures et des ensembles soudes."** *Welding in the World/le Soudage dans le Monde,* **WDWRA, 12(3/4):61-68. 1974. (in English and French);** *Soudage et Techniques Connexes,* **SOTCA, 28(11/12):506-510. Nov-Dec 1974. (in French); "Possibilita di impiego dell'emissione acoustica per il controllo non distruttivo delle saldature e delle strutture saldate."** *Rivista Italiana della Saldatura,* **RISAA, 26(6):340-346. Nov-Dec 1974. (in Italian).**

The study accompanied by bibliographical references is a survey of the present possibilities of acoustic emission based on its essential characteristics (sources of emission, nature of the signals, conditions of application). The equipment (transducer, amplifier, and signal processing equipment) is considered in relation to the difficulties arising from interference of various kinds. The applications of the processes are still limited, although perceptible progress has been made in specific cases.

724 **C. E. Hartbower. "Acoustic–Emission Measurements in the Ocean Engineering Program." Aerojet Solid Propulsion Company, Sacramento, California. May 15, 1971. report prepared at the request of the Naval Civil Engineering Laboratory, Port Hueneme, California, during a 14–day ACDUTRA period.**

The objective of this study was to make recommendations regarding the application of acoustic emission as a nondestructive inspection method in the ocean engineering program. Acoustic (or stress-wave) emission is the result of elastic energy released when a material undergoes cracking and/or plastic deformation. The report first presents a brief history of the acoustic-emission technology, with 45 references, together with a review of proof-testing concepts as affected by subcritical crack growth. With the ability to detect crack growth in real time, the utility of the proof test is greatly enhanced. Material certification of hyperbaric and other ocean-simulation facilities is an important application for acoustic emission. Selected pressure vessels are discussed including the 72-inch-ID test facility at NCEL, the 10-ft-ID *A* tank at Annapolis and the man-rated hyperbaric facility being installed at Panama City. With acoustic emission one can not only detect the presence of a growing crack and locate the crack by triangulation, but also have warning if a crack is approaching critical size. Also, incipient leaking of a pressure vessel is readily detected and located by the use of acoustic emission techniques. Various material-certification concepts are discussed including a simple, single-transducer, continuously monitored system and a multiple-transducer, computer-based, analysis-and-triangulation system. Details of a real-time, acoustic-emission, analysis-and-triangulation system are presented. Acoustic emission also can be an important tool in deep-submergence-simulation testing in that it will permit the detection of a crack at its beginning and then follow the crack-growth process to final fracture. There are numerous potential applications in deep-submergence-simulation testing; acoustic-emission should provide a highly sensitive measure of cracking in such test programs as window design, concrete pressure-vessel design, and stress-corrosion cracking studies. It is concluded that there are numerous applications for an acoustic-emission detection system in the NCEL Ocean Engineering Program. These applications include, in particular, the material certification of pressure vessels used to test devices in a simulated ocean environment. Because of the high pressures involved in such testing, after long and repeated usage these facilities become a potential hazard. Acoustic-emission technology offers a method for periodic (or even continuous) material certification to assure the safety of personnel who operate a given facility and who occupy a hyperbaric chamber.

725 **C. E. Hartbower. "Detection and Determination of Flaw Size by Acoustic Emission." Section VII. C, pp 418–473 in *Fracture Mechanics of Aircraft Structures.* H. Liebowitz, Editor. AGARD-AG-176. North Atlantic Treaty Organization, Advisory Group for Aerospace Research and Development. Jan 1974. [NTIS: AD-777-641]. [this monograph is a special publication of the Fracture Mechanics Working Group of the AGARD Structures and Materials Panel].**

However defect-free a structure may appear to be when it enters proof test and service, if there are active mechanisms for subcritical crack growth, such as fatigue, stress corrosion, hydrogen embrittlement, and/or strain aging, the life of the structure will be shortened. How much it is shortened will depend upon the rate of subcritical crack growth and the critical crack size of the material as determined by its fracture toughness. Proof testing provides a measure of maximum defect that can be present at the start of service. The difference between the critical crack size and the start of service is the amount of slow crack growth that can be tolerated. Without acoustic emission as a nondestructive inspection procedure, structures are needlessly lost in proof test and service. When a crack approaches critical size, at the onset of crack instability, the increase in crack-growth rate unmistakably heralds imminent fracture through the generation of an abrupt increase in acoustic emission count. Available acoustic emission instrumentation systems are so sensitive that it is possible to detect each stage of the failure process starting with deformation (dislocation pileups), crack propagation, and finally, the onset of instability—all in real time. Moreover, in the second and third stage of the failure process, as described above, it is possible by triangulation to locate the source of the signal. In some applications, the practical limitation for use of acoustic emission as a nondestructive inspection method has been and will continue to be extraneous noise. However, most noise problems are solved by the effective use of band-pass filters or special isolation techniques involving computer solutions.

726 **C. E. Hartbower. "Technical Note: Application of SWAT to the Nondestructive Inspection of Welds." *Welding Journal,* WEJUA, 49(2):54–s, 60–s, and 82–s. Feb 1970.**

The objective of the experiment described in this note was evaluation of the feasibility of the stress wave analysis technique (SWAT) as a method for obtaining data on delayed weld cracking. Some HY-80 steel bars were welded to form a test plate using a tungsten arc process. The delayed cracking process is described by data for stress-wave amplitude counts obtained at different periods following weld completion.

727 **C. E. Hartbower, F. J. Climent, C. Morais, and P. P. Crimmins. "Stress–Wave Analysis Technique Study on Thick-Walled Type A302B Steel Pressure Vessels." NASA CR-101887. Aerojet–General Corporation, Sacramento, California. July 1969. [NTIS: N69-36192].**

The results of investigations to determine relationships between stress-wave-emission characteristics and subcritical crack growth in Type A302, Grade-B alloy steel are presented. Stress-wave-emission characteristics which can be used as precursors for instability (failure) for unhydrogenated and hydrogenated material tested in air and 3% NaCl-water environments are discussed. The fracture behavior of unhydrogenated and hydrogenated A302B steel in these environments is also described. The results of background noise measurements at the Pad A and VAB Bottle Fields at the Kennedy Space Flight Center are presented and the feasibility of monitoring subcritical crack growth in Type A302B steel determined by correlating these results with the laboratory stress wave emission-fracture test data. Stress-wave-emission attenuation data are also presented and employed to determine sensor patterns for the pressure vessels at Pad A and the VAB Bottle Field.

728 **C. E. Hartbower and P. P. Crimmins. "Fracture of Structural Metals as Related to Pressure-Vessel Integrity and In-Scierivce Monitoring." Aerojet-General Corporation, Sacramento, California. July 1968; AIAA Paper 68-501. American Institute of Aeronautics and Astronautics, New York. 1968; paper presented at the Interagency Chemical Rocket Propulsion Group and American Institute of Aeronautics and Astronautics 3rd Solid Propulsion Conference, Atlantic City, New Jersey. June 4-6, 1968; ASME Publication 71-PVP-60. American Society of Mechanical Engineers, New York. 1971; paper presented at the American Society of Mechanical Engineers Pressure Vessel and Piping Conference, San Francisco, California. May 10-12, 1971.**

Consideration is given to problems of determining the maximum flaw that exists as a given pressure vessel structure goes into service, and of establishing an inspection method to ensure that the largest preexisting flaw will not grow to critical size before the end of a useful life-span. Selected case histories of proof-test failures are presented. Fracture-mechanics concepts in proof testing are discussed, as well as applications of the stress wave analysis technique (SWAT) for detecting crack growth during the proof test. Advanced SWAT systems and their potential applications are described.

729 **C. E. Hartbower and P. P. Crimmins. "Stress-Wave Emissions from Slow Crack Growth: A Review of Aerojet-General Research." Aerojet-General Corporation, Sacramento, California. 1970.**

The objective of the paper is to review the research conducted at Aerojet-General Corporation using the stress-wave analysis technique as a method for the detection of slow crack growth in metals. The research has been directed to three areas of concern: microcrack sources of stress-wave emissions; causes of slow crack growth, particularly fatigue, hydrogen embrittlement, stress-corrosion cracking, and spontaneous strain aging; and the use of stress-wave emission as a presage of unstable crack propagation. The research has shown that crack growth resulting in a sudden release of elastic energy can be detected by an accelerometer that converts the elastic wave into an equivalent electrical signal. Crack growth associated with fatigue, hydrogen embrittlement, stress-corrosion cracking, and spontaneous strain-aging failure mechanisms has been shown to produce discontinuous elastic-energy releases. The stress-wave analysis technique constitutes a unique nondestructive inspection method in that the material defect, when propagating, transmits its own signal, with the sensor acting as the receiver. Thus, with SWAT one can not only determine if there is a growing flaw, but also locate the flaw by a triangulation technique. Moreover, in most structural metals, as the crack increases in length, there is an increase in the rate and amplitude of stress-wave emission that serves as a presage of failure.

730 **C. E. Hartbower, W. W. Gerberich, and P. P. Crimmins. "Characterization of Fatigue-Crack Growth by Stress-Wave Emission." NASA CR-66303. Final Report. Aerojet-General Corporation, Sacramento, California. June 7, 1966. [NTIS: N67-19864].**

The report describes an investigation of fatigue crack propagation in two structural alloys using the elastic waves emanating from the propagating crack front to monitor cycle-by-cycle growth. The materials were 18% nickel maraging steel and D6aC low-alloy steel. Relationships were established between the stress-wave emission associated with fatigue crack growth and several metallurgical and loading variables. A tentative model for the mechanism of fatigue crack growth in an oblique shear mode and relationships between stress-wave emission and crack growth rate were also established.

731 **C. E. Hartbower, W. W. Gerberich, and P. P. Crimmins. "Mechanisms of Slow Crack Growth in High-Strength Steels." AFML-TR-67-26, Volume 1. Aerojet-General Corporation, Sacramento, California. Feb 1967. [see Reference 741 for Volume 2].**

A program was conducted to investigate the metallurgical nature of slow crack growth in 18% nickel maraging steel and D6aC low-alloy steel using the stress-wave-analysis technique (SWAT) as a new tool for monitoring incremental crack growth. The testing techniques and fracture-data analysis were based on linear-elastic fracture-mechanics concepts. The two heats of maraging steel investigated were found to be insensitive to hydrogen and aqueous environment for the sustained-load periods of time investigated. The D6aC, on the other hand, heat-treated to produce a variety of microstructure and strength levels, was susceptible to slow crack growth under all conditions investigated involving elevated temperature, hydrogenation, and aqueous environment. An inverse relationship was found between secondary-incubation time and crack-growth rate. At elevated temperature, the secondary-incubation time was found to increase at first and then reach a steady-state condition. For those conditions where there was a relatively small amount of carbon in interstitial solid solution, the steady-state secondary-incubation time was relatively long and independent of microstructure. For that condition where a relatively larger amount of carbon was in solid solution, the secondary-incubation time was relatively short and spontaneous strain-aging embrittlement was most apparent. In the latter case, the kinetics of both secondary-incubation time and crack-growth rate suggested the embrittling mechanism to be carbide precipitation. The D6aC subjected to prior hydrogenation and water environment produced a large amount of stress-wave emission. Based upon experimental secondary-incubation times, a theoretical interpretation of hydrogen embrittlement indicated that gross diffusion of hydrogen from one crack site to the next was the controlling mechanism. With both spontaneous strain-aging and hydrogen embrittlement, each crack step appeared to be approximately equal to the theoretical crack-tip displacement.

732 **C. E. Hartbower, W. W. Gerberich, and P. P. Crimmins. "Monitoring Subcritical Crack Growth by an Acoustic Technique." pp 371–380 in *Weld Imperfections,* A. R. Pfluger and R. E. Lewis, Editors. Addison-Wesley Publishing Company, Reading, Massachusetts. 1968; paper presented at the Symposium on Weld Imperfections, Palo Alto, California. Sep 19–21, 1966.**

The salient feature of the stress-wave analysis technique (SWAT) is that it permits the detection of elastic waves emanating from a crack *as it grows.* Thus, with SWAT, only those flaws that are growing are detected and under surveillance. It is the purpose of this brief paper to (a) acquaint the reader with some research applications of this technique, and (b) suggest its ultimate application as a supplementary nondestructive test method. There are several mechanisms by which a subcritical flaw may grow in service. For example, fatigue and stress-corrosion cracking are better known causes of slow crack growth. Hydrogen embrittlement and spontaneous strain aging are additional potential causes of subcritical crack growth. All of these mechanisms are under investigation at Aerojet-General using this technique. Each of the slow-crack-growth mechanisms involves discontinuous crack extension which may vary in both rate and magnitude as the cracking progresses. Research of these mechanisms is conducted by means of fracture test specimens containing fatigue precracks. Crack growth is monitored both by crack-opening displacement gage and by SWAT. The crack-opening displacement gage provides data on the gross length of the crack at any given instant, and SWAT monitors the increment-by-increment crack growth. Playback of tape-recorded data can provide a time expansion of as much as 16000:1; thus permitting a detailed step-by-step evaluation of the fracture process with precise measurements of crack incubation time. [reprinted by permission of Addison-Wesley Publishing Company. Copyright 1968].

733 **C. E. Hartbower, W. W. Gerberich, and P. P. Crimmins. "Monitoring Subcritical Crack Growth by Detection of Elastic Stress Waves."** ***Welding Journal,*** **WEJUA, 47(1):1s–18s. Jan 1968; paper presented at the American Welding Society National Fall Meeting, Houston, Texas. Oct 2–5, 1967.**

The paper describes a study of slow crack growth in D6aC low-alloy steel as influenced by hydrogen, stress corrosion, and fatigue using stress waves as a new technique for monitoring incremental crack growth. In low-cycle high-stress-intensity fatigue of D6aC heat treated to high toughness, each cycle involved a number of discontinuous crack jumps to form an increment of fatigue-crack growth. With an increasing stress intensity factor, both the stress-wave amplitude and the integrated value of stress wave emission increased. A linear relationship was found between fatigue crack growth rate and summation of stress-wave amplitudes. When heat treated to low toughness, the stress wave indicated intermittent cracking, with dormant periods of little or no crack growth ranging from two to three cycles to hundreds of cycles, followed by macroscopic growth in the next cycle. The D6aC steel was susceptible to slow crack growth both when hydrogenated and tested in distilled water. Over a given period of slow crack growth, there was a direct relationship between the number of stress waves and crack growth rate and, alternately, there was an inverse relationship between secondary-incubation time and crack-growth rate. As the crack lengthened under constant load, the stress intensity increased and the rate of stress-wave emission increased at a relatively gradual rate until a sudden increase in rate occurred corresponding to the start of rapid crack growth. Crack growth was found to occur in discontinuous steps for all test conditions investigated. A synergistic effect was noted when hydrogenated D6aC steel was tested in water. Hydrogen diffusion appeared to be the controlling mechanism in both types of embrittlement.

734 **C. E. Hartbower, W. W. Gerberich, and H. Liebowitz. "Investigation of Crack-Growth Stress-Wave Relationships."** ***Engineering Fracture Mechanics,*** **EFMEA, 1(2):291–308. Aug 1968; paper presented at the 1st National Symposium on Fracture Mechanics, Lehigh University, Bathlehem, Pennsylvania. June 19–21, 1967.**

Crack growth that results in a sudden release of elastic energy can be detected by use of a piezoelectric crystal that will convert the elastic stress wave into an electrical signal. Data on two widely different alloys, 7075-T6 aluminum and HY-80 steel, are presented in this paper. The objective was to determine (1) if, in these materials there is some recognizable, reproducible stress-wave characteristic(s) which predicts the onset of crack instability under plane-strain and plane-stress conditions; (2) if such characteristic varies from one material to another; and (3) if such characteristic is affected by specimen configuration. Single-edge-notch tensile tests involving a range of widths, thicknesses and crack depths all showed essentially the same stress-wave behavior. Plane-stress instability was marked in both materials by a number of large amplitude stress waves. In 7075-T6, K_{Ic} was marked by an order-of-magnitude increase in stress-wave amplitude; whereas, in HY-80, large amounts of crack growth occurred at stress intensity levels approaching the yield stress with stress-wave emission so small as to be barely detectable at the sensitivity level used. Fracture in 7075-T6 was normally in a flat, plane-strain mode; whereas, in HY-80, it was almost entirely in oblique shear by microvoid coalescence. [reprinted by permission of Pergamon Press Ltd].

735 **C. E. Hartbower, W. W. Gerberich, W. G. Reuter, and P. P. Crimmins. "Stress-Wave Characteristics of Fracture Instability in Constructional Alloys." U.S. Navy Contract N0014–66–C0340. Final Report. Aerojet–General Corporation, Sacramento, California. July 1968. [NTIS: AD–674–881]; IIW IX–700–70. International Institute of Welding. 1970.**

The objective of this research was to determine (1) if there is some recognizable, reproducible stress-wave characteristic or characteristics which predict the onset of crack instability under plane-strain and plane-stress conditions; (2) if such a characteristic varies from one material to another; and (3) if such a characteristic is affected by specimen configuration.

736 **C. E. Hartbower, A. T. Green, and P. P. Crimmins. "Correlation of Stress-Wave-Emission Characteristics with Fracture in Aluminum Alloys." NASA CR-102562, AGC Report 1246-Q-5. Aerojet-General Corporation, Sacramento, California. Dec 1969. [NTIS: N70-23323].**

The report describes the progress during the fifth quarter of a program to investigate the correlation of stress-wave-emission characteristics with fracture in selected aluminum alloys. During this reporting period, the cryogenic testing of part-through-cracked (PTC) tension specimens was completed, the plane-strain fracture toughness (K_{Ic}) was determined by computer, and the stress-wave data were analyzed.

737 **C. E. Hartbower, A. T. Green, G. A. Webber, and P. P. Crimmins. "Correlation of Stress-Wave-Emission Characteristics with Fracture in Aluminum Alloys." Report 1246-Q-4. Aerojet-General Corporation, Sacramento, California. Oct 1969.**

The report describes the progress during the fourth quarter of a program to investigate the correlation of stress-wave-emission characteristics with fracture in selected aluminum alloys. During this reporting period, part-through-crack tension specimens were tested at 75 and −320 °F. The report describes the instrumentation system employed for these tests and the stress-wave emission fracture characterization data obtained.

738 **C. E. Hartbower, C. F. Morais, W. G. Reuter, and P. P. Crimmins. "Acoustic Emission from Low-Cycle High-Stress-Intensity Fatigue."** ***Engineering Fracture Mechanics,*** **EFMEA, 5(3):765-789. Sep 1973; paper presented at Symposium on Fracture and Fatigue, George Washington University, Washington, D.C. May 3-5, 1972.**

The paper presents the findings of an Advanced Research Projects Agency (ARPA) study, the overall objective of which was to develop a nondestructive testing technique to determine flaw criticality based on acoustic emission. The research included an evaluation of sensors and instrumentation systems, using several materials and material conditions loaded in low-cycle, high-stress-intensity fatigue. The materials used for the study were D6aC tempered at 600 and 1100 °F, annealed and solution-treated-and-aged 6 Al - 4 V titanium and 7075-T6 aluminum. The test specimen was the precracked, single-edge-notch tension specimen; macrocracking was detected by crack-opening-displacement (COD) gage and microcracking by acoustic emission. The acoustic-emission system utilized 400 and 1000 kHz band-pass filtering at 100 db gain. The output signals of the totalizer and the COD gage were recorded on a single strip chart using a dual-pen recorder. The specimens were subjected to low-cycle, high-stress-intensity fatigue at 6 cycles per minute. In some tests, cycling was begun in air and finished in water. Acoustic emission was demonstrated to be highly effective as a nondestructive test method for following crack growth in low-cycle high-stress-intensity fatigue; acoustic emission confirmed the existence of periods of dormancy punctuated by periods of active fatigue crack growth. Using a dual-pen, strip-chart recorder displaying both crack-opening displacement and stress-wave count on the same chart, it was a simple matter not only to observe if there were crack growth in each individual cycle but also where in the cycle it occurred. Moreover, the process of stress-corrosion cracking during low-cycle, high-stress-intensity fatigue was readily detected by a marked increase in the stress-wave count rate. The utility of acoustic emission as a precursor of imminent failure was demonstrated for low-cycle, high-stress-intensity fatigue as well as for the case of environmentally assisted fatigue. Plots of cumulative stress-wave count versus cycle number consistently showed a marked increase in count rate several (10 to 20 or more) cycles before fracture. [reprinted by permission of Pergamon Press Ltd from *Engineering Fracture Mechanics*].

739 **C. E. Hartbower, C. F. Morais, W. G. Reuter, and P. P. Crimmins. "Detection of Weld Cracking in HY-80, HY-100 and HY-130 Steels by Stress-Wave Emission." Naval Ships System Command Contract N00024-70-C5215. Final Report. Aerojet Solid Propulsion Company, Sacramento, California. Mar 1971. [NTIS: AD-881-990L].**

A study was made to determine (1) the feasibility of detecting weld cracking in submarine hull construction, and (2) the duration of such cracking by stress-wave (acoustic) emission. The findings showed that stress-wave emission can be employed as a highly sensitive nondestructive inspection technique. Bursts of stress-wave emission indicative of delayed cracking were found to occur intermittently in some weldments for over 400 hours (approximately 17 days) after welding was completed. Two of the 33 weldments tested were sectioned and examined metallographically; the extent of cracking was found to correlate with cumulative stress-wave count. Thus, it appears that the present shipyard practice of a 7-day interval between welding and nondestructive inspection is unconservative.

740 **C. E. Hartbower, C. F. Morais, W. G. Reuter, and P. P. Crimmins. "Development of a Nondestructive Testing Technique to Determine Flaw Criticality." AFML-TR-71-218. Aerojet Solid Propulsion Company, Sacramento, California. Jan 1972. [NTIS: AD-747-225].**

The report presents the findings of a study to develop a nondestructive testing technique to determine flaw criticality based upon stress-wave emission. The research included an evaluation of various sensors and instrumentation systems, and several materials and material conditions tested to failure in rising load and low-cycle, high-stress-intensity fatigue. The fracture testing utilized a linear-elastic fracture-mechanics approach. Acoustic emission was the basis of the nondestructive inspection technique. The stress-wave-analysis technique (SWAT) was shown to be a highly sensitive nondestructive inspection (NDI) method, capable of detecting growing defects at least an order of magnitude smaller than any other known NDI method and, with a computerized system, capable of locating one or more defects in real time. Thus, the stress-wave-analysis technique (SWAT) can be used for in-service, continuous, NDI. When defects emit stress-waves and yet are too small to be confirmed by conventional nondestructive inspection, the suspect area is placed under surveillance using an array of transducers. When the defect approaches critical crack size, the acoustic emission count rate increases markedly, providing a precursor of crack instability and failure.

741 **C. E. Hartbower, W. G. Reuter, and P. P. Crimmins. "Mechanisms of Slow Crack Growth in High-Strength Steels and Titanium." AFML-TR-67-26, Volume 2. Aerojet-General Corporation, Sacramento, California. June 1969. [see Reference 731 for Volume 1].**

The metallurgical nature of subcritical crack growth in D6aC low-alloy steel, Grade-250, 18% nickel maraging steel, 6 Al - 4 V alpha-beta titanium and 3 Al - 13 V - 11 Cr beta titanium has been investigated using the stress-wave-analysis technique (SWAT) as a new tool for monitoring incremental crack growth. Three mechanisms of subcritical crack growth were investigated; vis., stress-corrosion cracking, hydrogen embrittlement, and spontaneous strain aging. The research is reported in two volumes. The earlier volume (1) dealt primarily with D6aC steel exposed to evaluate temperature, 70 °F aqueous environment, and hydrogenation. The objectives of Volume 2 were to better define the mechanism of spontaneous strain aging in D6aC steel, and to investigate the effect of hydrogenation or aqueous environment on incubation time and crack-growth rate in equiaxed- and platelet-alpha 6 Al - 4 V titanium, two aging conditions of 3 Al - 13 V - 11 Cr titanium, air-melt and vacuum-arc-remelt, grade-250, 18% nickel maraging steel, and 1750 °F austenitized, 1100 °F aged, D6aC steel, tested under sustained load. The crack-growth process was followed in greater detail than heretofore possible using an electronic-counter, stress-wave-detection system with the trigger level set just above the continuous emission produced by plastic deformation. The automation provided by the electronic-counting system and crack-opening-displacement strip chart permitted tests of up to 20 hours duration. The fact that primary crack incubation occurred after several hours of exposure in some materials and test conditions, suggested that much of the published threshold stress-intensity (K_{Iscc}) data that are based on time-to-failure tests rather than crack-growth-rate measurements are unconservative. Furthermore, tests in 165 °F circulating, distilled water produced marked changes in primary incubation time as well as changes in crack-growth mechanism in some of the material conditions investigated. Thus, an increase of less than 100 °F in water temperature was found to have a marked effect on subcritical crack growth. Intermediate water temperatures (between 70 and 165 °F) were not investigated.

742 **C. E. Hartbower, W. G. Reuter, C. F. Morais, and P. P. Crimmins. "Correlation of Stress-Wave-Emission Characteristics with Fracture in Aluminum Alloys." NASA CR-2072. Final Report. Aerojet Solid Propulsion Company, Sacramento, California. July 1972. [NTIS: N72-30470].**

The report presents the findings of a study to correlate stress-wave-emission characteristics with fracture in welded and unwelded aluminum alloys tested at room and cryogenic temperatures. The stress-wave-emission characteristics investigated were those which serve to presage crack instability; viz., a marked increase in (1) signal amplitude, (2) signal repetition rate, and (3) the slope of cumulative count plotted versus load. The alloys were 7075-T73, 2219-T87, and 2014-T651, welded with MIG and TIG using 2319 and 4043 filler wire. The testing was done with both unnotched and part-through-crack (PTC) tension specimens and with 18-inch diameter subscale pressure vessels. In the latter testing, a real-time, acoustic emission, triangulation system was used to locate the source of each stress-wave emission. With such a system, multiple emissions from a given location were correlated with defects found by conventional nondestructive inspection.

743 **C. E. Hartbower, W. G. Reuter, C. F. Morais, and P. P. Crimmins. "Use of Acoustic Emission for the Detection of Weld and Stress-Corrosion Cracking." pp 187-221 in** ***Acoustic Emission, ASTM STP 505.*** **American Society for Testing and Materials, Philadelphia, Pennsylvania. 1972. proceedings of symposium held in Bal Harbour, Florida. Dec 7-8, 1971.**

The paper describes the utility of acoustic emission for the detection of weld and stress-corrosion cracking. A brief review of the history of acoustic emission is presented together with a selected bibliography. The findings of two researches are presented to show how acoustic emission is used as a method for detecting weld and stress-corrosion cracking. A 100-ksi yield-strength, constructional-alloy weldment was found to emit bursts of acoustic emission for over 400 hours after welding and the acoustic emission was correlated with the amount of cracking as determined by metallographic examination. Moreover, in stress-corrosion studies with acoustic emission instrumentation, a conservative threshold stress intensity for stress-corrosion cracking can be determined based on crack propagation rather than time to failure. A linear relationship is shown between acoustic emission count and stress-intensity factor.

744 **W. F. Hartman. "Acoustic Emission and Discontinuous Yielding." COO-2204-2. Johns Hopkins University, Department of Mechanics and Materials Science, Baltimore, Maryland. 1971. [NTIS]; paper presented at the 10th Acoustic Emission Working Group Meeting, Pasadena, California. Dec 6-8, 1972; pp 875-885 in** ***Developments in Mechanics, Volume 7. Proceedings of the 13th Midwestern Mechanics Conference.*** **University of Pittsburgh, School of Engineering, Pittsburgh, Pennsylvania. 1973. conference held at University of Pittsburgh, Pittsburgh, Pennsylvania. Aug 13-15, 1973.**

The discontinuous yielding during tensile loading of many crystalline solids results in a serrated or steplike stress-strain curve and is known as the Portevin-Le Chatelier effect. In this paper, the replicable features of the discontinuous yielding of dead-weight loaded annealed specimens of aluminum and brass are documented and compared with the acoustic emission accompanying such deformations. It is shown that one method of acoustic emission monitoring provides data which is consistent with the hypothesis that the elastic energy released during a yield increment is proportional to the elastic energy stored since the last yield increment. Another method of monitoring gives a better indication of the number of acoustical events occurring and permits assessment of the frequencies of the emissions. It is concluded that a major portion of the acoustic emission during the strain-loading of crystalline materials is primarily associated with elastic-energy release during the propagation of nonhomogeneous deformation, which is the genesis of the Portevin-Le Chatelier effect during dead-weight loading.

745 **W. F. Hartman. "Acoustic Emission and the Portevin–Le Chatelier Effect." COO–2204–1, Progress Report, Mar 1–Nov 1, 1972. Johns Hopkins University. Department of Mechanics and Materials Science, Baltimore, Maryland. 1972. [NTIS].**

The discontinuous yielding during tensile loading of many crystalline solids results in a serrated or steplike stress-strain curve and is known as the Portevin-Le Chatelier effect. Replicable features of the discontinuous yielding of dead-weight loaded annealed specimens of several face-centered-cubic materials have been documented and compared with the acoustic emission accompanying such deformations. It has been found that the energy of the acoustic emission during a yield increment correlates with the elastic energy stored since the last yield increment. The materials tested represent different purities, grain sizes, stacking fault energies, annealing procedures, and yield strengths. Yet, all exhibit several regions of parabolic strain-hardening. The probable coefficients for each of these regions are discretely related and when a hardening transition occurs it is usually accompanied by a distinct change in the acoustic emission. The research is continuing with the realization that acoustic emission monitoring can become a unique aid in studying metallurgical phenomena of strain hardening.

746 **W. F. Hartman. "Acoustic Emission and the Portevin–Le Chatelier Effect." COO–2204–4. Johns Hopkins University, Baltimore, Maryland. 1973;** *Experimental Mechanics,* **EXMCA, 14(1):19–23. Jan 1974; pp 107–111 in** *Proceedings of the Third International Congress on Experimental Mechanics.* **B. E. Ross, Editor. Society for Experimental Stress Analysis, Westport, Connecticut. 1974. meeting held in Los Angeles, California. May 13–18, 1973.**

Many metals and alloys which exhibit repeated discontinuous yielding (Portevin-Le Chatelier effect) also emit rather interesting acoustic energy during work-hardening. Both phenomena are dramatic in dead-weight extensions of annealed specimens of alpha-brass or aluminum. The acoustic emission from such specimens was monitored and correlated with the features of the Portevin-Le Chatelier effect. It is shown that, when the discontinuous yielding subsides in aluminum, so does the acoustic emission; in fact, smooth continuous flow can occur in these materials with no detectable acoustic emission. Data are presented which are consistent with the hypothesis that, at room temperature, elastic energy released during a yield increment is proportional to the elastic energy stored since the last yield increment. This is not observed at elevated temperatures. It is concluded that additional studies of the acoustic-emission phenomena associated with plastic deformation can aid in achieving a better knowledge of the strain-hardening process for crystalline solids.

747 **W. F. Hartman. "Acoustic Emission and the Portevin–Le Chatelier Effect." COO–2204–10. Progress Report, Nov 1, 1972–Nov 1, 1973. Johns Hopkins University, Department of Mechanics and Materials Science, Baltimore, Maryland. 1973. [NTIS].**

The Portevin-Le Chatelier effect is the phenomenon of discontinuous yielding that occurs during the tensile loading of many crystalline metals and alloys. It produces a step-like or serrated stress-strain curve. The acoustic emission from the solids during the strain increments are being monitored and correlated with strain-hardening behavior as a function of composition, grain size, and temperature. It has been found that transitions in parabolic strain-hardening are accompanied by increases in acoustic emission. This suggests that microstructural changes do occur during such transitions. At elevated temperatures, the Portevin-Le Chatelier effect, in brass, coexists with a steady uniform creep deformation. The energy released as acoustic emission differs from that released at room-temperature tests. This is being studied as it relates to creep mechanisms and grain size. Other developments related to the main research program include the implementation of a spark-technique for simulating acoustic emission in a tensile specimen, so that calibrations can be performed. Also, a technique has been designed for analyzing the frequency content of the transient acoustic-emission signal.

748 **W. F. Hartman. "Acoustic Emission as an Aid in Studying Strain-Hardening Phenomena." Johns Hopkins University, Department of Mechanics and Materials Science, Baltimore, Maryland. paper presented at American Society for Nondestructive Testing National Spring Conference, Los Angeles, California. Mar 12–15, 1973;** ***Materials Evaluation,*** **MAEVA, 31(11):237–240. Nov 1973;** ***Testing, Instruments and Controls*** **(Australia), TICOA, 11(11/12): 13–16. Nov–Dec 1974.**

Many ductile polycrystalline metals and alloys exhibit quantized parabolic strain-hardening. During tensile testing a transition from one parabolic region to another appears as a second-order phase change. The origin of such a change is microscopic and ordinary optical inspection techniques cannot be used without unwanted difficulty, including the destruction of the testing configuration. It is here shown that the monitoring of the acoustic emissions during tensile tests of such materials provides an independent, nondestructive evaluation of the occurrence of these strain-hardening transitions.

749 **W. F. Hartman. "Acoustic Emission Detection of Flaws in Butt-Welded Plates." COO-2204-8. Johns Hopkins University, Department of Mechanics and Materials Science, Baltimore, Maryland. 1973; paper presented at the American Society for Nondestructive Testing National Fall Conference, Chicago, Illinois. Oct 1-5, 1973;** ***Materials Evaluation,*** **MAEVA, 32(12): 259-261, 268. Dec 1974.**

Acoustic emissions were monitored during mild flexural deformation of butt-welded steel plates after cooling of the welds. Measurement of the difference in the arrival times at two transducers was used to locate predominant sources. These were then compared with the location of known flaws. The study demonstrates the feasibility of employing acoustic-emission monitoring as a method of evaluating the quality of welds. Refinements, which are appropriate to quality control on an assembly line, are discussed.

750 **W. F. Hartman. "Acoustic Emission during Strain-Hardening Transitions in Copper." COO-2204-7. Johns Hopkins University, Department of Mechanics and Materials Science, Baltimore, Maryland. 1973;** ***Scripta Metallurgica,*** **SCRMB, 8(4):435-438. Apr 1974.**

During dead-weight extension of OFHC Cu polycrystals, parabolic strain-hardening transitions are accompanied by distinct increases in acoustic emission. This observation lends credence to the overall validity of the generalized parabolic-hardening-mode perspective and suggests that microstructural changes occur during the transitions.

751 **W. F. Hartman. "Continuous, In-Service Acoustic Emission Monitoring of Nuclear Reactor Components/Surveillance permanente de l'emission d'ondes acoustiques dans les composants de reacteurs nucleaires en service." Trodyne Corporation, Teterboro, New Jersey; Paper 5C/11 in** ***Proceedings of the Eighth World Conference on Nondestructive Testing,*** **volume containing Sections: 5C, 3J, 1D. International Committee for Nondestructive Testing, Secretariat du Cofrend, Institut de Soudure, 32 Boulevard de la Chapelle, 75880 Paris Cedex 18, France. 1976. conference held in Cannes, France. Sep 6-10, 1976. (paper in English).**

Described are the technique and results of 8 months of continuous acoustic emission monitoring of boiling-water reactor components.

752 **W. F. Hartman. "Correlation of Mechanical and Metallurgical Parameters with the Acoustic Emission during Tensile Deformation of Several Metals and Alloys." COO-2204-14. Johns Hopkins University, Department of Mechanics and Materials Science, Baltimore, Maryland. Dec 1975. [NTIS].**

Acoustic emission was monitored during plastic deformation under dead-weight tensile loading of polycrystalline aluminum, brass, copper, and steel. The acoustic emission data, both threshold-type and frequency-analysis type, were correlated with yield strength, stored elastic energy, strain hardening parameters, loading histories, diffusion processes, slip-band motion, and mosaic cell size. Major results are presented under the headings: Discontinuous Yielding (Portevin-Le Chatelier Effect), Parabolic Strain Hardening, and Frequency Analysis. The principal experimental techniques are also summarized.

753 **W. F. Hartman. "Establishing Acoustic Emission Monitoring for Safety and Reliability in Energy Production, Transportation and Manufacturing."** *Materials Evaluation,* **MAEVA, 32(8):20A, 21A, 23A, 24A. Aug 1974; paper presented at the American Society for Nondestructive Testing National Spring Conference, Los Angeles, California. Mar 11–14, 1974.**

The energy crisis and the economy continue to dominate most new directions in industry. The national needs are clear in energy production, transportation, and the manufacturing of consumer products. All of these are providing motivations for establishing efficient innovative techniques of assuring safety and reliability. One such technique is acoustic emission monitoring (AEM). Although its implementation is not yet widespread, this nondestructive method clearly has great potential for application to each of the above areas. The purpose of this paper is to pose some of these applications, especially those receiving less attention, and to suggest that certain potential applications could be effectively investigated so as to establish the feasibility of successfully using AEM. Attention is restricted to examples of AEM pertaining to the areas of energy, transportation, or manufacturing, and show how AE techniques can be used in these fields.

754 **W. F. Hartman. "Experiments with a Portable Acoustic–Emission Simulator and Calibration Device." COO–2204–9. Johns Hopkins University, Department of Mechanics and Materials Science, Baltimore, Maryland, 1973; paper presented at the 12th Acoustic Emission Working Group Meeting, New York. Dec 5–7, 1973.**

A portable electromechanical sparking device that simulates acoustic emission is evaluated as a compact ultrasonic source that can be used for calibrating AE systems. The device consists of a piezoelectric transient-voltage source which generates a spark wave in air. The wave has low amplitude, high frequencies, and is reproducible. The device is intended primarily for use on platelike structures, that is, plates, tanks, pressure vessels, etc. Surface waves are generated and can be used to check-out system sensitivity and source location functions. The experiments which test the reliability and versatility of the device are discussed.

755 **W. F. Hartman. "Towards Correlating Acoustic Emission with Strain-Hardening Transitions." COO-2204-3. Johns Hopkins University, Department of Mechanics and Materials Science, Baltimore, Maryland. 1972.**

Brief report of some initial findings of current research into the interrelation of acoustic emission, discontinuous yielding, and strain hardening. Data are presented showing acoustic emission from discontinuous yielding during dead-weight loading of brass, aluminum, and copper specimens.

756 **W. F. Hartman. "Towards Standards for Acoustic Emission Technology." pp 138–145 in** *Nondestructive Testing Standards: A Review, ASTM STP 624.* **H. Berger, Editor. American Society for Testing and Materials, Philadelphia, Pennsylvania. 1977. proceedings of a symposium held at the National Bureau of Standards, Gaithersburg, Maryland. May 19–21, 1976.**

Five years ago, acoustic emission monitoring was recognized barely as an emerging nondestructive testing method; yet significant progress has been made with respect to the establishment of standards. The NDT experts in acoustic emission technology presently are developing standard terminology, standard measurement procedures, standard calibration techniques, and standard personnel qualifications. The major collective efforts are being coordinated by the American Society for Testing and Materials (ASTM), American Society of Mechanical Engineers (ASME), American Society for Nondestructive Testing (ASNT), and the National Bureau of Standards (NBS). Reviewed and discussed are the current status of the developing standards, and some idealized standards that have not been formulated popularly or even seriously proposed. Also shown are how some of the limitations and deficiencies of the present state-of-the-art can be removed by the evolving or proposed standards. [reprinted by permission of the American Society for Testing and Materials, Copyright].

757 **W. F. Hartman and F. C. Keledy. "Quality Control of Spot Welding by Acoustic Emission Monitoring." Session 8, p 27 in** *The Second Acoustic Emission Symposium.* **Japan Industrial Planning Association, Dobashi Building 5F, 3-17-8 Akasaka, Minato-ku, Tokyo, 107 Japan. 1974. proceedings of a symposium held in Tokyo, Japan. Sep 2–4, 1974. [abstract only, paper not published in proceedings].**

One of the most promising uses of AE monitoring in manufacturing processes is the assessment of weld quality. The use of spotwelding is often impeded by a lack of confidence in the ability to make consistent welds. This lack of confidence can result in either a decline in the use of spot welds or the use of many more welds than a design-with-quality would require. Both of these alternatives produce substantial increases in the cost of the manufacturing processes. It is shown here that AE monitoring during spot welding can provide reliable assurance of weld quality. Special instrumentation for monitoring AE during spot welding will be described and the results of several feasibility studies will be presented. During the production of a spot weld, acoustic emissions result from many things, including friction, plastic deformation, expulsion, solidification, phase transformations, and post-weld cracking. During current flow, amplitude distribution and time discrimination allow separation of the acoustic emission associated with nugget formation from that caused by expulsion. Measurement of the former can even be used in a closed loop arrangement to control the welding time, so as to assure a minimum nugget size. During cooling, the acoustic emission caused by cracking can be separately monitored and recorded. For a given nugget size, the post-weld cracking count is shown to give a good prediction of the fatigue life of the spot weld. Because the instrumentation is versatile, having variable gain, multiple detection levels and timing controls, and selected data output, it can be easily adapted to many spot-welding applications. Using the AE instrumentation, 97% reliable, assessments of spotweld strengths have been achieved. Results are presented for AC spotwelding of steel, aluminum, nickel, and for capacitance-discharge spot-welding of wires and foils.

758 **W. F. Hartman and R. A. Kline. "Variations in Frequency Content of Acoustic Emission during Extension of HF-1 Steel."** *Materials Evaluation,* **MAEVA, 35(7):47–51. July 1977.**

Acoustic emissions resulting from continuous, dead-weight, tensile loading of quenched and tempered HF-1 steel were analyzed for frequency content. Three distinct frequency spectra were observed. These correspond to microplasticity, plasticity, and deformation after recovery. It is shown that frequency analysis of acoustic emission can be a sensitive evaluation technique for assessing structural integrity.

759 **H. Hatano. "Acoustic Emission: Its Fundamentals and Applications."** *Oyo Butsuri* **(Journal of Applied Physics), OYBSA, 44(6):564–576. 1975. (in Japanese).**

Fundamental investigations on acoustic emission are carried out for its application in the fields of materials science and safety engineering. Firstly, acoustic emission caused by plastic deformation and fracture of various types of metals and alloys is observed to study its physical properties. Secondly, a quantitative measuring method of acoustic-emission power and spectra is worked out. This method is applied to tension tests of monocrystals and polycrystals. Quantitative results are discussed in connection with microscopic mechanisms such as dislocation kinetics in the material. Thirdly, a multi-channel acoustic-emission source-location system is developed and is applied to hydrostatic tests of pressure vessels. Acoustic emission is shown to have a potential value for nondestructive inspection and safety monitoring of structures.

760 **H. Hatano. "Acoustic Emission during High–Cycle Fatigue Testing."** *Journal of the Society of Materials Science (Zairyo),* **ZARYA, 24(265):974–979. Oct 1975. (in Japanese).**

Acoustic emission (or stress wave emission) is measured during high-cycle fatigue testing of pure aluminum specimens. Experimental results obtained are as follows: (1) Burst-type acoustic emission begins to radiate intermittently in the early stage of fatigue testing before the macroscopic crack growth is observed; (2) Smoothed specimen radiates acoustic emission of several mV peak. Notched specimen radiates much larger acoustic emission of scores of mV peak under the same cyclic load; (3) Most of the acoustic emission signals are detected around the instance that the cyclic load attains a maximum; and (4) It is clear that the acoustic emission measurement using a peak voltmeter is useful for the study of fatigue, particularly that of microscopic fatigue crack growth.

761 **H. Hatano. "An Experimental Study on Detection of Stress Wave Emission."** *Journal of N.D.I.* **(***Hi-Hakai Kensa,* **Japan), HIHKA, 20(11):597-607. Nov 1971. (in Japanese); RFP-Trans-106. translated for Dow Chemical U.S.A., Rocky Flats Division, Golden, Colorado. 1972. [NTIS].**

Several subjects on detection of stress wave emission are experimentally considered as a fundamental study for its application to a nondestructive inspection technique. A method for detecting stress wave emission in a steel plate under a tensile test is confirmed. As a result of preliminary experiments, uniformity in size and shape of the test plate and in position of the transducer proves necessary for comparative research. On the other hand, it has become clear that preloading of both ends of the test plate suppresses various noises emitted before the plate yields and performance of transducers has an important part in effective detection. The origin of the recorded signals is determined between the sensors, or in the middle part of the test plate, by their propagation time.

762 **H. Hatano. "Detection and Application of Stress Waves." M. A. Thesis. Tokyo University, Tokyo, Japan. Feb 22, 1971. (in Japanese); RFP-Trans-114. translated for Dow Chemical U.S.A., Rocky Flats Division, Golden, Colorado. Oct 1972. [NTIS: N73-21853].**

When solid bodies such as metallic materials are deformed or destructed, sound waves are emitted as one of the energy-relief behaviors. These waves are called stress waves (SW). In this paper, the author has made a fundmental study on how to apply stress wave analysis for the practical nondestructive inspection. First of all, it was inferred from the mechanism of earthquakes that stress waves are generated by the dislocation of a solid. Therefore, the paper discusses the outline of the dislocation characteristics. Next, the author performed preliminary experiments for the selection of transducers, the determination of test plate shape, etc., and attempted to detect stress waves during tensile testing. As a result, major stress waves were discerned in relation with the plastic deformation of metallic materials. Also, the author measured the amplitude distribution of SW, its change with time, and its dependence upon the rate of deformation. These stress waves were generated from steel, aluminum, and zinc-test plates. With regard to steel-test plates, the relationships between their crystalline grains and shapes and the SW were studied. These experimental results could well account for the hypothesis that dislocation shifting would be the SW source. In the meantime, the author performed simple experiments to grasp the concepts of SW frequency distribution or propagation properties. Stress waves were found to be Lamb waves, judging from the propagation rate. Finally, the detection of SW was attempted by a hydrostatic pressure test of a quasi-rocket chamber. Even the same standard chambers had SW at low pressures when the strain concentrated part was added. The SW source location was inferred by time difference in the signals detected from four transducers on a test plate. As a result, its source existed in the strain concentrated part. Although there still remain many problems to solve in the future, the nondestructive inspection using SW was recognized to be effective.

763 **H. Hatano. "Quantitative Analysis of Acoustic Emission."** *Journal of the Acoustical Society of Japan (Nippon Onkyo Gakkaishi),* **NIOGA, 31(1):3–10. Jan 1975. (in Japanese).**

A novel method for the quantitative measurement of acoustic emission is worked out. The performance of the method is analyzed with regard to transducer loss and ultrasonic attenuation in the specimen. The magnitude of the error in the measurement is shown to be negligible for purposes of the present discussion. This method is actually applied to tension tests of pure aluminum specimens. The power spectrum of the acoustic emission for plastic deformation of the specimen is obtained quantitatively over a wide range from 100 kHz to 4 MHz. The total acoustic emission power attains a peak of 5 pW at the beginning of plastic deformation and decreases to 0.3 pW as strain increases. The autocorrelation function, which is the Fourier transform of the power spectrum, is given as a monotonically decreasing function of tau (τ). This shows that the elastic energy of the acoustic emission is radiated not in oscillatory form but in the form of random pulses. The mean value of the pulse width is estimated to be about 0.6 μs in the early stages of deformation and is observed to decrease gradually to less than 0.2 μs with deformation. This value is considered to correspond to the average dwell time of the elementary source events producing the acoustic emission. The change of the dwell time is attributed to an increase of the density of dislocations in the material. It is concluded that this quantitative method is a powerful tool for research on acoustic emission mechanisms in connection with dislocation kinetics.

764 **H. Hatano. "Quantitative Measurements of Acoustic Emission Related to Its Microscopic Mechanisms."** *Journal of the Acoustical Society of America,* **JASMA, 57(3):639–645. Mar 1975.**

A method for the quantitative measurements of the acoustic emission is worked out. Its performance is analyzed with regard to the transducer loss and the ultrasonic attenuation in the specimen. This method is actually applied to the tension tests of pure aluminum specimens. The power spectrum of the acoustic emission is obtained over a wide range of 100 kHz to 4 MHz. The total power attains a peak of 5 pW at the beginning of the plastic deformation and decreases to 0.3 pW with the increase of the deformation. The autocorrelation function for the acoustic emission is given as a monotonically decreasing function. This result reveals that the elastic energy of the acoustic emission is radiated not oscillatorily but in the form of random pulses. Mean value of the pulse widths is estimated to change from 0.6 to 0.2 μsec in the early stages of the deformation. This change is shown to be attributable to the increase of the density of dislocations in the material.

765 H. Hatano. "Quantitative Measurements Related to Acoustic Emission Source Mechanisms." Session 5, pp 77–96 in *The Second Acoustic Emission Symposium.* Japan Industrial Planning Association, Dobashi Building 5F, 3-17-8 Akasaka, Minato-ku, Tokyo, 107 Japan. 1974. proceedings of a symposium held in Tokyo, Japan. Sep 2–4, 1974.

Acoustic emission has been practically applied to various fields of engineering for the past ten years. The author has developed a multichannel acoustic emission location system and applied it to the fatigue and the hydorstatic tests of several types of pressure vessels including the full-size M-3A rocket chambers. A simplified source location technique named *Face-centered arrangement* and new types of acoustic emission sensors suitable for practical use have also been devised. The usefulness of the acoustic emission technique for nonestructive inspection and safety monitoring of structures has been proved. However, insufficient knowledge on the fundamental characteristics of the acoustic emission has considerably impaired the potential value of the technique. Apart from its engineering applications, acoustic emission is expected to afford many interesting subjects to materials scientists and physical acousticians since it has strong correlation with dislocation kinetics in the material. Further researches on its microscopic mechanisms are regarded to be worth making. The lack of methods for its quantitative measurements has been one of the large barriers to the basic investigations. The most fundamental weaknesses in the conventional methods are as follows: (1) They only detect a small proportion of the wide range of frequencies inherent in the acoustic emission. They do not offer enough of the necessary data even for the qualitative discussions since the acoustic emission pattern is subject to vary with its detection frequencies; and (2) *Count rate curve* is too sensitive to the threshold level of the detection for detailed researches, and scarcely offers quantitative information on the acoustic emission energy. Previous attempts in the direction of quantification seem to be prevented by the difficulty in devising high-sensitivity wide-range transducers and in calibrating acoustic responses which the specimen makes to the acoustic emission. The author has worked out a novel method for the quantitative measurements of acoustic emission over a wide range of frequencies during materials testing. In this paper, several of quantitative results obtained by the method are shown, and the acoustic emission source mechanisms are discussed briefly in connection with the dislocation theory.

766 H. Hatano. "Strain-Rate Dependence of Acoustic-Emission Power and Spectra in Aluminum Alloys. *Journal of Applied Physics,* JAPIA, 47(9):3873–3876. Sep 1976.

Strain-rate dependence of acoustic-emission power and spectra for deformation of aluminum alloys is investigated over a wide frequency range of 100 kHz to 4 MHz. In the case of Al-Cu-Mg alloy (2024) and pure aluminum (1100), it is found that acoustic-emission power is proportional to the first power of the plastic strain rate times the gauge length of the specimen. Normalized acoustic-emission power spectra show no significant change at a given strain depending on the strain rate, though they vary with the strain in early stages of deformation. In the case of Al-Mg alloy (5056). repeated bursts of acoustic emission are recorded accompanied by repeated discontinuous yielding. The mode of inhomogeneous deformation and the pattern of acoustic emission vary with the strain rate.

767 **H. Hatano, R. Horiuchi, and N. Niwa. "Fundamental Research on Stress-Wave Emission Method for Pressure Vessels." pp 31–40 in ISAS Report No. 492 (Vol. 38, No. 3). Institute of Space and Aeronautical Science, University of Tokyo, Tokyo, Japan. May 1973. [NTIS: N73-27781].**

Fundamental research for the application of the stress wave emission method to nondestructive inspection and safety monitoring of structures is carried out experimentally. Firstly, stress-wave emission of a model pressure vessel is observed during hydrostatic testing. Time differences in stress-wave emission signals are measured between four sensors attached to the surface of the vessel wall. Stress-wave emission that is radiated at pressures lower than the proof pressure is due to the structural imperfection which is liable to cause a fracture. Secondly, low-cycle fatigue testing of a model pressure vessel is monitored using a single channel system simplified for practical use. During the first pressure cycle, a large amount of stress-wave emission is recorded at high pressures. From the second to about the 5300th cycle, scarcely any stress-wave emission is detected. The usefulness of this method to predict the fatigue failure of the vessels in service is shown. Thirdly, it is observed that a full-size titanium-alloy rocket chamber under hydrostatic testing emits large stress waves on account of twinning of the material. The stress-wave emission method is demonstrated to have a potential value for nondestructive testing and safety monitoring of pressure vessels.

768 **H. Hatano, R. Horiuchi, and N. Niwa. "Stress Wave Emission Technique for Structures' Nondestructive Testing."** *Bulletin of the Institute of Space and Aeronautical Science, University of Tokyo,* **TDUKA, 10(2B):328–342. 1974. (in Japanese).**

769 **H. Hatano, S. Mizutani, R. Horiuchi, N. Niwa, Y. Ohsho, and T. Tsukikawa. "Stress-Wave Emission from Steels for Pressure Vessels."** *Tetsu-To-Hagane (Journal of the Iron and Steel Institute of Japan),* **TEHAA, 61(3):399–403. 1975. (in Japanese).**

The stress-wave emission (SWE) in several alloy steels which are frequently used as structural materials is observed during tensile testing. As to steels SB42, A387-C, and A515, a great amount of SWE is detected in the stage of initial yielding of every specimen annealed even at comparative low-temperatures. This experimental result suggests the usefulness of the SWE technique for nondestructive testing and safety monitoring of structures. On the other hand, in case of stainless steel SUS 27 that is solution-treated, little SWE is recorded compared with the alloys above mentioned.

770 **H. Hatano and E. Mori. "Acoustic–Emission Transducer and Its Absolute Calibration."** *Journal of the Acoustical Society of America,* **JASMA, 59(2): 344–349. Feb 1976.**

A new method is developed for absolute calibration of acoustic-emission transducers using a reciprocity technique in a Rayleigh-wave sound field. It is applied to practical transducers, and their sensitivity is obtained over a frequency range of 100 kHz to 1 MHz. The result is compared with frequency spectrum of an actual acoustic emission signal from a large structure detected by an identical transducer. It is shown that practical characteristics of transducers can be estimated with ease and repeatability in an acoustical environment similar to that of structures such as pressure vessel walls.

771 **H. Hatano and N. Niwa. "A Fundamental Experiment on Location of Stress Wave Emission by Triangulation." Paper No. 286, p 288 in** *Annual Convention Record of the Institute of Electronics and Communication Engineers of Japan, 1971.* **Apr 1971. (in Japanese). [available from Professor N. Niwa, Department of Electronics, Institute of Space and Aeronautical Science, University of Tokyo, Komaba, Meguro–ku, Tokyo, Japan].**

Time difference in stress-wave emission signals detected by a pair of transducers was measured utilizing a magnetic tape recorder, a dual-beam oscilloscope, etc.; and its source points in a thin plate were roughly inferred.

772 **H. Hatano and N. Niwa. "Detection of Stress Wave Emission and its Propagation Characteristics." Paper EM–71–22 presented at the meeting of the Society for the Study of Electrical Measurement, Institute of the Electrical Engineers of Japan. Sep 1971. (in Japanese). [available from Professor N. Niwa, Department of Electronics, Institute of Space and Aeronautical Science, University of Tokyo, Komaba, Meguro–ku, Tokyo, Japan].**

Methods of detecting stress wave emission were considered from the standpoint of electronics, and its propagation characteristics including its velocities in metal plates of different thickness were roughly measured.

773 **H. Hatano and N. Niwa. "Stress–Wave–Emission Method for Evaluation of Material Properties." Research Laboratory of Precision Machinery and Electronics, Tokyo Institute of Technology, Oh–okayama, Meguro–ku, Tokyo, Japan and Institute of Space and Aeronautical Science, University of Tokyo, Komaba, Meguro–ku, Tokyo, Japan, respectively. 1973; Paper E–12 presented at the 7th International Conference on Nondestructive Testing, Warsaw, Poland. June 4–8, 1973.**

Stress-wave emission (or acoustic emission) for commercial pure Al, Al-Mg alloy, Al-Cu-Mg alloy, and mild steel is investigated during tensile and compression testing.

Experimental results are discussed briefly from the viewpoint of strength mechanism. Influence of testing method, heat treatment, etc., on the stress-wave-emission data are surveyed experimentally. The stress-wave-emission method is shown to have potential value for evaluation of material properties. [Abstract 22-0087. *Metals Abstracts,* Vol. 9, 1976].

774 **H. Hatano, N. Niwa, T. Kishi, and R. Horiuchi. "A Fundamental Research on Stress Wave Emission in Mild Steel and Aluminum Alloys."** *Journal of N.D.I. (Hi-Hakai Kensa,* **Japan), HIHKA, 21(4):219–225. Apr 1972. (in Japanese); RFP-Trans-104. translated for Dow Chemical U.S.A., Rocky Flats Division, Golden, Colorado. 1972. [NTIS: N73-21854].**

Stress wave emission in mild steel and aluminum alloys during tensile testing is observed. Emission shows various patterns according to the respective materials. Mild steel radiates groups of large emissions with the propagation of Luders band. The emission signal for commercially pure aluminum reaches a maximum just after yielding and decreases smoothly as strain increases. Under the same experimental condition, a maximum value of the signal is sensitive to grain size. In the case of Al–Mg alloy, small cyclic fluctuations and intermittent bursts of emission are observed to be accompanied by serration phenomena. The Al-Cu-Mg alloy scarcely emits waves in the neighborhood of the yield stress, and with an increase of deformation, the emission attains a maximum. These phenomena are discussed from the viewpoint of strength mechanism in dislocation theory.

775 **H. Hatano, N. Niwa, T. Kishi, and R. Horiuchi. "Fundamental Studies for Application of Stress Wave Emission Method to Material Researches and Safety Monitoring." Paper No. 10 in** *Proceedings of the U.S.–Japan Joint Symposium on Acoustic Emission.* **Japanese volume. Japan Industrial Planning Association, Kawamura-Building 6F, 3-21-6 Akasaka, Minato-ku, Tokyo, Japan 107. 1972. symposium held in Tokyo, Japan. July 4–6, 1972. (paper in Japanese, abstract available in English).**

Lack of basic studies of stress wave emission (SWE) has considerably impaired the intrinsic merits of the SWE method. A measuring system including an electronic computer for practical use has been developed, but it is not easy to give a clarifying analysis of the data obtained by the system. The mechanism of the SWE is said to be in close relation with the behavior of dislocations in solid material. It is desired that fundamental characteristics of the SWE are investigated microscopically prior to its application studies. The purpose of this paper is threefold: (1) to make an analysis in reference to dislocation theory; (2) to point up the problems to be solved for application of the SWE method to nondestructive inspection of pressure vessels; and (3) to determine whether or not the method is serviceable enough as an aid to ensure the safety of structures in service. Stress wave emission in mild steel and aluminum alloys during tensile testing was observed. The SWE shows various patterns according to the respective materials. Mild steel radiates groups of large emissions with the propagation of Luders band. Emission signal for commercially pure aluminum reaches a maximum just after yielding and decreases smoothly as strain increases. Under the same experimental conditions, the maximum value of the signal is sensitive to grain size of the specimen. In the case of Al-Mg alloy, small cyclic fluctuations and intermittent bursts of SWE are observed accompanied by serration phenonmena. The Al-Cu-Mg alloy scarcely emits SWE in the neighborhood of the yield stress, and with an increase of deformation SWE attains a maximum. The variations in the SWE signals from mild steel as a function of its grain size are shown. Normalized curves of pulse-height distribution of the SWE signals are measured by a pulse-height analyzer and also are shown. These results are discussed from the viewpoint of the strength mechanism in dislocation theory. SWE from a pressure vessel and its source location are discussed. Stress wave emission from a model pressure vessel during hydrostatic testing is recorded to inquire into the relation between the SWE and its structural defects. Time differences in the rising of the burst-type SWE signals detected respectively by 4 sensors attached to the surface of the vessel are measured one after another, and their source points are calculated using an electronic computer.

Experimental data demonstrate that the SWE from a vessel observed at pressures lower than its proof pressure is traceable to the stress intensified region which is liable to cause fracture. In order to detect structural imperfections of vessels, further investigation into this kind of the SWE should be pursued intensively. Monitoring of fatigue failure of a pressure vessel by SWE method is also discussed. Stress wave emission from a model pressure vessel under low-cycle fatigue testing is observed using simplified electronic instruments. During the first pressure cycle, large SWE are recorded. Scarcely any emission is detected from the second to about the 5,300th cycle, which is the result of so-called Kaiser effect. Subsequently, a small amount of the SWE caused by fatigue crack growth is observed at every pressure cycle until the crack penetrates the vessel wall at the 5,300th cycle. Experimental results suggest the usefulness of the SWE method to predict fatigue failure of vessels in service.

776 **H. Hatano, N. Niwa, S. Kurosawa, and Y. Ando. "Observation of Stress Wave Emission from a Model Pressure Vessel during Low-Cycle Fatigue Testing."** *Journal of N.D.I.* **(*Hi-Hakai Kensa,* Japan), HIHKA, 21(4):238-241. Apr 1972. (in Japanese); RFP-Trans-105. translated for Dow Chemical U.S.A., Rocky Flats Division, Golden, Colorado. 1972. [NTIS].**

Stress wave emission from a model pressure vessel during low-cycle fatigue testing is observed using simplified electronic instruments. During the first pressure cycle, large emission signal is recorded at pressures higher than 15 kg/cm^2. Scarcely an emission is detected from the second to about the 5,300th cycle, which is the result of so-called Kaiser effect. Subsequently, a small amount of the emission caused by fatigue crack growth is observed at every pressure cycle until a crack penetrated the vessel wall at the 5,375th cycle. Experimental results suggest the usefulness of stress wave emission method to predict a fatigue failure of vessels in service.

777 **H. Hatano, N. Niwa, S. Kurosawa, and Y. Ando. "Stress Wave Emission during Hydrostatic Tests of Model Pressure Vessels."** *Journal of N.D.I.* **(*Hi-Hakai Kensa,* Japan), HIHKA, 21(2):104-105. Feb 1972. (in Japanese).**

778 **H. Hatano and K. Ono. "Acoustic Emission during Fracture-Toughness Tests of a Nuclear Pressure Vessel Steel and Its Weldment." Paper 7-2, pp 475-491 in *The Third Acoustic Emission Symposium.* Japan Industrial Planning Association, Dobashi Building, 3-17-8, Akasaka, Minato-ku, Tokyo, Japan. 1976. proceedings of a symposium held in Tokyo, Japan. Sep 16-18, 1976. [NTIS: AD/A-030-188].**

Because of the relevance to nondestructive evaluation of pressure vessels and other engineering structures, numerous studies of acoustic emission (AE) characteristics of structural steels have been conducted. In assessing relative AE activities, AE counts have commonly been measured as a function of stress, strain, or stress intensity factor. Increasingly, the rms voltages of AE signals have been employed as a measure of relative intensity. In addition, power density (frequency) spectrum and amplitude distribution analyses have become readily accessible techniques of AE signal characterization. Except for a few attempts, however, the absolute calibration of AE measurements has rarely been realized. Lack of systematic efforts to correlate physical processes of deformation and fracture to AE observations has so far limited knowledge on the origins of various AE signals. This is especially true in fracture related AE phenomenon. In many tests, the characterization of either AE signals or fracture parameters has been grossly inadequate. In order to correlate acoustic emission behavior to fracture characteristics, AE during fracture toughness tests of an ASME SA533B nuclear pressure vessel steel and its weldment was investigated at 23°, −15°, and −35°C. One significant feature of this study arises from the extensive compiled-data base of fracture toughness and other mechanical properties of the specimen materials under the sponsorship of Electric Power Research Institute, Palo Alto, California. In this report, however, results on AE measurements are summarized and discussed in reference to fractographic observations. Two distinct types of AE signals were recognized by means of waveform, rms voltage changes, and amplitude distribution analysis. One type of burst emissions was associated with the formation of equiaxed dimples, while the other type appears to arise from microcracking, tearing, and possibly fracture of inclusions.

779 **H. Hatano, I. Sagehashi, and N. Niwa. "A Preliminary Experiment on Detection of Stress Wave Emission in Metals." Paper No. 2440, p 2826 in *Joint Convention Record of Four Institutes of Electrical Engineers.* Japan. April 1970. (in Japanese). [available from Professor N. Niwa, Department of Electronics, Institute of Space and Aeronautical Science, University of Tokyo, Komaba, Meguro-ku, Tokyo, Japan].**

Stress wave emission in a steel plate during tensile testing has been successfully detected utilizing a piezoelectric accelerometer and other electronic instruments.

780 **H. Hatano, I. Sagehashi, Y. Yoshida, and N. Niwa. "Stress Wave Emission from a Model Pressure Vessel and Its Source Location." *Journal of N.D.I.* (*Hi-Hakai Kensa,* Japan), HIHKA, 21(4):234-237. Apr 1972. (in Japanese); RFP-Trans-107. translated for Dow Chemical U.S.A., Rocky Flats Division, Golden, Colorado. 1972. [NTIS].**

As a fundamental study for application of the stress wave emission method to nondestructive inspection technique, the emission from a model pressure vessel during hydrostatic testing is observed. Time differences in the rising of the burst-type emission signals detected respectively by 4 sensors attached to the surface of the vessel are measured one after another, and their source points are calculated using an electronic computer. Experimental results demonstrate that the stress wave emission from a vessel observed at pressures lower than its proof pressure is traceable to the stress intensified region which is liable to cause a fracture. In order to detect a structural imperfection of vessels, further investigation into this kind of emission should be pursued intensively.

781 **H. Hatano, H. Tanaka, R. Horiuchi, and N. Niwa. "Stress Wave Emission during Plastic Deformation in Pure Aluminum."** *Journal of the Japan Institute of Metals (Nippon Kinzoku Gakkai-Si),* **NIKGA, 39(7):675–679. July 1975. (in Japanese).**

Tensile tests of pure aluminum were used to measure stress wave emission (SWE) at room temperature for a better understanding of SWE during plastic deformation. The results were as follows: (1) In the case of a polycrystalline specimen, SWE had a maximum at the onset of macroscopic yielding and then decreased to a stationary level. The maximum value decreased with increasing grain size, and no peak was observed in the case of a single crystal; (2) It was found that the SWE energy per unit time increased linearly with strain rate; and (3) The frequency spectrum of SWE, which was determined over the frequency range of 100 kHz to 2 MHz, showed a wide distribution and depended on the extent of deformation.

782 **H. Hatano, Y. Yoshida, I. Sagehashi, and N. Niwa. "Detection of Stress Wave Emission in a Quasi-Rocket Chamber and Calculation of Its Source Points." Paper HPI-AE-10-38 submitted to the 10th meeting of the Atomic Energy Committee, High Pressure Institute of Japan. June 29, 1971. (in Japanese); RFP-Trans-94. translated for Dow Chemical U.S.A., Rocky Flats Division, Golden, Colorado. 1972. [NTIS].**

A considerable amount of stress wave emission in a small quasi-rocket chamber with a notch was observed during hydraulic static testing, and most of its source points were estimated to be near the notch by triangulation. Less emission in an identical one without notches was recorded at the same pressure.

783 **I. Hattori, M. Kishigami, and T. Nakajima. "Acoustic Emission in Fatigue and Fracture Test of Multi-Layered Cylindrical Pressure Vessel." Research Institute, Ishikawajima-Harima Heavy Industries Company, Ltd., Tokyo, Japan. July 1973.**

784 **D. R. Hay. "Potential Applications of Acoustic Emission in the North."** *Canadian Mining and Metallurgical Bulletin,* **CMMBA, 65(724):19. Aug 1972. [abstract only, paper was never published].**

In northern environments where many materials used in construction are susceptible to low-temperature embrittlement, acoustic emission techniques offer promise for the evaluation of metallurgical integrity and safety of structural components. Acoustic emission is a transient elastic wave produced as a result of the rapid release of energy in a material. In evaluation of structures, it is most useful in the detection of slow crack growth in materials of construction by monitoring the emissions released during crack extension. Applications include monitoring of pressure and storage vessels, bridges and similar structures, and construction equipment. Advantages of acoustic emission include increased sensitivity over conventional nondestructive testing techniques, the possibility of continuous monitoring in service, remote monitoring, and flaw location by triangulation. This paper will review the priniciples of acoustic emission, techniques of application, and theoretical correlations to fracture toughness as well as analysis of the temporal development of microcracking in composite solids. There is increasing interest toward interpretation of the emission activity in terms of the metallurgy of deformation. Results showing the correlation of emission characteristics to deformation mode in various states of stress will be presented.

785 **D. R. Hay and M. N. Bassim. "Detection of Incipient Brittle Fracture by Acoustic Emission." Failure Control International, Lachute, Quebec, Canada. Apr 1976.**

Acoustic emission is a powerful new supplementary nondestructive testing technique. This paper reviews the basic technical and application aspects of acoustic emission and describes recent laboratory and field observations. The fracture mechanics basis for the detection of *active* flaws in the subcritical region is outlined. The general features of acoustic emission and procedures for its application are described. Approaches to inspection via conventional nondestructive testing and acoustic emistion are compared. Theoretical and experimental correlations between acoustic emission and fracture mechanics developed by the authors as well as examples of application in their field work are described.

786 **D. R. Hay, M. N. Bassim, and D. Dunbar. "In-Service Acoustic Emission Monitoring of a Steam-Methane Reformer Furnace." Ecole Polytechnique, Department of Metallurgical Engineering, Montreal, Canada. 1975; paper presented at the 14th Acoustic Emission Working Group Meeting. San Antonio, Texas. Apr 28-30, 1975;** *Materials Evaluation,* **MAEVA, 34(1):20-24. Jan 1976.**

The technique of acoustic emission was applied to monitoring the start-up and steady state operation of a component in a steam-methane reformer furnace. This particular structure is characterized by the presence of high level extraneous noise and high skin temperature. Four independent transducers were mounted on waveguides which were welded to reference pins on the structure. Analysis of the emitted signals involved determination of the event rate and total counts and, at a later time, an analysis of some typical signals recorded during different phases of the start-up operation. The results demonstrate the feasibility of using acoustic emission for real-time monitoring of such complex structures.

787 **D. R. Hay, E. J. Ruzauskas, and N. Pollack. "Acoustic Emission in Metal–Matrix Composites." Drexel University, Metallurgical Engineering Department, Philadelphia, Pennsylvania. 1972; paper presented at the Refractory Composites Working Group Meeting, Houston, Texas. Feb 1–2, 1972.**

Acoustic emission is a transient elastic wave produced as a result of the rapid release of energy in a material. During deformation of a composite material, there exists a multiplicity of microfracture modes, each of which can give rise to acoustic emission. Characterization of the emissions produced by a material can be useful in two general regards, for fundamental investigation of the mechanisms and propagation of fracture as well as a technique for nondestructive surveillance of structure in service. In this study, emission from aluminum-tungsten and aluminum-stainless steel composites were investigated during mechanical testing in various states of stress.

788 **W. Heide. "Gerauschanalyse am Primarkreis von Kernreaktoren (Noise Analysis at the Primary System of Nuclear Power Plants)." Battelle-Institut eV, Frankfort am Main, Germany. 1973. (in German); paper presented at the 6th Meeting of the Working Group on Noise Analysis, Hannover, Germany. Feb 1973.**

Noise analysis at nuclear power plants intends to get information about dynamical response, critical conditions, and irregularities in the primary system. By a new ultrasonic noise analysis at reactors (one of the basic conditions for application of acoustic emission on nuclear pressure vessels), it is possible to extend and improve the surveillance of operational parameters such as pump-flow and control-rod-motion. At a pressure water reactor in Obrigheim, loose metal particles were detected.

789 **C. R. Heiple and R. O. Adams. "Acoustic Emission from Beryllium." RFP-2504. Rockwell International, Atomic International Division, Rocky Flats Plant, Golden, Colorado. July 1976; Paper 5-2, pp 342–359 in *The Third Acoustic Emission Symposium.* Japan Industrial Planning Association, Dobashi Building, 3-17-8, Akasaka, Minato-ku, Tokyo, Japan. 1976. proceedings of a symposium held in Tokyo, Japan. Sep 16–18, 1976.**

The acoustic emission from both powder and ingot-source beryllium was measured as a function of strain and prior heat treatment. Most measurements were made during tensile deformation, but a limited number of compression tests were also performed. The acoustic emission observed was of the burst type, with little or no contribution from continuous emission. The emission was characterized by the variation of burst rate and average energy per burst as a function of strain. The tensile behavior was qualitatively similar for all the materials tested. Burst rate maxima centered roughly at 0.1 percent and one percent plastic strain were observed. The magnitude but not the strain at the low strain burst rate peak was sensitive to prior thermal treatment, while the higher strain burst rate peak was insensitive to prior heat treatment. An energy per burst maximum was observed between the two burst rate peaks, the magnitude of which was moderately sensitive to heat treatment. The Kaiser effect is observed in the material studied. Emission during compression was similar to that observed in tension. The acoustic emission observed is attributed to dislocation motion, as proposed by James and Carpenter of LiF, NaCl, and Zn. Metallographic studies of the beryllium at various strains have ruled out microcracking and twin formation as major contributors to the acoustic emission.

790 C. R. Heiple, R. J. Merlini, and R. O. Adams. "Investigation of Fracture in Pressurized Gas Metal Arc Welded Beryllium." RFP-2463. Rockwell International, Atomics International Division, Rocky Flats Plant, Golden, Colorado. May 20, 1976.

Premature failures during proof testing of pressurized-gas-metal-arc (PGMA) welded beryllium assemblies were investigated. The failures were almost entirely within the beryllium (a forming grade, similar to HP-10 or S-240), close to and parallel to the weld interface. The aluminum-silicon weld filler metal deposit was not centered in the weld groove in the failed assemblies, and failure occurred on the side of the weld opposite the bias in the weld deposit. Tensile tests of welded samples demonstrated that the failures were unrelated to residual machining damage from cutting the weld groove, and indicated small lack-of-fusion areas near the weld start to be the most likely origin of the failures. Acoustic emission was monitored during tensile tests of the welds. The majority of acoustic emission was probably from crack propagation through the weld filler metal. Tensile bars cut from the region of the weld started to behave differently; they failed at lower loads and exhibited an acoustic emission behavior believed to be from cracking in the weld metal-beryllium interface. Improvement in the quality of these and similar beryllium welds can therefore most likely be made by centering the weld deposit and reducing the size of the weld start defect.

791 H. K. Helfrich. "Selbstregistrierende mikroseismiche Messungen (Automically Recorded Microseismic Observations)." ***Felsmechanik und Ingenieurgeologie,*** **FEINA, 4(1):1–13. 1966. (abstract in German, English, and French; article in German).**

Assisted by the Swedish Mining Research Foundation, the practical applicability of the method of recorded microseismic observations was tried out at the mines of the Boliden Mining Company with a device developed by the firm ADELTA AB,

Stockholm. American and Russian experiences in this field were used as a basis for the investigations. Basically, the device consists of a geophone, an amplifier, and headphones. It serves for the detection of locally appearing rock noise. The interpretation of the received sound impulses depends greatly on the observer and is hence subjective. If the basic apparatus is complemented by an impulse counter and by a recording device, it is possible to make objective observations and intrepretations. Such an automatic recorder may be used for occasional or periodic checking of rock movements. An additional multi-channel tape recorder permits the pinpointing of zones of imminent rock destruction. The principal of the microseismic method of observation is as follows: If a disturbance is created in a rock mass, for instance through excavation, a change in the stress flow will occur causing rock deformations. These deformations will affect existing discontinuities within the grain structure of the rock, initiating the formation of embryonic shear planes and fissures. Existing fissures will be enlarged. The deformation processes will cause rock noise which will be received, amplified, and recorded by above described apparatus. The noise impulses appear at different levels of magnitude and are registered in accordance with their frequency of occurance. A tendency towards an increase of the number of impulses per unit of time indicates a progressive deformation and distruction process and thus danger to collapse. There is a great number of areas of application for the microseismic method of observation. It may be used for checking the stability of tunnel crowns or of rock pilasters, especially when these are subjected to additional loads, for the pinpointing of danger zones, for the checking of mountain slopes when blasting, and for many more problems. Principally this method indicates the development of zones of destruction of rock masses and may be considered, therefore, as one of many possible methods for the checking of rock stability. Fully automatic devices may be used for continuously controlling existing or imminent danger zones of mountain slopes or of underground rock structures.

792 **E. G. Henneke, II and C. T. Herakovich. "Acoustic Emission Investigation of Composite Reinforced Metals." VPI-E-73-27. Virginia Polytechnic Institute and State University, Department of Engineering Science and Mechanics, Blacksburg, Virginia. Aug 1973.**

Acoustic emission count rates and total counts are presented as a function of strain for unidirectional boron-epoxy and for metals reinforced with unidirectional boron-epoxy. The loading is tensile and is applied parallel or perpendicular to the fiber direction. The acoustic emission results show that composite reinforced metal specimens which fail early exhibit a sharp increase in AE count rate at the knee of the bilinear stress-strain diagram. It is also shown that machining the edges of coupon specimens significantly increases the total counts prior to failure. The characteristic AE pattern from boron-epoxy is shown to be different for different prepreg materials. Ultrasonic pulse echo attenuation is shown to increase significantly due to damage during loading.

793 **E. G. Henneke II, C. T. Herakovich, G. L. Jones, and M. P. Renieri. "Acoustic Emission from Composite Reinforced Metals." Paper 2257. Virginia Polytechnic Institute and State University, Department of Engineering Mechanics,**

Blacksburg, Virginia. 1973; Paper No. 11 presented at the 1973 Society for Experimental Stress Analysis Fall Meeting, Indianapolis, Indiana. Oct 16–19, 1973; ***Experimental Mechanics,*** **EXMCA, 15(1):10–16. Jan 1975.**

Acoustic emission count rates are presented for tensile loading of unidirectional boron/epoxy and for aluminum sheets reinforced with unidirectional boron/epoxy. It is shown that different prepeg materials have different characteristic AE patterns. Results from composite reinforced metal specimens show that early failures are accompanied by a sharp increase in AE count rate at the knee of the bilinear stress-strain diagram. It is further shown that the count rates are a function of specimen fabrication and that higher total counts do not necessarily correspond to early failures.

794 **E. G. Henneke, II and H. W. Herring. "Spectrum Analysis of Acoustic Emissions from Boron–Aluminum Composites." pp 202–214 in** ***Composite Reliability, ASTM STP 580.*** **American Society for Testing and Materials, Philadelphia, Pennsylvania. 1975.**

Acoustic emissions were monitored from unidirectional tension specimens of boron-aluminum composites. It is shown that a sudden increase in count rate presages final failure. This sudden increase, however, occurs at a variable point relative to final failure load. Spectrum analysis in the range to 1 to 100 kHz was performed on the emissions with the result that no immediately obvious recognition pattern exists to distinguish between types of failure mechanisms. Thus, it is suggested that either acoustic emissions cannot be used to distinguish different failure modes in boron-aluminum composites in the frequency range studied or else the specimens used here had only one failure mode operating with sufficient energy release to be detected. The spectrum analysis did show, however, that the acoustic emissions had frequency components much higher than the frequencies for the fundamental modes of natural vibration of the specimen. [reprinted by permission of the American Society for Testing and Materials, Copyright].

795 **R. E. Herzog. "Forecasting Failures with Acoustic Emission."** ***Machine Design,*** **MADEA, 45(14):132–137. June 14, 1973.**

Editorial survey article on acoustic emission, describing what it is, its advantages and disadvantages as a nondestructive testing tool, and its applications.

796 **R. W. Higgs. "Sonic Signature Analysis for Arc Furnace Diagnostics and Control." pp 653–663 in** ***1974 Ultrasonics Symposium Proceedings.*** **Catalog No. 74 CHO 896-1SU. Institute of Electrical and Electronics Engineers, Inc., New York. 1974. symposium held in Milwaukee, Wisconsin. Nov 11–14, 1974.**

A computerized Fourier analysis system has been used to study the noise from scrap movement in a 50-ton electric arc melter. As part of this study, the melter roof has been opened during the heat to observe the condition of the scrap in the furnace. Preliminary data had lead to the hypothesis that the first major noise during the melt-down period after an elapsed energy consumption of 6 MWh would indicate scrap fall uncovering a wall area and leading to refractory lining damage by the arc. This hypothesis was proved untrue by visual observation of the scrap condition after the Fourier system indicated the occurrence of a major noise in the frequency bands most sensitive to scrap falling. More indicative would be the midpoint of the major noises during melt-down, a point which cannot be found in real-time monitoring. Difficulty was encountered also in detecting the quiet period after melt-down because the furnace is sometimes quiet between major scrap falls. The computerized Fourier system is useful for studying the characteristics of arc furnaces and their operating variables on a statistical, *a posteriori* basis.

797 **R. Hill. "The Use of Acoustic Emission for Characterising Adhesive Joint Failure."** *NDT International,* **NDITD, 10(2):63–72. Apr 1977.**

Acoustic emission measurements have been made on fracturing lap joints and it is shown that with some knowledge of the stresses involved information about the mode of fracture of the joint can be deduced. The mode of fracture is different for different adhesives. Tests on adhesives subjected to high temperatures indicate that acoustic emission might be used to prevent failure in adhesive joints accidentally exposed to a hostile environment. Adhesive joints tested suggest the Kaiser effect is only obeyed well near the fracture stress.

798 **R. Hill and R. W. B. Stephens. "Acoustic Emission: Evaluation of Transducers and Measuring Systems." pp 177–182 in** *Ultrasonics International 1973 Conference Proceedings.* **IPC Science and Technology Press Ltd, Guildford, Surrey, England. 1973; conference held at Imperial College, London, England. Mar 27–29, 1973.**

Acoustic emission produced by dynamic defects is related to the motion of these defects and the basic aim of acoustic emission measurement is to try and monitor these defect movements. However, in practice, this has proved to be difficult. The need to maintain a good signal-to-noise ratio within the measuring system has meant the signal distortion produced by resonant detectors has had to be tolerated. The exact details of the propagation and detection of acoustic emission are a problem for long-term study and the usual approach has been to take a simple transducer signal parameter and obtain a long time correlation between this and the mechanical behavior of the material. Some of the problems of the present acoustic emission detection systems and the problems still to be investigated, if the acoustic emission technique is to be fully effective, are briefly discussed. Results of electrical impedance measurements on resonant acoustic emission transducers are presented using various methods of mounting. The nature of the fundamental transducer vibration mode is discussed. Mounting the transducer on the acoustic emission test specimen using a solid material is to be avoided, for the reasons indicated.

799 **R. Hill and R. W. B. Stephens. "Acoustic Emission from Adhesive Joints." University of Aston, Birmingham, England and Chelsea College, London, England; paper presented at the Institute of Physics Conference on Acoustic Emission, Imperial College, London, England. Mar 14, 1972. [proceedings were never published].**

Paper reports on the development and evaluation of acoustic emission transducers and their use in determining if the adhesive joints tested obey the Kaiser effect. Electrical impedance measurements have been made to determine the resonances of the transducers and to investigate the various transducer mounting methods. These results will be reported. Pulse propagation studies have also been carried out to investigate the response of the measuring system to mechanical excitation. The resulting acoustic emission from one-half inch lap joints measured by these transducers, and particularly how well these adhesive materials obey the Kaiser effect, will be reported.

800 **R. Hill and R. W. B. Stephens. "Simple Theory of Acoustic Emission: A Consideration of Measurement Parameters."** *Acustica,* **ACUSA, 31(4): 224–230. Oct 1974.**

Acoustic emission can, potentially, yield details of defect motion within materials. The problem is proving difficult because of the complexities associated with the generation, propagation, and detection of acoustic emission. Here relationships are suggested between the acoustic emission stress wave and measurement parameters such as count, burst amplitude, and energy which may aid interpretation of acoustic emission measurement. Burst amplitude seems to be most simply related to the stress wave parameters, but other parameters may be suitable depending on the experimental conditions prevailing.

801 **R. Hill and R. W. B. Stephens. "Sonic Emission during Deformation of Solids."** *Archiwum Akustyki* **(Warszawa), ARAKB, 6(1):45–58. 1971. (in English).**

Deformation of material leading to a displacement of one region with respect of another could result in a stress wave, i.e., a sound wave. Stress wave emission provides information about the actual formation of disturbances within the specimen when by usual ultrasonic test methods the energy is injected into the body to detect the presence of defect. Stress wave emission has been observed in a diversity of materials such as gold, ice, and plastics. The onset of SWE is brought about when the deformation mechanism works sufficiently fast and it includes waves of different types propagating with different velocities. The frequency response of a specimen depends on its elastic properties, geometry, and restricting forces. The interpretation of the acoustic data obtained with SWE requires knowledge of the properties of the material. The parameters which are used to describe the nature of the SWE are the frequency content, the amplitude distribution, and the energy content. The energy of the SWE is low. The two main areas of application of the SWE are structural testing and material research. The important feature of testing by SWE is that it permits the location of the damage, in bodies of reasonable size.

802 **M. Hoff. "Acoustic Emission Source Location using Single and Multiple Transducer Arrays." pp 8–12 in *1972 Ultrasonics Symposium Proceedings.* Catalog No. 72 CHO 708–8SU. Institute of Electrical and Electronics Engineers, Inc., New York. 1972. symposium sponsored by Institute of Electrical and Electronics Engineers Group on Sonics and Ultrasonics, held in Boston, Massachusetts. Oct 4–7, 1972.**

Discrete stress wave emissions generated by the propagation of flaws in metals have been detected and processed to determine their source location. The accuracy of the required time of arrival measurements is compromised by the presence of extraneous acoustic sources and is complicated by the large number of signals generated. A computerized processing system has been developed that uses geometrically resolved statistical enhancement techniques to reject insignificant signals and to increase the amount of time between flaw detection and specimen failure. The feasibility of this location technique was tested using an array of four transducers mounted on eight-inch-diameter diaphragms of various alloys. Source location data were obtained when the diaphragms were pheumatically loaded to failure. The computer graphics display of these processed data dramatically illustrates flaw geometry and growth during the loading cycle. The application of this technique to the monitoring of larger specimens have been accomplished using two concentric four transducer arrays mounted on 36-inch diameter diaphragms. Two modes of operation allow switching between full area scanning using all 8 transducers, and high resolution tracking using 6 possible combinations of 4 transducers. Initial results indicate that 2 such arrays can be used to monitor even larger areas with little loss of location accuracy and with only a small increase in software complexity.

803 **K. G. Hoge, B. A. Kuhn, and V. L. Reshenk. "Flow Behavior of Mulberry Uranium." UCRL–51346. Lawrence Livermore Laboratory, Livermore, California. Feb 13, 1973.**

The mechanical properties of Mulberry uranium were studied at strain rates from 10^{-4} to 5×10^{3} sec^{-1} and at temperatures from 78 to 770 K. The nonlinear loading and off-loading characteristics of this alloy received particular attention. This nonlinear behavior is the result of a γ_S to γ_0 phase transformation and a reversible, diffusion-controlled ordering effect. A test series was conducted to determine the acoustic emission from phase transformation during tensile loading on Type-C alloy.

804 **J. T. Hoggatt. "Thin–Metal Lined PRD 49–III Composite Vessels." NASA CR–134555. Boeing Aerospace Company, Research and Engineering Division, Seattle, Washington. Jan 1974. [NTIS].**

In this program, a total of 22 eight-inch diameter PRD-49-III filament wound vessels were evaluated for burst strength and fatigue performance. Eighteen vessels contained a 0.003-inch (0.0762-mm) thick 321 stainless steel liner for fluid containment while the remaining four had elastomeric liners. Six of the metal-lined tanks were burst tested at 70 °F (21 °C). The final vessel design with a metallic liner gave an average burst pressure of 2560 psi (17.7×10^6 N/m^2) and an average performance factor (PV/W) of 1.75×10^6 inches (4.45×10^6 cm). Twelve metal-lined vessels were cycled at strain levels equivalent to 50% and 75% of ultimate at 70 °F (21 °C). All these vessels leaked in a relatively few cycles (20-60 cycles) with failure occurring in all cases in the metallic liner. The thin liner would de-bond from the composite and buckle during depressurization. No composite failures or indications of impending composite failures were obtained in the metal-lined vessels. An elastomeric lined vessel was successfully cycled for 2493 pressure excursions at 75% of ultimate strain prior to composite failure in the dome. The tests concluded that the filament wound composite vessels have excellent fatigue capabilities, however significant improvements are required in thin-metal liner technology.

805 **J. Holt. "Analysis of Acoustic Emission Signals in Compact Tension Specimens." Laboratory Note RD/L/N 107/72. Central Electricity Research Laboratories, Leatherhead, Surry, England. May 16, 1972.**

The overall shape of the acoustic emission signal and the total number of counts obtained using the *Dunegan mode* of counting in steel compact tension specimens are shown to be dominated by multiple reflections within the sample. Damping of the sample has resulted in overall signal decay times comparable with the ringdown times of the transducers used in acoustic emission work. A detailed examination of a typical emission from a damped sample has shown that information about the duration of the basic pulse reaching the transducer can be obtained by comparing the signal shape with that produced by pulse excitation of a simple analog circuit.

806 **J. Holt. "Electronic Noise in Acoustic–Emission Measurements."** ***Non–Destructive Testing,*** **NDETA, 9(2):66–70. Apr 1976.**

The factors controlling the background electronic noise level and the signal-to-noise performance of a typical single channel laboratory acoustic-emission system are analysed and the results are supported by measurements on a commercial apparatus. It is concluded that for such a system the possible improvement in signal-to-noise ratio obtainable by using improved electronics whilst retaining the present type of transducer is limited to less than 1 decibel.

807 **J. Holt. "Pulse Height Analysis in Acoustic Emission." Laboratory Note RD/L/N 106/72. Central Electricity Research Laboratories, Leatherhead, Surrey, England. May 16, 1972.**

A simple analog circuit is used to analyze the effect of the input signal duration on the output from an acoustic emission detection system. The results of the analysis are supported by measurements on the analog circuit and indicate that the peak amplitude of the output signal is a complex function of the input pulse duration. The limitations thereby imposed on the use of pulse height analysis to evaluate energy release rates from growing defects are discussed with reference to the range of pulse durations which could occur in an acoustic emission test.

808 **J. Holt. "The Use of Plastic Bases on Acoustic Emission Transducers." Laboratory Note No. RD/L/N 6/73. Central Electricity Research Laboratories, Leatherhead, Surrey, England. Mar 19, 1973.**

Commercial acoustic emission transducers are fitted with an epoxy plastic base to aid mounting on curved test surfaces and to provide electrical isolation. The effect of such bases on transducer performance is analyzed for the case of a large steel test sample. It is shown that useful gains may be obtained by optimizing the base thickness.

809 **J. Holt and T. W. M. Evans. "Measurement and Interpretation of the Amplitude Distribution of Acoustic Emission Signals." Laboratory Note No. RD/L/N 40/76. Central Electricity Research Laboratories, Leatherhead, Surrey, England. May 1976.**

The distribution of signal amplitudes obtained in an acoustic emission test determines the number of signals detectable above a given level of background noise. A knowledge of this distribution is therefore necessary if the result of laboratory measurements are to be used to predict the probability of defect detection in real situations with differing levels of background noise and signal attenuation. The complexity introduced into the signals by the effects of multiple reflections in the test piece raises difficulties in the measurement of amplitude distributions by means of conventional amplitude sorters; the design, calibration, and testing of equipment which overcomes these problems are described. This apparatus has been used to demonstrate that, despite the effects of multiple reflections, the measured amplitude distribution is a good representation of the distribution of stress wave amplitudes in the sample. This result establishes the technique as a means of studying fundamental characteristics of acoustic emission sources. Examples are presented of amplitude spectra from a variety of acoustic emission sources. It is shown that a basic log-normal distribution can be used to fit all the examples so far obtained, including the ubiquitous power-law spectra.

810 **J. Holt and D. J. Goddard. "Assessment of an Acoustic Emission Transducer for Waveguide Application." Laboratory Note No. RD/L/N 3/76. Central Electricity Research Laboratories, Leatherhead, Surrey, England. Mar 1976.**

In acoustic emission tests in hostile environments, e.g., the pressure vessel facility at MEL, it is necessary to mount the detecting transducers on extremely long acoustic waveguides. Ideally, the transducer diameter should not exceed that of the waveguide. The allowable waveguide diameter is however often limited by engineering considerations and the usual approach is to modify the end of the guide to match the diameter of the transducer. Modification of the transducer geometry affects not only the resonant frequency but the degree of signal dispersion introduced by the waveguide and also the intrinsic electrical noise generated by the transducer. An investigation of these conflicting effects has been made by using a model transducer matched in diameter to a practical waveguide. A laser pulse technique has been introduced to simulate acoustic emission signals, allowing the relevant waveguide modes to be identified and comparisons to be made with the transducer-waveguide combination used at present. Electrical impedance measurements have been used to estimate the effective electrical noise source of the transducer-waveguide combination. The results obtained indicate that an improvement of up to 16 db over the signal/noise ratio of the present system can be obtained.

811 **J. Holt and I. G. Palmer. "Correlation of Counting and Energy Measurements in Acoustic Emission Tests." Laboratory Note No. RD/L/N 52/76. Central Electricity Research Laboratories, Leatherhead, Surrey, England. May 1976.**

A variety of signal parameters have been used to assess the activity of a growing defect in an acoustic emission test. For example, work at the Central Electricity Research Laboratories has established the use of ringdown counting as a means of assessing plastic deformation in certain low alloy pressure vessel steels. There is now an increasing interest in monitoring the energy of the signal in the expectation that this will provide a better assessment of damage. In several instances however, including the case of low alloy steels, the results have been found to be essentially the same as those obtained by the counting methods; in other cases the relationships observed between total ringdown count and total energy have not been explained. Herein the problem of summation of the total ringdown count and total energy is analysed in detail. It is shown that, provided the correct summation procedure is used together with available amplitude spectrum results, these problems can be resolved. Reasonably satisfactory fits of the particular experimental data are also obtained leading to the suggestion that in certain cases the simultaneous summation of energy and counts can be used to provide an indication of the approach of failure.

812 **J. Holt and I. G. Palmer. "The Use of Acoustic Emission for Monitoring the Integrity of Steel Pressure Vessels." Paper C80/74, pp 46–51 in** ***Periodic Inspection of Pressurized Components.*** **Conference Publication 8, 1974. published by Mechanical Engineering Publications Limited for the Institution of Mechanical Engineers, London, England. 1974. conference held in London, England. June 4–6, 1974.**

The present status of the application of acoustic emission to the location and assessment of defects in steel pressure vessels are reviewed. Particular attention is given to the results and implications of a materials evaluation program on reactor pressure vessel steel. The further laboratory and model pressure vessel tests required to fully develop and establish the results of this program are discussed.

813 **J. Holt, I. G. Palmer, A. G. Glover, and J. A. Williams. "Some Examples of Laboratory Application and Assessment of Acoustic Emission in the United Kingdom. Part I: Acoustic Emission Behaviour of Low Alloy Steels." pp 35–50 in *Acoustic Emission.* Edited by R. W. Nichols. Applied Science Publishers Ltd, Ripple Road, Barking, Essex, England. 1976. proceedings of a colloquium on acoustic emission held during the 1975 Annual Assembly of the International Institute of Welding, Tel Aviv, Israel. July 7-13, 1975. [paper presented by B. J. L. Darlaston].**

This two-part paper summarises the work of a number of establishments in the United Kingdom (see Reference 1638 for Part II). The application of acoustic emission techniques to large steel structures requires a thorough laboratory investigation of the material parameters which govern acoustic emission in the particular steel. The first part of the paper deals with a low alloy pressure vessel steel and a high temperature creep-resistant steel. The second part considers the location of cracks in structures using acoustic emission techniques. A number of examples are discussed. The basic principles of acoustic emission are not discussed in this paper as the topic is fully covered by another paper in the colloquium. Although there is experience in the application of acoustic emission to pressure plant this, too, is omitted from the discussion as the general topic is also the subject of another paper.

814 **J. Holt, I. G. Palmer, and D. J. Goddard. "The Interpretation of Acoustic Emission Signals from the Deformation of Low Alloy Steels." pp 24–44 in *Schallemission: Anwendung bei der Untersuchung, Prufung und Uberwachung metallischer Werkstoffe.* Deutsche Gesellschaft fur Metallkunde eV, 637 Oberursel/Taunus bei Frankfurt/Main, Adenauerallee 21, Germany. 1974. proceedings of symposium held in Munich, Germany. Apr 18–19, 1974.**

The general features of acoustic emission tests on low alloy pressure vessel steels are reviewed and the particular effects of pearlite cracking, matrix deformation, and oxide cracking are described. Possible correlations of the acoustic emission signals and material parameters are analysed and their application to metallurgical studies is discussed.

815 **J. Holt, I. G. Palmer, and D. J. Goddard. "The Use of Acoustic Emission for the Detection of Cracks in Steel." Chapter 5, pp 22–27 in *The Detection and Measurement of Cracks.* The Welding Institute, Abington Hall, Abington, Cambridge, England. 1976.**

During defect extension in brittle materials, large amplitude acoustic emission signals are obtained from the effects of macroscopic crack extension and the fracture of large brittle inclusions. In ductile low alloy steels, however, gross failure occurs by ductile tearing caused by void coalescence which is an acoustically quiet process. Acoustic emission from such steels is consequently restricted to the effects associated with plastic deformation prior to gross crack extension. The signals obtained are thus strongly dependent on the precise metallurgical state of the steel. This sensitivity to microstructure raises severe problems in the application of acoustic emission to large structures, where the variation in heat treatment which inevitably occurs on an industrial scale is sufficient to drastically affect the local acoustic activity of the material. A laboratory investigation of the factors affecting the acoustic activity is therefore necessary to specify the realiability of the technique for a particular steel specification. However, even when this has been done, real defects may well differ from those introduced in laboratory specimens owing to the presence of oxide films, rust, etc., and also because real defects are most likely to occur in regions which differ in some way, e.g., a weld slag inclusion, from the parent plate or weld metal. For these reasons, caution must be exercised in extrapolating laboratory results to real situations, especially when attempting to assess defect size or proximity to failure. In this chapter the results are discussed of a laboratory study of acoustic emission from a low alloy pressure vessel steel (0.15% C-0.9% Mn-0.19% Cr-0.12% Ni), including the effects of oxidation, and the limitations they imply for the detectability of flaws in the as-specified material are considered.

816 **C. R. Horak (to Grumman Corporation). "Acoustic Emission Process and System for Improved Flaw Source Location." U.S. Patent 3,985,024. Oct 12, 1976.**

An acoustic emission process and system for improved flaw source location employing spatial filtration. Master-slive discrimination, rise time discrimination, and/or coincidence detection are integrated into a multichannel flaw source location system to reject unwanted signals. A computer provides calculation of the flaw source coordinates from the valid signals.

817 **C. R. Horak. "Structural Integrity Acoustic Emission Monitoring." ADN Report 09-01-72.2. Grumman Aerospace Corporation, Advanced Development Program, Bethpage, New York. Aug 1972.**

Note describes the evaluation of a prototype multichannel acoustic emission system used for monitoring structural integrity of complex structures. The system was laboratory-tested to establish output levels and basic characteristics. Interface problems with a multichannel high frequency magnetic tape recorder were resolved. Simulated flaw signals were recorded and played back into the system with satisfactory results. Structural integrity was monitored during F-14 wing-center section fatigue testing and recorded on magnetic tape for off-line analysis. This single test, monitoring a highly complex structure, provided valuable application experience and data because of the numerous geometries involved. Circuit modifications for improved system performance and a more versatile sensor-mounting method were developed during the course of testing. The practical application experience gained and the analysis of the magnetic tape recordings indicate the feasibility of monitoring complex structures for structural integrity. However, in order to meet reliability requirements, further optimization of the system is required.

818 **C. R. Horak and A. F. Weyhreter. "Acoustic Emission System for Monitoring Aircraft Components and Structures in a Severe Fatigue Environment." summary of paper in** *Paper Summaries: ASNT Spring Conference, March 28–30, 1977, Phoenix, Arizona.* **American Society for Nondestructive Testing, Columbus, Ohio. 1977.**

The need for a practical acoustic emission system which would effectively detect and locate fatigue cracks within the high nosie environment of aircraft ground and flight testing prompted the initiation of a design development program. The resulting system integrated known noise-discrimination techniques with computerized source location equipment. These discrimination techniques known as master-slave, coincidence, and rise time detection were integrated with a source location system and used on progressively more difficult aircraft elements of materials, components and structures. From 95 to 99% of all noises and vibrations external to the transducer array were eliminated by the hardware prior to computer source location processing. Crack initiation and propagation were detected and located in high strength steel landing-gear parts, titanium-welded preflawed plates, high strength steel elements, aluminum, and nickel-alloy structural bulkheads. The detection and location of crack initiation and propagation would have been virtually impossible in these fatigue environments had noise discrimination not been utilized. The basic concept for discrimination requires the system to look between the noise or vibrations for valid acoustic emission signals as opposed to the general concept of looking above a predetermined threshold. Cracks as small as 0.010 inches (0.0254 cm) long by 0.002 inches (0.005 cm) deep were detected and located by acoustic emission monitoring. These flaws were verified by conventional nondestructive testing methods in some cases after the parts were stressed to reopen the tightly closed fatigue cracks.

819 **C. R. Horak and A. F. Weyhreter. "Acoustic Emission System for Monitoring Components and Structures in a Severe Fatigue Noise Environment."** *Materials Evaluation,* **MAEVA, 35(5):59–63, 68. May 1977.**

The high noise environment caused by fatiguing material elements and aircraft components has severely limited the application of acoustic emission monitoring to detect and locate crack initiation and propagation. The need for a practical acoustic emission NDT system prompted the design and testing of a real-time system which integrated known noise discrimination techniques with computerized source location equipment. These discrimination techniques, previously used with count rate systems only, are known as master-slave, coincidence, and rise time detection. The application of this integrated system on progressively more difficult aircraft elements of materials and components, determined 95 to 99% of all noise external to the transducer array was rejected by the hardware prior to computer processing. Crack initiation and propagation were detected in high strength steel landing gear parts, titanium welded preflawed plates, and high strength steel elements undergoing severe fatigue testing. The detection and location of crack initiation and propagation would have been virtually impossible in this environment without noise discrimination. Cracks in threads as small as 0.010 inches (0.0254 cm) long by 0.002 inches (0.0050 cm) deep were detected and located by acoustic emission. The thread cracks required 80% maximum fatigue load to be applied and held prior to verification by magnetic particle inspection at 5X magnification.

820 **J. R. Houghton and P. F. Packman. "Pulse Shape of Acoustic Emission Signals." Report JRH-3-2360-02-75. Tennessee State University, School of Engineering and Technology, Nashville, Tennessee. Oct 1975; paper presented at the American Society for Nondestructive Testing National Fall Conference, Atlanta, Georgia. Oct 13-16, 1975.**

The methodology of signature analysis for acoustic emission pulses is outlined, using Fourier frequency spectrum, Fourier transform, shock spectrum, and shock spectrum ratios. The procedure compares the relative merits of each analysis technique to characterize the pulse received from the transducer in an acoustic emission test. Analytical and experimental pulses are modeled as passing through a high-pass filter and an acoustic emission transducer. Signatures of computer-modeled pulses are presented to show the relative sensitivity of each technique to small changes in the pulse-time-domain shape. This was done to optimize the acoustic emission transducer response and filter pole location so that potential differences between critical defects and benign acoustic emissions could be identified.

821 **J. R. Houghton, P. F. Packman, and M. A. Townsend. "A Technique for the Deconvolution of the Pulse Shape of Acoustic Emission Signals Back to the Generating Defect Source/Une technique pour la deconvolution de l'image des impulsions de l'emission acoustique, au defaut generateur." Tennessee State University and Vanderbilt University, Nashville, Tennessee; Paper 3K/2 in** *Proceedings of the Eighth World Conference on Nondestructive Testing.* **volume containing Sections 3K1, 3H1, 3G. International Committee for Nondestructive Testing, Secretariat du Cofrend, Institut de Soudure, 32 Boulevard de la Chapelle, 75880 Paris Cedex 18, France. 1976. conference held in Cannes, France. Sep 6-10, 1976. (paper in English). [NTIS: INIS-mf-3543].**

Acoustic emission signals recorded after passage through the instrumentation system can be deconvoluted to produce signal traces indicative of those at the generating source, and these traces can be used to identify characteristics of the source.

822 **J. R. Houghton, M. A. Townsend, and P. F. Packman. "Optimal Design and Evaluation Criteria for Acoustic Emission Pulse Signature Analysis."** *Journal of the Acoustical Society of America,* **JASMA, 61(3):859-871. Mar 1977.**

Design of pulse-recording systems and evaluation criteria (method of analyzing the pulse signature) are investigated with the objective of defining optimal approaches to pulse signal analysis. A representative situation which this paper addresses is the modeling of acoustic emission pulse analysis as a nondestructive means of failure detection in which pulse density counting is presently the most common evaluative criterion. The instrumentation is molded and selected analytical pulses are passed through the system. Two alternatives are considered here: frequency spectrum analysis and time domain reconstruction of the pulse or pulse train (deconvolution). The pulse recording-analysis problem is modeled, and the various analysis techniques are considered. Within practical constraint, optimal system designs are defined. It is shown that deconvolution of a pulse is a superior approach for transient pulse analysis. Reshaping of a transducer output back of the original input pulse is possible and gives an accurate representation of the generating pulse in the time domain. Frequency spectrum analysis methods and AE measurement systems have potential for segregation of different pulse shapes. Using deconvolution, in principle any definable transducer and filter system can be used to reconstruct pulse characteristics, i.e., to generate time domain signature. A number of major conclusions about the selection of design variables are made for general usage. In addition there are families of optimal selections for transducer and filter system parameters which give superior deconvoluted output signals. These may be significant when working with actual hardware and unknown characteristics. The optimal design results are new and possibly different from the expectations of engineers accustomed to working in the field of signal conditioning. In obtaining these results, modern rational design techniques on a rather complex design problem are demonstrated.

823 **E. Howells. "Acoustic Emissions from Stressed Dielectric Liquids." McGraw-Edison Power Systems Division, Research and Development Department, Franksville, Wisconsin. 1974; Reprint ISA ASI 74219(63–66). Instrument Society of America, Pittsburgh, Pennsylvania. 1974; paper presented at the Instrument Society of America 20th International Symposium, Albuquerque, New Mexico. May 21–23, 1974.**

A dielectric liquid when stressed to a level approaching its breakdown point, usually exhibits a phenomenon commonly known as *corona.* At this time partial discharges take place in a somewhat random fashion. They are pulse-like in nature and may be detected by monitoring the current supplied to the system to offset the discharges. A number of instruments are commercially available for this purpose and are capable of resolving discharges in the pico-coulomb range. Although the energy dissipated in these discharges is small, some portion of it must be converted into mechanical energy and appear as pressure pulses within the liquid. Some evidence of this can be found in the various reports documenting acoustic and transonic (20 to 50 kHz) observations associated with *corona* in devices such as oil filled transformers. The understanding of prebreakdown phenomena is limited. It was therefore decided that an investigation of these mechanical stress pulses (or acoustic emissions) might provide data which would complement the electrical information now available, if such acoustic emission did in fact exist.

824 **D. Hums and P. Jax. "Ermittlung beginnender Rissbildung in keramischen Werkstoffen bei thermischer Belastung mittels Schallemissionsanalyse**

(Acoustic Emission as an Aid to Detect Crack Formation in Ceramic Materials under Thermal Load)." Battelle-Institut eV, Frankfurt am Main, Germany. 1973. (in German); paper presented at the 7th International Conference on Science of Ceramics, Antibes-Juan-les-Pins, France. Sep 25, 1973.

Thermal loading experiments on ceramic materials have shown that acoustic emission is an aid to the detection and monitoring of fracture processes in the micro- and macro-ranges. The investigations were conducted with various types of ceramic materials. Acoustic emission was continuously measured during the heating and cooling phases of the firing process and under thermal shock load. It was found that critical phases during cooling, in which incipient micro- and macrocracking or fracture occurs, are sensitively recorded by AE: Microcrack formation in Al_2O_3 or quartz procelain in the transition range of the glass phase, microcrack formation in quartz procelain at the inversion point of quartz, i.e., 570 °C, and incipient microcracking or fracture in the temperature range below 400 °C. Quantitative determination of the extent of incipient damage, i.e., the number and mean length of the cracks, appears to be possible. Acoustic emission measurements on a stoneware body under thermal shock load also yielded valuable information. Microcracking in the transition range of the phase during cooling was observed in all temperature cycles, i.e., was found to be reversible; irreversible microcrack formation at the inversion point of quartz, however, occurred only within the first temperature cycle. In addition, macrocracking was detected in the temperature below 400 °C, especially during heating. Considering this detailed information on the kinetics of cracking obtained by it, acoustic emission is a useful aid to the development, production, and evaluation of ceramic materials. Its industrial application, e.g., for process control, should not present any great problems.

825 **D. O. Hunter, P. H. Hutton, and H. N. Pedersen. "Crack Detection in Pressure Piping by Acoustic Emission." pp 3.1–3.11 in "Nuclear Safety Quarterly Report—February, March, April, 1969 for Nuclear Safety Branch of USAEC Division of Reactor Development and Technology." BNWL-1084. Battelle-Northwest, Richland, Washington. June 1969. [NTIS].**

Work is in progress to adapt shear-wave detection of acoustic emission signals to practical field application requirements. This effort was indicated by previous study of wave propagation characteristics in flat plates. Use of wedge mountings to couple the signal from the plate surface to the transducer enhances signal detection. Acoustic emission detected during a recent test of carbon steel material showed evidence of a distinction between deformation and fracture generated signals. This correlates with an earlier observation from a pipe burst test conducted at the Battelle-Columbus test site. A thickness poled, surface mounted sensor with the fundamental resonance at about 1.2 MHz has been fabricated in an effort to enhance the detection of acoustic emission in the low MHz range. Although there is some loss of sensitivity compared to similar lower frequency sensors, initial tests show it to be well within the usable range. Local application of acoustic emission to investigations of potential slow crack growth by fatigue in process piping is continuing. Three of four joints monitored have shown limited indication of crack activity. Use of an acoustic emission system to monitor known flaws in the heads of the Containment System Experiment simulator vessel during blowdown tests has continued. Only limited indications of flaw movement have been detected.

826 **P. H. Hutton. "Acceptance Testing Welded Ammunition Belt Links Using Acoustic Emission." BNWL-SA-4524. Battelle-Northwest, Richland, Washington. Apr 1973; pp 402-415 in** *Proceedings of the Ninth Symposium on Nondestructive Evaluation.* **Southwest Research Institute, San Antonio, Texas. July 1973. symposium held in San Antonio, Texas. Apr 25-27, 1973; pp 107-121 in** *Monitoring Structural Integrity by Acoustic Emission, ASTM STP 571.* **Edited by J. C. Spanner and J. W. McElroy. American Society for Testing and Materials, Philadelphia, Pennsylvania. 1975. proceedings of a symposium held in Ft. Lauderdale, Florida. Jan 17-18, 1974.**

The objective of the work described in this paper was to provide a reliable method for 100% acceptance testing resistance welds used in the fabrication of 40-mm automatic weapon ammunition belt links. Four projection resistance welds (two on each side) are used to assemble the link. Since several variables must all be properly controlled simultaneously to produce a sound resistance weld, weld quality can vary widely. Links contain rounds of ammunition and are coupled together to form a belt which is automatically fed into the weapon. If one of the links in a belt fails, the immediate result is a jammed weapon. Conventional nondestructive test techniques cannot distinguish a sound resistance weld from a weak one. This work has shown, however, that by applying a moderate proof load to the weld joint, weak welds can be detected by acoustic emission. The technique is being adapted to production application with an expected inspection rate of 1200 links per hour in its initial version. This work is representative of many useful applications of acoustic emission that can be accomplished within existing state-of-the-art technology. [reprinted by permission of the American Society for Testing and Materials, Copyright].

827 **P. H. Hutton. "Acoustic Emission: A New Technique for Examining the Integrity of Welded Pressure Vessels." BNWL-SA-3710. Battelle-Northwest, Richland, Washington. 1971; paper presented at the 15th International Welding Congress, Stary Smokovec-High Tatras, Czechoslovakia. Oct 5-7, 1971; "Akusticka emisia: nova metoda kontroly celistvosti zvaranych tlakovych nadob."** *Zvaranie,* **ZVARA, 20(9/11):341-347. Sep-Nov 1971. (in Slovak).**

Acoustic emission which is basically elastic waves propagated in a solid material by energy released as the material deforms or fractures has potential for providing structural integrity surveillance in a variety of ways not attainable by conventional nondestructive testing methods. In the case of welded pressure vessels, acoustic emission surveillance could be beneficially applied during fabrication, acceptance testing, and in-service use to enhance vessel safety. The application can start with testing of vessel material samples to help evaluate tensile and fatigue properties.

During vessel fabrication, acoustic emission provides a sensitive tool for detecting and locating flaws formed in the welds as they cool. This is particularly beneficial in the case of heavy section vessels requiring many passes to complete a weld. When the vessel is completed, acoustic emission can again be applied during hydrostatic testing to identify incipient failure sites either in the weld zones or in the body of the plates by detecting localized deformation or cracking around flaws. Finally, the technique could be applied in most cases to vessels in-service on a continuous surveillance basis to detect vessel deterioration from fatigue, stress corrosion cracking, etc., and warn of impending brittle or ductile fracture. An inherent limitation of the acoustic emission technique is the fact that it will not detect a static flaw. Deformation or fracture or both at least on a localized basis must be occurring to generate the acoustic emission signals. This lecture discusses briefly the method, the indicators, and the state-of-the-art for the acoustic emission applications described.

828 **P. H. Hutton. "Acoustic Emission: A New Tool for Evaluating Structural Soundness." BNWL-SA-2449. Battelle-Northwest, Richland, Washington. Aug 1969; pp 165–171 in** ***Proceedings of the 7th Symposium on Nondestructive Evaluation of Components and Materials in Aerospace, Weapons Systems and Nuclear Applications.*** **Western Periodicals Company, North Hollywood, California. July 1969. symposium held in San Antonio, Texas. Apr 23–25, 1969.**

Acoustic emission is defined as the elastic waves produced in a solid material by energy released as the material is deformed or fractured. These waves can be detected at the surface of the material with high sensitivity transducers. This provides the basis for a new technique for monitoring metal structures, pipes, pressure vessels, graphite shapes, etc., to detect structural deterioration before failure occurs. Flaw detection resolution using acoustic emission exceeds that of any of the conventional nondestructive techniques such as ultrasonics, radiography, etc. It is, however, limited to detection of an active strain or flaw area—it will not detect static flaws. This paper discusses development work conducted at Battelle-Northwest to apply acoustic emission to detect flaw growth in nuclear reactor pressure systems. It also deals with applications of acoustic emission to detect formation of cracks in welds, metal fatigue, and use as a surveillance tool in studying pressure vessel fracture modes. Preliminary work on application to graphite is included.

829 **P. H. Hutton. "Acoustic Emission Allows Detection of Crack Development Before Failure Occurs."** ***SAE Journal of Automotive Engineering*****, SJAUA, 79(8):33–37. Aug 1971.**

Incipient failures due to metal fatigue and spontaneous cracking in welds can be detected by acoustic emission techniques. Resolution exceeds that of ultrasonics and radiography. Only active strain or flaw areas are detected. The fatigue-detection system is described. Acoustic emission from low-cycle tension-compression fatigue testing of a high Ni alloy at 1000 °F (538 °C) is summarized, as are tests of center-notch panel specimens of Al. Cycles-to-failure ranged from 362,000 to 391,000. Three out of four tests showed a consistent exponential relationship between emission count and stress intensity. A sharp increase in acoustic emission preceding visible crack development was found due to macrocrack nucleation. Acoustic emission can detect crack formation in welds even while welding is in progress and can detect weld porosity induced by flux interruption. [Abstract 22-1283. *Metals Abstracts,* Vol. 4, 1971].

830 **P. H. Hutton. "Acoustic Emission Applied Outside of the Laboratory." BNWL-SA-3775. Battelle-Northwest, Richland, Washington. 1971; pp114-128 in** *Acoustic Emission, ASTM STP 505.* **American Society for Testing and Materials, Philadelphia, Pennsylvania. 1972. proceedings of symposium held in Bal Harbour, Florida. Dec 7-8, 1871.**

The paper presents a cross section of beneficial applications which illustrate the wide variety of present day uses of acoustic emission surveillance. These uses range from detection of flaws in pressure vessels during hydrostatic testing to inspection of building floors for detecting weak points, detection of impending failure in localized segments of rock structure in mining, and process control in metal-forming work. While these examples are not all inclusive, they do show that acoustic emission is not just a laboratory novelty and that beneficial use of the phenomenon is beginning to be realized.

831 **P. H. Hutton. "Acoustic Emission Applied to Determination of Structural Integrity." BNWL-SA-3147; Battelle-Northwest, Richland, Washington. Mar 1970. [NTIS: BNWL-SA-3147, AD-752-794]; pp 28-29 in** *Proceedings of Mechanical Failures Prevention Group, Eleventh Meeting.* **W. T. Sawyer. Chesapeake College, Wye Mills, Maryland. Apr 13, 1971. conference held in Williamsburg, Virginia. Apr 7-8, 1970. [NTIS: AD-724-745].**

Acoustic emission is defined as the elastic waves produced in a solid material by energy released as the material is deformed or fractured. These waves can be detected at the surface of the material with high sensitivity transducers. This provides the basis for a new technique for monitoring metal structures, pipes, pressure vessels, graphite shapes, etc., to detect structural deterioration before failure occurs. Flaw detection resolution using acoustic emission exceeds that of any of the conventional nondestructive techniques such as ultrasonics, radiography, etc. It is, however, limited to detection of an active strain or flaw area—it will not detect static flaws. This presentation discusses development work conducted at Battelle-Northwest to apply acoustic emission to detect flaw growth in nuclear reactor pressure systems, formation of cracks in welds, metal fatigue, and thermally induced cracking in ceramics.

832 **P. H. Hutton. "Acoustic Emission Detection in the Presence of Hydraulic Noise." BNWL-933. Battelle-Northwest, Richland, Washington. Dec 1968. [NTIS].**

When use of *acoustic emission*, produced as metals deform and fracture, is considered as a means of continuously monitoring the integrity of nuclear reactor pressure boundaries, segregation of emission data from background noise must be evaluated. Testing has been performed to demonstrate detection of acoustic emission in the presence of turbulent flow and cavitation hydraulic noise. The results show that an acoustic emission sensing system operating in the 1.5 to 2.5 MHz frequency range with shear sensors is essentially insensitive to either turbulent flow or cavitation noise. At the same time, this system will detect acoustic emission signals clearly at a distance of at least 3 feet from the signal source. Follow-on work is necessary to optimize the sensing technique, adapt it to routine field application, and establish the maximum distance at which an acoustic emission signal can be reliably detected.

833 **P. H. Hutton. "Acoustic Emission in Metal: A New NDT Tool." BNWL-SA-1430. Battelle-Northwest, Richland, Washington. Sep 1967; paper presented at the Institute of Electrical and Electronics Engineers 1967 Symposium on Sonics and Ultrasonics, Vancouver, British Columbia, Canada. Oct 5, 1967.**

834 **P. H. Hutton. "Acoustic Emission in Metals as a NDT Tool." BNWL-SA-1262. Battelle-Northwest, Richland, Washington. Sep 1967. [NTIS]; *Materials Evaluation*, MAEVA, 26(7):125-129. July 1968; paper presented at the American Society for Nondestructive Testing National Fall Conference, Cleveland, Ohio. Oct 18, 1967.**

The potential of acoustic emission as nondestructive testing and metallurgy research tools is discussed. Acoustic emission is the pressure wave produced by energy released by deformation and fracture. With adequate sensors and signal amplification, these pressure waves can be detected at the surface which is deforming or cracking. Applications include monitoring flaw growth in reactor pressure tubes; remote and continuous surveillance of operational systems, such as natural gas pipe lines, and aircraft structures to detect flaw formation and growth; monitoring welds for cracking during cooling; monitoring hydrostatic testing of missile fuel tanks; tool for fracture research in metallics and nonmetallics; and monitoring space vehicles for micrometeorite penetration.

835 **P. H. Hutton. "Acoustic Emission Monitoring for Continuous Crack Detection in Nuclear Reactor Pressure Boundaries." BNWL- 1597. Battelle-Northwest, Richland, Washington. Nov 1971. [NTIS].**

A program to develop application of acoustic emission techniques to reactor system integrity surveillance has been sponsored at Battelle-Northwest by the U. S. Atomic Energy Commission, Division of Reactor Development and Technology, from February 1966 to July 1970. This report covers the final portion of the program (January 1969 to July 1970). Full discussion of the first period of the program was previously reported in BNWL-997 (see Reference 841). Capabilities resulting from this program include use of acoustic emission monitoring to testing, flaw formation during welding, incipient fatigue failure points in metal structures, and stress corrosion cracking in metals. Better definition of the acoustic emission signal and propagation was achieved, as well as development of specialized detection instrumentation and suitable techniques for data analysis (including real-time source location techniques).

836 **P. H. Hutton. "Acoustic Emission-What It is and Its Application to Evaluate Structural Soundness of Solids." BNWL-SA-2983. Battelle-Northwest, Richland, Washington. Feb 1970; paper presented to the Central Ohio Section of American Society for Nondestructive Testing, Columbus, Ohio. Feb 17, 1970, and to the Chicago Section of American Society for Nondestructive Testing, Chicago, Illinois. Feb 19, 1970.**

Acoustic emission is defined as the elastic waves produced in a solid material by energy released as the material is deformed or fractured. These waves can be detected at the surface of the material with high sensitivity transducers. This provides the basis for a new technique for monitoring metal structures, pipes, pressure vessels, graphite shapes, etc., to detect structural deterioration before failure occurs. Flaw detection resolution using acoustic emission exceeds that of any of the conventional nondestructive techniques such as ultrasonics, radiography, etc. It is, however, limited to detection of an active strain or flaw area—it will not detect static flaws. This presentation discusses development work conducted at Battelle-Northwest to apply acoustic emission to detect flaw growth in nuclear reactor pressure systems. It also deals with applications of acoustic emission to detect formation of cracks in welds, metal fatigue, and use as a surveillance tool in studying pressure vessel fracture modes.

837 **P. H. Hutton. "Application of Acoustic Emission to Detect Cracking in Welds and to Monitor a 22-Inch Diameter Tee Pressurized to Failure." Y-40662. Battelle-Northwest, Richland, Washington. May 1969.**

Demonstration of acoustic emission techniques for detection of weld cracking and for detection of flaw formation and growth during pressure test of a 22-inch diameter pipe tee to failure was performed. The purpose of the subject test work was to: (1) Demonstrate detection of crack formation during submerged arc welding of heavy section steel by interrogation of acoustic emission; and (2) Demonstrate detection of acoustic emission under normal pressure test conditions, and determine correlation between emission response and the mechanical-metallurgical phases of pressurizing a 22-inch diameter pipe tee to failure. The test specimen used in weld monitor tests consisted of two plates, 2 inches thick by 12 inches wide by 24 inches long, of A533 carbon steel joined with a U-groove weld along the 24-inch dimension by the submerged arc machine welding process. Induced weld cracking was readily detected by the acoustic emission monitor system both during and after welding. Noise from the welding process did not obscure indications of crack formation. Two phases of pressure testing a 22-inch diameter, 1.5-inch wall, SA-212-B carbon steel pipe tee were monitored for acoustic emission response. These were a proportional limit test where the tee was pressurized to an indicated 2.0% maximum plastic strain, and a burst test of the same tee where it was pressurized until a through-wall fracture developed. Acoustic emission response showed four distinct phases in the proportional limit test. Interpretation of these phases requires correlation with mechanical data compiled by Combustion Engineering. Attainment of a new maximum pressure level in the step-wise pressurization sequence was readily identifiable in the acoustic emission pattern. Emission detected during the burst test indicated cracking occurring in both tee crotches prior to final failure on one side. There was evidence of crack activity starting at 5200 psig pressure. Each new maximum pressure level was detectable in terms of acoustic emission and the amount of emission followed a rising curve as pressure increased toward failure. A distinction in type of emission signal was observed and is felt to be related to plastic deformation as opposed to crack formation. Water leaking around a fitting was detectable with the acoustic emission system well before it became evident from rate of specimen pressurization.

838 **P. H. Hutton. "Detecting Acoustic Emission in the Presence of Hydraulic Noise."** *Non-Destructive Testing,* **NDETA, 2(2):111–115. May 1969.**

The acoustic emission produced as metals deform and fracture can be used to monitor the integrity of nuclear reactor pressure boundaries, as long as the emission data can be separated from background noise. Tests have shown that a sensing system using shear-wave sensors in the frequency range 1.5 to 2.5 MHz can detect acoustic emission in the presence of turbulent flow and cavitation noise, and at distances of at least 3 feet (1 metre) from the signal source. Further work is needed to improve the sensing technique, establish the maximum distance at which an emission signal can be reliably detected, and adapt the technique for practical use.

839 **P. H. Hutton. "Detecting Acoustic Emission in the Presence of Nuclear Reactor Hydraulic Noise." BNWL-SA-2366. Battelle-Northwest, Richland, Washington. June 1969; paper presented at the American Nuclear Society 15th Annual Meeting, Seattle, Washington. June 15–19, 1969.**

When a solid material undergoes plastic deformation it produces signals, or what is known as acoustic emission, which, in the case of a nuclear reactor, can be used to monitor the integrity of the reactor's pressure boundaries. One of the major questions concerning the use of acoustic emission has been the ability to distinguish emission signals from background noise. The work to be described in this paper was a test to demonstrate how acoustic emission could be detected in the midst of turbulent flow and cavitation hydraulic noise. The results show that an acoustic emission sensing system operating in the 1.5 to 2.5 MHz frequency range with shear mode sensors is essentially insensitive to either turbulent flow or cavitation noise, while at the same time it can detect acoustic emission distinctly.

840 **P. H. Hutton. "Detection of Flaw Growth in Pressure Piping by Sensing Acoustic Emission." BNWL-SA-1076. Battelle-Northwest, Richland, Washington. 1966; paper presented at a Meeting of the Atomic Energy Commission Advisory Committee on Reactor Safeguards, Washington, D.C. Dec 1966.**

The development of a system for detection of flaw growth in reactor pressure piping by sensing attendant acoustic emission is discussed. A general discussion on methods for measuring this acoustic emission is given. Results from deformation tests, ductility tests, and brittle failure tests are presented. Sensor development is briefly considered. Methods of analysis are discussed.

841 **P. H. Hutton. "Detection of Incipient Failure in Nuclear Reactor Pressure Systems Using Acoustic Emission." BNWL-997. Battelle-Northwest, Richland, Washington. Apr 1969. [NTIS].**

Development work to apply acoustic emission detection as a means of continuously monitoring nuclear reactor pressure boundaries for flaw formation or growth is summarized. The nature of the acoustic emission signal has been clarified, specialized detection techniques have been developed, identification of acoustic emission in the presence of hydraulic noise has been demonstrated, and the technique has been beneficially applied to selected problems both in the field and in the laboratory.

842 **P. H. Hutton. "Flaw Detection by Acoustic Emission." BNWL-SA-4576. Battelle-Northwest, Richland, Washington. 1972. [NTIS]; paper presented at Oregon State University Seminar, Corvallis, Oregon. Jan 31, 1973.**

Research on the acoustic emission phenomenon is reviewed, methods for detecting acoustic emission are discussed, and uses of the acoustic emission testing method to detect flaw growth are described. Principle applications include the monitoring of pressure vessels for flaw growth during hydrostatic proof testing and while in service; fatigue crack growth monitoring of in-service materials; and surveillance of welds for flaw formation. It is concluded that acoustic emission technology has reached the point where it can be effectively applied to material characterization in the laboratory, detection and location of flaws during welding and proof testing, and continuous in-service monitoring. Conventional nondestructive testing must, however, be utilized to establish the size of a detected flaw. Significant development work remains to realize acoustic emission testing's full potential.

843 **P. H. Hutton. "Identification of Worn Bullet Forming Dies by Acoustic Signature in Process." BNWL-SA-3813. Battelle-Northwest, Richland, Washington. 1971; pp 59-73 in** ***Proceedings of the Eighth Symposium on Nondestructive Evaluation in Aerospace, Weapons Systems, and Nuclear***

Applications. **Western Periodicals Company, North Hollywood, California. 1971. symposium held in San Antonio, Texas. Apr 21–23, 1971.**

The results of a feasibility study of the acoustic signature approach applied to a conventional six-station bullet-forming press are discussed. Data were obtained from the second pointing operation during normal operation of the press. The results show a distinct difference in spectral content of the acoustic information from a *good* and *bad* die-punch set. The ultimate objective of this study would be to monitor all operations in the process to rapidly identify when a particular die and punch needs to be changed to maintain the product within tolerance.

844 **P. H. Hutton. "Integrity Surveillance of Pressure Systems by Means of Acoustic Emission." BNWL-SA-2194. Battelle-Northwest, Richland, Washington. Oct. 1969; pp 1017–1032 in** *Pressure Vessel Technology. Part 2: Materials and Fabrication.* **American Society of Mechanical Engineers, New York. 1969. proceedings of the 1st International Conference on Pressure Vessel Technology, Delft, Netherlands. Sep 29–Oct 2, 1969.**

Acoustic emission is defined as the elastic waves produced in a solid material by energy released as the material is deformed or fractured. These waves can be detected at the surface of the material with high sensitivity transducers. This provides the basis for a new technique for monitoring metal structures, pipes, pressure vessels, etc., both during fabrication and during system operation to detect flaw formation and growth before failure occurs. The basic phenomena was first studied in the United States and Germany in late 1940 and early 1950. The first practical application of acoustic emission as a surveillance technique was in the United States in 1964 when it was successfully used to detect flaw growth in rocket motor casings during hydrostatic proof testing. The monitor system permitted depressurization of faulty casings for repair prior to failure under test. Extending the application to continuous surveillance of complex systems such as a reactor primary pressure system in the field during operation, however, imposes significantly more demanding requirements. Flaw-detection resolution potential using acoustic emission far exceeds that of any of the more conventional nondestructive techniques such as ultrasonics, radiography, etc. It is, however, limited to detection of an active strain or flaw area; it will not detect static flaws. The paper discusses development work conducted at Battelle-Northwest directed toward applying acoustic emission to detect flaw growth in nuclear reactor pressure systems during reactor operation. It also deals with application of acoustic emission to detect formation of cracks in welds as they cool, metal fatigue, stress corrosion cracking, and as a surveillance tool in studying pressure vessel fracture modes.

845 **P. H. Hutton. "Investigate Feasibility of Utilizing Acoustic Emission Methods to Evaluate the Integrity of a Prototype ADAPTS Oil Storage Container." Report No. CG-D-37-76. Battelle-Northwest, Richland, Washington. Dec 1975. [NTIS: AD/A-025-910].**

The report discusses acoustic emission characteristics of nylon fiber-polyurethane material used in the U.S. Coast Guard's Prototype Oil Storage Container, a subsystem of the Air Deliverable Anti-Pollution Transfer System (ADAPTS). Characteristics were determined by laboratory testing. Results are considered in the interest of potential use of acoustic emission methods to evaluate the physical integrity of items manufactured with this material. Generation of acoustic emission and its relationship to material loading were favorable to the intended application. The specific application is concluded to be impractical, however, due to high attenuation of acoustic emission signals in the material. Acoustic emission does have potential application to this type of material where a low attenuation signal transmission path such as metal sheathing is available.

846 **P. H. Hutton. "Monitoring High Temperature Fatigue using Acoustic Emission Techniques." BNWL-SA-3989. Battelle-Northwest, Richland, Washington. Dec 1971. [NTIS]; paper presented at the American Society for Testing and Materials Symposium on Fatigue at Elevated Temperatures, held at the University of Connecticut, Storrs, Connecticut. June 18-22, 1972.**

The paper presents a unique method for detecting the onset and progress of fatigue damage at elevated temperatures. The technique utilizes acoustic emission information which can now be measured in a high temperature environment. Acoustic emission profiles of the fatigue process in metals show distinct and reproducible features which can be related to microcracking and the nucleation of a macrocrack. These results have been observed with a variety of metals in environments ranging from room temperature to 1200 °F. Application of this detection method is still in the development stages.

847 **P. H. Hutton. "Nuclear Reactor Background Noise vs Flaw Detection by Acoustic Emission." BNWL-SA-3820. Battelle-Northwest, Richland, Washington. 1971; pp 1-16 in** *Proceedings of the Eighth Symposium on Nondestructive Evaluation in Aerospace, Weapons Systems, and Nuclear Applications.* **Western Periodicals Company, North Hollywood, California. 1971. symposium held in San Antonio, Texas. Apr 21-23, 1971.**

The paper discusses results obtained from current investigation of the character of reactor background noise at various stages of operation of the San Onofre PWR at San Clemente, California and its relation to acoustic emission detection. The results are encouraging to *clean* detection of acoustic emission in the 1 to 2 MHz region and recorded data samples are being used to investigate means of identifying acoustic emission signals within the general noise.

848 **P. H. Hutton. "Nuclear Reactor System Noise Analysis, Dresden I Reactor, Commonwealth Edison Company." BNWL-867. Battelle-Northwest, Richland, Washington. Oct 1968.**

Work is in progress to utilize *acoustic emission* produced as metals deform and fracture, as a means of continuously monitoring the integrity of nuclear reactor pressure boundaries. To help accomplish this, a definition of the noise field is needed because it must be considered in detecting and analyzing acoustic emission during reactor operation. In-plant noise measurements on several reactor systems will be necessary to provide an adequate definition. The first of these has recently been completed at the Dresden I Nuclear Power Reactor at Morris, Illinois, and the results are encouraging. For this reactor, the noise field above 500 kHz, is low. This provides a region nominally from 750 kHz to about 3 MHz where signal attenuation with increasing frequency rises sharply for ferrous materials which is conducive to reliable detection of acoustic emission signals. Detector response to acoustic emission has been demonstrated within and well beyond this region. Although the results of this first study can not be considered universally applicable to reactor systems, they do provide strong encouragement for continued monitor system development. They, also, define a region for beneficial effort on improved detection techniques.

849 **P. H. Hutton. "Performance of Noise Analysis Study." Task 9, Project IV, pp 104–120 in "In-Service Inspection Program for Nuclear Reactor Vessels." R. D. Wylie. SwRI Project 17-2440, EEI Project RP-79. Biannual Progress Report No. 5. Southwest Research Institute, San Antonio, Texas. May 28, 1971. [NTIS: PB-200-518].**

The objective of this task is to characterize the noise background in operating nuclear reactors to determine both spectral content and amplitude data. Such information is needed for final specification of an acoustic emission surveillance system for detecting the noise emitted by a growing flaw in the presence of background hydraulic flow noise. During this period, the background noise of the San Onofre pressurized water reactor was measured. The results are encouraging to *clean* detection of acoustic emission in the 1 to 2 MHz region and recorded data samples are being used to investigate means of identifying acoustic emission signals within the general noise.

850 **P. H. Hutton. "Performance of Noise Analysis Study." Task 9, Project IV, pp 97–133 in "In-Service Inspection Program for Nuclear Reactor Vessels." R. D. Wylie. SwRI Project 17-2440, EEI Project RP-79. Biannual Progress Report No. 6, Volume 1. Southwest Research Institute, San Antonio, Texas. Jan 7, 1972. [NTIS: PB-207-591].**

For this task, the background noise of the San Onofre pressurized water reactor was measured. The results are encouraging to *clean* detection of acoustic emission in the 1 to 2 MHz region. Recorded data samples were used to investigate means of identifying acoustic emission signals within the general noise. The final results of the above evaluation are reported herein.

851 **P. H. Hutton. "Predicting Metal Fatigue Life by Acoustic Emission." Battelle-Northwest, Richland, Washington. 1968.**

852 **P. H. Hutton. "Progress Report March, April, May, June, 1966, 04 Nuclear Safety, Pressure Piping Crack Monitoring Detection of Metal Overstress by Acoustic Emission." BNWL-CC-763. Battelle-Northwest, Richland, Washington. Aug 2, 1966. [NTIS].**

Results of tests to determine feasibility of monitoring for and locating crack initiation or growth in reactor pressure piping by means of transducer initiated acoustic waves are described. Factors discussed include types of transducers and other equipment, acoustic emission signatures, mechanisms of signal propagation, location of origin of acoustic emission in piping, effect of extraneous noise, and correlation and evaluation of acoustic emission data. This is first report on this project. Subsequent reports are BNWL-CC-900 (Reference 853) and NDTMD-MEMO-67-2 (Reference 854).

853 **P. H. Hutton. "Progress Report July, August, September, 1966, 04 Nuclear Safety, Pressure Piping Crack Monitoring Detection of Metal Overstress by Acoustic Emission." BNWL-CC-900. Battelle-Northwest, Richland, Washington, Oct 28, 1966. [Official Use Only].**

The purpose of this report is to summarize first quarter FY-67 accomplishments on Pacific Northwest Laboratory's program to develop a monitor system to detect crack formation or growth in reactor pressure piping by sensing the accompanying acoustic emission. The program is being sponsored by the Division of Reactor Development and Technology, U.S. Atomic Energy Commission, Washington, D.C. Details and background of the work were described in the previous progress report, BNWL-CC-763 (see Reference 852). For the sake of brevity and conciseness, this report will simply expand on those areas where work has been done during the intervening period. The three main areas of effort have been: (1) definition of the general acoustic response pattern related to the gross aspects of forming and extending a crack in various materials; (2) development of a monitor system prototype concept exclusive of transducers; and (3) development of a suitable, high temperature transducer. Tests using double cantilever beam (DCB) specimens of various materials to establish conditions of crack formation and growth have indicated that material ductility is a major controlling factor in the acoustic response pattern. It appears to effect both acoustic emission intensity and the point in the crack formation-growth sequence at which the main emission occurs. A concept has been developed for the prototype of a full scale monitor system. Hardware development is being limited to the analyzer portion of the system at this time because it is the part most significant to demonstrating feasibility of the intended application. Signal level and signal rate are both being investigated as possible parameters for evaluating acoustic emission data. Of the various transducers discussed in the previous progress report for potential high temperature application, the capacitive or electrostatic transducer now looks most promising. A significant improvement in sensitivity has been achieved and a trial model used during recent tests produced generally satisfactory data. The sequence of effort on the program is being adjusted somewhat from earlier plans. Some of the more detailed investigative phases will receive only moderate attention, temporarily, in favor of first demonstrating the basic feasibility of detecting acoustic emission and making a meaningful analysis under postulated service conditions.

854 **P. H. Hutton. "Progress Report October, November, December, 1966, 04 Nuclear Safety, Crack Detection in Pressure Piping by Acoustic Emission." NDTMD–MEMO–67–2. Battelle–Northwest, Richland, Washington. Feb 17, 1967. [Official Use Only].**

Research efforts continued on a program to develop a surveillance system for detecting crack formation or growth in reactor pressure piping by sensing the attendant acoustic emission. Program study work this quarter has been directed toward analysis of acoustic emission signals from a selected material subjected to different modes of failure. Types A-212-B carbon steel chosen for the test work exhibited distinctly different emission both in pattern and signal content for conditions of gross plastic deformation, ductile fracture, and brittle fracture. Evidence begins to indicate that acoustic emission signal characteristics may permit a degree of definition which will distinguish between shear and cleavage fracture. This would be significant in analyzing local stress levels and strain rates. Current hardware development on the analyzer portion of the prototype monitor system has utilized a count rate approach which is producing promising results. Summaries of test data prepared with this equipment to date show good consistency and definition. Low noise charge amplifiers fabricated locally using field-effect transistors (FETs) enhance the application of the count rate analyzer by improving the signal to noise ratio of the data. The scanner portion of the prototype monitor system concept has been eliminated in favor of a continuous monitor of the transducers by groups. This latter approach can better satisfy the signal source-location function required for this system. Development work on a high temperature, high sensitivity, electrostatic transducer has produced encouraging results. Laboratory test models have shown good sensitivity and an extremely broad frequency-response range. Although the mechanical strength of the dielectric must be improved, this does not appear to be an insurmountable problem. Overall proof of this transducer is being expedited in an effort to avoid the necessity of developing a specialized room-temperature transducer for detailed study of acoustic emission characteristics.

855 **P. H. Hutton. "Proof Testing Ammunition Belt Links by Acoustic Emission." BNWL–SA–4522. Battelle–Northwest, Richland, Washington. Mar 1973; paper presented at the American Society for Nondestructive Testing National Spring Conference, Los Angeles, California. Mar 12–15, 1973.**

The work discussed in this paper was directed to the objective of providing a reliable method for 100% proof testing of the welded section of 40-mm automatic weapon ammunition belt links. The belt links of concern are fabricated from approximately 1/32-inch thick sheet steel strips. The strips are preformed into link halves which are joined by four projection welds, two on each side. If these welds are not sound, the lind can at least partially separate in service. This results in a jammed weapon. The work has demonstrated that by using acoustic emission techniques during proof loading to a level which will not damage a sound link, weak welds can be detected. No other presently available nondestructive testing technique is capable of detecting these faulty welds. The paper discusses feasibility determination and adaptation for production application.

856 P. H. Hutton. "Reactor System Noise Analysis Dresden I Reactor, Commonwealth Edison Company." NDT-MEMO-68-8, Preliminary Report. Battelle-Northwest, Richland, Washington. 1968.

Development of new and more discerning techniques for nondestructively testing reactor pressure boundaries to assess their integrity is of general interest to the nuclear industry for further enhancement of nuclear safety. One such technique which shows considerable promise is based on detection of *acoustic emission* or elastic waves generated in metal by the energy released as it plastically deforms and fractures. This technique has potential application for integrity surveillance during both hydrostatic testing of a reactor system and on-line operation. It is however, limited to detection of a dynamic condition, i.e., it will not detect a static flaw or discontinuity. The U.S. Atomic Energy Commission, Division of Reactor Development and Technology in Washington, D.C., has sponsored development work at Battelle's Pacific Northwest Laboratory for the past two and one-half years directed toward development of the technology and equipment needed to apply this new surveillance technique routinely to reactor systems. Since acoustic emission is an acoustic phenomenon, one of the obvious questions to be resolved is that of determining whether normal reactor background noise would mask or confuse acoustic emission information to the extent that it could not be reliably identified. The recent outage and subsequent start-up of Commonwealth Edison Company's Dresden I Nuclear Power Reactor offered one of the first opportunities to obtain definitive information on background noise from a boiling water power reactor. It should be noted that the cooperation extended by Commonwealth Edison employees in support of this study has been outstanding and contributed much to obtaining data important to the acoustic emission program.

Preliminary evaluation of noise data measured at various locations on the Dresden I reactor start-up and power level increase to 100 MW electrical (approximately half power) indicates that application of acoustic emission surveillance during reactor operation is feasible on this reactor. Significant components of the reactor noise appear to be below 750 kHz and especially below 300 kHz. An electronic filter is effective in attenuating the noise components observed. From 750 kHz to 5 MHz (upper limit of the monitor system), the level of signal observed was essentially that of the normal electronic noise of the monitor system. Some of the frequency components in this region may have been contributed by the reactor system (this can be resolved by laboratory investigation); however, the significant facet is that the total level was within that normally produced by the electronic system alone. It thus appears that a frequency *window* between 750 kHz and about 2.5 MHz (the point at which signal attenuation versus increasing frequency starts to rise sharply in ferrous metals) is available on the Dresden I reactor for detection of acoustic emission during reactor operation. Acoustic emission signal analysis work in progress at Battelle-Northwest shows significant frequency components up to at least 5 MHz. Transducers with a relatively flat response into the low MHz range similar to those used in the Dresden I study will be necessary to detect acoustic emission under operating conditions. With the strong noise components indicated in the 0 to 300 kHz range, it would be difficult, if not impossible, to identify acoustic emission using a transducer with a major resonance in that region. The results of the Dresden I noise study together with current signal analysis information are encouraging for the development of a practical acoustic emission monitor system.

857 **P. H. Hutton. "Standards and Calibration Procedures." Task 5, Project IV, pp 145–148 in "In-Service Inspection Program for Nuclear Reactor Vessels." R. D. Wylie. SwRI Project 17–2440, EEI Project RP–79. Biannual Progress Report No. 7. Southwest Research Institute, San Antonio, Texas. Aug 25, 1972. [NTIS: PB–212–144].**

The objective of this task is to provide standards and calibration procedures to be applied to acoustic emission monitor systems for use in continuous surveillance of nuclear reactor pressure boundaries. Preamplifiers and sensors used to perform a noise analysis study on San Onofre reactor have been examined in place, removed, and returned to BNW for examination after approximately a 1-year exposure to a reactor environment. Noise data obtained from San Onofre reactor have been used to calibrate the artificial noise source installed on the system evaluation facility (EBOR) at the National Reactor Test Site, Idaho Falls, Idaho.

858 **P. H. Hutton. "Use of Acoustic Emission to Study Failure Mechanisms in Metals." BNWL–SA–1865. Battelle–Northwest, Richland, Washington. Nov 1968; ASME Publication 69–MET–8. American Society of Mechanical Engineers, New York. 1969; paper presented at the American Society of Mechanical Engineers Pressure Vessels and Piping Conference, Washington, D.C. Apr 1969.**

The author, defining acoustic emission as the elastic wave produced in metal by energy imparted as the metal deforms and fractures, declares that use of acoustic emission gives a much greater degree of resolution in surveillance of metal deformation and cracking than any other available technique. He discusses the nature of acoustic emission, its detection, and use of the information for flaw detection and evaluation, as well as applications of the technique for detection of stress-corrosion cracking, weld cracking, metal fatigue, and flaw growth in pressure vessels.

859 **P. H. Hutton, C. E. Fitch, and H. N. Pedersen. "Crack Detection in Pressure Piping by Acoustic Emission." pp 3.1–3.32 in "Nuclear Safety Quarterly Report–August, September, October, 1968 for Nuclear Safety Branch of USAEC Division of Reactor Development and Technology." BNWL–926. Battelle–Northwest, Richland, Washington Dec 1968. [NTIS].**

Basic wave propagation study for the purpose of improving acoustic emission detection techniques has produced, to date, evidence that shear wave detection offers distinct advantages in signal-to-noise ratio and signal resolution. High cycle, tension-tension fatigue testing of 304L stainless steel has shown the onset of acoustic emission activity at about 20% of the total number of cycles required to produce a visible surface crack. Design has been completed for an environmental test of transducer crystal materials to determine the combined effect of high temperature and nuclear radiation. A circuit to provide a continuous record of accumulated number of acoustic emission events has been satisfactorily tested. Testing has been performed to demonstrate detection of acoustic emission in the presence of turbulent flow and cavitation hydraulic noise.

The results show that an acoustic emission sensing system operating in the 1.5 to 2.5 MHz frequency range with shear sensors is essentially insensitive to either turbulent flow or cavitation noise. At the same time, this system will detect acoustic emisison signals clearly at a distance of at least three feet from the signal source. An acoustic emission monitor system installed on the Containment Systems Experiment simulator vessel has been operated in conjunction with integrity evaluation tests on the vessel with satisfactory results.

860 P. H. Hutton, W. D. Jolly, and J. B. Vetrano. "Acoustic Emission for Periodic and Continuous Flaw Detection in Pressure Vessels." BNWL-SA-4248. Battelle-Northwest, Richland, Washington. May 1972; pp 47-170 in ***Proceedings of the U.S.-Japan Joint Symposium on Acoustic Emission.*** **English volume. Japan Industrial Planning Association, Kawamura-Building 6F, 3-21-6 Akasaka Minato-Ku, Tokyo, Japan 107. 1972. symposium held in Tokyo, Japan. July 4-6, 1972;** ***Pressure Engineering, Journal of Japan High Pressure Institute, (Kotsu Gasu),*** **KOGAA, 11(2):70-91, 69. 1973. (in Japanese).**

The primary topic of this paper concerns detection of flaws in pressure vessels and piping. With pressure vessels, acoustic emission can be beneficially applied during fabrication, acceptance testing, and in-service use to enhance vessel safety. The application can start with testing of vessel material samples to help evaluate properties such as tensile and fatigue strength. During vessel fabrication, acoustic emission is a tool for detecting and locating flaws formed in welds. When the vessel is completed, acoustic emission can be applied during hydrostatic testing to detect incipient failure points in either the weld zones or in the body of the plates. Finally, the technique can be applied to vessels in service on a continuous surveillance basis to detect vessel deterioration from fatigue, stress corrosion cracking, etc., and warn of impending failure.

861 P. H. Hutton and R. N. Ord. "Acoustic Emission." Chapter 1, pp 1-30 in ***Research Techniques in Nondestructive Testing.*** **R. S. Sharpe, Editor. Academic Press, New York. 1970.**

Discussion follows of the use of acoustic emission as a device for monitoring the structural integrity of physical assemblies of solid materials. Acoustic emission is a phenomenon arising from energy released as a solid material undergoes plastic deformation and fracture. As an integrity monitoring device, the use of acoustic emission makes it possible to remotely detect crack formation and movement at the time it occurs. Also, simultaneous interrogation of three or more sensors provides the necessary information to locate the source of the signals by triangulation. Significant potential applications include the detection of incipient fatigue failure in pressure vessels (e.g., nuclear reactors) and airframes, penetration of space vehicles by micrometeorites, and onset of stress corrosion cracking in susceptible structures. A state-of-the-art summary is followed by a review of the theoretical aspects and physical principles involved. Sensor design, signal conditioning, and data analysis and readout are then discussed in detail, along with current applications and areas of further development. It is shown that acoustic emission constitutes a powerful tool for both structural integrity surveillance and for basic study of material failure mechanisms. Full utilization of the technique does require considerable additional development work. In the meantime, the current state-of-the-art permits reliable and beneficial application to many selected industrial and laboratory problems.

862 **P. H. Hutton, R. N. Ord, C. E. Fitch, H. N. Pedersen, and J. C. Spanner. "Crack Detection in Pressure Piping by Acoustic Emission." pp 3.1–3.31 in "Nuclear Safety Quarterly Report–May, June, July, 1968 for Nuclear Safety Branch of USAEC Division of Reactor Development and Technology." BNWL-894. Battelle-Northwest, Richland, Washington. Oct 1968. [NTIS].**

Detailed study of acoustic emission wave propagation in reactor materials has continued. Emphasis is being placed on study in the frequency range of 0.5 to 3.0 MHz and on the use of critical-angle transducer orientation. Some success has been achieved in detecting signal variations between brittle and ductile material. There are also indications that what was earlier thought to be a variation in signal frequency content, as observed by lower frequency sensors, is actually a reflection of acoustic emission repetition rate or signal density for increments of deformation and crack growth. It appears that critical angle detection in the higher frequency range (low MHz) will improve the signal-to-noise ratio and provide a sharper wave front for improved source location resolution. One MHz, PZT-5A transducers have been fabricated by using a ceramic mounting and backing material in place of the previously used epoxy to improve their high temperature longevity. These are currently in the testing process. A low-noise, wide-band preamplifier has been fabricated and satisfactorily field-tested for acoustic emission detection. The performance is superior to previously used commercial preamplifiers, and the cost is about one third that of the commercial models. Background-noise spectra measurements have been made on the Dresden I Nuclear Power Reactor during start-up from cold flow to full power operation. For this reactor, it appears that an acoustic emission monitor system operating in the range of 750 kHz to 3 MHz would be relatively unhampered by reactor system noise interference.

863 **P. H. Hutton, R. N. Ord, H. N. Pedersen, and J. C. Spanner. "Crack Detection in Pressure Piping by Acoustic Emission." pp 3.1–3.13 in "Nuclear Safety Quarterly Report–July, August, September, October, 1967 for Nuclear Safety Branch of USAEC Division of Reactor Development and Technology." BNWL-754. Battelle-Northwest, Richland, Washington. June 1968. [NTIS].**

Acoustic emission signals can be tentatively categorized on the basis of the mode of metal deterioration and signal frequency. Plastic deformation produces a *continuous* signal with an approximate 20 kHz sporadic burst signal overlayed. Ductile fracture generates a burst signal of approximately 50 kHz. Brittle fracture shows a burst signal with frequency ranging up to at least 250 kHz. Acoustic emission has been shown to correlate well with mechanical parameters of crack growth (crack velocity and change in crack area). It has also been demonstrated that crack growth in a 0.25-inch thick section of steel is readily detectable at crack velocities less than 0.001 inches per minute. Broad band (350 kHz) PZT-5A transducers have been fabricated and used for a period of 12 hours in a temperature environment of 650 to 675 °F with no detectable deterioration of ability to sense acoustic emission. A lithium niobate ($LiNbO_3$) transducer has been fabricated and tested in comparison with PZT-5 transducers currently being used for detection of acoustic emission. The sensitivity of the $LiNbO_3$ is comparable to that of the PZT-5. This transducer has a potential service temperature of 1000 °C. A system analysis study on application of acoustic emission to detect crack growth in pressure piping is in progress. The objective is to define significant parameters and point out areas requiring concentrated research and development effort. An integrated, solid-state acoustic emission monitor system is being fabricated for use in study work. This incorporates improvements over the currently used assemblage of laboratory instruments and will promote greater consistency of measurements and reduce preparation time for monitoring tests. Acoustic emission program work was reviewed with Phillips Petroleum Company personnel at a meeting in Idaho Falls, Idaho on October 20, 1967, and with Atomic Energy Commission personnel on November 3, 1967. Acoustic emission generated during elevated temperature burst testing of a 24-inch diameter, 1.64-inch wall, Type A-106-B steel pipe specimen with a through-wall flaw was monitored at the Battelle-Columbus test site. Flaw propagation was detected by acoustic emission 10.5 minutes before a surface crack appeared and 11 minutes before specimen failure. The data also indicate which end of the flaw was moving at a given time. Spectrum analysis of noise generated by turbulent water flow, and water boiling shows the significant frequency content to be below 20 kHz. Adaptation of acoustic emission as a means of detecting incipient failure in reactor pressure boundaries (principally piping) is the objective of a development program in progress at Battelle-Northwest. The USAEC Division of Reactor Development and Technology is sponsoring this program which has been in progress since February 1966. The total program effort has been designated under six task headings to aid in maintaining a balanced effort toward the objective of a reliable application technique. Discussion of progress is presented by task designation. Essentially all of the study work to date has been performed on low alloy, high strength carbon steel (A-212-B) representative of material used extensively in nuclear reactor pressure systems.

864 P. H. Hutton, R. N. Ord, H. N. Pedersen, and J. C. Spanner. "Crack Detection in Pressure Piping by Acoustic Emission." pp 3.1–3.14 in "Nuclear Safety Quarterly Report—November, December, 1967, January, 1968 for Nuclear Safety Branch of USAEC Division of Reactor Development and Technology." BNWL-816. Battelle-Northwest, Richland, Washington. Sep 1968. [NTIS].

Assurance of the integrity of nuclear reactor pressure boundaries is of vital importance to public safety. A development program in progress at Battelle-Northwest to detect incipient failure in metals by measuring acoustic emission supports this premise. As a metal deforms and fractures, elastic waves are generated and can be detected at the metal surface. It is believed that a reliable surveillance system can be developed using this phenomenon. The scope and duration of the program is currently being reviewed in light of three levels of sophistication: (1) detect and locate flaws during hydrostatic test of reactor components or the assembled reactor system; (2) detect and locate flaws on a continuous basis on an operating reactor; and (3) detect, locate, and measure flaws on a continuous basis on an operating reactor. It is anticipated that future program scope will be resolved during the coming quarter. As in previous reports, the progress for this quarter is discussed under six task headings which describe the total program effort. Low alloy carbon steel has continued to be the primary test material. Acoustic emission data from five single edge notch (SEN) specimens have been analyzed for emission rate versus load on the Battelle SEL 840A computer. The type of flaw and the material condition both have a significant effect on the shape of the curve. A 3 by 6 by 42-inch double cantilever bean of A-302-B steel tested below the nil-ductility transition temperature (–160 °F) by the Metallurgy Department for study of heavy section steel properties has been monitored for acoustic emission. The intent of this test was to correlate acoustic emission with fracture mechanics information. Preliminary tests of applying acoustic emission to detect flaw formation as a weld cools showed that generation of cracks can be readily detected. This has the potential for early beneficial application to weld surveillance during reactor pressure system construction. A much simplified version of the capacitive transducer has been designed and applied in an acoustic emission test with good results. A formal report on the initial system analysis study has been compiled and submitted for publication. A solid-state acoustic emission signal conditioning system has been fabricated and successfully applied to acoustic emission test work. Availability of this sytem results in both time and cost savings in performing tests, as well as enhancing uniformity of data acquisition. A high temperature PZT-5 transducer was supplied to Phillips Petroleum Company for test application in their SM-1A pressure vessel annealing studies. Two burst tests involving 24-inch diameter, 1.64-inch wall, Type A-106-B steel pipe with part-through-wall flaws were monitored for acoustic emission at the Battelle-Columbus test site. Deformation emission was observed to be a low level and distinguishable from that generated by fracture. Data analysis is not yet complete.

865 **P. H. Hutton, R. N. Ord, H. N. Pedersen, and J. C. Spanner. "Crack Detection in Pressure Piping by Acoustic Emission." pp 3.1–3.29 in "Nuclear Safety Quarterly Report–February, March, April, 1968 for Nuclear Safety Branch of USAEC Division of Reactor Development and Technology." BNWL–885. Battelle-Northwest, Richland, Washington. Oct 1968. [NTIS].**

Enhancement of public safety through development of improved techniques for integrity surveillance of nuclear reactor pressure boundaries is an area of considerable merit. Detection of acoustic emission or elastic waves produced in metal as it deforms and fractures is the basis for such a development program at Battelle-Northwest. The technique would provide a warning of incipient failure in time to take necessary remedial action. The program began in February 1966 under the sponsorship of the USAEC Division of Reactor Development and Technology. A coordinated effort evolved between this program and acoustic emission work being done by Phillips Petroleum Company is aimed at increased overall efficiency in developing a tested integrity surveillance system utilizing acoustic emission. The program effort for this quarter, as in previous reports is discussed under six task headings. Low alloy carbon steel was the primary test material. However, some work also has been done with Type 304L stainless steel. Investigation of the effects of the transmission media and specimen geometry on propagation of acoustic emission signals is in progress. Evidence indicates that there is a short duration packet of information at the leading edge of the detected acoustic emission signal which is not affected by specimen geometry, but is characteristic of the source mechanism. It also appears that selected mode detection, i.e., shear wave or Rayleigh wave by critical angle selection, would be advantageous to signal detection and to source location functions. The PZT-5A transducer assembly used for acoustic emission study work was modified to achieve about a five-fold increase in sensitivity. A high temperature (600 °F) life test of this assembly is in progress with no indication of crystal deterioration after 500 hours of continuous testing. A lithium niobate piezoelectric transducer with improved sensitivity is being assembled for testing at elevated temperature (>700 °F).

A program prepared jointly with Phillips Petroleum Company (PPCo) has been directed toward developing three successive levels of refinement in acoustic emission monitor systems for integrity surveillance of reactor pressure boundaries. In general, Battelle-Northwest is responsible for R&D work and PPCo is responsible for plant application test work under this program. Acoustic emission data obtained during burst tests of two 24-inch diameter, 1.64-inch wall, Type A-106-B steel pipe specimens at the Battelle-Columbus test site have been analyzed. Both tests involved part-through-wall falws. Low level signals attributed to metal deformation controlled the shape of the emission rate curve. The curve showed a distinctive preliminary peak in both tests about 4 minutes prior to failure. A small amount of slow crack growth was indicated in both specimens. Burst signals signifying this growth contained a unique packet of high frequency energy at the leading edge which appeared to be independent of geometry effects. Arrangements have been made to install instrumentation on Commonwealth Edison's Dresden I Nuclear Power Reactor to examine the background noise spectrum during startup and operation.

866 P. H. Hutton and D. L. Parry. "Assessment of Structural Integrity by Acoustic Emission." ***Materials Research and Standards,*** **MTRSA, 11(3):25-32. Mar 1971.**

The article discusses the application philosophy and methods of the acoustic emission technique. Cases in which the technique is applied to measure structural integrity outside the laboratory also are presented. These range from the monitoring of welds for crack formation to the surveillance of buildings and earth structures for structural deterioration.

867 P. H. Hutton and H. N. Pedersen. "Crack Detection in Pressure Piping by Acoustic Emission." pp 3.1-3.18 in "Nuclear Safety Quarterly Report—May, June, July, 1969 for USAEC Division of Reactor Development and Technology." BNWL-1187. Battelle-Northwest, Richland, Washington. Sep 1969. [NTIS].

Investigation of wedge mountings for coupling the shear component of acoustic emission signals from a plate surface to a sensor has been extended to the study of heavy section 4-inch thick steel plate. Work to date, using PZT-5 sensors, shows imporved sensitivity for the wedge mount versus surface mount up to 1.5 MHz, but the signal front obtained through the current wedges does not have good definition. Maximum acoustic emission detection range in 4-inch thick steel using various sensors is being evaluated. Thickness-poled, surface-mounted PZT-5 sensors appear to be limited to 4 to 5 feet for reliable detection. However, edge-mounted PZT-5 shear sensors receive well up to at least 8 feet (maximum dimension of the specimen). Detection frequencies used were 500 to 700 kHz. A simple method for obtaining relative calibration of sensor response characteristics has been developed using a small air abrasive unit to provide a broad band excitation source. Lithium niobate used in the shear mode shows good sensitivity to acoustic emission and offers definite potential for use in a high temperature sensor. This material is a piezoelectric with a service temperature range >1500 °F. The PZT-5A peizoelectric crystals exposed to a total of 4.9×10^8 roentgens of gamma radiation showed no detectable effect. Tuned preamplifiers of the type used by Battelle-Northwest for acoustic emission detection have been tested for resistance to temperature damage. Results show them capable of withstanding 185 °F for an extended period. Long-term service at 200 °F is attainable by using special high temperature wire and would meet anticipated temperature requirements for continuous surveillance of thermal power reactors. A high pass filter incorporating 36 db per octave attenuation has been developed and tested. This filter will enhance ability to detect low level acoustic emission signals in the presence of lower frequency background noise. On the basis of total engineering information, sections of process piping in local reactors monitored for fatigue crack growth by acoustic emission detection are to be replaced. Arrangements have been made to salvage for examination those sections producing limited acoustic emission indications of crack growth. This examination will contribute to establishing a criterion for evaluating acoustic emission indications. Increasing the monitor frequency from >200 kHz to >600 kHz on the acoustic emission monitor applied to the Containment System Experiment simulator vessel was found to reduce noise interference without degrading the data.

868 **P. H. Hutton, H. N. Pedersen, and J. C. Spanner. "Crack Detection in Pressure Piping by Acoustic Emission." pp 4.1–4.10 in "Nuclear Safety Quarterly Report–January, February, March, 1967 for Nuclear Safety Branch of USAEC Division of Reactor Development and Technology." BNWL–433. Battelle-Northwest, Richland, Washington. July 1967. [NTIS].**

A double cantilever beam specimen of A-302-B carbon steel was tested in cooperation with the Metallurgy Department to study the brittle fracture characteristics and the attendant acoustic emission. Limited acoustic emission detected was in the form of 100 to 120 kHz bursts of about 50-microsecond duration. The test temperature (−140 °C) introduced detection problems which must be overcome before adequate data can be obtained from low temperature tests. Static burst tests have been performed using 3-inch diameter by 0.25-inch wall A-106-B carbon steel pipe. These tests were performed as the initial phase of the acoustic emission field-application problem study, and also to study basic signal characteristics. Acoustic emission in the form of burst type signals was evident early in the test and increased abruptly in density and amplitude as the initial flaw began to propagate in a ductile manner. Triangulation verified the flaw as the signal source. Analysis of hydraulic noise data from an operating reactor showed essentially all of the frequency content to be below 10 kHz. This demonstrated that the higher frequency components of acoustic emission above 10 kHz must be utilized in a practical crack monitor system. Fabrication of experimental, multi-layer electrostatic transducers by vapor deposition has been contracted with Thermionics, Inc., Hayward, California, to expedite evaluation of this technique. Three units, one each of WO_2, ZrO_2, and SiO_2 dielectric with tantalum electrodes, have been received and partially tested. The WO_2 has been determined to be unsuitable for this application. Both the ZrO_2 and SiO_2 still look promising; however, a problem of shorting between successive layers is being experienced.

869 **P. H. Hutton, H. N. Pedersen, and J. C. Spanner. "Crack Detection in Pressure Piping by Acoustic Emission." pp 4.1–4.11 in "Nuclear Safety Quarterly Report–April, May, June, 1967 for Nuclear Safety Branch of USAEC Division of Reactor Development and Technology." F. W. Albaugh, J. J. Fuquay, H. Harty, E. E. Voiland, and D. C. Worlton, BNWL-537. Battelle-Northwest, Richland, Washington. Dec 1967. [NTIS].**

Burst testing of specimens fabricated from 3-inch diameter by 0.25-inch wall, A-106-B carbon steel pipe has provided additional insight into the nature of acoustic emission from this material. The outer surface condition of the pipe is a mjaor factor in producing continuous emission at a readily detactable level as the metal deforms. With an *as received* oxide layer in place, the continuous emission from metal deformation is readily detectable; with the oxide removed, the continuous signal is barely detectable using gains of 10^4. The continuous emission is apparently transmitted as a surface wave. All tests produced an abundance of transient or burst type signals with frequency content as high as 250 kHz. Acoustic emission test data analyzed with a digital analyzer through the range of 20 kHz to 180 kHz showed a spectrum similar to that previously derived from selected data using an analog analyzer. The significant difference is the signal density in the later data which was sufficiently low so that an analog analyzer showed no response.

870 **P. H. Hutton and J. R. Skorpik. "A Simplified Approach to Continuous AE Monitoring Using Digital Memory Storage." BN-SA-574. Battelle-Northwest, Richland, Washington. 1976; Paper 2-6, pp 152-166 in *The Third Acoustic Emission Symposium.* Japan Industrial Planning Association, Dobashi Building, 3-17-8, Akasaka, Minato-ku, Tokyo, Japan. 1976. proceedings of a symposium held in Tokyo, Japan. Sep 16-18, 1976.**

The basic theory of acoustic emission monitoring is established and the technology is being applied to evaluation of a wide range of material experiments and structural analysis programs. In the course of this work, the general trend in acoustic emission

(AE) monitoring systems has been one of increasing complexity. This is particularly true of systems designed for continuous monitoring of a structure. These systems are usually multichannel (in the vicinity of 20 to 40 or more channels) and require a dedicated computer for data processing and storage. Such systems may cost from $100,000 to $500,000. Some applications require this degree of sophistication. An example is simultaneous monitoring of a large material volume such as a nuclear reactor pressure vessel during hydrostatic testing to detect and locate defects. There are many structural monitoring applications of AE, however, where the system complexity just described is not needed. In a structure such as a pressure vessel or a bridge, once initial inspection has been accomplished, a large portion of the material volume is of no real concern to continued structural integrity. There are selected areas such as known subcritical defects, locations which experience has shown tend to develop defects (e.g., welds), locations with known high design stresses (e.g., nozzle penetrations of a pressure vessel) or portions of a structure which are especially critical to the integrity of the total structure (e.g., certain suspension members in a bridge). To apply a complex AE system to continuously monitor such selected areas over a long period of time would not only be very expensive but also may be impractical because an operator with the required skill may not be routinely available. A unique concept for continuous AE monitoring in selected applications such as those described above has been developed at Battelle-Northwest. It reverses the trend to increasing complexity while retaining such important requirements as reliability, permanent data record, and easy retrieval of data for analysis. The original concept grew out of a special need in a program being performed for the U.S. Federal Highway Administration to apply AE to bridge monitoring.

871 **P. H. Hutton and J. R. Skorpik. "AE Monitoring Simplified Using Digital Memory Storage and Source Isolation." BN-SA-668. Battelle-Northwest, Richland, Washington. Mar 1977; summary of paper in** *Paper Summaries: ASNT Spring Conference, March 28–30, 1977, Phoenix, Arizona.* **American Society for Nondestructive Testing, Columbus, Ohio. 1977;** *Materials Evaluation,* **MAEVA, 35(11):55–60. Nov 1977.**

The general trend in acoustic emission monitoring systems has been one of increasing complexity. This is particularly true in systems for continuous monitoring which are usually multichannel (perhaps 20 to 40) and incorporate a dedicated minicomputer. A unique concept which reverses this trend for selected applications has been developed at Battelle-Northwest. The concept originated in work performed for the U.S. Federal Highway Administration. This concept uses solid-state digital memories to store acquired data in a permanent form which is easily retrieved. It also uses a fundamental method to accept AE data only from a selected area. The digital memory system is designed for short-term or long-term (months) monitoring. It has been successfully applied in laboratory testing such as fatigue crack growth studies, as well as field monitoring on bridges and piping to detect crack growth. The features of simplicity, versatility, and low cost contribute to expanded practical application of acoustic emission technology.

872 **P. H. Hutton and J. C. Spanner. "Detection of Metal Overstress by Acoustic Emission." BNWL-SA-438. Battelle-Northwest, Richland, Washington. Dec 27, 1965; paper presented to an U.S. Atomic Energy Commission Advisory Committee on Reactor Safeguards, Washington, D.C. Nov 1965.**

Experimental work conducted indicates a tremendous potential for application of acoustic emission as a nondestructive surveillance tool to assess structural integrity of operating systems such as reactor pressure systems. Possible applications for the study of fracture mechanics, brittle fracture, and ductility changes due to nuclear irradiation are also indicated. The method and its advantages are described.

873 **P. H. Hutton and J. B. Vetrano. "Crack Detection in Pressure Piping by Acoustic Emission." pp 3.1–3.15 in "Nuclear Safety Quarterly Report—August, September, October, 1969 for USAEC Division of Reactor Development and Technology." BNWL-1266. Battelle-Northwest, Richland, Washington. Dec 1969. [NTIS].**

Extensive investigation has been made of acoustic emission signal propagation characteristics in thick plate (4 inches) and in heavy wall piping (0.25 to 1 inch). Data analysis to date shows that surface waves are not dependable for acoustic emission detection and that although the surface-mounted thickness poled sensor is not responsive to all types of shear, its diversity of mode sensitivity is an advantage for emission detection. This can, however, be a problem in establishing a single propagation velocity value for triangulation of source location. The wedge-mounted shear sensor offers a possible solution to the problem of ambiguity of propagation velocity. Also, in light of the fact that the Lamb wave propagation mode appears to be active, use of a narrow band monitoring frequency will permit prediction of the most prominent wave velocity for a given material thickness. The room temperature portion of an investigation of acoustic emission signal attenuation as a function of metal temperature and emission monitoring frequency has been completed. Based on measured attenuation, these initial results indicate that a given acoustic emission signal in room temperature material would be detectable about 2.5 times as far away at a 400 kHz detection frequency as at 1.2 MHz. Study of the effect of temperature on generation of acoustic emission in stainless steel is in progress. The room temperature phase has been completed to form a reference base for subsequent tests at 600 and 900 °F. A significant additional feature of the data from specimens that failed in fatigue is the prominent identification of initial macrocrack formation. A prototypic high temperature (650 °F) sensor that has been fabricated uses injection molding of high temperature plastic to form a case around a PZT-5A crystal. This first unit shows good response characteristics for acoustic emission detection. The technique reduces sensor fabrication cost by a factor of three and also improves sensor uniformity.

I

874 **P. H. Hutton and J. B. Vetrano. "Crack Detection in Pressure Piping by Acoustic Emission." pp 3.1–3.15 in "Nuclear Safety Quarterly Report–November, December, 1969, January 1970 for USAEC Division of Reactor Development and Technology." BNWL-1315-1. Battelle-Northwest, Richland, Washington. Mar 1970. [NTIS].**

Investigation of acoustic emission signal propagation in pipe specimens has shown that propagation characteristics in pipes can be unique from that in flat plates. Wall thickness, wall thickness-to-diameter ratio, and helical propagation paths are significant factors. It appears that plate wave theory will permit selecting a detection arrangement for a given configuration which will minimize propagation velocity ambiguities and thus enhance general source location accuracy. Limited additional work to investigate the effect of temperature on generation of detectable emission in stainless steel has shown no discernible effect up to 600 °F.

875 **P. H. Hutton and J. B. Vetrano. "Crack Detection in Pressure Piping by Acoustic Emission." pp 3.1–3.9 in "Nuclear Safety Quarterly Report–February, March, April, 1970 for USAEC Division of Reactor Development and Technology." BNWL-1315-2. Battelle-Northwest, Richland, Washington. May 1970. [NTIS].**

Study of acoustic emission produced by the fatigue failure process in metal has resulted in tangible evidence that the nucleation of a macrocrack is identifiable by its acoustic emission. The same work also produced evidence of the flaw size resolution attainable in carbon steel by acoustic emission, i.e., microcracking produced readily detectable emission.

876 **I. Yu. Ievlev. "Systematic Analysis of the Frequency Spectrum of Acoustic Emission in the Acoustic Frequency Range."** *Trudy Uralskogo Politekhnicheskogo Instituta,* **TUPIA, (215):149–151. 1973. (in Russian).**

A method of frequency analysis at $\leqslant$11 kHz is proposed, and an example of its application is given in studying the kinetics of the thermoelastic martensitic transformation, i.e., the phase transformation $\beta_1 \rightleftharpoons \gamma'$ in a Cu-14 Al-1.6 Ni alloy with the martensite temperature M_S 60°. The specially prepared specimen was cooled at 10° per minute, and the frequency spectrum was recorded at 60 to 30°. The method is extremely sensitive and selective. [Abstract 125446d reprinted by permission of the American Chemical Society from *Chemical Abstracts,* Vol. 81, 1974].

877 **I. Yu. Ievlev, V. P. Melekhin, and V. V. Bukhalenkov. "Detection of Single Acoustic Emission Signal."** *Trudy Uralskogo Politekhnicheskogo Instituta,* **TUPIA, (215):146–148. 1973. (in Russian).**

878 **I. Yu. Ievlev, V. P. Melekhin, R. I. Mints, and V. M. Segal. "Some Features of the Emission of Sound during a Thermoelastic Martensitic Reaction.** *Soviet Physics–Solid State,* **SPSSA, 15(9):1761–1763. Mar 1974; translated from** *Fizika Tverdogo Tela,* **FTVTA, 15(9):2647–2650. Sep 1973. (in Russian).**

The results are given of an investigation of the parameters (intensity, frequency range, and amplitude spectrum) of the emission of sound which accompanied a thermoelastic martensite reaction: $\beta_1 \rightleftharpoons \gamma'$ in the Cu-14.5% Al-1.6% Ni alloy. It was found that the sound emission is due to the inhomogeneity of the shear deformation in the system formed by a martensitic crystal and the matrix. The intensity of the sound emission reflected the kinetics and completeness of the martensitic reaction. The frequency range of the acoustic signal was related to the reaction mechanism and could be used to determine the density of the effective barriers in the path of a growing crystal. The amplitude spectrum of sound was governed by the thermodynamic features of the forward and reverse martensitic reactions.

879 **K. Iida, M. Onoe, Y. Takahashi, H. Yamada, and C. Lee. "Acoustic Emission during Bending Fatigue of Aluminum Alloy." Session 6, pp 15–30 in** *The Second Acoustic Emission Symposium.* **Japan Industrial Planning Association, Dobashi Building 5F, 3-17-8 Akasaka, Minato-ku, Tokyo, 107 Japan. 1974. proceedings of a symposium held in Tokyo, Japan. Sep 2–4, 1974.**

Recent papers on acoustic emission during fatigue loading have demonstrated the utility of acoustic emission as a valuable measure in the investigation of fatigue behaviour of a material. And it is not long since the dependency of acoustic emission on one of fatigue-crack propagation parameters, the propagation rate, the range of stress intensity factor, and the maximum value of stress intensity factor had obtained general recognition. Some papers showed that the use of properly designed devices to eliminate noise from sources other than those desired and the application of the short-time correlation method afford the possibility of continuous detection of acoustic emission resulted from fatigue loading of a material. Furthermore, in experimental evidence that the acoustic emission during fatigue can be divided into two categories, acoustic emission resulted from crack extension and one generated from crack closure phenomenon, tells the considerable utility of acoustic emission technique for the elucidation of mechanisms of fatigue crack initiation and propagation. For example, the hopeful possibility of acoustic detection of the fatigue crack initiation should be helpful in discussion of fatigue-crack initiation problems, and besides detection of crack closure phenomenon by acoustic emission must promote the analysis of such behaviours, that are affected by fatigue-crack closure phenomenon, as fatigue-crack propagation rate, residual stresses along the edges of the propagated crack. Two of the present authors, Onoe and Yamada reported in a previous paper a successful monitoring of crack initiation and propagation by acoustic emission signal in steel specimen, and also mentioned that acoustic emission was generated during tension phase in the beginning stage and during both tension and compression phases just prior to final failure. The purpose of the work described was to study the acoustic emission behaviour during high-cycle fatigue crack propagation in an aluminum alloy and to attempt correlation with the crack growth rate of a wide range.

880 K. Iida, Y. Sakamoto, H. Yamaguchi, K. Fujisawa, and T. Okazawa. "Acoustic Emission during Low Cycle Fatigue Test of Hemispherical Pressure Vessel Models." Paper 1–3, pp 47–64 in *The Third Acoustic Emission Symposium.* Japan Industrial Planning Association, Dobashi Building, 3–17–8, Akasaka, Minato–ku, Tokyo, Japan. 1976. proceedings of a symposium held in Tokyo, Japan. Sep 16–18, 1976.

In the design of a pressure vessel, it is important to evaluate the reliability for safety to prevent failures. For the purpose of getting available data for these evaluations, several investigations were carried out by the Research Committee for Safety of High Tensile Strength Steel organized in the Institution for Safety of High Pressure Gas Engineering under auspices of Ministry of Internal Trade and Industry, Japan. In one of these investigations, the low-cycle fatigue behavior of a hemispherical pressure vessel model with misaligned welded joint was investigated at Wakayama Works of Sumitomo Metal Industries, Ltd. At that time, AE measurement was carried out for the purpose of detecting fatigue cracks. In the present paper, AE measurements during a low-cycle fatigue test of hemispherical pressure vessel models of 1000 mm inner diameter with a misaligned welded joint are described. In these measurements, the determination of AE locations was made using four transducers, and two methods were used to eliminate the influence of mechnical noises.

881 T. Ikoma, M. Ogura, and Y. Adachi. "Acoustic Emission from Single Crystals of Gallium Arsenide." Paper 5–1, pp 329–341 in *The Third Acoustic Emission Symposium.* Japan Industrial Planning Association, Dobashi Building, 3–17–8, Akasaka, Minato–ku, Tokyo, Japan. 1976. proceedings of a symposium held in Tokyo, Japan. Sep 16–18, 1976.

Crystal defects in a semiconducting single crystal have a great influence on electronic characteristics of semiconductor devices. The perfect crystal technology has developed dislocation-free silicon crystals, which led to large-area devices of silicon such as a thyristor and LSI. In compound semiconductors as GaAs and GaP, crystal defects are more easily introduced during a fabrication process and/or a long time operation of a device than in silicon. In a semiconductor double-hetero laser diode, dark line defects (DLD) play an important role in determining the lifetime for a continuous wave-operation. The DLDs were easily observed by photoluminescence and cathode-luminescence patterns because the luminescence efficiency is low in this region. The DLDs were closely observed by the transmission electron microscope and were found to be a bunch of dislocations. During a degradation of laser diodes, DLDs are growing up by glide and/or climb motions in an active region. The lifetime of a continuous wave double-hetero laser diode is in many cases less than an hour, although more than 10,000 hours are reported for a high quality diode. The observation of DLDs by the scanning electron microscope and photoluminescence patterns together with such a short lifetime suggests that DLDs grow rapidly to make a large pileup of dislocations. Another example which shows the importance of dislocations is a GaAs microwave device. A thermo-compression bonding generates dislocations and deteriorates the microwave performance of a Gunn-diode. The conditions of a thermo-compression bonding have been investigated in relation to the reliability and noise performance of a Gunn Oscillator. In compound semiconductor devices, the generation and movement of dislocations are much more important than in silicon devices. Therefore, a nondestructive inspection of dynamic behavior of crystal defects is needed from the standpoint of a device reliability and a process control. The acoustic emission technique (AET) has attracted increasing interest as a real-time inspection of large structures, and also as a tool of fatigue study of materials. The microscopic nature of AE has been studied in metals (Cu, Zn) and ionic crystals (KCl, LiF). From the view point of crystal perfection, semiconducting crystals such as Si, Ge, and GaAs are more advantageous than metal and ionic crystals. The AE was studied in LiF in relation to the emission of dislocation loops from a Frank-Read source. It was proven that the emission of dislocation loops generates acoustic waves in an ionic single crystal. Since a GaAs double-hetero laser diode encounters the problem of quick degradation due to a generation of DLDs, a careful control of fabrication process is required for III-V compound semiconductor devices. The AET is attractive as a method of a real-time inspection of generation and multiplication of defects in semiconductor crystals and devices. A preliminary study has been done to detect AE signals from a GaAs single crystal which is deformed by compression and heating. The purpose of this paper is to show that the AET can be applied to the detection of dislocation generations in GaAs crystals and devices by presenting preliminary experimental results about the correlation between AE and dislocation patterns.

882 **S. M. Ikonopisov and A. A. Girginov. "Analysis of Methods of Indicating Electrical Breakdown in the Anodic Oxidation of Aluminum."** *Doklady Bolgarskoi Akademii Nauk,* **DBANA, 28(2):257–260. 1975. (in Russian).**

The relative validity of a number of methods of indicating the development of electrical breakdown during the anodic oxidation of Al films and plates is discussed. Preferred methods include visible sparks (recorded subjectively or objectively), acoustic emission, and the incidence of low frequency oscillations; these all develop simultaneously and may be used as reliable criteria for the purpose in question. Other proposed criteria (e.g., high frequency oscillations) are less reliable and only apply to a limited range of surface finishes or electrolytes. [Abstract 22-1480. *Metals Abstracts,* Vol. 9, 1976].

883 **H. Imaeda, H. Kusanagi, H. Kimura, and H. Nakasa. "Acoustic Emission Characteristics during Tensile Tests of Various Metals with Several Types of Crystal Structures." Paper 7-3, pp 492–512 in** ***The Third Acoustic Emission Symposium.*** **Japan Industrial Planning Association, Dobashi Building, 3-17-8, Akasaka, Minato-ku, Tokyo, Japan. 1976. proceedings of a symposium held in Tokyo, Japan. Sep 16–18, 1976.**

Acoustic emission is defined to be the phenomenon that stress waves generate internally during dynamic processes in various materials. The dynamic processes such as plastic deformation, phase transition, and cracking occur as the result of externally applied stress or some other unstable situation in materials. These external and/or internal conditions give a remarkable effect on the AE characteristics. Techniques to utilize the AE characteristics have two important areas of application. The first is the use as a nondestructive technique due to the ability of the AE techniques to detect the growth of flaws or cracks under an applied stress. The second is the use as a basic research tool for describing the mechanisms by which materials deform under different conditions. It is expected that the AE test results are related to the microstructural features which influence mechanical properties of a material. To predict the AE behavior of structural materials in practical use and to evaluate the structural integrity, it is necessary to get the basic background information of the AE characteristics in tensile specimens of the structural materials and more basically of pure metals. The purpose of this work are to get the standard AE characteristics in typical metals available and to investigate the effect of crystal structures on AE characteristics. Main parameters used for representing AE characteristics are wave-form, frequency spectrum, count rate, rms value, and amplitude distribution. These parameters measured in the same conditions are presented for 14 metals which are classified into 4 types of crystal structures.

884 **J. Imamasa and T. Miki. "Experimental Verification for Application of Fracture Mechanics to Failure of Pressure Vessel." Paper VIII-441 in** *Dritte Internationale Tagung uber den Bruch. Zusammenstellung der Tagungsberichte, Teil IX: Technische Anwendungen.* **Verein Deutscher Eisenhuttenleute, Dusseldorf, Germany. 1973. volume of preprints of the 3rd International Congress on Fracture, held in Munich, Germany. Apr 8-13, 1973. [preprints out of stock from publisher and proceedings never published].**

Results of two series of experiments, which were conducted to verify the application of fracture mechanics to the catastrophic failure of a pressure vessel, were reported in this paper.

885 **T. Imanaka, K. Sano, and M. Shimizu. "Acoustic Emission: A Critical Review."** *Nippon Kinzoku Gakkai Kaiho (Bulletin of the Japan Institute of Metals),* **NKZKA, 12(12):871-881. Dec 1973. (in Japanese); UCRL-Trans-10747. translated for Lawrence Livermore Laboratory, Livermore, California. 1974. [NTIS].**

A review emphasizing recent research on transducer design, elimination of noise, preamplification, and audiomonitor systems. The metallurgical significance of the quantities measured and recent applications to the fracture in 18% Ni maraging steel and plastic deformation of Cu single crystals are discussed. [Abstract 22-0766. *Metals Abstracts,* Vol. 7, 1974].

886 **T. Imanaka, K. Sano, and M. Shimizu. "Application of Acoustic Emission."** *Kawasaki Seitetsu Giho* **(Kawasaki Steel Technical Report), KWSGB, 5(1): 1-18. Jan 1973. (in Japanese).**

A brief introduction to the field of acoustic emission and the acoustic emission monitoring system which has been used in the laboratories are presented. A consideration to the basic mechanisms of acoustic emission is briefly discribed. Acoustic emission characteristics of the various steels and some applications to material research are also presented. The feasibility of acoustic emission technique to the study of the dynamic feature of dislocation is discussed in relation to ultrasonic attenuation measurements.

887 **T. Imanaka, K. Sano, and M. Shimizu. "Dislocation Attenuation and Acoustic Emission during Deformation in Copper Single Crystal."** *Crystal Lattice Defects,* **CLADA, 4(1):57–64. Jan 1973.**

The early stages of plastic deformation in copper single crystals have been studied by simultaneous measurements of stress, strain, acoustic emission, and attenuation of 27-MHz longitudinal ultrasonic waves, which were recorded continuously during deformation. The overdamped resonance at 9 to 225 MHz has also been measured at various steps of deformation. The frequency dependence of attenuation agrees well with that calculated by the Granato-Lucke theory even at the presence of large bias stress. The attenuation has increased through two stages, the boundary of which is a macroscopic yield point. These observations are reasonably interpreted in terms of segment length change and multiplication of mobile dislocations which participate in plastic flow. The fundamental feature of acoustic emission and its correlation to the ultrasonic attenuation have been also studied, and feasibility of acoustic emission technique to the study of dynamic feature of dislocation is demonstrated. The results were compared with those in polycrystalline aluminum.

888 **T. Ingham and D. G. Dawson. "The Application of Acoustic Emission Measurements on Laboratory Testpieces to Large Scale Pressure Vessel Monitoring." Paper G3/8 in** *Structural Mechanics in Reactor Technology.* **Compiled by T. A. Jaeger. Volume 3, Part G–H. North-Holland Publishing Company, Oxford, England. 1975. proceedings of the 3rd International Conference on Structural Mechanics in Reactor Technology held in London, England. Sep 1–5, 1975. [distributed in the U.S. and Canada by American Elsevier Publishing Company, Inc., New York].**

The development of laboratory acoustic emission testing at REML (Risley Engineering Materials Laboratory) is outlined. Tests using single sensors have indicated that emission activity in pressure vessel steels could be related to toughness and strength; however, emission data from this type of test could not be related to structural monitoring. A more detailed series of measurements using an emission location technique was made on BS1501/151/Grade 28A mild steel to relate specimen activity with that of large structures. A test pressure vessel containing 4 artificial defects was monitored for emission whilst pressure cycling to failure. Test pieces cut from both the failed vessel and from as-rolled plate material were tested in the laboratory. No emission was reliably detected from the defects in the pressure vessel whereas emission was easily detected from both types of test piece. A marked difference in emission characteristics was observed between plate and vessel test pieces. Activity from vessel material was virtually constant after general yield and emission amplitudes were low. Plate test pieces showed maximum activity at general yield and more frequent high amplitude emissions. An attempt has been made to compare the system sensitivities between the pressure vessel test and laboratory tests. In the absence of an absolute calibration device, system sensitivities were estimated using dummy signals generated by the excitation of an emission sensor. The measurements have shown an overall difference in sensitivity between vessel and laboratory tests of approximately 25 db. The reduced sensitivity in the vessel test is attributed to a combination of differences in sensors, acoustic couplant, attenuation, and dispersion relative to laboratory tests and the relative significance of these factors is discussed. Signal amplitude analysis of the emissions monitored from laboratory test pieces showed that, with losses of the order of 25 to 30 db, few emissions would be detected from the pressure vessel test. It is concluded that no reliable prediction of acoustic behaviour of a structure may be made from laboratory tests unless test pieces of the actual structural material are used. A considerable improvement in detection sensitivity is also required for reliable detection of defects in low-strength ductile materials and an absolute method of system calibration is required between tests.

889 **T. Ingham, A. L. Stott, and A. Cowan. "Acoustic Emission Characteristics of Steels–Part 1: Acoustic Measurements from Tensile Tests."** *International Journal of Pressure Vessels and Piping,* **PRVPA, 2(1):31–50. Jan 1974.**

Acoustic emission measurements have been made in conjunction with ambient temperature tensile tests on nine structural steels with yield strengths ranging from 247 to 1606 MN/m^2 (16 to 104 t/in^2). The acoustic activity is shown to be associated with initial yielding of the steels. The relative amounts of emission monitored during the tests have permitted an *emission* rating of the steels to be made. Acoustic emission has been shown to increase with increasing strength and decreasing ductility of the steels. Tests to check *recovery* phenomena, i.e., the Kaiser effect, have confirmed the absence of significant acoustic emission on second loading until the initial load has been attained.

890 **T. Ingham, A. L. Stott, and A. Cowan. "Acoustic Emission Characteristics from Steels—Part 2: Acoustic Measurements from Fracture Toughness Tests."** *International Journal of Pressure Vessels and Piping,* **PRVPA, 3(4):267–293. Oct 1975. [see Reference 889 for Part 1].**

Acoustic emission measurements have been made in conjunction with fracture toughness tests on nine steels of yield strengths from 247 to 1606-MNm^{-2}. Tests made under conditions where failure would be analysed in terms of linear elastic fracture mechanics (brittle behavior) or general yielding fracture mechanics (ductile behavior) have shown the emission characteristics at brittle and ductile failure to be significantly different. The applicability of relationships of the form [Acoustic Emission = Constant x (Fracture Toughness)^n] has been investigated and the effect of test piece size examined. The relative amounts of emission monitored during the tests have permitted an acoustic *rating* of the steels to be made. The validity of the Kaiser effect has been demonstrated but a detrimental influence of possible corrosion-oxidation at the crack faces has been noted. The significance of the results has been assessed in terms of the application of acoustic emission to provide 100% coverage during the periodic inspection of pressure vessels.

891 **D. R. Ireland, R. A. Wullaert, A. S. Tetelman, and H. L. Dunegan. "Acoustic Emission Characteristics of Irradiated Reactor Pressure Vessel Steels." Task 8, Project IV, pp 46–63 in "In-Service Inspection Program for Nuclear Reactor Vessels." R. D. Wylie. SwRI Project 17-2440, EEI Project RP-79. Biannual Progress Report No. 4. Southwest Research Institute, San Antonio, Texas. Jan 7, 1971. [NTIS: PB-198-274].**

A short investigation was undertaken to determine if there were any differences in acoustic activity from flaw initiation or propagation in irradiated reactor structural materials versus unirradiated materials. The program was conducted as part of the Connecticut Yankee reactor pressure vessel surveillance program wherein data were being obtained on the effects of irradiation on mechanical properties. The results of the program indicated that through a combination of changes in yield strength and stress intensity factor the acoustic emission activity of A302B steel was significantly changed by irradiation. These changes are in the direction to enhance the acoustic emission as a result of irradiation.

892 **K. Ishihara, K. Yamada, Y. Otsuka, and T. Fuji. "AE Monitoring of the Fatigue Test of Spherical Pressure Vessel." Paper 1-2, pp 31–46 in** ***The Third Acoustic Emission Symposium.*** **Japan Industrial Planning Association, Dobashi Building, 3-17-8, Akasaka, Minato-ku, Tokyo, Japan. 1976. proceedings of a symposium held in Tokyo, Japan. Sep 16–18, 1976.**

At present, there are more than 2,000 spherical pressure vessels made of high tensile strength steel in Japan and in the large majority is in Kanagawa Prefecture. According to the results of an investigation performed by the Industry Safety Section of Kanagawa Prefecture in 1969, cracks in the welding beads were found in nearly 30% of the spherical pressure vessels, even in those vessels which contents were not supposed to bring about stress corrosion cracking. This necessitated close study of the causes of cracking. With this background, the Industry Safety Section of Kanagawa Prefecture organized the committee for the security and conservation of the spherical pressure vessels in 1970. This committee is composed of scientists, users, fabrication specialists, and material makers. Under the sponsorship of the committee, the Technical Research Center of Nippon Kokan K.K. constructed a model spherical pressure vessel of 3 metres in diameter and 40 mm in thickness made of high tensile strength steel, and has conducted a series of hydraulic fatigue tests to find out the causes of initiation and propagation of crack since 1970. In this test, together with the nondestructive testing methods such as ultrasonic testing and radiographic testing, a new monitoring method was used, that is, acoustic emission technique. This paper relates to the seventh test in the series.

893 **K. Ishikawa and H. C. Kim. "Stress Wave Emission and Plastic Work of Notched Specimens."** *Journal of Materials Science,* **JMTSA, 9(5):737–743. May 1974.**

The stress-wave energy released from notched specimens of structural steel was measured in order to compare it with the recently proposed J-integral which takes account of the effect of large plastic deformation around the crack tip in ductile materials. Close agreement was observed between the J-integral and the differential stress-wave energy released. This suggests that the increment of the stress-wave energy released is proportional to the decrement of the work done on the specimen during tensile testing under the plane stress condition. This result, combined with information obtained from linear elastic fracture mechanics, leads to a relationship between the differential stress-wave energy released and the stress intensity factor K, $[\Delta(SWER)/\Delta a] \propto K^2$. It was also found that in the region before general yielding, the stress-wave energy release was proportional to the development of plastic zone size. A larger portion of the accumulated stress-wave energy released at the catastrophic crack growth point reached virtually the same value for each specimen, independent of the initial crack length. This implies that void formation and coalescence are not influenced by the initial crack length, but by the geometry of the crack tip. [reprinted by permission of Chapman & Hall Ltd].

894 **K. Ishikawa and K. Tsuya. "Acoustic Emission and Hardness of Metals."** *Journal of the Japan Institute of Metals (Nippon Kinzoku Gakkai–Si),* **NIKGA, 38(7):673–674. July 1974. (in Japanese).**

Letter report presents acoustic emission data for several metals and alloys. The data show a linear relationship between accumulated counts and hardness.

895 **E. Isono. "History of the Acoustic Emission Method: A New Tool for NDT."** *Journal of N.D.I. (Hi–Hakai Kensa,* **Japan), HIHKA, 19(11):579–598. Nov 1970. (in Japanese); RFP–Trans–79. translated for Dow Chemical U.S.A., Rocky Flats Division, Golden, Colorado. 1971. [NTIS].**

A history of the acoustic emission method is presented and use of the method for monitoring pressure tests for a missile fuel tank or a nuclear pressure vessel, studying destructive mechanisms (plastic deformation, brittle fracture, fatigue, corrosion stress cracks, etc.), and studying or monitoring weld cracks are also discussed.

896 **E. Isono, N. Taniguchi, and T. Udagawa. "An Application of Acoustic Emission to the Weld Cracking Test." pp IB7.1–IB7.10 in** *Cracking and Fracture in Welds.* **Japan Welding Society, 1–11 Kanda Sakuma–cho, Chiyoda–ku, Tokyo, Japan. 1972. proceedings of the First International Symposium on the Precaution of Cracking in Welded Structures Based on**

Recent Theoretical and Practical Knowledge, Tokyo, Japan. Nov 8–10, 1971. [paper presented in Sub-Session I-B: "Assesment of Weld Cracking." Discussion in "Special Volume" section of proceedings, pp 33–35].

Acoustic emission gives the best way to investigate crack initiation and its chronological behavior. Acoustic emision technique developed by the authors have been examined extensively for several crack tests. One tentative test was performed on Y-groove cracking test (JIS Z 3158) using this technique. Brief results are as follows: (1) Time for crack initiation delays with increasing preheating temperature, (2) Crack initiation and propagation, if occurs, finish within about one hour.

897 **E. Isono and T. Udagawa. "Propagation Characteristics of AE Waves in Plates and Pipes."** *Journal of N.D.I.* **(***Hi–Hakai Kensa,* **Japan), HIHKA, 22(7):387–393. July 1973. (in Japanese).**

Acoustic emission wave characteristics were investigated on 1 to 25-mm thick plates and 420-mm diameter steel pipe using pseudo-AE source made out of step wave. Observation of the fluctuation of receiving wave shapes and amplitudes with changing distance and direction of AE source and receiving sensor, measurements of propagation velocity, and consideration of propagating paths were studied. Both with water and without water as containing media were tested in the case of pipe. The wave which arrives first after propagating through steel plates is S_0 mode of Lamb waves, their sound velocity being 5500 m/sec in each case of plate thickness. Secondary arriving wave varies its sound velocity with different thicknesses and it ranges from 1900 to 3500 m/sec. The utilization of the first arriving S_0 mode wave for the location is a most reliable method. However, since the wave has small amplitude in general, the threshold level setting influences greatly the result of location.

898 **E. Isono, T. Udagawa, and M. Ogasawara. "Recovery of the Kaiser Effect." Paper No. 12 in** *Proceedings of the U.S.–Japan Joint Symposium on Acoustic Emission.* **Japanese volume. Japan Industrial Planning Association, Kawamura-Building 6F, 3–21–6 Akasaka, Minato–ku, Tokyo, Japan 107. 1972. symposium held in Tokyo, Japan. July 4–6, 1972. (paper in Japanese, abstract available in English).**

Recently there is an increasing effort to develop acoustic emission as a practical tool to detect crack initiation or propagation. One of the most important problems in pressure testing of vessels is how to evaluate the effect prestressing has on acoustic emission, i.e., the so-called *Kaiser effect.* Pressure vessels are first pressure-tested at the time they are constructed, and also after they are repaired as a result of periodic inspection. Acoustic emission data from the first pressure test provides a base with which to compare the acoustic emission data after subsequent repairs. Therefore, the Kaiser effect has to be taken into account in the practical application of acoustic emission phenomena. Under investigation in the laboratory is the application of acoustic emission techniques to pressure test monitoring of welded steel pipes. The recovery of the Kaiser effect on the pressure test is examined. Observations were made in pressure tests and flattening tests of pipe, and also in tensile tests to obtain the basic data to clarify this phenomena. Results obtained are reported.

899 E. Isono, T. Udagawa, and N. Taniguchi. "A Study of Cracking Phenomena by Acoustic Emission Technique." *Journal of N.D.I.* **(***Hi-Hakai Kensa,* **Japan), HIHKA, 21(4):226–233. Apr 1972. (in Japanese).**

An acoustic emission technique, which has come to be utilized recently in various fields, can exhibit its best capability when applied as a monitor of tracing phenomena in process of time, such as initiation and propagation of weld cracks or delayed failure cracks. The authors who have been interested in this method and have attempted the development of a suitable apparatus necessary for the measurement of the acoustic emission have reported its successful results with low temperature notched-bend test. The paper describes influences of steel material and preheating temperature on weld cracking behavior in Y-groove cracking test specimens using the apparatus mentioned. Brief summaries are as follows: There is a great difference in the cracking behavior between JIS–SM class structural steel and high tensile strength steel. The former develops cracks intensely right after welding, but the cracking mostly ceases within several minutes, while the latter has a high emission rate at 30 to 60 minutes after welding which lasts several hours with sporadic acoustic emission. Also, preheating has an effect of delaying the weld crack initiation. The results clearly indicate that the acoustic emission technique can easily give detailed information which has been difficult or impossible to obtain with the conventional testing techniques.

900 V. S. Ivanova, Z. G. Fridman, and E. G. Smirnov. "Application of Stress Wave Emission to Study of Fracture Mechanism." *Russian Metallurgy (Metally),* **RMLYA, (6):121–123. 1974; translated from** *Izvestiya Akademii Nauk SSSR, Metally,* **IZNMA, (6):149–152. Nov–Dec 1974. (in Russian).**

The propagation of cracks and the onset of final rupture in Mo single crystals as the result of a sharp blow were studied by a method based on an analysis of the acoustic emission (stress waves) occurring at various stages in the fracture process, and the results were compared with data relating to W, V, and other materials. A sharp increase in the rate of acoustic emission at the instant of fracture in samples undergoing severely localized plastic deformation was attributed to a ductile crack in polycrystalline aggregates. The increase in signal amplitude occurring at the instant of fracture indicated a transition to brittle crack propagation. [Abstract 22-0443. *Metals Abstracts,* Vol. 8, 1975].

901 M. Iwamoto, M. Tomizawa, S. Deguchi, E. Yamamoto, and R. Motegi. "Several Applications of Acoustic Emission on the Japanese National Railway." Session 8, pp 46–74 in *The Second Acoustic Emission Symposium.*

Japan Industrial Planning Association, Dobashi Building 5F, 3-17-8 Akasaka, Minato-ku, Tokyo, 107 Japan. 1974. proceedings of a symposium held in Tokyo, Japan. Sep 2-4, 1974.

The paper discusses acoustic emission occurring when fatigue bending test is performed on rails. In a fatigue test, the rail specimen is tested until it is crushed while repeatedly increasing or decreasing the amount of AE wave generated when the rail is subjected to plastic deformation. In this process of the test, when cracks occur on the rail the amount of AE wave is seemingly decreased due to concentration of stress. However, the amount of AE wave increases because of sporadic propagation in the plastic zone and cracks develop. After that an outstandingly great amount of AE wave occurs. Immediately after the occurrence of the AE wave, the rail fails. It may be considered that the pattern of processes from occurrence of AE wave to failing includes a noise generating process up to the stage when the test piece is broken in with the testing machine, AE wave-generating process on the basis of plastic deformation, grouped AE wave-generating process, fatigue crack occurrence process, process of AE wave decreasing due to concentration of stress, and sporadic occurrence of AE wave at a high level. In addition to the above processes, a study was made also of the difference in wave mode of AE wave propagating in the rail. In the study, it was confirmed that the AE waves detected at the rail head, bottom, and in a rail section were the longitudinal waves while those detected at the rail web were the guided waves with a shear-wave component.

J

902 **D. R. James. "An Investigation of the Relationship between Acoustic Emission and Dislocation Kinetics in Crystalline Solids." Ph.D. Thesis. University of Denver, Denver, Colorado. 1970; University Microfilms 71-8029.**

The sounds emitted from single crystals of irradiated LiF, annealed LiF, NaCl, and zinc during constant strain-rate compressive deformation were recorded and analyzed. The acoustic emission consisted of discrete pulses of varying amplitudes, and the rate of pulsing changed with total strain during the continuous deformation experiments. The acoustic pulse rate versus total strain profile for the annealed crystals was found to initiate before the yield point, attain a maximum after yield, and decline in the far plastic region. The pulse rate versus total strain profile for irradiated LiF crystals was observed to initiate and attain a maximum in the macroscopic elastic range and then decline to a minimum at yield. A mathematical relationship between the acoustic emission pulse rate and the kinetics of some dislocation process was sought. Absolutely no correlation was found between the pulse rate and the rate of change of total dislocation density, as verified by a series of etch-pit experiments on LiF at various values of total strain. Therefore, the operation of dislocation multiplication sources could not have produced the observed emission.

The pulse rate was next assumed to be proportional to the mobile dislocation density as suggested in the recent literature on acoustic emission research. This assumption, although not subject to direct examination, was explored. Computer programs were utilized to reveal the implied dislocation velocity behavior when the pulse rate was assumed to be proportional to N_m. This approach resulted in totally unreasonable dislocation velocity versus total strain behavior. Therefore, it was concluded that acoustic emission pulse rate was not directly related to the mobile dislocation density. The third assumption examined for correlation was that the pulse rate was proportional to the rate of change of mobile dislocation density. By using the indirect method of predicting dislocation velocity behavior as described above, reasonable behavior was found. From the data obtained, a reasonable correlation between acoustic emission pulse rate and the rate of change of mobile dislocation density was determined. Based on this determination, it was proposed that the source of the acoustic pulsing was dislocation breakaway from pinning points. The breakaway was associated with yield in the annealed crystals and with the breakaway from minor pinning points in the presence of stronger pinning points in irradiated LiF. The observed pulse amplitudes were used to estimate that some 100 to 1000 cm of dislocation line length had been involved in the generation of an individual acoustic pulse. From the previously determined relationship between pulse rate and rate of change of mobile dislocation density, it was suggested that approximately 10^4 segments of dislocation line participated in the creation of a pulse. It was suggested that several breakaway events may have occurred nearly simultaneously, and their combined force pulse may have stimulated other segments along the slip plane to breakaway, thus accumulating the force which was measured. The concept of stimulated dislocation breakaway from pinning points appeared to be a reasonable model to explain the pulse amplitudes and rate observed in this study, and may provide the basic concept for explaining discontinuous flow, jerky dislocation motion, slip-line broadening, and other features of plasticity.

903 **D. R. James. "Overview of Acoustic Emission and Dislocation Kinetics." Air Force Weapons Laboratory, Kirtland AFB, New Mexico. 1971; hand out material for lecture presented at a short course in Acoustic Emission Technology, sponsored by Dunegan Research Corporation, held in Livermore, California. July 19-23, 1971.**

904 **D. R. James, D. A. Benson, and R. E. Belland. "MTS Modification for Acoustic Emission Testing."** *Review of Scientific Instruments,* **RSINA, 43(4):694-695. Apr 1972.**

A simple, inexpensive modification to a closed loop servo-hydraulic testing machine was made which eliminated virtually all machine noise with respect to acoustic emission monitoring. [reprinted by permission of the American Institute of Physics from *Review of Scientific Instruments*].

905 **D. R. James and S. H. Carpenter. "Acoustic Emission Test Facility."** *Review of Scientific Instruments,* **RSINA, 42(8):1131-1136. Aug 1971.**

An acoustic emission test facility was designed and constructed to allow constant strain-rate compressive deformation of samples up to an applied load of approximately 180 kg. Strain rates of from 1.6×10^{-3} to 3.2×10^{-6} seconds^{-1} were obtained with total system noise limited to less than 10 microvolts peak-to-peak referred to the input of the preamplifier. Overall signal amplification was 9900 ±100 over a band-width of 200 Hz to 200 kHz. [reprinted by permission of the American Institute of Physics from *Review of Scientific Instruments*].

906 **D. R. James and S. H. Carpenter. "Relationship between Acoustic Emission and Dislocation Kinetics in Crystalline Solids."** *Journal of Applied Physics,* **JAPIA, 42(12):4685-4697. Nov 1971.**

The sounds emitted from single crystals of irradiated LiF, annealed LiF, NaCl, and zinc during constant strain-rate compressive deformation were recorded and analyzed. The acoustic emission consisted of discrete pulses of varying amplitudes, and the rate of pulsing changed with total strain during the continuous deformation experiments. A reasonable correlation between acoustic emission pulse rate and the rate of change of mobile dislocation density was found and led to the conclusion that the pulses were generated by dislocation breakaway from pinning points. The measured dislocation densities were used to estimate that several thousand centimeters of dislocation line length had been involved in the generation of an individual acoustic pulse. It was suggested that some 10^5 to 10^6 segments of dislocation line were involved in a cumulative fashion due to the stimulating effect of the moving acoustic wave on bowed dislocation segments. The concept of stimulated dislocation breakaway from pinning points appeared to be a reasonable model to explain the acoustic results of this study and may provide the fundamental basis for explaining discontinuous plastic flow, jerky dislocation motion, slip line broadening, and other features or crystal plasticity. [reprinted by permission of the American Institute of Physics from *Journal of Applied Physics*].

907 **D. R. James and S. H. Carpenter. "The Effect of Sample Size on the Acoustic Emission Generated during Tensile Deformation of 7075 Aluminum Alloy."** *Scripta Metallurgica,* **SCRMB, 10(8):779–781. Aug 1976.**

The investigation reported was undertaken to determine the effects of sample size on the acoustic emission generated during the tensile deformation of 7075-T651 aluminum alloy. Previous work described in the literature gives conflicting data on the effects of sample size. Early work suggested that the acoustic emission generated was dependent on the surface area of the test specimens. Later, work by Fisher and Lally indicated that the acoustic emission was proportional to the sample length, but was insensitive to changes in sample thickness. In order to identify actual sources of emission and to develop workable theoretical models, an accurate knowledge of how the acoustic emission depends on sample size is necessary. At present at least one theoretical model, assuming that the amount of acoustic emission is proportional to the volume of deforming material, has been developed to explain certain experimental results. [reprinted by permission of Pergamon Press Ltd].

908 **L. C. James. "Experiments on Noise as an Aid to Reactor and Plant Operation."** *Nuclear Engineering,* **NUENA, 10(104):18–22. Jan 1965.**

Noise is probably one of the most frequently used means of fault diagnosis in engineering circles; often used naturally and unconsciously. The general accepted maximum range of the audible spectrum is 20 to 20,000 cycles per second. With such a range, it is clear the noise received by the human ear in the vicinity of an engineering installation is an integration of many frequency components. In any installation if a particular frequency of sound can be identified and it has not previously occurred in the normal running background noise, it is highly probable that the signal can be identified with a particular defect behavior. With amplification above background, the hearing or recording of sound at that particular frequency in future will then indicate malfunctioning and enable action to be taken before failure occurs. It is also possible to convert heat energy to sonic energy and therefore to indicate temperatures or heat flux conditions provided again the device can be designed to generate sound at a known frequency which does not occur in the normal background. In a similar way, a sound-generating device can be constructed so that the frequency is modified by a mechanical movement and hence, by previous calibration, it becomes possible to detect such things as liquid level changes, pressure changes, etc. The response time of the detecting system, including the time constants of actual recording and sorting apparatus, is governed by the propagation of sound waves in the medium. The velocity of sound in air at normal atmospheric pressure is approximately 332 metres per second (m/s), in steel it is approximately 5000 m/s, while in water, depending on conditions, it is around 1500 m/s. These are the media which one could expect to encounter in a typical engineering installation. The delay between a disturbance causing a noise and the indication reaching ear or recording equipment is therefore negligible over the distance encountered in such installations.

909 **R. James. "Predictive Maintenance System Improved at Exxon Chemical Plant."** *Oil and Gas Journal,* **OIGJA, 74(5):59–64. Feb 2, 1976.**

Computerized analysis of acoustic signals from remote sensors located near critical components of essential equipment at Exxon's Baytown refinery permits early warning of potential catastrophic damage. About 800 sensing channels monitor equipment on the 250-acre plant. Decisions on shut down are made by operating personnel. Sensors are attached to such equipment as gears, pumps, and bearings which monitor in sequence at intervals while potential cracks are monitored continuously. One example of success was detection of bearing race corrosion in a 705 rpm vertical motor. Additional surveillance equipment is contemplated and development of smaller systems is under way.

910 **R. James, B. Reber, B. Baird, and W. Neale. "Application of High Frequency Acoustic Techniques for Predictive Maintenance in a Petrochemical Plant." Exxon Chemical USA, Baytown, Texas. 1973; paper presented at the 28th Annual Petroleum Mechanical Engineering Conference, Los Angeles, California. Sep 17-20, 1973.**

Lead times, provided by the use of high frequency acoustic instrumentation in detecting incipient mechanical failure, establish this instrumentation technique as an effective tool for performing predictive maintenance. Applications where high frequency acoustic monitoring has provided predictive maintenance data in a typical petrochemical plant and the potential economic impact on the plant's operation are described.

911 **C. S. Jan, R. B. Pond, Sr., and R. E. Green, Jr. "Acoustic Emission during Deformation of Lead-Tin Alloys." AFOSR-TR-75-1066. Johns Hopkins University, Department of Mechanics and Materials Science, Baltimore, Maryland. Dec 1974. [NTIS: AD/A-012-958]; paper presented at the American Society for Metals Materials Testing and Quality Control Conference, Detroit, Michigan. Oct 21-24, 1974.**

Acoustic emission measurements were made on lead-tin binary alloys of different compositions during constant-strain-rate tensile tests. It was found that different compositions possess different microstructures and have different characteristic structure sensitive acoustic emission patterns. Most of the acoustic emissions occurred within the first few percent of strain. Immediate reloading of a number of the test specimens which had been previously loaded and unloaded showed that although the acoustic emission activity was reduced on the second loading, emission commenced at the beginning of reloading at load levels much lower than the maximum level reached on the first loading. These results give direct experimental evidence against the often quoted *Kaiser effect.* A series of measurements on specimens of constant alloy composition solidified at different rates show different acoustic emission activities. The results of the present investigation show that for lead-tin alloys the chemical composition, microstructure, degree of deformation, and prior mechanical history all affect the acoustic emission to a marked degree. Therefore the deformation mechanisms active in these alloys cannot be simply characterized by analysis of the acoustic emission counts alone. When a solid material undergoes a dynamic process such as phase transformation, plastic deformation, interfacial debonding, or fracture, stored energy may be released in the form of acoustic emissions. These emissions, whether in the form of discrete pulses or in the form of a continuous wave train, are characteristic of the microstructural process, which is, in turn, structure dependent. The acoustic emission produced by tin was early noted by tinsmiths as *tin cry* and was studied by Mason *et al.*, Kaiser, and Portevin. Lead also exhibits acoustic emission as reported by Kaiser, Schofield, and Siegel. When lead and tin are combined to form an alloy, their original individual physical properties will change according to the chemistry of the alloy, and therefore alterations may be expected in the acoustic emissions. Tatro has reported that a 40-percent tin, 60-percent lead solder produced acoustic emission when subjected to an external load. In the present work, results are given of acoustic emission measurements on lead-tin binary alloys of different compositions during tensile deformation.

912 P. Jax. "Acoustic Emission as an Aid to Detecting and Monitoring Crack Formation and Crack Growth in Ceramic Materials." pp 55–86 in "EWGAE-ISPRA: Proceedings of the Third Meeting of the European Working Group on Acoustic Emission." A. C. Lucia, Editor. Report EUR-5513e. Commission of the European Communities, Directorate General Scientific and Technical Information and Information Management, Luxembourg, Belgium. 1976. meeting held at the Joint Nuclear Research Center, Ispra, Italy. Sep 25–26, 1974.

Acoustic emission was applied to detect incipient crack formation and macroscopic crack growth during thermal stressing of ceramics. Different ceramic materials were exposed to rapid cooling conditions after a firing process or to thermal shock. Acoustic emission was continuously measured throughout these tests by means of a waveguide device. The method proved to be sensitive in detecting critical phases of precracking processes in the microscopic range. For example, during cooling after firing at 1300 °C in the transition range of the glass phase, microcracks are formed in quartz, porcelain, and alumina; for quartz, porcelain, and a stone ware microcracking at the quartz transition point at 570 °C can give a further critical damage of the ceramic body. Below 400 °C macroscopic cracks were formed as indicated by AE and visual observation. It seems that the extent of macroscopic crack formation can be quantitatively predicted by AE. Continuously measuring AE signals during the firing process of alumina seems to be a qualitative method to give an early warning of the final fracture of the ceramic body. The measurements during thermal shock of a stoneware gave interesting results on the kinetics of crack formation and crack growth. Whereas microcracking in the transition range of the glass phase was proved to be reversible in all temperature cycles, irreversible cracking at the quartz transition point appeared only during the first cooling period. Macroscopic crack formation was critical in the temperature range of 20 to 400 °C, especially during the heating period. With these results AE offers itself for services in materials research, surveillance of ceramic products during fabrication, or materials testing.

913 **P. Jax. "Amplitude Distribution of Acoustic Emission." Battelle-Institut eV, Frankfurt am Main, Germany. 1973; paper presented at the Second Annual Meeting of the European Stress Wave Emission Working Group, Frankfurt am Main, Germany. Sep 1973.**

The amplitude distribution of burst emission and the continuous type of acoustic emission during the palstic deformation of various metals and alloys were measured and analyzed. The use of this property of acoustic emission is generally discussed. Similar relationships as have been found during plastic deformation will also hold for the following processes: crack growth, hydraulic noises, and electrical background noise. The pulse rate was found to depend on the discriminator threshold in specific ways. Using these empirical laws, amplitude distribution functions for burst and continuous emission were calculated. The underlying statistics are discussed. It is shown theoretically and experimentally that this statistic is caused by the statistics of the primary stress pulses generated in the specimen. That means that multiple reflections within the specimen and the damped oscillations of the transducer do not influence these statistics.

914 **P. Jax. "Evaluation of Cracks Formed during Heat Treatment for Stress Relief by Acoustic Emission." Paper G6/4 in** *Structural Mechanics in Reactor Technology, Preprints of the 2nd International Conference.* **Compiled by T. A. Jaeger. Volume 3, Parts G–H. Commission of the European Communities, Brussels, Belgium. 1973. conference held in Berlin, Germany. Sep 10–14, 1973. [NTIS: CONF-730942-P3].**

A new nondestructive testing method—the acoustic emission analysis (AEA)—was applied to the problem of detecting undercladding cracks in reactor steel A508 Class II with overlay welds. This was made by measuring acoustic emission during the heat treatment for stress relief of clad blocks. The results of the AEA were compared with those of other NDT methods and micrographs. It was found that AEA is a sensitive method for detecting cracks. The extent of crack formation can be qualitatively predicted by an acoustic emission parameter, the number of total emission counts during the heating period. By this method it is possible to classify the cracks roughly into different sizes. Especially a reasonable value of the acoustic emission parameter for industrial quality control can be given. Due to the principles of the method, the AEA is able to follow crack formation during the whole heat treatment and to give detailed information of the kinetics of cracking in dependence of time and temperature: Undercladding cracks in reactor steel are not formed in the period of heating-up the blocks, but at the final temperature of the heat treatment. The velocity of crack formation in this stage gradually increases from zero to a maximum, which is reached after an incubation time of 2 to 4 hours at 600 °C, and decreases then continuously—also during further heat treatments. These results are in good correspondence to existing theories about the cause of undercladding cracks. To study in more detail the mechanism of crack formation, acoustic emission was measured during creep tests of specimens, which had a similar heat treatment and structure as the coarse-grained heat affected zone of the clad blocks. According to the AEA and to photographs, microcracks are already formed after times of less than 30% of rupture lives. The rupture of the specimen is announced by an increase of acoustic emission activity, that means in this case a range of macroscopic crack formation, which takes nearly 7% of the total rupture life. It was found by acoustic emission analysis that crack propagation along a grain boundary cannot be spontaneous, but must proceed in steps.

915 **P. Jax. "Feststellung von Unterplattierungsrissen mit Hilfe der Schallemissionsanalyse (Determination of Subplating Cracks with the Help of Acoustic Emission Analysis)." Interim Report. Battelle-Institut eV, Frankfurt am Main, Germany. Oct 1973. (in German). [NTIS: NP-20015].**

The results obtained on the use of sound emission analysis for the detection of subplate weld cracks in 22 Ni-Mo-Cr 37 steel are reported and compared with the theoretical predictions. The formation of subplate cracks was also detected in BHW38 steel (a high-V variation of the original steel). Noise sound analysis showed that these crack formations, in contrast to that in 22 Ni-Mo-Cr 37 steel, occurred during the heating of the sample. Possible reasons for this are discussed.

916 **P. Jax. "Feststellung von Unterplattierungsrissen mit Hilfe der Schallemissionsanalyse (Determination of Subplating Cracks with the Help of Acoustic Emission Analysis)." Final Report. Battelle-Institut eV, Frankfurt am Main, Germany. Apr 1973. (in German). [NTIS: NP-20016].**

Laboratory studies on low-stress anneal of plate reactor steel 22 Ni-Mo-Cr 37 have shown that noise emission analysis is suitable as a nondestructive method for the detection of subplate cracks. The extent of the crack formation can be determined qualitatively with a noise emission parameter that sums the pulse rate over the anneal duration. The method therefore permits a coarse classification of the fissures by size and frequency. The studies showed that a minimum length of the subplate crack of 1 mm corresponds to a control value of 4×10^6 for the sum of the pulse rates. Sound emission analysis shows the kinetics of the crack formation as a function of the anneal duration and the anneal temperature. The formation or growth rate of the cracks reaches a maximum at a certain incubation period and then continuously decreases even with subsequent anneal. These results agree with theoretical results on the appearance of subplate cracks. The origin and the mechanism for the formation of subplate cracks were determined.

917 **P. Jax. "Feststellung von Unterplattierungsrissen mit Hilfe der Schallemissionsanalyse (Determination of Cracks under Claddings with Acoustic Emission)."** ***Materialprufung,*** **MTPRA, 18(2):44–45. Feb 1976. (in German).**

918 **P. Jax. "Schallemission bei plastischer Verformung von Metallen (Acoustic Emission during Plastic Deformation of Metals)." pp 59–117 in** ***Schallemission: Anwendung bei der Untersuchung, Prufung und Uberwachung metallischer***

Werkstoffe. **Deutsche Gesellschaft fur Metallkunde eV, 637 Oberursel/Taunus bei Frankfurt/Main, Adenauerallee 21, Germany. 1974. proceedings of symposium held in Munich, Germany. Apr 18–19, 1974. (paper in German).**

919 **P. Jax and J. Eisenblatter. "Acoustic Emission Measurements during Plastic Deformation of Metals." pp 2–8 in** ***Battelle Information 15,*** **Battelle-Institut eV, Frankurt am Main, Germany. Apr 1973.**

Acoustic pulses in a frequency range which extends far into the ultrasonic region are emitted in metals during plastic deformation. The intensity and characteristics of this *acoustic emission* were measured during the deformation of various steels and face-centered cubic metals. The results provide interesting information on the kinetics of slip processes taking place during plastic deformation. Acoustic emission is the basis of a new nondestructive method of evaluation—acoustic emission analysis.

920 **P. Jax and D. Hums. "Feststellung und Uberwachung beginnender Rissbildung und des Risswachstums in keramischen Werkstoffen mit Hilfe der Schall-Emissionsanalyse (Determination and Surveillance of Formation and Growth of Cracks in Ceramic Materials by Means of Acoustic Emission Analysis)."** ***Keramische Zeitschrift,*** **KERZA, 26(11):625–630. 1974. (in German).**

A new nondestructive testing method, acoustic emission, was applied to detect incipient crack formation and macroscopic crack growth during thermal stressing of ceramics. The method is based on the phenomenon that local elastic energy is suddenly released during cracking processes giving rise to AE signals picked up by sensitive piezoelectric transducers. The AE was measured during thermal shock and cooling of different ceramic materials. The method proved to be sensitive in detecting critical phases of precracking processes in the microscopic range. In the cooling period after the firing process the transition range of the glass phase is such a critical phase; for quartz, porcelain and a stoneware microcracking of the quartz transition point at 570 °C can give a further critical damage of the ceramic body. Below 400 °C macroscopic cracks were formed as indicated by AE and visual observation. It seems that the extent of macroscopic crack formation can be quantitatively indicated by AE. Continuous measuring of AE-signals seems to be a qualitative method to predict the final fracture of the ceramic body. With these first results AE offers its services in materials research, surveillance of ceramic products during fabrication, and materials testing.

921 **W. M. Jensen, A. C. Knoell, and C. Zweben. "Development of Boron Epoxy Rocket Motor Chambers." Jet Propulsion Laboratory, Pasadena, California and Materials Sciences Corporation, Blue Bell, Pennsylvania. 1972; paper presented at the Reinforced Plastics/Composites Institute of the Society of the Plastics Industry 27th Annual Technical Conference. Feb 8–11, 1972.**

A 71-cm diameter 74-cm length boron-epoxy composite rocket motor chamber was designed based on the geometric configuration of the JPL Application Technology Satellite titanium alloy apogee motor chamber. Because analyses showed large stress concentrations in the domes, the configuration was modified using the same basic constraints for openings and attachments. The rocket motor chamber was then fabricated by filament winding with boron-epoxy tape and hydrostatically tested to failure at 264 N/cm^2, 57.2 N/cm^2 above the design value. Two more rocket motor chambers were fabricated with the same basic constraints, but shortened to 57.6 cm for a smaller propellant load. The first of these short chambers failed in proof because of filament winding fabrication difficulties. The second chamber was successfully fabricated and passed the hydrostatic proof test.

922 **E. C. Jessen, H. Spanheimer, and A. J. De Herrera. "Prediction of Composite Pressure Vessel Performance by Application of the Kaiser Effect in Acoustic Emissions." H300-12-2-037. Hercules Incorporated, Systems Division, Magna, Utah and Thiokol Corporation, Wasatch Division, Brigham City, Utah. June 1975.**

Chambers for solid-propellant rocket motors are often fabricated from continuous fibers embedded in an epoxy matrix. Quality acceptance testing of the chambers consists of a hydrotest at the maximum expected or greater operating pressure for the action time of the motor. However, such testing does not define the burst strength of the chamber, nor whether it could withstand a second pressurization at the same conditions. In addition, as the factor of safety is reduced for high performance chambers, such testing could cause fatigue of the composite. Evaluation of acoustic emissions during testing could provide assurance of chamber integrity, predict burst performance, and define reliability from testing conducted at pressure levels lower than the maximum expected operating pressure. Acoustic emission testing is being used to evaluate the integrity of metal pressure vessels, but the technique has not advanced to predicting burst pressure. A large amount of work has been conducted on small laboratory composite specimens, and these results have been applied to evaluation of large cases in the Trident-I (C-4) Program. In laboratory specimens the construction is simple, and the location of flaws is well-defined. Actual motor cases have complex construction, and the flaw location is unknown. The methods of data acquisition are similar to those used for metal cases. Results have shown that (1) chambers have an identifiable acoustic emission output denoted as the signature, (2) acoustic emission in good quality cases are minimal during periods of constant pressure, and (3) in successive pressurizations, the acoustic emissions are minimal until the previously achieved maximum pressure is obtained. This last effect is known as the Kaiser effect. These effects have been used to generally identify the flawed region, to associate acoustic emissions with events during testing, and to define chamber quality. The evaluation of data from more than 60 large chambers and the method of correlation of the data with the above effects to define chamber performance are presented.

923 **A. F. Joffe. *The Physics of Crystals.* McGraw-Hill Book Company, New York. 1928. p 50.**

The author gives the following statement about acoustic emission: "While the quantitative investigation which is still in progress has not furnished definite results as yet, some qualitative results have been obtained which throw light on the phenomenon of plasticity. The investigation was first carried on by Prof. P. Ehrenfest and myself, and then continued by Miss Classen. We noticed that the shear of heated rock salt and zinc progresses by small jumps, each accompanied by a noise like a tick of a clock. In a quiet room those ticks are rather loud and follow at regular time intervals. Many hundreds of ticks may be noticed, their frequency depending upon the load applied."

924 **W. D. Jolly. "Acoustic Emission during Welding Indicates Defect Formation." BNWL-SA-2486A. Battelle-Northwest, Richland, Washington. 1969; paper presented at the American Society for Metals Materials Engineering Congress, Philadelphia, Pennsylvania. Oct 13-16, 1969. [abstract only; see Reference 925 for content of presentation].**

Acoustic emissions, pulses of high frequency sonic energy, result from the formation of weld defects as the weld metal cools. Cracks, porosities, and inclusions are detected as an increase in acoustic emission rate. Artificially induced weld defects in steel, aluminum, titanium, and stainless steel cuase a significant change in acoustic emission rate within one minute after welding. Acoustic emission rate is related to defect temperature. Experimental results from automatic and manual gas tungsten arc, automatic submerged arc, electron beam, and resistance welding are presented. A correlation is shown between acoustic emission rate, and metallographic data from welds in stainless steel. Methods for locating weld defects and practical applications are discussed.

925 **W. D. Jolly. "Acoustic Emission Exposes Cracks during Welding Process." BNWL-SA-1967. Battelle-Northwest, Richland, Washington. July 10, 1968;** ***Welding Journal,*** **WEJUA, 48(1):21-27. Jan 1969.**

Acoustic emission was observed from a series of welds in stainless steel plate. A direct correlation was established between acoustic emission rate and radiographic analysis of single pass machine welds in 0.125-inch 304L stainless steel plate containing induced defects at specific locations. Acoustic emission rate from multipass welds in 0.5-inch 304L and 316 stainless steel plate was greater than the emission rate from single pass welds by about an order of magnitude, and the same direct correlation between radiographic analysis and acoustic emission rate was found. The onset of acoustic emission occurs 20 to 45 seconds after the defect region begins to solidify. Acoustic emission rate from induced defect regions increases as the defect region continues to cool, reaching a peak at about 400 °C.

926 **W. D. Jolly. "An In Situ Weld Defect Detector: Acoustic Emission." BNWL-817. Battelle-Northwest, Richland, Washington. Sep 1968. [NTIS].**

The report describes an investigation of the application of acoustic emission to the detection of weld defects during and immediately following the welding operation. Both manual and machine welds fabricated by the gas-tungsten-arc process produce acoustic emission. Acoustic emission from welds in 304L and 316 stainless steel is examined in detail. A correlation is made between acoustic emission rate and radiographic analysis data. A relationship between acoustic emission rate and weld defect region temperature is demonstrated. Acoustic emission pulse characteristics and spectral power distribution are presented. Problems involved in locating weld defects by means of acoustic emission signals are discussed.

927 **W. D. Jolly. "Develop and Evaluate the Coinicident Detection–Location System." Task 10, Project IV. pp 79–90 in "In–Service Inspection Program for Nuclear Reactor Vessels." E. R. Reinhart. SwRI Project 17–2440, EEI Project RP–79. Progress Report No. 8. Final Technical Report. Southwest Research Institute, San Antonio, Texas. May 21, 1973. [NTIS: PB–222–701].**

The coincident detection-location (CDL) technique was evaluated on the EBOR pressure vessel using both a three-channel coincidence detection system and a four-channel coincidence detection system. The four-channel system (CDL-4) reduced simulated full power reactor noise count by a factor of 7.5 over the three channel system (CDL-3). A growing fatigue crack was located by the CDL-3 system in the presence of simulated full-power reactor noise, and a correlation was established between acoustic emission rate and crack growth rate. The practical limit of signal-to-noise amplitude ratio was S/N = 0.25 for both systems. Fatigue cracks, monitored by both CDL systems, emitted acoustic emissions of up to 1500 microvolt peak-to-peak amplitude at the sensor output. Acoustic emission rate signature of fatigue cracks was found to depend upon the magnitude of cyclic load applied.

928 **W. D. Jolly. "Die Anwendung von Schallemission fur die Schweissnahtkontrolle (The Use of Acoustic Emission as a Weld Quality Monitor)."** *Zeitschrift fur Werkstofftechnik/Journal of Materials Technology*, **ZWKTA, 2(2):97–101. 1971. (in German).**

This is a German translation of the author's paper BNWL-SA-2727. See Reference 942.

929 **W. D. Jolly. "Evaluation of the Acoustic Emission System." Task 3, Project IV. pp 54–63 in "In–Service Inspection Program for Nuclear Reactor Vessels" R. D. Wylie. SwRI Project 17–2440, EEI Project RP–79. Biannual Progress Report No. 2. Southwest Research Institute, San Antonio, Texas. Dec 2, 1969. [NTIS: PB–188–347].**

The purpose of this task is to identify and correct functional limitations of the procured acoustic emission system with respect to the requirements for long-term surveillance of an operating reactor pressure boundary. Toward this objective, a program was outlined and preparation of a laboratory test facility was initiated. Studies were initiated to establish test criteria and evaluate anticipated functional limitations of the procured system. Significant preliminary results were obtained in three areas of study: (1) PZT-5A (lead zirconate titanate) sensors can tolerate the radiation level at a reactor pressure boundary but appear to suffer serious loss of sensitivity at temperatures above 300 °C (572 °F). Preliminary experiments indicate that the sensitivity of PZT-5A at 100 °C (212 °F) can be maintained up to 360 °C (680 °F) by the application of a bias potential; (2) A basic limitation on flaw location accuracy is the ability to electronically measure the time between signal arrivals at 4 sensors. Initial measurements on a flat plate, using a two-sensor acoustic emission system, indicate that time between signal arrivals can be electronically measured to ±0.2 microseconds; and (3) Experience is being gained on the application of acoustic emission monitoring techniques to heavy section components. In field testing, stress corrosion cracking, welding, and hydrostatic testing of thick section steel structures have all been successfully monitored.

930 **W. D. Jolly. "Evaluation of the Acoustic Emission System." Task 3, Project IV, pp 72–97 in "In-Service Inspection Program for Nuclear Reactor Vessels." R. D. Wylie. SwRI Project 17-2440, EEI Project RP-79, Biannual Progress Report No. 3. Southwest Research Institute, San Antonio, Texas. July 30, 1970. [NTIS: PB-193-433].**

The laboratory evaluation of the prototype acoustic emission monitor system was started upon completion of the system February 2, 1970. A functional check-out of the system was completed in February. The laboratory evaluation facility, a 96 by 48 by 4-inch mild steel plate with reactor noise simulator was completed and ultrasonic characteristics of the plate were determined. This test plate was used extensively in the investigation of three new approaches to the system–reactor interface problem: (1) an acoustic coupler-waveguide technique which may permit the removal of sensor and preamplifier from the hostile environment at the pressure boundary; (2) a *donut* sensor configuration with low output impedance and improved shear wave response; and (3) a magnetic sensor with low output impedance and operating temperature up to 500 °C.

931 **W. D. Jolly. "Evaluation of the Acoustic Emission System." Task 3, Project IV, pp 25–45 in "In-Service Inspection Program for Nuclear Reactor Vessels." R. D. Wylie. SwRI Project 17-2440, EEI Project RP-79. Biannual Progress Report No. 4. Southwest Research Institute, San Antonio, Texas. Jan 7, 1971. [NTIS: PB-198-274].**

In support of the acoustic emission system evaluation, a study of the acoustic coupler-waveguide approach to the reactor-surveillance system interface problem was continued. Measurements of signal loss due to length, bending, restraining clamps, and coupling pressure were made on Zircaloy-2 waveguides. Results indicate waveguide length of 40 feet is practical. Incentives for using a modular system of close-spaced four sensor arrays were found to be outweighed by the magnification of arrival time measurement error for sources located outside the array. One emission source yielded only 5% of emission bursts within 3 inches of the actual location. An alternative circular configuration (three radial, one central sensor), analyzed by computer simulation, offered some improvement over the rhombic array but arrival time measurement-error magnification in the centrally located sensor was prohibitive. Triangulation range-accuracy measurements at 1.0 and 0.5 MHz demonstrated that the close-spaced sensor array was not satisfactory for the reactor surveillance system.

932 **W. D. Jolly. "Evaluation of the Acoustic Emission System." Task 3, Project IV, pp 84–103 in "In-Service Inspection Program for Nuclear Reactor Vessels." R. D. Wylie. SwRI Project 17–2440, EEI Project RP–79. Biannual Progress Report No. 5. Southwest Research Institute, San Antonio, Texas. May 28, 1971. [NTIS: PB–200–518].**

The EEI Prototype acoustic emission monitor system was modified to incorporate the successive approximation method of signal processing and source location. A wired logic breadboard signal processor was designed and evaluated on the Laboratory Test Plate. The wired logic processor performed the complete source location operation independent of a computer but reevaluation of the design resulted in a signal processor-computer program design which was incorporated in the modified system. The modified system (Prototype II) was designed for a hexagonal sensor pattern in modules of 7 sensors spaced 60 inches apart. The system includes 10 sensors arranged in two overlapping hexagons. In preparation for the field evaluation on EBOR the attachment and routing of waveguides on an actual operating reactor was examined. Waveguides may be bent to a radius of 3 inches, clamped by standoffs using a low density bushing and routed through the primary reactor shield. A new method of source location, coincident detection, appears to offer the possibility of reactor surveillance at frequencies below 1.0 MHz. Experimental results indicate that acoustic emissions well below the reactor background noise level may be detected and located using digital delay lines and background subtraction techniques.

933 **W. D. Jolly. "Field Evaluation of a Prototype Acoustic Emission Reactor Surveillance System." Southwest Research Institute, San Antonio, Texas. 1973; paper presented at the American Society for Nondestructive Testing National Spring Conference, Los Angeles, California. Mar 12–15, 1973.**

As a part of a research project sponsored by the Edison Electric Institute and managed by Southwest Research Institute, Battelle Northwest developed a prototype acoustic emission system for evaluation under simulated reactor operating conditions. Experimental results of operation of the prototype system on the EBOR reactor vessel at the National Reactor Test Station, Idaho, at elevated temperature and in the presence of simulated reactor coolant flow noise indicate that on-line reactor surveillance is feasible. A successive approximation triangulation technique and a coincident detection location technique were evaluated. The coincident detection location technique was used to locate simulated emission sources at a signal-to-noise amplitude ratio of −12 dbV. Fatigue cracks grown from the inner and outer surface of the EBOR pressure vessel exhibited similar acoustic emission response.

934 W. D. Jolly. "Field Evaluation of the Acoustic Emission System." Task 4, Project IV, pp 91–96 in "In-Service Inspection Program for Nuclear Reactor Vessels." R. D. Wylie. SwRI Project 17–2440, EEI Project RP–79. Biannual Progress Report No. 6, Volume 1. Southwest Research Institute, San Antonio, Texas. Jan 7, 1972. [NTIS: PB–207–591].

The field evaluation task was initiated in April 1971. The completed Prototype Reactor Surveillance System (PRSSII) was functionally tested and shipped to the EBOR Facility August 25, 1971. The test series scheduled for the EBOR Facility includes 3 series of functional tests of the Nondestructive Test Facility and the PRSSII. Three test series are scheduled to evaluate the effect of temperature, pressure, and reactor coolant flow noise on the ability to detect and locate growing cracks. The last of the 6 test series includes cracks artificially grown in the vessel wall.

935 W. D. Jolly. "Field Evaluation of the Acoustic Emission System." Task 4, Project IV. pp 57–71 in "In-Service Inspection Program for Nuclear Reactor Vessels." E. R. Reinhart. SwRI Project 17–2440, EEI Project RP–79. Progress Report No. 8. Final Technical Report. Southwest Research Institute, San Antonio, Texas. May 21, 1973. [NTIS: PB–222–701].

Evaluation of the prototype reactor surveillance system on the nondestructive test facility (EBOR Pressure Vessel) at the National Reactor Test Station in Idaho was completed in July 1972. Acoustic emission from a fatigue bar and from a voluntary source was detected at a wall temperature of 507 °F on the EBOR pressure vessel and the emission sources correctly located. Acoustic emission from a fatigue crack grown directly in the 6-inch vessel wall was detected in the presence of simulated full-power reactor coolant flow noise. The fatigue crack was located within ±6 inches and a correlation was established between acoustic emission total count and crack growth. A growing fatigue crack on the inner surface of the EBOR pressure vessel was detected and located by the prototype reactor surveillance system. The acoustic waveguide system, used to protect sensors and preamplifiers from the high temperature at the vessel surface, operated without degradation through two cycles to vessel wall temperature above 450 °F. Experiments completed in the field evaluation of the Prototype Reactor Surveillance System demonstrate that an on-line reactor surveillance system for continuous monitoring of crack growth and location of defects are feasible.

936 W. D. Jolly. "General Guidelines for Acoustic Emission Systems." Tasks 6 and 7, Project IV. pp 91–97 in "In-Service Inspection Program for Nuclear Reactor Vessels." E. R. Reinhart. SwRI Project 17–2440, EEI Project RP–79. Progress Report No. 8. Final Technical Report. Southwest Research Institute, San Antonio, Texas. May 21, 1973. [NTIS: PB–222–701].

Acoustic emission monitor systems may be used to advantage on nuclear steam supply systems. Hydrostat monitor systems detect discontinuities in a pressure boundary during hydrostatic pressurization and may warn of incipient failure. On-line surveillance systems must operate reliably for several years while processing data at 1,000 to 10,000 data sets/sec to extract acoustic emissions from coolant flow noise. An acoustic waveguide system removes the transducer and electronics from the region of high temperature and radiation. On-line systems may be used to detect leaks or loose parts in addition to acoustic emissions. A limited system may be used to monitor a known discontinuity or isolated areas where flaw growth is credible.

937 **W. D. Jolly. "Nondestructive Testing: Advanced NDT Methods." pp 17–25 in "Pacific Northwest Laboratory Division of Isotope Development Programs Quarterly Report: November 1968–January 1969." K. Drumheller, Editor. BNWL-1010. Battelle-Northwest, Richland, Washington. Feb 1969. [NTIS].**

Investigation of the longitudinal and shear components of acoustic emissions to improve acoustic-emission monitoring techniques was concluded. The development of a portable acoustic-emission monitoring system was initiated. A feasibility experiment was conducted to identify the problems involved in monitoring acoustic emission from a specimen at high temperature.

938 **W. D. Jolly. "Nondestructive Testing: Advanced NDT Methods and Concepts." pp 21–24 in "Pacific Northwest Laboratory Division of Isotope Development Programs Quarterly Report: February–April 1969." W. F. Sheely, Editor. BNWL-1085. Battelle-Northwest, Richland, Washington. June 1969. [NTIS].**

Acoustic emission studies of the submerged arc welding process indicate that slag inclusions can be detected during improper bead formation. Radiographic studies of weld beads laid on various metals and alloys confirmed that acoustic emissions can indicate the presence of cracks, porosity, and inclusions in many materials, but not in some. Acoustic emission appeared also to be sensitive to lack of penetration in electron-beam weld. Efforts are under way to evaluate a technique of locating the flaw that causes detected acoustical emission.

939 **W. D. Jolly. "Nondestructive Testing: Development of Advanced NDT Methods." pp 44–59 in "Pacific Northwest Laboratory Division of Isotope Development Programs Quarterly Report: May–July, 1968." K. Drumheller, Editor. BNWL-877. Battelle-Northwest, Richland, Washington. Sep 1968. [NTIS].**

The application of acoustic emission techniques to the detection of weld defects during the welding process has been investigated. Single-pass tungsten-inert-gas welds in 0.125-inch 304L stainless steel plate, and resistance welds in 6061 aluminum were studied.

940 **W. D. Jolly. "Pressure Vessel Investigations: Acoustic Emission Analysis of the Second Large Tensile Specimen at Southwest Research Institute." pp 44–50 in "Heavy-Section Steel Technology Program Semiannual Progress Report for Period Ending August 31, 1970." F. J. Witt, Program Director. ORNL–4653. Oak Ridge National Laboratory, Oak Ridge, Tennessee. Apr 1971. [NTIS].**

Emphasis during the reporting period has been placed upon the simulated service tests activity, although effort was continued under the complex stress state studies and the specific safety research studies. Activities reported include the development of a fatiguing procedure for flaw sharpening of the simulated service test specimens, evaluation of the acoustic emission data obtained during the testing of the second, large flawed tensile specimen at approximately 210 °F at Southwest Research Institute, testing of the third large tensile specimen and a one-sixth scale model at 50 °F at Southwest Research Institute, completion of a parametric study for investigating geometry effects of the SwRI large tensile specimen, and continued expediting of procurement of six intermediate test vessels from Taylor Forge Company. Acoustic emission test results indicated yielding and crack extension at loads below those at which these behaviors were observed experimentally and tended to support the predictive capabilities of the method.

941 **W. D. Jolly. "The Application of Acoustic Emission to In–Process Weld Inspection." BNWL–SA–2212. Battelle–Northwest, Richland, Washington, 1969. [NTIS]; *Materials Evaluation,* MAEVA, 28(6):135–139,144. June 1970; paper presented at American Society for Nondestructive Testing National Spring Conference, Los Angeles, California. Mar 10–13, 1969.**

Experimental results show that cracks and porosity in the welds (austenitic stainless steel, 6061 Al alloy) emit acoustic signals during and after joint fusion; that some weld defects not detected by radiography were detected by specially designed transducers sensitive to the acoustic signals emitted by the weld defects, the verification being by metallographic sectioning; that the acoustic emission rate from weld defects is a function of the weld defect temperature; that there is a relationship between the acoustic emission count and the nugget diameter of resistance welds; and that the weld-bead conformation on submerged-arc welds is related to the changes in the acoustic emission rate recorded during welding. [Abstract 22–0865. *Metals Abstracts,* Vol. 3, 1970].

942 **W. D. Jolly. "The Use of Acoustic Emission as a Weld Quality Monitor." BNWL–SA–2727. Battelle–Northwest, Richland, Washington. Sep 1969. [NTIS]; paper presented at the American Society of Mechanical Engineers 24th Annual Petroleum Engineering Conference, Tulsa, Oklahoma. Sep 21-25, 1969.**

Acoustic emission monitoring is discussed to provide a general understanding of this technique used as a weld quality monitor. The most important feature of the acoustic emission technique as opposed to other nondestructive testing methods is that the data is real-time; thus, one can more effectively determine the causes of the weld defects. The wide-ranging applications of this technique are illustrated by examples of real-time data from submerged-arc, gas tungsten arc, and spot welding. Acoustic emission can be used to detect defective welds and roughly indicate the extent of the defects; moreover, crack growth can be related to other dynamic conditions of the weld. As a supplement to established weld inspection methods, acoustic emission monitoring can result in reduction of rework costs and improvment of weld quality.

943 **W. D. Jolly and M. R. Compton. "Procure an Acoustic Emission System." Task 2, Project IV, pp 81–83 in "In-Service Inspection Program for Nuclear Reactor Vessels." R. D. Wylie. SwRI Project 17-2440, EEI Project RP-79. Biannual Progress Report No. 5. Southwest Research Institute, San Antonio, Texas. May 28, 1971. [NTIS: PB-200-518].**

The procurement task was held open in anticipation of modifications to the prototype acoustic emission system resulting from the laboratory evaluation of the system. Modifications were made to expand the system from 8 to 10 channels and incorporate a new source location method, as well as to increase the data file capability of the system. The new system, designated Prototype II, locates emission sources with a hexagonal array of sensors in modules of 7 by the successive approximation method and files up to 100 separate locations. Completion of these modifications concludes the procurement task.

944 **W. D. Jolly, T. J. Davis, and M. R. Compton. "Evaluation of the Acoustic Emission System." Task 3, Project IV, pp 47–90 in "In-Service Inspection Program for Nuclear Reactor Vessels." R. D. Wylie. SwRI Project 17–2440, EEI Project RP–79. Biannual Progress Report No. 6, Volume 1. Southwest Research Institute, San Antonio, Texas. Jan 7, 1972. [NTIS: PB-207-591].**

The laboratory evaluation task is completed. The Prototype Reactor Surveillance System, Model II (PRSSII) consists of 10 signal acquisition channels interconnected in 2 hexagonal modules. Data generating circuits perform the logic and arrival time difference measurement to form a data set acceptable to the computer. Acoustic emission source designation and data collation are performed by computer software and a periodic record is printed. The report is a detailed description of circuits and functions of the PRSSII.

945 **W. D. Jolly, T. J. Davis, and J. R. Sheff. "Develop and Evaluate the Coincident Detection-Location System." Task 10, Project IV, pp 149–195 in "In-Service Inspection Program for Nuclear Reactor Vessels." R. D. Wylie. SwRI Project 17–2440, EEI Project RP–79. Biannual Progress Report No. 7. Southwest Research Institute, San Antonio, Texas. Aug 25, 1972. [NTIS: PB-212-144].**

The coincident detection-location technique uses digital delay lines to perform continuous correlation of signal arrivals at three transducers. Processing of both coolant flow noise and acoustic emissions is performed by hard-wired logic. Events are segregated into 64 resolution elements. Statistical analysis indicates that false locations due to coolant flow noise will be distributed over more resolution elements than will acoustic emission from discrete locations. Experimental results indicate that location of a discrete source is possible when signal amplitude is 6 to 12 dbV below peak coolant flow noise amplitude. Acoustic emissions from fatigue bars on the EBOR vessel could not be detected in the presence of simulated reactor full power noise.

946 **W. D. Jolly and D. L. Parry. "Field Evaluation of the Acoustic Emission System." Task 4, Project IV, pp 97–143 in "In-Service Inspection Program for Nuclear Reactor Vessels." R. D. Wylie. SwRI Project 17-2440, EEI Project RP-79. Biannual Progress Report No. 7. Southwest Research Institute, San Antonio, Texas. Aug 25, 1972. [NTIS: PB-212-144].**

Work on the field evaluation task began in September 1971 following delivery of the prototype reactor surveillance system to the EBOR facility and completion of modifications to the EBOR pressure vessel. The prototype reactor surveillance system was evaluated using both direct mounted transducers for low temperature operation and 40 inches long aluminum waveguides for high temperature operation. The ability to detect and locate acoustic emission sources originating within the 10-transducer array was demonstrated using both externally grown cracks and during the initial hydrostatic pressurization of the EBOR vessel. Acoustic integrity analysis of the EBOR vessel during the initial hydrostatic test performed by Jersey Nuclear Company indicated that no significant flaws existed in the EBOR vessel at that time. The reactor coolant flow simulator designed and installed by Southwest Research Institute was calibrated to produce a noise spectrum similar to that of the San Onofre Reactor. Acoustic emissions from externally grown fatigue cracks were not of sufficient amplitude or repetition rate to be detected in the presence of this simulated flow noise. However, initial data from a flaw grown directly in the vessel wall inidicate that during crack growth acoustic bursts are of sufficient amplitude to be detected above the simulated reactor flow noise. The prototype reactor system was evaluated at vessel wall temperatures from ambient to 400 °F. No significant change was detected in either the acoustic properties of the vessel wall or the sensitivity of the waveguide system used for signal acquisition with the prototype reactor surveillance system.

947 **W. D. Jolly, E. R. Reinhart, and S. P. Ying. "Acoustic Emission Experiments at EBOR." pp 109–135 in "A Joint Industry/Utility Program to Develop NDT Techniques for Nuclear Reactor Inspection: An Addendum Report to Steering Committee EEI Project RP-79." Compiled by E. R. Reinhart. Southwest Research Institute, San Antonio, Texas. May 21, 1973. [NTIS: PB-224-698/1].**

The work reported herein covers the experiments performed by Southwest Research Institute at the EEI RP 79 Nondestructive Test Evaluation Facility (NDTF) in Idaho.

The SwRI was one of four contributing industrial participants in the Joint Industry/ Utility Program to Develop NDT Techniques for Nuclear Reactor Inspection. The primary objectives of the SwRI experiments were to evaluate the technique of acoustic emission as applied to the inspection of the primary coolant system of a nuclear reactor and to use the tests to develop acoustic system components for possible areas of applications to this field. Results presented cover experiments in hydrostatic as well as on-line monitoring applications of acoustic emission.

948 **G. L. Jones. "Acoustic Emission Studies of Composite and Composite Reinforced Metals." Ph.D. Thesis. Virginia Polytechnic Institute and State University, Blacksburg, Virginia. Mar 1974; Xerox University Microfilms 74-20,371.**

Acoustic emission results for tensile tests of unidirectional boron-epoxy and aluminum and titanium reinforced with unidirectional boron-epoxy are shown. Both the emission count rate and the total emission count are presented for each type of specimen tested. Two analytical models for predicting fiber breakage from the emission results are discussed. One of these models is used to find the percentage of fibers broken at various strain levels and the results are presented also. Use of these findings to develop a technique for predicting imminent failure of composites is discussed. Suggestions for improving the testing procedure are also made.

949 **M. H. Jones and W. F. Brown, Jr. "Acoustic Detection of Crack Initiation in Sharply Notched Specimens." NASA-RP-180. National Aeronautics and Space Administration, Lewis Research Center, Cleveland, Ohio. 1964. [NTIS: N64-21304]; *Materials Research and Standards,* MTRSA, 4(3):120-129. Mar 1964.**

Of the several techniques employed for pop-in load determination, acoustic detection appears to offer the greatest flexibility under a wide variety of testing conditions. This paper describes an acoustic system that uses an ordinary phonograph pickup and a tape recorder. Both sound and load are recorded on the tape. Results obtained with this technique are described for tests on several sheet alloys, using different specimen types provided with both sharp machined notches and fatigue cracks. In addition, acoustic pop-in data for 7075-T6 aluminum alloy are compared with data obtained with a compliance gage.

950 **M. H. Jones and W. F. Brown, Jr. "An Acoustic Method for Determining the Onset of Crack Propagation in Sharply Notched Specimens." *Materials Research and Standards,* MTRSA, 3(5):384. May 1963. [abstract only]; paper presented at the American Society for Testing and Materials 66th Annual Meeting, Atlantic City, New Jersey. June 23-28, 1963.**

The paper describes a system for detecting and recording the sounds of cracks propagating in laboratory specimens designed for fracture toughness testing. Using this technique, the characteristic plane strain fracture toughness for heavy sections (Ki_C) may be determined from relatively small sheet samples. Plane-strain fracture toughness data obtained by the acoustic method are presented for several alloys. These data are compared with results obtained with compliance gages and tests of sharply notched cylindrical specimens. [reprinted by permission of the American Society for Testing and Materials, Copyright].

951 **H. Jost. "Schallemissionsmessungen an bruchmechanischen Proben wahrend wechselnder und zugiger Belastung (Acoustic Emission Measurements on Fracture Specimens during Fatigue and Uninterrupted Loading)." pp 149-169 in** *Schallemission: Anwendung bei der Untersuchung, Prufung und Uberwachung metallischer Werkstoffe.* **Deutsche Gesellschaft fur Metallkunde eV, 637 Oberursel/Taunus bei Frankfurt/Main, Adenauerallee 21, Germany. 1974. proceedings of symposium held in Munich, Germany. Apr 18-19, 1974. (paper in German).**

The article reports on the improvement of the sound emission analysis of changing and tensile loads and shows that crack initiation in steels is always indicated by an increase of the acoustic emission pulse rate. This acoustic emission is not only due to the growth of the plastic region in front of the crack, but also to friction of the crack sides.

952 **H. Jost and J. Eisenblatter. "Using Acoustic Emission Analysis in the Determination of Cracking during Welding." pp 45-48 in** *Battelle Information 14,* **Battelle-Institut eV, Frankfurt am Main, Germany. 1972.**

953 **J. F. Judge. "Special Report: The Crisis in Non-Destructive Testing."** *Aerospace Technology,* **AETEB, 21:28-36. Aug 28, 1967.**

954 **J. H. Jurenka. "Acoustic Emission Monitoring during Creep Tests of Steel Specimens." General Electric Company, Schenectady, New York. Apr 1977; summary of paper in** *Paper Summaries: ASNT Spring Conference, March 28-30, 1977, Phoenix, Arizona.* **American Society for Nondestructive Testing, Columbus, Ohio. 1977.**

It is desirable that an engineering material is able to sustain the design load for a significant fraction of its life after the formation of an easily detected crack. This property allows the repair or replacement of a component after the warning crack appears, but before the component becomes disabled. Since this is an important property of materials, many past tests have reported the proportion of life from appearance of the first visible crack to final rupture. This is relatively simple from room temperature tests. However, during tests made in high-temperature media, such as creep tests, the tests are difficult. It would be desirable to have a means of monitoring for cracks appearing in a specimen within a furnace. Furthermore, this means would be more valuable if the monitoring device would operate unattended. While the bending method of detecting crack propagation appears to be simple and reliable, it is severely limited in that it can be applied only to tensile-loaded, axi-symmetrical specimens, and is applicable to rupture tests of omni-axial specimens or components. As another means of detecting equivalent-sized cracks, an acoustic emission monitoring method seemed the most promising since it would impose relatively few restrictions on the size, shape, or state of stress of the specimen or component being tested. The acoustic emission transducers presently available (PZT-5) are limited with respect to the temperature. For this reason lead metaniobate with Curie point of 550 °C and lithium niobate with highest Curie temperature of 1210 °C have been receiving increased attention as potential transducers for high-temperature operations. However, bonding of the transducer for high temperature is a difficult problem. The problem of monitoring acoustic emission at high temperature has been overcome with the use of a waveguide technique. The rational design and optimum use of transmission lines, be they straight or curved, requires an accurate governing theory for describing their vibration and wave-propagation characteristics. The application of the waveguide technique is demonstrated. The acoustic emission monitoring technique for high temperatures is outlined. The monitoring of structural degradation of specimens operating at high temperature and high load is demonstrated. Acoustic emission generated by crack initiation and crack propagation until rupture was recorded.

K

955 **I. M. Kaganova and M. I. Kaganov. "Cerenkov Emission of Sound by a Particle Moving in Metal."** *Soviet Physics–Solid State,* **SPSSA, 15(7): 1410–1413. Jan 1974; translated from** *Fizika Tverdogo Tela,* **FTVTA, 15:2119–2125. July 1973. (in Russian).**

The frequency distribution of emitted acoustic energy is calculated. The polarization of acoustic waves is discussed and it is shown that mainly transverse waves are emitted at moderate frequencies although the total energy of longitudinal waves is higher than that of the transverse waves. It is assumed that the mechanism leading to the emission of acoustic waves can be described as follows: A particle generates electromagnetic waves which are transformed to acoustic waves by the conduction electrons.

956 **J. Kaiser. "Erkenntnisse und Folgerungen aus der Messung von Gerauschen bei Zugbeanspruchung von Metallischen Werkstoffen (Conclusions and Results from Sound Measurement in the Tensile Stressing of Metals)."** *Archiv fur das Eisenhuttenwesen,* **AREIA, 24(1–2):43–45. Jan–Feb 1953. (in German).**

Metals other than tin emit noises during deformation requiring sensitive electro-acoustic instruments with considerable amplification to detect them. The sound emission is irreversible, i.e., when a certain stress has been applied and released, no sound is detected on subsequent loading until the same stress level is reached. In this way, the level of plastic strain retained in a specimen can be determined. The relation between the tensile stress, on the one hand, and the amplitude or frequency of the concomitant sound, on the other, is characteristic for each metal and is influenced by such factors as yield point, ultimate tensile strength, the level of internal stresses, surface layer effects, etc. It is suggested that a valuable experimental technique for the study of relaxation problems might be developed on this basis.

957 **J. Kaiser. "Materialprufverfahren (Material Testing Procedure)." German Patent No. 852,771. Oct 20, 1952. (in German); UCRL–Trans–10185. translated for Lawrence Radiation Laboratory, Livermore, California. June 1967. [NTIS].**

A material testing procedure is described in which the frequency or amplitude or number of sound effects occurring when the specimen is stressed are determined and evaluated to indicate previous stressing, macroscopic internal stresses, annealing after previous stressing and stress limits for elastic limit, rupture limit, tearing limit and yield point. The equipment consists of an indicator, preferably a peizocrystal to convert sound effects produced by stress into electrical vibrations, a voltage amplifier, and an indicator.

958 **J. Kaiser. "Uber das Auftreten von Gerauschen beim Schmelzen und Erstarren von Metallen (Occurrence of Noises during Melting and Solidification of Metals)."** *Forschung auf dem Gebiete des Ingenieurwesens,* **FGIAA, 23(1–2):38–42. 1957. (in German).**

Investigation of the acoustic emission of several pure metals shows a correlation between the occurence of sound emissions and conventional methods in determining structural changes due to thermal and mechanical processes. Materials studied included tin, zinc, lead, aluminum, and their alloys, and Wood's metal.

959 **J. Kaiser. "Untersuchungen uber das Auftreten Geraushen beim Zugversuch (An Investigation into the Occurence of Noises in Tensile Tests, or A Study of Acoustic Phenomena in Tensile Tests)." Ph.D. Thesis. Technische Hochschule Munchen, Munich, Germany. 1950. (in German); UCRL-Trans-1082(L). translated for Lawrence Radiation Laboratory, Livermore, California. June 1964. [NTIS and John Crearar Library, Chicago, Illinois].**

This work is considered the first comprehensive study of acoustic emission as related to metals. Kaiser observed the acoustic emission phenomenon in steel, aluminum, copper, brass, lead, tin, wood, and zinc. The onset of emission occurred before the macroscopic yield point was reached. He observed that each material produced a characteristic spectra of frequency and amplitude which were related to stress level; however, these observations have not been confirmed by other investigators. He estimated that the intensities of this phenomenon were two or four orders of magnitude below the audible threshold; therefore, suitable isolation and amplification was required for the detection of such noises. Kaiser also reported the irreversibility effect of acoustic emission. Upon reloading a specimen, little significant noise is generated until the previous maximum stress is reached. Working with polycrystalline specimens, Kaiser theorized that the acoustic emissions were associated with the slippages between the grain boundaries as a result of the applied stress.

960 **W. J. Kalb. "SWAT Stamps Out Material Flaws."** *Iron Age,* **IRAGA, 202:58. Dec 5, 1968.**

961 **A. Kamio. "On the Relation of a Recorded Value of Acoustic Emission to the Sizes of the Cracks."** *Journal of N.D.I.* **(***Hi-Hakai Kensa***, Japan), HIHKA, 20(8):368-369. Aug 1971. (in Japanese); RFP-Trans-118. translated for Dow Chemical U.S.A., Rocky Flats Division, Golden, Colorado. 1972. [NTIS].**

Measurement of acoustic emission for the detection of cracks produced during or after welding or for the detection of dangerous parts when a given structure is under load, has been practiced many times recently. Recorded values of AE, however, differ depending upon the device and the recording method used. For the proper evaluation of these values, a comparison with some standard AE source will be necessary. The author took a drop test of a steel ball as a constant AE source and with this standard, examined the characteristics of AE measuring and recording devices. Next he attempted to approximate the area of a crack to a falling steel ball by its AE, which was considered to be equal to the AE of the crack. Acoustic emission equipment used was a Nortec NDT-200.

962 **H. W. Kamm and I. R. Kraska. "A Review of Nondestructive Methods for the Detection of Concealed Cracks." AFML-TR-71-120. General American Research Division, General American Transportation Corporation, Niles, Illinois. July 1971.**

The report is a discussion of NDT methods for detecting concealed cracks with emphasis on crack detection under fasteners and painted and plated surfaces. Many methods are considered. They are classified as: recommended, potentially applicable, and inapplicable. The recommended methods are ultrasonic and eddy current which are now field applicable. Potentially applicable methods are acoustic emission, acoustic impact, and nonfilm imaging systems. Acoustic emission is chosen as the best of these three since it is potentially the quickest. An experimental program is necessary to test this method on actual or simulated structures and evaluate its ability to expose potential and actual structural failure. This will determine whether it is an alternate to the simpler ultrasonic and eddy current methods or complementary to them. Inapplicable methods are given to holography, visual, liquid penetrant, magnetic, thermal, and radiography.

963 **T. Kanagawa, M. Hayashi, and H. Nakasa. "Estimation of Spatial Geo-Stress Components in Rock Samples Using the Kaiser Effect of Acoustic Emission." Paper 3-4, pp 229-248 in *The Third Acoustic Emission Symposium.* Japan Industrial Planning Association, Dobashi Building, 3-17-8, Akasaka, Minato-ku, Tokyo, Japan. 1976. proceedings of a symposium held in Tokyo, Japan. Sep 16-18, 1976.**

In the field of rock mechanics, acoustic emission has been to date applied mainly to monitoring the rock burst in tunnels and to evaluating the stability of pillars in mines. This paper reports a new attempt to estimate anisotropic stresses in the ground, which is based on the Kaiser effect of AE signals emitted in the laboratory compressive tests of boring cores. The resulting geo-stress was compared with the one obtained in the field test *in situ* by the over-coring method already developed. It is emphasized that special effort has been made to eliminate some improper AE signals caused by the stress concentration at the corner of a rock specimen. This effort was successfully completed by attaching a suitable adhesive hunch made of epoxy resin-cement mixture on both loading ends of the rock specimen. It will be shown that the anisotropic geo-stress components can be estimated in the laboratory compressive tests on boring cores sampled in several directions.

964 **T. Kanazawa, K. Iida, and K. Okumi. "Acoustic Emission in Brittle Fracture Accompanied by Fibrous Crack." Paper 7-4, pp 513-530 in *The Third Acoustic Emission Symposium.* Japan Industrial Planning Association, Dobashi Building, 3-17-8, Akasaka, Minato-ku, Tokyo, Japan. 1976. proceedings of a symposium held in Tokyo, Japan. Sep 16-18, 1976.**

Acoustic emission is closely related to the basic process of deformation and fracture and is generally regarded as an important clue to solve their mechanisms. However, there are at present only a few reports on AE in the testing of brittle fracture under the moderate low temperatures because of technical difficulties in cut-off of noise caused by coolant and others and in obtaining an AE sensor that is available in cooling environment. In this paper, the application of AE technique for material evaluation is examined, using four points bending tests, especially taking notice of the test range between ductility and brittleness, and several criteria in fracture mechanics are discussed in the relation to AE characteristics.

965 **A. Kanno. "Utilization of Standard A. E. Source in Crack A. E. Measurement." Paper No. 13 in** ***Proceedings of the U. S.-Japan Joint Symposium on Acoustic Emission.*** **Japanese volume. Japan Industrial Planning Association, Kawamura-Building 6F, 3-21-6 Akasaka, Minato-Ku, Tokyo, Japan 107. 1972. symposium held in Tokyo, Japan. July 4-6, 1972. (paper in Japanese, abstract available in English).**

Concerning the safety evaluation of weld structures such as pressure vessels by the acoustic emission method, it will be required that the AE values obtained by various apparatuses are able to be compared. Moreover it is desired that measured AE values are correlated quantitatively with the critical phenomena propagating. These requirements will result in the use of a standard AE source. In this paper, the dropping of a 2-mm diameter ball bearing was adopted as the standard AE source, considering the frequency characteristics in the measuring system. The mathematical relation of measured voltage of an emission signal to various factors is discussed. If the experimental conditions are kept the same as in field testing, test results can be evaluated by comparing emission voltage with the results from the standard AE source. This paper examines the relation between AE energy and crack area in the tearing test of a weld part.

966 **A. Kanno and M. Sakaki. "Correlation between Acoustic Emission and Crack Size Calibrated by Standard AE Source Generated by Dropped Steel Ball." Session 3, pp 67-79 in** ***The Second Acoustic Emission Symposium.*** **Japan Industrial Planning Association, Dobashi Building 5F, 3-17-8 Akasaka, Minato-ku, Tokyo, 107 Japan. 1974; proceedings of a symposium held in Tokyo, Japan. Sep 2-4, 1974; Report No. 48-NS-S01. Ship Research Institute, Tokyo, Japan. 1972. [NTIS: JPNRSR-28].**

The quantitative treatment of AE signals relating to the defective phenomena is one of urgent requirements in the application of the AE method for the investigation of integrity of structures under loading. This paper presents a basic study of quantitative treatment of AE signals, in which AE signals were calibrated by the standard AE generated by a dropped steel ball. This method enables the calibration of AE measuring systems of different measuring apparatus, probe location, recording apparatus, and sample size. The summation of each calibrated AE source energy was correlated with the crack area in tearing test of tack weld part by TIG method. Following the paper published in the proceedings of the first symposium (see Reference 965), this paper includes some theoretical considerations about the characteristics of the standard AE signals. The experimental results of larger specimens and bombe testing were included.

967 A. Kanno, M. Sakaki, and K. Katsumata. "Correlation between Acoustic Emission and Crack Size Calibrated by Standard AE Source Generated by Dropped Steel Ball." *Journal of N.D.I.* (*Hi-Hakai Kensa*, Japan), HIHKA, 23(7):357–363. July 1974. (in Japanese).

A linear correlation in log-log plot between energy of acoustic emission and area of crack was obtained in a tearing test of TIG welded specimen. Energy of AE was calibrated in comparison with a standard AE source which was generated by a steel ball dropped on the tested specimen from a known height. As for the standard AE, experiments show that: (1) ringing frequency of standard AE signal is approximately inversely proportional to the diameter of a ball and independent of the dropping height and the thickness of plate, and (2) the amplitude of standard AE signal is proportional to the 2/3 power of the dropping height. These results were well-explained by a theory on rebound height of dropped ball.

968 A. Kanno, M. Sakaki, M. Shimada, and K. Katsumata. "Relation between Acoustic Emission and Crack Growth in Mild Steel." Paper 5-3, pp 360–375 in *The Third Acoustic Emission Symposium.* Japan Industrial Planning Association, Dobashi Building, 3-17-8, Akasaka, Minato-ku, Tokyo, Japan. 1976. proceedings of a symposium held in Tokyo, Japan. Sep 16–18, 1976.

The acoustic emission method has been applied to the pressure testing of pressure vessels in preservice or inservice test in Japan. In these applications, the sensitivity levels, which are set to be large enough over the environmental noise such as mechanical and electrical ones, seem to be rather higher comparing with the acoustic emission obtained by a small size test specimen. Although there exist some deviations between the acoustic emission in large structure and that of small specimen, the investigation of amplitude of acoustic emission in small test specimen will be important to evaluate the test results of large-scale application of acoustic emission method. This paper presents the following studies; (1) the acoustic emission of bending fatigue crack growth of mild steel, and (2) the acoustic emission of static bend test. The elaborated efforts were made to eliminate the mechanical noise in the testings.

969 P. Karpiniuk and Z. Pawlowski. "Acoustic Emission during Deformation and Fracture Processes." IPPT-PAN, Warsaw, Poland. 1973; Paper E-10 presented at the 7th International Conference on Nondestructive Testing, Warsaw, Poland. June 4–8, 1973.

Acoustic emission technique has grown in the last ten years to a new tool in nondestructive testing of materials, components, and assemblies. It is a technique that allows to record the elastic waves generated due to a spontaneous energy release in material. Various mechanisms connected with deformation and fracture processes are involved.

The sources of acoustic emission may be in general: dislocation motion, break-away from dislocation pileups, vacancy generation, microcrack formation, crack initiation, and propagation. The acoustic emission technique makes it possible to distinguish between propagating and nonpropagating cracks, or speaking more generally between *active* and *passive* material flaws. The aim of this paper is to discuss some aspects of elastic waves generation in ductile and brittle materials (grey iron, steel, AlMg Si alloy, and duralumin), as well as in fibre-glass reinforced plastics subjected to tensile stresses and fatigue.

970 **B. S. Kasatkin, A. K. Tsaryuk, V. F. Musiyachenko, L. I. Mikhodui, B. N. Kolomiichuk, and V. V Volkov. "The Acoustic Method of Observing the Delayed Breakdown of Welded Joints."** *Automatic Welding,* **AUWEA, 25(10):33–36. Oct 1972; translated from** *Avtomaticheskaya Svarka*, **AVSVA, (10):32–35. Oct 1972. (in Russian).**

The determination of the kinetics of retarded failure in welds (Fe-Si and Cr-Mn steel) is discussed. The method is based on acoustic detection, which relies on the fact that as cracks develop energy is released in the form of sound waves, and this may be converted into electrical signals. An analysis of weld failure by this method emphasizes the major part played by local regions of martensite in retarded failure. [Abstract 55-0584. *Metals Abstracts,* Vol. 6, 1973].

971 **W. T. Kattner. "Nondestructive Testing Capabilities Needed in the Aerospace Industry: A Technological Forecast." FTFP–3. General Dynamics, Convair Aerospace Division, Fort Worth, Texas. July 21, 1971.**

The nondestructive testing requirements of the aerospace industry in the 1970-1985 period were identified. The approximate dates when the necessary capabilities to perform these tests in the factory, in the maintenance shop, and in flight were determined by means of the Delphi technique of technological forecasting. The results of this study are summarized in several tables. Some of the extensive and varied uses are suggested. Several means of evaluating the content of the forecast are presented. It is suggested that where certain nondestructive testing capabilities are needed before the forecast dates, consideration should be given to increasing the financial commitment to the development of one or more of the more promising methods identified in this investigation. This forecast was made following the conventional Delphi technological forecasting technique with three differences: (1) The initial statements of the testing capabilities needed in the aerospace industry were obtained from a group of concerned engineers, quality control personnel, and scientists; (2) Two groups of participants were used as independent Delphi panels to obtain data relative to the reproducibility of Delphi forecasts; and (3) In the fourth round, the participants were invited to indicate the technique which seemed most likely to provide each needed capability. This report is concerned with the content of the forecasts of when the nondestructive testing capabilities will be available for use; a second report is planned dealing with the reproducibility of Delphi panel results. The forecasts of both panels are given to aid the concerned reader in evaluating the relative validity of the forecasts. Other indications of the approach to consensus are also discussed.

972 **Y. Katz. "Micro-Mechanical Approach to Stress Corrosion Cracking in Titanium Alloys." Ph.D. Thesis. University of California, Berkeley, California. Sep 1969; published as UCRL-19046. Lawrence Radiation Laboratory, Berkeley, California. Sep 1969. [NTIS: PB-187957].**

Stress corrosion cracking (SCC) in T-aqueous 35% NaCl systems were investigated on precracked specimens under constant load. Materials tested were commercially pure Ti-75A, alpha-beta Ti - 8 Al - 1 Mo - 1 V and Ti - 5 Al - 2.5 Sn, and all beta Ti - 13 V - 11 Cr - 3 Al alloys. Contoured specimens, with stress intensity (K) independent of crack length, allowed a determination of the crack propagation rate dependence on K. Such a relationship was utilized for evaluating the total time to failure in a single-edge notch specimen since K = f(c) was available. An acoustic emission technique was designed for investigating the micromechanical nature of SCC. Acoustic emission results, as well as fractographic evidence, show the process to be discontinuous slow crack growth. Corralation between stress-wave emission activity and crack velocity was shown with a higher velocity being associated with more frequent activity and larger wave amplitudes. Transgranular cleavage dominated the fracture surface under active environmental conditions. In the Ti - 13 - 11 - 3 the cleavage was on the {100} planes. Tests performed at different temperature show SCC to be a thermally activated process. The apparent activation energy was about 9.5 kcal/mol for the Ti - 75A, 4.5 kcal/mol for the Ti - 5 - 2.5, and ranged from 2.3 to 3.5 kcal/mol for the Ti - 8 - 1 - 1. In the latter the apparent activation energy was affected by the applied stress intensity factor. Consequently a thermally activated triggering mechanism, combined with a mechanically controlled crack jump was proposed. The initiation and reinitation of the crack step were attributed to a diffusion controlled process while the crack step was attributed to a mechanical driving force. Introducing a mechanical term in the rate equation both the low apparent activation energy observed, as well as the high crack velocity, could be explained. Hydrogen embrittlement was hypothesized to be a possible mechanism for the thermally activated process.

973 **Y. Katz. "The Role of the Mechanical Factor in Stress Corrosion Cracking." Paper IV-322 in** ***Dritte Internationale Tagung uber den Bruch. Zusammenstellung der Tagungsberichte, Teil V: Einfluss von Korrosion und Umgebung auf Bruchvorgange.*** **Verein Deutscher Eisenhuttenleute, Dusseldorf, Germany. 1973. volume of preprints of the 3rd International Congress on Fracture, held in Munich, Germany. Apr 8-13, 1973. [preprints out of stock from publisher and proceedings never published].**

Stress corrosion cracking in Ti-aqueous NaCl system were studied. Materials included, alpha, beta, and alpha-beta Ti-alloys. Information regarding the slow crack growth region, in precracked fatigue specimens was accumulated. In addition the micromechanics of the crack growth was investigated by acoustic emission techniques. The combined experimental results, together with observed thermal effects, stimulated discussion on the mechanical driving force involved in stress corrosion cracking.

974 **Y. Katz and W. W. Gerberich. "On the Discontinuous Nature of Stress Corrosion Cracking in Titanium Alloys."** *International Journal of Fracture Mechanics,* **IJFMA, 6:219–221. 1970.**

The complexity of SCC (stress corrosion cracking) behavior arises from the fact that the phenomena involve mechanical (due to the stress field) and electrochemical (due to the environment) driving forces together with their interactions. Separation of these two major influences is difficult, still critical for the evaluation of proposed models. Specifically, this report is concerned about the use of the acoustic emission technique during a SCC investigation of titanium alloys in 3.5% NaCl. In principle, the basic contribution that might be expected from the acoustic technique relates to the question of whether or not the transgranular cleavage process involved continuous growth. That is, any *brittle crack* steps during the slow crack growth process would emit elastic stress waves which could be acoustically detected. Experiments are being performed on Ti-8-1-1, Ti-5-2.5 alloys (α/β structure) and Ti-13-11-3 alloys (all β structure) in an annealed condition. Fatigue pre-cracked notch specimens with a thickness of about 0.125 inches are being tested under constant load. Specimen geometries also include the contoured double-cantilever beam specimen ($K \simeq$ constant) instrumented with a crack-opening displacement gage and a piezoelectric transducer. These provide slow-crack rate determinations and stress wave recordings. The amplification set-up for the stress-wave emission (SWE) was of about 60,000 gain and a pass-bank frequency between 30 and 48 kHz was used to filter out mechanical noise. The experimental results showed that environmental crack growth was accompanied by stress-wave emission. Furthermore, the stress wave activity had a clear correlation with the slow crack growth rate.

975 **Y. Katz and M. D. Merz. "On the Relationship between a Duplex Microstructure and the Fracture Behaviour of a Low-Carbon Alloy Steel."** *Journal of the Iron and Steel Institute,* **JISIA, 210 (Part 12):911–915. Dec 1972.**

The fracture behaviour of a low-carbon alloy steel (Cr,Mo,V) was investigated. Various microstructures were obtained by thermomechanical processing of metastable austenite. High fracture toughness occurred when a structure of tempered martensite and tempered bainite was present. The presence of these two distinct constituents produced high resistance to crack propagation due to cracking of one constituent ahead of the major crack. An acoustic emission technique, as well as metallography and fractography, was utilized in order to study the micromechanical nature of the fracture event. Stress relaxation and enhancement of plastic flow resulted in large-scale yielding, with crack propagation. With respect to this behaviour, the relationship between fracture toughness and microstructure was discussed, particularly with regard to the problem of an extended plastic zone at the crack front.

976 **U. Kayser. "Die induzierte Phononenemission als Aktivierungsmechanismus bei der athermischen Martensitbildung im Stahl (Induced Phonon Emission as the Activation Mechanism in the Athermal Martensite Formation in Steels)."** *Blech, Rohre und Profile,* **BLCHA, 19(5):191–194. 1972. (in German).**

Thermal and acoustic wave processes in the metal lattice structure and the activation mechanism of athermal martensite formation in steel were examined. The austenite-martensite transformation of supercooled austenite crystals is activated by acoustic microwaves formed by induced phonon emission. The phonons are emitted within the range of frequency of the lattice atoms. The austenite-martensite transformation can occur only when the austenite is supercooled enough so that phonons are emitted. The athermal martensite formation is explosion-like and is an acoustic maser effect. This is followed by intensive sound emission of the transformation. The impact hardening of Mn steel and martensite formation due to surface pressure by rubbing pressure can be explained by the acoustic activation mechanism of this phase change. [Abstract 117022g, reprinted by permission of the American Chemical Society from *Chemical Abstracts,* Vol. 77, 1972].

977 U. Kayser. "Induzierte Schallemission bei der Stahlhartung (Industrial Acoustic Emission in Steel Tempering)." ***Umschau,*** **UMSCA, 72(18):594–595. 1972. (in German).**

The analysis of sound emission is used to elucidate the formation of martensite in industrial steel. The athermal martensitic transformation in steel is activated by stimulated emission of acoustic phonons. The martensitic *burst* transformation is regarded as a phonon-maser effect. [Abstract 19246s reprinted by permission of the American Chemical Society from *Chemical Abstracts,* Vol. 78, 1973].

978 B. J. Keene and G. D. Sims. "Design and Assessment of a Portable Stress Wave Emission System." NPL Report IMS–21. National Physical Laboratory, Division of Inorganic and Metallic Structures, Teddington, Middlesex, England. Sep 1973. [NTIS: N74–16118].

A portable stress-wave emission system was constucted and found acceptable for laboratory analysis of failures. Some indication of methods of applying the equipment is given. Stress waves produced by failure processes are detected by piezoelectric transducers in good acoustic contact with the specimen. The transducer converts the pressure wave it receives into an electrical signal. The portable equipment provides collection, amplification, and processing of these signals. This type of equipment could, for the first time, allow the evaluation of the potential of stress-wave emission techniques in a wide range of material and structural problems, and allow the use of these techniques in quality control applications.

979 H. D. Keiser. "Tin." p 640 in ***The Encyclopedia Americana, International Edition.*** **Volume 26. Americana Corporation, New York. 1970.**

"If a bar of the metal is bent, it emits a creaking sound known as the *cry* of tin, which results from mechanical twinning of the crystals during bending of the metal."

980 **F. C. Keledy and K. R. Notvest (to Trodyne Corporation). "Acoustic Emission Crack Monitor." U.S. Patent 3,713,127. Jan 23, 1973.**

A system for monitoring crack growth in a structure in which acoustic energy released as a result of a crack event is sensed and integrated to provide a binary signal representative of the number and intensity of individual crack events. The binary crack energy data signals are accumulated and a warning device is activated when the stored crack data exceeds a predetermined level. Means may be provided to discriminate crack event signals from similar acoustic signals resulting from noise and impact or vibration.

981 **M. P. Kelly. "Acoustic Emission Monitoring of Vessel V-9 by Dunegan/ Endevco." Section 5.1.1, pp 60–67 in "Quarterly Progress Report on Reactor Safety Programs Sponsored by the NRC Division of Reactor Safety Research for October–December 1974, II. Heavy–Section Steel Technology Program." G. D. Whitman. ORNL-TM-4805, Vol. II. Oak Ridge National Laboratory, Oak Ridge, Tennessee. Mar 1975. [NTIS for entire report; Section 5.1 of Pressure Vessel Investigation available as D/E Report V-9 from Dunegan/ Endevco, San Juan Capistrano, California]. [see also Reference 223].**

Intermediate test vessel V-9 was monitored for acoustic emission during test. The high incidence of emissions from a weldment remote from the prepared flaw led to some doubt as to where the ultimate failure would occur. Failure was at the prepared flaw. Subsequently the weldment in question was examined carefully ultrasonically. The comparison of locations of ultrasonic indications with the distribution of acoustic emission events shows a definite correlation. Further examination, including sectioning of the weldment, is planned.

982 **M. P. Kelly. "AE Monitoring of Natural and Artificial Defects in the KEMA Vessel." Technical Report DE 74-7. Dunegan/Endevco, San Juan Capistrano, California. June 1974.**

The application of a 10-channel acoustic emission source location system to the detection and location of natural and artificial defects in a low-alloy steel pressure vessel is presented. Test results demonstrate that cracking of artificial defects during pressurization can be accurately located and categorized in real time. Additional results indicate excellent correlation between acoustic emission and ultrasonics in assessing various types of defects in natural welds.

983 **M. P. Kelly and R. L. Bell. "Detection and Location of Flaw Growth in the EBOR Nuclear Reactor Vessel." Technical Report DE 73-4. Dunegan/ Endevco, San Juan Capistrano, California. 1973.**

Recognizing the rapid growth in both the size and number of nuclear power plants and the importance of safety to the industry, a research and development program for in-service inspection of nuclear reactor vessels (EEI RP79) was initiated by the Edison Electric Institute and the Tennessee Valley Authority. Project IV of this program pertained to the investigation of acoustic emission techniques for continuous surveillance of in-service reactors. EBOR (experimental beryllium oxide reactor), a nonoperating modified reactor located at the National Reactor Test Station in Idaho was made available by the AEC for EEI RP79. Along with several other interested organizations, Dunegan/Endevco participated in Project IV from January 1972 to the final hydrotest on October 13, 1972. The objectives of Dunegan/ Endevco's participation in the EEI-TVA Project RP79 were: (1) Determine the capabilities of acoustic emission to ascertain the structural integrity of nuclear pressure vessels in proof tests and on-line, and (2) Develop and evaluate systems capable of detecting acoustic emision on-line in an operating power plant environment. A program was defined to meet these objectives. It consisted of three work phases. The first was an initial scoping. In this phase, studies were completed which defined the physical environment, the *noise* environment, wave propagation characteristics of the EBOR vessel, requirements for the transducer, and locational accuracy, With the information obtained in the scoping, the second phase–system development and fabrication–was begun. Two systems were produced for testing on the EBOR vessel. Using prior hydrotest experience and data from the scoping, the Model 1010 was built. It consists of ten channels of instrumentation and was designed specifically for the hydrotest type of environment. Operating in the 100 to 300 kHz bandpass it was capable of detecting and locating acoustic emission sources with a transducer spacing of more than 10 feet. For the on-line evaluation an entirely different system was constructed. It incorporated a transducer capable of driving 1000 feet of cable before preamplification with a special frequency response. An entirely different hardware and software system was designed to manipulate, store, analyze, and display the information for on-line monitoring. This was required because of the special high noise environment found in nuclear power plants. Phase three consisted of evaluating the hydrotest and on-line systems on the EBOR vessel. The on-line system was used to monitor several hydrotests and fatigue crack growth at room temperature and at 450 °F. The effect of the simulated San Onofre operating noise was also investigated in conjunction with the fatigue crack growth. The Model 1010 acoustic emission location system was used to monitor a hydrotest of the EBOR vessel to 1875 psi. A flaw indication was detected, but the acoustic emission indications were so minute that the structural integrity was judged to be good. When crack growth was promoted by crack pressurization, excellent data were obtained. The crack area was located within several inches and showed significant crack growth. In the final high temperature test, crack growth was detected, located, and quantified with the on-line monitoring system. In addition, tests were performed at the same time which showed the capability of the system to operate in the simulated San Onofre noise environment with only minor loss of resolution.

984 **M. P. Kelly, D. O. Harris, and A. A. Pollock. "Detection and Location of Flaw Growth in Metallic and Composite Structures." pp 221-240 in** *Monitoring Structural Integrity by Acoustic Emission, ASTM STP 571.* **Edited by J. C. Spanner and J. W. McElroy. American Society for Testing and Materials, Philadelphia, Pennsylvania. 1975. proceedings of a symposium held in Ft. Lauderdale, Florida. Jan 17-18, 1974.**

The application of a multichannel real-time acoustic emission location system to the detection and location of structural defects in three different materials and geometries is presented. Test results from welded steel pressure vessels demonstrate that incremental crack growth during pressurization can be accurately located and assessed. Additional results from graphite-epoxy honeycomb structures indicate that acoustic emission is also a viable tool in detecting and locating impending failure in composite materials. [reprinted by permission of the American Society for Testing and Materials, Copyright].

985 M. P. Kelly and R. J. Schlamp. "Acoustic Emission Preservice Inspection of Calvert Cliffs Unit 2 Nuclear Power Plant." Technical Report DE 76-7. Dunegan/Endevco, San Juan Capistrano, California. May 1976.

Acoustic emission is the term applied to the spontaneously generated elastic waves produced within a material under stress. Plastic deformation and the nucleation and growth of cracks are the primary sources of acoustic emission in metals. Unlike most other nondestructive test (NDT) methods, the energy comes from the material itself. As a result, AE is more sensitive to growing defects and potentially less reliant on operator interpretation than conventional NDT techniques. Acoustic emission offers a further advantage in the inspection of complex structures containing inaccessible areas. Normally just a few stationary transducers can detect, locate, and qualify defects over a large area, including regions inaccessible by alternate methods. Acoustic emission techniques have been successfully applied to the evaluation of structural integrity in bridges, nuclear and petrochemical pressure vessels, pipeline, wind tunnels, rocket motor cases, composite materials, and aircraft. Acoustic emission has already been incorporated into production line testing in one Nuclear Regulatory Commission facility. Although acoustic emission has not specifically been incorporated into the American Society of Mechanical Engineers (ASME) Boiler and Pressure Vessel Codes, there does exist a proposed ASME standard covering the application of the technique. Several acoustic emission examinations have already been performed in accordance with this document. This report presents the acoustic emission results obtained during the preservice inspection of the Baltimore Gas and Electric Company Calvert Cliffs Unit 2 nuclear reactor and primary coolant system. The primary coolant system consists of four reactor coolant pumps, two steam generators, one pressurizer, and associated piping. The purpose of the test was twofold: (1) To locate and quantify any defects found to be growing during initial hydrostatic testing; and (2) To establish baseline records which should be useful in follow-on testing throughout the useful life of the plant. Two Dunegan/Endevco (D/E) Model 1032 Source Location Systems were utilized in the inspection which was performed in conjunction with the system primary hydrostatic proof test on January 18 and 19, 1976.

986 M. P. Kelly and R. J. Schlamp. "Detecting Structural Degradation by Acoustic Emission." summary of paper in ***Paper Summaries: ASNT Spring Conference, March 28–30, 1977, Phoenix, Arizona.*** **American Society for Nondestructive Testing, Columbus, Ohio. 1977.**

Acoustic emission is more sensitive to growing defects and potentially less reliant on operator interpretation than conventional NDT techniques. It has a distinct advantage in the inspection of complex structures which include inaccessible areas. Normally just a few transducers can detect, locate, and qualify defects over a large area. Some examples of the detection of subcritical crack growth during periodic proof testing of structures will be reported. Advantages of using AE during proof testing as opposed to continuous monitoring of structures are discussed.

987 **C. R. Kennedy, D. S. Kupperman, and B. J. Wrona. "Acoustic Emission from Thermal-Gradient Cracks in UO_2." Argonne National Laboratory, Materials Science Division, Argonne, Illinois. Aug 1975; paper presented at the American Society for Nondestructive Testing 35th National Fall Conference, Atlanta, Georgia. Oct 13-16, 1975. [NTIS: CONF-751016-3]; *Materials Evaluation,* MAEVA, 34(4):91-96. Apr 1976.**

A feasibility study has been conducted to evaluate the potential use of acoustic emission to monitor thermal-shock damage in direct electrical heating of UO_2 pellets. In the apparatus used for the present tests, two acoustic-emission sensors were placed on extensions of the upper and lower electrical feedthroughs. Commercially available equipment was used to accumulate acoustic-emission data. The accumulation of events displayed on a cathode-ray-tube screen indicates the total number of acoustic-emission events at a particular location within the pellet stack. The results of the preliminary study have led to the following conclusions. (1) Acoustic-emission events can be detected with commercially available equipment with sufficient resolution to estimate the extent of cracking in an individual pellet; (2) Summation of acoustic events is an indication of the amount of cracking that has taken place in the pellets. These tests have indicated that acoustic emission can be used to monitor thermal-shock damage in UO_2 pellets subjected to direct-electrical heating. Acoustic-emission monitoring could prove valuable in other experiments involving direct-electrical-heating techniques in which the results may be expected to be influenced by pellet cracking.

988 **J. N. Kerawalla. "An Investigation of the Acoustic Emission from Commercial Ferrous Materials Subjected to Cyclic Tensile Loading." Ph.D. Thesis. University of Michigan, Ann Arbor, Michigan. 1965; University Microfilms 66-6632.**

The purpose of this study was to observe the difference in the acoustic emission as well as surface behavior of commercial ferrous materials when they are subjected to cyclic tensile loading above and below the failure limit. The term acoustic emission here refers to the elastic pulses generated within a material while it is under the influence of externally applied stress. The materials used for the study were 1018 and 1042 cold-rolled, and 1020 and 1045 hot-rolled steels. Four specimens were prepared out of each of the four steels selected. The first specimen was subjected to a static tensile test, the second specimen was used to obtain microstrain data relating to the plastic deformation in the material, the third specimen was subjected to fatigue above the fatigue limit, while the fourth specimen was subjected to fatigue below the fatigue limit. The specimens undergoing fatigue were subjected to surface as well as acoustic observations in their virgin state, and after every 10^3, 10^4, 10^5, 10^6, and 10^7 cycles of fatigue. The surface observations were performed by examining the specimens under an optical microscope and by examination of carbon-collodion replicas of the specimen under an electron microscope. The surface observations reveal that slip lines do occur near the surface of those specimens which are stressed below the fatigue limit but these lines are diffuse, less frequent, and can be observed by replication only after etching. The specimens stressed above the fatigue limit show intense slip in localized regions called fatigue slip bands. These bands are not parallel to the primary slip lines. These primary slip lines were observed only in the case of 1020 hot-rolled steel specimens which were loaded beyond the upper yield point. The fatigue slip bands usually first appear in a middle of a grain and grow in a longitudinal direction until they encounter an inclusion crater or a grain boundary, and then they begin to widen. Then cracks begin to nucleate at inclusion craters or grain boundaries where there is a conglomeration of slip bands. These cracks propagate via cracks formed in adjacent grains.

The acoustic emission studies show that there are two distinct types of emission, the burst type and the high frequency type. The high frequency type of emission occurs only in the case of the hot-rolled (recrystallized) steels when they are undergoing the abrupt elongation after the upper yield point. Contrary to the results reported by Kaiser, acoustic emission can be observed at stresses less than a previously applied stress. The high frequency emission happens to be nonrepeatable, and occurs only when the dislocations are unpinned from the solute atoms at the upper yield stress. The burst type emission occurs during the loading as well as the unloading of the specimen. The burst type emission occurring on unloading has been studied in great detail for the first time and it is found that the total emission on unload (which is the sum total of all the pulses counted by the electronic detection system during the unloading operation) is a function of the maximum stress reached during the load-unload cycle. The total emission on unload increases with fatigue or strain rate, but it decreases with aging or plastic deformation. Hence it can be useful in detecting the abrupt increase in plastic deformation which occurs when a specimen is stressed beyond the fatigue limit. It is hypothesized that the burst type emission observed on unloading is due to the acoustic energy released by nonlinear but elastic motion of dislocation lines or tangles as they return to their respective positions of minimum potential energy. [reprinted by permission of University Microfilms International from *Dissertation Abstracts International,* Vol. 27, No. 2B, 1966. Copyright 1966 by University Microfilms International].

989 L. A. Kerridge. "The RPE Stress Wave Emission Monitoring Facility." RPE Memorandum No. 684. Rocket Propulsion Establishment, Westcott, Aylesbury, Bucks, England. July 1976. [copy of report, designated RPE Memo 684, BR-56288, can be obtained from the Defence Research Information Centre, Station Square House, St. Mary Cray, Orpington, Kent BR5 3RE, England].

Described is the Rocket Propulsion Establishment facility for monitoring acoustic emission during laboratory testing of materials and hydraulic pressure testing of, in particular, carbon fiber-reinforced plastic, rocket-motor components. Typical records obtained are shown.

990 **A. W. Khair. "A Study of Acoustic Emission during Laboratory Fatigue Tests on Tennessee Sandstone." pp 57–86 in** ***Proceedings First Conference on Acoustic Emission/Microseismic Activity in Geologic Structures and Materials.*** **H. R. Hardy, Jr. and F. W. Leighton, Editors. Trans Tech Publications, Clausthal, West Germany. 1977. conference held at Pennsylvania State University, University Park, Pennsylvania. June 9–11, 1975.**

In these studies cylindrical specimens of Tennessee sandstone were cyclically loaded under uniaxial compressive stress. A triangular loading cycle (frequency 0.25 Hz for most tests) was applied using an MTS electrohydraulic loading system. Fatigue characteristics of the test material were studied by carrying out a series of tests at different maximum stress levels. Examination of stress-strain-time data indicated the presence of creep which changes character as the number of cycles increase. The acoustic emission pattern for subsequent cycles appeared approximately similar, however, it was interesting to note that the peak acoustic emission rate was not coincident with the peak stress. Observation of a large number of cycles indicated a slowly varying (modulated) acoustic emission pattern. Besides the cyclic tests, a number of specimens were loaded directly to failure at a constant rate of loading. Here the acoustic emission rate was found to correlate well with the observed deformation, and the peak acoustic emission rate was found to occur at a stress level near, but somewhat lower, than the failure stress.

991 **G. K. Kharchenko, B. A. Zaderii, and S. S. Kotenko. "The Acoustic Method of Locating Cracks in Electron–Beam Welds between Niobium and Molybdenum."** ***Automatic Welding,*** **AUWEA, 26(2):70–71. Feb 1973; translated from** ***Avtomaticheskaya Svarka,*** **AVSVA, (2):72–73. 1973. (in Russian).**

Electron-beam butt-welding of molybdenum and niobium alloy sheet material was monitored for acoustic emission during the welding process and post-weld cooldown. It was found that cracks formed in the 50 to 600 °C temperature range.

992 **N. Kiesewetter. "Acoustic Emission from Moving Dislocations."** ***Scripta Metallurgica,*** **SCRMB, 8(3):249–252. Mar 1974.**

During the yielding of a crystal, dislocations move in their slip planes and give rise to the radiation of sound waves. The movement of a dislocation and thus its acoustic emission depends on the different types of interaction with other dislocations, defects, grain boundaries, and surfaces. Shown here is the calculation of the acoustic emission of screw dislocations, which are generated continuously by a source near the surface. Driven by the applied stress and by an image force, these dislocations move towards the surface and are swept out of the material. The resulting formulae for this type of acoustic emission can also be applied to the case, where in the interior of the material, continuously generated screw dislocations of different sign run against each other and are annihilated. The method presented here is based on Eshelby's work on the sound radiation of harmonically oscillating screw dislocations. [reprinted by permission of Pergamon Press Ltd].

993 **N. Kiesewetter and P. Schiller. "The Acoustic Emission from Moving Dislocations in Aluminum."** *Physica Status Solidi (a),* **PSSAB, 38:569–576. 1976.**

The acoustic emission during tensile tests is investigated for several types of aluminum alloys. Findings show that the mean square amplitude of the continuous emission is proportional to strain rate and sample length. It is assumed that the continuous emission is generated by dislocation loops, which are created continuously at Frank-Read sources. According to this model, the mean square amplitude of the continuous emission U^2 is proportional to the total length of the mobile dislocations, L, and the size of the loops l: $\overline{U^2} \sim Ll$. Information about l was taken from Kuhlmann and Wilsdorf's observations of slip lines at the surface of deformed aluminum single crystals.

994 **Y. Kikuta, S. Ochiai, and S. Asano. "Acoustic Emission Detection of Crack Propagation Modes in Hydrogen Delayed Cracking."** *Technology Reports of the Osaka University* **(Japan), TROUA, 24(Report 1162):65–74. Mar 1974; IIW IX-837-73 and IIW IIA-327-73. International Institute of Welding. 1973.**

Acoustic emission detection method was applied to research hydrogen-delayed cracking propagation modes. Here the specimens of high temperature-hydrogen occulsion under various hydrogen partial pressures were delayed-cracked by various applied stress levels. Crack propagation velocity was found to increase exponentially by the increase of stress level at crack tip (stress intensity factor K_I), when initial hydrogen level is fixed. Under a fixed stress intensity factor K_I, the velocity is drastically multiplied by the increase of initial hydrogen level. There exists a definite relationship between crack propagation velocity and acoustic emission rate (emissions per minute), and the logical deduction is that the size of subcritical cracks decreases when applied stress level is boosted. Fractographic observations confirm this unique fact of hydrogen-delayed cracking also.

995 **Y. Kikuta, S. Ochiai, and S. Asano. "Detection of Crack Propagation in Hydrogen-Delayed Cracking by Use of Acoustic Emission." HB No. 9256. Henry Brutcher Technical Translations, Altadena, California. 1974; translated from** *Tetsu-To-Hagane* (*Journal of the Iron and Steel Institute of Japan*), **TEHAA, 59(4), Lecture 269:S271. 1973. (in Japanese).**

Paper discusses the investigation of acoustic emission from the initiation and propagation of hydrogen-delayed cracking. Also discussed are the preparation of Charpy V-notch and linear compliance test specimens, the experimental procedure used to detect delayed cracking in a three-point bend test, acoustic emission from crack opening displacement (COD) specimen, and delayed cracking of a linear-compliance specimen.

996 **H. C. Kim. "Characterization of Mechanical Properties by Acoustic Emission using an Energy Criterion." pp 591–596 in *1975 Ultrasonics Symposium Proceedings.* Catalog No. 75 CHO 994–4SU. Institute of Electrical and Electronics Engineers, Inc., New York. 1975. symposium held in Los Angeles, California. Sep 22–24, 1975.**

The paper presents an energy approach to the acoustic emission from solids undergoing deformation. A new way of describing the acoustic emission during three stages of work hardening in single crystals has been proposed taking into consideration currently acceptable theories of work hardening. The effect of size of specimen on the acoustic emission energy has been investigated for two rectangular prisms having different volumes but using the same strain rate. In cyclic deformation at constant stress amplitude three stages of acoustic emission energy release were observed and corresponded to stages of initial cyclic hardening, saturation of hardening and lastly failure. An overall physical picture of acoustic emission associates it with the incremental change of stored energy of the specimen caused by the mechanical deformation and is independent of the material being necessarily crystalline.

997 **H. C. Kim. "Dependence of Stress Wave Emission and Modes of Fracture on the Void Content in Carbon Fibre Composite (Unidirectional)." pp 3–1 to 3–2 in "Newsletter on the Non-Destructive Testing of Composite Materials and Adhesive Bonded Joints." Issue Number 3. Royal Aircraft Establishment, Farnborough, Hants, England. June 1974.**

Stress wave energy release (E_S) and modes of failure are investigated for carbon fibre-epoxy composite as a function of void content. Observations have shown that the modes of failure markedly depend on the void content. A low void content results in catastrophic failure, while high void content leads to shear failure.

998 **H. C. Kim. "Stress Wave Emission in Carbon Fibre/Epoxy Composite and Epoxy Specimens." Imperial College, London, England; paper presented at the Institute of Physics Conference on Acoustic Emission, Imperial College, London, England. Mar 14, 1972. [proceedings were never published].**

The stress-wave emission technique has been applied to carbon fiber epoxy composites and epoxy specimen while they are undergoing tensile, microstrain, cyclic, double-notch interlaminar shear, and flexural tests. Also an attempt has been made to correlate their behavior with the energy released, in the form of acoustic emission, during deformation. Results of measurements of frequencies have been presented, together with details of the observation of wave forms. The amplitudes of acoustic pulses were analyzed according to a newly proposed energy criterion. The acoustic energy released increased with increasing load up to fracture; and at the proportional limits, detected by strain gage transducer, higher levels of acoustic emission activity were detected. One of the important properties exhibited by carbon fiber epoxy composite was that acoustic emission was detected on unloading; also on reloading up to the same stress level, the acoustic emission activity was reduced but not removed. The observation is in sharp contrast to metals where appreciable acoustic emission is not generated until the stress level reaches the previous stress on reloading (Kaiser effect).

999 **H. C. Kim, W. G. B. Britton, and R. W. B. Stephens. "Chelsea College, University of London." pp 4-1 to 4-3 in "Newsletter on the Non-Destructive Testing of Composite Materials and Adhesive Bonded Joints." Issue Number 2. Royal Aircraft Establishment, Farnborough, Hants, England. July 1973.**

Acoustic emission techniques have been used to investigate the progress of damage accumulated in unidirectional CFRP on subjecting the specimen to simple tension and constant stress cyclic deformation. This work has led to the introduction of an energy criterion to correlate the stress wave energy released with the mechanical work done, and encouraging agreement has been found between these parameters. Microstrain experiments have been conducted at various stress levels and observed hysteresis behavior is in good agreement with measured stress wave energy released. The void contents have been controlled effectively by varying the dwell time, and subsequent interlaminar shear tests showed a significant correlation of measured strength with total stress wave energy released and its energy release rates.

1000 **H. C. Kim, A. P. Ripper Neto, and R. W. B. Stephens. "Reply by Kim, Ripper Neto, and Stephens" (to Stone and Dingwall).** *Nature Physical Science,* **NPSCA, 241:69-70. Jan 15, 1973.**

1001 **H. C. Kim, A. P. Ripper Neto, and R. W. B. Stephens. "Some Observations on Acoustic Emission during Continuous Tensile Cycling of a Carbon Fibre/Epoxy Composite."** *Nature Physical Science,* **NPSCA, 237:78-80. May 29, 1972.**

Reported are preliminary experiments on the acoustic emission from a composite when subjected to sequential cycles of tensile stressing. Two conditions of cycling were followed: (1) the maximum cycling stress was gradually increased with successive cycles, and (2) all cycles were of the same stress amplitude.

1002 **E. J. Kimble, Jr. "Development of Mobile Microseismic Field Equipment." pp 339-356 in** *Proceedings First Conference on Acoustic Emission/Microseismic Activity in Geologic Structures and Materials.* **H. R. Hardy, Jr. and F. W. Leighton, Editors. Trans Tech Publications, Clausthal, West Germany. 1977. conference held at Pennsylvania State University, University Park, Pennsylvania. June 9-11, 1975.**

The paper describes briefly the design and development of a mobile microseismic monitoring facility developed for use in a number of field studies under way by the Pennsylvania State Rock Mechanics Laboratory. The system allows for seven channel monitoring of field microseismic data. Data are recorded on magnetic tape and associated amplifiers, filters, as well as the tape recorder and other equipment are housed in a small van. Power for operation of the system is supplied by a 5 kW motor generator unit. Various difficulties encountered in development and use of the monitoring facility are discussed, as well as special problems associated with monitoring microseismic activity in areas of high background noise.

1003 **I. Kimpara and M. Takehana. "Analysis of Acoustic Emission from Internal Failure of Glass Fiber Reinforced Plastics Composites." Session 9, pp 2–20 in** ***The Second Acoustic Emission Symposium.*** **Japan Industrial Planning Association, Dobashi Building 5F, 3-17-8 Akasaka, Minato-ku, Tokyo, 107 Japan. 1974. proceedings of a symposium held in Tokyo, Japan. Sep 2–4, 1974.**

Glass-fiber reinforced plastics (GFRP) have now become to be used as some of the most successful man-made composite materials in many application fields. The GFRP have homogeneous or inhomogeneous anisotropy in strength and rigidity and have various modes of fracture strength. They exhibit different natures from those in conventional isotropic homogeneous materials. Developing suitable experimental methods to detect internal failure of composite materials during deformation would be important in order to evaluate internal structural integrity of composite materials such as GFRP. It has been well-known that GFRP emit clear and audible noise during deformation. This acoustic emission seems to involve much useful information on internal failure of GFRP. In this report, a tensile failure process of woven-roving laminates is discussed on the basis of observed AE properties. The Kaiser effect of AE under repeated loading is also discussed. Internal failure behavior at lower stress level around the first knee are analyzed. It is shown that the AE method requires a sensitive sensor to evaluate the failure mechanisms and the structural integrity of GFRP.

1004 **V. K. Kinra and H. Kolsky. "The Interaction between Bending Fractures and the Emitted Stress Waves."** ***Engineering Fracture Mechanics,*** **EFMEA, 9(2):423–432. 1977.**

Described are an experimental investigation of the interaction between the brittle fractures, which are observed in glass beams in flexure, and the stress waves generated by these fractures. A number of different experimental techniques have been employed. These are (a) The use of resistance strain gages to detect and record the waves generated; (b) High speed photography with a Kerr-Cell framing camera. The camera was also used to take high speed photoelastic patterns of the stress distributions set up during fracture; and (c) Direct measurements of the duration of the fracture process by measuring the interval between the breaking of two electrical contacts on opposite sides of a glass specimen. It was found that the fracture started at some point on the side of the specimen which was under tension, and rapidly traversed about 70% of the cross section. It then slowed down as it further penetrated the zone which was initially under compression. When the bar was sufficiently long, i.e., when there was no interference from reflected pulses, the crack was found to take a sharp turn and travel slowly (i.e., at velocities much lower than the wave velocities in glass) in a direction parallel to the axis of the beam. Fracture was finally completed when the compressive longitudinal pulses generated by the fracture returned from the ends of the beam as pulses of tension, and reached the fracture plane. This last stage could be lengthened either by using long rods or by suitably changing the acoustic conditions at the ends of a rod. It is shown that the shapes of emitted pulses which are observed some distance from the fracture plane can be adequately predicted by the assumption used earlier for tensile fractures, namely that as fracture progresses. The emitted wave corresponds to that expected from the relief of the stress present in the fracture cross section before fracture started. [reprinted by permission of Pergamon Press Ltd].

1005 N. Kirby and P. G. Bentley. "A Note on Acoustic Emission Measurements at REML." pp 507–524 in "Non-Destructive Testing for Reactor Core Components and Pressure Vessels." IAEA-145. International Atomic Energy Agency, Vienna, Austria. 1972. report of a panel meeting held in Vienna, Austria. Nov 29–Dec 3, 1971. [NTIS: CONF-711114]; IIW-IX-802-72 and IIW-X-672-72. International Institute of Welding. 1972.

The development of high sensitivity instrumentation for detecting and processing acoustic emission signals is outlined. Details are given of tests on fracture toughness test pieces of low strength steels and the benefits and limitations of such data are discussed. A summary of development work to establish location techniques is given.

1006 N. Kirby and P. G. Bentley. "A Note on Acoustic Emission Measurements from Low and Medium Strength Pressure Vessel Steels." *Australian Welding Journal,* **AUWJA, 17(3):23–28. May–June 1973; "Note sur les measures d'emission acoustique dans le cas d'acier de resistances ordinaire ou moyenne pour recipients a pression."** *Soudage et Techniques Connexes,* **SOTCA, 27(5/6):235–242. May–June 1973. (in French);** *Welding in the World/le Soudage dans le Monde,* **WDWRA, 11(1/2):11–20. 1973. (in English and French).**

A description is given of the following aspects of the UKAEA acoustic emission program: (1) development of high-sensitivity instrumentation for detecting and processing emission signals; (2) development of instrumentation and analysis for defect location; and (3) characterization of emissions and patterns of emission for low- and medium-strength pressure vessel steels.

1007 **N. Kirby and P. G. Bentley. "Pressure Vessel Inspection by Acoustic Emission."** *British Chemical Engineering and Process Technology,* **BCPTA, 16(10):912-913. Oct 1971.**

The principles of stress wave emission noise produced during deformation processes and their application to the inspection of pressure vessels for defects, are outlined. The method is compared with other nondestructive test methods. [Abstract 22-0165. *Metals Abstracts,* Vol. 5, 1972].

1008 **T. Kishi, Y. Obata, H. Tanaka, Y. Sakakibara, R. Horiuchi, and K. Aoki. "Acoustic Emission Peak under Cyclic Deformation."** *Nippon Kinzoku Gakkaishi (Journal of Japan Institute of Metals),* **NIKGA, 40(5):492-498. May 1976. (in Japanese).**

Acoustic emission behavior under cyclic deformation was studied with the use of pure aluminum. An acoustic emission peak accompanied by the Bauschinger effect was newly found, and named the *Bauschinger peak.* The characteristic and mechanism of this peak were as follows: (1) This peak was observed with the appearance of irreversible reverse plastic strain, and the peak height decreased with increase in the number of cycles and took a constant at the saturation stress; (2) The peak height and the reverse plastic strain at which a maximum of the peak was attained increased in the strain amplitude; (3) The peak height increased with increase in strain rate, but its strain rate dependence was not so simple as that of simple tensile deformation. This result shows, that there exists an increment of acoustic emission signals due to the Bauschinger effect; and (4) From the above results in addition to the observation of microstructure, it is likely that this peak is strongly related with the dislocations easily mobile in cells during stress reversal, with the aid of internal stress, which may annihilate each other in this region.

1009 **T. Kishi, S. Saito, Y. Mishima, and R. Horiuchi. "Acoustic Emission Signals of Ti Alloy and Maraging Steel during Plane Strain Fracture Toughness Testing." Paper 7-1, pp 455-474 in** *The Third Acoustic Emission Symposium.* **Japan Industrial Planning Association, Dobashi Building, 3-17-8, Akasaka, Minato-ku, Tokyo, Japan. 1976. proceedings of a symposium held in Tokyo, Japan. Sep 16-18, 1976.**

In recent years, the acoustic emission technique has been increasingly applied for detecting flaws and cracks in materials either under testing or in service. In fracture-toughness testing, fractography and electric resistivity methods are used to detect the initiation and propagation of slow crack growth. On the other hand, the acoustic emission has a possibility of becoming a more efficient method of detecting flaws, if the mechanical and electrical noises can be distinguished from the original emissions of crack propagation and the relationship between emissions due to plastic deformation and those due to crack growth can be made clearer. The purpose of the paper is to detect the initial point of slow crack growth and more importantly to study the microscopic mechanism of crack propagation during fracture toughness testing, by means of the acoustic emission technique.

1010 **T. Kishi, Y. Sakakibara, H. Tanaka, R. Horiuchi, and Y. Obata. "The Application of Acoustic Emission Technique to the Study of Strain Hardening and Fatigue Hardening." Paper 4–4, pp 305–325 in** ***The Third Acoustic Emission Symposium.*** **Japan Industrial Planning Association, Dobashi Building, 3–17–8, Akasaka, Minato-ku, Tokyo, Japan. 1976. proceedings of a symposium held Tokyo, Japan. Sep 16–18, 1976.**

In recent studies, of acoustic emission, there is a general agreement that acoustic emission during plastic deformation is related to dislocation motion. However, the relationship between acoustic emission and metallurgical factors in plasticity is not yet clear. Sedgwick and James and Carpenter have proposed a relationship between acoustic emission and mobile dislocation miltiplication, while Siegel, Gillis, and Hamstad, as an extended treatment, have applied the concept of inhomogeneous deformation to the behaviours of acoustic emission during plastic deformation. From these aspects, the authors studied the AE behaviours of Luders deformation and polycrystalline yielding and explained the results from the concepts of inhomogeneous deformation of Luders band and grain size, and as an advanced study of inhomogeneous deformation acoustic emission behaviour due to Bauschinger effect was studied, because backward deformation is controlled by the inhomogeneous deformation of the units of cell or substructure. Sanker has already measured the unloading emission, however, this acoustic emission behaviour has not been reported. Under these situations, a new AE peak accompanied by Bauschinger effect was found out already by the authors, and the characteristic of this peak was discussed in detail. In this paper, a new method to obtain a back stress component in flow stress, which controlls strain hardening and fatigue hardening, is proposed, by analysing this Bauschinger AE peak, and as an extended study, Kaiser effect and fatigue strength are discussed from the view point of obtained back stress.

1011 **T. Kishi, H. Tanaka, and R. Horiuchi. "Acoustic Emission Peak of Bauschinger Effect."** ***Scripta Metallurgica,*** **SCRMB, 9(10):1023–1026. Oct 1975.**

Tension and compression tests were used to obtain acoustic emission data from pure Al polycrystal specimens taken from extruded rods. Results show the existence of a new acoustic emission peak caused by the Bauschinger effect. The results of measurements of strain rate and amplitude effects on peak behaviour show that the peak is strongly influenced by the behaviour of dislocations which are easily mobile during stress reversal and which may annihilate each other in this region. [Abstract 32-0375. *Metals Abstracts,* Vol. 9, 1976].

1012 **P. J. Klass. "Non-Destructive Evaluation Effort Grows."** ***Aviation Week & Space Technology,*** **AWSTA, 104(7):59–65. Feb 16, 1976.**

1013 **M. V. Klassen-Neklyudova.** ***Mechanical Twinning of Crystals.*** **Consultants Bureau, New York. 1964. pp 5, 59, 62 (Figure 81), and 79. translated from Russian by J. E. S. Bradley. original Russian text published by the Press of the Academy of Sciences of the USSR for the Institute of Crystallography in Moscow, Russia. 1960.**

Mention is made to the sound produced by tin, *tin cry*, and zinc during twinning and elastic stretching. Reference is made to the work of Forster and Scheil (Reference 487), Mason, McSkimin, and Shockley (Reference 1139), and Thompson and Millard (Reference 1757) in their observations of acoustic emission.

1014 **R. A. Kline. "The Spectral Analysis of Acoustic Emission Signals." M. S. Engineering Thesis. Johns Hopkins University, Baltimore, Maryland. 1975.**

The ability to associate the frequency spectra of acoustic emission signals with the deformation process responsible for their generation may prove to be a valuable research and nondestructive testing technique. Towards this end, AE signals resulting from the tensile loading of 70-30 yellow brass and HF-1 steel were analyzed for frequency content and correlated with the mechanical behavior of the specimen. The results of these tests were used as the basis for a model of AE generation.

1015 **R. A. Kline and W. F. Hartman. "Frequency Analysis of Acoustic Emission Signals." pp 1631-1635 in** *Proceedings of the Second International Conference on Mechanical Behavior of Materials.* **American Society for Metals, Metals Park, Ohio. 1976. conference held in Boston, Massachusetts. Aug 16-20, 1976.**

In the past twenty years, acoustic emission techniques have been used to investigate a wide variety of material phenomena including slip, twinning, phase transformations, creep, crack nucleation and propagation, etc. Virtually all of these experimental programs relied upon the numerical counting of AE amplitudes or events (or both) as a measure of the extent of the acoustical activity present. Although the usefulness of these methods is beyond question, the fact remains that simple counting techniques are insensitive to the mechanisms of AE generation. Thus, despite the numerous investigations which have established AE as a viable NDT technique, the precise mechanisms producing stress waves in a deforming body remain unknown. In order to establish the relation between the deformation and the sensed AE event, a more complete knowledge of the nature of the acoustic emission signal is necessary. This requires a complete characterization of the rise time, amplitude distribution, energy density, decay characteristics, and frequency spectrum of the signals, as well as the rate of their occurrence and the initial amplitudes of the events which have been observed extensively. It is proposed that measurement of the frequency spectra of AE events is a sensitive technique of studying the relation between specimen deformation and the resultant AE. It provides basic information about the times and distances fundamental to the production of stress waves in a deforming material.

1016 **J. L. Knill, J. A. Franklin, and A. W. Malone. "A Study of Acoustic Emission from Stressed Rock."** *International Journal of Rock Mechanics and Mining Sciences,* **IJRMA, 5:87-121. 1968.**

Acoustic emission is generated in stressed materials as a result of internal deformation and may, therefore, be used to examine the failure mechanisms of materials within the laboratory and *in situ*. The literature relating to acoustic emission is reviewed, particular attention being given to experimental systems for the detection, recording and analysis of such emissions, the interpretation of the acoustic record, and the practical application of acoustic techniques in rock mechanics. An experimental acoustic system, developed for use both in the laboratory and the field, and the ancillary loading equipment are described. The methods of processing and analysing the acoustic record are illustrated by reference to a series of laboratory tests on a variety of rock samples. Various parameters are described, including pulse rate, signal energy content, and amplitude distribution; the variation of these parameters during uniaxial loading of rock specimens is described. [reprinted by permission of Pergamon Press Ltd].

1017 G. C. Knollman and J. L. Weaver. "Evaluation of an Acoustic Emission Monitor for On-Line Quality Control in Spotwelding of Electronic Components." Paper 6-1, pp 413-427 in ***The Third Acoustic Emission Symposium.*** **Japan Industrial Planning Association, Dobashi Building, 3-17-8, Akasaka, Minato-ku, Tokyo, Japan, 1976. proceedings of a symposium held in Tokyo, Japan. Sep 16-18, 1976.**

The widespread use of resistance welding for assembly of high-reliability components has generated the need for a method of monitoring the quality of small resistance welds. Failure of a single spotweld in a complex electronic subassembly can cause the failure of an expensive device or system. Reduction of failure rate by use of redundant spotwelds (as employed in structural components) is not economically feasible in the manufacture of electronic components. The quality of each weld should be monitored at the time it is made (on-line) to avoid costly salvage or subsequent failure. A simple approach to control of spotweld quality would involve monitoring of the weld current or the voltage drop across the weldment during the current pulse. While this method provides an indication of grossly abnormal weld conditions, it is not directly related to the flow and expulsion processes which determine weld strength. However, these processes generate sound waves which propagate from the weld and into the weld electrodes. This acoustic emission (largely ultrasonic) can be detected by a piezoelectric transducer and can be used to monitor the formation of a resistance weld provided that the weld emits sound waves that are detectable above the ambient noise. This paper describes the performance of a commercially available acoustic emission monitor as applied to quality control in spotwelding of electronic components.

1018 L. Knopoff and F. Gilbert. "First Motions from Seismic Sources." ***Bulletin of the Seismological Society of America,*** **BSSAA, 50(1):117-134. Jan 1960.**

An application of dynamic dislocation theory gives the elastodynamic radiation resulting from the sudden occurrence of an earthquake due to faulting. The fault plane is visualized as a geometrical discontinuity across which there exists a sudden discontinuity in either one component of the strain tensor or one component of the displacement vector. It is shown that there are eight independent models, if unilateral faulting is assumed; and an argument is presented to demonstrate the likelihood that unilateral faulting does not exist in nature. For bilateral faulting, the eight independent models are reduced in number to five. Of these five, two are more likely to occur in nature than the others. One of these, the displacement dislocation model, has a first-motion radiation pattern formally identical with that of a double couple in an unfaulted medium. The second, the shear-strain dislocation model, has a first-motion radiation pattern formally identical with that of an isolated force in an unfaulted medium. The latter type of mechanism may occur in deep-focus earthquakes. Another type of radiation, corresponding to the single couple in an unfaulted medium, results from the sudden release of shear strain in a laminar region.

1019 **R. M. Koerner and A. E. Lord, Jr. "Acoustic Emissions in Medium Plasticity Clayey Silt."** *ASCE Journal of the Soil Mechanics and Foundations Division*, **JSFEA, 98(SM-1):161–165. Jan 1972.**

The technique of acoustic emission monitoring of soil under stress appears to have potential. Utilization of such a technique may provide an alternative to measuring soil strains, which in the *in situ* state is difficult. This is evidenced by the stress-acoustic emission curves which show general similarity to stress-strain curves of the soil in conventional testing. It is possible that the technique may offer a non-destructive testing method for the monitoring of a wide variety of soils under stress.

1020 **R. M. Koerner and A. E. Lord, Jr. "Acoustic Emissions in Stressed Soil Samples."** *Journal of the Acoustical Society of America*, **JASMA, 56(6): 1924–1927. Dec 1974.**

Although acoustic emission techniques have been used in a wide variety of materials, scarcely any attention has been given to particulate systems such as soils. Yet, it is in such particulate materials where the concept of strain is fictitious and an alternate parameter is needed. The results of this study suggest that acoustic emission might be such a parameter. Included herein is a laboratory study of the acoustic-emission response of a series of soil samples at different water contents and under different stress states. The stress states are hydrostatic compression and triaxial shear compression tests. In each type of test, the soil samples are laterally confined by pressure which simulates the soil in its natural state. The results show that high confinement pressure and high water content both produce high levels of acoustic emission. The emissions begin instantaneously upon the application of the stress. In all cases, the emissions cease when the soil samples reach an equilibrium position under the applied stress state. Some insight as to the wide range of problems which might be solved using acoustic-emission techniques is also noted.

1021 **R. M. Koerner and A. E. Lord. "Acoustic Emission Monitoring of Earth Dam Stability."** *Water Power & Dam Construction,* **WAPOA, 28(4):45–49. Apr 1976.**

While it has long been recognized that earth masses generate audible sounds during deformation, little practical use has been made of this phenomenon. Such sounds are most apparent immediately prior to and during failure. The paper explains the techniques used for the monitoring of soil masses, and discusses some case histories.

1022 **R. M. Koerner, A. E. Lord, [Jr.], and J. N. Deisher. "Acoustic Emission Detection of Underground Gasoline Storage Tank Leaks." summary of paper in** *Paper Summaries: ASNT Spring Conference, March 28–30, 1977, Phoenix, Arizona.* **American Society for Nondestructive Testing, Columbus, Ohio. 1977.**

Many of the 4,000,000 underground gasoline tanks in service in the country are losing various amounts of their product through leaks. This leads to economic losses and wastage of an increasingly scarce resource. It also has serious environmental consequences, such as ground water pollution. An AE technique has been devised as a competitor to the Kent-Moore Tank Systems Tightness Tester. It has been laboratory tested and is ready for field testing and evaluation. Design of the leak detection system, its sensitivity level, and the relationship between AE-like activity and leak size will be discussed.

1023 **R. M. Koerner, A. E. Lord, Jr., and J. N. Deisher. "Acoustic Emission Stress and Leak Monitoring to Prevent Spills from Buried Pipelines." Drexel University, Philadelphia, Pennsylvania. 1976; paper presented at the 1976 National Conference on Control of Hazardous Material Spills, sponsored by the U.S. Environmental Protection Agency and the Oil Spill Control Association of America, held in New Orleans, Louisiana. Apr 25–28, 1976.**

Presented in this paper is a series of laboratory test results of pipelines under stress and containing leaks using acoustic emission monitoring techniques. Such acoustic emissions are generated by the pipeline when it goes into a plastic state in any one of a number of stress modes, or when it develops a leak of as little as 0.013 inches in diameter. In the latter series of tests air, water, and oil were used under pressure up to 200 psi. In addition to an explanation of the procedure used in acoustic emission monitoring, and the equipment which is required, the paper also contains a section on recommendations for further research.

1024 **R. M. Koerner, A. E. Lord, Jr., and W. M. McCabe. "Acoustic Emission Studies of Soil Masses in the Laboratory and Field." pp 243–256 in** *Proceedings First Conference on Acoustic Emission/Microseismic Activity in Geologic Structures and Materials.* **H. R. Hardy, Jr., and F. W. Leighton, Editors. Trans Tech Publications, Clausthal, West Germany. 1977. conference held at Pennsylvania State University, University Park, Pennsylvania. June 9–11, 1975.**

Utilizing the laboratory studies described earlier by Lord *et al.* (see Reference 1098), it was decided that field sites could be economically monitored as long as the internal emissions from the soil mass were brought to the ground surface by metal rod waveguides. A study was conducted which showed that the emissions could be transmitted for considerable distances using reinforcing rods, instrument pipes, bailing wire, etc. Using these waveguides, six sites have been, or are being, monitored at the present time. On a qualitative basis we have found the following: (1) Sites where no emissions are recorded are not deforming; (2) Sites where relatively low emission levels are recorded are deforming; and (3) Sites where high emission levels are recorded are unsafe. The ultimate goal of the project is quantification of acoustic emissions to the imposed stress level and in this regard, at least one of these sites has given excellent agreement of soil emissions to ground settlement.

1025 J. Kolar. "Raziskave akusticne emisije pri utrujanju snovi (Research of Acoustic Emission due to Fatigue of a Material)." Diplomsko delo st. 2256. Fakulteta za strojnistvo, Univerza v Ljubljani, Ljubljana, Yugoslavia. 1975. (in Slovenian).

1026 H. Kolsky. "The Stress Pulses Propagated as a Result of the Rapid Growth of Brittle Fracture." *Engineering Fracture Mechanics,* **EFMEA, 5(3):513–522. Sep 1973.**

Stress pulses are generated whenever a brittle fracture occurs, and this paper first describes the phenomena observed when a Hertzian fracture is produced by the impact of a spherical indentor on a block of glass. The block is sufficiently large to be regarded as a half space, and the wave propagated over the surface of the glass is diverging from the region of impact and consists of a combination of dilatation, distortional, and Rayleigh surface waves. At large distances of travel, the last of these predominates since it diverges cylindrically, while the others are diverging spherically. Satisfactory agreement was observed between the wave shapes predicted theoretically and those observed experimentally. The next problem was concerned with the fracture of glass rods in simple tension, here the fracture was initiated at a scratch on the surface of the rod and diverged cylindrically across the rod at high velocity from the point of initiation. The growth of the fracture results in the propagation of two pulses, one longitudinal in nature and the other flexural. The shape of these pulses was measured by means of strain gages fixed to opposite sides of the surface of the rod parallel to the diameter which contained the point of initiation. In order to account for the shapes of these pulses theoretically, some assumption had to be made about the stress distribution over the cross section during the fracture process. The stress on the fractured portion clearly vanished, and it was assumed that since the fracture growth was so rapid it would be reasonable to assume that the stress in the unfractured portion retained the value it had before fracture commenced. On this assumption the shapes of the longitudinal pulse and the flexural pulse could be predicted, and from these the records of the strain pulses to be expected from the two strain gages could be determined. It was assumed that, the longitudinal pulse was propagated without change in form, while the propagation of the flexural pulse was calculated by the use of the Timoshenko theory. Excellent agreement between theory and experiment was obtained. Experiments were next carried out on the fracture of thin rods in flexure, and it was found that the observed pulses had much longer durations. The type of treatment used for extensional fracture could here clearly not account for the whole phenomenon, since it would appear to predict that fracture started at the surface, ran to the neutral axis, and then stayed there permanently. In fact it was found that with short specimens, the fracture rapidly approached the neutral axis and stayed in the neighborhood of this axis until an unloading wave arrived from the end of the rod. As longer and longer rods were employed the lengths of the pulses increased, and where long rods were used, it could be seen that the stress was continuing to decrease even in the absence of reflected pulses from the ends. The program of experimental and theoretical work on the problem is continuing. [reprinted by permission of Pergamon Press Ltd].

1027 H. Kolsky. "Wave Propagation Effects and Fracture." pp 199–214 in *Mechanical Properties at High Rates of Strain,* J. Harding, Editor. Conference Series No. 21. Institute of Physics, London, England. 1974. proceedings of a conference held in Oxford, England. Apr 1974. [reprint of paper available from Professor H. Kolsky, Division of Applied Mathematics, Brown University, Providence, Rhode Island 02912].

The first part of this paper reviews the research that has been carried out during the last hundred years on the fractures produced by the propagation, reflection, and diffraction of stress pulses of large amplitude; in particular the phenomena of *spall* fractures and the fractures produced by the reinforcements of reflected pulses are considered. Two other problems related to wave propagation, namely the terminal speed of running cracks and the *dynamic* tensile strength of brittle solids are next considered. The second half of the paper discusses some recent investigations carried out in the author's laboratory and elsewhere on the stress pulses generated when brittle fractures occur. Three distinct problems are considered. These are (a) the stress pulses generated in Hertzian impact experiments when fractures are produced by the impact of steel balls on glass blocks; (b) the stress pulses set up when a rod is broken in a simple tensile test; and (c) the stress pulses set up by the fractures produced in flexure experiments on glass rods and bars. The information that these results give on the fracture process itself is then discussed. [reprinted by permission of The Institute of Physics, copyright 1974].

1028 E. G. Konovalov and V. S. Galkov. "Acoustic Emission Recording as a Means of Investigating Solid-State Welding Processes." ***Welding Production*, WEPRA, 19(6):82–85. June 1972; translated from *Svarochnoe Proizvodstvo,* SVAPA, (6):51–52. June 1972. (in Russian).**

Described are the acoustic emission testing techniques and results obtained during solid-state microwelding of gold and aluminum wires 40 micrometres in diameter to silicon coated with 1 micrometre of aluminum or nickel. Frequency spectra generated during the welding process are presented for investigations in the frequency range 0.02 to 30 MHz. Experimental data indicate that the microwelding techniques that provide the highest joint quality are associated with acoustic emission spectra that include increased high-frequency peaks.

1029 **E. G. Konovalov, V. S. Galkov, and N. S. Ivanov. "Acoustic Emission in the Cutting of Metals."** *Doklady Akademii Nauk Belorusskoi SSR,* **DBLRA, 15(9):796–798. Sep 1971. (in Russian).**

The spectra of the acoustic emission (3 to 30 MHz) during the cutting of foils (up to 100 micrometres) of Al, Cu, Ni, Nb, Ti, and W were recorded.

1030 **E. G. Konovalov and E. N. Rudenok. "Deformation Stages of Material and Carrier Frequency of Acoustic Emission."** *Doklady Akademii Nauk Belorusskoi SSR,* **DBLRA, 18(7):616–617. 1974. (in Russian).**

Reported are the results of an investigation to study the variation in acoustic emission frequency from copper and iron strained in tension. Monitoring the frequency of vibration gives a measure of deformation and an indication of the yield stress.

1031 **E. G. Konovalov and E. N. Rudenok. "Stressed State of Material and Acoustic Emission."** *Doklady Akademii Nauk Belorusskoi SSR,* **DBLRA, 17(4):323–324. Apr 1973. (in Russian).**

The mechanism underlying the spontaneous generation of sound from metallic and nonmetallic materials subjected to concentrated stress is discussed in the light of some special experiments made with copper and Armco iron samples. The acoustic emission began shortly after the onset of loading, but in the early stages it was difficult to measure owing to the distorting influence of the clamps in the stress-producing apparatus. With increasing degree of deformation the energy of the sound emission also increased. However, after reaching the yield point of the material it passed through the maximum and decreased gradually. These results were in agreement with existing data. The generation was attributed to crack and pore formation. [Abstract 32-0909. *Metals Abstracts,* Vol. 6, 1973].

1032 **A. G. Konstantinova. "The Connection between the Energy of Elastic Pulses Generated in the Destruction of Solids and the Stress and Dimensions of the Ruptures."** *Bulletin of the Academy of Sciences of the USSR, Geophysics Series,* **BASRA, 2:135–137. May 1962; translated from** *Izvestiya Akademii Nauk SSSR, Seriya Geofizicheskaya,* **IAGFA, 2:194–198. Feb 1962. (in Russian).**

Griffith's theory of fracture is used in a discussion of the energy emitted in the form of stress wave emission from rock specimens designed to fracture along their length. Conventional instrumentation is used, the pulses being recorded on magnetic tape, the relative energy of the pulses being calculated from the oscillogram traces produced after replaying the magnetic tape recording into an oscilloscope.

1033 **A. G. Konstantinova. "The Shape of Elastic Pulses Accompanying Rock Breaking."** *Bulletin of the Academy of Sciences of the USSR, Geophysics Series,* **BASRA, pp 421–426. 1959; translated from** *Izvestiya Akademii Nauk SSSR, Seriya Geofizicheskaya,* **IAGFA, 1959. (in Russian).**

Stress pulses generated by samples of rock under uniaxial compressive loading are displayed using conventional electrical instrumentation in conjunction with accelerometers for pulse detection. Discusses attempt to damp out interfering pulses in the experimental set-up.

1034 **A. G. Konstantinova. "Time Distribution of Elastic Pulse Energy during the Destruction of Rock."** *Bulletin of the Academy of Sciences of the USSR, Geophysics Series,* **BASRA, pp 1056–1061. 1960; translated from** *Izvestiya Akademii Nauk SSSR, Seriya Geofizicheskaya,* **IAGFA, 1960. (in Russian).**

The paper presents an analysis of the time distribution of elastic pulse energy to reveal certain aspects of the fracturing processes during failure of rock specimens under uniaxial compressive loading.

1035 **J. L. Koury. "Solid Strand Burn Rate Technique for Predicting Fullscale Motor Performance." AFRPL–TR–73–49. Air Force Rocket Propulsion Laboratory/MKMB, Edwards AFB, California. Oct 1973.**

The degree of success or failure for meeting the space and ballistic mission requirements of rocket-launched vehicles depends greatly on the ability to accurately predict propulsion performance. Solid rocket motor-performance predictions are possible by establishing a correlation between full-scale motor performance, and small ballistic test motors, liquid strand burn-rate tests or solid strand burn-rate tests (or both). The accuracy of these predictions will be maintained if the correlation between motor performance and any of the test techniques remains constant. On two operational motors, a shift in the correlation between liquid strand burn rate (LSBR) and large motors was noticed by SAMSO and the cause of the shift was suspected to be related to some subtle change in the propellant raw material, perhaps the ammonium perchlorate oxidizer. An investigation was initiated to re-establish the correlation between full-scale motor performance and liquid strand burn-rate tests and to determine the cause of the shift. The AFRPL approach toward establishing this correlation was to perform solid strand burn-rate tests on propellant taken from cartons which were cast during motor manufacture. This paper describes the results of the AFRPL solid strand burn-rate tests on Titan III propellant. During the program, an improved monitoring system to detect the combustion events of solid propellant ignition and burnout was developed and demonstrated. The paper includes discussion of: (1) Solid strand burn-rate test method; (2) Acoustic emission detection system; (3) Burn rate variability of solid strands; and (4) Correlation between full-scale motor performance, and small test motors, and liquid and solid strand burn-rate tests. The results of this effort determined the applicability of using the solid strand burn-rate technique for predicting full-scale motor performance for current and future Air Force systems. The cost savings associated with using strand burn rate-test techniques instead of small motor tests are discussed.

1036 **Y. Krampfner, A. Kawamoto, K. Ono, and A. T. Green. "Acoustic Emission Characteristics of Copper Alloys under Low-Cycle Fatigue Conditions." NASA CR-134766, UCLA-ENG-7529. University of California, School of Engineering and Applied Science, Los Angeles, California, and Acoustic Emission Technology Corporation, Sacramento, California. Apr 1975. [NTIS: N75-27137].**

The acoustic emission characteristics of pure copper, zirconium-copper, NARloy-Z, Glid Cop Al 10, and NASA 1-1 have been determined in order to aid the development of nondestructive evaluation schemes of thrust chambers via AE techniques. The AE counts, rms voltages, frequency spectrum, and amplitude distribution analysis were employed to evaluate the AE behavior under fatigue loading conditions. The results of AE testing were interpreted with the detailed evaluation of wave forms and crack propagation characteristics, as well as scanning electron fractographs of fatigue tested samples. In the cold-worked condition, crack propagation in NARloy-Z produces AE signals that can be utilized for its detection. In other alloys, with a possible exception of NASA 1-1, AE signals are too weak to reliably detect the propagation of a crack. A sample of annealed alloys produces continuous type AE signals at the beginning of a fatigue test. However, as the sample work-hardens, the AE behavior becomes similar to that of a cold-worked sample. When a sample of zirconium containing alloys is annealed repeatedly after each fatigue loading cycle, numerous surface cracks are produced during the subsequent fatigue cycle, emitting strong burst type AE signals. While frequency spectrum analysis of the AE signals does not readily identify the type of AE signals or their origins, amplitude distribution analysis exhibits responses that are characteristic of certain types of AE signals. The latter can be incorporated in nondestructive evaluation schemes of thrust chambers using AE techniques.

1037 **N. Krauland and K. Westerberg. "Experiences with AE-Measurement in Boliden Mines." pp 195-203 in *Proceedings First Conference on Acoustic Emission/Microseismic Activity in Geologic Structures and Materials.* H. R. Hardy, Jr., and F. W. Leighton, Editors. Trans Tech Publications, Clausthal, West Germany. 1977. conference held at Pennsylvania State University, University Park, Pennsylvania. June 9-11, 1975.**

Measurement of acoustic emission was applied to underground mining in some of the Boliden Company's underground mines. The work was conducted under the auspices of the Swedish Mining Research Association. The line of approach was to use the

AE noise rate as a measure of the intensity of fracture processes in the rock masses, indicating a possible impending instability of a stope. A short review is given of instrumentation developed. Attempts are being made to more clearly define the physical nature of acoustic emission pulses with a view of distinguishing pulses generated by different sources as, e.g., drilling, machinery, ventilation, etc., and hopefully rock fracturing. A further point of interest is the transmission characteristics of acoustic emission through rock (attenuation as a function of frequency). Requirements for an efficient AE measuring system are considered. These are outlined both with regard to the necessary knowledge of the physical nature of AE as well as with regard to the aspects of measuring systems to be used as practical tools in mining.

1038 **K. G. Kreider, Rapporteur. "Rapporteur's Report, Session 7: Testing of Composite Materials." pp 967–992 in** ***Proceedings of the 1975 International Conference on Composite Materials. Volume 2.*** **Edited by E. Scala, E. Anderson, I. Toth, and B. R. Noton. Metallurgical Society of AIME, New York. 1976. conference cosponsored by the Society for the Advancement of Material and Process Engineering, Society of the Plastics Industry, Society of Plastics Engineers, Institute of Aeronautics and Astronautics, and American Society for Metals, held in Geneva, Switzerland. Apr 7–11, 1975 and in Boston Massachusetts. Apr 14–18, 1975.**

A review is given of thirteen papers submitted for presentation in Session 7: *Testing of Composite Materials.* The author (rapporteur) has briefly summarized each paper and has included several illustrations from each paper. The conclusions of each contributing author have been included together with the rapporteur's comments. For the complete text of the two papers concerning acoustic emission see References 602 and 644.

1039 **R. J. Kroll. "Stress Waves in Test Specimens due to Simulated Acoustic Emission." Ph.D. Thesis. Michigan State University, East Lansing, Michigan. 1962; University Microfilms 63–1740.**

When a test specimen is loaded, acoustic emissions occur at various strain levels. These acoustic emissions create small stress waves in the test specimen. These stress waves are detected using a piezoelectric crystal at the end of the test specimen. However, the signal from the piezoelectric crystal is complicated and difficult to interpret. The purpose of the investigation is to determine an optimum system design, which consists of the specimen, the electronic instrumentation, and the experimental technique, for the study of acoustic emissions. Natural acoustic emissions create extremely small amplitude stress waves. Therefore, in this study the acoustic emission is simulated by burying a small piezoelectric crystal in an enlarged version of the test specimen. The simulated acoustic emission is similar in character to the natural emission, but its amplitude is much larger. The nature of the stress wave propagation through the specimen is determined analytically and verified experimentally. The knowledge of the stress wave propagation allows an optimum specimen design to be postulated. An experimental investigation verifies that this specimen design cleans up the signal from the crystal. The optimum specimen design is used in a system in which the electronic instrumentation and the experimental techniques are varied. Several different systems are found which produce an acceptable output signal for acoustic emission research. This output consists of a relatively large amplitude and short duration pulse, followed by lower amplitude *hash.* A recommendation of an optimum system design for the study of acoustic emissions is given. Analytical and experimental studies are conducted to establish the validity of using a simulated acoustic emission sourse and an enlarged test specimen.

1040 **R. J. Kroll and C. A. Tatro. "Stress-Wave Propagation in Axiallysymmetric Test Specimens."** *Experimental Mechanics,* **EXMCA, 4(5):129–134. May 1964; paper presented at the Society for Experimental Stress Analysis Annual Meeting, Boston, Massachusetts. Nov 6–8, 1963.**

When a body is stressed, dislocations tend to pile up against an obstacle. If they suddenly break through the obstacle, then a small stress wave, known as an acoustic emission, is generated. Acoustic-emission research is attempting to correlate this phenomenon to the onset of yielding of a tensile-test specimen. The reliable detection of the acoustic emissions depend on the proper design of the system consisting of the test specimen, the instrumentation and the experimental techniques. In turn, the design must be based on a fundamental knowledge of the stress-wave propagation through the specimen. In this paper, a theoretical and experimental investigation of the stress-wave propagation through a test specimen due to a stress-wave source, similar to an acoustic emission, is presented. This stress-wave propagation is shown to be typical for more general types of excitation of any similar type of an axiallysymmetric body. The method of producing the stress-wave source is described. The instrumentation system for detection of the stress wave must contain a variable electronic filter to clean up the signal and eliminate noise. Several different size and differently proportioned test specimens are used to determine scale effects and to optimize the specimen design. Stress waves are also mechanically introduced into separate components of the test specimen to confirm the theoretical predictions for the modes of stress-wave propagation through these components. The theoretical predictions of the stress-wave propagation through the entire test specimen are verified by the experimental investigations. An optimum design of the entire system is presented.

1041 **B. A. Kuhn. "Acoustic Emission Monitoring of Ring-Cracking in Glass and Alumina." UCID-16076. Lawrence Livermore Laboratory, Livermore, California. July 7, 1972. [NTIS].**

Experiments were conducted to determine whether acoustic emission monitoring of ring cracking in glass and alumina is feasible. Pyrex 7740 was selected for the preliminary investigation to allow a visual correlation to be made, along with the acoustic emission data. The test results on the glass showed an excellent correlation between the visual ring-crack propagation and the rapid increase in the acoustic emission rate.

Tests were also conducted on alumina. The acoustic emission data from the alumina provided the same type of rapid increase in acoustic emission activity, at the onset of ring cracking, that was observed in the glass. A microscopic inspection of the fracture surface on the alumina showed that a ring crack did form, providing sufficient evidence that acoustic emission could be used to detect ring cracking in an opaque material.

1042 A. Kumar. "Application of Acoustic Emission to Weld Monitoring." Interim Report M-117. Construction Engineering Research Laboratory, Champaign, Illinois. Dec 1974. [NTIS: AD/A-003-992].

This report discusses the feasibility of applying acoustic emission to real-time monitoring of gas metal arc welding. A state-of-the-art survey indicated that investigators had some success with detecting cracks, porosity, and slag inclusion during the welding of steels by monitoring acoustic emission. The presence of cracks was most easily detected in gas tungsten arc welding of stainless steels. Slag inclusions associated with roped bead in submerged arc welding can also be detected by acoustic emission. In the present work, cracks were introduced in a restrained weldment by hydrogen additions to the shielding gas. Selective filtering was used to increase the signal-to-noise ratio. Results show that the acoustic emission rate from cracks is much higher than that from the welding arc noise. A real-time weld monitor for detection of cracks in steel is therefore technically feasible. However, when porosity was introduced, the acoustic emission rate was about the same order as that from the welding arc noise. Thus, porosity could not be detected by the acoustic emission in a real-time weld monitoring situation.

1043 K. Kuribayashi, H. Tanaka, T. Kishi, R. Horiuchi, and A. Kato. "The Fundamental Study of Acoustic Emission in the Plastic Deformation of Metals and and Alloys." Paper 4-2, pp 271-287 in *The Third Acoustic Emission Symposium.* Japan Industrial Planning Association, Dobashi Building, 3-17-8, Akasaka, Minato-ku, Tokyo, Japan. 1976. proceedings of a symposium held in Tokyo, Japan. Sep 16-18, 1976.

An acoustic emission is a transient elastic wave generated by the rapid release of an elastic energy within a material. This definition is broad enough to contain the various events such as a final catastrophic fracture of a structure or the microscopic diffusion of one atom by one lattice point. Most AE technology and its applications have been developed mainly in the field of the nondestructive testing, including the location of the AE source. Fundamental studies of the plastic deformation by means of AE utilizing its dynamic character have not so developed compared with the numerous number of the experiments of the location of AE source or the applications as the nondestructive testing. Indeed about the mechanism of the generation of AE the correspondence between the burst-type of AE and the discontinuity of the stress-strain relation proposed by Gillis and Hamstad gives fairly good explanation to AE of the inhomogeneous deformation such as the serration and the development of the Luders band. Multiplication of mobile dislocations and the unlocking of dislocations from the obstacles can provide rather sophisticated interpretation to the plausible mechanism of the generation of AE. The singularity of AE affected with the various metallurgical factors, such as the concentration of solute atoms, temperature, and the stacking fault energy, can not thoroughly be explained by these mechanisms. Then in the present paper, the dynamic behaviors of AE in various oriented single crystals of fcc pure Al and Al-Mg solid solution were precisely discussed. The effect of the various metallurgical factors which affect the AE behavior was also developed from the point of the absolute magnitude of AE and the distribution of the AE amplitudes.

1044 **F. N. Kusenberger, J. Lankford, Jr., P. H. Francis, and J. R. Barton." Nondestructive Evaluation of Metal Fatigue." AFOSR-TR-72-1167. SwRI Project No. 15-2800. Southwest Research Institute, San Antonio, Texas. Apr 1972. [NTIS: AD-743-312].**

The important role of nondestructive evaluation methods in complementing the tools of fracture mechanics for preventing catastrophic fatigue failures in engineering structures is discussed. Magnetic perturbation signatures from surface and subsurface fatigue cracks in AISI 4340 steel rod-type axially loaded specimens are illustrated, compared, and considered in terms of the plastic zone and residual stresses in the vicinity of the fatigue cracks. Also, magnetic records from drilled holes and fatigue cracks in 4340 and maraging steels are presented which show significant differences in response from the two materials. Observations concerning fatigue crack initiation events in the vicinity of nonmetallic inclusions in 4340 steel using electron replication microscopy are described. A model for crack initiation in high strength steel is proposed, and the significance of the initiation events on the earlier detection of fatigue damage is also discussed. An analytical approach for predicting the stresses and strains, effects of interfacial debonding, etc., in the vicinity of surface and near surface inclusions are described, as well as the status of development indicated. Finally, preliminary acoustic emission investigations on rod-type specimens containing surface fatigue cracks are reviewed and the results compared with those from single edge-notched flat specimens.

1045 **J. Kvapil. V. Kment, and C. Barta. "Application of Acoustic Thermal Analysis to a Corundum Single Crystal."** ***Growth of Crystals,*** **GRCRA, 5B:157-160. 1963; translated from** ***Rost Kristallov,*** **RKANA, 5B:157-160. 1963. (in Russian). proceedings of the 3rd Moscow (Russia) Conference.**

Solid-state theory shows that the thermal modes of a lattice are closely related to acoustic waves. It has been shown that even sudden lattice changes (caused by melting, reaction, or change of allotropic form) can give rise to acoustic waves; in these cases use was made of the decrepitation method employed in mineralogy to find the temperature of formation of hydrothermal minerals. The specimen is heated in an electric oven, and the sound produced by rupture of microcavities is transmitted via a sound guide to a microphone. These sounds are relatively loud, and there is no difficulty in recording them. Difficulties appear when the method is applied to changes in the structure itself; the sounds are weak, and high gain is necessary to record them, which leads to amplification of noise from the surroundings, and forces one to resort to acoustic insulation of the apparatus. The transfer via the air is as good as in the decrepitation method, probably because the frequencies are not too high. The method has been applied to single crystals. As heating is used, this procedure is termed as acoustic thermal analysis; it has yielded interesting results with some crystalline materials of low hardness, so testing was done on corundum single crystals.

L

1046 **R. J. Landy. "Acoustic Emission: Application to Fracture Mechanics." B.S. Thesis. University of Queensland, Department of Mechanical Engineering, St. Lucia, Queensland, Australia. Oct 1976.**

The use of fracture mechanics enables a critical defect size in a structure to be specified at a given operating stress provided that the fracture toughness and yield properties of the material are known. Conversely, for a given defect size it is possible to predict the failure stress. The principal objective of acoustic emission monitoring is to relate the acoustic emission characteristics to the deformation ahead of a crack and in this way enable one of the variables, defect size or applied stress, to be replaced by acoustic emission data. It has been common practice to relate the total count recorded to stress intensity factor which is a function of flaw size, and the stress in the structure and flaw location with respect to the principal stresses. The acoustic emission variable used is not always total count and in particular, the workers in this field use the count rate when investigating fatigue problems. These fatigue tests not only include strictly controlled experiments under the laboratory conditions, but also actual on site monitoring of pressure vessels, etc. The nature of acoustic emission signals is advantageous in the location of flaws in a test structure. In the tests completed in service, not only has the extent of crack growth been evaluated and hence the structure's condition in relation to failure in service assessed, but the location of the most critical flaw has been achieved with some accuracy. The basic principles and equipment are given in Chapter 1. The results of experiments undertaken in plane stress, plane strain, and elevated temperature conditions are discussed in the following chapters. Application of acoustic emission testing to actual structures is analyzed and the problems that have arisen are discussed in relation to possible solutions. Included in the last chapter, problems with the results and experimental set up are presented and in conclusion, recommendations and areas for future research are listed.

1047 **E. A. La Roche, Jr. "Acoustical Methods for Predicting Thermal Shock Resistance." M. S. Thesis. Clemson University, Clemson, South Carolina. 1970.**

It was desired to develop a reliable test for evaluation of the thermal shock resistance of refractory bricks. Such a test should predict the relative thermal shock resistance of refractories and allow engineers to choose the best refractory for a specific application. Determination of the most severe thermal conditions which can be endured by a refractory should be provided. The application of sonic testing methods to evaluating thermal shock resistance was studied. The technique of listening to bricks cracking during thermal shock was improved to the point of practical application. The effects of cooling rates, temperature gradients, and cyclic heating and cooling were successfully evaluated. The relative thermal shock resistance of several refractory compositions was reliably determined. A decaying amplitude sonic test was used to measure internal damping within refractory bricks. The internal damping appeared to be related to the structural integrity of the refractories. Tests performed on a series of refractories indicated that thermal shock damage in the bricks was accompanied by an increase in internal damping. It appeared that the decaying amplitude sonic test could be developed into a reliable method of evaluating relative thermal shock resistance.

1048 G. Latham, M. Ewing, J. Dorman, D. Lammlein, F. Press, N. Toksoz, G. Sutton, F. Duennebier, and Y. Nakamura. "Moonquakes." ***Science,*** **SCIEA, 174(4010):687–692. Nov 12, 1971.**

Although the average rate of seismic energy release within the moon appears to be far below that of the earth, over 100 events believed to be moonquakes have been recorded by the two seismic stations installed on the lunar surface during Apollo Missions 12 and 14. With few exceptions, the moonquakes occur at monthly intervals near times of perigee and apogee and show correlations with the longer-term (7-month) lunar gravity variations. The repeating moonquakes are believed to occur at not less than 10 different locations. However, a single focal zone accounts for 80 percent of the total seismic energy detected. This active zone appears to be 600 kilometres south-southwest of the Apollo 12 and 14 sites and deep within the moon. Each focal zone must be small (less than 10 kilometres in linear dimension) and fixed in location over a 14-month period. Cumulative strain at each location is inferred. Thus, the moonquakes appear to be releasing internal strain of unknown origin, the release being triggered by tidal stresses. [Copyright 1971 by the American Association for the Advancement of Science].

1049 T. Lauhoff (to Interatom Internationale Atomreaktorbau GmbH, Germany). "Elektroakustischer Wandler mit einem magnetostriktiven Element (Electroacoustic Transducer with a Magnetostrictive Element)." German Patent 2,333,482. Jan 16, 1975. (in German).

An electroacoustic converter, and in particular a receiver for body and liquid sound waves, is described. With small dimensions, it is simple and robust, can be used at high temperatures, and also registers small values of acoustic pressure.

1050 P. A. Laura, H. Vanderveldt, and P. Gaffney. "Acoustic Detection of Structural Failure of Mechanical Cables." ***Journal of the Acoustical Society of America,*** **JASMA, 45(3):791–793. Mar 1969.**

The letter describes an investigation of the sounds emitted when cables are loaded to failure. Detectable stress waves are emitted well before reaching the failure load, with individual wires failing first, followed by the failure of a strand. This process continues until ultimately the entire cable fails.

1051 **P. A. Laura, H. H. Vanderveldt, and P. G. Gaffney. "Mechanical Behavior of Stranded Wire Rope and Feasibility of Detection of Cable Failure."** *Marine Technology Society Journal*, **MTSJB, 4(3):19–32. May–June 1970.**

Experimental results on detection of wire-rope failure by means of monitoring stress wave emissions in the cable system are reported. In addition, data are presented as a result of experimentation conducted on a variety of steel wire ropes. The specimens, varying in length, material, core, and geometry were subjected to a tensile load until total failure occurred. Mechanical properties, such as load versus elongation, load versus change in diameter, initial breaking strength, reserve strength, modified Poisson's ratio, and modified modulus of elasticity are presented.

1052 **C. E. Lautzenheiser, A. S. Greer, Jr., W. D. Jolly, and S. P. Ying. "Evaluation of Heavy Section Vessels Using Acoustic Techniques." Paper C117/74, pp 146–152 in** *Periodic Inspection of Pressurized Components.* **Conference Publication 8, 1974. published by Mechanical Engineering Publications Limited for the Institution of Mechanical Engineers, London, England. 1975. conference held in London, England. June 4–6, 1974.**

During the past five years, Southwest Research Insitute has maintained an active research and development program designed to meet the continuing need to improve inspection techniques. The Institute has been especially active in the areas of acoustic emission, acoustic holography, acoustic spectrometry, and advanced ultrasonics. Practical results of this research are presented along with some discussion on application of NDT technology to nuclear power systems.

1053 **F. J. Lavoie. "Nondestructive Testing."** *Machine Design,* **MADEA, 41:121–135. Sep 4, 1969.**

Comprehensive article on NDT. Included is a section on acoustic emission in which the author discusses the activities at Battelle-Northwest and Aerojet-General.

1054 **F. V. Lawrence, Jr., E. P. Cox, and E. M. Honig, Jr. "The Effect of Lack of Penetration Defects on Static Tensile Behavior of High-Strength Structural Steel Weldments." CERL-IR-M-123. Construction Engineering Research Laboratory, Champaign, Illinois and University of Illinois, Department of Civil Engineering, Urbana, Illinois. Feb 1975. [NTIS: AD/A-007-132].**

The paper investigates the effect of lack of penetration (LOP) in welded T-1 steel plates on tensile properties. Tensile tests of welded plates having varying sizes of LOP were conducted and monitored for acoustic emissions. Fracture surfaces were examined. Using a documented computer program, the J-integral of fracture mechanics was evaluated at several points for each specimen. An anticipated decrease in tensile strength and ductility with increasing LOP was observed. The fracture surface was found to pass in and out of the weld metal, the extent depending on LOP. The J-integral analysis showed that welded T-1 steel joints with LOP are well approximated by the rigid-plastic model. This permits a critical value of the J-integral to be computed from the load and deflection at peak load. There appears to be a critical LOP value below which the J-integral is not valid. Moreover, the fracture occurs in the base metal for LOP less than one critical value, and in the weld metal for LOP greater than a second critical value higher than the first. For intermediate LOP values, failure occurs in the weld metal, but gross elongation occurs in the base metal. By using base metal yield and tensile strength, a net section criterion, and a limit load stress, four modes of fracture behavior can be identified. These modes indicate three ranges of influence of LOP size on fracture. For LOP size less than a lower critical value, base metal dominates fracture. For LOP size greater than a larger critical value, LOP size controls fracture. For intermediate LOP size, the LOP dominates but does not fully control fracture. These results express quantitative relationships between LOP defect size and tensile properties. The range of allowable LOP defect sizes for which tensile strength may not be seriously affected is thus established.

1055 **A. M. Lazarev, I. V. Gulevskii, and L. Yu. Odnopozov. "Use of Standard Equipment for Fatigue Testing by the Acoustic-Emission Method."** *Soviet Journal of Nondestructive Testing,* **SJNTA, 11(3):383–385. May–June 1975; translated from** *Defektoskopiya,* **DEFKA, (3):133–135. May–June 1975. (in Russian).**

The principal difficulties in using the acoustic emission method for fatigue testing stem from noise interference, against the background of which it is difficult and often impossible to detect the AE signals. An experiment was conducted on standard testing machinery. In the testing of samples on a machine built by the Leningrad Shipbuilding Institute (LKI-1), findings show that the machine has a high intrinsic noise level, against the background of which it is impossible to detect AE signals. It was shown by observation of the noise signals on a cathode-ray oscilloscope and measurement of the noise level that the machine noise is a stationary random process. This fact enabled use of a spectral analyzer of the SK4-13 type.

1056 **H. H. Le. "An Electrical Model for a Transducer Used in the Study of Acoustic Emission from Materials under Tensile Stress." M.Sc. Thesis. University of Manitoba, Electrical Engineering Department, Winnipeg. Manitoba, Canada. May 1972.**

The thesis describes a method for modeling a transducer used in the study of acoustic emission from materials under tensile stress. The method involves the measurement of the input impedance of the transducer over a wide range of frequencies from which one derives the electrical elements of the model. The model is then used in the electrical analysis of the system. As an illustrative example, the present technique is applied to determine the electrical model of a commercial transducer used in the study of acoustic emission from Fe 3% Si specimens. The model together with the model for the preamplifier used in the actual system makes possible the computation of the impulse response which relates the output waveform and the output statistical properties to the input waveform and the input statistical properties.

1057 **J. B. Lean and J. Plateau. "Observation des ondes sonores prenant naissance au cours de la deformation plastique des metaux, en relation avec le probleme de la rupture fragile de l'acier doux (Observations of Sound Waves Originating during Plastic Deformation of Metals in Relation to the Problem of Brittle Fracture in Mild Steel)."** *Memoires Scientifiques de la Revue de Metallurgie,* **MRMTA, 56(1):91–99. June 1959. (in French).**

Sound effects during tensile testing at different temperatures (room, −196 °C and 200 °C) of mild carbon steel (0.07%), 4% Si steel, and Duralumin were recorded and correlated with formation of Piobert-Luders lines and mechanism of rupture. [abstract reprinted from *Engineering Index,* p 1326, 1959].

1058 **J. B. Lean, J. Plateau, C. Bachet, and C. Crussard. "Sur la formation d'ondes sonores, au cours d'essais de traction, dans des eprouvettes metalliques (On Formation of Sound Waves during Traction Tests in Metal Specimens)."** *Comptes Rendus Hebdomadaires des Seances de l'Academie des Sciences,* **COREA, 246:2845–2848. May 19, 1958. (in French); UCRL-Trans-905(L). translated for Lawrence Radiation Laboratory, Livermore, California. Jan 1963.**

Tension-tests were carried out on specimens of extra-mild steel at temperatures ranging from +200 to −196 °C. Sound waves are emitted during formation of Luders lines or waves of plastic deformation corresponding to Portevin-LeChatelier effect. Brittle fracture at low temperature, without appreciable elongation, is preceded by identical sound waves. [Abstract 975-Q. *Metal Abstracts,* 1958].

1059 **J. B. Lean, J. Plateau, and C. Crussard. "Contribution a l'etude du mecanisme de la rupture fragile de l'acier doux (Study of Mechanism of Brittle Fracture of Mild Steel)."** *Memoires Scientifiques de la Revue de Metallurgie,* **MRMTA, 56(4):427–452. Sep 1959. (in French).**

Tensile impact tests on smooth and notched bars at 150 to −210 °C and with rates of load application of 1.4×10^{-5} to 3.8×10^{-1} per sec; statistical analysis of results; study of formation of Luders-Piobert bands by sonic methods, and modern theory lead to hypothesis on mechanism of brittle fracture. [abstract reprinted from *Engineering Index,* p 1326, 1959].

1060 **L. S. S. Lee and F. L. Sharp. "Correlations of Acoustic Emission and Fracture Mechanics of Pressure Cycled Pipes Containing Longitudinal Surface Flaws." ASME Publication 74-PVP-22. American Society of Mechanical Engineers, New York. 1974; paper presented at the American Society of Mechanical Engineers Pressure Vessels and Piping Conference, Miami Beach, Florida. June 24–28, 1974.**

Pressure cycling tests on 304 stainless steel tubing containing simulated longitudinal surface cracks were conducted in the acoustic emission test program. Two different patterns of the summation of emission during the period of fatigue test were observed; one was approximately linearly increasing while the other approximated a step increase at some particular fraction of the fatigue life. The step increase in emission coincided with a marked increase in the crack-opening displacement and is believed to be associated with yielding at the crack root and/or subsequent crack extension. Acoustic emission is correlated with loading parameters using fracture mechanics. Two types of fracture mechanics formulation are considered; namely, the stress intensity factor K_I and the effective stress σ^* approach. The occurrence of a step increase in emission in the second pattern mentioned above can be associated with the attainment of particular values of stress intensity factor K_{IAE} or effective stress σ^*_{AE}. The first pattern arises when the initial loading values of K_I or σ^* are equal or greater than these threshold values. The effective stress threshold σ^*_{AE} lies between the elastic limit strength and the 0.2% offset yield strength. An example problem is given to illustrate the application of the concept of the acoustic emission stress threshold, K_{IAE} or σ^*_{AE}, for setting up a nondestructive inspection scheme.

1061 **S. S. Lee and J. H. Williams, Jr. "Acoustic Emission during Unloading of Elastically Stressed Magnesium Alloy." NASA CR-2840. Massachusetts Institute of Technology, Cambridge, Massachusetts. June 1977.**

A magnesium alloy was quasistatically cycled elastically between zero load and tension. Both loading and unloading stress delays were found and the unloading stress delay was further studied. An analytical expression was written for the unloading stress delay which is an elastic constitutive parameter. The potential use of these results for the acoustic emission monitoring of elastic stress states is discussed.

1062 **F. Leighton and W. Blake. "Rock Noise Source Location Techniques." Report of Investigations 7432. U.S. Bureau of Mines, Washington D.C. Sep 1970.**

The Bureau of Mines has developed rock noise source-location techniques that may be used to obtain precise and practical source locations from microseismic field data. These techniques enable one to monitor all or any part of a rock structure and map progressive stress buildup or rock movement. Thus, the microseismic method of determining instability in rock structures is becoming an invaluable engineering tool for the mining or civil engineer engaged in rock excavation and construction.

1063 **F. Leighton and W. I. Duvall. "A Least Squares Method for Improving the Source Location of Rock Noise." Report of Investigations 7626. U.S. Bureau of Mines, Washington D.C. 1972.**

The least squares method of rock noise source location presented is shown to be more accurate and reliable than standard direct solution methods. This method is particularly effective in improving solutions from rock noise locations outside of geophone array boundaries. A least squares solution for a medium in which the seismic velocity is the same in all directions is also presented. In this special case, one can solve for both the source coordinates and the velocity.

1064 F. W. Leighton and B. J. Steblay. "Applications of Microseismics in Coal Mines." pp 205–229 in *Proceedings First Conference on Acoustic Emission/Microseismic Activity in Geologic Structures and Materials.* H. R. Hardy, Jr., and F. W. Leighton, Editors. Trans Tech Publications, Clausthal, West Germany. 1977. conference held at Pennsylvania State University, University Park, Pennsylvania. June 9–11, 1975.

The paper describes the instrumentation, techniques, and results of current microseismic investigations by the Bureau of Mines in coal mines. Rock noises were monitored in a deep western Colorado coal mine for the purpose of studying the phenomena of coal-mine bumps and bounces (seismic energy releases large enough to be felt and heard by personnel or to damage the structure). Data preceding several bounces of varying magnitude were recorded and analyzed. Techniques were developed to locate and delineate bounce areas within the mine structure and to estimate the stability of individual suspect areas. These data indicate that microseismic techniques may be used to establish where and when coal-mine bounces will occur. Microseismic equipment, sensitive to frequencies in the ultrasonic range, was also used to monitor a controlled roof-fall experiment at the Bureau's experimental mine at Bruceton, Pennsylvania. Results of this experiment indicate that this equipment may be ideal for use in an early warning system to predict roof falls.

1065 M. G. Le Levier. "Utilisation des Techniques Acoustiques pour l'Inspection et le Controle des Reacteurs (Acoustic Techniques in Reactors Surveillance)." CEA-BIB-205. Centre d'Etudes Nucleaires de Saclay, Gif-sur-Yvette, France. Oct 1973. (in French). [NTIS].

An annotated bibliography of literature on acoustic, mainly ultrasonic techniques used in the inspection and evaluation of nuclear reactors. Included in the 125 references are 14 on acoustic emission and 9 on boiling detection. The bibliography is arranged in five sections: reviews, reports, conferences, patents, and standards, and is cross-referenced with author index and subject index.

1066 W. C. Leschek (to Westinghouse Electric Corporation). "Acoustic Emission Transducer and Monitoring System." U.S. Patent 3,855,847. Dec 24, 1974.

An improved acoustic emission transducer and transmission system for the on-line monitoring of acoustic signals generated by a metal when under mechanical stress. The transducer structure includes a tubular housing having a continuous side wall, a bottom wall, and an open top. A piezoceramic element is fastened to the bottom of the housing by means of a conductive high temperature adhesive. A positive electrode is fixed to the piezoceramic element by the adhesive, and both the piezoceramic element and the positive electrode are cushioned and insulated from the side wall of the housing by means of an inner tubular housing of high temperature insulating material. The inner housing assembly is sealed to enclose the piezoceramic element, adhesive, and electrode—the sealant preventing moisture from penetrating the assembly and shorting the crystal. The upper end of the positive electrode passes through the sealant and is connected to an electrical cable which transmits any signal picked up by the transducer to a remote location. The cable is attached directly to a transformer which is used to couple maximum power from the transducer to an output transmission line. The transformer is selected to match the impedance of the transducer to the transmission line.

1067 **W. C. Leschek. "Acoustic Emission Transducer Calibrator."** ***Materials Evaluation,*** **MAEVA, 33(2):41–48. Feb 1975.**

A method for calibrating an acoustic emission transducer is described. Calibration is performed by comparing the output signal of the acoustic emission transducer against that of a primary standard, while both transducers are acoustically loaded by a medium which is representative of a large metal structure, such as a nuclear reactor pressure vessel wall, and are exposed to the same diffuse random noise sound field. The primary standard transducer is calibrated independently, using a reciprocity technique in a gated continuous-wave sound field.

1068 **W. C. Leschek. "Severe Environment Acoustic Emission Detection Apparatus." Westinghouse Research Laboratories, Pittsburgh, Pennsylvania. 1975.**

A severe environment acoustic emission detection apparatus has been developed for use on operating commercial nuclear reactors. Its function is to detect acoustic emissions (elastic stress waves) as they strike the outer surface of the primary pressure boundary, and to relay this information to a remotely located computerized source location and monitoring system. The apparatus features a severe environment acoustic-emission transducer and utilizes a matched and balanced transducer-cable-amplifier subsystem. This subsystem permits the piezoelectric sensor's low level electrical output signal to travel with relatively little attenuation or interference over a long distance without the aid of a preamplifier, thereby allowing all active electronics to be removed from the immediate reactor environment which only the transducer and cable need survive.

1069 **W. C. Leschek (to Westinghouse Electric Corporation). "Transmetteur d'emission acoustique et systeme d'ecoute pour controle (An Acoustic Emission Transmitter and Controlled Listening System)." French Patent 2,189,720. Jan 25, 1974. (in French).**

1070 W. C. Leschek and P. E. Carpenter (to Westinghouse Electric Corporation). "Transducteur electro-acoustique perfectionne (Improved Electro-Acoustic Transducer)." French Patent 2,262,469. Sep 19, 1975. (in French).

The device comprises a tabular metal enclosure, a window and a piezoelectric member, an elongated damping member in contact with said piezoelectric member, an insulating pad against the damping member and, therefore, against the piezoelectric member, and electrical conductors connected to one of the surfaces of said piezoelectric member and to said damping member, respectively. The device can be used, in particular, for inspecting the defects, and flaws being formed in the wall of pressurized containers, e.g., nuclear reactors.

1071 G. I. Leskov, E. N. Trunov, and L. I. Zhivaga. "Acoustic Emission Transducers for the Inspection of Weld Quality." *Welding Production,* **WEPRO, 21(7): 66–68. July 1974; translated from** *Svarochnoe Proizvodstvo,* **SVAPA, (7):38–39. July 1974. (in Russian).**

Article describes the construction of both a piezoelectric and an electromagnetic transducer for use in the detection of cracking during and after welding by acoustic emission.

1072 R. G. L'Hermite. "Volume Changes of Concrete." pp 659–694 in *Chemistry of Cement: Proceedings of the 4th International Symposium, Washington 1960.* **National Bureau of Standards Monograph 43, Volume 2. U.S. Government Printing Office, Washington, D.C. Sep 27, 1962.**

The report presents an up-to-date account of the present state of knowledge on the deformation of concrete. The following phenomena have been examined; hygrometric, thermal, chemical, and dynamic phenomena (action of forces) in relation to the nature of cement, of aggregates, and the composition of the concrete. Included is a brief description of the use of acoustic emission to detect plastic deformation and cracking in concrete under stress. Placed on a fairly slender prismatic test specimen is a microphone connected to an amplifier and a device capable of recording the sound characteristics produced in the concrete under the action of loading. Also set up is an apparatus for measuring the speed of shock pulses. As the specimen is loaded to a load T_1 the frequency, intensity, and speed of sound emission is fairly constant. Above T_1 the frequency and intensity of emission increase to the breaking point of the specimen, while the speed of sound drops beyond T_1. The emissions are indications of internal ruptures that disturb the continuity of the structure. Their length is a few millimeters and their thickness a few microns. The increase in the load augments the density of this cracking and enlarges the cracks, which, as they join, lead to rupture. The report has appended to it a specially prepared bibliography containing 230 references.

1073 **R. L'Hermite. "What Do We Know About the Plastic Deformation and Creep of Concrete?"** *RILEM (Reunion Internationale des Laboratoires d'Essais et de Recherches sur les Constructions) Bulletin,* **RILBA, (1):21–51. Mar 1959.**

The author has analysed, from his abundant documentation, a hundred French and foreign publications which have appeared during the past forty years. As a result of his own considerable work during the past twenty years, still in progress, he has succeeded in crystallizing an easily understood synthesis on the present state of the important and difficult problem of the plastic deformation and creep of concrete. The author's conclusions are of direct use by engineers or builders, although a systematic study of the problem may not be possible. The effects of the composition of the concrete, of the method of conservation, of the volume under stress, the age of the concrete, the importance and duration of the loading, and unloading are also considered. The creep under different stress conditions is also examined. The creep is an hydro-constrictive phenomenon dependent upon the movements of the water; it is covariant with the shrinkage. The state of deformation tends towards a value proportional to the stress with a minimum corresponding to the shrinkage. Testing and measuring apparatus are described and numerous experimental results are presented which give to this publication an essentially reliable character. A theory is proposed, after dealing with other theories and is presented in an impartial and objective manner, rather to satisfy the demands of those engaged in research and in experiment than to give a theoretical conclusion on the problem.

1074 **S. T. Li and W. J. McGonnagle. "Some Potentials and Frontiers of Nondestructive Testing and Evaluation, Part 1."** *International Journal of Nondestructive Testing,* **IJNTA, 4(3):215–229. Dec 1972.**

Article discusses the potentials and frontiers of acoustical and ultrasonic techniques currently being used and under development for nondestructive testing. Acoustic emission testing is one of the subjects covered. Other ultrasonic techniques discussed include spectroscopic analysis, imaging, critical angle reflectivity, and holography.

1075 **S. T. Li, V. Ramakrishnan, and J. E. Russell. "Advances in Nondestructive Testing of Concrete." pp 1–11 in** *Highway Research Record, No. 378, Nondestructive Testing of Concrete.* **Highway Research Board, National Academy of Sciences, Washington, D.C. 1972. paper presented at the 50th Annual Meeting of the Highway Research Board, Washington, D.C. Jan 18–22, 1971.**

The paper reviews recent advances in the following nondestructive methods of testing concrete: ultrasonic, resonance, radioactive, electrical, initial surface absorption, chemical analysis, and hardness. A brief review is given of the research work in progress at the South Dakota School of Mines and Technology (Rapid City). There is a great potential in the concrete industry for nondestructive methods of testing concrete. Two major fields where those methods could prove to be superior to traditional methods are quality control in the construction of structural members, both precast and cast-in-place, and monitoring strength development to determine acceptable times for the removal of form work or transfer of prestressing force to concrete.

1076 S. T. Li, V. Ramakrishnan, and J. E. Russell. "Where Stands Nondestructive Testing of Concrete and Whither?" *International Journal of Nondestructive Testing*, **IJNTA, 2:281–300. 1970.**

This article presents an up-to-date status of nondestructive testing of concrete and indicates the promising direction in which it may evolve in the near future. A brief description on acoustic emission is included.

1077 T. C. Lindley, I. G. Palmer, and C. E. Richards. "Acoustic Emission Monitoring of Fatigue Crack Growth." Laboratory Note No. RD/L/N 134/76. Central Electricity Research Laboratories, Leatherhead, Surrey, England. Feb 1977.

The acoustic emission technique has been used to monitor fatigue crack growth. Emissions occurring near the peak load of the fatigue cycle were separated from those occurring near the minimum load by gating techniques. The peak load emissions were attributed to both crack extension and to deformation and fracture events occurring within the crack-tip plastic zone, and these emissions have been correlated with the energy released during crack growth. The effect of material and testing variables on these emissions was also examined. The minimum load emissions were attributed to crack closure processes; they could be important in crack closure studies and in detecting fatigue cracks in materials *quiet* at peak load. Single overloads resulted in fatigue crack growth retardation, accompanied by a sharp fall in peak and minimum load acoustic emission. This could be useful for the practical application of acoustic emission monitoring to fatigue situations.

1078 P. G. Lingenfelder. "Acoustic Emission Monitoring of Graphite/Epoxy Composite Structures." Final Report. Lockheed Missiles and Space Company, Inc., Sunnyvale, California. Aug 1974.

Acoustic emission monitoring was proposed as a factory acceptance test for the Trident-I composite equipment section in order to gain greater confidence in the strength of the equipment section and warn of failures which would occur below the proof-load level. The report describes the development work which was performed in attempting to achieve that goal. A survey was made of the acoustic emission services available commercially and the equipment available for purchase. With this knowledge of the commercial state-of-the-art, the testing and acoustic emission monitoring of composite parts began covering graphite/epoxy tape and cloth structures both subscale and fullscale. An acoustic emission monitoring technique was developed on graphite/epoxy tape structures to predict failure within 5% and to locate the weak areas and the probable failure site. However, on graphite/epoxy cloth structures the test results to date indicate that a correlation between acoustic emission and failure load cannot be established. As a result acoustic emission monitoring is not recommended for cloth structures, but is highly recommended for tape structure where knowledge of the failure load and initiation site are required.

1079 **N. Lions, J. Baron, H. Buis, R. Berger, and A. Bret. "Dispositifs de visualisation par ultra-sons dans le sodium (visus) et de detection acoustique dans le Reacteur Phenix (Ultrasonic Visualization in Sodium System (Visus) and Acoustic Detection in the Fast Reactor Phenix)." CEA-CONF-2647. CEA Centre d'Etudes Nucleaires de Cadarache, 13-Saint-Paul-les-Durance, France and Societe d'Etudes Techniques et d'Entreprises Generales (SODETEG), 92-Plessis-Robinson, France. 1973. (in French); paper presented at the Congress of the Societe des Electriciens des Electroniciens et des Radioelectriciens, Vittel, France. Sep 11-15, 1973; ANL-Trans-1052 (CONF-730983-36). translated for Argonne National Laboratory, Argonne, Illinois. June 1975. [NTIS].**

Two acoustic instruments, developed at the fast neutron department of the CEN/ Cadarache will be presented. They are in use as part of the instrumentation of the reactor Phenix. The first one is of the active type, the second of the passive type. The active instrument operates like a sonar and visualises objects which are in the reactor vessel, in particular the fuel-assembly beads during handling operations. The passive instrument, the acoustic detector, supervises the reactor core by analysing the acoustic noise produced in the reactor core. The electro-acoustic converters are, in both cases, situated outside the reactor. The acoustic vibrations are conducted by means of waveguides. Both instruments are designed to operate in a hostile medium, high temperature liquid metal, high neutron flux, and strong gamma field.

1080 **R. G. Liptai. "A Proposal to Initiate a Program in Acoustic Emission from Frictional and Fracture Processes in Rocks." Lawrence Radiation Laboratory, Livermore, California. Oct 1969.**

1081 **R. G. Liptai. "Acoustic Emission from Composite Materials." UCRL-72657. Lawrence Radiation Laboratory, Livermore, California. Feb 3, 1971. [NTIS: UCRL-72657, CONF-710403-1]; pp 285-298 in *Composite Materials: Testing and Design, ASTM STP 497.* American Society for Testing and Materials, Philadelphia, Pennsylvania. 1972. proceedings of the 2nd ASTM Symposium on Composite Materials, held in Anaheim, California. Apr 20-22, 1971.**

A review of the application of acoustic emission analysis to the evaluation of material properties and defect structure is presented, including a brief discussion of experimental techniques. Sources of acoustic emission in composite materials under strain are discussed. Particular attention is directed toward fiber-reinforced composites; and experimental results on filament-wound NOL rings and vessels (glass-epoxy) are reported. It is concluded that acoustic emission data are helpful in establishing the mechanisms governing operative fracture modes and assessing structural integrity.

1082 **R. G. Liptai. "An Investigation of the Acoustic Emission Phenomenon." Ph.D. Thesis. Michigan State University, East Lansing, Michigan. 1963; University Microfilms 64–4986.**

Acoustic emissions are the sounds produced during the straining of a metal. The present effort was directed toward establishing the mechanisms governing the emission source. Oriented single crystal specimens of 99.99+% aluminum were used. A special device for silent loading was necessary since the sound level associated with the emission process is low. Strains up to 2% were produced. An ADP piezoelectric crystal placed on the end of the specimen detected the emission pulses, and the frequency range between 2 and 20 kilocycles was monitored. A complete description of the electronic detection and analysis systems is given. It is concluded that the micromechanisms associated with the emission source are the same as those controlling slip band formation. The pulses seem to originate as a result of dislocation avalanche at the surface. These breakaways dissipate strain energy by producing new surface, heat, and elastic stress waves. Electron micrographs of oxide replicas support the hypothesis that emissions pulses result from slip-band formation. The micrographs show that a physical interruption of the oxide coating occurs due to plastic deformation. Anodic coatings were found to be effective in controlling slip-band formation, acting as barriers to dislocation motion. The thickness of these coatings was the quantity which appeared to be the most appropriate in analyzing emission data. Thicker films, of the order of 1.0×10^{-3} inches, greatly increased emission activity. Also, films of this magnitude shifted the most active period of emission to higher strain values. It is shown that a semilogarithmic relationship exists between oxide thickness and average strain.

1083 **R. G. Liptai, H. L. Dunegan, and C. A. Tatro. "Acoustic Emission Generated during Phase Transformations in Metals and Alloys." UCRL–50525. Lawrence Radiation Laboratory, Livermore, California. Sep 6, 1968. [NTIS]; *International Journal of Nondestructive Testing*, IJNTA, 1:213–221. 1969; pp 183–191 in *Physics and Nondestructive Testing*. W. J. McGonnagle, Editor. Volume 2. Gordon and Breach Science Publishers, New York. 1971.**

Phase transformations in metals and alloys usually are grouped in one of two general categories: nucleation-and-growth, or martensitic (shear) transformations. The choice depends on the atomic mechanism involved in the transformation. Unlike the nucleation-and-growth process, which involves thermal activation and diffusion, martensitic transformations are diffusionless reactions characterized by cooperative movements of atoms through small distances relative to neighboring atoms. Acoustic emissions are the stress waves or lattice vibrations that are generated during these shear-like reactions. Emission analysis has been used for material evaluation and for characterization of martensitic transformations. Experiments were conducted with gold-47.5 atomic % cadium, indium-22 at % thallium, cobalt, and plutonium.

The use of emission analysis and the experimental results will be discussed. Metallurgical processes other than martensitic phase transformations can generate acoustic emissions. Emission analysis of these processes for material evaluation will also be discussed. An electrosonic technique is used to detect acoustic emission in the frequency range between 90 to 200 kHz. The electronic equipment (for example, lead-zirconate-titanate sensing transducer, amplification, and recording apparatus) and the experimental techniques will be described.

1084 **R. G. Liptai and D. O. Harris. "Acoustic Emission – An Introductory Review."** *Materials Research and Standards*, **MTRSA, 11(3):8–10,44. Mar 1971.**

A brief introduction to the field of acoustic emission is presented. The basic mechanisms of acoustic emission, a brief historical survey of the field, and some of its applications to materials research and structural integrity evaluation are discussed. Areas of future development also are presented.

1085 **R. G. Liptai, D. O Harris, R. B. Engle, and C. A. Tatro. "Acoustic Emission Techniques in Materials Research." UCRL-72582. Lawrence Radiation Laboratory, Livermore, California. July 9, 1970. [NTIS: N71-15064]; paper presented at the Symposium on Advanced Experimental Techniques in the Mechanics of Materials, San Antonio, Texas. Sep 9–11, 1970;** *International Journal of Nondestructive Testing*, **IJNTA, 3:215–275. Dec 1971; pp 163–223 in** *Advanced Experimental Techniques in the Mechanics of Materials*. **P. H. Francis and U. S. Lindholm, Editors. Gordon and Breach Science Publishers, New York. 1973.**

A review of the application of emission analysis to evaluate materials properties and defect structure is presented. Topics discussed include fracture toughness and crack propagation, fatigue, plastic deformation, and creep processes in metals, composites, and rock materials. The status of emission techniques as applied to the evaluation of structural integrity is reported. A complete discussion of experimental techniques and data acquisition and processing systems is given. The authors conclude that acoustic emission techniques have wide applicability to experimental studies in materials research and to evaluation analysis of structural integrity. Directions of future developments and applications are discussed.

1086 **R. G. Liptai, D. O. Harris, and C. A. Tatro. "An Introduction to Acoustic Emission." UCRL-73209. Lawrence Radiation Laboratory, Livermore, California. May 27, 1971. [NTIS]; pp 3–10 in** *Acoustic Emission, ASTM STP 505*. **American Society for Testing and Materials, Philadelphia, Pennsylvania. 1972. proceedings of symposium held in Bal Harbour, Florida. Dec 7–8, 1971.**

A brief introduction to the field of acoustic emission is presented. The basic mechanisms of acoustic emission and a brief historical survey of the technology, along with some of its application to materials research and structural evaluation programs, are discussed.

1087 **R. G. Liptai, D. O. Harris, and C. A. Tatro. "Foreword." UCRL-73676. Lawrence Livermore Laboratory, Livermore, California. 1971.**

This is the *Foreward* for *Acoustic Emission, ASTM STP 505.* (see Reference 1907).

1088 **R. G. Liptai and C. A. Tatro. "Acoustic Emission: A Surface Phenomenon." pp 287-341 in** *Proceedings of the 4th Annual Symposium on Nondestructive Testing of Aircraft and Missile Components.* **Southwest Research Institute, San Antonio, Texas. Feb 1963.**

Acoustic emission is concerned with the emission of sounds, the generation and propagation of which are a result of the straining and deformation of a metal. The level of sound associated with acoustic emission is low, at least an order of magnitude smaller than that produced by the stroking of a specimen with the end of a frayed cotton thread. Sounds such as *tin cry*, reorientation of grain boundaries, metallurgical transformations as that of martensite, Portevin-LeChatelier saw tooth slip, and others have been observed and studied for years; however, these may also be considered as sources of acoustic vibrations. From these studies it is apparent that sounds have been noted in metal deformation for many years; however, the specific nature of the sounds associated with basic deformation processes has been in a state of uncertainty. The authors opine that acoustic emissions are basic in nature and related to the basic deformation processes, that is, related to slip. In this work, aluminum was chosen for a specimen material because of its copious emission behavior and its ability to be surface-coated with various types of oxide surfaces. Orientated single crystal specimens were used and a critical examination of surface conditions was conducted using light and electron microscopy. Theoretical considerations such as microcrack formation, slip-band formation, slip at surface barriers, and dislocation pile up at surface barriers are given with conclusions based on a combination of experimental and theoretical deliberations. Suggested topics of acoustic emission investigation and application are also provided.

1089 **R. G. Liptai and C. A. Tatro. "Acoustic-Emission Inspection." pp 234-243 in** *Metals Handbook, 8th Edition, Volume 11: Nondestructive Inspection and Quality Control.* **American Society for Metals, Metals Park, Ohio. 1976.**

1090 **R. G. Liptai and C. A. Tatro. "Acoustic Emission Techniques in Nondestructive Testing." UCRL-74549. Lawrence Livermore Laboratroy, Livermore,**

California. Feb 6, 1973. [NTIS]; Paper E-11 presented at the Seventh International Conference on Nondestructive Testing, Warsaw, Poland. June 4–8, 1973.

Acoustic emissions are defined as transient elastic waves that are generated by the rapid release of energy within a material as a result of the application of external energy sources. Common mechanisms for the generation of acoustic emission activity include plastic deformation, crack initiation, and crack propagation. In one sense, acoustic emission techniques for nondestructive testing and evaluation are different from those usually associated with the field. The basis for most conventional NDT methods is interpretation of the interaction of a defect or discontinuity with some sort of energy (X-rays, ultrasonics, thermal energy, microwaves). In acoustic emission, the defect or discontinuity itself provides a source of energy, the release of which generates the acoustic emission. In this sense, the defect plays an active role in preference to the passive role it usually plays during the evaluation by conventional NDT methods. A review is given of the application of emission analysis to evaluate materials properties and defect structure. Emphasis is placed on a discussion of emission techniques as applied to the evaluation of structural integrity of materials, components, and structures. The characteristics of acoustic emission techniques as a unique NDT method are presented. Examples of the current status of the application of acoustic emission for NDT are reported and future applications and developments are discussed.

1091 D. J. Lloyd and K. Tangri. "Acoustic Emission from Al_2O_3-Mo Fibre Composites." *Journal of Materials Science,* **JMTSA, 9(3):482–486. Mar 1974;** *Proceedings of the British Ceramic Society,* **PBRCA, (25): 169–178. 1975.**

Acoustic emission has been used to investigate the fracture process in Al_2O_3 and an Al_2O_3-Mo fiber composite. The acoustic emission data indicate extensive microfracturing at loads considerably less than that required to cause fracture in Al_2O_3. The acoustic emission emanating from the composite in the early stages of fracture is similar to that observed in Al_2O_3. However as final fracture is approached, the acoustic emission profiles differ considerably. The acoustic emission profiles are discussed in terms of the fracture mechanisms. [reprinted by permission of Chapman & Hall Ltd].

1092 C. S. Lockman and A. T. Green. "Accelerometer Techniques as Applied on Polaris First Stage A3 Production Chamber Hydrotests." Final Report on AGC Task No. 75–466 and Lockheed Missile and Space Division Task No. 7. Aerojet-General Corporation, Sacramento, California. July 15, 1964.

The objective of the task was to evaluate the accelerometer data acquired on the first-stage Polaris A3 filament-wound production chamber hydrotests, and to correlate the results with the data obtained from the experimental chamber hydrotests. Using an accelerometer technique developed during the hydrotests of first-stage Polaris A3 experimental chambers, data recordings were made during hydrostatic testing of production chambers. Analysis of the hydrotest accelerometer recordings show a relationship between the acceleration impulses emanating from the filament-wound chamber during hydrotest and the structural integrity of the chamber after the hydrotest.

1093 **D. Lockner and J. D. Byerlee. "Acoustic Emission and Fault Location in Rocks." pp 99–107 in** ***Proceedings First Conference on Acoustic Emission/ Microseismic Activity in Geologic Structures and Materials.*** **H. R. Hardy, Jr. and F. W. Leighton, Editors. Trans Tech Publications, Clausthal, West Germany. 1977. conference held at Pennsylvania State University, University Park, Pennsylvania. June 9–11, 1975.**

1094 **A. E. Lord, Jr. "Acoustic Emission." Chapter 6, pp 289–353 in** ***Physical Acoustics: Principles and Methods.*** **W. P. Mason and R. N. Thurston, Editors. Volume XI. Academic Press, New York. 1975.**

This is a review article which covers the acoustic emission work in metals and rocks and more specifically in soils, wood, and magnetic processes. The article includes a historical review of microseismic studies in mines and acoustic emission in metals. An extensive bibliography is included.

1095 **A. E. Lord, Jr., J. N. Deisher, and R. M. Koerner. "Attenuation of Elastic Waves in Pipelines as Applied to Acoustic Emission Leak Detection." summary of paper in** ***Paper Summaries: ASNT Spring Conference, March 28–30, 1977, Phoenix, Arizona.*** **American Society for Nondestructive Testing, Columbus, Ohio. 1977;** ***Materials Evaluation,*** **MAEVA, 35(11):49–54, 60. Nov 1977.**

The American pipeline and terminal system is the largest in the world and is currently experiencing a large expansion effort. One of the major reasons for the expansion is the lack of visibility of this mode of transporation. While this is desirable for a number of social, political, and economic reasons, it does present a problem for surveillance and monitoring insofar as leak detection is concerned. It appears that acoustic emission monitoring can fill this need. Design studies of an AE monitoring system, capable of detecting a pipeline leak and locating its source, will be described. Attenuation of the AE signal, choice of wave propagation mode to be detected, background noise, sensor frequency, and the leak source location procedure will be discussed.

1096 **A. E. Lord, Jr., and R. M. Koerner. "Acoustic Emissions in Soils and their Use in Assessing Earth Dam Stability.** ***Journal of the Acoustical Society of America,*** **JASMA, 57(2):516–519. Feb 1975.**

Although considerable effort has been expanded into acoustic-emission research and related applications, scarcely any has been in the direction of particulate materials. One such material is soils where numerous earthen structures present challenging problems insofar as their stability and structural integrity are concerned. Such structures involving soils in whole or in part are earth dams, embankments, retaining walls, etc., where both fundamental and field problems exist. This paper is a brief review into our studies on the acoustic-emission response of soils. The majority of the work has been on laboratory-size soil samples, although some large-scale model work and field instrumentation are also reported. Parallel studies as to the fundamentals of the emissions in soils are also noted. Thus far it appears that acoustic-emission monitoring techniques can provide the nucleus of a field monitoring technique to assess stability of earthen structures.

1097 **A. E. Lord, Jr. and R. M. Koerner. "Acoustic Emission Response of Dry Soils."** *Journal of Testing and Evaluation,* **JTEVA, 2(3):159–162. May 1974.**

An investigation into the acoustic emission response of a series of axially stressed dry-soil samples is presented. The correlation between acoustic emissions and strain is noted and strongly suggests that acoustic emission monitoring of soil masses may lead to a nondestructive field evaluation technique. Preliminary insight into the fundamental mechanism of acoustic emissions in dry soils is presented. It is seen that granular soils yield higher accumulated emission counts than fine grained soils. The summary curves which are presented show this clearly defined trend and can be regarded as laboratory calibration curves for the soils which were tested. [reprinted by permission of the American Society for Testing and Materials, Copyright].

1098 **A. E. Lord, Jr., R. M. Koerner, and J. W. Curran. "Fundamental Studies of Acoustic Emission in Soils." pp 135–148 in** *Proceedings First Conference on Acoustic Emission/Microseismic Activity in Geologic Structures and Materials.* **H. R. Hardy, Jr. and F. W. Leighton, Editors. Trans Tech Publications, Clausthal, West Germany. 1977. conference held at Pennsylvania State University, University Park, Pennsylvania. June 9–11, 1975.**

Acoustic emissions from soil samples undergoing laboratory testing were monitored under a number of different stress states. These stress states included unconfined compression, isostatic compression, and triaxial shear compression. Tests were carried out on various soil types at different water contents and densities. General findings are that the gross acoustic emissions increase with decreasing water content, increasing percentages of the granular phase, increasing amount of confining stress, and increasing shear stress. Subsequent studies were aimed at a frequency decomposition of these total emissions. It was found that soil emissions are in the 1 to 80 kHz region with a shift from the lower to the higher value with increasing confining pressure around the soil specimen. This laboratory work has given insight to proper equipment selection for the field monitoring of soil masses described in a companion paper by Koerner *et al.,* presented at this conference (see Reference 1024).

1099 **A. E. Lord, Jr., R. M. Koerner, and W. M. McCabe. "Acoustic Emission Behavior of Sand as Used in Foundation Bearing Capacity."** *Materials Evaluation,* **MAEVA, 34(5):103–108. May 1976; paper presented at the American Society for Nondestructive Testing National Fall Conference, Atlanta, Georgia. Oct 13–16, 1975.**

Based upon recent research of acoustic emission monitoring in various soil systems, i.e., earth dam monitoring, an extension has been made into the procedures for monitoring soil masses beneath structural foundations. Since such soils are often cohesionless sands and gravels, some fundamental work as to the effects of physical characteristics on acoustic emission generation has been included. It has been found that soils consisting of angular particle shapes which are well-graded generate large acoustic emission counts under stress application. It was also found that particle size, per se, does not have a significant influence on the acoustic emission generation. A large-scale laboratory bearing capacity experiment was performed showing a strong correspondence between load and acoustic emission counts. In this experiment, failure was clearly defined, a distinct improvement over existing procedures in which failure is not so easily observed and must be estimated. A potential use of the technique in monitoring a building foundation in a mine subsidence area is discussed where preliminary acoustic emission rates are high and a complete monitoring scheme has been proposed to the owner.

1100 **A. E. Lord, Jr., R. Usatschew, and M. Robinson. "Acoustic Emissions Associated with Changes of Magnetization in Thin Nickel Rods."** *Letters in Applied and Engineering Sciences,* **LAESB, 2(1):1–9. Mar 1974.**

Acoustic emissions have been observed during magnetization of thin nickel rods. The acoustic emissions are consistent with the familiar Barkhausen magnetic effect. The emissions are much more numerous in as-received (presumably cold-worked) material than in annealed material, and occur only at the steep portion of the hysteresis loop. The acoustic emission count is a strong function of the rate of application of the magnetic field. The count varies from zero at a rate of 15 oersteds per second to 1350 at 90 oersteds per second. [reprinted by permission of Pergamon Press Ltd].

1101 **G. K. Lucey, Jr. "Resonant and Antiresonant Frequencies of Thick Disks and Thick Rods."** *Journal of the Acoustical Society of America,* **JASMA, 43(6): 1324–1328. June 1968.**

Correction factors are derived that show the effect of thickness on the fundamental resonant-frequency equations for cylindrical disks and rods and an equation is given that defines a boundary between a thick disk and a thick rod. These corrections are verified for isotropic and planar isotropic materials using the published frequency data of Shaw and McMahon. Direct comparisons are made with the correction factors of Moseley, Rayleigh, Onoe and Tiersten, Bancroft, and Baerwald. Experimental evidence is provided to show that upon substitution of appropriate frequency constants, the corrections apply to the antiresonant frequencies of thick disks and thick rods.

1102 **A. C. Lucia, R. Brunnhuber, M. Franchi, and M. Galli. "A New Methodology for the Application of Acoustic Emission." pp 1641–1645 in** *Proceedings of the Second International Conference on Mechanical Behavior of Materials.* **American Society for Metals, Metals Park, Ohio. 1976. conference held in Boston, Massachusetts. Aug 16–20, 1976.**

One of the most peculiar properties of acoustic emission is its definite relation to the dynamics of material behavior on both macro- and microscopic scale. For this reason, the AE makes it possible to monitor in real time the evolution of a process of progressive damage which may be taking place in a structure or in a specimen, with a sensitivity superior to that of any other technique. As an NDT technique, AE has recently had a rapid development both in the field of inspecting the integrity of structural materials used in mechanical constructions, and in the physico-metallurgical field of detecting and examining processes arising in materials (plastic deformations, phase transformations, hydrogen embrittlement, twinning, etc.). It is obvious to foresee the potential of AE for a wider and more skillful application both with traditional materials, for production process control, and for determining the properties of the materials themselves, and for the designing and realization of non-traditional materials of prespecified performance. The information content of the AE signal is rich; the problem is that of extracting such information from a non-stationary (or, within certain limits, locally stationary) random signal, with a frequency content which can range from a few kHz to 1 or 2 MHz.

1103 **A. C. Lucia, R. Brunnhuber, M. Franchi, and M. Galli. "Field Applications of Acoustic Emission: Experiments and Methodology." Paper 3-3, pp 212–228 in** *The Third Acoustic Emission Symposium.* **Japan Industrial Planning Association, Dobashi Building, 3-17-8, Akasaka, Minato-ku, Tokyo, Japan. 1976. proceedings of a symposium held in Tokyo, Japan. Sep 16–18, 1976.**

A vast range of instruments specially designed for AE is today available, as well as a know-how of a number of possibilities of practical applications. However, the industrial fall-out of AE technology is still rather modest especially with respect to the sensitivity, potentiality, and characteristics which make AE the newest and most versatile among the NDT techniques. One reason for the lack of confidence in AE could be the fact that, if the methodology of application is not sufficiently strict, the results are equivocal and difficult to interpret. By methodology of application is intended the precautions that have to be taken and how a measurement of AE has to be organized for it to be significant. It entails a series of preliminary tests being carried out on the structure examined and on the material of which it is made in order to facilitate later the interpretation of the results. These preliminary investigations (treatments to which the material has been submitted, stresses already applied on the structure and, if possible, finger printing or signature of the main types of defects possible on the structure) enable the field of the interpretation of the results to be restricted and their use is much greater in that AE has still not reached the last stage of its development. It is, in fact, necessary to acknowledge that, at an application level, there are still various problems to be solved and that, since AE is still a technique in evolution, all is not clear or of immediate and univocal interpretation.

1104 **A. C. Lucia, R. Brunnhuber, and G. Redondi. "Stress Wave Emission Source Location and Characterization by Digital Methods." Session 2, pp 42–66 in** *The Second Acoustic Emission Symposium.* **Japan Industrial Planning Association, Dobashi Building 5F, 3–17–8 Akasaka, Minato-ku, Tokyo, 107 Japan. 1974. proceedings of a symposium held in Tokyo, Japan. Sep 2–4, 1974.**

The two main problems to be solved in the use of acoustic emission as a means of control for pressure vessels or, more generally, for plants, are the location of sources and their characterization, where characterization means the extraction from the signals of informations on nature and entity of the flaw causing the signals themselves. In the field of location, much work has been carried out by different laboratories throughout the world, and techniques and equipments have been developed. In all the solutions proposed, special-purpose systems have been developed which range from the electronic circuit for the measurement of differences in signal detection times by the sensors to the memories enabling a storage, and consequently a statistic of the delay values to be made, to a final computer used for the actual triangulation. On the other hand, the source characterization problem is still far from a satisfactory solution. Energetic and probabilistic or spectral type analyses have been tried, together with the classical measurements of count-rate and total counts, but a methodology enabling a certain and reliable diagnosis on the state of the plant, has not yet been developed. Also in this case, the analyses are usually accomplished by means of special-purpose equipments, each enabling the experimenter to do only one type of analysis, and without the possibility of wide-range variations in measurement parameters. The purpose of our work is to reduce as much as possible the utilization of special hardwares in source identification (source location and characterization), by accomplishing all the elaborations via software. The final aim is to effect the identification using for example the terminal of a large computer. One great advantage of this methodology is the possibility of combining and connecting the two phases of the identification, and mainly of executing all sorts of signal analyses and comparisons in order to extract informations about the state of the plant under test. It is obvious that, before this last stage can be reached, two intermediate stages have to be successfully overcome: (a) preparation of a location code carrying out also the determination of the time-of-arrival differences of the signals at the transducers, and (b) elaboration of a reliable methodology for source characterization.

1105 **A. C. Lucia, E. Ohlmer, D. Schwalm, and G. Volta. "Incipient Failure Identification and Location." pp 255–262 in Report EUR-5060. Euratom J.R.C., Ispra, Italy. 1973.**

The safety and the availability of a nuclear power plant are goals that can be achieved by incorporating adequate precautions and facilities into the design and into the preoperational and operational stages of the plant. Surveillance methods during operation are based on the realistic assumption that failures and malfunctions are the result of progressive degradations in time, consuming the initial safety margins. Such degradations are, in general, sources of various signal or parameter fluctuations which, when accurately detected, serve to prevent failures and damages. The incipient failure diagnosis involves two main technical problems. First: the identification of the *signature* of a failing state among the various signals emitted by a complex system. Second: the location of the source and then of the failing sub-system. At Ispra, both problems were examined. As possible failure *signatures,* many years have been spent investigating the neutron noise and more recently, stress wave emission. As a location method, a classical triangulation technique, suitable for all types of signals, has been primarily developed. An alternative technique, suitable only for stationary signal sources, has also been outlined.

1106 A. C. Lucia and G. Redondi. "Acoustic Emission Signal Interpretation." pp 147–174 in "EWGAE–ISPRA: Proceedings of the Third Meeting of the European Working Group on Acoustic Emission." A. C. Lucia, Editor. Report EUR–5513e. Commission of the European Communities, Directorate General Scientific and Technical Information and Information Management, Luxembourg, Belgium. 1976. meeting held at the Joint Nuclear Research Center Center, Ispra, Italy. Sep 25–26, 1974.

1107 A. C. Lucia and G. Redondi. "On the Interpretation of the Acoustic Emission Signals." ASME Paper No. 75–WA/PVP–20. American Society of Mechanical Engineers, New York. 1975; paper presented at the American Society of Mechanical Engineers Winter Annual Meeting, Houston, Texas. Nov 30–Dec 5, 1975; *Transactions of the ASME, Journal of Pressure Vessel Technology,* **JPVTA, 98(3):199–207. Aug 1976.**

The AE signals are of a nonstationary random type, the nonstationarity being connected with the evolution of the source, whose characterization is here pursued by the development of methodologies of statistical analysis of nonstationary signals. The AE signal is hypothesized to be constituted by a combination of a stationary random component and a deterministic one, whose behavior represents the evolution of the source. The determination of this latter component is carried out for signals from tensile tests on compact tension specimens. A second approach to the signal interpretation problem is also presented. It is based on the representation of the AE signal as random pulses of varying duration and amplitude.

1108 A. C. Lucia, G. Volta, R. Brunnhuber, M. Franchi, M. Galli, and B. Ganss. "Ultra–Acoustic Wave Propagation and Source Location Experiments." Report EUR–5396e. Euratom J.R.C., Technology Division, Ispra, Italy. 1976.

The report deals with the location of acoustic emission sources and with some problems connected therewith: propagation of ultrasonic waves and the use of waveguides. Some results of experiments carried out in this field are given. Furthermore, a chapter is devoted to the problem of the determination of the frequency response of the transducers.

1109 **R. F. Lumb and R. J. Hudgell. "Acoustic Emission Characteristics of Structures Welded in Mild Steel/Characteristiques des emissions acoustiques des constructions soudees." British Gas Engineering Research Station, Newcastle Upon Tyne, England; Paper 3K/17 in** *Proceedings of the Eighth World Conference on Nondestructive Testing.* **volume containing Sections 3K2, 3H2, 5A. International Committee for Nondestructive Testing, Secretariat du Cofrend, Institut de Soudure, 32 Boulevard de la Chapelle, 75880 Paris Cedex 18, France. 1976. conference held in Cannes, France. Sep 6–10, 1976. (paper in English). [NTIS: INIS-mf-3456].**

Information has been published previously on the emission characteristics of different mild steels. The paper shows that significantly different characteristics are obtained from the same structure depending on whether the source is clean or dirty plate; submerged arc, semi-automatic CO_2, or manual metal arc weld deposit; or weld heat-affected zone.

1110 **Yu. I. Lykov, L. A. Maslov, and V. I. Panin. "Determining the Duration of a Crack** ***Step*** **by Measuring the Spectrum of Acoustic Emission Signals."** *Soviet Journal of Nondestructive Testing,* **SJNTA, 10(6):687–689. Nov–Dec 1974; translated from** *Defektoskopiya,* **DEFKA, 10(6):75–77. Nov–Dec 1974. (in Russian).**

An arrangement is described for measuring the spectrum of acoustic emission signals from single *steps* in a crack. For a step of 1 mm in a crack, the first minimum of the spectral curve occurs at a frequency of 200 kHz. The results obtained are compared with theoretical values.

1111 **L. C. Lynnworth. "High Temperature Acoustics–Sixties' Research Spawns Seventies' Applications."** *Naval Research Reviews*, **NARRA, 23:1–22,32. July 1970.**

Discussed is the use of high temperature acoustic waves to measure properties of solids (moduli, strength, etc.), and of fluids (viscosity, thermal conductivity, etc.). The method can also characterize the state of a system, measure or monitor (or both) integrity of welds, inspect and test materials nondestructively, and measure processes nondestructively. The nature of ultrasonic waves is outlined, and ultrasonic measuring systems for use at high temperature are described. Applications of the procedure are extensively discussed, including acoustic emission.

1112 **L. C. Lynnworth. "Industrial Applications of Ultrasound: A Review, II. Measurements, Tests, and Process Control Using Low-Intensity Ultrasound."** *IEEE Transactions on Sonics and Ultrasonics,* **IESUA, SU-22(2):71–101. Mar 1975.**

The following applications are reviewed: ultrasonic measurement of flow, temperature, density, porosity, pressure, viscosity and other transport properties, level, position, phase, thickness, composition, anisotropy and texture, grain size, stress and strain, elastic properties, bubble, particle and leak detection, nondestructive testing, acoustic emission, imaging and holography, and combinations of these. Principles, techniques, equipment, and application data are summarized for these areas. Most of the measurements utilize approaches designed to respond primarily to sound speed, but some depend on attenuation effects. Most equipment in use involves intrusive probes, but noninvasive, externally-mounted transducers are being promoted in several areas. Both pulse and resonance techniques are widely used. Limitations due to the influence of unwanted variables are identified in some cases. A bibliography and list of vendors provide sources for further information.

1113 **L. C. Lynnworth and J. E. Bradshaw. "Magnetostriction Transducers for Acoustic Emission, Impulse, Vibration, and Noise Analysis."** *Materials Research and Standards*, **MTRSA, 11(3):33-35. Mar 1971.**

Magnetostriction transducers respond to frequencies up to approximately 1 MHz. Bandwidth is controlled through choice of geometry. Mechanically, magnetostrictive metals are much more rugged than piezoelectric ceramics. Magnetostriction transducer characteristics include transduction in the presence of high temperature and nuclear radiation, remote sensing via wire waveguides, ruggedness, and magnetic coupling to rotating structures.

1114 **L. C. Lynnworth and E. P. Papadakis. "Measurement of Elastic Moduli of Materials at Elevated Temperature." Office of Naval Research Contract N00014-73-C-0023, I.D. No. NR384-320/04-27-72/468. Panametrics, Inc., Waltham, Massachusetts. May 31, 1974. [NTIS: AD-780-231].**

Ultrasonic measurement of elastic properties of solids at elevated temperature comprises the central theme in this report of some twelve years of research and development under the support of the Office of Naval Research Acoustics Branch. The principal methods utilized were momentary contact with pressure coupling, for bulk specimens of cross-sectional dimensions large compared to wavelength, and continuous contact, with welded joints, for wires and rods of cross-sectional dimensions small compared to wavelength. The bulk specimens were tested with broadband pulses or RF bursts having frequencies typically above 1 MHz. The wires were tested with pulses having 0.1 MHz as a typical center frequency. Attenuation data were obtained in metals up to ~2500 Kelvin, and sound velocity data, to over 3500 K. In steel and several refractory metals and alloys, correlation of the slope change in the moduli-temperature function, a peak in the attenuation-temperature function, and microstructural or grain boundary phenomena, have been accomplished. Other topics include: transducers, velocimetry, spectrum analysis, infrasonic measurements of moduli in plastic tensile test specimens, measurement of polymerization, and applications of ultrasonics such as nondestructive testing, acoustic emission, thermometry, flowmetry, and process control. The above topics are covered in nine chapters.

The major portion of the material consists of the authors' publications of work supported at least in part since 1962 by ONR contracts Nonr-3918(00), N00014-71-C-0050 and N00014-73-C-0023.

1115 **L. C. Lynnworth, E. P. Papadakis, D. R. Patch, K. A. Fowler, and R. L. Shepard. "Nuclear Reactor Applications of New Ultrasonic Transducers."** *IEEE Transactions on Nuclear Science*, **IETNA, NS-18(Part-1):351-362. Feb 1971; paper presented at Institute of Electrical and Electronics Engineers 2nd Nuclear Power Systems Symposium. New York. Nov 4–6, 1970.**

New high temperature and broadband magnetostrictive and piezoelectric ultrasonic transducers have been developed for testing, measurement, and control. Potential reactor applications include measuring fuel centerline temperature, sodium temperature profile (ten points with only one wire), occurence and location of bubbles in liquid coolants, liquid level, flow gas pressure, nondestructive testing, and acoustic emission surveillance.

M

1116 **E. S. Machlin and M. Cohen. "Burst Phenomenon in the Martensitic Transformation."** *Journal of Metals,* **JOMTA, 3:746–754. Sep 1951. (Same as** *Transactions of the AIME,* **TAIMA, 191:746–754. Sep 1951).**

The martensite reaction in single crystals and polycrystals of 70% Fe-30% Ni alloys is shown to be autocatalytic in nature, producing bursts of transformation during cooling. The temperature of the first burst of transformation, called M_b occurs below M_s in these alloys. Experiments were devised to test the athermal embryo and strain embryo theories of martensite nucleation. The results indicate that internal strains, either within the virgin austenite or around existing martensitic plates, control the nucleation process in these alloys. Furthermore, the growth of martensitic plates is not limited by the attainment of an elastic balance with the austenitic matrix, but by the occurrence of plastic deformation at the martensite boundaries which interferes with the propagation mechanism.

1117 **N. J. Magnani. "Acoustic Emission and Stress-Corrosion Cracking of U-4½ wt % Nb." SLA-73-5195. Sandia Laboratories, Albuquerque, New Mexico. Feb 1973; paper presented at the Society for Experimental Stress Analysis 3rd International Congress on Experimental Mechanics, Los Angeles, California. May 13–18, 1973;** *Experimental Mechanics,* **EXMCA, 13(12): 526–530. Dec 1973.**

Stress corrosion crack propagation in uranium 4.5-weight percent niobium has been monitored with crack-opening displacement and acoustic emission techniques. Excellent correlation was obtained between the direct crack velocity measurements and the emission count rate at all stress intensities. The data also show that the emission rate is related to the crack velocity and that as the cracking velocity increases the counts per unit area of new crack surface increase. The results of the investigation show the feasibility of acoustic emission detection techniques as a useful stress corrosion testing tool.

1118 **N. J. Magnani, K. D. Boultinghouse, and C. J. Miglionico. "The stress Corrosion Cracking Behavior of U-4.5 wt % Nb in Laboratory Air." SC-RR-710426. Sandia Laboratories, Albuquerque, New Mexico. Mar 1972**

The stress corrosion cracking behavior of U-4.5 wt % Nb has been determined in dry (~10% RH) air using side-grooved tapered double-cantilevered-beam specimens. Crack propagation was monitored using crack opening displacement gauges and acoustic emission detection techniques. The threshold stress intensity for stress corrosion cracking was determined to be ~20 ksi·inch$^{1/2}$ (22 MN·m$^{-3/2}$) and the plane strain fracture toughness ~50 ksi·inch$^{1/2}$ (55 MN·m$^{-3/2}$).

1119 **R. Mah. "Acoustic Emission from Deformed Metals." M. S. Thesis. University of Illinois, Urbana, Illinois. 1970.**

The first task of this investigation was to develop equipment and test procedures for the detection of acoustic emission. The relationship between acoustic emission and the stress and strain history for different materials was then studied for both notched and unnotched specimens. The new experimental methods developed were used for proving the cause of acoustic emission and also extended to fatigue tests where the same equipment was used to detect the onset of crack propagation.

1120 **R. Mah, R. L. Kochen, and J. M. Macki. "Corrosion-Induced Acoustic Emissions from Uranium 4.5-Weight Percent Niobium Binary Alloy." RFP-2351. Rockwell International, Rocky Flats Plant, Golden, Colorado. Dec 15, 1975. [NTIS].**

The ability to monitor corrosion and stress-corrosion cracking of uranium 4.5-weight percent niobium was investigated. The acoustic emissions from stress-corrosion cracking were monitored using smooth four-point bend specimens, which were immersed in oxygen-saturated water containing chloride ions. The acoustic emissions from corrosion were monitored from rectangular corrosion coupons. These latter specimens were exposed to nitrogen-saturated and oxygen-saturated solutions of distilled water and distilled water with chloride ions. Dilute solutions of hydrochloric acid were also used. Findings show that stress-corrosion crack monitoring of the alloy was successful. Corrosion monitoring of the alloy showed a direct correlation between corrosion rate (based on weight loss) and total acoustic emissions. For a specific range, one can calculate the corrosion rate or uranium alloys in a corrosive solution from the total acoustic emissions generated. Studies to determine the source of corrosion-induced acoustic emissions were inconclusive.

1121 **M. J. P. Maigret. "Contribution de l'analyse acoustique pour l'etude des mecanismes de ruptures des materiaux composites (Contribution of Acoustic Analysis to the Study of Fracture in Composite Materials)."** *Mecanique-Materiaux-Electricite,* **MMXEA, (304-305):31-33. Apr-May 1975. (in French).**

The application of acoustic emission and the experimental technique used in studying composite materials are discussed. Fiber composites with varying resin content were studied.

1122 **K. Malen and L. Bolin. "A Theoretical Estimate of Acoustic-Emission Stress Amplitudes."** *Physica Status Solidi (B) Basic Research,* **PSSBB, 61(2):637-645. Feb 1974.**

A theoretical estimate is made of the stress amplitude for acoustic emission due to the sudden local change of an inelastic strain. The source may thus be, e.g., dislocation motion, crack propagation, twinning or martensitic transformation. The dynamic aspect of the acoustic emission event is emphasized. The estimates have been compared with some experimental results reported in the literature and a reasonable agreement is found.

1123 **A. W. Malone. "An Experimental System for the Laboratory Study of Acoustic Emission from Stressed Rock." M.Sc. Thesis. Imperial College, Department of Geology, London, England. 1965.**

1124 **G. A. Malone and R. A. Duscha. "Nondestructive Evaluation of Electroformed Outer Shells for Regeneratively Cooled Thrust Chambers." Textron Bell Aerospace Company, Buffalo, New York and National Aeronautics and Space Administration, Lewis Research Center, Cleveland, Ohio. 1974; paper presented at the American Electroplaters Society Symposium, Atlanta, Georgia. Dec 9–10, 1974.**

Electroforming is presently a prime technique for producing the outer jackets of regeneratively cooled thrust chambers for rocket engines. Good bond integrity of the electroformed joints adjacent to the coolant passages is important to the useful service life of these aerospace systems. This investigation demonstrated that nondestructive evaluation techniques such as ultrasonic *C*-scan, holography, and acoustic emission are capable of detecting low integrity bonds which could lead to premature failures in the subject hardware.

1125 **G. A. Malone, R. Stauffis, and R. Wood. "Nondestructive Tests of Regenerative Chambers." NASA CR–120980, BAC Report No. 8654–953003. Bell Aerospace Company, Buffalo, New York. July 1972. [NTIS: N72–32502]. [This report consists of two volumes: the report and an appendix, both published under the same report number].**

Flat test panels simulating internally cooled regenerative thrust chamber walls were fabricated by electroforming, brazing, and diffusion bonding to evaluate the feasibility of nondestructive evaluation techniques to detect bonds of various strength integrities. Ultrasonics, holography, and acoustic emission were investigated and found to yield useful and informative data regarding the presence of bond defects in these structures.

1126 **G. A. Malone, L. Vecchies, and R. Wood. "Nondestructive Tests of Regenerative Chambers." NASA CR–134656, BAC Report No. 8654–953005, Final Report. Bell Aerospace Company, Buffalo, New York. June 1974.**

The objective of the program was to define the capabilities and limitations of nondestructive evaluation methods to detect and locate bond deficiencies in regeneratively cooled thrust chambers for rocket engines. Nondestructive evaluation methods used were those of demonstrated capability from previous work under Contract NAS 3-14376, NASA Report CR-120980 (see Reference 1125). Under the contract, flat test panels and a cylinder were produced to simulate regeneratively cooled thrust chamber walls. Planned defects with various bond integrities were produced in the panels to evaluate the sensitivity, accuracy, and limitations of nondestructive methods to define and locate bond anomalies. Holography, acoustic emission, and ultrasonic *C*-scan were found to yield sufficient data to discern bond quality when used in combination and in selected sequences. Bonding techniques included electroforming and brazing. Materials of construction included electroformed nickel bonded to Nickel-200 and OFHC copper, electroformed copper bonded to OFHC copper, and 300 series stainless steel brazed to OFHC copper. Variations in outer wall strength, wall thickness, and defect size were evaluated for nondestructive test response.

1127 **G. A. Malone, L. Vecchies, and R. Wood. "Nondestructive Tests of Regenerative Chambers – Appendix." NASA CR-134657, BAC Report No. 8654-953005, Final Report. Bell Aerospace Company, Buffalo, New York. June 1974.**

This volume of a two-volume report contains the appendix to report NASA CR-134656 (Reference 1126).

1128 **J. C. Manaranche. "L'Emission Acoustique (Acoustic Emission)."** *Measures, Regulation, Automatisme,* **MRAUA, 38(11):81–84. Nov 1973. (article, in French), 38(10):79. Oct 1973. (bibliography).**

Industrial applications of the method are presented with particular attention being given to the control of pressurized vessels in the nuclear industry.

1129 **C. L. Mantell.** *Tin: Its Mining, Production, Technology, and Applications.* **American Chemical Society Monograph Series. Book Department, The Chemical Catalog Company, Inc., New York. 1929. pp 13 and 16.**

In the chapter on the history of tin, Mantell quotes the Arab chemist Geber (or Jabir) in his *Summa Perfectionis Magisterii,* which was written in the eighth century: "I intimate to the sons of learning that tin is a metallic substance which is white, but not pure white; it has a little ring, and emits a creaking sound." In describing the physical and chemical properties of tin Mantell states: "When a bar of tin is bent, it emits a characteristic creaking sound called the *cry* of tin, resulting from the grinding of the crystals one against the other during the bending of the metal."

1130 **J. Marini. "Surveillance of Evolution of Defects in Stainless Steel Pipings Subject to Fatigue Cycles in Temperature/Surveillance de l'evolution de defauts dans des Tuyauteries en acier inoxydable soumises a des cyclages de**

fatigue en temperature." Framatome, Courbevoie, France; Paper 5C/9 in *Proceedings of the Eighth World Conference on Nondestructive Testing.* volume containing Sections: 5C, 3J, 1D. International Committee for Nondestructive Testing, Secretariat du Confrend, Institut de Soudure, 32 Boulevard de la Chapelle, 75880 Paris Cedex 18, France. 1976. conference held in Cannes, France. Sep 6–10, 1976. (paper in French).

The surveillance of internal crack growth in austenitic ICL-167 CN steel is possible by using ultrasonic techniques. The fracture mechanics allows the prediction of the evolution of these cracks under fatigue loading.

1131 **G. G. Martin. "Acoustic Emission from the Martensitic Transformation." Note ARL/Mat. 110. Department of Defence, Australian Defence Scientific Service, Aeronautical Research Laboratories, Melbourne, Australia. May 1975. [NTIS: AD/B-009-881].**

The paper outlines the principles of acoustic emission and briefly lists some of its users. It also reviews the literature on the martensitic transformation with special attention to those transformations studied using acoustic emission. Results are given of further tests carried out to assess: (a) the suitability of an Fe-Ni alloy as a reference for acoustic emission monitoring systems; and (b) the usefulness of acoustic emission as a research tool in the study of phase transformations. The high sensitivity to the martensitic start temperature (M_S) shown by acoustic emission requires a more accurate definition of M_S.

1132 **G. G. Martin. "Bibliography on Stress Wave Emission: 1950–1972." ARL/MET.1. Department of Supply, Australian Defence Scientific Service, Aeronautical Research Laboratories, Melbourne, Australia. Feb 1973. [NTIS: ARL-MET-1 or N73-30500].**

The bibliography lists 250 known sources of information relating to acoustic emission. References are listed chronologically under two headings–articles readily available in Australia and those available elsewhere. The author has classified each paper in the first section into one of five categories: Introduction, Theory, Evaluation, Instrumentation, and Application. Following this are listed materials discussed, type of test performed, and descriptors of the main features of the paper.

1133 **G. G. Martin. "Bibliography on Stress Wave Emission: 1950–1973." ARL/MET.2. Department of Supply, Australian Defence Scientific Service, Aeronautical Research Laboratories, Melbourne, Australia. Feb 1974. [NTIS].**

This is the first revision of ARL/MET.1 (see Reference 1132). The author has added 106 new references to the Main Index making a total of 234 references of papers readily available in Australia as of December 1973. To the Supplementary Index, 86 new references were added to make a total of 199 references. Thus the bibliography lists 433 references.

1134 R. L. Martin. "Detection of Ball Bearing Malfunctions." ***Instruments and Control Systems*****, INCSA, 43(12):79–82. Dec 1970.**

Bearing malfunction is generally detected only when a component fails completely. Inability to predict incipient failures may lead to extensive damage or to periodic replacement of satisfactory components. Possible bearing failure modes include excessive contact stress, misaligned loads, material flaws, lubricant failure or contamination, and electrical discharge between balls and races. In any case the malfunction is manifest as a defect in a race or a ball. Such failure can be detected by frequency analysis of a signal obtained from the bearing.

1135 T. Maruyama, H. Togawa, and S. Tominaga. "Acoustic Emission Initiated in the Fracture Toughness Testing of High Strength Steel." Session 5, pp 58–76 in ***The Second Acoustic Emission Symposium.*** **Japan Industrial Planning Association, Dobashi Building 5F, 3-17-8 Akasaka, Minato-ku, Tokyo, 107 Japan. 1974. proceedings of a symposium held in Tokyo, Japan. Sep 2–4, 1974.**

Acoustic emission is noticeable as an effective technique to detect in real time the initiation and propagation of cracks during the hydraulic test and/or operation of the pressure vessel. The study of this technique is now under way. However, influences of various factors on the emission characteristics are not sufficiently investigated, and many of them still remain uninvestigated. It is considered that the emission characteristics are greatly affected by factors: namely, (1) type of crack involving fatigue crack during operation, welding cracking, stress corrosion cracking, shape discontinuity, notch, etc.; (2) fracture mode as represented by cleavage, shear and intergranular fracturing, etc.; (3) size effect in terms of thickness, etc.; (4) where to set a level for detecting AE; and (5) electric and mechanical noises. There are several reports on the influence of the factors from (3) to (5), but the influences of the factors from (1) and (2) are hardly investigated. In experiments described in this report, the relations between emission characteristics and type of crack and/or fracture mode were investigated by the fracture toughness test using single-edge notched bend specimens, thus establishing the fundamental characteristics regarding the prediction of failure by means of relating the emission characteristics to the deforming-fracture mode at the crack tip and to stress intensity factor and fracture toughness as the parameters used in fracture mechanics.

1136 W. P. Mason. "Dislocation Drag Mechanisms and Their Effects on Dislocation Velocities." pp 487–502 in ***Dislocation Dynamics.*** **A. R. Rosenfield, G. T. Hahn, A. L. Bement, Jr., and R. I. Jaffee, Editors. McGraw-Hill Book Company, New York. 1968. proceedings of the 2nd Battelle Institute Materials Science Colloquia, held in Seattle, Washington and Harrison, British Columbia. May 1–6, 1967.**

Briefly mentioned is the experiment of detecting twinning in tin by Mason, McSkimin, and Shockley which is reported in Reference 1139.

1137 W. P. Mason. "Internal Friction, Acoustic Emission and Fatigue in Metals for High Amplitude Ultrasonic Frequencies." *Engineering Fracture Mechanics,* **EFMEA, 8(1):89–101. 1976.**

High amplitudes (strains up to 3×10^{-3}) have been used in studying internal friction, acoustic emission, and fatigue in metals. The internal friction in the high amplitude range is due to breakaway of dislocations from pinning points and by the generation of Frank-Read loops which results in plastic strain. There are two plastic regions, one when the added loops are held up by the grain boundaries, and the second when the loops break through the grain boundaries. In the second region, slip bands are produced in the metal. As the amplitude increases slip bands can join up and fatigue in the metal occurs. Ultrasonic frequencies are useful for studying fatigue since a large number of cycles can occur in a reasonable time. Acoustic emission–that is, a noise in the sample associated with the dislocation motion–can be studied by putting transducers on the sample. It is shown that the emission is closely associated with the internal friction since it goes through two regions in the plastic range. [reprinted by permission of Pergamon Press Ltd].

1138 W. P. Mason, D. N. Beshers, M. C. Jon, and J. T. Kuo. "Acoustic Emission Caused by Large Alternating Strains." *Ultrasonics,* **ULTRA, 13(3):128–131. May 1975. [NTIS: AD/A-016-087].**

By vibrating specimens at high strains, acoustic emission has been observed under controlled conditions. Continuous emission, in the form of strong harmonic generation, has been observed for polycrystalline copper and brass. In single crystal copper, a non-harmonic, burst type, response has also been observed. For brass, the internal friction and the third harmonic were analyzed in terms of the dislocation models of Granato-Lucke and Peguin *et al.* A dynamic dislocation multiplication mechanism is suggested for the burst-type emissions.

1139 W. P. Mason, H. J. McSkimin, and W. Shockley. "Ultrasonic Observation of Twinning in Tin." *Physical Review,* **PHRVA, 73(10):1213–1214. May 1948.**

Experiments have been performed with a view to observing the effects of individual moving dislocations. If a dislocation moves across a test specimen with a speed approaching that of sound, a displacement of the order of the lattice constant d should occur in a time of order of diameter of specimen divided by the speed of sound. This requires instruments capable of measuring displacements of about 10^{-8} cm occurring in times of 10^{-6} sec., values which lie in ranges already exploited in connection with ultrasonic delay lines. The method of experimentation consists of pressing the metal specimen directly against a quartz crystal with a liquid seal, applying stresses which deform the metal, and observing the resulting motion imparted to the quartz. The specimen is prepared from 99.9 percent pure tin. Results are in general agreement with passage at the speed of sound of twinning dislocations from side to side in the specimen. However, the magnitude of the displacement is about ten times larger than would be expected from the twinning of one pair of planes. The net volume contraction of the tin as viewed from the quartz is about 10^{-8} cm^3. Somewhat similar experiments carried out with large aluminum crystals failed to give detectable signals.

1140 **J. Masounave, J. Lanteigne, M. N. Bassim, and D. R. Hay. "Acoustic Emission and Fracture of Ductile Materials."** *Engineering Fracture Mechanics,* **EFMEA, 8(4):701–709. 1976.**

Acoustic emission, radius of the contraction zone on the side of the specimen, and load-load point displacement were measured during fracture of single-edge notched specimens containing fatigue cracks of different lengths. It was found that there exists a three-fold correlation between formation of a contraction zone, noticeable acoustic emission, and deviation from linearity on the load-load point displacement curves. The stress intensity factor K was evaluated from the extent of contraction on the specimen surface and related to the acoustic total count. A correction for K was made to account for the considerable plasticity at the crack tip. Finally, a model for the formation and development of the plastic and contraction zones is formulated. [reprinted by permission of Pergamon Press Ltd].

1141 **J. R. Mastandrea and M. V. Scherb. "Electret Polymer Transducer." NT Control Number LAR-11239. McDonnell Douglas Corporation, Huntington Beach, California under contract to Langley Research Center. 1972.**

Traditionally, impact transducers have been piezoelectric ceramics such as lead zirconate titanate or quartz. Recently, however, a new class of piezoelectric materials has received considerable attention. Polymers, such as tetrafluoroethylene (TFE) and fluoronated ethylenepropylene resins (FEP), exhibit a limited piezoelectric effect in their natural states, and this effect is heightened considerably if the polymers are permanently polarized. The permanently-polarized, or electret, state is produced by heating the polymer and placing it in an electric field, where it is allowed to cool. This material is much less expensive than ceramic piezoelectrics. Furthermore, small thin-film polymer transducers can be used with integrated-circuit amplifiers, to produce higher output voltages and frequency responses. The electret-sensing element is metallized on both sides, and one side is bonded to a conductive housing with conductive epoxy. The other side is similarly bonded to a thin disk that is wired to a recorder. For a higher level of output signal, an amplifier may be added. The front surface of the transducer is the sensing surface. The output voltage is generated between the wire to the recorder and the external housing of the transducer. With an impedance-converter amplifier, an electret transducer produces an output signal of 220 millivolts, across a resistance of 1 megohm, and a capacitance of 100 picofarads, in response to an acoustic pulse. The more expensive equivalent ceramic transducer produces approximately the same output signal.

1142 **W. Matley, P. Howser, and D. Brock. "Acoustic-Emission Detection System."** *Non-Destructive Testing,* **NDETA, 7(5):257–258. Oct 1974.**

The authors describe a system for the detection and processing of acoustic emission. The system was designed principally for carbon-fiber reinforced plastics. A compact, low-cost instrument which detects and processes acoustic-emission signals over a wide bandwidth has been produced by the Department of Electronic and Electrical Engineering at the University of Surrey in collaboration with the Royal Aircraft Establishment, Farnborough. The instrument divides the emissions into four levels according to their amplitude. Unwanted emissions or noise arising from the testing machine can be eliminated by a guard-transducer system. This equipment is designed to provide high gain and low basic noise over a wide bandwidth (70 kHz to 1 MHz) so that the contributions of a broad spectrum of emissions can be processed, and meaningful frequency analysis performed when required. Special damped transducers are used with the equipment to provide a wide frequency coverage. To accommodate the wide dynamic range, the signals are sorted into four levels of amplitude each separated by a factor of X10 from its neighbor. Thus the most sensitive threshold detector L4 provides an output pulse when the level of signal at the transducer exceeds 20 μV peak: a gain of 94 db. The L3 provides a pulse of 200 μV peak threshold and so on. A stepped attenuator allows the overall gain to be reduced by 24 db. The analogue signal is available after the preamplifier (gain X 50) for recording or Fourier analysis.

1143 B. W. Maxfield. "Discussion Summary: Electromagnetic Detection of Acoustic Emission Signals." pp 332–333 in *Acoustic Emission, ASTM STP 505*. American Society for Testing and Materials, Philadelphia, Pennsylvania. 1972.

The discussion summarizes the work reported by Maxfield and Cochran in Report COO-3150-1, "Electromagnetic Detection of Acoustic Emission from a Martensitic Transformation." (see Reference 1145).

1144 B. W. Maxfield. "Electromagnetic Generation and Detection of Acoustic Waves." NYO-2150–73. Cornell University, Laboratory of Atomic and Solid State Physics, Ithaca, New York. June 1971.

In the last few years, much experimental and theoretical effort has been expended in the study of a purely electromagnetic means of generating ultrasound in conducting materials. The basic physics behind this phenomena is now well understood in essentially all materials above liquid nitrogen temperature (77 K). In this regime, a magnetic field of a few kilogauss is necessary to produce an acoustic wave of reasonable amplitude. The major difference between this and other means of acoustic wave generation is the complete absence of any mechanical bond or other physical contact to the material under study. This feature can be important for certain practical applications. For instance, mechanical bonds do not cycle over a wide temperature range. Electromagnetic generation (EG), being a contactless method, can cycle as rapidly as the material can be colled or heated. Continuous operation at high or cryogenic temperature is possible; acoustic waves can also be generated in liquid metals. The EG method should prove useful in a variety of applications; for example, the continuous inspection of rotating objects (shafts, etc.), moving sheet, tube, rod and strip stock of many commercial materials, and porous materials (compacted powders). Acoustic waves can also be generated in nonconductors by plating them with a conducting foil or film. Another important feature of EG is that the polarization of the acoustic wave is determined by the direction of the magnetic field and not the cut of a crystal. Both longitudinal and transverse waves can be generated with comparable efficiency. Small permanent magnets, electromagnets, and pulsed field magnets are available, thus in many applications the necessity of a magnetic field will not complicate the measurement procedure. Each potential application should be accessed on an individual basis; the purpose of this paper is to provide the necessary background to permit the reader to evaluate any possible application, including its use in the study of acoustic emission (stress wave emission).

1145 **B. W. Maxfield and R. Cochran. "Electromagnetic Detection of Acoustic Emission from a Martensitic Transformation." COO-3150-1. Cornell University, Laboratory of Atomic and Solid State Physics, Ithaca, New York. Mar 1972. [NTIS]; *Materials Evaluation*, MAEVA, 31(2):17-20. Feb 1973.**

The acoustic emission resulting from the martensitic phase transformation in a gold-47.5 at % cadmium alloy has been observed using a purely electromagnetic means to detect the acoustic signals. This detection method eliminates transducer-related resonances which usually make a determination of the frequency spectrum associated with acoustic emission difficult. Both the temporal and frequency responses of individual acoustic emission events have been determined. For the small specimen used in this investigation, the frequency response is dominated by acoustic cavity resonance effect in the specimen. This work shows that electromagnetic detection of acoustic emission may be especially useful when attempting to determine the frequency spectrum of various sources of acoustic emission.

1146 **V. F. Mazzio, R. L. Mehan, and J. V. Mullin. "Basic Failure Mechanisms in Advanced Composites." NASA CR-134525. General Electric Company, Space Sciences Laboratory, Philadelphia, Pennsylvania. June 3, 1973.**

The purpose of this research effort is to establish the fundamental failure mechanisms which result from the interaction of thermal cycling and mechanical loading of carbon-epoxy composites. The work is confined to epoxy resin unidirectionally reinforced with HTS carbon fibers. The approach consists of first identifying local fiber, matrix, and interface failure mechanisms using the model composite specimen containing a small number of fibers so that optical techniques can be used for characterization. After the local fracture process has been established for both mechanical loading and thermal cycling, engineering composite properties and gross fracture modes are then examined to determine how the local events contribute to real composite performance. In this way, the critical events in the fracture process can be identified and a rational basis for optimizing the properties of these materials established. Much of the effort in model specimen studies has been directed toward the tensile and compressive behavior under fast and slow thermal cycling between 220 K and 422 K (−65 °F and +300 °F). Parameters such as resin modification (to alter the load carrying capabilities of the resin under mechanical and thermal cycling conditions), oxidative surface treatment of the fiber, variations in fiber spacing, and the effects of thermal cycling rates on polymer molecular structure and behavior have been considered. The use of acoustic emission analysis in studying the failure process and mechanical behavior in both model specimens and engineering specimens of carbon-epoxy composites, as well as equipment and techniques, has also been given consideration.

1147 F. M. Mazzolai and H. C. Kim. "Orientation Dependance of Acoustic Emission in Copper Single Crystals." pp 110–118 in ***Journees d'Etudes sur l'Emission Acoustique.*** **P. F. Gobin, R. Goutte, and C. Guillaud, Organizers. Institut National des Sciences Appliquees de Lyon, Avenue Albert Einstein, 69621 Villeurbanne, Lyon, France. Mar 1975. proceedings of a conference held in Lyon, France. Mar 17, 1975. (paper in English).**

Acoustic emission during plastic deformation of polycrystalline materials has been investigated extensively and it has been suggested that the release of acoustic emission may be related to dislocation breakaway from pinning points, breakaway of dislocation pileups, movement of dislocation tangles, operation of Frank-Read sources, and dislocation microload drop associated with dislocation avalanche. One common point in these proposed source mechanisms is that the release of acoustic emission may be related to the same fundamental process occurring during deformation. However, most workers have failed to identify the source of emission other than loosely associating it with dislocation movement. This failure may be caused by two factors, firstly, the *ring-down count* and *ring-down count rate* parameters, criterions currently used are too crude, and, secondly, they overlooked the difference in dislocation dynamics during yielding and work hardening stages. The present work was undertaken to investigate the acoustic emission energy release during three stages of work hardening and the energy spectrum of emission pulses in copper single crystals during compression testing.

1148 S. L. McBride and T. S. Hutchison. "Helium Gas Jet Spectral Calibration of Acoustic Emission Transducers and Systems." Royal Military College of Canada, Kingston, Ontario, Canada. 1976; ***Canadian Journal of Physics,*** **CJPHA, 54(17):1824–1830. Sep 1, 1976.**

The excitation and spectral calibration of acoustic emission systems using a helium gas jet is shown to produce system independent acoustic emission spectral information. The calibration source makes no mechanical contact with the specimen or structure under investigation and can be conveniently used for specimen or structural response mapping. [reprinted by permission of the National Research Council of Canada].

1149 S. L. McBride, T. S. Hutchison, and R. Murphy. "Acoustic Emission and the Semiconductor–Metal Transition in Vanadium Dioxide Thin Films." ***Scripta Metallurgica,*** **SCRMB, 8(4):431–434. Apr 1974.**

The electrical resistance of the vanadium dioxide film was measured concurrently with the acoustic emissions using the four-lead method. The data show all the main features observed during several hundred cycles of the sample through the transition temperature range. Little variation in the behavior was observed provided the thermal history of the sample was not irregular. The temperature was cycled between 50 and 90 °C.

1150 **B. O. McCauley. "Development and Testing of a Wide-Temperature Acoustic Emission Sensor." ERR-FW-1308. General Dynamics, Convair Aerospace Division, Fort Worth, Texas. May 1, 1972.**

An acoustic emission sensor has been developed that will function in an altering temperature over a range of -73 °C to + 176 °C. A high sensitivity, a high frequency response, and a noise-free operation under temperature variations are accomplished by coupling the transducer element directly to the test article with minimal backloading. The transducer element is enclosed in a brass housing and the latter is mounted on the test article with rubber-based, wide-temperature adhesive which prevents direct contact between the brass housing and the test article. Effective acoustic coupling between the test article and the sensor element is accomplished through a mixture of wide temperature bearing grease and a dry lubricant.

1151 **B. O. McCauley, Y. Nakamura, and C. L. Veach. "Study of Acoustic Emission during Mechanical Tests of Large Flight Weight Tank Structure." pp 233-257 in** *Proceedings of the Ninth Symposium on Nondestructive Evaluation.* **Southwest Research Institute, San Antonio, Texas. July 1973. symposium held in San Antonio, Texas. Apr 25-27, 1973.**

A PPO-insulated, flight-weight, subscale, aluminum tank was monitored for acoustic emissions during a proof test and during 100 cycles of environmental test simulating space flights. The use of a combination of frequency filtering and appropriate spatial filtering to reduce background noise was found to be sufficient to detect acoustic emission signals of relatively small intensity expected from subcritical crack growth in the structure. Several emission source locations were identified, including the one where a flaw was detected by post-test X-ray inspections. For most source locations, however, post-test inspections did not detect flaws; this was partially attributed to the higher sensitivity of the acoustic emission technique than any other currently available NDT method for detecting flaws. For these nonverifiable emission sources, a problem still remains in correctly interpreting observed emission signals.

1152 **M. L. McCauley. "Monitoring Slope Stability with Acoustic Emission." pp 257-269 in** *Proceedings First Conference on Acoustic Emission/Microseismic Activity in Geologic Structures and Materials.* **H. R. Hardy, Jr., and F. W. Leighton, Editors. Trans Tech Publications, Clausthal, West Germany. 1977. conference held at Pennsylvania State University, University Park, Pennsylvania. June 9-11, 1975.**

Acoustic emission is being used by the California Department of Transportation to monitor slope stability. A study of the use of acoustic emission to monitor stability, done by the University of California for the Department in 1962, demonstrated the feasibility of the technique. The second phase of study, done by the Transportation Laboratory, indicated the practicality of the method. The third phase evaluated various types of equipment that could improve the quality of the acoustic records. Several case histories illustrate the capabilities of acoustic emission. The technique has been used to evaluate the stability of landslides, cut slopes, and the effectiveness of corrective measures used on unstable slopes. There is an inverse relationship between the acoustic emissions and slope stability. Acoustic emission data can be correlated with information from other types of monitoring.

1153 **M. L. McCauley. "The Use of Sub-Audible Rock Noise (SARN) to Monitor Slope Stability."** *Engineering Geology,* **EGGOA, 2:1–8, 1965.**

1154 **R. W. McClung. "Nondestructive Testing Methods."** *Annual Review of Materials Science,* **ARMSC, 4:1–20. 1974.**

The article presents some of the recent advances in the standard nondestructive testing methods and discusses newer technologies being applied, including thermal techniques, holography, acoustic emission, microwaves, Mossbauer effect, and exo-electron emission.

1155 **R. W. McClung. "On Nondestructive Testing in the U.S.A. of Prestressed Concrete Pressure Vessels for Nuclear Reactors." pp 527–540 in "Non-Destructive Testing for Reactor Core Components and Pressure Vessels." IAEA–145. International Atomic Energy Agency, Vienna, Austria. 1972. report of a panel meeting held in Vienna, Austria. Nov 29–Dec 3, 1971. [NTIS: CONF-711114].**

Many nondestructive testing techniques are used to assure the integrity of prestressed concrete reactor vessels in the U.S. Metallic members are examined by radiographic, ultrasonic, magnetic particle, and penetrant techniques to meet the requirements of ASME Boiler and Pressure Vessel Code, Section III. The concrete is fabricated to meet the requirements of the American Concrete Institute. After fabrication and during proof testing and operation, the test methods include measurement of stress on the prestressing tendons, strain on both concrete and metallic members, temperature, helium permeation, and moisture distribution in the concrete, leakage in the liner and penetrations, and ultrasonic examination of the liners for corrosion thinning. Acoustic emission is being investigated for possible application.

1156 **R. W. McClung. "Remote Inspection of Welded Joints." ORNL–TM–3561. Oak Ridge National Laboratory, Oak Ridge, Tennessee. Sep 1971. [NTIS].**

Maintenance and repair of reactor components by remote welding will require development of adequate nondestructive examination techniques to assure the integrity of the weld. High temperature and radiation levels will require remote inspections and add to the complexity of the performance. Recommended developments include improved closed-circuit television for visual examination, evaluation of new high-temperature penetrants, study of neutron radiography, improved transducers for ultrasonics, and application of acoustic emission and eddy currents. Sophisticated mechanical positioning, manipulation, and scanning equipment will be required for most of the techniques.

1157 **J. C. McClure, Jr., S. Bhattacharya, and K. Schroder. "Correlation of Barkhausen Effect Type Measurements with Acoustic Emission in Fatigue Crack Growth Studies."** *IEEE Transactions on Magnetics,* **IEMGA, MAG-10(3):913-915. Sep 1974.**

Discontinuous changes in the magnetization of a ferromagnetic sample similar to the rapid changes in magnetization which take place in the Barkhausen effect are caused by the propagation of a fatigue crack. These changes were detected through induced voltages on a pick-up coil surrounding the specimen, which can be quantitatively related to the crack growth. Simultaneous acoustic emission measurements confirmed that the magnetization changes were due to the fatigue crack. In some cases the resolution of the Barkhausen technique is superior to the resolution of acoustic emission measurements.

1158 **N. J. McCormick, R. C. Erdmann, and C. A. Stevens. "On the Efficiency of Acoustic Emission Testing."** *Transactions of the American Nuclear Society,* **TANSA, 21:373. 1975.**

Techniques for acoustic emission testing of nuclear reactor components have been discussed previously. Here a measure from information theory is used to assess the benefits of improving the efficiency of such schemes for identifying structural flaws.

1159 **B. B. McCullough. "An Experimental Method for the Investigation of Acoustic Emission." M.S. Thesis. Tulane University, New Orleans, Louisiana. 1965.**

Acoustic emissions are sound pulses that are emitted from crystalline materials when subjected to an applied stress. Previous investigators have determined that surface finish and strain rates are parameters which affect acoustic emission in metals. The purpose of this investigation is to determine and evaluate an experimental method that would enable investigators to study acoustic emission and the parameters which affect this phenomenon. Sound levels associated with acoustic emission are low and can be detected only by highly sensitive devices with high amplification. The transducer employed for the detection of sound pulses is an ammonium dihydrogen phosphate piezoelectric crystal. The sensitivity of this transducer requires that the loading device be completely silent and free of any mechanical or electrical vibrations which might be confused with the actual acoustic emission signals. The system used in this investigation utilizes the buoyancy principle with load being applied as water is drained from a loading tank. This loading technique enables a wide range of loading rates so that the effect of strain rates on acoustic emission may be evaluated. Magnetic shielding and an anechoic chamber were used to reduce external mechanical and electrical vibrations. Surface effects on acoustic emission necessitates special care in the design and preparation of specimens to be investigated. Results are presented in the form of histograms with acoustic emission per ten-second intervals versus specimen strain. The results present ample proof that the experimental method used is capable of detecting and evaluating acoustic emission; however, an evaluation with suggested improvements is presented for consideration. The results obtained were in agreement with those by previous investigators, and the experimental method discussed should be helpful in determining the relation between acoustic emission and metallurgical properties of metal.

1160 **J. W. McElroy. "Development of Acoustic Emission Testing for the Inspection of Gas Distribution Pipelines." pp 59–79 in** *Monitoring Structural Integrity by Acoustic Emission, ASTM STP 571. Edite*d **by J. C. Spanner and J. W. McElroy. American Society for Testing and Materials, Philadelphia, Pennsylvania. 1975. proceedings of a symposium held in Ft. Lauderdale, Florida. Jan 17–18, 1974.**

The paper describes how a utility became interested in acoustic emission testing for the survey inspection of its buried gas pipelines. Several weld failures emphasized the need to develop inspection techniques to monitor these pipelines in order to locate weld flaws in high-stress regions to permit prompt weld repairs and assure system integrity. It was found that a weld must have a certain metallurgical flaw and must be located in a high-stress region in order to precipitate an inservice oxyacetylene weld failure in a pipeline. It was found that metallurgical flaws in high-stress regions can be located by acoustic emission tests. The paper describes the field development of the acoustic emission testing technique and the correlation of the field results with the laboratory results. It was found that the stress levels imposed by either internal pressurization or by external loading in the field tests are sufficient to cause critical flaws to emit and be located. Then replacement or reparation of the critical flaws located in this manner would restore the integrity of the pipline. Discussed in the paper are the other nondestructive correlations such as radiography and correlations through tension tests, hydrostatic tests, bend tests, and fatigue tests. [reprinted by permission of the American Society for Testing and Materials, Copyright].

1161 **K. R. McKinney, H. H. Chaskelis, and S. W. Freiman. "Prediction of Flaw Sizes from Acoustic Emission Measurements in Ceramics."** *Journal of the American Ceramic Society,* **JACTA, 59(7–8):369–370. July–Aug 1976.**

Recent work of Freiman *et al.*, and Evans and Linzer has shown that the acoustic emission count rate in ceramics is a function of crack velocity. The present note shows that acoustic emission count rate (normalized for crack width) for polycrystalline ceramics can be expressed as a function of crack velocity alone (no dependence on the stress intensity factor, K_I) and that acoustic emissions provide a technique by which the size of the largest propagating flaw in a structure giving rise to emissions can be determined.

1162 **L. McLain. "Sounding Out a Way to Discover Faults in Material Acoustically."** *Engineer,* **ENGIA, 241(6232):33. Aug 21, 1975.**

Stored elastic energy releases stress waves of a frequency within audio range, when a controlled load induces local deformation. The *cry* of Sn is quoted as an example. Gains of ~80 to 100 db are used from an electronic noise level of ~20 μV. Usually amplification is needed. Apparatus is mentioned which can pinpoint a source of stress wave emitted from twins, slip bands, phase change, and movement of a dislocations. The method can be used when defects are too small for conventional nondestructive testing. [Abstract 22-0405. *Metals Abstracts,* Vol. 9, 1976].

1163 **W. McMahon and E. S. Scherba. "Quick Release Acoustic Sensor Holding Fixture." NASA Tech Brief B72-10076. National Technical Information Service, Springfield, Virginia. 1972.**

1164 **R. L. Mehan and J. V. Mullin. "Analysis of Composite Failure Mechanisms Using Acoustic Emissions."** *Journal of Composite Materials*, **JCOMB, 5:266-269. Apr 1971.**

Advantages of the local heterogeneity of advanced composites in acoustic emission analysis are pointed out. Investigations in which the acoustic signals were detected by a low-mass high-frequency accelerometer bonded to the surface of the specimen are discussed. The signals were amplified and stored on magnetic tape. Typical photographs of individual fiber fractures of boron filaments in a three-filament composite are shown together with the acoustic signature typical of such events. Similar pictures are presented for carbon fiber fracture and debonding in novolac matrix.

1165 **P. Mehdizadeh. "Assessment of Corrosion Fatigue Damage by Acoustic Emission and Periodic Proof Tests." pp 286-302 in** *Proceedings of the Tenth Symposium on Nondestructive Evaluation.* **Southwest Research Institute, San Antonio, Texas. June 1975. symposium held in San Antonio, Texas. Apr 23-25, 1975;** *Materials Evaluation,* **MAEVA, 34(3):55-63. Mar 1976; paper presented at the American Society for Nondestructive Testing National Fall Conference. Atlanta, Georgia. Oct 13-16, 1975.**

Acoustic emissions generated by corrosion fatigue specimens during fatigue cycling and periodic proof tests were monitored to follow the progress of corrosion fatigue damage. Acoustic emission montoring of periodic proof tests was used to detect the initiation of corrosion fatigue damage. This method can be utilized to establish the initiation of incipient fatigue cracks after as little as 5% of the total fatigue life. A qualitative method based on the plateauing of acoustic emission during periodic proof loading was used to predict the remaining fatigue life. This scheme can predict the remaining fatigue life up to 70% of the actual values. The present investigation shows that periodic overloading has no adverse effect on fatigue performance in salt water. In fact, the overloading due to periodic proof tests improved the corrosion fatigue life under the testing conditions.

1166 P. Mehdizadeh (to Continental Oil Company). "Detecting Incipient Fatigue Damage in Metal." U.S. Patent 3,911,734. Oct 14, 1975.

A method of detecting incipient fatigue damage in a metal specimen by periodic proof tests wherein the metal specimen is removed from service, subjected to a load and the acoustic emission characteristics determined during application of the load.

1167 R. H. Meister. "Schallemissionsmessung: ein neues zerstorungsfreies Prufverfahren (Acoustic Emission Measurement: A New Nondestructive Test Method)." *Technica,* **TCHNA, 22(20):1956–1958. Sep 21, 1973. (in German).**

Acoustic emission testing is concerned with the causes, propagation, and, in particular, the assesment of acoustic waves, which are propagated at the velocity of sound, through plastically deformed solids and solids in which cracks are formed. The causes of the smallest cracks by overstressing, fatigue, stress corrosion, or embrittlement, and their subsequent development can be determined and their position accurately located. The technique can be applied to high and low alloy steels and many other industrial metals. Relationships between plastic deformation and acoustic emission can be determined. [Abstract 22-0248. *Metals Abstracts,* Vol. 7, 1974].

1168 R. P. Meister and M. J. Golis. "NDE of Hydrogen Reforming Furnace Tubes." pp 35–37 in *Paper Summaries: ASNT 36th National Fall Conference, September 27–30, 1976, Houston, Texas.* **American Society for Nondestructive Testing, Columbus, Ohio. 1976.**

This paper presents a review of the possible applicable NDT techniques for the inspection of hydrogen reforming tubes for cracks, internal carburization, and random corrosion. In view of typical discontinuities, and the state-of-the-art, such techniques as multifrequency eddy current, acoustic emission, and ultrasonics will be considered. The possibility of utilizing experimental systems and techniques will be reviewed.

1169 **R. P. Meister, M. D. Randall, D. K. Mitchell, L. P. Williams, and H. E. Pattee. "Summary of Nondestructive Testing Theory and Practice." NASA CR-2120. Battelle Columbus Laboratories, Columbus, Ohio. Oct 1972. [NTIS: N72-33482].**

The ability to fabricate design critical and man-rated aerospace structures using materials near the limits of their capabilities requires a comprehensive and dependable quality assurance program. The quality assurance program must rely heavily on nondestructive testing methods for thorough inspection to assess properties and quality of hardware items. A survey of nondestructive testing methods is presented to provide space program managers, supervisors, and engineers who are unfamiliar with this technical area with appropriate insight into the commonly accepted nondestructive testing methods available, their interrelationships, use, advantages, and limitations. Primary emphasis is placed on the most common methods as liquid penetrant, magnetic particle, radiography, ultrasonics, and eddy current. A number of the newer test techniques including thermal, acoustic emission, holography, microwaves, eddy-sonic and exo-electron emission, which are beginning to be used in applications of interest to NASA, are also discussed briefly.

1170 **V. P. Melekhin, R. I. Mints, and A. M. Kugler. "Effects of the Elastic Deformation of Zinc on Acoustic Emission."** *Izvestiya Vysshikh Uchebnykh Zavedenii, Tsvetnaya Metallurgiya*, **IVUTA, 14(3):128–131. 1971. (in Russian); RFP-Trans-123. translated for Dow Chemical U.S.A., Rocky Flats Division, Golden, Colorado, Jan 1973. [NTIS].**

The relation between the mechanism underlying the elementary acts of plastic deformation in metals (slip, twinning, etc.) and the amount of energy transformed into acoustic (as distinct from thermal, electric, and light) emission are discussed in the light of special experiments conducted on Zn crystals. The amplitude, intensity, and duration of the acoustic signals following such acts were studied in a specially developed electronic detection apparatus. Acoustic emission arose largely from the immediate vicinity of regions subject to plastic yield. The mean amplitude of the acoustic pulses accompanying twinning was an order of magnitude greater than that accompanying simple slip. [Abstract 32-0988. *Metals Abstract,* Vol. 4, 1971].

1171 **R. H. Merrill. "Design of Underground Mine Openings, Oil-Shale Mine, Rifle, Colo." Report of Investigations 5089. U.S. Bureau of Mines, Washington D.C. Dec 1954.**

The report contains a summary of the various studies made to determine the structural design of the production mine area of the Oil-Shale mine, Rifle, Colorado. Prior to this investigation, it had been determined from a study of the geology, the structure of the deposit, and mining costs that the deposit should be mined by open-stoping. The problem remained to determine the maximum safe roof span, the optimum extraction ratio, and the size of the pillars. These factors were determined from a study of the geologic structure of the deposit, the physical properties of the oil shale, and the magnitude and distribution of the stress around proposed underground openings. The latter was determined from both theoretical calculations and model studies. Safe roof spans, pillar sizes, pillar contours, and optimum extraction ratios were calculated from elastic theory and the physical properties of the mine rock. Safe roof spans, pillar sizes, and time effects were determined by Wright and Bucky by testing mine models in a centrifuge. The models were made from rock specimens taken from the proposed mine roof and pillars. The results from both studies agreed. The roof-span determinations were checked by tests made in a full-scale experimental stope, referred to as the testroom. The room was instrumented to measure roof sag, separations in the roof strata, and incipient failures. The dimensions of the testroom before instrumentation were 50 by 100 feet. The room was widened in successive increments to 80 by 200 feet. Between widening operations, periods up to 14 months were allowed to provide a time test of the roof spans.

The results of the sag measurements showed that the roof sag was greater than predicted by theory during the periods in which the room was being widened. Also, the roof continued to sag during periods when the dimensions of the room were unchanged. This anomalous sag was attributed to an elastic or plastic effect and is referred to as time sag. The sag measurements included a comparison of the sag at different elevations in the roof, referred to as differential sag. The differential-sag measurements did not disclose the development of any significant spearations within the 8-foot thick roof layer. However, stratascope examinations made in the roof disclosed some separations but none that appeared at the same elevations in all bore holes. The microseismic method of determining incipient rock failure was used concurrently with the other instrumentation. Loose slabs and other small-scale rock failures were detected and located by this method. Also, forewarning data were recorded prior to the failure of a 20-inch slab that occurred in August 1951. Since the roof sagged more than predicted by theory and numerous planes of weakness appeared at various roof elevations, it was concluded that the roof is made up of many layers or strata and that these strata were connected from point to point within the roof by bonds of varying strength. Some of these bonds may be ruptured by the stresses induced by time sag. This effect, called time deterioration, occurs at a slow rate. An 80- by 100-foot section of the room stood for 3 years before failure of the 20-inch slab. The computed safe extraction ratio was 82 percent. The selected mining program required only a 75-percent extraction ratio; thus, the selected areal size of the pillar, for 60-foot roof spans, was 60 feet square. The safety factors for pillars, which vary with the depth of overburden, were computed from elastic theory and tested by Wright and Bucky by mine-model studies. Again, essentially the same results were obtained. It was also determined, in the model studies, that the pillar rock would suffer a time effect but that the effect was negligible if the pillar were loaded below 75 percent of the compressive strength of the rock.

1172 R. H. Merrill and T. A. Morgan. "Method of Determining the Strength of a Mine Roof." Report of Investigations 5406. U.S. Bureau of Mines, Washington, D.C. 1958.

A method of determining the strength of a mine roof layer in place is described. The method was tested on a limestone roof layer 50 feet wide, 100 feet long, and 20 inches thick. In-place measurements were accomplished by introducing air under pressure into a separation between the layer and the overlying roof, and as pressure was increased, the corresponding strain and deflection of the roof layer were measured until the roof failed. The ultimate strength of the roof agreed with the tensile strength of the roof rock (modulus of rupture) as determined in the laboratory; the modulus of elasticity as measured in place did not agree with the laboratory-determined values; the deflection and tensile strain in the layer in place showed good agreement with the theory of a body-loaded beam clamped at both ends. This procedure for determining the strength of a mine-roof layer is safer and less costly than successively widening a roof to failure. For example, it would have been necessary to mine the room to a width of 114 feet and a length of 228 feet to produce failure in this roof layer. The method is not necessarily limited to limestone; any thinly bedded rock type can be tested by this procedure when there is an open parting over the mined area.

1173 T. E. Michaels. "Acoustic Emission of Zircaloy-2 during Tensile and Fatigue Loading." BNWL-727. Battelle-Northwest, Richland, Washington. Aug 1968. [NTIS].

The acoustic emission of Zircaloy-2 has been studied during tensile and fatigue loading at room temperature. A considerable amount of signal was received during these loading tests and metallographic evidence relating acoustic emission to deformation processes was acquired.

1174 T. E. Michaels and M. C. Fraser. "Acoustic Emission Behavior of Zircaloy-2 Pressure Tubing under Applied Stress." BNWL-545. Battelle-Northwest, Richland, Washington. Sep 1967. [NTIS].

The acoustic emission of Zircaloy-2 pressure tube material was studied during burst tests. Three material conditions were studied; as-received pressure tube material, as-received with addition of 275 ppm hydrogen, and irradiated material.

1175 D. J. Millard. "Twinning in Single Crystals of Cadmium." Ph.D. Thesis. University of Bristol, Bristol, England. 1950.

A study was made of twinning in single crystal wires of cadmium. Twinning was detected by a sensitive piezoelectric device that was specially developed for this work. It was found that providing twinning occurred before a specimen yielded in slip, the start of twinning both at room temperature and at -183 °C was governed by a resolved shear stress law. It was also found that at room temperature, creep by twinning took place. A model of a stepped matrix-twin boundary was made. Such a step was seen to have many of the properties of a dislocation and, since its motion converts a region of matrix into a region of twin, it has been called a twinning dislocation. An attempt has been made to explain the observed phenomena in terms of the movement of such twinning dislocations. The investigation of the start of twinning is believed to have been more thorough than any previous investigation of this nature. No previous study of creep by twinning is known to the author. Although the possibility that twinning dislocations are responsible for twin growth has previously been suggested, the detailed theories given in the discussion are new.

1176 **D. G. Miller. "Sounds Generated by Wood when under Stress."** *Research News* **(Canadian Department of Forestry, Forest Product Research Branch), 6(5): 6–7. 1963. [copy available from Library, Eastern Forest Products Laboratory, Department of the Environment, Montreal Road, Ottawa, Ontario, K1A OW5, Canada].**

1177 **A. R. Milne. "Statistical Description of Noise under Shore–Fast Sea Ice in Winter."** *Journal of the Acoustical Society of America,* **JASMA, 39(6):1174–1182. June 1966.**

Ambient noise under shore-fast sea ice in midwinter is generated at the surface of the ice by wind action as well as by mechanical cracking caused by declining air temperatures. An attempt has been made to relate field measurements of underwater noise to the sources of the noise by means of a geometrical model of the environment. The statistical properties and the spectra of the under-ice noise can be related to the surface density and the waveform of the source pressure pulses by means of the model.

1178 **A. R. Milne. "Thermal Tension Cracking in Sea Ice: A source of Underice Noise."** *Journal of Geophysical Research*, **JGREA, 77(12):2177–2192. Apr 20, 1972.**

A summary is given of the important observations that relate the rheological properties of sea ice to the occurrence of surface tension cracks and explain how the noises produced by myriad surface cracks are observed under the ice. Ice-cracking noise is expected where the surface of the sea ice is cold and has a sufficiently low salinity to approximate a brittle solid. This condition is more likely to occur in seas covered by a high percentage of multiyear ice and where wind mixing in leads was absent when the seas congealed. The acoustic water waves observed from crack-excited plate waves are described and two examples of single-frequency Crary wave arrivals are shown. A mechanism is postulated to explain the tendency for regularity in the spacing of pressure ridging in the Artic Ocean. Thermal pre-stressing at the ice surface, added to flexural stresses caused by wind-induced flexural-gravity standing waves, can produce significant tensile stress maximums at antinode spacings. For 3 metres of ice, the predicted antinode spacings are 132 metres where a high probability of through-the-ice fractures can occur. Flexural-gravity waves that propagate in the presence of thermal pre-stressing should produce a correlation of wind speed with ice-cracking noise. [reprinted by permission of author and *Journal of Geophysical Research,* copyright by American Geophysical Union, 1972].

1179 **A. R. Milne and J. H. Ganton. "Ambient Noise under Artic Sea Ice."** *Journal of the Acoustical Society of America,* **JASMA, 36(5):855–863. May 1964.**

Underwater ambient-noise spectra and amplitude distributions are described for data acquired during field experiments made in the spring, late summer, and winter within the Canadian Artic Archipelago. Measurements in each case were made using a bottom-mounted hydrophone. The ice cover was fast to the shore during the spring and winter experiments. In the summer experiment, the ice cover, although 10/10, was free to move. The ambient noise was generally impulsive and at times highly non-Gaussian. The significant noise was the result of mechanical acitivity associated with the ice cover. For shore-fast spring and winter ice, surface cracks as a result of thermal stresses were important; for late summer ice, relative motion of the floes was important.

1180 **A. R. Milne and J. H. Ganton. "Diurnal Variations in Underwater Noise beneath Springtime Sea Ice."** *Nature,* **NATUA, 221:851–852. Mar 1, 1969.**

Underwater ambient noise in ice-covered seas can show a dependence on air temperature. This temperature dependence is particularly striking in spring in the latitudes of the Canadian Artic Archipelago where the air temperatures are, in turn, dependent on the altitude of the Sun.

1181 **A. R. Milne and J. H. Ganton. "Noise beneath Sea Ice and Its Dependence on Environmental Mechanisms."** *U. S. Naval Journal of Underwater Acoustics,* **21(1):69–81. Jan 1971.**

The state of the sea-ice and its reaction to the atmosphere results in the production of underwater acoustic noise from a variety of mechanisms. Moving floes in the summer and fall produce noise with an intensity dependent on the mean daily wind. Under fast-ice, noise is produced by other mechanisms. One is thermal-noise which is caused by cracks as a result of surface cooling; another is wind-noise caused by the flow of air over the crystalline snow or ice surface. The latter two mechanisms are surface phenomena, so that the resultant underwater noise spectrum depends on the response of the ice and snow acting as a layered medium.

1182 **A. R. Milne, J. H. Ganton, and D. J. McMillin. "Ambient Noise under Sea Ice and Further Measurements of Wind and Temperature Dependence."** *Journal of the Acoustical Society of America,* **JASMA, 41(2):525–528. Feb 1967.**

Measurements of underwater-pressure spectral density from M'Clure Strait (2-5 May 1966), the Beaufort Sea (31 July and 16 August 1965), and the Canada Deep (14 April 1965) are presented. The M'Clure Strait measurements, made under shore-fast sea ice, exhibited a variability related to air-temperature and wind-speed changes of a similar nature to those reported previously for midwinter under-ice noise. The great differences in the magnitude and variation with frequency of the ambient noise observed under sea ice is again illustrated with three sets of measurements obtained during 1965-1966. The noise observed under the shore-fast ice of M'Clure Strait in the springtime appeared to be governed by air-temperature and wind-speed changes in a similar manner to noise observed previously in the midwinter [J. H. Ganton and A. R. Milne, *Journal of the Acoustical Society of America,* 38:406-411. 1965]. On the other hand, measurements made under the moving ice cover of the Canada Deep in the spring-time and of the Beaufort Sea in the summer appeared to be less amenable to description in terms of easily observable physical variables.

1183 **Yu. V. Miloserdin, V. M. Baranov, and K. I. Molodtsov. "Recording of Acoustic Emission Signals in Construction Elements."** *Soviet Atomic Energy*, **SATEA, 32(4):376–377. Apr 1972; translated from** *Atomnaya Energiya,* **AENGA, 32(4):330–331. Apr 1972. (in Russian).**

Sonic and ultrasonic generation during the deformation of materials (acoustic emission) may be employed in the defectoscopic investigation of reactor elements. Acoustic warning signals are produced when cracks arise in structural members. The detecting apparatus may consist of a pulse detector, and either an oscillosgraph or other recorder. More elaborate equipment for registering microcracks is described in block diagram form, and includes an acoustic transducer, piezoelectric gage, preamplifier, amplifier, and recorder.

1184 **R. I. Mints, V. S. Kortov, and V. P. Melekhin. "Effect of the Mechanisms of Plastic Deformation on the Acoustic and Exo-Electron Emission [of Metals]."** *Akademiya Nauk Ukrayins'koyi SSR, Metallofizika,* **MFIZA, (44):93–96. 1973. (in Russian); Translation RTS–9680. British Library, Lending Division, Boston Spa, Wetherby, West Yorks, England. Feb 1975. (in English).**

The mechanisms underlying the acoustic and exo-electron emission of metals are discussed. The results indicate that transient emission of these classes is an almost unvarying accompaniment to the plastic deformation of metal and the intensity and general characteristic of the emission depend on the nature of the plastic deformation. In Zn crystals, twinning and slip processes produce emissions of different intensities, as may be appreciated on analyzing the amplitude distribution of the sound pulses. The parameters of the emitted signals are related to the mobility of the defects created in the surface layers of the metals. [Abstract 33-1867. *Metals Abstracts,* Vol. 6, 1973].

1185 **R. I. Mints, V. P. Melekhin, I. Yu. Ievlev, and V. V. Bukhalenkov. "Acoustic Emission in a Thermoelastic Martensite Reaction."** *Soviet Physics–Solid State,* **SPSSA, 14(5):1368–1369. Nov 1972; translated from** *Fizika Tverdogo Tela,* **FTVTA, 14(5):1582–1583. May 1972. (in Russian).**

The relationship between the parameters of the acoustic emission attending the nucleation, growth, and disappearance of the martensite crystal in the thermoelastic martensite transformation was investigated. The object of the study was the alloy Cu + 14% Al + 1.6% Ni (by charge), which was quenched from 1200 K in water and in which the thermoelastic martensite transformation $\beta_1 = \gamma'$ is realized. The results confirmed the identical nature of the dislocation mechanisms governing the twinning and martensite transformation processes. [Abstract 3518. *Physics Abstracts*, 76(958). 1973].

1186 **S. R. Mintzer and R. Pascual. "Acoustic Emission and Microstructure during Plastic Deformation of an Aluminum Alloy."** *Materials Science and Engineering,* **MSCEA, 19(2):289-290. June 1975.**

This is a *Short Communication* on a research program to study the acoustic emission response of 2024 aluminum with different degrees of aging during tensile deformation. All specimens exhibited a maximum count rate. It is reported that as the aging time was increased the height of the count-rate curve decreased and the width of the peak and strain at which it occurred increased. The authors state there is an obvious relationship between the shape of the count rate versus strain curves and the microstructure resulting from aging, however, they have reported detailed information on this elsewhere (see Reference 1376).

1187 **M. Mirabile. "Acoustic Emission Energy and Mechanisms of Plastic Deformation and Fracture." pp 189-221 in** *Schallemission: Anwendung bei der Untersuchung, Prufung und Uberwachung metallischer Werkstoffe,* **Deutsche Gesellschaft fur Metallkunde eV, 637 Oberursel/Taunus bei Frankfurt/Main, Adenauerallee 21, Germany. 1974. proceedings of symposium held in Munich, Germany. Apr 18-19, 1974;** *Non-Destructive Testing,* **NDETA, 8(2):77-85. Apr 1975.**

The application of acoustic emission to industrial NDT is hindered by a lack of exact knowledge of the significance of parameters such as total count, to characterize the emissions. These parameters, and any novel parameters which can be derived, must be critically re-examined for their relation to metallurgical, mechanical, and geometrical characteristics of structures if progress is to be achieved. The author describes experimental and theoretical work to this end on emissions from plastic deformation and fracture of various test pieces. The author concludes from his findings that the problems of noise and transducer sensitivity seem to be surmountable.

1188 **M. Mirabile. "Application de l'emission acoustique a l'etude de la rupture des aciers (Application of Acoustic Emission to the Study of the Fracture of Steel)."** *Memoires Scientifiques de la Revue de Metallurgie,* **MRMTA, 71(9): 495-507. Sep 1974. (in French); BISI-13314. translated by the British Industrial and Scientific International Translation Service, The Metals Society, London, England. Sep 1975. [available from the American Society for Metals, Metals Park, Ohio].**

Acoustic emission is a comparatively new technique of nondestructive testing, but at present it does not produce a correct interpretation of the experimental data. Some theoretical models are proposed, making it possible to interpret the origin of acoustic emission. The results obtained on low-strength steels are linked with the different fracture mechanisms. Possible effects of these results on the practical application of the method are demonstrated. [Abstract 22-0416. *Metals Abstracts,* Vol. 8, 1975].

1189 **M. Mirabile. "Plastic Deformation and Fracture Mechanisms Investigation by Means of the Stress-Wave-Emission Technique." Session 4, pp 66–91 in** *The Second Acoustic Emission Symposium.* **Japan Industrial Planning Association, Dobashi Building 5F, 3–17–8 Akasaka, Minato-ku, Tokyo, 107 Japan. 1974. proceedings of a symposium held in Tokyo, Japan. Sep 2–4, 1974.**

Stress-wave emission is a technique which has found, in the recent past, many uses and applications, both as a fundamental research tool and as a nondestructive test. However, its use at an industrial level is as yet limited because of many unresolved problems regarding the quality and the reliability of the information provided by the emitted signals. It is not well-established, for instance, whether it is sufficient for practical nondestructive checks to use the total count N_T alone in order to ascertain where a crack is growing, how far it has grown, and why this is occurring. The question thus arises of how it is possible to distinguish a plastic deformation process from a fracture process by means of the total count and/or its time derivative alone. Furthermore, while K_{Ic} is a decisive parameter in brittle fracture prevention, it is of little interest in the study and characterization of ductile materials at present of greater interest, where the conditions of crack instability are controlled by the crack-opening displacement. Thus for the SWE technique to develop towards its industrial applications under the large variety of conditions in which a nondestructive control is possible and required, the significance of the old parameters must be critically revised. New quantities need to be found and correlated with the metallurgical, mechanical, and geometrical characteristics of the structure. The aim of the present paper is to make some contribution to these problems. In the next section, material properties, experimental procedures which allow new quantities to be measured, and revised forms of the traditional parameters are illustrated. Results are reported which relate SWE activity to plastic deformation processes occurring during tensile testing of unnotched testpieces, as well as to fracture processes occurring in notched testpieces during fracture mechanics tests. An attempt is made to explain such results according to the plastic deformation processes in differently strengthened materials, to Cottrell's model of brittle crack growth and to matrix-inclusion decohesion processes during ductile fracture.

1190 **M. Mirabile. "Power Spectral Density in the SWE Techniques and a Possible Theoretical Explanation." Centro Sperimentale Metallurgico, Rome, Italy; paper presented at the Institute of Physics Conference on Acoustic Emission, Imperial College, London, England. Mar 14, 1972. [proceedings were never published].**

In order to investigate by help of SWE the various mechanisms of fracture, hydrogen embrittlement and so on, the knowledge of three probability functions seems very important: P_k (t), probability of k emissions in the time interval 0-t; P (ϵ, t) probability for a pulse of energy ϵ to be emitted in the same interval; and P (f,t) the probability that a pulse emitted in this interval contains the frequency f. While literature gives many informations about the function P_k(t) and few about the P (ϵ, t), the function P (f,t) is almost unknown. Its knowledge both by theoretical and experimental investigation could give many informations in the dislocation dynamics, from which a better explanation might be reached of the mechanism of SWE itself. In fact some results obtained suggest a model for the SWE as follows: if a dislocation moves in its slip plane with a nonrelativistic speed, it brings along with itself its displacement field as an electric charge does for its electromagnetic field. Just as in this case, it is possible to define a Poynting's vector for a moving dislocation. In practice, a dislocation can be accelerated for example by an external driving force which varies its state of motion, or by the breakaway of piled-up dislocations. In this frame, it can be assumed that the emitted energy depends on the acceleration of dislocation, and that the Fourier spectrum of a single pulse is the Fourier transform of some function of its Poynting's vector. These considerations extended to the fracture processes permit predictions of different power spectral density (PSD) for brittle or ductile fracture. On bending specimens of 38 NCD 4 Steel, some measurements have been carried out of both frequency spectrum of single pulses and PSD for all pulses emitted, in the 20 to 20,000 Hz range. Accounting for the influence of transducer transfer function and the Eigen functions of a machine-specimen system, it can be concluded that the PSD under brittle is different from the PSD under ductile fracture conditions. These results can be explained by the theory of radiation from moving dislocations; futhermore they allow to evaluate for both brittle and ductile fracture the order of magnitude of the energy lost as acoustic emission, as related to the total energy supplied by the machine.

1191 **M. Mirabile and E. Palombi. "Dependence of Stress Wave Emission upon Brittle and Ductile Fracture Mechanisms." Paper II–234 in** *Dritte Internationale Tagung uber den Bruch. Zusammenstellung der Tagungsberichte, Tiel III: Bruchmechanik und Prufverfahren.* **Verein Deutscher Eisenhuttenleute, Dusseldorf, Germany. 1973. volume of preprints of the 3rd International Congress on Fracture, held in Munich, Germany. Apr 8–13, 1973. [preprints out of stock from publisher and proceedings never published].**

In recent years, a number of investigations have been carried out in order to clarify the relation between stress wave emission and mechanical variables in fracture mechanism tests. However, the problem of the dependence of SWE upon fracture mechanism has been scarcely investigated. The authors have studied this dependence. Some typical results are discussed and qualitatively interpreted in terms of dislocations dynamics.

1192 **J. R. Mitchell. "Acoustic Emission for Monitoring High Cycle Fatigue Crack Propagation." M.S. Thesis, University of Oklahoma, School of Aerospace, Mechanical and Nuclear Engineering, Norman, Oklahoma. 1973.**

The study of high-cycle fatigue, utilizing acoustic emission techniques, involves all the difficulties encountered in the past and adds to this list the traditional problems associated with fatigue investigation. Thus, until the present time, the study of crack growth rate has been confined to constant load situations and low-cycle fatigue. In these investigations, the crack growth rate has been greater than 10^{-4} inches per cycle which is above the rate encountered in most engineering applications. In this investigation, the crack growth rates are between 10^{-6} and 10^{-4} inches per cycle which is a realistic situation. The detection of these slowly growing cracks has been made possible by the use of specimen isolation techniques and modifications to the acoustic emission detecting system. It has been the aim of this project to perform a realistic investigation of high-cycle fatigue crack growth utilizing acoustic emission. Included in this study are: the layout of an experimental procedure and calibration system which eliminates some of the shortcomings of previous acoustic emission research; an examination of crack growth rate versus acoustic emission rate for three metals; and a detailed explanation of the micromechanisms, as identified by acoustic emission, which lead to fatigue failure.

1193 **J. R. Mitchell and D. M. Egle. "Acoustic Emission Monitoring of Fatigue Crack Propagation." summary of paper in** ***Paper Summaries: ASNT Spring Conference, March 28–30, 1977, Phoenix, Arizona.*** **American Society for Nondestructive Testing, Columbus, Ohio. 1977.**

A series of fatigue experiments were conducted to investigate the relationship between acoustic emission activity and crack growth rate for 7075-T6, 2024–T3 aluminum and mild steel. Cantilevered specimens, notched to promote crack growth, were cycled under tension-tension loading using controlled displacements. Acoustic emission was monitored in the conventional manner utilizing a narrow-band filtered resonant transducer. A pseudo-emission source was used to maintain constant detection sensitivity for all experiments. Two emission types could be identified: a rate dependent emission occurring at points of maximum velocity (and strain rate) in the displacement cycle, and a rate independent burst type emission occurring at the points of maximum and minimum deflection as well as points of maximum velocity. The rate dependent emission was associated with crack nucleation and the burst emission with crack extension. Although the count rate-crack growth rate relationship showed considerable scatter, distinctive trends among the three materials were evident.

1194 **J. R. Mitchell, D. M. Egle, and F. J. Appl. "Detecting Fatigue Cracks with Acoustic Emission." University of Oklahoma, School of Aerospace, Mechanical and Nuclear Engineering, Norman, Oklahoma. 1972; pp 121–126 in** ***Proceedings of the Oklahoma Academy of Science.*** **Volume 53. Oklahoma Western State College, Weatherford, Oklahoma. 1973. meeting held in Weatherford, Oklahoma. Nov 17, 1972.**

Acoustic emission has emerged as a technique for detecting crack growth in a wide range of materials and has been useful in proof testing structures susceptible to fatigue damage. This proof testing involves removing the structure from service and loading it in an ideal environment. Results of the present study, which is directed towards the use of the acoustic emission technique in an in-service crack monitoring system, indicate that continuously monitored acoustic emission can detect crack growth and failure in a specimen undergoing low-cycle fatigue. Tests with typical aircraft bolted structures show mechanical slippage at the joint. Preliminary results of high-cycle fatigue tests indicate that acoustic emission from crack growth may be detected in a high background noise environment.

1195 J. S. Mitchell and J. E. Borhaug. "Modern Methods and Instrumentation for On–Stream Predictive Analysis of Rotating Machinery." Dow Industrial Service, Pleasant Hill, California. 1972; paper presented at the American Society for Nondestructive Testing National Fall Conference, Cleveland, Ohio. Oct 16–19, 1972.

Vibration from operating machinery contains a vast amount of information which defines mechanical performance. It has been customary to record levels and take oscilloscope photographs of time domain waveforms and shaft orbits, but these methods restrict the amount of information which can be assimilated and analyzed. Transforming the vibration signal from the time to the frequency domain with a wave or frequency analyzer is one method of presenting a complex vibration signal in an easily interpretable form. The plots thus obtained may be analyzed on their own merits to detect abnormalities or more importantly can be compared over a period of time to detect small trends before they reach a level of severity which would force a shutdown. Several case histories illustrate the use of this type of analysis.

1196 L. D. Mitchell. "An Investigation of the Correlation of the Acoustic Emission Phenomenon with the Scatter in Fatigue Data." Ph.D. Thesis. University of Michigan, Ann Arbor, Michigan. 1965; University Microfilms 66–6657.

In the past, sounds have been heard coming from materials such as magnesium and tin during bending. Recent observations show that many materials emit sounds when stressed, but these sounds are generally below the threshold of hearing. Such sounds have been termed acoustic emissions. Two principal types of emissions can be identified: (1) a burst type in which a sudden energy release produces an exponentially decaying vibration of the specimen, and (2) a *high frequency* type which will be referred to as high rate emission and which appears to be a continuous wave emitted from the specimen. The source of the high rate emission may be a high rate impulse excitation or, less likely, a periodic forcing function. The principal effort in this investigation has been to develop a nondestructive testing technique using the characteristics of the acoustic emission as predictive variables. A literature review completed in the areas of acoustic emission and fatigue showed that identical models have been proposed to explain both the emission and fatigue behavior of materials. Because of this similarity a correlation has been sought between fatigue life and the acoustic emission characteristics of AISI E4340 steel. In addition to the emission variables other variables that have been thought to affect fatigue life were included as possible predictions. The correlation experiment had a totally randomized design.

This helped to insure that important correlations would not be obliterated by unknown data trends. Two statistical programs were used to analyze the data. The first was the Multiple Classification Analysis program; the second, the Stepwise Regression with Simple Learning program. The former provided information concerning the importance of each predictive variable and its relationship to fatigue life. The relationships derived from this program were supplied to the regression program along with other standard relationships. The computer then developed a least squares mathematical model of fatigue life behavior based upon the predictors and functions of the predictors. Both programs rejected all the acoustic emission characteristics as well as functions of them as possible predictors of fatigue life because they lacked significance in the prediction. Moreover, the programs showed that other variables such as rms surface finish are of little importance in predicting fatigue life and thus they, also, were rejected as predictors. The predictions obtained from the regression program showed that the engineering stress beyond the upper yield stress was the best predictor of fatigue life.

It is concluded that there is little reason to believe that the acoustic emission characteristics as determined in this work are correlated with the finite fatigue life of AISI E4340 steel. Also, it seems that burst type emission is due to an elastic mechanism whereas high rate emission is due to a plastic deformation process occurring at or near the upper-lower yield point. This research uncovered a number of previously unreported acoustic phenomena which may prove helpful in studying material damping, mechanical hysteresis effects, locking and frictional stresses, and diffusion rates. The following phenomena are examples of what has been observed. A new type of emission which appears as a continuous low frequency wave was observed to occur during both the loading and unloading at stresses of from 4,000 to 20,000 psi but not outside of this range. This emission has a frequency in the vicinity of 250 kHz. Strain aging was found to change the stress level at which the rate of load emission is greatest. Reproducible emission phenomenon was observed during the unloading of the specimens. A range of stress, ΔS, was observed at the beginning of the unload portion of the load-unload cycle. During this stress range, no emission occurs. Moreover, it was found that ΔS increases as the maximum applied stress, S_{max}, increases. [reprinted by permission of University Microfilms International from *Dissertation Abstracts International,* Vol. 27, No. 2B, 1966. Copyright 1966 by University Microfilms International].

1197 **L. D. Mitchell and J. R. Frederick. "Plastic Stress and Acoustic Emission as Predictors of Fatigue Life."** *Experimental Mechanics*, **EXMCA, 10(8):328–332. Aug 1970; paper presented at 1970 Society for Experimental Stress Analysis Spring Meeting. Huntsville, Alabama. May 19–22, 1970.**

The high rate of acoustic emission from AISI-E4340 annealed steel when it is stressed can be used to accurately indicate the yield stress. Predictive equations for the fatigue life of AISI-E4340 annealed steel have been developed by the use of *Multiple Classification Analysis* and *Stepwise Regression* computer programs based on an experimental study of 175 specimens. The acoustic emission from the fatigue specimens was investigated prior to fatiguing, but no parameter of the emission was found to correlate significantly with fatigue life. The high-rate emission which occurs at yield is found to be useful in determining the value of the yield stress. The value of the stress into the plastic range determined in this manner was defined by both statistical approaches as the best predictor of fatigue life. The fatigue stress level was rejected as a lesser predictor. Surface finish in the range of 90 to 350 microinches was found to have little correlation with fatigue life of this material.

1198 **M. Miyairi, M. Tomizawa, E. Yamamoto, and H. Hisakawa. "Several Applications of Acoustic Emission on the Japanese National Railway." Paper No. 5 in** *Proceedings of the U.S.-Japan Joint Symposium on Acoustic Emission.* **Japanese volume. Japan Industrial Planning Association, Kawamura-Building 6F, 3-21-6 Akasaka, Minato-ku, Tokyo, Japan 107. 1972. symposium held in Tokyo, Japan. July 4-6, 1972. (paper in Japanese, abstract available in English).**

For the purpose of monitoring fatigue failure of track and vehicular parts laboratory tests were conducted on welded rail and bar specimens. These included bend, fatigue bend, and fatigue tension testing to failure of welded rail and bar specimens, and a ball-drop test on an 18-foot long rail. Also a frequency analysis was made of a vehicle moving at a constant speed of 20 km/hour. Preliminary test results indicate acoustic emission looks promising as a practical inspection technique for maintaining high reliability in track and vehicles.

1199 **M. E. Mizell and W. T. Lundy, Jr. "In-Flight Crack Detection System for the C-135 Lower Center Wing Skin." ISA Reprint 76240. Instrument Society of America, Pittsburgh, Pennsylvania. 1976; paper presented at the 22nd International Instrumentation Symposium, San Diego, California. May 25-27, 1976. [copy of reprint also available as Technical Report DE 76-2 from Dunegan/Endevco, San Juan Capistrano, California].**

A system has been developed by Oklahoma City ALC to detect the occurrence of unstable crack growth in C/KC-135 lower center wing skin panels in flight. Wing skin material currently used on C/KC-135 aircraft is a strong, but brittle, aluminum. As part of a program to assure structural integrity pending major modification accomplishment, a wing center section crack detection system was needed to insure *fail safe* operation. Acoustic emission transducers are used to pick up crack signals. A signal processor discriminates between crack and noncrack signals and provides a warning light and a permanent record when a crack is detected.

1200 **Mobile Oil Corporation. "Method and Apparatus for Detecting Particulate Material in Flow Stream." British Patent 1,435,972. May 19, 1976.**

1201 **J. A. Mock. "Special Report: A Guide to Nondestructive Testing."** *Materials Engineering,* **MAENB, 69(6):60-70. June 1969.**

1202 **I. L. Mogford. "Acoustic Emission and Fracture in Steels." Central Electricity Research Laboratory. Leatherhead, Surrey, England; paper presented at the Institute of Physics Conference on Acoustic Emission, Imperial College, London, England. Mar 14, 1972. [proceedings were never published].**

The work has been concerned with interpretation of the acoustic emission response of steels during deformation and subsequent fracture. The aim is to establish if the technique can be used to assure structural integrity, as well as a means of defect location. The classical behavior of emission rate rising smoothly to a maximum coincident with fast fracture, as originally reported for high strength steel, is not found universally. Data are presented showing how the acoustic emission characteristics of mild steel and low alloy steel are related to particular failure modes over a wide temperature range. The emission data have been analyzed in terms of signal amplitude distribution, as well as rate of emission. Finally the correspondence between the experimental data and the behavior of model and real pressure vessels is discussed.

1203 **K. Mogi. "Acoustic Emission from Stressed Rocks." Paper No. 3 in** *Proceedings of the U. S.-Japan Joint Symposium on Acoustic Emission.* **Japanese volume. Japan Industrial Planning Association, Kawamura-Building 6F, 3-21-6 Akasaka, Minato-ku, Tokyo, Japan 107. 1972. symposium held in Tokyo, Japan. July 4-6, 1972. (paper in Japanese, abstract available in English).**

Most rock materials are brittle and inhomogeneous. In such inhomogeneous brittle rocks, a number of elastic waves of shock-type (acoustic emission) are generated by microfractures during deformation. Acoustic emission has been used to examine the mechanics of deformation and failure of rocks. A large increase of acoustic emission activity before fracture suggests a method for predicting earthquakes, which may be regarded as a large-scale example of rock fracture. In this paper, some fundamental features of acoustic emission from stressed rocks are discussed mainly based on our recent studies.

1204 **K. Mogi. "Magnitude Frequency Relation for Elastic Shocks Accompanying Fractures of Various Materials and Some Related Problems in Earthquakes."** *Bulletin of the Earthquake Research Institute, University of Tokyo,* **TDJKA, 40:831-853. 1962.**

1205 **K. Mogi. "Rock Fracture."** *Annual Review of Earth and Planetary Sciences,* **AREPC, 1:63-84. 1973.**

Studies of rock fracture have been carried out in solid earth science, which deals with properties and processes of the earth, and in civil, mining, and petroleum engineering. This area of study is relatively new, and knowledge has increased rapidly in the past two decades, primarily because of recent developments in experimental equipment and techniques. In earth science particularly, rock fracture has recently been given considerable attention due to the growing interest in earthquakes and other tectonic phenomena. As a branch of materials science, studies of rock fracture have some special considerations. First, most rocks are brittle and have structural and mechanical inhomogeneities ranging from microscopic to macroscopic scale. Second, the failure behavior of rock is greatly affected by varying pressures and high temperatures which may characterize the natural conditions of deformation. Third, the stress may sometimes have to be applied at a slow rate to simulate tectonic conditions. And, it is important to recall that rock fracture takes place in an infinite medium in nature. Recently, many reviews have appeared on this subject, some containing extensive bibliographies. In this chapter, recent studies on some fundamental problems in rock fracture are outlined and critically reviewed.

1206 **K. Mogi. "Some Discussions on Aftershocks, Foreshocks, and Earthquake Swarms–The Fracture of a Semi-Infinite Body Caused by an Inner Stress Origin and Its Relation to the Earthquake Phenomena." (3rd paper).** *Bulletin of the Earthquake Research Institute, University of Tokyo,* **TDJKA, 41:615–618. 1963.**

1207 **K. Mogi. "Study of Elastic Shocks Caused by the Fracture of Heterogeneous Materials and Its Relations to Earthquake Phenomena."** *Bulletin of the Earthquake Research Institute, University of Tokyo,* **TDJKA, 40:125–173. 1962.**

1208 **K. Mogi. "The Fracture of a Semi-Infinite Body Caused by an Inner Stress Origin and Its Relation to the Earthquake Phenomena." (1st paper).** *Bulletin of the Earthquake Research Institute, University of Tokyo,* **TDJKA, 40:815–829. 1962.**

1209 **K. Mogi. "The Fracture of a Semi-Infinite Body Caused by an Inner Stress Origin and Its Relation to the Earthquake Phenomena. 2nd paper: Case of the Materials Having some Heterogeneous Structures."** *Bulletin of the Earthquake Research Institute, University of Tokyo,* **TDJKA, 41:595–614. 1963.**

1210 **J. A. Moore. "Acoustic Emission Monitoring of Advanced Aerospace Components." summary of paper in** *Paper Summaries: ASNT Spring Conference, March 28–30, 1977, Phoenix, Arizona.* **American Society for Nondestructive Testing, Columbus, Ohio. 1977.**

The paper summarizes the efforts at Vought to monitor the acoustic emissions from advanced aerospace components during static and fatigue testing. The acoustic emission plots derived during fatigue testing of a multilaminated fuselage fitting, fatigue testing of a conventional flap attach fitting, and static testing of the attach lug of a steering component are reviewed. Plots were obtained during fatigue testing and static loading to failure of graphite composite outer wing panel test articles. These data were used to evaluate plots for production wings that are scheduled for extensive in-flight testing. Also reviewed are the results of monitoring pressure gas welds during the welding operation and during proof load, plus the results of monitoring flash-welded control rods during proof load. Finally, the results from monitoring carbon-carbon materials (Space Shuttle leading edge) during static loading are presented.

1211 **J. F. Moore. "The Early Detection of Fatigue Damage." NA-70-640. Third Semiannual Report, 1 June 1970–30 November 1970. North American Rockwell Corporation, Los Angeles, California. Nov 30, 1970.**

The report describes the continuing effort to develop nondestructive test methods capable of determining the extent of fatigue damage and providing a means of predicting the future safe life of aerospace materials and structures. During this report period, the fatigue process in 1100-0 aluminum was studied by means of exoelectron emission and acoustic emission measurements. The exoelectron observations were by means of measuring the current produced by fatigue damage in air; the acoustic emission measurement was based on the number of acoustic events occurring in selected time intervals during fatigue. It appears that there is a trend indicative of a stress-independent relationship between the change of exoelectron emission current and the precentage of the spent fatigue life based on the current after a selected number of cycles. The intensity of acoustic emission was shown to increase or decrease by orders of magnitude during the fatigue life. The first significant change of acoustic emission was apparently correlatable to 8 to 20 percent of the specimen fatigue life. The continuing metallographic examination confirmed that surface slip striations appear early in the fatigue process and appear to directly correlate with the exoelectron emission phenomenon.

1212 **J. F. Moore and S. Tsang. "The Early Detection of Fatigue Damage by Exoelectron Emission and Acoustic Emission." pp 143–157 in** ***Testing for Prediction of Material Performance in Structures and Components, ASTM STP 515.*** **American Society for Testing and Materials, Philadelphia, Pennsylvania. 1972.**

The paper describes the development of nondestructive test methods capable of determining the extent of fatigue damage and providing a means of predicting the future safe life of aerospace materials and structures. The fatigue process of 1100-0 aluminum was studied by means of exoelectron emission and acoustic emission measurements. The exoelectron observations were made in both vacuum and air. The exoelectron emission measurement in vacuum was accomplished by counting the emission events for selected time intervals. The measurements in air were performed using a specially developed current-measuring system with a resolution of 10^{-15} amperes. The acoustic emission measurement in air was based on the number of acoustic events occurring in selected time intervals during fatigue. The exoelectron emission curves exhibited an early rapid rise and fall of emission intensity, and again a rapid rise near the end of the test. The initial emission decrease coincided with metallurgically observed changes of surface slipbands. There was also a trend indicative of a stress-independent relationship between the change of exoelectron emission current and the percentage of life after a selected number of fatigue cycles. Analysis of the acoustic emission data showed a marked increase in emission rate in less than 50% of the fatigue life of the material.

1213 **J. F. Moore, S. Tsang, and G. Martin. "The Early Detection of Fatigue Damage." AFML-TR-71-185. North American Rockwell Corporation, Los Angeles, California. Sep 1971. [NTIS: AD-730-348].**

This is the final technical report for a program directed at the development of nondestructive test methods for the detection of early fatigue and fracture damage in metals and alloys. The program is based on an interdisciplinary approach designed to interrelate the factors of early fatigue damage with measurable physical phenomena. The program initially concentrated on a comprehensive study of the existing knowledge of fatigue phenomena in metals, and the results of the study are described in terms of fatigue and fatigue-associated phenomena, metallurgical structure, effect of interrelating fatigue phenomena on physical properties, and the availability of appropriate measurement techniques and equipment. Next, the program developed a series of controlled fatigue experiments to measure quantitatively the fatigue effects in selected metal specimens. These tests also included a systematic metallographic evaluation to determine the actual depth and character of the surface layer affected by progressive fatigue, particularly in the early stages of fatigue. Finally, NDT methods were evaluated in terms of their potential detection and measurement capability of the observed fatigue-related effects and damage as determined by the study and fatigue evaluation tests. Based on this analysis, the following NDT methods were selected for detailed evaluation for static and in-process measurement of fatigue damage: Exoelectron emission, acoustic emission, and ultrasonic surface wave attenuation and velocity. The change of exoelectron emission current was found to have a relation to the spent fatigue life. A marked change of acoustic emission intensity could also be related to the percentage of fatigue life. A correlation existed between the attenuation of ultrasonic surface wave and the length of fatigue time early in the process. Characterization of fatigue-damaged surface was studied by measurements of surface potential difference and ellipsometric parameters. There is a relationship between surface potential difference and the length of fatigue time. Values of ellipsometric parameters also undergo a change in fatigue process. Based on the promising potential of the exoelectron emission and ultrasonic surface wave test methods, further development is recommended using actual air vehicle components and service test conditions.

1214 **J. F. Moore, S. Tsang, D. O. Thompson, and S. A. Hoenig. "Nondestructive Detection of Structural Damage uniquely Associated with Fatigue." AFML-TR-74-131. Rockwell International Corporation, Los Angeles Aircraft Division, Los Angeles, California. July 1974. [NTIS: AD/A-005-357].**

This is the final technical report for a program directed at the development of reliable nondestructive testing and inspection methods to locate and assess fatigue damage in Air Force aircraft structures. The primary nondestructive test approach is the application of exoelectron emission to the detection of fatigue damage. The program initially conducted a series of controlled fatigue tests in an ultrahigh vacuum chamber equipped with devices to measure the photo-stimulated exoelectron current and to fully characterize the surface state of the specimen prior to, during, and after fatigue measurement. The stimulated exoelectron current was found to increase with increase in applied strain amplitude and decrease with surface oxide thickness. A new exoelectron emission mechanism, based upon the roughness-enhanced, optical-photoyield coupling phenomena, was developed. Next, the program was directed to perform a series of fatigue tests in air on specimens of different shapes with and without discontinuities under various loading and environmental conditions, in order to develop a practical technique for applying exoelectron emission and surface potential differnece measurements to assess fatigue damage, as well as to predict final failure sites. Based upon the nondestructive measurements, a quantitative-statistical method was developed to facilitate the analysis and comparison of exoelectron emission data gathered during fatigue test for the prediction of the failure site. Another method of analyzing both exoelectron emission and surface potential difference data obtained during several interruptions of the fatigue cycling was also developed. Finally a prototype system design for the automatic scanning and exoelectron emission testing of aircraft-type structures was developed.

1215 **V. W. Moore, Jr. "Acoustical Methods of Evaluating Thermal Shock Resistance." M. S. Thesis. Clemson University, Clemson, South Carolina. 1972.**

A study was made to perfect a system of measuring the acoustic emissions generated by refractories during exposure to rapidly changing temperatures. An equipment arrangement was developed which successfully measured these sounds without pick-up of extraneous noise. Tests were conducted which demonstrated that the acoustic emission system would predict the thermal shock resistance of commercial refractories. It was found suitable for magnesia as well as alumina-silica refractories. The test method not only ranked refractories as to their thermal shock resistance, but also indicated the temperature intervals during which thermal damage occurred. The system proved useful in designing optimum firing schedules for the production of refractories and other ceramics. A resonant decay time test was evaluated for measuring the extent of thermal shock damage. A test method was developed and it was demonstrated that a decrease in resonant frequency and a decrease in decay time occurs after exposure of samples to a thermal shock test. The test did not appear to be suitable for quantitative ranking of the extent of damage between different types of refractories. Instead it shows only whether or not damage has occurred. The test did appear to be a superior method for determining the resonant frequency of refractories to support calculations of Young's Modulus of Elasticity.

1216 **C. F. Morais and A. T. Green. "Establishing Structrual Integrity Using Acoustic Emission." pp 184–199 in** *Monitoring Structural Integrity by Acoustic Emission, ASTM STP 571.* **Edited by J. C. Spanner and J. W. McElroy. American Society for Testing and Materials, Philadelphia, Pennsylvania. 1975. proceedings of a symposium held in Ft. Lauderdale, Florida. Jan 17–18, 1974.**

Acoustic emission techniques have advanced from laboratory use to applications, such as establishing the quality of pressure containment and many other structures. The authors discuss the statistical relationship between acoustic emission data and the structural integrity of a series of production pressure vessels. In this application, acoustic emission data obtained from the first 25 percent of a required hydroproof test determined the bursting strength of the vessel. Additional information contained in the data is shown to establish the mode of structural degradation. At least seven ways of utilizing acoustic emission data to establish the quality of pressure vessels are presented. Use of acoustic emission techniques to detect and locate a growing defect in a structure, while under a relatively severe environmental condition, is presented. Various methods of data processing are described and the relations to parameters which may indicate the degree of quality of the structure are shown. Data from materials such as glass-reinforced plastics, concrete, glass, and metallics are presented. References to less well-known efforts show the universal nature of the technology. [reprinted by permission of the American Society for Testing and Materials, Copyright].

1217 **V. P. Morozov. "Theoretical Analysis of the Acoustic Emission from Cavitation Line Vortices."** *Soviet Physics–Acoustics,* **SOPAA, 19(5):468–471. Mar–Apr 1974; translated from** *Akusticheskii Zhurnal,* **AKZHA, 19:736–742. Sep–Oct 1973. (in Russian).**

Vibrations of cavitation vortices are investigated, first in an incompressible liquid and then with allowance for the compressibility of the medium. Expressions are derived for the vortex resonance frequencies; these expressions differ from the Kelvin equations in that they allow for the influence of surface tension. It is shown that for a cavitation void of finite length an acoustic emission component occurs due to incomplete pressure compensation near the ends of the vibrating vortex.

1218 **T. M. Morton, R. M. Harrington, and J. G. Bjeletich. "Acoustic Emissions of Fatigue Crack Growth." Lockheed Palo Alto Research Laboratory, Lockheed Missiles and Space Company, Inc., Palo Alto, California. July 1972;** *Engineering Fracture Mechanics,* **EFMEA, 5(3):691–697. Sep 1973.**

Acoustic emissions of fatigue crack growth have been monitored and quantitatively correlated with growth rate and the applied range of stress intensity for high cycle fatigue of 2024-T851 aluminum alloy. The data suggest a more cogent relationship for acoustic emissions and the applied range of stress intensity rather than between acoustic emissions and the average crack growth rate. Since nearly all crack growth is expected during the maximum load portion of the fatigue cycles, only the emissions from the acoustic events in the vicinity of the peak load were incorporated in correlations with da/dn and ΔK. Large amplitude emissions in the proximity of the minimum cyclic load were also detected. Because of their characteristics, these emissions are attributed to crack surface interference and, consequently, were not included in the correlation analyses. [reprinted by permission of Pergamon Press Ltd from *Engineering Fracture Mechanics*].

1219 **T. M. Morton, R. E. Lewis, and R. M. Harrington. "Acoustic Emission Research - 1971." LMSC-D266144. Lockheed Missiles and Space Company, Inc., Sunnyvale, California. 1972.**

1220 **T. M. Morton, S. Smith, and R. M. Harrington. "Effect of Loading Variables on the Acoustic Emissions of Fatigue-Crack Growth." Lockheed Palo Alto Research Laboratory, Lockheed Missiles and Space Company, Palo Alto, California. 1973; paper presented at the Society for Experimental Stress Analysis 3rd International Congress on Experimental Mechanics, Los Angeles, California. May 13-18, 1973;** ***Experimental Mechanics,*** **EXMCA, 14(5):208-213. May 1974.**

A series of experiments has been conducted to determine the effect of loading variables such as cyclic frequency, load ratio, and material on acoustic emission from fatigue crack propagation. It is shown that the applied stress intensity range (ΔK) is a controlling parameter for all materials studied, while the other parameters have lesser effects. Two potential methods for engineering application of acoustic emission during fatigue loading are described.

1221 **F. J. Moskal and J. D. Fageol. "Acoustic-Emission Signal-Processing Analog Unit for Locating Flaws in Large Tanks." NASA Tech Brief B73-10325. NTIS, Springfield, Virginia. Sep 1973.**

(Request to Technology Utilization Officer, Marshall Space Flight Center, Code A&PS-TU, Marshall Space Flight Center, Alabama 35812 for further information on this Tech Brief will result in a copy of disclosure T.U. No. SID-94518, NT Control No. MFS-24424.) The reportable item describes a well-known nondestructive test method, that of acoustic emission and triangulation for the purpose of flaw detection and location in large tanks. Triangularly located sensor arrays pick up the emission to be fed into the computer. It is held that even though the technique itself is known within the state-of-the-art, the improvement in processing equipment is substantial and new. Moreover, the acoustic signal locating mode of this system yields a certain acceptable repeatability (25 parts per 1000), a further step towards standardizing NDT equipment. Vast areas of potential applications are open to this instrumentation such as atomic reactors and power generation, marine and aircraft vessels, and structural frames and bridges. An analog processing unit has been developed to process signals from an array of acoustic emission sensors on a large tank to permit more precise location of flaws in the tank structure. The unit establishes the location of the detected flaw within a small subtriangle area and then converts the data to digital form for further processing by an on-line computer. A more complete description is found in report SD71-264 (see Reference 1708).

1222 **A. Q. Mowbray. "Assault on Fracture Pays Off."** *Materials Research and Standards,* **MTRSA, 4(3):103–106 and 5A. Mar 1964.**

The ASTM Special Committee on Fracture Testing of High-Strength Materials has provided a focal point for interest, talent, and information in the field of fracture; enforced decisions in national aerospace programs which might not otherwise have been made; created widespread appreciation of the need for sharp-notch tests in materials evaluation programs; and evolved a national approach to the analysis of fracture. This article gives a brief history of the committee's work and describes present and future programs. Under present and future programs, mention is made to the acoustic emission work of M. H. Jones and W. F. Brown, Jr. at NASA-Lewis Research Center (see Reference 949). [reprinted by permission of the American Society for Testing and Materials, Copyright].

1223 **G. L. Mowrey. "Computer Processing and Analysis of Microseismic Data." pp 427–443 in** *Proceedings First Conference on Acoustic Emission/Microseismic Activity in Geologic Structures and Materials.* **H. R. Hardy, Jr., and F. W. Leighton, Editors. Trans Tech Publications, Clausthal, West Germany. 1977. conference held at Pennsylvania State University, University Park, Pennsylvania. June 9–11, 1975.**

Computer programs designed to rapidly and accurately process and analyse micro-seismic data, which have been previously recorded on a multichannel analog magnetic tape recorder, are discussed. Currently at Pennsylvania State, there exists a suite of computer programs which provide the user with a means of digitizing multichannel analog data, performing mathematical operations upon the digitized data, displaying the digitized data both before and after mathematical processing, and providing hard copy output of the observed display by way of an x–y plotter or line printer. These programs are run on a hybrid computer facility consisting of a time-sharing DEC system-10 computer interfaced to an EAI-680 analog computer. A computer program designed to locate the source of microseismic activity by the method of least squares reduction of seismic arrival times is also discussed and can be run on the batch system IBM-370 Model-168 computer.

1224 **M. W. Moyer. "Acoustic Emission for Nondestructive Testing and Certification." Y-DA-2855. Union Carbide Corporation, Y-12 Plant, Oak Ridge, Tennessee. Feb 20, 1969. [NTIS]; paper presented at the 3rd Acoustic Emission Working Group Meeting, New Orleans, Louisiana. Feb 20–21, 1969.**

The equipment employed in using acoustic emission testing as a tool for materials testing and certification is described. Tests performed on 8 aluminum alloys, an aluminum cantilever beam, and stress corrosion of uranium alloys are discussed. Future plans for the examination of welds and stress corrosion are given.

1225 **J. Mozina and I. Grabec. "Acoustic Emission during Corrosion/Emission acoustique durant la corrosion." University of Ljubljana, Ljubljana, Yugoslavia; Paper 3K/18 in** *Proceedings of the Eighth World Conference on Nondestructive Testing.* **volume containing Sections 3K2, 3H2, 5A. International Committee for Nondestructive Testing, Secretariat du Cofrend, Institut de Soudure, 32 Boulevard de la Chapelle, 75880 Paris Cedex 18, France. 1976. conference held in Cannes, France. Sep 6–10, 1976. (paper in English).**

Description of the acoustic emission caused by corrosion of steel and aluminum alloys in HCl solution, and possibilities of the application of this phenomenon for NDT of corrosion resistance are discussed.

1226 **R. A. Muenow. "Large Scale Applications of Acoustic Emission." Law Engineering Testing Company, Atlanta, Georgia. 1972; paper presented at the 1972 Institute of Electrical and Electronics Engineers Ultrasonic Symposium. Boston, Massachusetts. Oct 4–7, 1972. [paper not published in proceedings];** *IEEE Transactions on Sonics and Ultrasonics*, **IESUA, SU-20:48. Jan 1973, [abstract only].**

Acoustic emission techniques have been developed to a point where simple but effective use can be made of them in the construction industry. Applications include the in-service monitoring of bridges (steel, concrete, and wood), cranes, buildings, amusement rides, rotating equipment, and elevated water tanks. The methods of test are used only for indications and serve as a warning to employ other more detailed inspection techniques. Acoustic emission has also been used as a quality control tool for the uniform mixing of concrete and the grading of aggregate. Several of the applications will be discussed along with numerous slides showing installations. [reprinted from *IEEE Transactions on Sonics and Ultrasonics*].

1227 **R. A. Muenow. "Sonic Emission Nondestructive Testing." Law Engineering Testing Company, Atlanta, Georgia. 1971; paper published in** *Loss Prevention* **(a publication for coal and ore dock operators and their employees by Bridge Operators Engineering & Inspection Service, Minneapolis, Minnesota), 45(1): 28–35. July 1971; paper presented at the 50th Annual Meeting of Coal and Ore Dock Superintendents, Green Bay, Wisconsin. Mar 25–26, 1971.**

1228 **R. A. Muenow. "Uses of Acoustic Emission in the Construction Industry." Session 9, p 1 in** *The Second Acoustic Emission Symposium.* **Japan Industrial**

Planning Association, Dobashi Building 5F, 3-17-8 Akasaka, Minato-ku, Tokyo, 107 Japan. 1974. proceedings of a symposium held in Tokyo, Japan. Sep 2-4, 1974. [abstract only, paper not published in proceedings].

Acoustic emission techniques will be described as applied to structural analysis of old railroad bridges, new interstate bridges, airport runways, pile-driving equipment, cranes, and prestressed or post tensioned concrete. Case histories will include the analysis of railroad bridges which are both riveted and welded over 60 years old as opposed to new interstate bridges, where complete NDT inspection is performed and acoustic emissions are measured as early as in the fabrication shop. Airport runways have been monitored with AE techniques to develop crack information in such locations as turn offs and touch-down areas. Data will be presented illustrating crack propagation versus repetitive loadings and aircraft weight. Data will be shown which could be used as an early warning system as to loss of subgrade. Pile-driving equipment has been monitored with AE to produce data relative to pile breakage and loading characteristics. Data and slides will illustrate the correlation between blow weight, number of blows, and resistance to penetration as measured by AE. Numerous illustrations of long-time studies on various types of cranes using AE to detect cracks or loose bolts will be given. A typical 4-year study will be shown correlating AE activity to actual repairs. All AE data will be presented on prestressing and post-tensioning systems to record slippage or breakage of wire strands during the stressing operation, as well as during erection and use.

1229 **E. A. W. Muller. "Zerstorungsfreie Materialprufung (Nondestructive Testing of Materials)."** *Archiv fur Technisches Messen und Industrielle Messtechnik,* **ATMIA, 455(12):229-232. Dec 1973. (in German).**

A brief but comprehensive survey is given on X-ray radiography, gamma radiography, radiometric gages, neutron radiography, proton radiography, ultrasonic testing, acoustic emission testing, eddy current testing, magnetic flux testing, potential probe method, microwaves, exo-electron emission, IR thermography, holography, polarized lasers, and penetrant inspection.

1230 **J. V. Mullin, V. F. Mazzio, and R. L. Mehan. "Basic Failure Mechanisms in Advanced Composites." NASA CR-121000. General Electric Company, Space Sciences Laboratory, Valley Forge, Pennsylvania. Mar 1972. [NTIS].**

The purpose of this effort was to identify failure mechanisms in carbon-epoxy composites as a basis for more reliable prediction of the performance of these materials. The approach involved both the study of local fracture events in model specimens containing small groups of filaments and fractographic examination of high fiber content engineering composites. A good deal of emphasis has been placed on the correlation of model specimen observations with gross fracture modes. The effects of fiber surface treatment, resin modification and fiber content were studied and acoustic emission methods were applied. Some effort was devoted to analysis of the failure process in composite-metal specimens.

1231 **J. V. Mullin and R. L. Mehan. "Evaluation of Composite Failures through Fracture Signal Analysis."** *Journal of Testing and Evaluation,* **JTEVA, 1(3):215–219. May 1973.**

The paper describes the use of acoustic monitoring techniques to identify specific failure mechanisms in composite materials. Because of their inherent complexity, composites exhibit a number of fracture modes depending on loading conditions and the properties and volume ratios of the constituents. The technique of fracture signature analysis attempts to use the frequency content and nature of the signals as a means for identifying specific critical events in the fracture process. Some of the difficulties associated with this technique are described together with a description of the techniques applied and results obtained with boron-epoxy. [reprinted by permission of the American Society for Testing and Materials, Copyright].

1232 **C. W. Musser. "NDT Systems for Establishing Weld Integrity of Space Vehicles."** *Materials Evaluation,* **MAEVA, 27(2):42–48. Feb 1969; paper presented at the American Society for Nondestructive Testing/American Welding Society Symposium on NDT of Welding, Honolulu, Hawaii, Mar 18–19, 1968.**

Along with advances in the technology of processes and new alloys, nondestructive testing methods have been upgraded and implemented to keep pace with the requirements for producing stronger, and lighter-weight space vehicles. However, in spite of the new developments in materials and processes, welding in one form or another is still utilized to a great extent and is expected to continue to be used for many years to come. No welding process has been developed to date which does not require inspection evaluation to ensure that the assembled components have the required design integrity. Penetrating radiation has been improved through the development of new filters, new energy sources and filmless readout capabilities. However, the basic technique is still much the same and is still the primary method used for determining weld integrity. Filmless readout methods have reduced the time required for evaluation; however, in welds greater than 0.250 inches in thickness, loss of sensitivity and resolution are encountered with this method. This capability, however, is expected to be improved within the next decade. Large space vehicles such as the Saturn V have thousands of feet of welds which must be radiographed; so mechanized radiographic systems have been developed and will be further improved. Supplementary techniques and methods such as ultrasonics, eddy current, and infrared are presently being used to provide earlier evaluation restuls. These methods also complement radiography since they can be used to ensure weld integrity in areas or in designs where it is difficult or impossible for radiography to be utilized. Battery-operated portable systems have also been developed for use in areas where access is available from only one side of an assembly. Lightweight versions of these systems will also make them available for use in outer space.

The NDT systems developed and implemented for the Saturn V Program, along with new techniques and methods still in the development stages, are described in this paper. Slides will be utilized to show the equipment and handling systems being described. These systems represent the latest developments in NDT and are expected to be utilized for all weld evaluations required through the presently projected space programs. A mechanized radiographic system used on the Saturn V S-1C Stage with automatic film reel units and motorized camera and tube drives will be described and shown. An X-ray vidicon-TV monitor readout for a filmless method will also be shown and discussed. In addition, ultrasonic methods for inspecting large components and raw materials will be shown, and a newly developed ultrasonic scanning system for evaluating fusion spot welds will be discussed and displayed. Other ultrasonic methods developed for weld inspection will be discussed. Eddy current and latest penetrant methods utilized for surface defect detection will be shown and described for applications on the S-1C. Extremely large components and assemblies required modification of standard techniques to preclude excessive handling and expensive tooling. Infrared methods developed as an aid to weld evaluation are capable of determining depth of penetration for double-sided welds where lack of penetration could not be detected by other NDT methods. The paper will show the necessity for custom-designed integrated NDT systems to ensure reliability of welded assemblies required for space vehicles.

1233 **J. Myl and J. Kvapil. "Ausnutzung von Schalleffekten zur Indizierung von Veranderungen des Kristallgitters I. Verfolgen der Verandungen bei Chemischen Vorgangen."** *Collection of Czechoslovak Chemical Communication,* **CCCCA, 27:1517–1520. 1962. (in German).**

1234 **J. Myl and J. Kvapil. "Verwendung Akustischer Effekte zur Indizierung von Anderungen des Kristallgitters (Application of Acoustic Effects to the Identification of Changes in the Crystal Lattice)."** *Collection of Czechoslovak Chemical Communications,* **CCCCA, 27:1099–1104. 1962. (in German).**

Sudden lattice changes are accompanied by acoustic effects which are usually below the range of audible intensities. A simple apparatus was constructed to record these effects in the course of heating the crystals. It is possible to determine the temperatures at which the phase transformation or sudden changes of the thermal expansion take place. The analysis requires only small samples (0.1 g) and the accuracy is to within ± 10°. [Abstract 9307h reprinted by permission of the American Chemical Society from *Chemical Abstracts,* Vol. 57, 1962].

N

1235 T. Nagata, K. Kobayashi, A. Imazu, and R. Miura. "An Application of the AE Technique to Low-Cycle Fatigue Test of Piping Components." Paper 1-1, pp 11-30 in *The Third Acoustic Emission Symposium.* Japan Industrial Planning Association, Dobashi Building, 3-17-8, Akasaka, Minato-ku, Tokyo, Japan. 1976. proceedings of a symposium held in Tokyo, Japan. Sep 16-18, 1976.

In recent years, exellent advances have been achieved in the acoustic emission technology, and many experimental and practical works have been reported in this field. On the other hand, social needs to the high reliability of the reactor power plants seem to be becoming more and more enthusiastic. The in-service inspection of the reactor plants, as well as the high reliability of the structural design, has been recognized to be significantly important. There are being developed or improved various in-service inspection methods. Especially the AE method is one of these that are expected most, because it has a possibility to monitor the plants in their operating conditions. Several works have been reported already concerning the application of the AE technology to the actual reactor plants. Power Reactor and Nuclear Fuel Development Corporation (PNC) is developing two types of reactors as a national project in Japan. One is Advanced Thermal Reactor (ATR) and the other is Fast Breeder Reactor (FBR). In the case of the latter, it is operated at relatively higher temperatues than the conventional Light Water Reactors (BWR and PWR), and there are various differences in the reactor characteristics between them. So it was decided to perform a series of tests of the structural components in the primary coolant piping system of the FBR at Oarai Engineering Center of PNC. They are called as piping structural tests, and many actual size or scale-down models of the piping structural components of Type 304 austenitic stainless steel have been tested under monotonic or cyclic loadings. There are measured a variety of values, and at the same time AE characteristics of the structures are measured and analyzed to evaluate the applicability of the AE method to the actual FBR plants. This report shows results of the AE measurement in a low-cycle fatigue test of the N-8 specimen that was conducted last year. The N-8 test specimen was a model of the inlet nozzle to the reactor vessel of the primary coolant circuit. It was of Type 304 austenitic stainless steel, and cyclic out-of plane bending moment loadings were applied at room temperature until induced fatigue cracks grew through the wall and then became cracks of relatively large size. The AE signals were measured in the form of various values, and after the test there were applied some analyses from several kinds of view point.

1236 Y. Nakamura. "Acoustic Emission Monitoring of Thompson's D6aC Compression Fatigue Test I." General Dynamics, Convair Aerospace Division, Fort Worth, Texas. Apr 14, 1969.

The compression fatigue test on Thompson's specimen No. 12FTJ19742-1, conducted from April 3 till April 12, 1969 was monitored continuously by an acoustic emission monitoring system. The purpose of the monitoring was to detect initiation and growth of cracks in the specimen. The monitoring system used was similar to those used for the previous tests on Humphries specimens and FW No. 1, Phase II. Several modifications were made, however, in order to improve the signal-to-noise ratio, achieving almost complete rejection of test-fixture noise. The system detected the initiation of two different types of cracks and their growth in relation to the cyclic loading. The report briefly describes the monitoring system used for the test, and the results of the monitoring.

1237 **Y. Nakamura. "Acoustic Emission Monitoring of Thompson's D6aC Compression Fatigue Test II." General Dynamics, Convair Aerospace Division, Fort Worth, Texas. Apr 30, 1969.**

The compression fatigue test on Thompson's specimen No. 12FTF19742-2, conducted from April 14 until April 22, 1969 was monitored continuously by an acoustic emission monitoring system. The purpose of the monitoring was to detect initiation and growth of cracks in the specimen during the fatigue cycling. The specimen was loaded under uniaxial compression and cycled between 28,000 and 140,000 psi levels (average stress across the least cross section) at a rate of 10 to 12 cycles per minute. The test was terminated after 10^5 cycles. The report describes the results of the monitoring.

1238 **Y. Nakamura. "Acoustic Emission Monitoring on Humphries Specimen Z-7." Preliminary Report. General Dynamics, Convair Aerospace Division, Fort Worth, Texas. Jan 16, 1969.**

The fatigue test on the Humphries Specimen Z-7 conducted from January 11 until January 15, 1969 was monitored by a passive ultrasonic detection system. The purpose of the monitoring was to detect acoustic emissions associated with initiation and growth of cracks within two high-stressed areas in the specimen. The detection system used for the monitoring has been described in the previous report on Humphries Specimen Z-6. Two changes in the test setup and procedures from the last test were made: (1) two, instead of one, master sensors were used in each high-stressed area in order to obtain a wider coverage of monitoring; and (2) a gate circuit was employed to count only those signals which were sensed near the maximum load of each loading cycle. Acoustic emissions originating in the area around bolt holes 1 to 9 are clearly identified. This preliminary report briefly described the results of the test. A final report will follow.

1239 **Y. Nakamura. "Acoustic Emission Monitoring System for Detection of Cracks in a Complex Structure."** *Materials Evaluation*, **MAEVA, 29(1):8–12. Jan 1971; paper presented at the American Society for Nondestructive Testing National Spring Conference, Los Angeles, California. Mar 9–12, 1970.**

An acoustic emission monitoring system designed to detect, in real time, initiation and growth of cracks in a complex structure during static as well as fatigue testing has been developed. The system consists of arrays of acoustic sensors, logic circuits, and output devices. An effective spatial- and frequency-filter combination permits the use of the system in the presence of the heavy background noise ordinarily encountered in the testing of a large, complex structure. The system is being used successfully in static and fatigue testing of full-scale, as well as small-scale specimens of aircraft components and structures.

1240 **Y. Nakamura. "Detection and Analysis of Acoustic Emission Signals." pp 445–457 in** ***Proceedings First Conference on Acoustic Emission/Microseismic Activity in Geologic Structures and Materials.*** **H. R. Hardy, Jr., and F. W. Leighton, Editors. Trans Tech Publications, Clausthal, West Germany. 1977. conference held at Pennsylvania State University, University Park, Pennsylvania. June 9–11, 1975.**

A concise review of techniques used in detection and analysis of acoustic emission signals for application in materials research and intensity surveillance of nongeologic structures is presented. Under-damped piezoelectric transducers have been used almost exclusively for detection of AE signals to achieve high sensitivity of signal detection. Arrays of such detectors are often used for source locations and for elimination of unwanted signals. Detected AE signals are analyzed for source locations and for the significance of detected signals. For the latter, frequently used techniques include analyses in terms of event counts, various forms of weighted signal counts, distribution of signal amplitudes, distribution of signal durations, and frequency spectra.

Y. Nakamura. "Instruction Manual for ARL–6 Acoustic Emission Monitoring System." FZM–5420. General Dynamics, Convair Aerospace Division, Fort Worth, Texas. Aug 19, 1969. [report contains proprietary information and has not been released to public].

1242 **Y. Nakamura. "Location of Acoustic Emission Source." pp 315–316 in "1970 Applied Research Program." Research Summary ARR–15. General Dynamics, Convair Aerospace Division, Fort Worth, Texas. [unpublished].**

1243 **Y. Nakamura (to General Dynamics Corporation). "Method and Apparatus for Acoustic Emission Non–Destructive Testing." U.S. Patent 3,858,439. Jan 7, 1975.**

Described are a method and apparatus for the detection of incipient structural failures by acoustic emission monitoring in the presence of background noise. One or more *master* sensors are placed on the structure to be tested and are surrounded by a plurality of *slave* sensors. Each sensor detects acoustic emission in the structure and sends a corresponding signal to an Acoustic Emission Monitor (AEM). The AEM includes electronic circuitry which accepts only signals which reach the master sensors before they reach the slave sensors, thus excluding signals originating outside the area covered by the sensor array. The AEM also contains frequency filters excluding signals outside the desired frequency range, to exclude most background *noise.* These techniques permit the detection of acoustic emissions from flaws and propagating cracks in structures of complex shape in an acoustically noisy environment.

1244 **Y. Nakamura. "Passive Ultrasonic Monitoring on Humphries Specimen Z-6." General Dynamics, Convair Aerospace Division, Fort Worth, Texas. Dec 1968.**

The fatigue test on the Humphries Specimen Z-6, conducted from November 27 till December 6, 1968, was monitored by a passive ultrasonic detection system. The purpose of the monitoring was to detect, if possible, initiation and growth of cracks within two high-stressed areas in the specimen. The detection system was based in part on the stress wave analysis technique (SWAT) developed at Aerojet-General Corporation. Several modifications and improvements had been made based on previous experience with the testing on FW No. 1 carry-through box, laboratory experiments, and theoretical considerations. The results of the monitoring clearly indicate that the newly developed detection system can detect small, growing fatigue cracks in D6aC steel well in advance of the failure of the specimen, but that careful attention must be given to the placement of the transducers when operating in the normal noise environment of the test facilities. The report briefly describes the passive ultrasonic monitoring system used for the test on the Humphries Specimen Z-6, and the results of the test. Some recommendations for future tests are also made.

1245 **Y. Nakamura. "Spatial Filtration in AE Detection." Session 2, pp 19–41 in** ***The Second Acoustic Emission Symposium.*** **Japan Industrial Planning Association, Dobashi Building 5F, 3–17–8 Akasaka, Minato-ku, Tokyo, 107 Japan. 1974. proceedings of a symposium held in Tokyo, Japan. Sep 2–4, 1974; "Spatial Filtration in Detection of Acoustic Emission."** ***Journal of N.D.I.*** **(*Hi-Hakai Kensa*, Japan), HIHKA, 24(7):400–404. July 1975. (in Japanese).**

In practical applications of acoustic emission, consideration must first be given to detection of acoustic emission signals of interest. In certain applications, such as a monitoring of ordinary pressure vessels, this poses no problem because acoustic emission signals of intensity sufficiently greater than background noise are normally expected. However in many other applications, the signal detection itself is a problem either because the background noise level is too high or because the acoustic emission signals to be detected are too weak. The former situation arises in such applications as monitoring of a complex structure, where background noises originated by relative movements of various joints in the structure are several orders of magnitude greater, both in number and in intensity, then AE signals to be detected; while the latter situation occurs frequently in testing of materials that intrinsically emit only weak AE signals, where even normal laboratory environment may turn out to be too noisy.

In either case, it becomes necessary to devise a special scheme so that AE signals buried in the background noise may selectively be detected. Frequency filtering is used nearly universally in these and other situations, and a significant improvement in signal-to-noise ratio is often achieved, provided that the AE signals of interest have distinctly different frequency content from most background noises. In many cases, however, the frequency filtering only gives a partial solution to the problem because, after a frequency filtering, a significant amount of noise energy may still remain in the same frequency band as that of AE signals. Various forms of spatial filtration techniques have been developed in the past several years to overcome this difficulty. This paper describes basic principles and some variations of the technique, presents some examples of applications from published data, and discusses problems associated with use.

1246 Y. Nakamura. J. W. Hagemeyer, and C. L. Veach. "Acoustic Emission Observations of Delayed Cracking in Steel Specimens Containing Interference–Fit Fasteners: A Preliminary Study." ERR–FW–1185. General Dynamics, Convair Aerospace Division, Fort Worth, Texas. Aug 15, 1971.

Acoustic emission monitoring was found to be a valuable technique for observing delayed cracking in stressed steel specimens. The D6aC steel specimens, stressed by interference fit bolts, gave emission signals at rates which were directly related to their stress state and the amount of hydrogen charging they had previously received. A short-term, high-rate phase immediately after stressing, followed by a long-term, low-rate phase was observed for all charged specimens but not an uncharged control. Approximate calculations of the relative energy of cracking implied by the observed acoustic signals correlate well with the amount of charging given each specimen and with the obvious crack growth. The addition of water to a stressed but uncharged specimen produced a pattern of emission signals similar to that of the hydrogen-charged specimens suggesting the initiation of stress-corrosion cracking. It is concluded from this preliminary study that the acoustic monitoring technique used here, because of its high sensitivity and ability to give immediate indications of crack growth, would be a useful tool for studying delayed cracking.

1247 Y. Nakamura, B. O. McCauley, A. H. Gardner, J. C. Redmond, J. W. Hagemeyer, and G. M. Burton. "Development of an Acoustic Emission Monitoring System." ERR–FW–901. General Dynamics, Convair Aerospace Division, Fort Worth, Texas. Sep 12, 1969.

A system for monitoring acoustic emissions from cracks in a specimen during a fatigue test has been in development since September 1968. The primary purpose of the task is to develop a monitoring system capable of detecting the initiation and growth of cracks in aircraft components and structures during fatigue cycling. After a few preliminary tests, which indicated that cracking could be detected prior to the failure in specimens using equipment available in the laboratory, the development of the monitoring system progressed through a series of tests on both small-scale and full-size specimens of aircraft components and structures. The early models were not totally successful because of the extremely heavy background noise from test fixtures. Therefore, the development effort was concentrated on how to eliminate the background noise. The developed system, considered to be a unique, production-type equipment, utilized a combination of spatial, time-domain, and frequency filtering to eliminate test-fixture noise. Spatial filtering is accomplished by the use of arrays of acoustic sensors and a series of logic circuits. The system is effective in detecting cracks in the noisy environments of fatigue test. The developed system, however, still has several limitations and further research and development is necessary.

1248 **Y. Nakamura, B. O. McCauley, and C. L. Veach. "Study of Acoustic Emission during Mechanical Tests of Large Flight Weight Tank Structure." NASA CR-115761, MSC-04800, FZK-390. General Dynamics, Convair Aerospace Division, Fort Worth, Texas. Mar 15, 1972. [NTIS: N72-31667]. [see also Reference 1151 for publication].**

A PPO-insulated, flight-weight, subscale, aluminum tank was monitored for a acoustic emissions during a proof test and during 100 cycles of environmental test simulating space flights. The use of a combination of frequency filtering and appropriate spatial filtering to reduce background noise was found to be sufficient to detect acoustic emission signals of relatively small intensity expected from subcritical crack growth in the structure. Several emission source locations were indentified, including the one where a flaw was detected by post-test X-ray inspections. For most source locations, however, post-test inspections did not detect flaws; this was partially attributed to the higher sensitivity of the acoustic emission technique than any other currently available NDT method for detecting flaws. For these nonverifiable emission sources, a problem still remains in correctly interpreting observed emission signals.

1249 **Y. Nakamura and C. L. Veach (to General Dynamics Corporation). "Method and Apparatus for Detecting Progressive Cracks in Materials." U.S. Patent 3,919,883. Nov 18, 1975.**

An acoustic emission monitoring technique for detecting the growth of formation of cracks or other flaws in structures is disclosed. A growing crack emits acoustic signals which can be detected by an appropriate sensor on the structure surface. It has been found that the rate of amplitude buildup decreases as distance from a sensor to the site of the acoustic event increases. Thus it is possible by slecting only acoustic signals which have at least a selected build-up rate, to eliminate most background noise and other signals originating further than selected distance from the sensor. Thus, the formation and growth of cracks in the selected area can be accurately monitored in an acoustically noisy environment.

1250 **Y. Nakamura, C. L. Veach, and B. O. McCauley. "Amplitude Distribution of Acoustic Emission Signals." ERR-FW-1176. General Dynamics, Convair Aerospace Division, Fort Worth, Texas. June 15, 1971; pp 164-186 in**

Acoustic Emission, ASTM STP 505. **American Soceity for Testing and Materials, Philadelphia, Pennsylvania. 1972. proceedings of symposium held in Bal Harbour, Florida. Dec 7–8, 1971.**

Acoustic emission signals from slowly growing cracks have been studied experimentally using precracked compact tension specimens of D6aC steel heat treated to several different fracture toughness values but with the same yield strength. The observed signal amplitude is distributed over a range of more than 60 db, reflecting the statistical nature of the fracture process in a microscopically heterogeneous medium. A clear correlation between the crack growth behavior and the emission amplitude distribution is found. A shift in the amplitude distribution with stress intensity level at a crack tip and a definite difference between the amplitude distribution of emission signals and that of rubbing noise may have direct bearing on the application of the acoustic emission technique as a nondestructive inspection method for materials as well as structural tests.

1251 **Y. Nakamura, C. L. Veach, and B. O. McCauley. "Study of Acoustic Emission Signals from Static Loading of Precracked Specimens." pp 316–317 in "1970 Applied Research Program." Research Summary ARR-15. General Dynamics, Convair Aerospace Division, Fort Worth, Texas. [unpublished].**

1252 **H. Nakasa. "Acoustic Emission Techniques for Material Diagnostics."** *Nippon Kikai Gakkai Shi (Journal of the Japan Society of Mechanical Engineers).* **NKGKA, 77(668):803–807. July 1974. (in Japanese).**

1253 **H. Nakasa. "Application of A. E. Techniques to Material Diagnostics." Paper No. 8 in** *Proceedings of the U.S.–Japan Joint Symposium on Acoustic Emission.* **Japanese volume. Japan Industrial Planning Association, Kawamura-Building 6F, 3–21–6 Akasaka, Minato-ku, Tokyo, Japan 107. 1972. symposium held in Tokyo, Japan. July 4–6, 1972. (paper in Japanese, abstract available in English).**

Characteristics of AE signals emitted from solid materials in deformation and fracture are determined by frequency-, amplitude-, and emission-rate analysis. Material diagnostics, such as inspection of integrity, monitoring of degradation, anticipation of fracture and so on, could be performed by analyzing AE characteristics precisely. This report presents a measuring system of AE signals for the purpose of material diagnostics, and shows typical test results of measuring AE characteristics of various materials in tensile testing, degraded by work-hardening, thermal deterioration, and neutron irradiation embrittlement. AE characteristics of such sound materials as stainless steel, carbon steel, Al-alloy, and FRP are presented. The AE characteristics of V-alloy and pure Cu which have been irradiated by fast neutrons to 2×10^{17} nvt (>180 keV) are presented. Variations of AE characteristics by work-hardening (carbon steel), thermal deterioration (FRP and high Cr steel), and irradiation embrittlement (pure V and pure Al) are also presented.

1254 **H. Nakasa. "Application of Acoustic Emission Techniques to Material Diagnostics." Report No. IAEA-SM/168/D-5. Central Research Institute of Electric Power Industry, Tokyo, Japan. 1973; pp 461–478 in** *Nuclear Power Plant Control and Insturmentation 1973.* **Publication STI/PUB/341. International Atomic Energy Agency, Vienna, Austria. 1973. proceedings of a symposium (CONF-730108) held in Prague, Czechoslovakia. Jan 22–26, 1973.**

The paper discusses an instrumentation system developed for detecting and anlayzing acoustic emission signals, and deals with applications of acoustic emission techniques to the material diagnostics for various tensile-test specimens degraded by neutron-irradiation or prestraining and for model pipings of a fast breeder reactor.

1255 **H. Nakasa. "Defect Estimation and Safety of Steel Structures: Part 4. Acoustic Emission and Examination of Safety of Structures."** *Journal of Steel Structures C,* **JSSCD, 11(112):21-31. 1975. (in Japanese).**

1256 **H. Nakasa. "Measurement of Acoustic Emissions in Tensile Tests of Several Materials." CRIEPI Technical Report 71069. Central Research Institute of Electric Power Industry, Tokyo, Japan. 1972. (in Japanese).**

Acoustic emission characteristics of carbon steel, stainless steel, high-Cr steel, Al, Cu, FRP, mica, and glass were measured in the process of tensile tests by the Instron universal testing instrument. Previously reported measuring systems were utilized in order to investigate frequency-, amplitude- and emission-rate characteristics of acoustic emission signals. Test results demonstrated availability of acoustic emission techniques for material diagnostics, such as determining integrity of materials, monitoring of degradation, anticipation of fracture, etc.

1257 **H. Nakasa, H. Kusanagi, and H. Kimura. "Instrumentation System for Parametric Analysis of Acoustic Emission Characteristics and Its Application to Integrity Assessment of Structural Materials." Paper 2-5, pp 131-151 in** *The Third Acoustic Emission Symposium.* **Japan Industrial Planning Association, Dobashi Building, 3-17-8, Akasaka, Minato-ku, Tokyo, Japan. 1976. proceedings of a symposium held in Tokyo, Japan. Sep 16-18, 1976.**

Acoustic emission techniques have recently received the wide application in many industrial areas due to the great potential as the structural integrity monitor. On the other hand, the instrumentation system for such purposes has come to necessitate more and more complicated and sophisticated analysis on AE characteristics. The AE signals show that when, where, how often, and what kinds of AE events occur in a structural material which is deformed to fracture under an applied stress. The main parameters which define the AE characteristics are as follows: (1) waveform, peak amplitude, and frequency content of AE signals, which show the nature of the

AE event such as energy and duration time; (2) emission rate and energy distribution of AE signals detected by a transducer, which show dynamic behavior of microfracture or deformation (or both); and (3) relative signal arrival times at several transducers and the consequent source location, which present spatial identification of defective areas. Large-scale AE instrumentation systems available at present or in the near future can be classified into the following types: (1) the system which has a great number of signal detection channels and lays stress on signal location for hydrostatic test monitoring of large scale structures; (2) the system which has a capability of acquisition of every AE parameter in the first arrival signal for each individual discrete AE event; and (3) the system which can statistically analyse all independent AE parameters cited above. This paper deals with a mini-computer-based instrumentation system of the above Type 3 and its applications to integrity assessment of structural materials. The availability of this system has been evaluated in various kinds of applications to materials testing, accelerated-failure-mode testing of model structures, and on-line monitoring of the structural integrity for the large scale structures in practical use. The main works to be applied are as follows: (1) basic material tests for metal, concrete and rock, and the subsequent application of the Kaiser effect to estimation of residual stress; (2) monitoring and estimation of geostress in an under-sea tunnel; (3) various failure-mode tests for model structures such as steel pressure vessels, piping components, and concrete vessels for radioactive waste disposal; and (4) pre-service integrity monitoring for an oil-storage tank, a nuclear reactor component, etc. On the basis of the above experiences, it will be shown that the parametric analysis of AE characteristics is useful for the evaluation of the structural integrity.

1258 **H. Nakasa and T. Nagata. "Application of Acoustic Emission Techniques to Fatigue Tests of Piping Components."** *Pressure Engineering, Journal of Japan High Pressure Institute (Kotsu Gasu),* **KOGAA, 13(2):72–79. 1975. (in Japanese).**

It is important to know the progress of fatigue cracks with proper techniques. Some experimental works on the precise understanding of the dynamic behavior of fatigue cracks due to cyclic stress have been carried out by acoustic emission. The acoustic emission is a phenomenon that solid material releases strain energy in the course of deformation to fracture and emits elastic waves, therefore dynamic and minute deformation and fracture phenomena within solid material can be known by AE methods with high sensitivity and real-time response. The AE method was applied to know the fatigue damage of piping elements, such as branch pipes, bend pipes, and nozzle pipes, for the primary coolant piping system of FBR. Instrument system for the AE method consists of (1) AE detecting system, (2) signal processing system (3) analyzing system, and (4) signal recording system. The experiment by the AE method was performed on a bend pipe of 318, 5-mm outer diameter, 6.5-mm wall thickness, and 457.2-mm radius of curvature by the application of displacement-controlled cyclic load, and crack propagation mode was investigated. The AE accumulation curve showed strong correlation with crack propagation. Also AE will be widely applied for the diagnosis of structural soundness and deterioration monitoring.

1259 **H. Nakasa and Y. Tomoda. "On the Method of Measuring Acoustic Emission Signals Emitted in the Process of Deformation and Fracture of Solid Materials." CRIEPI Technical Report 71064. Central Research Institute of Electric Power Industry, Tokyo, Japan. 1972. (in Japanese).**

Characteristics of acoustic signals emitted from solid materials in deformation and fracture are determined by frequency-, amplitude- and emission-rate analysis. Nondestructive and dynamic testing of materials, such as diagnostics of integrity of materials, monitoring of degradation, anticipation of fracture, etc., could be performed by analyzing acoustic emission characteristics precisely. The report presents a measuring system of acoustic emission signals for the purpose of diagnostics of various solid materials. In this system, a multichannel pulse-height analyzer for gamma-ray spectrometry is utilized in order to investigate the amplitude- and emission-rate characteristics of acoustic signals in detail. Availability of this measuring system was demonstrated by test results of analyzing acoustic emission characteristics of glass disks in fracture and steel plates under tension.

1260 **H. Nakasa, Y. Tomoda, and H. Takaku. "Application of Acoustic Emission Techniques to Evaluating Degrees of Embrittlement in Steel." Report No. 73027. Central Research Institute of Electric Power Industry (CRIEPI), Technical Laboratory No. 1, Komae-shi, Tokyo, Japan. Oct 1973. (in Japanese).**

The assessment of material integrity is one of the most important problems for assuring safety of the reactor pressure vessel. For the purpose of evaluating degrees of neutron-irradiation embrittlement as part of reactor vessel surveillance, acoustic emission characteristics in tensile tests have been investigated for several metal specimens degraded by neutron irradiation and for mild-steel specimens degraded by strain aging. The linear relationship between the logarithms of the AE amplitude, a, and the number of emissions, n(a), [i.e., $\log n(a) = -m \log a + \log n_0$] holds in the metal specimen as well as in the rock specimen and earthquakes. The change of the slope, m, is sensitive to degrees of embrittlement, and the AE nondimensional parameter can be an indicator to estimate the degree of material degradation. As the m-valve can be obtained easily from the relationship between AE amplitude and the number of emissions near the yield stress in the tensile test, AE techniques utilizing the m-valve is applicable to diagnostics for irradiation embrittlement in reactor vessel surveillance.

1261 **H. Nakasa, S. Ueda, and T. Nagata. "Acoustic Emission Analysis on Fatigue Failure of Fast Breeder Reactor Piping Components." Session 9, pp 21–43 in *The Second Acoustic Emission Symposium.* Japan Industrial Planning Association, Dobashi Building 5F, 3-17-8 Akasaka, Minato-ku, 107 Japan. 1974. proceedings of a symposium held in Tokyo, Japan. Sep 2–4, 1974.**

The assessment of structural integrity of the primary coolant system is one of the most important problems for the safe operation of the liquid-metal-cooled fast breeder reactor (LMFBR). To prevent the fatigue failure of the primary piping, it is necessary to monitor any deterioration and to detect incipient flaw formation and growth before catastrophic failure occurs. A research and development work on the application of the acoustic emission techniques to monitoring fatigue failure of LMFBR piping components has been performed since March 1972 as a co-operative research project by the Power Reactor and Nuclear Fuel Development Corporation and the Central Research Institute of Electric Power Industry. The objectives of the research program are as follows: (1) Development of the AE instrumentation system available for the fatigue tests of piping components, (2) Investigation of AE characteristics in the fatigue processes, and (3) Evaluation on the availability of the AE techniques to the practical LMFBR primary coolant piping. To evaluate the low-cycle fatigue strength and investigate AE characteristics, 13 bending-fatigue tests for the Type 304 stainless-steel piping components such as elbow, branch, reducer, and nozzle connections, were carried out under various test conditions in 1972 and 1973. Furthermore, 4 elbow pipings and 2 nozzle models were tested up to now (in July of 1974). The paper summarizes typical test results which were mainly obtained up to August 1973, and discusses the AE characteristics in the process of fatigue failure of the piping components.

1262 **C. A. Nanney. "Evidence for Correlations between Microseisms and Earthquakes." NRL Report 5237. U.S. Naval Research Laboratory, Washington, D.C. Jan 13, 1959.**

A study was made to determine whether any correlations exist between the occurrence of earthquakes and the properties of microseisms. Correlations were found between (a) microseism frequency and the occurrence of large earthquakes, (b) microseism amplitudes and the occurrence of large and deep earthquakes, (c) smaller shocks and a local increase of microseism amplitudes, (d) long-period (>10 seconds) microseisms and the occurrence of deep earthquakes, and (e) microseism propagation directions and the great circle path to the location of significant earthquakes. The preceding correlations were all found to test significantly at a 0.01 statistical significance level or greater.

1263 **C. A. Nanney. "Possible Correlations between Earthquakes and Microseisms."** ***Nature,*** **NATUA, 181(4612):802–803. Mar 22, 1958.**

1264 **V. D. Natsik. "Radiation of Sound by a Dislocation Emerging to the Surface of a Crystal."** ***JETP Letters,*** **JTPLA, 8(6):198–202. Sep 20, 1968; translated from** ***Zhurnal Eksperimental'noi i Teoreticheskoi Fiziki, Pis'ma v Redaktsiyu,*** **ZFPRA, 8(6):324–329. Sep 20, 1968. (in Russian).**

The mechanism of sound radiation produced when a dislocation passes through an elastic modulus discontinuity plane, for example through grain boundaries in a poly crystal, on emerging to the surface of a crystal, etc., is discussed in analog with electrodynamics, this radiation mechanism is called transition.

1265 **V. D. Natsik and A. N. Burkanov. "Radiation of Rayleigh Waves by Edge Dislocations Emerging at the Surface of a Crystal."** *Soviet Physics–Solid State,* **SPSSA, 14(5):1111–1116. Nov 1972; translated from** *Fizika Tverdogo Tela,* **FTVTA, 14(5):1289–1296. May 1972. (in Russian).**

It is shown that the emergence of an edge dislocation at the surface of a crystal is accompanied by emission of a sound pulse which propagates along the surface from the point of emergence. The spectral characteristics and shape of the pulse are analyzed and its intensity is estimated.

1266 **V. D. Natsik and K. A. Chishko. "Acoustic Emission Associated with the Annihilation of Dislocations."** *Soviet Physics–Solid State,* **SPSSA, 14(11): 2678–2682. May 1973; translated from** *Fizika Tverdogo Tela,* **FTVTA, 14(11):3126–3132. Nov 1972. (in Russian).**

It is shown that the collision and subsequent annihilation of dislocations of opposite sign are accompanied by the emission of acoustic pulses. This effect is analyzed in the example of the annihilation of two linear edge and screw dislocations. In connection with the stated problem, general expressions are derived for the emission fields of a system of straight dislocations executing arbitrary motion. The acoustic emission of converging dislocations of opposite sign consists of two parts. The first part is caused by the nonstationary character of the dislocation motion, and its intensity is determined by the acceleration at precollision times. The second part represents an acoustic pulse generated at the instant of collision. The intensity of this pulse depends on the relative velocity of the dislocations at the instant of collision. Also analyzed is the spectral composition of the emission, and the total energy radiated by a pair of straight dislocations of opposite sign is determined.

1267 **J. R. Neel. "Die Monitor** *Listens* **for Press Malfunctions."** *Metal Stamping,* **MLSMA, 10(8):16–18. Aug 1976.**

Article describes an acoustical die monitor built by the Westinghouse Electric Control Products Division in Beaver, Pennsylvania. The device utilizes audible signals from pressing operations by means of the vibrations that cause them and then responds immediately to the signals.

1268 **L. X. Nepomuceno and H. Onusic. "Emissao Acustica na Inspecao Nao Destrutiva (Acoustic Emission in Nondestructive Inspection)." Report 7001/E-013. Laboratorio de Acustica e Sonica, Sao Paulo, Brazil. 1969. [NTIS: N70-40401, in Portuguese]; NASA-TT-F-13643. NASA Technical Translation, Washington, D. C. June 1971. [NTIS: N71-28914, in English].**

The development of procedures for using acoustical emissions for nondestructive testing is investigated. Described also are the phenomenon of acoustical emission and a theoretical model which explains the experimental data; the methods for detecting fractures and deformation in materials; examples of the analysis of fractures; and the determination of critical loads for structures.

1269 **E. V. Nesmashnyi, B. V. Skiba, and V. N. Sosedov. "Wideband Piezoelectric Receiver for the Investigation of Acoustic Emission Signals."** *Soviet Journal of Nondestructive Testing,* **SJNTA, 11(5):631–637. Sep–Oct 1975; translated from** *Defektoskopiya,* **DEFKA, (5):107–115. Sep–Oct 1975. (in Russian).**

The theoretical amplitude-frequency characteristics of multilayered piezoelectric transducers are analyzed to determine the configuration of an acoustic emission receiver containing a piezoceramic plate with resonance frequency in the megahertz range, which is the optimum in terms of bandwidth and sensitivity. It is shown that a receiver configuration incorporating a damper mechanically impedance-matched with a piezoelectric plate operating close to the electrical no-load regime is the most acceptable version. A procedure is discussed for estimating the bandwidth of the receiver by comparison of the theoretical and experimental impulse oscillograms for impact of a ball against the protective baseplate of the receiver. A specific wideband receiver design is described, and the distortion of the impact impulse by the given receiver is estimated.

1270 **D. R. Newman. "Development of Standards and Calibration Procedures." Task 5, Project IV. pp 73–77 in "In-Service Inspection Program for Nuclear Reactor Vessels." E. R. Reinhart. SwRI Project 17–2440, EEI Project RP–79. Progress Report No. 8. Final Technical Report, Southwest Research Institute, San Antonio, Texas. May 21, 1973. [NTIS: PB–222–701].**

In order to insure that the acoustic emission system is operating at the proper sensitivity levels, the development of standards and calibration procedures was investigated in this task. There are three desirable items for checking out any acoustic system used for the monitoring of reactor pressure vessels: (1) a repeatable, inexpensive calibration source for checking out the response of individual transducers and coupling materials or methods, (2) a means of checking out the transducer array installed on the reactor vessel, and (3) a method of checking out the electronic portion of the acoustic emission system independently of the transducers. This task concerned itself primarily with Items 1 and 2 because of the need for devices of this nature and because of the many different types of electronic systems used on the various acoustic emission systems.

1271 **D. R. Newman and J. C. Crowe. "Recent Developments in Nondestructive Evaluation Applicable to Encapsulated Isotopes." pp 138–165 in** *Proceedings: Third International Symposium on Packaging and Transportation of Radioactive Materials.* **Volume 1. symposium sponsored by U.S. Atomic Energy Commission and seven Hanford contractors, held in Richland, Washington. Aug 16–20, 1971. [NTIS: CONF–710801 (Vol. 1)].**

Nondestructive evaluation techniques have been applied recently at Battelle-Northwest to primary containers and capsules of radioisotopes to ensure the highest possible integrity of these containers. New techniques have been developed in the ultrasonic and acoustic emission areas. These techniques, along with a philosophy of bringing NDE personnel into the design of the container before the design is finalized so suggestions can be made as to how to make the design *testable*, have proven to be an important asset in maintaining container integrity.

1272 **R. H. Nichols, Jr., "Acoustic Emission."** *Journal of the Acoustical Society of America*, **JASMA, 50(2, Part 1):436. Aug 1971.**

Presented are descriptive excerpts taken from a paper, "Acoustic Emission—An Introductory Review," by R. G. Liptai and D. O. Harris in the March 1971 issue of *Materials Research and Standards,* (see Reference 1084).

1273 **R. H. Nichols. "Quality Control by the Cry of Materials under Stress."** *Journal of the Acoustical Society of America,* **JASMA, 58(6):1343–1344. Dec 1975.**

Brief note on the application of acoustic emission techniques used by Western Electric to detect microcracks produced during the bonding process of silicon integrated circuits to ceramic mounts.

1274 **R. W. Nichols, Editor.** *Acoustic Emission.* **Applied Science Publishers Ltd, Ripple Road, Barking, Essex, England. 1976.**

The book contains seven papers concerned with the application of acoustic emission to the inspection of pressure vessels at various stages, including inspection during fabrication, during proof testing, and surveillance during operation. These papers are the proceedings of a colloquium on acoustic emission held during the 1975 Annual Assembly of the International Institute of Welding in Tel Aviv, Israel, July 7–13, 1975. For individual papers see References 722, 813, 1284, 1332, 1462, 1638, and 1729.

1275 **R. W. Nichols. "Some Developments in Testing and Evaluation Techniques for the Prevention of Fracture in Welded Structures." pp 491–504 in** *Mechanical Behavior of Materials.* **Volume V: Composites, Testing and Evaluation. Society of Materials Science, Japan, Kyoto, Japan. 1972. proceedings of the International Conference on Mechanical Behavior of Materials, held in Kyoto, Japan. Aug 15–20, 1971.**

An important recent development in the design of metal structures arises from the practical realisation that materials and structures always contain defects and that such defects change the property requirements for a chosen strength. This has led to the development of fracture mechanics methods relating critical size of defect, material properties, and failure stress. Different forms of fracture mechanics are applicable to different materials and an assessment is made of the available approaches–simulative tests, model tests, and tests based on analytical models especially linear elastic fracture mechanics and general yield fracture mechanics. The implications of such approaches to material evaluation are discussed highlighting the need for critical inspection for defects and the potential of acoustic emission methods as an overall and ultimate inspection tool.

1276 **R. W. Nichols. "The Interaction between NDT and Materials Technology."** *British Journal of Non-Destructive Testing,* **BJNTA, 15(3):65–75. May 1973.**

The interaction on nondestructive testing, acceptance codes for defects, and fracture mechanics is discussed. The ability of modern methods of NDT to detect small defects is surveyed, but it is suggested that this sometimes leads to unnecessary rejections. For example, fracture mechanics has shown that much larger quantities of porosity and slag inclusions can be tolerated than most present specifications permit. The implications of fracture mechanics for preservice and in-service inspection are detailed. The significance of recent developments in ultrasonic testing and stress-wave emission techniques are discussed.

1277 **G. D. Nicholson and J. Becker. "Acoustic Emission Evaluations." Appendix D, pp D–1 to D–28 in "Examination and Evaluation of Capsule–F for the Connecticut Yankee Reactor Pressure-Vessel Surveillance Program. J. S. Perrin, J. W. Sheckherd, and V. G. Scotti. Final Report on work performed under the Nuclear Utilities Technical Assistance Plan (NUTAP). Battelle Columbus Laboratories, Columbus, Ohio. Mar 30, 1972.**

A considerable amount of supplementary data was obtained in this program. The results of these evaluations are presented in Part B of the report. The work included fracture-mechanics evaluations of base-metal WOL specimens, acoustic emission studies of the WOL specimens, and instrumented impact test evaluation of the Charpy specimens. During the program, six irradiated base-metal WOL fracture-toughness specimens were loaded to failure and monitored with acoustic emission instrumentation. The acoustic emission analog signals were recorded on magnetic tape and replayed with optimized gain settings into the signal-processing equipment. The processed acoustic emission data were analyzed by computer to determine the degree of fit to the model, $\Sigma N = AK^m$. The values of the m exponents for the least-squares lines varied from 2 to 5. In this examination, as in the previous one, more acoustic emission activity was recorded for the lower temperature tests (−106 °F). An unexpected behavior in the acoustic emission characteristic was observed; near failure, the acoustic emission count rate sharply decreased.

1278 **J. Nicolas. "Courants de Foucault multi-frequences et Emission acoustique (Multifrequency Eddy Currents and Acoustic Emission)."** *Qualite, Revue Pratique de Controle Industriel,* **QRPCA, 14(67):34, 53. June 1974. (in French).**

1279 **A. Nielsen. "Acoustic Emission Detection during Straining of Steel." RISO-M-1290. Danish Atomic Energy Commission, Research Establishment RISO, Roskilde, Denmark. Sep 1970.**

The report was published as an internal report only and is not available to the public. The work covered here is contained in formal report TRG Report 1983 (C), published by the United Kingdom Atomic Energy Authority, (see Reference 1285).

1280 **A. Nielsen. "Acoustic Emission Investigations during Pressure Vessel Testing at Reactor Materials Laboratory, Culcheth." in** *Practical Fracture Mechanics for Structural Steel.* **M. O. Dobson, Editor. United Kingdom Atomic Energy Authority, in association with Chapman and Hall Ltd., London, England. 1969. proceedings of the Symposium on Fracture Toughness Concepts for Weldable Structural Steel, Risley, England. Apr 29-30, 1969.**

Acoustic emission during the pressurisation of steel pressure vessels has been investigated using transducers located on the vessel surface. Emissions were detected from the start of pressurisation. After rising to a maximum at relatively low pressure, they subsequently tended to decrease and decay rapidly when pressure was held constant. In repeated pressurisations, there was virtually no acoustic activity until the maximum recorded previous pressure was exceeded. Introduction of artificial defects in the vessels raised the level of acoustic activity but the general pattern of behavior was unchanged. No significant rise in activity was detected during the immediate approach to failure.

1281 **A. Nielsen. "Acoustic Emission Surveillance Methods." RISO-M-1429. Danish Atomic Energy Commission, Research Establishment RISO, Roskilde, Denmark. Sep 1971.**

The report was prepared for the Committee on Reactor Safety Technology's Working Group: viz., Material and mechanical problems related to the safety aspects of steel components in nuclear plants. The purpose was to present the status of acoustic emission with respect to quality control and assurance, in fabrication and operation, of steel pressure vessels. Acoustic emission techniques are currently used during testing of pressure vessels to point out locations to be subjected to further nondestructive testing, thus increasing the efficiency of the control of the vessels. The question of predicting failure by acoustic emission surveillance during testing is still open and subject to extensive experimental investigations. For in-service acoustic emission surveillance the efficiency in securing structural integrity is not clear, but information is currently being collected from practical applications.

1282 A. Nielsen. "Acoustic Emission Surveillance Methods." RISO-M-1491. Danish Atomic Energy Comission, Research Establishment RISO, Roskilde, Denmark. May 1972.

The report was prepared for the Committee on Reactor Safety Technology Working Group: viz., Materials and mechanical problems related to the safety aspects of steel components in nuclear plants. It continues the report under RISO-M-1429 (see Reference 1281). The current report is based on information received recently from fruitful collaboration within the Working Group, from the literature, and from personal experience and communications. The purpose is to present the status of the acoustic emission technique with respect to its application during proof testing, recurrent testing, and service, of steel pressure vessels for nuclear plants. During proof testing, it is evident that the location of acoustic emission sources can be carried out reliably. The problem of warning against catastrophic failure cannot be given a clear answer, although an experienced operator can make an approach to the problem and promising investigations are in progress. During recurrent inspection, it should be possible to gain further information on the state of the steel structure by applying the acoustic emission technique if the structure is accessible. Not much experience has been gained of acoustic emission surveillance during service in terms of Signature Analysis, but promising investigations are in progress to solve the various problems, by using the acoustic emission technique in the hostile environment of a nuclear reactor.

1283 A. Nielsen. "Acoustic Emission Surveillance Methods." Riso Report No. 277. Danish Atomic Energy Commission, Research Establishment Riso, Roskilde, Denmark. Dec 1972. [NTIS]; IIW X-703-73. International Institute of Welding. 1973.

The status of the acoustic emission technique with respect to its application during proof testing, recurrent testing, and service of steel pressure vessels for nuclear plants is discussed. During proof testing it is evident that the location of acoustic emission sources can be carried out reliably. The problem of warning against catastrophic failure cannot at present be given a clear answer, although an experienced operator can make an approach to the problem and promising investigations are in process. During recurrent inspection it should be possible to gain further information on the state of the steel structure by applying the acoustic emission technique if the structure is accessible. Not much experience has been gained of acoustic emission surveillance during service in terms of signature analysis, but promising investigations are in progress to solve the various problems by using the acoustic emission technique in the hostile environment of a nuclear reactor.

1284 **A. Nielsen. "The Use of Acoustic Emission for Detection of Defects as They Arise during Fabrication." pp 11-33 in *Acoustic Emission.* Edited by R. W. Nichols. Applied Science Publishers Ltd, Ripple Road, Barking, Essex, England. 1976. proceedings of a colloquium on acoustic emission held during the 1975 Annual Assembly of the International Institute of Welding, Tel Aviv, Israel. July 7-13, 1975.**

Three different applications of acoustic emission technique are considered: (1) Surveillance during welding to detect defects as they are formed. The welding processes considered are arc welding, electroslag welding, electron-beam welding, resistance spot welding, and a few other welding methods; (2) Surveillance immediately after welding to detect delayed cracking; and (3) Surveillance during stress relief to detect reheat cracking. Work done within this field indicates that the technical difficulties in the application of the AE technique can be handled, and that advantages could be gained with respect to production rate and quality.

1285 **A. Nielsen, F. G. Latham, and N. Kirby. "Acoustic Emission from Steel Pressure Vessels." TRG Report 1983 (C). United Kingdom Atomic Energy Authority, Risley, England. Apr 1970. [NTIS]; IIW IX-684-70 and IIW X-575-70. International Institute of Welding. 1970.**

Preliminary investigations were made of acoustic emission from mild steel pressure vessels. Using piezoelectric accelerometers of 60 and 120 kHz natural frequency, the acoustic activity was found to be high in the initial stages of pressurization and to decrease at higher pressures. A similar pattern was followed by vessels containing large defects although the rate of emission was greater. There was no significant increase in the rate of emission as the vessels approached failure. Repeated pressurization of vessels showed reduced acoustic activity on second and subsequent pressure cycles, especially when the rest period between cycles was short. Further investigation is necessary to relate acoustic signals (in terms of amplitude, rate, and frequency) with metallurgical changes (deformation, cracking, tearing, etc.) in the structure.

1286 **M. Nikolic. "Raziskava akusticne emisije rezalnega procesa (Research of Acoustic Emission of Cutting Process)." Diplomsko delo st. 2058. Fakulteta za strojnistvo, Univerza v Ljubljani, Ljubljana, Yugoslavia. 1974. (in Slovenian).**

The work describes some properties of acoustic emission generated by a cutting process of steel and aluminum on a turning lathe.

1287 **K. Nishida. "Development of an Acoustic Emission Monitoring Capability in the U.S. Navy." Session 2, pp 1–18 in** ***The Second Acoustic Emission Symposium.*** **Japan Industrial Planning Association, Dobashi Building 5F, 3-17-8 Akasaka, Minato-ku, Tokyo, 107 Japan. 1974. proceedings of a symposium held in Tokyo, Japan. Sep 2–4, 1974.**

The U.S. Navy effort to develop an acoustic emission monitoring capability is aimed at reducing the time and cost required for surveillance of naval structures during fabrication, test, and service life. It includes a broad range of materials, loading conditions, and structures. The approach is (1) to design and construct a laboratory acoustic emission system with post and real time capability and (2) to conduct parallel tests in the laboratory to develop background in the technique. Specific applications of interest to the Navy include materials behavior, weld cracking, corrosion and fatigue, evaluation of inaccessible areas in structures, evaluation of advanced structures which incorporate new materials and concepts, and monitoring of prototype pressure vessels. This summary discusses some of the results obtained from tests conducted in the past few years together with the problems encountered. These include the hydrostatic behavior of a high strength syntactic foam used as buoyancy material, the weld behavior of high strength steel, the hydrostatic behavior of the prototype titanium pressure hull for the submersible ALVIN, and the fatigue of an internally pressurized vessel. It was concluded on the basis of the many tests performed that static tests generally do not appear to present major difficulties for acoustic emission monitoring. Cyclic conditions, however, can be extremely troublesome because of hydraulic noise, and monitoring is presently being accomplished on a sampling basis. Although major problems remain to be solved before the potential of acoustic emission can be realized, the technique is considered valuable in assessing the behavior of structures.

1288 **K. Nishida. "Structural Studies of Massive Glass, Sept 1968–May 1969." Report 3180. Naval Ship Research and Development Center, Washington, D.C. Aug 1969.**

The report summarizes structural studies of massive glass conducted at the Naval Ship Research and Development Center over the past few years. The accumulated data are directed towards improving the understanding of the response of glass structures to hydrostatic pressure. Results of analytical and experimental studies are presented and discussed. The experimental portion of the program is centered on the development of reliable compression joints in glass structures. The analytical studies involve primarily the use of computer techniques to analyze and evaluate structural details in glass structures. Since much of the technological effort in these studies are still in progress, the data presented herein are not conclusive in many cases.

1289 **N. Niwa. "An Epitome of Stress Wave Emission."** ***Journal of N.D.I.*** **(*Hi–Hakai Kensa,* JAPAN), HIHKA, 18(9):416–417. Sep 1969. (in Japanese).**

The situation of studies of stress wave emission inspection by the author on his tour in the United States is outlined, and the importance of stress wave emission technique is accentuated as an epoch-making nondestructive testing tool.

1290 **N. Niwa. "For the Special Issue on** *AE.*" *Journal of N.D.I.* (*Hi-Hakai Kensa,* **Japan). HIHKA, 21(4):205-206. Apr 1972. (in Japanese).**

1291 **N. Niwa, I. Sagehashi, and H. Hatano. "Heat Treatment of Steel and Stress Wave Emission."** *Journal of N.D.I.* (*Hi-Hakai Kensa*, **Japan), HIHKA, 20(2):98. Feb 1971. (in Japanese).**

Influence of heat treatment on stress wave emission in a hot-rolled steel plate was investigated during tensile testing. Under the same experimental conditions, full-annealed steel emitted much more signals than raw steel. On the other hand, a large amount of emission was detected from deformation of the plate in the weld area.

1292 **N. Niwa, I. Sagehashi, and H. Hatano. "Preliminary Experiments on Stress Wave Emission Detection." Paper HPI-AE-5-16 submitted to the 5th meeting of the Atomic Energy Committee, High Pressure Institute of Japan. June 30, 1970. (in Japanese); RFP-Trans-95. translated for Dow Chemical U.S.A., Rocky Flats Division, Golden, Colorado. 1972. [NTIS].**

Main characteristics of stress wave emission in a steel plate during tensile testing were surveyed carefully. It was made clear that emissions are detected mainly immediately after the test plate begins to yield, and the so-called nonreversibility of stress wave emission was also confirmed.

1293 **M. J. Noone and R. L. Mehan. "Observation of Crack Propagation in Polycrystalline Ceramics and Its Relationship to Acoustic Emission." pp 201-229 in** *Fracture Mechanics of Ceramics.* **R. C. Bradt, D. P. H. Hasselman, and F. F. Lange, Editors. Volume 1: Concepts, Flaws, and Fractography. Plenum Press, New York. 1974. proceedings of a symposium held at Pennsylvania State University, University Park, Pennsylvania. July 11-13, 1973.**

The paper describes the detection and observation of discrete events taking place during loading of ceramics and which culminate in the development of a critical fracture crack. Ceramic materials with a high degree of optical transparency greatly facilitate the direct observation of crack movement within the bulk material away from surface disturbances. Polycrystalline ceramics with potential for applications as window materials resistant to ballistic impacts were used in most of the work to be discussed. A simple technique is described in which acoustic emission monitoring was used to facilitate detection of subcritical crack growth in these materials so that direct correlation of visual and acoustic events could be attempted. The data obtained from acoustic monitoring were a valuable supplement to the usual information obtained during mechanical testing of some ceramics and provided a greater understanding of the fracture process. The objective was to identify the source of the acoustic events in the transparent ceramics so that data gathered during testing of opaque structural ceramics might be correctly interpreted. The equipment and techniques used were developed for testing of fiber-reinforced composite materials in the author's laboratory where correlation of acoustic and visual data obtained in the testing of model (transparent matrix) systems has been used to interpret failure processes in structural composites.

1294 **P. J. Noronha and E. C. McKannan. "A Review of the State-of-the-Art of the Non-Destructive Testing of Flight Pressure Vessels." pp 188–203 in** *Prevention of Structural Failure: The Role of Quantitative Nondestructive Evaluation.* **T. D. Cooper, P. F. Packman, and B. G. W. Yee, Editors. Materials/Metalworking Technology Series, Number 5. American Society for Metals, Metals Park, Ohio. 1975.**

The paper reports on the state-of-the-art of the NDE of space vehicle pressure vessels. These vessels differ from conventional ground-based pressure vessels in the low safety factors used in order to permit maximum weight reduction. The design of flight vessels is based on a nominal stress requirement and a fracture mechanics approach, and optimization of the weight of the vessel is based on the smallest size defect that can be detected with a high degree of confidence. In addition for future space vessels, consideration of proof testing for several cycles of use will be required. The wide variety of metals used for fabrication, and the different defects that may be present in them at every stage, up to completion of the vessel are described. Techniques currently being used for NDE are described along with their advantages, limitations, and limits of detectability at high levels of confidence. Techniques being considered for use in the future to improve the limits of the minimum flaw size that can currently be detected include the Delta scan and acoustic emission techniques and these are described. The construction of space vessels for use in the future has been modified to reduce the presence of critical defects and so to improve the cost effectiveness to projected NDE requirements.

1295 **K. R. Notvest. "Acoustic Emission Monitoring of a Submerged-Arc Weld." Curtiss-Wright Corporation, Woodridge, New Jersey. Sep 18, 1975.**

1296 **K. R. Notvest (to Trodyne Corporation). "Acoustic Emission Spot Welding Controller." U.S. Patent 3,824,377. July 16, 1974.**

A system for detecting the acoustic emission emanating from a spot weld in which the emission from spot-weld nugget formation, weld expulsion, and post-weld cracking are separately detected and converted to binary signals. The binary signals are accumulated and compared with predetermined count levels. Means are provided to turn off the spot-welding machine when the accumulated count indicates that the spot-weld nugget has reached the desired volume. Means may also be provided to activate alarm signals if the weld nugget fails to reach the desired volume, or if the weld nugget suffers excessive expulsion or post-weld cracking.

1297 **K. R. Notvest. "Control of Resistance Spot Welding with Acoustic Emission." Trodyne Corporation, Teterboro, New Jersey. 1972; paper presented at the 9th Acoustic Emission Working Group Meeting, Denver, Colorado. June 14–16, 1972.**

Spot welding has always had the problem of control and evaluation. Welds are made extremely rapidly and the finished spot weld is buried between two sheets, making weld inspection difficult and expensive. The practice of running test strips for destructive testing may be misleading if the test material does not exactly match the work in composition, geometry, and fit-up. The principal parameter of a spot weld is an expression of nugget size, such as nugget diameter, penetration, or lap shear strength. The quantity of expulsion or spitting is a secondary consideration. Post-weld cracking is an acceptance criteria more prevalent in complex alloys. All the above are physical events that generate acoustic emission. If the aggregate of acoustic emission can be detected, reasonably free of interference, the various emission patterns can be separated and associated with the causative source. Properly implemented, such an acoustic emission detection system can automatically control and evaluate a spot welding operation.

1298 **K. Notvest. "Effect of Thermal Cycles in Welding D6aC Steel."** *Welding Journal,* **WEJUA, 45(4):173s–177s. Apr 1966.**

Crack susceptibility of D6aC ultra-high-strength steel welds was studied in relation to the thermal cycles experienced in welding and heat treating. Good correlation was obtained between resistance to cracking and weld ductility for any given weld thermal cycle. Optimum as-welded ductility was obtained when preheat, weld-heat input and postheat parameters were adjusted so that the weld transformed at 650 to 530 °F from 3 to 30 minutes after welding. Transformation times and temperatures were found to be lower for weld metal compared to wrought base metal because of complete solution of alloying elements. Physical properties of tempered welds were found to vary considerably compared to quench and tempered wrought material. Post-heating or stress relieving at temperatures around 800 and 1225 °F resulted in a sharp drop in weld ductility, both at temperature and below. Stress relieving at 1325 °F provided a stable, ductile structure which was not affected by subsequent exposure to 800 or 1225 °F. Production experience on large size rocket-motor components has verified that lower bainite weld structures post-heated at 530 to 600 °F and stress relieved at 1325 °F will have considerably greater ductility. Welds so treated will survive subsequent repair, fabrication, and final heat treatment without cracking.

1299 **K. R. Notvest (to Trodyne Corporation). "Vorrichtung zum Steuern von Schweissvorgangen (Device for the Control of the Welding Process)." German Patent 2,227,527. Nov 22, 1973. (in German). [see U.S. Patent 3,824,377, Reference 1296].**

1300 **N. V. Novikov, S. I. Likhatskii, and A. L. Maistrenko. "Determination of the Moment of Breakaway of a Crack by the Acoustic Method in Eccentric Tensile Testing of Notched Specimens."** ***Strength of Materials,*** **SMTLB, 5(9):1049–1053. Sep 1973; translated from** ***Problemy Prochnosti,*** **PPCNB, (9):21–25. Sep 1973. (in Russian).**

An acoustic method of indicating the beginning of failure of a material is described. Its application significantly widens the possibilities of testing materials for failure viscosity. The basic element of the acoustic system is a piezoceramic sensor attached to the surface of a flat sample near the notch. Samples of the AMg6 alloy and steel 65G were tested. [Abstract 045919. *Engineering Index,* Vol. 73, 1974].

1301 **G. Nystrom. "Some Experiences from Field Application of Acoustic Emission at TRC/Quelques experiences de l'application d'emission acoustique pratiquees sur le terrain." Tekniska Rontgencentralen AB, Stockholm, Sweden; Paper 3K/11 in** ***Proceedings of the Eighth World Conference on Nondestructive Testing.*** **volume containing Sections 3K2, 3H2, 5A. International Committee for Nondestructive Testing, Secretariat du Cofrend, Institut de Soudure, 32 Boulevard de la Chapelle, 75880 Paris Cedex 18, France. 1976. conference held in Cannes, France. Sep 6–10, 1976. (paper in English).**

A description of the AE equipment and data handling applied to the detection of loose parts and leakage of tubing inside the containment of nuclear reactor power stations is given. An experiment with AE during hydrotest of 105-mm shells is also presented.

O

1302 A. E. Oaks. "Nondestructive Testing of Scout Rocket Motors." NASA CR-2013. General Electric Company, Philadelphia, Pennsylvania. Apr 1972. [NTIS: N72-24526].

The nondestructive tests applied to Scout rocket motors were reviewed and appraised. Analytical techniques were developed to evaluate the capabilities of the radiographic and ultrasonic procedures used. Major problem areas found were the inadequacy of high voltage radiography for detecting unbonds and propellant cracks having narrow widths, the inability to relate the ultrasonic signals received from flat-bottomed holes in standards to those received from real defects, and in the general area of the specification of acceptance criteria and how these were to be met. To counter the deficiencies noted, analyses were conducted to the potential utility of radiometric, acoustic, holographic, and thermographic techniques for motor and nozzle-bond inspection, a new approach to qualifying magnetic particle inspection and the application of acoustic emission analysis to the evaluation of proof and leak-test data.

1303 L. Obert. "Measurement of Pressures on Rock Pillars in Underground Mines, Part I." Report of Investigations 3444. U.S. Bureau of Mines, Washington, D.C. Apr 1939.

The report presents a relatively simple laboratory method for determining the variation of the velocity of sound with pressure in small columns of stone. The results of the investigations show that the velocity of the rocks studied, except limestone, changes sufficiently with pressure to warrant anticipation of a high degree of success when the method is applied to actual mine pillars. The samples studied were of uniform material which was free from cracks or large variations in composition. In actual mine structures the composition seldom, if ever, reaches this degree of uniformity, and a study of the variations due to these differences will form part of the next investigation. As such variations and discontinuities tend to make the material less elastic in general, it is anticipated that the variation of velocity with pressure will be at least as much as that found for the homogeneous samples studied in the laboratory.

1304 L. Obert. "Measurement of Pressures on Rock Pillars in Underground Mines, Part II." Report of Investigations 3521. U.S. Bureau of Mines, Washington, D.C. June 1940.

The report presents a method of measuring the velocity of sound in rock mine pillars *in situ* and describes the apparatus required to make such measurements. In a series of experiments, the method has been applied to rock pillars in the Leadwood Mine of the St. Joseph Lead Company, from which conclusions are drawn as to the possibilities of the method. Further measurements on concrete replacement pillars in the same mine are reported, together with a laboratory study of the corresponding grades of concrete from which the accuracy of the method is indicated. The St. Joseph Lead

Company has developed a static method of determining pressures exerted by the roof on the concrete pillars, which requires a knowledge of Young's modulus of elasticity for the pillars. Sonic measurements of the elastic modulus of these pillars were made *in situ* to assist in this determination. The report is concluded with a summary of the progress to date and an outline of future activities.

1305 **L. Obert. "The Microseismic Method: Discovery and Early History." pp 11–12 in** ***Proceedings First Conference on Acoustic Emission/Microseismic Activity in Geologic Structures and Materials.*** **H. R. Hardy, Jr. and F. W. Leighton, Editors. Trans Tech Publications, Clausthal, West Germany. 1977. conference held at Pennsylvania State University, University Park, Pennsylvania. June 9–11, 1975.**

The paper provides a brief historical review of the development of microseismic techniques since their initial investigation some forty years ago.

1306 **L. Obert. "Use of Subaudible Noise for Prediction of Rock Bursts." Report of Investigations 3555. U.S. Bureau of Mines, Washington D.C. Jan 1941.**

The occasional rock noises in a mine caused by breaking and ripping of the back, settling, and other causes are well-known to all mining engineers. These noises often precede certain types of rock failure and as such are used to a certain extent as warnings. Besides these audible rock noises there are, however, an even larger number of subaudible rock noises, which must be picked up with some type of vibration microphone and amplified in order that they may be heard. It was suggested in an earlier Bureau of Mines publication that a study of these subaudible noises might be useful in determining rock movements preceding failure. During the course of some seismic investigations at the Ahmeek Mine of the Calumet and Hecla Consolidated Copper Company, it was noted that the ground, which was normally *quiet* became *noisy* (by noisy is meant amplified subaudible noise) and that the noise continued to increase for about 15 minutes until it terminated with a rock burst that occurred not over 50 feet from the point of observation. It is evident that, if there is always an increase in the noise immediately before a rock burst, it could be used to predict rock bursts. This incident started a program of research, and this report covers a study on the use of subaudible noises for predicting rock bursts. The investigation is part of a comprehensive program of research being carried out by the Bureau of Mines on problems resulting from pressure in underground mines.

1307 **L. Obert and W. I. Duvall. "Micro-Seismic Method of Determining the Stability of Underground Openings." Bulletin 573. U.S. Bureau of Mines, Washington, D.C. 1957.**

The cracking and popping of rock, often heard by men working underground, for years has been interpreted as a warning of danger and unstable ground. However, not until 1939 was it discovered that rock under stress generates subaudible and sometimes audible seismic disturbances (referred to as *microseismims*), which can be detected with the use of a suitable seismic apparatus. Since that date, laboratory and field investigations have been in progress to ascertain to what extent this phenomenon can be utilized to detect, localize, and determine the magnitude of areas of instability in and around mine openings and to predict if and when failure will occur. These investigations have included a determination of the types of problems to which the method is applicable and the effect of rock type, geological defects, such as joints, fractures, faults, and other uncontrollable factors, on the reliability of results. Six reports have been prepared covering the investigations conducted between 1940 and 1952. These reports describe some of the characteristics of microseismims, as determined from both laboratory and field tests; the equipment used to pick up and record microseismims; the general test procedure and method of analyzing results; and applications of the method to specific mining problems. This report contains a summary of the investigations conducted between 1940 and 1955. The characteristics of microseismims and the microseismic equipment are briefly reviewed. Generalizations regarding experimental procedure and analysis of results are presented. Examples of the various types of mining problems to which the microseismic method is applicable are cited, and the results of a few specific investigations are discussed.

1308 L. Obert and W. Duvall. "Microseismic Method of Predicting Rock Failure in Underground Mining. Part I: General Method." Report of Investigations 3797. U.S. Bureau of Mines, Washington D.C. Feb 1945.

The report is the first of four to be presented on the microseismic method of predicting rock failure in underground mining. The application of the microseismic method to one specific case, namely, that of predicting rock bursts in deep mining, has been described in two previous reports. Since the publication of these reports, the method has been extended to other fields of mining, and it is the purpose of this report (Part I) to describe the general microseismic method, giving the empirical facts upon which it is based, the different testing procedures, the method of recording and analyzing the data, and the interpretations that can be obtained therefrom. Part II (Report of Investigations 3803) will describe a series of laboratory experiments that were performed to give information on the origin of microseismims and the microseismic properties of various rocks. Part III will cover application of the method to specific mining problems together with a brief description of a number of investigations performed during the last three years. Part IV will describe in detail the microseismic equipment used in the various phases of these investigations. In underground mining, there are visible and sometimes audible indications of excessive rock pressure, such as cracks, pinchoidal fracturing in drifts and working faces, hourglassing of pillars, squeezes on sets and cribs, etc., which are familiar to all mining engineers. Although these indications often precede failure and, to a certain extent, are used as a warning, they do not provide in themselves adequate means of determining the existing pressure condition. The need for a quantitative or quasi-quantitative method of determining the stress distribution in underground mine structures, such as pillars, arches, and the rock overlying mine openings, as a means of improving the safety of working conditions and the economic operation of the mine is self-evident. The problem of developing such a method was undertaken by the Bureau of Mines in 1936. One seismic method for determining the pressure, applicable only to large pillars of uniform ore or rock, was developed and first reported in 1939. This report describes a second and more general method applicable in all types of hard-rock mining.

1309 L. Obert and W. I. Duvall. "Seismic Methods of Detecting and Delineating Subsurface Subsidence." Report of Investigations 5882. U.S. Bureau of Mines, Washington, D.C. 1961.

Subsidence to surface occurs in a large proportion of underground mines. Preliminary to any indication of subsidence on the surface, there is usually a breaking up and sloughing of the cover into the mine opening which, in a period of time ranging from minutes to many years, progresses to the surface. This process is referred to as subsurface subsidence. In many mining operations, information regarding the process of subsurface subsidence is important. For example, if there are buildings or other improvements on the surface, or if the direction of the subsurface subsidence might threaten other underground workings, such as shafts, or if the subsidence might be progressing in a direction which would fall outside the anticipated subsidence limits, the problem could assume special significance. In the past some information regarding subsurface subsidence limits has been obtained by driving exploratory dirfts or by diamond drilling. However, for the amount of information that can be obtained from one drift or from one diamond drill hole, these methods are costly. Because of the need for less costly and time-consuming procedures, this investigation was undertaken by the Bureau of Mines to determine the possibilities of developing or employing seismic (geophysical) methods. To date, four seismic methods have been considered, the traveltime, microseismic, traveltime difference, and seismic reflection methods. In this report, the principal features of these methods and the necessary instrumentation are discussed. Some experimental data are presented to indicate the type of analysis and the accuracy that may be expected from these procedures.

1310 L. Obert and W. Duvall. "The Microseismic Method of Predicting Rock Failure in Underground Mining. Part II: Laboratory Experiments." Report of Investigations 3803. U.S. Bureau of Mines, Washington D.C. Mar 1945.

The report is the second of four on the microseismic method of predicting rock failure in underground mining in which the results of a number of laboratory experiments designed to give information on the origin of microseismims (small-scale seismic disturbances) and microseismic properties of various mine rocks are presented. Part I of this report (Report of Investigations 3797) discussed the results of field experiments, the development of the microseismic method, the test procedure as applied to generalized problems, the method of analyzing microseismic data, and the conclusions that can be drawn therefrom. Part III will describe the application of the method to a number of specific mining problems investigated during the last 3 years, and Part IV will describe in detail the microseismic apparatus employed in the various phases of this work. In the course of the development of this method and in applying it to mining problems, there has always been some conjecture regarding the origin of microseismims. It was known from observations in many mines that microseismims originated in rock under pressure and that the microseismic rate, that is, the number of microseismims recorded per unit time, increased as the time of failure approached. It was also known that microseismims were produced at the time when the rock was visibly cracked or fractured; however, it was observed that the microseismims often preceded any indications of cracking or farcturing of recent origin. This raised the question: Did the microseismims unaccounted for by this cracking or fracturing originate by intermovements along fissures, seams, or fractures of geological origin, or did the microseismims originate in uniform pieces of homogeneous rock without occasioning any visible indications of failure? The answer to this question might have considerable significance when the application of the method to mining problems is considered. For example, if microseismims originate only along seams, joints, fractures, etc., it would be impossible to employ the method in uniform homogeneous formations; whereas, if they do originate in homogeneous rock the method would have general application. Part of this report discusses a series of laboratory experiments designed to give information on the origin of microseismims; the remainder is devoted to a study of the microseismic properties of various mine rocks. In the course of applying the method to a specific mining problem, a series of preliminary experiments has always been necessary to determine the microseismic properties of the rock. These preliminary experiments involved microseismic measurements in rock that was known to be under pressure and, wherever possible, following a sequence of mining that eventually led to an induced failure (such as that caused by the removal of pillars, remnants, timber, or other support). These preliminary experiments often were difficult to effect and always time-consuming. It is the purpose of this part of the investigation to develop a laboratory method for determining the microseismic properties of rock that would replace at least part of the preliminary testing that heretofore had to be done in the mine.

1311 L. Obert and W. Duvall. "Use of Subaudible Noises for the Prediction of Rock Bursts, Part II." Report of Investigations 3654. U.S. Bureau of Mines, Washington D.C. July 1942.

The investigation reported herein is part of a comprehensive research program being conducted by the Metal Mining Research Section, Mining Division, Bureau of Mines, on pressure problems in underground mines. This program was initiated in 1937 and for the first 3 years was concerned primarily with measurement of the pressure on mine pillars and supports and their strength. During this period, investigations were conducted in both the laboratory and the field in the comparatively shallow lead and zinc mines of Missouri and Oklahoma. As a result, a seismic method of determining the pressure on mine pillars was developed. Before this research program was initiated, Chas. F. Jackson, Chief of the Mining Division, suggested that rock under pressure might give rise to noises inaudible to the unaided ear, that is, subaudible noises. As a result of this suggestion, listening tests were made concurrently with the routine pressure measurements to see if any rock noises were present; with one exception, it was found there were no rock noises in any of these shallow mines. At the same time, pressure measurements disclosed that, with one exception, pillars were not under pressure amounting to more than 10 percent of the crushing strength of the rock. The one exception was a pillar in the Big John Mine of the Eagle Picher Mining and Smelting Company in the Tri-State district, which showed some physical evidence of failure, such as small cracks and spalling; it was reported that the pillar had been heard to crack or snap at intervals. Listening tests in this pillar disclosed that subaudible noises of comparatively small intensity were present and occurred at 1- to 2-minute intervals, on the average. Unfortunately, owing to the shortness of this pillar it was impossible to make any pressure measurements.

In 1940, it was decided to attempt pressure measurements in deep mining. The Ahmeek Mine of the Calumet and Hecla Consolidated Mining Company, Ahmeek, Michigan, was selected, and tests were made in small rib pillars in the stopes at a vertical depth of approximately 3800 feet below the surface. During these tests, the pillars showed repeated evidences of pressure, such as cracking, spalling and hourglassing, and the back became displaced with respect to the footwall. Moreover, several small rock bursts occurred in the stope adjacent to these pillars. It was not surprising that the pressure measurements showed these pillars to be under a pressure approximately equal to that of the crushing strength of the rock. Throughout these experiments, a number of listening tests were made, and it was found that subaudible noises were continuously present, at times occurring with such frequency that they could not be counted. Listening tests at another section in the mine disclosed that this area, which normally was comparatively quiet (that is, with few subaudible noises present), became noisy just before a rock burst. On the strength of these observations, a series of tests of a preliminary nature was conducted over a period of approximately 1 month to determine something about the nature of these subaudible noises and whether they could be used for predicting rock bursts. The tests disclosed that several types of rock noises could be classified, in addition to the noises due to rock drilling and machinery. The latter were found to be so intense that they obscured the rock noises, making it impossible to make observations except when the mine was inoperative. It was concluded that the occurrence of subaudible rock noises could be correlated with the occurrence of rock bursts; but, owing to the statistical nature of these tests, observations over an extended period with the proper apparatus would be required before it could be determined whether these correlations were accurate enough to permit prediction of rock bursts. This report describes a series of tests over a 2-month period at the Ahmeek Mine with a new apparatus designed particularly for this purpose. The results of these tests are discussed both in respect to the nature of the subaudible noises and to the ability to predict rock bursts. The report also describes preliminary investigations at the Sunshine Mine of the Sunshine Mining Company, Kellogg, Idaho; the Frood Mine of the International Nickel Company, Ltd., Copper Cliff, Ontario; and the Lake Shore Mine of the Lake Shore Mines, Ltd., Kirkland Lake, Ontario, Canada.

1312 **J. Obraz. "Akusticka emise a jeji vyuziti v defektoskopii (Acoustic Emission and Its Application in Crack Fault Detection)."** *Jemna Mechanika a Optika,* **JMKOA, 18(12):325–326. Dec 1973. (in Czech).**

1313 **D. T. O'Connor and D. W. Prine (to Magnaflux Corporation). "Detection of Paster Welds with Acoustic Emission." U.S. Patent 3,782,183. Jan 1, 1974.**

Given are a method and apparatus for the nondestructive detection of paster welds wherein the test piece is repeatedly flexed at the weld to be tested, and then any acoustic emission produced by such flexing, indicative of the presence of a paster weld, is detected. The invention provides for apparatus to convert the acoustic emission to electrical pulses which may then be integrated to provide a signal dependent upon the amplitude and rate of such acoustic emission.

1314 **M. Ogasawara, T. Udagawa, H. Imai, E. Isono, and M. Suzuki. "Application of Acoustic Emission Technique to the Identification of Weld Defects of Welded Pipe." Session 8, pp 28–45 in** *The Second Acoustic Emission Symposium.* **Japan Industrial Planning Association, Dobashi Building 5F, 3-17-8 Akasaka, Minato-ku, Tokyo, 107 Japan. 1974. proceedings of a symposium held in Tokyo, Japan. Sep 2–4, 1974.**

Soundness of weld seam of pipe is usually guaranteed by application of combined nondestructive inspections (NDI), such as X-ray, ultrasonic, magnetic particles, and eddy current as well as pressurizing test. However, so-called *cold weld* in electric resistance welded (ERW) pipes are considered to be undetectable by the conventional NDI, though it can be the cause of failure during pressurizing test or in service. Further assurance of quality of weld is obtained by mechanical testings such as the flattening test and guide-bend test. As are the characteristics of coupon test, they can not assure the soundness of whole line of weld seam. Improvement of NDI system could be expected if acoustic emission technique (AE) might be applied, incorporated with pressurizing test, to evaluate the quality of weld seam. It is especially interesting to know whether cold weld can be detected by AE method. In the experiment, thin rubbers were put between the sample and the machine to shut noise generated due to mechanical movement of the machine. The pressurizing velocity varied greatly from sample to sample due to somewhat arbitrary manual operation of a pressure valve of the ring expander. The AE was recorded with NSC type instrument and Dunegan's preamplifier and sensor. The AE data were analyzed in terms of energy rate. In addition, instead of energy rate method, pulse height analysis was applied. After the ring expansion test, the ring samples were cut, flattened and notched on the welding line by pressing a wedge. Then, the sample was borken by the large size Charpy-type testing machine. In most cases, the fractured surface thus obtained coincided to the plane of welded seam. Defects were certificated by the direct observation of the fracture surfaces. This certification of defects was used as an evaluation standard of the results of each NDI. In this report, detectability of cold weld is investigated by means of AE technique during a ring expansion test which was intended to simulate the pressurizing test of pipes on the production lines.

1315 **N. Ogura. "Cracking at the Toe of Welded Joint and Acoustic Emission." pp 62–64 in** *Significance of Defects in Welded Structures.* **T. Kanazawa and A. S. Kobayashi, Editors. University of Tokyo Press, Bunkyoku, Tokyo, Japan. 1974. proceedings of the Japan-U.S. Seminar on the Significance of Defects in Welded Structures, held in Tokyo, Japan. Oct 15–19, 1973.**

Three recent catastrophic failures during hydraulic testing in Japan, two spherical pressure vessels (1968), and one forged head of a multiple layered vessel (1970), have stimulated study of the condition under which brittle fracture can be initiated. A major point of interest in these three failures was that failure occurred at the toe and ran along the heat affected zone of a welded joint. Failure was induced by delayed hydrogen cracking at the toe with the crack growing to a critical size before triggering the brittle fracture. The role of dissolved hydrogen in promoting toe crack of a welded joint is thus the object of this investigation.

1316 **H. Okada, K. I. Yukawa, and H. Tamura. "Application of Acoustic Emission Technique to the Study of Stress Corrosion Cracking in Distinguishing between Active Path Corrosion and Hydrogen Embrittlement."** *Corrosion,* **CORRA, 30(7):253-255. July 1974.**

Numerous studies have been conducted on the effect of anodic or cathodic polarization on time to failure as a tool to differentiate active path corrosion (APC) from hydrogen embrittlement (HE). By acoustic emission technique, it is possible to detect high amplitude acoustic signals accompanying hydrogen induced microcracking at the advancing crack tip, in contrast to APC where cracks propagate by anodic dissolution along with preferential paths and significant acoustic emission cannot be expected. The purpose of this study is to differentiate the two mechanisms (APC and HE) in cracking of high strength and stainless steels by using an acoustic emission technique with or without galvanostatic polarization.

1317 **H. Okada, K. Yukawa, and H. Tamura. "Transition of Cracking Mechanisms from Active Path Corrosion to Hydrogen Embrittlement in High Strength Steels in a Boiling Nitrate Solution."** *Corrosion,* **CORRA, 32(5):201-203. May 1976.**

The mechanism of nitrate cracking of plain carbon steels and low alloy steels has been confirmed as active path corrosion (APC). It should be recalled that acidification of the corrodent at the crack front due to hydrolysis of the dissolved metal ions enhances hydrogen entry into steels, even in nitrate solutions. Hydrogen embrittlement (HE), therefore, could occur if the strength level of steels exceeds the critical value for a given hydrogen content. The purpose of this study is to distinguish between two mechanisms, APC and HE, in nitrate cracking of low alloy steels with the same chemical composition, but heat treated to three different strength levels, by the simultaneous use of high amplitude acoustic emission (AE) and crack growth measurements as previously reported.

1318 **R. O'Neil and G. M. Jordan. "Safety and Reliability Requirements for Periodic Inspection of Pressure Vessels in Nuclear Industry." Paper C48/72, pp 140-146 in** *Periodic Inspection of Pressure Vessels.* **Institution of**

Mechanical Engineers, London, England. 1972. proceedings of conference arranged by the Applied Mechanics Group of the Institution of Mechanical Engineers, held in London, England. May 9–11, 1972.

The use of pressure vessels to contain the primary coolant in nuclear reactors introduces the need for stringent safety and reliability criteria. Expressed in terms of failure probability per year, these criteria imply conviction of failure rates considerably below that currently achieved in conventional pressure vessels, even if such a comparison was valid. This paper, after briefly deriving standards required, considers the factors used in building up the safety and reliability cases, and their relative importance. These include design, manufacturing, construction, inspection, and testing standards, as well as the fracture mechanics methods involved in specifying tolerable defect sizes and inspection sensitivity. New inspection and testing techniques soon to become available are also outlined. Finally, the paper indicates, by way of example, the standard of safety and reliability likely to be achieved in the nuclear industry by pressure vessels which are built, inspected, tested, and repeat-tested, in accordance with the recommended procedures.

1319 **H. Ono, H. Nakasa, T. Machida, M. Fukusima, M. Hori, S. Ueda, A. Imazu, T. Nagata, M. Kikuchi, K. Doi, and K. Okabayashi. "Acoustic Emission Analysis on Fatigue Failure of LMFBR Primary Circuit Piping (Second Report)." PNC-J256 74-03, CRIEPI-74002. Central Research Institute of Electric Power Industry, Tokyo, Japan. May 1974. (in Japanese). [work performed in cooperation with the Power Reactor and Nuclear Fuel Development Corporation].**

The cooperative work by Power Reactor and Nuclear Fuel Development Corporation and Central Research Institute of Electric Power Industry on the measurement of acoustic emissions in the fatigue test of LMFBR piping components has been performed since March in 1972. Acoustic emission analysis on five test runs from March to October in 1972 were previously reported in January of 1973. This report summarises the results of eight test runs which were conducted from October 1972 to August 1973. Acoustic emission characteristics, such as amplitude, rate and summation of emissions, wave shape, frequency content, and time difference for signal location, have varied corresponding to the fatigue progress. These results have demonstrated that the acoustic emission techniques are useful for inspection of integrity, monitoring of degradation, and anticipation of fracture in LMFBR piping components.

1320 **H. Ono, H. Nakasa, T. Machida, Y. Tomoda, M. Fukusima, K. Takigawa, M. Hori, S. Ueda, T. Nagata, M. Kikuchi, and K. Okabayashi. "Acoustic Emission Analysis on Fatigue Failure of LMFBR Primary Circuit Piping (First Report)." PNC-J256 73-02, CRIEPI-72001. Central Research Institute of Electric Power Industry, Tokyo, Japan. Jan 1973. (in Japanese). [work performed in cooperation with the Power Reactor and Nuclear Fuel Development Corporation]. [English translation available as report JAPFNR-240 from NTIS].**

The Power Reactor and Nuclear Fuel Development Corporation and the Central Research Institute of Electric Power Industry determined to perform the cooperative work on the measurement of acoustic emissions in the fatigue test of LMFBR piping components. The following items were considered as important: (1) The development of the instrumentation systems available for the test of piping components; (2) To obtain the characteristics of acoustic emissions in the piping components; and (3) Evaluation on the avilability of acoustic emission method for practical LMFBR primary circuit piping. The report summarises the test results which were conducted from March to October 1972, and includes: (1) The instrumentation systems developed and shown in this report; (2) The amplitude and emission rate characteristics of acoustic emission signals investigated in the test of piping components; and (3) The test program planned for the future.

1321 **K. Ono. "Acoustic Emission and Microscopic Deformation and Fracture Processes." UCLA-ENG-7465. University of California, School of Engineering and Applied Science, Los Angeles, California. Sep 1974. [NTIS: AD-787-086]; Session 4, pp 1–63 in** *The Second Acoustic Emission Symposium.* **Japan Industrial Planning Association, Dobashi Building 5F, 3-17-8 Akasaka, Minato-ku, Tokyo, 107 Japan. 1974. proceedings of a symposium held in Tokyo, Japan. Sep 2–4, 1974.**

The paper reviews diverse observations of acoustic emission from materials and attempts to interpret them on the basis of current knowledge of plastic deformation and fracture. The results of a recent study on these topics will be introduced together with those from other groups. In order to correlate a variety of experimental results, important aspects of measurement techniques will be examined. This section discusses the nature of emission processes, vibration mode, transducer characteristics, threshold-crossing counts, signal intensity measurements, and frequency spectrum and amplitude distribution analyses techniques. In particular, the merits and drawbacks of various frequency spectrum analysis methods are critically discussed. In the next section, acoustic emission behavior during plastic deformation of solids is presented and discussed. An attempt is made to unify diverse and often conflicting findings on the basis of dislocation dynamics of glide and twinning. Observed acoustic emission characteristics have been classified into five categories, which arise from heterogeneous and homogeneous yielding, dynamic strain aging, twinning, and other heterogeneous deformation; and strain-induced transformation of acoustic emission during plastic deformation are examined in relation to these metallurgical aspects as well as the mechanical vibration viewpoint. For example, tensile deformation of magnesium is mainly via twinning, which generates shear mode acoustic emission. On the other hand, the yielding of a high strength low alloy steel sample excites longitudinal resonances along the length of the sample. It is shown that the microyielding concept with suitable modifications best explains these acoustic emission behaviors observed. At the same time, however, it must be noted that current understanding is still qualitative and needs further experimental and theoretical research efforts. Fracture processes generally produce high levels of acoustic emission, except for fracture in low-strength ductile materials. Various fracture mechanisms found in different classes of materials are correlated to observed acoustic emission behavior. These include ductile fracture via void coalescence, crack nucleation via second-phase fracture, and slow crack growth via tearing, cleavage, and and quasi-cleavage crack propagation. Relationships between threshold-crossing counts and stress-intensity factor or crack-tip opening displacement are discussed. It is shown that the consideration of amplitude distribution effects can substantially alter such relationships and simple analyses previously reported by others will lead to an erroneous conclusions. More elaborate analyses of acoustic emission from fracture processes are presented, mainly based on recent works on aluminum alloys that show the importance of the mode of vibration and transducer excitation. The results suggest the possibility of identifying the mechanisms of acoustic emission through frequency spectrum and vibration mode analyses.

1322 **K. Ono. "Amplitude Distribution Analysis of Acoustic Emission Signals." UCLA-ENG-7569. University of California, School of Engineering and Applied Science, Los Angeles, California. Aug 1975. [NTIS: AD/A-017-004];** ***Materials Evaluation,*** **MAEVA, 34(8):177-181, 184. Aug 1976.**

Acoustic emission signal characterization via amplitude distribution analysis is discussed from theoretical and experimental viewpoints. Fundamental amplitude distribution characteristics of continuous- and burst-type acoustic emission signals are derived and compared with experiment. Regardless of the signal bandwidth, continuous-type signals exhibit the Rayleigh-type peak distribution. Relationship between the amplitude distribution of acoustic emission counts and the peak distribution of burst signals is discussed, considering three model-peak distribution functions. Proposed in this paper are new calibration method of acoustic emission instrumentation and a means to separate the contributions of continuous- and burst-type acoustic emission signals.

1323 **K. Ono. "Recent Developments in Acoustic Emission as Applied to Welding Defects." UCLA-ENG-7375. University of California, School of Engineering and Applied Science, Los Angeles, California. Sep 1973. [NTIS: AD-768-396]; pp 48-58 in** ***Significance of Defects in Welded Structures.*** **T. Kanazawa and A. S. Kobayashi, Editors. University of Tokyo Press, Bunkyoku, Tokyo, Japan. 1974. proceedings of the Japan–U.S. Seminar on the Significance of Defects in Welded Structures, held in Tokyo, Japan. Oct 15-19, 1973.**

Recent developments in acoustic emission testing as related to welded structures are presented, focusing the attention to signal characterization, fracture induced acoustic emission, and large scale testing to locate flaws. Successful applications as well as the areas in need of further research are identified.

1324 **K. Ono and A. T. Green. "Acoustic Emission Testing for Structural Applications." University of California, School of Engineering and Applied Science, Los Angeles, California and Acoustic Emission Technology Corporation, Sacramento, California. 1976. [NTIS: AD/A-024-064]; paper presented**

at the American Society of Civil Engineers Specialty Conference on Dynamic Response of Structures: Instrumentation, Testing Methods and System Identification, held at University of California at Los Angeles Extension, Los Angeles, California. Mar 30–31, 1976.

Acoustic emission is a class of dynamic response of structures in microscopic scale. It is the transient elastic energy released when materials undergo plastic deformation or fracture (or both) under various types of loading. Significant advances in the detection of acoustic emission activities have brought about a new nondestructive testing method, which has been applied successfully for the evaluation of structural integrity of aerospace tanks, rocket motor cases, pressure vessels and piping systems, etc. It is also a valuable inservice method for nuclear pressure vessels. More than ten nuclear reactors have been instrumented and monitored in recent years. Except for the applications to pressure boundaries, use of acoustic emission testing has been sparse in large-scale structures. Since many buildings and structures may contain potentially hazardous flaws, wider utilization of acoustic emission testing should be considered. This paper describes acoustic emission phenomena and provides a representative coverage of reported applications.

1325 **K. Ono, G. Huang, and H. Hatano. "Acoustic Emission Characteristics of Structural Steels/Comportement des emissions acoustiques dans les aciers." University of California, Los Angeles, California. June 1976. [NTIS: AD/A-026-603]; Paper 3K/3 in** ***Proceedings of the Eighth World Conference on Nondestructive Testing.*** **volume containing Sections 3K1, 3H1, 3G. International Committee for Nondestructive Testing, Secretariat du Cofrend, Institut de Soudure, 32 Boulevard de la Chapelle, 75880 Paris Cedex 18, France. 1976. conference held in Cannes, France. Sep 6–10, 1976. (paper in English). [NTIS: INIS-mf-3451].**

The paper describes acoustic emission behavior of C-Mn steels, 4130 steel, and SA533B steels during tensile and fracture toughness tests. Nonmetallic inclusions were found to be the primary source of burst-type emission.

1326 **K. Ono, T. C. Schlotthauer, and T. J. Koppenaal. "Acoustic Emission from Ferrous Martensites." UCLA-ENG-7334. University of California, School of Engineering and Applied Science, Los Angeles, California. May 1973. [NTIS: AD-761-220]; pp 386–401 in** ***Proceedings of the Ninth Symposium on Nondestructive Evaluation,*** **Southwest Research Institute, San Antonio, Texas. July 1973. symposium held in San Antonio, Texas. Apr 25–27, 1973; Technical Report No. 4. Philco-Ford Corporation, Aeronutronic Division, Newport Beach, California. July 1973. [NTIS: AD-763-443].**

Acoustic emission during martensitic transformation of two Fe-Ni-C alloys (31.4 Ni-0.03 C and 22 Ni-0.44 C) has been investigated as a function of their thermomechanical history. Starting conditions examined include (1) as annealed equiaxed austenite (γ); (2) 20, 50, and 80% rolling of Item 1 at room temperature; (3) transformed to martensite (a') and then rapidly reverted to γ; (4) reverted γ from 20, 50, and 80% deformed alpha prime; and (5) reverted γ after five thermal cyclings with or without prior deformation. The AE total counts (N) were monitored as a function of temperature (T). While specimen-to-specimen variation in N was occasionally as high as 50%, extrapolated M_S temperatures for a given alloy condition could be established to within ±5 °C. The shape of N versus T curves as well as the magnitude of N changed systematically with microstructures of the alloys. For example, 20% or 50% deformation of the low carbon γ greatly increased N immediately below M_S which was almost unchanged. Further deformation to 80% changed the shape of the N-T curve completely and the M_S temperature was depressed. These and other findings were correlated to eddy current testing results for volume fraction transformed and microstructure observation via transmission electron microscopy. The interpretation of the AE results is discussed. In the second part of the paper, the use of AE technique for M_S temperature determination in alloy steels is discussed. This study was a part of an alloy development program and demonstrated the merit of AE technique for this application. In all, 15 different steel compositions were examined. The shape of N-T curves was varied, but M_S measurement was simple and rapid. In certain cases, M_S temperature could not be assigned unambiguously. This was traced to alloy segregation as revealed by optical metallography. Possible causes of the acoustic emission behavior are discussed. Experimental results as well as their implications are presented.

1327 **K. Ono, R. Stern, and M. Long, Jr. "Application of Correlation Analysis to Acoustic Emission." UCLA-ENG-7202. University of California, School of Engineering and Applied Science, Los Angeles, California. Jan 1972. [NTIS: AD-736-965]; pp 152-163 in *Acoustic Emission, ASTM STP 505.* American Soceity for Testing and Materials, Philadelphia, Pennsylvania. 1972. proceedings of symposium held in Bal Harbour, Florida. Dec 7-8, 1971.**

Correlation analysis techniques have been used to characterize acoustic emission pulses from metals. A brief description of correlation functions and the method of their determination are presented. The time-varying spectra of a magnesium sample have been obtained using the present techniques, and then compared with the more common methods of correlation analysis in acoustic emission studies are included.

1328 **K. Ono and H. Ucisik. "Acoustic Emission Behavior of Aluminum Alloys." UCLA-ENG-7514. University of California, School of Engineering and Applied Science, Los Angeles, California. Feb 1975. [NTIS: AD/A-008-204]; *Materials Evaluation,* MAEVA, 34(2):32-44. Feb 1976.**

Acoustic emission studies of high strength aluminum alloys (7001-T75, 7049-T6, 7049-T73, 7075-T651, and 7075-T7651) are described. The following acoustic emission parameters, i.e., threshold crossing counts, their dependence on the threshold level, rms voltages, and frequency spectra, were measured during fracture toughness tests and correlated to stress intensity factor. Three types of transducers were used and their effects were evaluated. In the T7 conditions, acoustic emission interpreted to result from plastic zone expansion was initially detected followed by that due to slow crack growth. Such transition was established through the analysis of frequency spectra, the mode of vibration, as well as the count versus stress intensity factor relationship. In the T6 conditions, no clear transition was present, indicating that slow crack growth initiates at a low level of stress intensity factor. These and other findings on acoustic emission behavior are discussed in detail.

1329 **M. Onoe. "Acoustic Emission Research in Japan for Flaw Detection in Weld." pp 22–47 in *Significance of Defects in Welded Structures.* T. Kanazawa and A. S. Kobayashi, Editors. University of Tokyo Press, Bunkyoku, Tokyo, Japan. 1974. proceedings of the Japan–U.S. Seminar on the Significance of Defects in Welded Structures, held in Tokyo, Japan. Oct 15–19, 1973; IIW X-753-74. International Institute of Welding. 1974.**

The paper reviews research effort in acoustic emission in Japan. A brief outline of the formation and activities of the Japanese Committee on Acoustic Emission (JCAE) is presented. The paper reviews some of the research work presented at recent JCAE meetings, however, primary discussion is on the application of acoustic emission techniques to the detection of delayed cracking in welds in high tensile strength steel weldments.

1330 **M. Onoe. "Development of Acoustic Emission Techniques for Pressure Vessel Inspection–TAB-AE(IRI) Project." Paper 2-4, pp 118–130 in *The Third Acoustic Emission Symposium.* Japan Industrial Planning Association, Dobashi Building, 3-17-8, Akasaka, Minato-ku, Tokyo, Japan. 1976. proceedings of a symposium held in Tokyo, Japan. Sep 16–18, 1976.**

Acoustic emission due to plastic deformation, crack initiation and propagation within a structure can be detected by sensors put on the surface. The location of sources can be determined from differences of arrival times of an AE wave to several sensors. Hence there has been much interest to apply AE techniques to detect locate, and identify defects in a pressure vessel, especially at a hydrostatic test. Before the AE inspection of pressure vessels will be widely accepted as reliable and economical means for the integrity assurance of a vessel, it is essential to establish a procedure for conducting a test upon the consensus among owners, fabricators, and inspectors. The recent ASME proposal provides some, but not all, a common basis with the above aim in mind. The Industrial Research Institute has run a study program on AE from 1974, referred to as the TAB-AE project. The project is partially supported by a grant from Japan Bicycle Association. The steering committee of the project consists of experts from industries, national laboratories, and universities. An outline of the project is given and discussions of some of the problems encountered during experiments.

1331 **M. Onoe. "Instrumentation for Measuring Acoustic Emission."** *Journal of N.D.I. (Hi–Hakai Kensa*, **Japan), HIHKA, 21(4):207–218. Apr 1972. (in Japanese).**

1332 **M. Onoe. "Japanese Experience in Laboratory and Practial Applications of Acoustic Emission to Welded Structures." pp 75–101 in** *Acoustic Emission.* **Edited by R. W. Nichols. Applied Science Publishers Ltd, Ripple Road, Barking, Essex, England. 1976. proceedings of a colloquium on acoustic emission held during the 1975 Annual Assembly of the International Institute of Welding, Tel Aviv, Israel. July 7–13, 1975.**

The review describes the activities of the Japanese Committee on Acoustic Emission, and then basic studies on AE caused by cracking due to hydrogen embrittlement, stress corrosion, and solidification. In the third section, AE caused by delayed cracking and lamellar tearing in weld are discussed. The fourth section describes the present state of uses of multichannel AE location systems in hydrostatic proof tests of pressure vessels. Efforts towards the establishment of a standard procedure for conducting AE inspection are mentioned. It is also shown that AE inspection is more effective than ultrasonic inspection for detecting flaws commonly called *cold weld* in electric resistance welded pipes. In the last section, there are descriptions of applications of the AE technique to tensile and fatigue tests of welded structures, including nuclear reactor components.

1333 **M. Onoe. "Report of the Second Acoustic Emission Symposium."** *Journal of N.D.I. (Hi–Hakai Kensa*, **Japan), HIHKA, 23(12):663–676. Dec 1974. (in Japanese).**

1334 **M. Onoe. "Special Issue on Acoustic Emission."** *Journal of N.D.I. (Hi–Hakai Kensa*, **Japan), HIHKA, 21(4):203–204. Apr 1972. (in Japanese).**

This is the introductory article in the special issue on acoustic emission of the *Journal of N.D.I.* (Japan).

1335 **M. Onoe. "U.S.–Japan Joint Symposium on Acoustic Emission."** *Pressure Engineering, Journal of Japan High Pressure Institute (Kotsu Gasu),* **KOGAA, 11(2):65–69. 1973. (in Japanese).**

Summary of the first acoustic emission symposium, sponsored by the High Pressure Institute of Japan, the Japanese Society for Non-Destructive Inspection, and the Japan Industrial Planning Association (secretariat), and held in Tokyo, Japan, July 4-6, 1972.

1336 **M. Onoe, K. Yamaguchi, H. Ichikawa, and T. Shimada. "A Multichannel Acoustic Emission Source Location System and Its Application to Fatigue Test of a Reactor Vessel."** *Electrical Engineering in Japan,* **EENJA, 95(3): 129–137. May–June 1975; translated from** *Denki Gakkai Ronbunshi, C,* **DGRCA, 95C(6):139–146. June 1975. (in Japanese).**

The characteristics of AE signals and transducers have been studied, resulting in the development of multichannel on-line AE location system which had been used successfully in a fatigue test of a reactor vessel. This paper describes features of the system such as hardware and software for noise reduction, prevention of erroneous location, and fast data processing. The results of fatigue test which was conducted in cooperation with the Central Research Laboratory of Electric Power Industry and Japan Atomic Energy Research Institute are also reported. [reprinted by permission of Scripta Publishing Company].

1337 **M. Onoe, K. Yamaguchi, H. Ichikawa, T. Shimada, and Y. Shimozuma. "Multichannel AE Source Location System and Its Application to Fatigue Test of Model Reactor Vessel." Session 2, pp 82–102 in** *The Second Acoustic Emission Symposium.* **Japan Industrial Planning Association, Dobashi Building 5F, 3-17-8 Akasaka, Minato-ku, Tokyo, 107 Japan. 1974. proceedings of a symposium held in Tokyo, Japan. Sep 2–4, 1974.**

Interest in acoustic emission has been generated for several years. After a few years' survey and basic research, the potential and importance of acoustic emission was considered for assuring integrity of large structures, and the development of a multichannel, real-time acoustic emission source location system was begun in early 1972. The system was designed mainly as a digital system because of its dependability, flexibility, and expandability. It was considered that the location system could be used, if possible, for many purposes such as fundamental laboratory experiments, fatigue test and field experiments, and be useful for fundamental development of in-service monitoring in which there was interest also. Consequently, it was to have high gain amplifiers, selectable bandwidth filters, software and hardware functions of noise or error rejection, and fast locating and sorting capacity. The system was successfully used in a fatigue test of model nuclear reactor vessel conducted jointly by Japan Atomic Energy Research Institute, Central Research Institute of Electric Power Industry, and Institute of Industrial Science of the University of Tokyo in July and September, 1973. The first half part of this paper reports the outline of the system structure and its features, that is, analog equipments, time difference detector, and software function. The latter half part is about the test of model reactor vessel and its location results.

1338 **A. Ookawa and K. Yazu. "The Energy Radiated from a Dislocation by an Accelerated Motion through Impurity Fields."** *Journal of the Physical Society of Japan,* **JUPSA, 18(Supplement I):36–43. 1963. proceedings of the International Conference on Crystal Lattice Defects. 1962.**

When a segment of a dislocation under otherwise uniform motion is accelerated and decelerated successively by passing through localized internal strain field, elastic waves are radiated outwards. In consequence of this type of radiation loss, the dislocation motion is impeded, giving rise to an apparent excess frictional stress for plastic deformation. In crystal provided with many local fields, the dislocation motion changes its character from the quasi-static case to the dynamical case with increasing average velocity, such as to show for moderate dislocation velocity a peak value of the radiation loss and the frictional stress. The estimated maximum frictional stress, $\sim 0.5\ kg/mm^2$ for Cu-10%-Zn, is rather small but of the same order as the lower yield stress of the alloy, and the radiation loss is suggested as a mechanism giving rise to the lower yield stress.

1339 **C. N. Owston. "Carbon Fibre Reinforced Polymers and Non-Destructive Testing."** *British Journal of Non-Destructive Testing,* **BJNTA, 15(1):2–11. Jan 1973.**

A brief description of carbon-fiber reinforced polymers is followed by a discussion of the philosophy of NDT as applied to these materials. An outline of the flexural fatigue properties of CFRP and a description of several NDT techniques leads to a discussion of the correlation between NDT and fatigue performance.

1340 **C. N. Owston. "Non-Destructive Testing and Fibre Reinforced Composites at Cranfield Institute of Technology." p 5-1 in "Newsletter on the Non-Destructive Testing of Composite Materials and Adhesive Bonded Joints." Issue Number 2. Royal Aircraft Establishment, Farnborough, Hants, England. July 1973.**

For a variety of reasons work on the above field was not possible for about 12 months, but the situation has now changed and two projects are under way. Most of the previous work at Cranfield on CFRP was concerned with flexural fatigue strength. Although not a parameter of general application, it was appropriate to the CFRP compressor blade program with which the work at Cranfield was associated. Even in metals, it is difficult, if not impossible to predict fatigue life from the initial state of a specimen. In fibre-reinforced materials, fatigue performance is more variable, specimen to specimen, than in most metals and it was a surprising achievement when found that fatigue performance could be predicted from a nondestructive examination of specimens before testing. A combination of an eddy-current test to measure volume fraction and ultrasonic attenuation to characterise void content and acoustic emission to evaluate internal activity gave a failure parameter which correlated well with fatigue strength. The first of the new projects is a continuation and extension of the flexural fatigue work except that the current effort relates to push-pull fatigue, a more generally applicable system. The aim is to find if the nondestructive test failure criterion, derived for flexural fatigue, will work in push-pull; and also to improve basic knowledge of the justification of the failure criterion and extend its compass to include additional types of defect. To assist in this work, construction has begun on a hydraulically powered loading frame, capable of ±10 tons and of cycling at up to 20 Hz about any mean load. The frame has an ultimate capability of 20 tons for future development. The machine will permit study of relatively large specimens containing a variety of defects and their relative importance can be determined. The second project is not primarily NDT. It is concerned with the effect of defects on the compression creep performance of glass reinforced structures. Nevertheless, it is anticipated that this program will contribute useful results to NDT work.

1341 **C. N. Owston. "Non-Destructive Testing of Carbon Fibre Reinforced Polymers." pp 1-21 in** *Non-Destructive Testing and Inspection Applied to Composite Materials and Structures.* **C. N. Owston and E. H. Jaffe. AGARD-R-590. North Atlantic Treaty Organization Advisory Group for Aerospace Research and Development. Feb 1972. proceedings of the 32nd AGARD Structures and Materials Panel Meeting, London, England. Mar 31, 1971. [NTIS: AD-739-780].**

The paper describes test techniques such as eddy current, ultrasonic testing, acoustic emission, and thermal methods applied to specimens made of carbon-fibre reinforced polymers. Results obtained from fatigued specimens are discussed. The report contains a second paper entitled "Non-Destructive Inspection Practices Used in Production of Composite Airframe Structures" by E. H. Jaffee, and describes NDI techniques as a tool for quality control during the manufacturing process of composite airframe structures.

P

1342 **P. F. Packman. "Metal Defects Make Themselves Heard."** *Iron Age,* **IRAGA, 212(9):39–41. Aug 30, 1973.**

This is a question-and-answer article resulting from an interview by *Iron Age* and the author in which he outlines the whys and wherefores of acoustic emission as a promising new nondestructive testing technique.

1343 **M. Page. "Ultrasonics Spearhead Today's Developments in Non-Destructive Testing."** *Metals and Materials,* **MEMTA, pp 20–22. Mar 1977.**

Survey article briefly describes several recent developments, including research in stress-wave emission at the NDT Centre in Harwell and The Welding Institute. Also the Seventh International Engineering and Quality Control Exhibition is previewed.

1344 **I. G. Palmer. "Acoustic Emission Measurements on Reactor Pressure Vessel Steel." Laboratory Note No. RD/L/N 133/72. Central Electricity Research Laboratories, Leatherhead, Surrey, England. June 9, 1972;** *Materials Science and Engineering*, **MSCEA, 11(4):227–236. Apr 1973.**

Acoustic emission measurements have been made on laboratory specimens of reactor-pressure vessel steel and the results correlated with crack-tip yielding and crack extension. It has been shown that most of the emissions arise from plastic zone expansion, and that ductile crack growth is relatively quiet. The emission behaviour is structure sensitive, and is markedly affected by prestrain. The effects of annealing on the recovery of the emission behaviour are described, together with the effects of oxidation and corrosion.

1345 **I. G. Palmer. "The Use of Acoustic–Emission Techniques and Liquid Penetrants to Detect Surface Cracks."** *Non-Destructive Testing.* **NDETA, 7(1):25–27. Feb 1974.**

A method is described in infiltrating surface defects with adhesive liquids. After curing, high amplitude acoustic emissions are obtained from cracking of the adhesive when the defect opens under load. Thus the defects can be found.

1346 **I. G. Palmer, B. J. Brindley, and R. P. Harrison. "The Relationship between Acoustic Emission and Crack Opening Displacement Measurements."** *Materials Science and Engineering,* **MSCEA, 14(1):3–6. Apr 1974.**

The acoustic emission from crack-tip yielding in reactor pressure vessel steel is characterized in terms of the crack-opening displacement. It is found that in laboratory specimens, the experimental results deviate from the theoretical curve when general yielding occurs. In pressure vessels, the theoretical correlation is expected to be followed.

1347 **I. G. Palmer and P. T. Heald. "The Application of Acoustic Emission Measurements to Fracture Mechanics." Laboratory Note No. RD/L/N 110/72. Central Electricity Research Laboratories, Leatherhead, Surrey, England. May 17, 1972;** ***Materials Science and Engineering*****, MSCEA, 11(4):181–184. Apr 1973.**

The acoustic emission from carbon-manganese pressure vessel steel is characterized in terms of the crack-tip plasticity. It is found that the total emission count is directly proportional to the plastic zone size, and it appears that the source of emission is the immediate vicinity of the elastic-plastic boundary. This relationship is used to calculate the total emission count before failure as a function of either the defect size or the applied stress.

1348 **I. G. Palmer, J. Holt, and D. J. Goddard. "Acoustic Emission Measurements on Type-316 Stainless Steel." pp 11–26 in "EWGAE–ISPRA: Proceedings of the Third Meeting of the European Working Group on Acoustic Emission." A. C. Lucia, Editor. Report EUR-5513e. Commission of the European Communities, Directorate General Scientific and Technical Information and Information Management, Luxembourg, Belgium. 1976. meeting held at the Joint Nuclear Research Center, Ispra, Italy. Sep 25–26, 1974.**

Acoustic emission measurements have been made on Type-316 stainless steel in the solution-treated condition, as part of a feasibility study for the monitoring of CFR components. The work involved testing both plain tensile specimens and precracked compact tension specimens in the temperature range 20 to 200 °C. At 20 °C plastic deformation was a quiet process but ductile crack growth was accompanied by high amplitude emissions capable of detection on plant. At 200 °C, both plastic deformation and ductile crack growth were quiet.

1349 **M. N. Palmer (to Gulf Research and Development Company). "Methods and Apparatus for Detecting Leaks." U. S. Patent 3,500,676. Mar 17, 1970.**

Describes an instrument to detect both the sound of an intravalve leak which is in the sonic range, and the sound of a leak to atmosphere which is in the ultrasonic range. The instrument is primarily used with pressurized fluid systems. An external touch probe is provided for ultrasonic sounds. A tunable amplifier is used so that steady leakage sounds can be distinguished from other extraneous noises, and a frequency divider is utilized to convert ultrasonic vibrations to the audible range.

1350 **T. Papa, D. Sette, L. Stagni, and A. Congiu Castellano. "Acoustic Emission Produced during Heating of Tin and Zinc Single Crystals."** *Journal of Testing and Evaluation,* **JTEVA, 3(1):48–50. Jan 1975.**

Acoustic emission activity has been detected during heating of tin (Sn) and zinc (Zn) single crystals from 0 to 130 °C. The emission of Sn occurs copiously only if the heating starts below a temperature of about 18 °C, and lasts up to a higher temperature between 60 and 80 °C. The results may be interpreted by relating acoustic emission to processes in the alpha ⇔ beta phase transformation. Although still qualitative, such a hypothesis gives useful information on the nucleation time of the α-phase and on the temperature hysteresis of the transformation. [reprinted by permission of the American Society for Testing and Materials, Copyright].

1351 **E. P. Papadakis. "Panel 4 Report on: Evaluation of Existing and Potential Techniques and Criteria for NDT and Materials Characterization." pp 463–480 in "Proceedings of the Interdisciplinary Workshop on Nondestructive Testing–Materials Characterization." D. O. Thompson, Program Manager. AFML-TR-73-69. North American Rockwell Science Center, Thousand Oaks, California. Apr 1973. [NTIS: AD-760-174]. Workshop held at North American Rockwell Science Center, Thousand Oaks, California. Nov 13–15, 1972.**

1352 **E. P. Papadakis. "Possible NDT Test Method for Ceramic Substrate Intergity Based upon Statistical Distribution of Acoustic Emission Count Totals under Torque." Ford Motor Company, Metal Melting and Processing Department, Detroit, Michigan. Aug 22, 1975.**

A nondestructive testing method utilizing acoustic emission has been devised to detect catalytic converter substrates in which cracks propagate excessively under stress. The method involves applying an axial load and then a torque about the axis, and *listening* with an instrument for the bursts of acoustic emission which accompany small increments of crack propagation. The best data come from the torque cycle which is torquing, holding a few seconds, and releasing the torque. The number of AE bursts during this cycle is recorded. Good specimens follow a normal frequency distribution on a log scale; specimens suspected of being weak give excess AE bursts and deviate from the normal curve. A durability test is proposed to verify that the specimens with excess acoustic emissions fail earlier in service. A durability test is a prerequisite to the establishment of a meaningful NDT test for substrate integrity.

1353 **E. P. Papadakis. "Sonic and Ultrasonic Methods for Nondestructive Testing of Catalytic Converter Substrates."** *Materials Evaluation,* **MAEVA, 34(2):25–31. Feb 1976.**

Sonic and ultrasonic measurements have been effective in defining new areas for the nondestructive testing of catalytic converter substrates of both the laid-up and the extruded type. For laid-up cylinders, the frequency of the lowest mechanical resonance peak is related to the degree of misalignment of the layers. Badly cracked pieces show high damping or extraneous resonances (or both). Acoustic emission under stress can be used to detect specimens in which cracks can propagate at the given level of applied stress, thus differentiaing between cracked and sound material. Ultrasonic wave velocity is inversely related to porosity in the ceramic material, and has been used to isolate other variables as well as porosity.

1354 W. J. Pardee. "Theoretical Analysis of Acoustic Emission Spectra." pp 667–679 in "Proceedings of the ARPA/AFML Review of Quantitative NDE." D. O. Thompson, Program Director. AFML–TR–75–212, SC595.10AR. Science Center, Rockwell International, Thousand Oaks, California. Jan 1976. [NTIS]. conference held at the Science Center, Rockwell International, Thousand Oaks, California. July 15–17, 1975.

The fracture of 20- to 40-micrometre silicon carbide grains on a steel slab produces high frequency and wide-band acoustic bursts that are similar to those observed during acoustic emission tests. The experimental detection and frequency analysis of these bursts have been used for a prototype of more complicated NDE acoustic emission studies and as a *white noise* standard. A conceptual separation is made of the acoustic signal generation process into two parts: a localized inelastic process which is the source of the burst, and the elastic propagation of this signal to a remote boundary where it is deleted. The theoretical determination of the first from measurements of the second will be demonstrated, and the implications and limitations discussed.

1355 W. J. Pardee and L. J. Graham. "Analysis of Acoustic Emission Frequency Spectra." pp 597–600 in *1975 Ultrasonics Symposium Proceedings.* Catalog No. 75 CHO 994–4SU. Institute of Electrical and Electronics Engineers, Inc., New York. 1975. symposium held in Los Angeles, California. Sep 22–24, 1975.

It has been suggested that the frequency analysis of an acoustic emission pulse can provide additional information about the nature and magnitude of its source. Presented are theoretical calculations and experimental measurements of two relatively simple and qualitatively different simulated acoustic emissions. The first is the fracture of small (20 to 40 micrometres) silicon carbide grains on a steel slab. This has been found useful as a practical measure of medium and transducer response. The second is the impact of a small elastic sphere on the same slab. The primary interest in both cases is to test the theoretical understanding of the medium response, and, consequently, to determine the extent to which we can determine the source spectrum from the measured spectrum.

1356 W. J. Pardee and L. J. Graham. "Frequency Analysis of Two Types of Simulated Acoustic Emissions." Science Center, Rockwell International, Thousand Oaks, California. 1977; *Journal of the Acoustical Society of America,* JASMA, 63(3):793–799. Mar 1978.

It has been suggested that the frequency spectrum of an acoustic emission burst may be a partial *signature,* identifying the source. To explore this possibility, a detailed study was made, theoretically and experimentally of two qualitatively different simulated acoustic emissions. The first is the fracture of small (20 to 40 micrometres) silicon carbide grains on a steel slab. The second is the impact of a small elastic sphere of various materials on the same slab. The primary objective has been to test current understanding of the medium response, and, consequently, the extent to which the source spectrum can be determined from the observed spectrum.

1357 R. W. Parkinson and C. T. Peters. "Acoustic Emission Activity during Accelerated Conversion of High Alumina Cement Pastes." *Journal of Materials Science,* JMTSA, 12(4):848–850. Apr 1977.

1358 D. L. Parry. "A Proposed Method of Applying Acoustic Emission Analysis to the Pre-Service and In-Service Inspection of Reactor Coolant Systems." Jersey Nuclear Company, Richland, Washington. 1970.

Described is Jersey Nuclear Company's plan to include acoustic emission testing in the pre-service and in-service inspections of nuclear reactor coolant systems. The essential feature of the plan is the application of unique acoustic emission analysis techniques to pre-service hydrostatic tests applied to the reactor system, and the correlation of the results with ultrasonic examinations of specified regions of the system. The resulting data establish a base from which acoustic emission procedures can be re-applied at in-service inspection intervals to obtain rapid and economic assesment of the coolant system integrity. As outlined in the proposal, the plan is implemented in full accordance with existing ASME XI codes on in-service inspection of reactor coolant systems.

1359 D. L. Parry. "Acoustic Analysis Technology for Incipient Failure Detection in Pressure Vessels." Idaho Nuclear Corporation, Idaho Falls, Idaho. 1969; presentation made to the American Society for Metals Materials Engineering Congress, Philadelphia, Pennsylvania. Oct 13–16, 1969. [paper never written].

Idaho Nuclear Corporation, under the auspices of Division of Reactor Development and Technology, Atomic Energy Commission, has developed, and is in the process of field testing and evaluating, a new nondestructive test technique for the 100% integrity analysis of the pressure vessels and primary system components of large nuclear power plants. The technique uses acoustic analysis technology to detect and locate flaws or cracks in the structure undergoing test. The results from engineering scale laboratory tests, and field application of the system during the acceptance hydrostatic testing of pressure vessels, has shown its potential as a promising and powerful new nondestructive test tool. In addition, tests have demonstrated the applicability of acoustic analysis technology for the measurement of structural metal condition, such as degree of embrittlement and transition temperature range.

1360 **D. L. Parry. "Acoustic Emission Analysis of Pressure Vessels and Piping." pp 17–32 in** *Proceedings of the Eighth Symposium on Nondestructive Evaluation in Aerospace, Weapons Systems, and Nuclear Applications.* **Western Periodicals Company, North Hollywood, California. 1971. symposium held in San Antonio, Texas. Apr 21–23, 1971.**

Acoustic analysis technology is now being applied as a commercially available nondestructive test tool. It has proven to be capable of rapidly detecting and locating multitudes of flaws in large complex structures under field test conditions. Analysis capabilities range from the detection and location of microcracking to the identification and location of significant crack growth. A recent acoustic analysis test of a nuclear pressure vessel resulted in the detection and accurate location of 13 weld slag and porosity regions. The defects were acoustically identified as being insignificant to the integrity of the vessel, but represented potential problem sites requiring close surveillance during planned future periodic in-service inspections. The defect sites were cross-checked and confirmed using standard nondestructive test techniques. Tests conducted on pressure piping have enabled the detection and location of fatigue-originated microcracking and stress-corrosion cracking.

1361 **D. L. Parry, "Acoustic Emission Integrity Analysis of Pressure Vessels and Piping." Jersey Nuclear Company, Richland, Washington. 1971; pp 553–568 in** *Proceedings of the 1st International Conference on Structural Mechanics in Reactor Technology.* **EUR-4820, Volume 4. Commission of the European Communities, Brussels, Belgium. 1972. conference held in Berlin, Germany. Sep 20–24, 1971.**

Acoustic emission analysis techniques are now being applied in industry for the detection and location of defects and leaks in commercial structures and pressure systems. It has proven to be capable of rapidly detecting and locating multitudes of flaws in large complex structures under field test conditions. Analysis capabilities range from the detection and location of microcracking to the identification and location of significant crack growth. The new nondestructive test tool utilizes the micro-acoustic signals that emanate from material flaws. It utilizes the high frequency noise generated by leaking gas or fluids to detect and locate leaks in pressure systems, such as vessels and piping. Acoustic analysis tests of nuclear pressure vessels have resulted in the detection and accurate location of flaws and defect regions. The defect sites were cross-checked and confirmed using standard nondestructive test techniques. Tests conducted on pressure and drill piping have enabled the detection and location of material flaws, corrosion, and leaks.

1362 **D. L. Parry. "Acoustic Integrity Analysis in the Oil and Chemical Industry." API Preprint No. 32–72. Jersey Nuclear Company, Richland, Washington. Apr 19, 1972; paper presented at the 37th Midyear Meeting of the American Petroleum Institute, Division of Refining. New York. May 9, 1972.**

The nondestructive test technology of acoustic emission analysis is now finding wide application in the chemical and petroleum industry. The technology has been demonstrated to be a powerful tool for the qualification of new pressure vessels and piping as well as the requalification of in-use pressure systems. It provides a means of guarding against vessel failure during hydrostatic acceptance tests. It also provides a means of quickly assessing a complete structure for defect significance and location. Computer technology is applied for analysis of the micro-emissions emitted from defects present in the structure undergoing testing. Relative significance of the defects to the imposed stress conditions is analyzed, and the location of these defects is computer-displayed on a map which is a geometrically accurate layout of the structure undergoing test. Large pressure vessels, storage tanks, valve casings, as well as thousands of feet of piping can be tested in a quick, accurate manner. Acoustic analysis techniques are now being developed for acoustically assessing the quality of welds as they are made in the manufacture of large vessels. The technology of acoustic emission analysis is fast approaching the status of a standard nondestructive test tool in the chemical and petroleum industry.

1363 **D. L. Parry. "Ensayos no destructivos por** ***analisis de emisiones acusticas*** **(Nondestructive Testing by Acoustic Emission Analysis)."** ***Revista de Soldadura,*** **RSLDB, 3(1):17–27. Jan–Mar 1973. (in Spanish).**

Described is the Exxon Nuclear Company's NDT-ACOUSTICS monitoring and source location system and its operation. Typical applications are discussed as well as the verification of acoustic emission results by ultrasonic, eddy current, and radiographic testing.

1364 **D. L. Parry. "Industrial Application of Acoustic Emission Analysis Technology." pp 150–183 in** ***Monitoring Structural Integrity by Acoustic Emission, ASTM STP 571.*** **Edited by J. C. Spanner and J. W. McElroy. American Society for Testing and Materials, Philadelphia, Pennsylvania. 1975. proceedings of a symposium held in Ft. Lauderdale, Florida. Jan 17–18, 1974.**

Acoustic emission analysis technology as a nondestructive testing tool is being applied in industry to a wide variety of structures. Its acceptance, however, has been more rapid in industries where it is applied to metallic pressure containment structures. These applications include pipelines, piping systems, storage vessels, pressure vessels, and even complex pressure containments such as the primary coolant systems of large nuclear power plants. The application techniques used and a description of typical results obtained during field applications provide additional insight into a new powerful testing tool that is rapidly gaining acceptance throughout the industry. [reprinted by permission of the American Society for Testing and Materials, Copyright].

1365 **D. L. Parry. "Inservice Inspection of Nuclear Power Plant Steam Supply System Using NDT-Acoustics." XN-310. Exxon Nuclear Company, Richland, Washington. Aug 8, 1975; pp 332-348 in** *Proceedings of the Tenth Symposium on Nondestructive Evaluation.* **Southwest Research Institute, San Antonio, Texas. June 1975. symposium held in San Antonio, Texas. Apr 23-25, 1975.**

Acoustic emission analysis technology is being utilized for the volumetric preservice and inservice inspection of nuclear-power plant primary-coolant systems. Typical tests results, derived from the NDT-ACOUSTICS inservice inspection of a large dual cycle nuclear power plant, are presented to demonstrate the accuracy and reliability of the technology. These results demonstrate that the technology, when supplemented by such standard nondestructive testing techniques as ultrasonics and radiography, provides a powerful tool that improves the probability of conducting 100-percent integrity analysis of nuclear-plant primary coolant systems.

1366 **D. L. Parry. "NDT-Acoustics: Acoustic Emission Analysis, Nondestructive Test Technology." JN-98. Jersey Nuclear Company, Richland, Washington. Dec 1972.**

In a few short years, acoustic emission analysis has progressed from the research laboratory to a worldwide nondestructive testing technology. It is applicable for the analysis of materials ranging from metals to woods and structures ranging from laboratory specimens to the primary pressure systems of large nuclear power plants. Acoustic emission analysis, which for convenience Jersey Nuclear calls NDT-ACOUSTICS, utilizes the microscopic signals generated in materials undergoing stress to analyze the physical integrity of structures and structural components. The technology is equally applicable for the detection and location of minute leaks in pressure systems. A presentation of the typical results obtained in the routine field application of NDT-ACOUSTICS demonstrates that this technology is now taking its place beside the standard nondestructive test tools of ultrasonics and radiography.

1367 **D. L. Parry. "NDT-Acoustics: New Nondestructive Test Technology Used in the Nuclear and Petrochemical Industries." Jersey Nuclear Company, Richland, Washington. 1973; Paper G6/2 presented at the 2nd International Conference on Structural Mechanics in Reactor Technology, held in Berlin, Germany. Sep 10-14, 1973. [paper not published in proceedings].**

1368 **D. L. Parry. "Nondestructive Flaw Detection by Use of Acoustic Emissions." IDO-17230. Phillips Petroleum Company, Idaho Falls, Idaho. May 1967. [NTIS].**

A requirement exists for a safeguards system that will ensure the integrity of the primary loop of a nuclear power installation. To meet this requirement, a technique of detecting the acoustic emissions of materials under stress has been applied as a nondestructive diagnostics tool for reactor pressure vessel and primary loop flaw analysis. The presence of a flaw in a metal under stress is evidenced by the low threshold of appearance for the acoustic signals, by an occurrence of burst signals, and by the increase in the continuous-emission-signal intensity with applied stress. Investigations into the feasibility of applying the acoustic phenomena as a safeguards system for ensuring primary loop integrity are being conducted as a test program. This test program has been developed around large-scale, complex reactor vessel testing, supplemented by laboratory-scale tensile-specimen testing. Acoustic signals have been successfully detected, recorded, and analyzed from two complex, embrittled pressure vessels tested under conditions of applied stress. The acoustic data were recorded on magnetic tape which was post-test analyzed by instrumentaiton systems designed to relate the detected acoustic signal characteristics with such test parameters as applied stress, stress rate, and material flaw size. Tensile specimen testing has been used for the detection, recording, and analysis of the continuous-emission-type acoustic signal. It has enabled the instrumentation of a detection and analysis system which allows the interpretation of the continuous-emission signal on a signal rate versus strain or stress basis. The continuous-emission-type signal has been shown to have essentially no threshold effect, appearing instantaneously with the first stress. It is statistically rich, and initial investigations into statistical time distribution analysis have indicated strongly that it may provide sufficient data for determining flaw size, growth rate, and prediction of the failure-stress value. Continuous-emission data and burst data have indicated the possibility of using the emission phenomena for determining metal characteristics such as ductility and nonstress.

1369 **D. L. Parry. "Nondestructive Flaw Detection in Nuclear Power Installations."** *Transactions of the American Nuclear Society,* **TANSA, 10(1):330–331. June 1967.**

Briefly describes how the detection of acoustic emissions from materials under stress has been applied as a NDT tool for reactor pressure vessel and primary-loop flaw analysis. The results of a test program to assess the suitability of the technique for monitoring the integrity of an embrittled reactor pressure vessel are reported.

1370 **D. L. Parry. "Nondestructive Flaw Detection in Nuclear Power Installations." pp 107–126 in** *Incipient Failure Diagnosis for Assuring Safety and Availability of Nuclear Power Plants.* **E. P. Epler and D. P. Roux, Coordinators. U.S. Atomic Energy Commission, Division of Technical Information. Jan 1968. [NTIS: CONF-671011]. proceedings of conference sponsored by Oak Ridge National Laboratory, held in Gatlinburg, Tennessee. Oct 30–Nov 1, 1967.**

Acoustic signals are known to be generated in metals under stress. These signals emanate from so called *healthy* or nonflawed materials as well as flawed materials. However, it has been observed that a certain characteristic *burst* signal is indicative of material flaw growth. The threshold of appearance of the acoustic *burst* signal, the increased occurence rate of the signal with increasing material stress, and the acoustic energy release signature are all considered defining parameters of material flaw characteristics. Investigations are being conducted into the feasibility of applying the acoustic phenomena to flaw diagnostics. These investigations are directed at developing a nondestructive technique for the detection and definition of flaws in the large complex pressure vessels and primary system of nuclear power installations. They consist of applying the technique of detecting and interpreting acoustic emissions emanating from materials under applied stress.

1371 **D. L. Parry. "Preservice Inspection of a Nuclear Power Plant Primary Pressure Containment System by Acoustic Emission Technology." pp 163–169 in** *Proceedings of the Eleventh Symposium on Nondestructive Evaluation.* **Southwest Research Institute, San Antonio, Texas. 1977. symposium held in San Antonio, Texas. Apr 2–22, 1977.**

Acoustic emission analysis technology applied as a volumetric nondestructive testing tool can provide a complete and accurate means of qualifying the integrity of a nuclear-plant primary-pressure containment system. The accuracy and reliability of the new technology are often spot-checked by use of standard nondestructive testing techniques such as ultrasonics and radiography. The checks are usually conducted on only selected source locations that have been indicated by the AE inspection to be predominant discontinuities; re-evaluation of all indicated discontinuity locations is seldom, if ever, carried out. However, following a recent AE inspection of the pressure containment system of an Austrian nuclear plant, a complete reinspection of all indicated discontinuity regions was conducted. This re-evaluation inspection was conducted by the Austrian Licensing Agency using ultrasonic techniques. The Licensing Agency reported results showed a 100% confirmation that included even the most minor discontinuities detected, analyzed, and reported by the AE inspection. These results represent the most thorough, independent confirmation of AE inspection ever obtained. The results are a resounding confirmation of the accuracy and reliability of AE as a nondestructive inspection technology.

1372 **D. L. Parry and D. L. Robinson. "Elk River Acoustic Leak Detection Tests." IN–1371. Idaho Nuclear Corporation, Idaho Falls, Idaho. May 1970. [NTIS].**

The report discusses laboratory tests and subsequent field tests conducted to show the feasibility of using acoustic techniques for detecting and locating leaks in the primary coolant system of water-cooled nuclear power reactors. The laboratory tests showed that acoustic bursts can be created in a crack when water is forced through the crack under temperature and pressure conditions necessary to cause the water to flash to steam within the length of the crack. Tests conducted on the Elk River reactor verified laboratory findings. Elk River tests also show the definite potential for using acoustics to detect and locate leaks in reactor pressure systems.

1373 **D. L. Parry and D. L. Robinson. "Incipient Failure Detection by Acoustic Emission: A Development and Status Report." IN–1398. Idaho Nuclear Corporation, Idaho Falls, Idaho. July 1970. [NTIS].**

A practical incipient failure detection system, based on the phenomenon of acoustic emissions from imperfections in structural material under stress, has been developed and tested on pressure vessels of various sizes and degrees of complexity. The acoustic emission technique has proven capable of rapidly detecting and locating flaws during hydrostatic acceptance and requalification testing of nuclear reactor pressure vessels. The technique, coupled with other standard nondestructive testing techniques, can provide improved capability for the integrity analysis of a wide variety of complex industrial structures.

1374 **D. L. Parry, L. D. Watson, and D. L. Robinson. "Preliminary Report, Dresden I Reactor Plant Noise (RPN) Analysis." Phillips Petroleum Company, Idaho Falls, Idaho. June 1968.**

Tests conducted by Phillips Petroleum Company with an acoustic analysis system during the last year have suggested that the reactor plant noise (RPN) spectrum generated by fluid flow in the primary system of a nuclear power installation may be a limiting factor in applying acoustic analysis techniques for flaw detection and location on an operating nuclear power system. This evidence was obtained from tests on reactors at the Nuclear Reactor Test Site and from in-plant test experience on the SM-1A nuclear power facility at Fort Greely, Alaska. During the SM-1A tests, nearly all acoustic analysis data were obscured when primary system flow was turned on during hydrostatic stressing of the vessel for acoustic test purposes. To avoid this noise interference, the SM-1A tests were performed without fluid flow. To fully assess the extent of reactor plant noise interference with in-plant, on-line acoustic analysis, there was a need to apply noise analysis techniques to typical commercial nuclear power installations for the purpose of studying their noise spectra. Battelle's Pacific Northwest Laboratories accordingly made arrangements in early 1968 to analyze the RPN spectrum of the Dresden I Nuclear Power Station. The Phillips Company was asked by the U.S. Atomic Energy Commission to participate in the performance of this analysis. A secondary objective was the comparison in techniques and equipment as employed by the two organizations in their respective acoustic activities.

1375 **E. U. Parsons. "Acoustic Emissions Detect Failures."** ***Pollution Engineering,*** **PLENB, 7(7):30–32. July 1975.**

Antifriction bearings play a large part in the operation of equipment that mines, transports, manufactures, and distributes the world's goods. There are billions of ball and roller bearings in use, and all are subject to failure resulting from the normal fatigue limits of the materials used and the severity of the service conditions. When a bearing is used under normal conditions of load and speed, it can be expected to give from 20,000 to 200,000 hours of service. Under severe conditions of high speed, heavy load, vibration, hostile environment, and lack of proper maintenance, it can fail in a few hours. Most failures are the result of: (1) Excessive speed or load, (2) Operation in a hostile environment, (3) Damage due to faulty installation, (4) Fatigue of the bearing materials, and (5) Improper maintenance. Any one or combination of these conditions can shorten the life expectancy of the bearing and cause damage to the equipment. Production is lost while the equipment is shut down for repairs. For bearings in critical service, early warning of incipient failure is important. Experience in the aircraft and aerospace industries produced data showing the direct relationship between increased acoustic emission and onset of bearing failure. Research in the bearing manufacturing industry showed relationships between certain types of emitted signals and specific types of failures.

1376 **R. Pascual. "Acoustic Emission and Dislocation Multiplication during Serrated Flow of an Aluminum Alloy."** *Scripta Metallurgica,* **SCRMB, 8(12):1461:1466. Dec 1974.**

Acoustic emission data were obtained during plastic deformation of Al 2024 tensile specimens. The data were plotted versus strain simultaneously with the stress-strain curve. The results indicate a strong correlation between acoustic emission and dislocation multiplication. The results are only qualitative. [Abstract 31-1083. *Metals Abstracts,* Vol. 8, 1975].

1377 **R. Pascual, M. Ahlers, R. Rapacioli, and W. Arneodo. "Acoustic Emission and the Martensitic Transformation of β Brass."** *Scripta Metallurgica,* **SCRMB, 9(1):79–84. Jan 1975.**

Acoustic emission is known to be a powerful, even if not widely used method to study martensitic transformations. Elastic waves generated during the transformation shear can be detected by suitable transducers. These data can give useful information about the processes involved in the different stages of the transformation. In this note are reported the results of a study of brass single crystals in which the martensitic transformation is induced either by cooling or by externally applied stresses above M_S. The results are complemented by optical observations and by following the electrical resistivity during the transformation. [reprinted by permission of Pergamon Press Ltd].

1378 **A. Pattnaik and A. Lawley. "A Study of Deformation and Fracture in Al-$CuAl_2$ Composites by Acoustic Emission." Paper 50, pp 250–254 in** ***The Microstructure and Design of Alloys.*** **Monograph Report Series No. 36. Institute of Metals, London, England. 1973. proceedings of the 3rd International Conference on the Strength of Metals and Alloys. Volume 1, held in Cambridge, England. Aug 20–25, 1973.**

In the present study, attention was directed to the effect of microstructure on acoustic emission during flow and fracture of directionally solidified Al-$CuAl_2$. The necessary characterization of structure and a correlation with mechanical behavior have been made by Pattnaik and Lawley. From these studies and other observations, it is clear that premature fracture of the brittle $CuAl_2$ θ-phase occurs on tensile deformation; this is the phenomenon of most interest vis-a-vis acoustic emission. Acoustic emission was monitored over the entire tensile stress-strain curve for a spectrum of microstructures ranging from lamellar to spheroidized.

1379 **A. Pattnaik and A. Lawley. "Role of Microstructure on Acoustic Emission in the Deformation of Al-$CuAl_2$ Composites." Research Project No. 336, Technical Report No. 10. Drexel University, Department of Metallurgical Engineering, Philadelphia, Pennsylvania. April 1973. [NTIS: AD-758-378].**

Acoustic emission has been monitored during the ambient temperature tensile deformation of directionally solidified Al-$CuAl_2$. Attention was directed to the effect of microstructure in that a range of structures was examined ranging from lamellar to spheroidized; coarsening was promoted by elevated temperature exposure at temperatures up to 0.98 $T_{eutectic}$ for times up to 100 hours. Burst-type acoustic emission is associated with premature failure of the brittle $CuAl_2$ (θ) phase in Stage II of the stress-strain curve i.e., θ elastic, α solid solution plastic. The marked increase in total acoustic emission or in the size of individual bursts with coarsening of the lamellar structure is rationalized in terms of an increase in the size of fracture events. Cracks can propagate from one θ-platelet to others via bridging, or through other interconnecting paths between θ-lamellae. Thus, acoustic emission identifies differences in the local fracture process as affected by composite structure. A variation in the number of cracks with location in a specimen, and from specimen to specimen, and the lack of a clear correlation between composite strain and the density of cracks, precluded the development of a quantitative correlation between acoustic emission and composite damage.

1380 **H. N. Pedersen and J. C. Spanner. "Detection, Location, and Characterization of Flaw Growth in Metals Using Acoustic Emission Methods." pp 163-164 in *Incipient Failure Diagnosis for Assuring Safety and Availability of Nuclear Power Plants.* E. P. Epler and D. P. Roux, Coordinators. U.S. Atomic Energy Commission, Division of Technical Information, Washington, D.C. Jan 1968. [NTIS: CONF-671011]. proceedings of conference sponsored by Oak Ridge National Laboratory, held in Gatlinburg, Tennessee. Oct 30-Nov 1, 1967.**

The Battelle-Northwest Laboratory in February 1966 initiated a program under AEC-DRDT sponsorship to nondestructively detect incipient failure of reactor pressure piping using the acoustic emission generated by fracture, cleavage, dislocation unpinning, twinning, etc., of single or multiple grains of the piping material. The purpose of the program is fourfold; namely (1) the detection of acoustic emission against the noise background of the operating reactor, (2) the location of the point of origin of the acoustic emission, (3) the characterization of the acoustic emission by the type of failure or flaw-growth mechanism in the metal, and (4) prediction of the time of failure.

1381 **P. J. Pekrul. "On-Line Vibration and Loose Parts Monitoring of Nuclear Power Stations as a Preventative Maintenance Tool."** *Materials Evaluation,* **MAEVA, 34(7):154–159, 164. July 1976; paper presented at the American Society for Nondestructive Testing National Fall Conference, Atlanta, Georgia. Oct 13–16, 1975.**

A large number of power station outages have been traced to vibration related causes. Currently in the nuclear power industry, a significant amount of preventative maintenance equipment is being installed to diagnose these vibration problems before serious damage results. A system for on-line vibration and loose parts monitoring is described as well as the results of some experience with these systems. A benefit-to-cost ratio justification for such a system is presented.

1382 **M. Perdrix, F. Lakestani, J. C. Baboux, and P. Fleischmann. "Impulse Responses of Piezoelectric Discs: Application to the Measurement of their Constants and to the Calibration of Receiving Transducers/Responses impulsionnelles de disques piezoelectriques: application a la mesure de leurs constantes et a l'etalonnage des recepteurs." Institut National des Sciences Appliquees de Lyon, Laboratoire d'Ultrasons, 69621 Villeurbanne, Lyon, France; Paper 3J/8 in** *Proceedings of the Eighth World Conference on Nondestructive Testing.* **volume containing Sections: 5C, 3J, 1D. International Committee for Nondestructive Testing, Secretariat du Cofrend, Institut de Soudure, 32 Boulevard de la Chapelle, 75880 Paris Cedex 18, France. 1976. conference held in Cannes, France. Sep 6–10, 1976. (paper in French).**

Two devices based on the study of the responses of a disc excited by a current impulse are given. The piezoelectric constant, the permittivity and the coupling factor are measured with the first; the receiving transducer sensitivity as a function of the ultrasonic wave frequency is obtained with the other.

1383 **M. Perosa. "Raziskave medsebojne povezave med trdoto in akusticno emisijo pri jeklu." Diplomsko delo st. 2021. Fakulteta za strojnistvo, Univerza v Ljubljani, Ljubljana, Yugoslavia. 1974. (in Slovenian).**

The work describes the correlation between acoustic emission and hardness of a material tested by Vickers and Brinell hardness tests.

1384 **C. T. Peters. "Now Hear This: Acoustic Emission (AE) is the Technical Name for** ***Noises*** **Emitted by Stress Materials."** *Plastics and Rubber Weekly,* **PLRWB, (631):40–41. May 7, 1976.**

1385 **C. T. Peters and C. Larson. "Acoustic Emission to Monitor Electrodeposition."** *NDT International,* **NDITD, 9(4):197–199. Aug 1976.**

Brittle electroplates are subject to the build-up of high internal stress which can lead to undesirable stress cracking during plating. The authors monitored acoustic emission during electrodeposition of hard chromium onto polished copper to see whether they could detect the onset of such cracking. They used commercial acoustic-emission equipment and found they could just detect emission due to evolution of hydrogen, as well as much stronger signals from stress cracking. They conclude that the method is a sensitive means of monitoring crack formation during electrodeposition and that further work could lead to determination of crack size and distribution, as well as the onset of cracking.

1386 **R. W. Peverley. "Acoustic Emission Data Systems on Nuclear Power Plants." Bechtel Power Corporation, Houston, Texas. 1974; paper presented at the 13th Acoustic Emission Working Group Meeting, Laguna Hills, California. Aug 19–21, 1974.**

The ASME Section XI Inservice Inspection Committee is currently working on an acoustic emission procedure. This procedure, when adopted, will provide for the use of acoustic emission techniques to conduct in-service inspection of nuclear power plant pressure-retaining components. The adequacy of the acoustic emission technique has, in fact, been demonstrated on several operating nuclear power plants. Because of the inherent economic advantages of this technique, its eventual approval will result in many utility companies including acoustic emission data acquisition and processing equipment in their plant inventory. Since the purchase of such equipment will require significant expenditures of capital funds, the full and effective utilization of this equipment must be assured. This paper explores the feasibility of incorporating a data system within a nuclear power plant that will not only be capable of weld inspection, but can also be used for loose-parts monitoring, rotating-machinery wearout maintenance detection, and seismic-vibrations monitoring. The equipment required for each system would be similar, i.e., transducers, preamplifiers, signal conditioners, tape recorders, spectrum analyzers, computer capability, and readout devices. Each of the four phenomena to be measured vary basically only in their frequency regime. Thus, the adaptation of the basic system to meet individual requirements would be minimum. This approach, if properly planned, would provide the utility companies with the capability of accomplishing four tasks with the same basic equipment. This paper describes the requirements for each data system and proposes an approach for combining these into a common system. The integration of this system into the nuclear power plant is also discussed.

1387 **R. W. Pfeil. "Acoustic Test for Detection of Flaws in Ball Bearings." pp 4–21 to 4–33 in "Proceedings of Aerospace Testing Seminars." Lockheed Missiles & Space Company, Inc., Palo Alto, California. May 1973.**

Presented are some methods by which ball bearings can be electronically tested. The most common defects found in bearings are pits, gouges, chips, and debris in the raceways. Each of these conditions will cause a bearing to run roughly with localized hot spots, which will cause degradation of the lubricant and early failure of the bearing.

Since the bearing will exhibit vibration due to the rolling elements impacting with the flaws, there are methods by which tests can be performed; namely, sensing the vibration or the acoustic phenomena. First, since the impact of the rolling elements with the flaws will shock-excite the races and balls at their natural frequencies, there is an indicator of bearing condition through sensing the vibration amplitude for a given element resonance. The larger the flaws, the greater the impact; therefore, bearing condition can be equated to the amplitude seen at resonance. An advantage of this technique is that the frequency sensed is not a function of rotational speed. A second method is to consider the rates at which the balls impact with the flaws. By knowing the rotational speed and the geometry of the bearing, one can calculate the repetition rates at which the balls will impact with the flaw. An advantage to this technique is that the location of the flaw can be predicted; that is, whether the flaw is in the inner race, outer race, or ball. A flaw located on the ball will contact both the inner and outer race during rotation; therefore, the flaw repetition frequency would be twice the ball-spin frequency.

1388 **J. W. Phillips. "Stress Pulses Produced during the Fracture of Brittle Tensile Specimens."** *International Journal of Solids and Structures*, **IJSOA, 6:1403–1412. 1970.**

A theoretical model similar to that used by Miklowitz is assumed for the fracture of uniaxial tensile specimens. For this model, both symmetric and antisymmetric pulses are propagated away from the fracture zone. The theoretical predictions for surface strain are found to agree remarkably well with experimental results. [reprinted by permission of Pergamon Press Ltd from *International Journal of Solids and Structures*].

1389 **J. W. Phillips. "The Stress Pulses Produced by Brittle Fracture of Rods." Technical Report No. 8 on National Science Foundation Grant NSF–GK–749. Brown University, Division of Applied Mathematics, Providence, Rhode Island. Dec 1968.**

Theoretical and experimental investigations of the unloading stress pulses produced in brittle circular rods when they fracture in uniaxial tension have been carried out. A simple fracture model is developed for the generation of longitudinal and flexural stress pulses within the fracture zone, and Fourier techniques are employed to predict the ever-changing shapes of these pulses as they propagate in both axial directions away from the fracture zone. Strain-gage detection of actual pulses produced in glass rods pulled slowly to failure shows the existence of both longitudinal and flexural pulses, and indeed remarkable agreement between theory and experiment is found. Additional experimental work on glass rods bent slowly to fracture is reported. Again it is found that both longitudinal and flexural pulses are produced during fracture. Although a theoretical analysis of this related problem has not been carried out, certain conclusions regarding the total time required for fracture can be drawn from a study of the strain-gage records.

1390 **A. G. Pickett. "Development of a Nondestructive Test Method Evaluation Facility." Project II. pp 25–48 in "In-Service Inspection Program for Nuclear Reactor Vessels." E. R. Reinhart. SwRI Project 17-2440, EEI Project RP-79. Progress Report No. 8. Final Technical Report. Southwest Research Institute, San Antonio, Texas. May 21, 1973. [NTIS: PB-222-701].**

A Nondestructive Test Method Evaluation Facility (NDTF) was constructed by modifying the Experimental Beryllium Oxide Reactor (EBOR), located at the National Reactor Test Site (NRTS), to accomplish objectives of the RP-79 Program. The accomplishments of the Program are documented by reports available from the National Technical Information Service (NTIS), Operations Division, Springfield, Virginia 22151. Final tests at the EBOR site, sponsored by the Edison Electric Institute (EEI) program, were completed in May 1972. During the course of tests at EBOR, various industrial groups participated in the test program in order to evaluate and develop nondestructive testing techniques. The initial experimental results were so promising that many of the participants wanted to continue the testing while the facility was still available. Joint funding was, therefore, received from the industrial participants, EEI, and the Tennessee Valley Authority (TVA), and testing was resumed and continued until November 1972. The major objective of this continued effort was to monitor flaw growth in the reactor pressure vessel with simulated plant noise and environment using prototype acoustic emission and acoustic holographic systems and conventional nondestructive inspection techniques. Southwest Research Institute (SwRI) provided program management and technical staff. Aerojet Nuclear Company operated the facility and performed all necessary modifications. The report further outlines the overall experimental program.

1391 **A. G. Pickett and E. R. Reinhart. "Development of a Nondestructive Test Method Evaluation Facility: Technical Support." Task 4, Project II, pp 24–33 in "In-Service Inspection Program for Nuclear Reactor Vessels." R. D. Wylie. SwRI Project 17-2440, EEI Project RP-79. Biannual Progress Report No. 5. Southwest Research Institute, San Antonio, Texas. May 28, 1971. [NTIS: PB-200-518].**

Removal of the EBOR vessel head and internals was completed. The SwRI conducted a nondestructive examination in October 1970 of the EBOR vessel to determine the baseline condition of the welds, prior to the addition of any artificial defects in the vessel. While at INC, SwRI representatives outlined instrumentation and access requirements needed to initiate and monitor artificial defects in the EBOR vessel. In addition to the above effort, SwRI has designed, built, and tested a number of 12 by 2 by 0.5 inch tabs that are designed to be used as acoustic emission noise sources attached to the EBOR outside wall. Each tab has a loading fixture which causes controlled crack growth in the tab. Ultrasonics is used to remotely monitor the crack growth. In addition to the crack tabs, a method is being developed to enable flaws to be initiated and grown in the EBOR vessel wall. The high pressure hydraulic facility needed for this study has been built and will shortly be tested.

1392 A. G. Pickett, E. R. Reinhart, and S. P. Ying. "Acoustic Emission from Irradiated Steels." pp 33–41 in *Proceedings of the Eighth Sumposium on Nondestructive Evaluation in Aerospace, Weapons Systems, and Nuclear Applications.* Western Periodicals Company, North Hollywood, California. 1971. symposium held in San Antonio, Texas. Apr 21–23, 1971.

Acoustic emission is a nondestructive testing method presently being developed for safety monitoring of nuclear reactor pressure retaining components. Since the material properties of reactor steels are known to be affected by irradiation, a short study was conducted to determine what effect, if any, irradiation may have upon acoustic emission signal characteristics. Acoustic emission from irradiated and nonirradiated steels was detected during tensile and wedge-opening load (WOL) tests. Specimens were fabricated from A302-B base metal and a representative grade of weld metal. The irradiated specimens were contained in the first capsule removed for evaluation in the San Onofre nuclear-reactor materials surveillance program. Tensile tests were conducted at room temperature and the WOL specimens were tested at liquid nitrogen temperatures. The yield points detected by acoustic emission in the tensile tests are higher for the irradiated specimens in comparison with the unirradiated steels. The accumulation of acoustic emissions for both kinds of tests all have higher counts from irradiated specimens. The accumulative counts of WOL tests have been fitted to a theoretical model. From this experiment, it can also be concluded that the acoustic emission technique is applicable for monitoring of materials irradiated by a nuclear reactor.

1393 M. Pikalo and M. Sipek. "The Use of Acoustic Emission Techniques in Metallurgical Research and Praxis/L'utilisation des techniques d'emission acoustique dans les essais et dans la pratique metallurgique." Zelezarna Ravne, Ravne na Koroskem, Yugoslavia; Paper 3K/15 in *Proceedings of the Eighth World Conference on Nondestructive Testing.* volume containing Sections 3K2, 3H2, 5A. International Committee for Nondestructive Testing, Secretariat du Cofrend, Institut de Soudure, 32 Boulevard de la Chapelle, 75880 Paris Cedex 18, France. 1976. conference held in Cannes, France. Sep 6–10, 1976. (paper in English).

Detection of acoustic emission from martensitic transformations and comparison between dilatation and acoustic emission are discussed. Also discussed is the monitoring of flake growth in high strength steel with acoustic emission techniques.

1394 H. F. L. Pinkney. "Surface Motion on a Solid Elastic Sphere Following a Brief Localized Disturbance." Ph.D. Thesis. Stanford University, Division of Engineering Mechanics, Stanford, California. May 1959. [available on inter-library loan from the Engineering Library, Stanford University].

1395 R. B. Pipes, N. J. Ballintyn, and W. R. Scott. "Acoustic Emission Response of Metal Matrix Composites." Report No. NADC-76082-30. University of Delaware, Department of Mechanical and Aerospace Engineering, Newark,

Delaware and Naval Air Development Center, Air Vehicle Technology Department, Warminster, Pennsylvania. Jan 23, 1976. [NTIS: AD/A-026-024].

The program investigates acoustic emission techniques for nondestructive evaluation of the integrity of a boron-aluminum composite material. The influence of biaxial states of stress upon the acoustic response of the composite material is examined through implementation of a test program which utilizes the off-axis tensile coupon specimen. Both mechanical property and acoustic emission test results are presented.

1396 **W. M. Pless and C. D. Bailey. "Detection of Large Crack Extensions in Aircraft Wing Structure by Acoustic Peak Detectors." pp 30–43 in** *Proceedings of the Tenth Symposium on Nondestructive Evaluation.* **Southwest Research Institute, San Antonio, Texas. June 1975. symposium held in San Antonio, Texas. Apr 23–25, 1975.**

As a major part of the Flight Structural Monitoring System (FSMS) program, tests were conducted to establish the detectability limits of FSMS to crack extension induced in a large aircraft wing structure through cyclic loading. The FSMS concept was developed at the Lockheed-Georgia Company to fulfill a need defined by the Air Force Special Projects Office to extend a flight crew's ability to detect and localize large crack extensions occurring during flight. An acoustic emission system was designed for on-board use featuring strategically located piezoelectric transducers, signal processing and display unit, and a warning light on the pilot's annunciator panel. The system detects and displays peak acoustic transients well above the aircraft noise background as determined by a flight noise survey. Detectability limits were determined through the use of a full-scale center wing box removed from an in-service C-130E aircraft. The wing box was installed in a fatigue test rig and instrumented alternately on the *upper* and *lower* surfaces with eight FSMS transducers bonded at close-in and remote locations with respect to the fracture site, along the same wing panel and across joints on adjacent panels. Large crack extensions were induced by tension-tension loadings at specified percentages of design limit load acting on crack-starter sawcuts made at well-defined locations on the specimen. Voltage responses produced by crack extensions were recorded on magnetic tape which were analyzed to provide sensitivity and acoustic attenuation data for use in optimizing transducer spacings in the later flight test phase of the program. In addition, results indicated the value in using a FSMS-like system not only to monitor flight critical structure on aircraft during flight, but to monitor large structural specimens during load tests to detect unstable crack extensions. Acoustic attenuation was seen to vary from 11 db per foot for close-in sensors to 2 db per foot at sensors 8 to 16 feet from the cracks. Sensitivity varied from 23,000 millivolts per square inch near the crack to 150 mV/in^2 at distances 16 feet from the crack for crack extensions which create crack surface areas ranging from 0.10 to 0.30 square inches.

1397 **W. M. Pless, C. D. Bailey, and J. M. Hamilton [Jr.]. "Acoustic Emission Detection of Fatigue Crack Growth in a Production Size Aircraft Wing Test Article under Simulated Flight Loads." Lockheed-Georgia Company,**

Marietta, Georgia. 1977; summary of paper in *Paper Summaries: ASNT Spring Conference, March 28–30, 1977, Phoenix, Arizona.* **American Society for Nondestructive Testing, Columbus, Ohio. 1977;** *Materials Evaluation,* **MAEVA, 36(5):41–48. Apr 1978.**

A structural article representative of a large full-production military aircraft wing was subjected to flight-by-flight load spectra to initiate and propagate fatigue cracks in 17 selected fastener holes in 8 test areas. One hundred unit flights were applied to the wing to initiate crack growth at sharp notch sawcuts and existing cracks, followed by 2000 unit flights to propagate the cracks. Seven test areas were monitored for acoustic emission using a 32-channel AE flaw locator system and triangulated sensor arrays. The AE was detected from 12 of the 15 test holes monitored which experienced crack growth. After the tests, the holes were removed from the structure for fractographic analysis of crack growth. Correlations between AE and crack growth for several of the test holes are discussed. Experimental positioning of sensor arrays to determine possible airborne AE system configurations are discussed.

1398 **W. M. Pless, C. D. Bailey, and J. M. Hamilton. "Acoustic Emission Monitoring of Crack Propagation on the Expedited Wing Fatigue Test Article (X993)." AFML-TR-76-214, LG77ER-0042. Lockheed-Georgia Company, Marietta, Georgia. Dec 1976.**

The final report describes work accomplished, under a U.S. Air Force contract, whose objective was to evaluate the ability of acoustic emission techniques to detect crack growth in a large full-production aircraft wing fatigue test article subjected to simulated flight spectra. Fatigue cracks were propagated at 17 selected predamaged fastener holes in 8 test areas on the test article. The flight load spectra consisted of 50 variable tension-tension gust cycles followed by one-half of a compressive ground cycle. Sixteen of the fastener holes were monitored with a 32-channel acoustic emission flaw locator system containing a computer, CRT monitor, input-output teleprinter console, and tape recorder. Linear and triangular sensor arrays were used to detect and locate cracks. Twelve fatigue cracks were detected at 15 fastener holes monitored where crack growth occurred. Acoustic emission data were correlated to crack growth data, determined through fractographic analysis, for several of the test cracks. Complex structural features which were accommodated successfully for detection of AE across joints are described.

1399 **I. Polajnar, I. Grabec, and V. Prosenc. "Akusticna emisija pri varjenju (Acoustic Emission in Welding)."** *Strojniski Vestnik/Mechanical Journal,* **STJVA, 21(11–12):244–246. Nov–Dec 1975. (in Slovenian).**

Preliminary results are given concerning investigations of acoustic emission in welding. The results indicate that the detection of acoustic emission can give a determined insight into phenomena in the weld and also an interpretation of the connection between acoustic emission and weld quality is possible. [Abstract 55-1270. *Metals Abstracts,* Vol. 9, 1976].

1400 A. A. Pollock. "A Computer Program to Generate Longitudinal Frequency Response Functions." *Acustica,* **ACUSA, 19(5):292–294. 1967/1968. [summary in English, German, and French, article in English].**

The paper describes the response of a thin dumb-bell shaped specimen to pulse excitation at a point. A computer program models the pulse transmission process and predicts the output spectrum of transducers mounted on the ends of the specimen. With the help of this program, the loading effect of transducers mounted on the specimen can be studied and the effects of varying the electro-acoustic matching conditions can be evaluated in detail. The method is readily applicable to determining the response of more complicated sectioned systems to first-mode longitudinal excitation.

1401 A. A. Pollock. "Acoustic Emission." *Engineering,* **ENGNA, 209(5433):639–642. June 1970.**

Acoustic emissions are defined. The form and the origins of emissions are described. It is shown how emissions may be detected, amplified, recorded, and analysed. The main application of acoustic emissions is a nondestructive testing technique in the monitoring of cracks and flaws in pressure vessels and welded structures.

1402 A. A. Pollock. "Acoustic Emission." pp 1–11 in *Journees d'Etudes sur l'Emission Acoustique.* **P. F. Gobin, R. Goutte, and C. Guillaud, Organizers. Institut National des Sciences Appliquees de Lyon, Avenue Albert Einstein, 69621 Villeurbanne, Lyon, France. Mar 1975. proceedings of a conference held in Lyon, France. Mar 17, 1975. (paper in English).**

Advantages and difficulties of acoustic emission as a nondestructive testing technique are reviewed. It is shown that acoustic emission opens a new dimension in the study of material deformation and that much remains to be learned in the field of the theory of the source and the theory of acoustic wave propagation and detection.

1403 A. A. Pollock. "Acoustic Emission: A Review of Recent Progress and Technical Aspects." Chapter 2, pp 49–84 in *Acoustics and Vibration Progress.* **R. W. B. Stephens and H. G. Leventhall, Editors. Volume 1. Chapman and Hall, London, England. 1974. [reprint of chapter available as Technical Report DE 74–UK from Dunegan/Endevco, San Juan Capistrano, California].**

The chapter touches on many aspects of acoustic emission technology. An introductory section describes some of the forces leading to the growing use of this method for failure prevention, defect identification, and quality assurance. The review section describes recent progress under headings of materials understanding, signal processing methods, mathematical models, experimental technique, instrumentation, and practical applications. The third section covers a miscellany of technical points, brought together in a discussion of the factors that influence emission data. Source behavior, the influence of load system and load rate, wave propagation effects, and transducer and system characteristics are considered. Of particular interest is a new mathematical model defining the first-approximation relationship between amplitude distribution, ringdown count, and system gain.

1404 A. A. Pollock. "Acoustic Emission–2: Acoustic Emission Amplitudes." *Non-Destructive Testing,* **NDETA, 6(5):264–269. Oct 1973. [This is the second in a series of articles on acoustic emission. See References 1716, 212, 336, and 712 for Parts 1, 3, 4, and 5, respectively].**

The author discusses the significance of the amplitude of acoustic emission and its usefulness as a source of information. He describes the theory and application of amplitude sorting. He also presents some experimental results and concludes that the method has a significant future role as part of a more complete characterization of emission signals.

1405 A. A. Pollock. "Acoustic Emission: Basic Considerations and Energy Transfer and some Structural Testing Results." Cambridge Consultants, Bar Hill, Cambridge, England; paper presented at the Institute of Physics Conference on Acoustic Emission, Imperial College, London, England. Mar 14, 1972. [proceedings were never published].

Fundamental considerations show that in order to satisfy system-boundary conditions the displacement waveform of an acoustic emission must be step-like and the stress waveform must be pulse-like. The energy spectrum is broadband extending down to zero frequency and up to a frequency that depends on the duration of the source event. The excitation of theoretically predicted structural resonances is confirmed by frequency analysis of emission waveforms. The energy flow when materials deform involves conversion of stored elastic energy into surface energy, energy of mechanical deformation, acoustic emission, and heat. For appreciable emission, an excess of elastic energy must be available over and above the energy required for mechanical deformation and creation of new surface. Fracture mechanical calculations for a medium-strength steel at two different temperatures show that the excess energy is greater for the more brittle condition; this supports experimental observations that over a given loading range a more brittle material gives more emission. Representation of the crack growth process on axes of stored energy against crack length permits representation of emission in comparison with other terms and shows that for larger crack jumps a greater fraction of the energy available appears in the acoustic emission. Engineering applications of acoustic emission must be supported by fundamental understanding. In trying to interpret emission from structures it is necessary to rely, for economical reasons, on laboratory testing for background information; and the application of laboratory results to large structures is one of the key areas of this developing technology. A case recently studied has been the proof testing of a military bridge. By comparing the amplitude distribution and other features of emission from the bridge, with emission from laboratory trials, some conclusions could be drawn about the nature of the deformation taking place in the bridge.

1406 **A. A. Pollock. "Acoustic Emission and Fracture Mechanics." pp 133–148 in** *Schallemission: Anwendung bei der Untersuchung, Prufung und Uberwachung metallischer Werkstoffe.* **Deutsche Gesellschaft fur Metallkunde eV, 637 Oberursel/Taunus bei Frankfurt/Main, Adenauerallee 21, Germany. 1974. proceedings of symposium held in Munich, Germany. Apr 18–19, 1974.**

1407 **A. A. Pollock. "Acoustic Emission from Filament Wound Pressure Vessels." pp 4–2 to 4–4 in "Newsletter on the Non-Destructive Testing of Composite Materials and Adhesive Bonded Joints." Issue Number 1. Royal Aircraft Establishment, Farnborough, Hants, England. Jan 1973.**

The report is a brief summary of tests conducted early in 1971 to investigate the applicability of acoustic emission to the testing of fibre reinforced plastic pressure vessels. This was a small scale investigation of a preliminary nature. Proof pressures conventionally used for testing these particular vessels are not far short of nominal failure pressure and there is a risk that a vessel might be significantly weakened by the proof test before it went into service. It would therefore be desirable to reduce the proof pressure if adequate information about vessel strength and integrity could be acquired by other means. It was conjectured that acoustic emission might provide such means. Hydrotesting to failure was conducted on 6 CFRP and 2 GRP spheroidal vessels while acoustic emission were monitored.

1408 **A. A. Pollock. "Acoustic Emission from Materials." pp 2–4 in** *Encyclopaedic Dictionary of Physics, Supplementary Volume 3.* **J. Thewles, Editor-in-Chief. Pergamon Press, Oxford, England. 1969.**

1409 **A. A. Pollock. "Acoustic Emission from Materials." Chapter 8, pp 63–70 in** *Non-Destructive Testing: Views, Reviews, Previews.* **H. B. Egerton, Editor. United Kingdom Atomic Energy Authority Research Group, Harwell Post-Graduate Series. Oxford University Press, London, England. 1969.**

The chapter presents a definition of acoustic emission and its sources. Following sections include a summary of past and present work on acoustic emission, techniques for observing acoustic emission, interpretation of results and possible applications, distribution of emission events in time, and emission waveform and frequency content.

1410 **A. A. Pollock. "Acoustic Emission from Solids undergoing Deformation." Ph.D. Thesis. University of London, London, England. 1970.**

Acoustic emissions are stress waves generated in solid materials by mechanical deformation. They carry information about the source distrubance which may be of interest to the materials scientist, to the nondestructive testing engineer, and to the geologist. Practical applications of acoustic emission in the aerospace and nuclear industries are well-advanced. In this thesis, the physical basis of acoustic emission is examined and the factors which influence emission generation are discussed. The process of emission propagation in a particular specimen geometry is studied, first through an analysis of normal modes and then by means of a computer model. The study shows how the observed frequency spectrum of the emissions is influenced by specimen geometry, and the main predictions are confirmed by experiments on adhesive bond specimens of appropriate geometry. Experiments on crack growth in titanium, steel, and aluminum are described. Brittle materials yielded more emission than ductile materials. In titanium and in steel at −60 °C, the emission rate depended on the stress intensity. In steel at −135 °C, no such dependence was observed. The different patterns of emission behaviour in the various materials are related to the different mechanisms of crack growth. In connection with these experiments, the concept of *emission signature* is discussed and methods for the systematic comparison of different materials are suggested. In the final chapter, tests on adhesive bonds are described. Emission from a bond depended on the surface condition and on the glue-line thickness, and could be used to diagnose poor adhesion or to predict the failure strength of the bond. The root mean square error of failure strength predictions, made at 40% of failure load, was 22% for a series of bonds of widely varying strength. Bond testing appears to be a promising area for the practical application of acoustic emission.

1411 **A. A. Pollock. "Acoustic Emission Methods of NDT."** *British Journal of Non-Destructive Testing*, **BJNTA, 13(3):85–89. May 1971.**

The potential and limitations of acoustic emission testing are reviewed. Actual and possible applications discussed include crack-growth assessment, in-process weld inspection, and recognition of deformation and failure mechanisms, and fatigue crack instability. An idealized representation of brittle crack growth in a double-cantilever beam, demonstrating a relation of acoustic emission to fracture quantities, is presented. [Abstract 22-1032. *Metals Abstracts,* Vol. 4, 1971].

1412 **A. A. Pollock. "AE Monitoring of Natural and Artificial Defects in the KEMA Vessel." Dunegan/Endevco, Royston, England. Sep 1974; Technical Report DE 74–7A. Dunegan/Endevco, San Juan Capistrano, California. Sep 1974.**

Following M. P. Kelly's report DE 74-7 of June, 1974 (see Reference 982), the purpose of this report is to provide further detailed information on certain aspects of the test data. In particular, this report is concerned with the correlation between AE source location data and the results of the ultrasonic examination carried out by RTD prior to the KEMA tests. Critical attention is also paid to the sources of error affecting the source location data acquired during this test. These two factors are considered highly pertinent in the present state-of-the-art of source location technology.

1413 **A. A. Pollock. "Application of Acoustic Emission as an On-Line Monitoring System for Nuclear Reactors." NUREG-0159, Biannual Progress Report to the U.S. Nuclear Regulatory Commission, Office of Nuclear Regulatory Research. Dunegan/Endevco, San Juan Capistrano, California. Jan 1977. [NTIS: PB-262-194; also available as Technical Report DE 76-9 from Dunegan/Endevco].**

Permission was received from Baltimore Gas and Electric to monitor the reactor vessel of Calvert Cliffs Unit No. 2. Transducer layout was planned and a 30-sensor, 8-pulsed configuration was defined for covering the nozzle regions, barrel section, and lower header. Practical details of equipment installation were resolved and materials were made ready for shipment. Laboratory tests to support the vessel monitoring program were planned and material samples procured. In parallel with this activity, transducers were installed on a by-pass line nozzle on Dresden III through the collaboration of Commonwealth Edison. Preparations for this installation included analysis to determine the optimum sensor positioning, taking dispersion and other factors into acccount. The Dresden system was calibrated, background noise levels were assessed, and electronics for area-of-interest inspection was specified.

1414 **A. A. Pollock. "Electrical Circuit Means for Use in Acoustic Emission Detecting and/or Recording Apparatus." U.S. Patent 3,822,586. July 9, 1974; British Patent 1,391,175. Apr 16, 1975.**

Electrical circuit means for use in acoustic emission detecting and/or recording apparatus comprise logic gates which in operation of the apparatus are connected to respective transducers so as to receive electrical signals representative in time of the distances of the transducers from a source of acoustic emission. Each gate is set prior to the receipt of a signal from the associated transducer to afford an output signal on receipt of a signal at its input. Further logic elements are provided which are connected with the logic gates and are adapted upon receipt by one of the gates by an input signal to reset for a predetermined time at least the remainder of the logic gates so as to prevent the appearance of an output signal on any of said remainder of the logic gates at said predetermined time.

1415 **A. A. Pollock. "Fatigue Monitoring by Acoustic Emission."** *Journal of the Society of Environmental Engineers,* **JSEEB, 13(2):3-6. June 1974.**

The article outlines the basic principles of acoustic emission. It is shown that acoustic emission is a useful technique for monitoring fatigue and for assessing fatigue damage in a variety of engineering structures.

1416 **A. A. Pollock. "Metals and Rocks: AE Physics and Technology in Common and in Contrast." pp 383-401 in** *Proceedings First Conference on Acoustic Emission/Microseismic Activity in Geologic Structures and Materials.*

H. R. Hardy, Jr. and F. W. Leighton, Editors. Trans Tech Publications, Clausthal, Germany. 1977. conference held at Pennsylvania State University, University Park, Pennsylvania. June 9–11, 1975.

Outside the geological disciplines acoustic emission has been extensively studied, both as a phenomenon and as a technique for the nondestructive testing of materials and engineering structures such as pressure vessels, pipes, bridges and aircraft. The purpose of this paper is to review the fundamental understanding and applications techniques that have developed over the past 20 years, insofar as they may be relevant to geological applications of the AE phenomenon. The paper includes consideration of emission generation, wave propagation, frequency spectra, and various distribution functions. An advantage of amplitude or energy distribution functions is that the characteristic b-value should be independent of the distance from source to detector, thus facilitating the application of laboratory to field situations.

1417 **A. A. Pollock. "Quantitative Evaluation of Acoustic Emission from Plastic Zone Growth." Technical Report DE 76-8. Dunegan/Endevco, San Juan Capistrano, California. Aug 1976.**

Tests on three compact tension specimens of A516 Grade-70 Steel are described. Emission from the growth of the plastic zone is analyzed in depth by reference to available models. Quantitative relationships between events, count, and system sensitivity for this material are demonstrated. These relationships involve the use of amplitude distributions and structural calibration methods. This approach shows the way to apply laboratory data, quantitatively, to the evaluation of defect severity in the field test environment. Ductile tearing in A516 Grade-70 parent plate is essentially quiet. In the field test situation, weldments will probably be the dominant source of the observed emission.

1418 **A. A. Pollock. "Ringdown versus Energy." pp 144–146 in "EWGAE–ISPRA: Proceedings of the Third Meeting of the European Working Group on Acoustic Emission." A. C. Lucia, Editor. Report EUR-5513e. Commission of the European Communities, Directorate General Scientific and Technical Information and Information Management, Luxembourg, Belgium. 1976. meeting held at the Joint Nuclear Research Center, Ispra, Italy. Sep 25–26, 1974.**

The method of ringdown counting has been subject to considerable criticism during the past two years and little has been said in its defense. The purpose of this paper is to clarify the distinctions between the different signal processing techniques and to indicate how ringdown counting may be used in a quantitatively meaningful way.

1419 **A. A. Pollock. "Sounds of Stress."** ***Interface,*** **5(6):3–5. 1971. [*Interface* is a house journal of the AIM Associates Cambridge Ltd group of companies and is published by Cambridge Consultants Ltd, Bar Hill, Cambridge, England.]**

1420 **A. A. Pollock. "Stress-Wave Emission: A New Tool for Industry."** *Ultrasonics,* **ULTRA, 6(2):88-92. Apr 1968.**

Acoustic emissions are noises made by materials as they deform. The alternative term *stress wave emission* (SWE) is preferable since it does not imply restriction to audible frequencies. Many common deformation mechanisms generate detectable emission. The effect has been recognized and studied with increasing interest in the last fifteen years: stress-wave analysis is now being used for large-scale nondestructive testing and also for materials research in the laboratory.

1421 **A. A. Pollock. "Stress-Wave Emission during Stress Corrosion Cracking of Titanium Alloys." Document D1-82-0658. Boeing Scientific Research Laboratories, Seattle, Washington. Oct 1967.**

The document describes a preliminary investigation to ascertain the value of the use of acoustic emission as a tool for the study of stress corrosion cracking in Ti:8-1-1 alloys. Emissions generated during crack propagation were detected and recorded. Their characteristics suggested that the *source processes* were mechanical, not chemical in nature. Emission parameters showed systematic variations with load, crack velocity, and stress intensity (K). The brittle cleavage of single grains or groups of grains is proposed as one source of acoustic emission which is consistent with the experimental data. The vibrational behavior of the specimen-transducer system was studied. These experiments led to an improved understanding of the fundamentals of the emission process, the general form of the source stress wave, its energy spectrum, and its dissipation within the specimen-transducer system. A procedure for the estimation of source event energies was developed.

1422 **A. A. Pollock. "Stress-Wave Emission in NDT."** *Non-Destructive Testing,* **NDETA, 2(4):178-182. Aug 1969.**

Acoustic emission (or *stress-wave emission*) is one of the most promising new techniques of nondestructive testing. It is currently in use in the aerospace and nuclear industries for monitoring the growth of cracks and flaws in pressure vessels. It is particularly useful for locating flaws which may be tens of feet from the detecting transducer. Failure strength prediction and metal condition diagnosis can be carried out by use of acoustic emission. It can also be used to predict the failure strength of adhesive bonds and to detect poor adhesion.

1423 **A. A. Pollock. "Trouble Talks: Acoustic Emission."** *Machine Design,* **MADEA, 48(8):72-76. Apr 8, 1976.**

1424 **A. A. Pollock and W. J. Cook. "Acoustic Emission Testing of Aerial Devices." Technical Report DE 76-10. Dunegan/Endevco, San Juan Capistrano,**

California. 1976; paper presented at the 1976 Annual Conference of the Southeastern Electric Exchange, Engineering and Operating Division, New Orleans, Louisiana. Apr 26–27, 1976.

Beginning with the first overhead power line construction, a need existed for a safe and economical means of elevating and supporting a lineman to perform aerial work. This continuing need has produced an evolution in overhead line work methods which included the use of personal climbers, work platforms, noninsulated aerial devices, and, in recent years, the fully insulated aerial device. The noninsulated aerial device was a safe and simple means of placing a lineman in the work position. However, protection to the lineman from energized conductors and equipment was no greater, and possibly less than he had while standing directly on a wood pole. Development of the fully insulated aerial device provided the possibility of greatly improved safety to the lineman, as well as the circuits on which he worked. In order to maintain the insulation qualities built into these fully insulated aerial devices, an adequate test program was a necessity. Many utility companies instituted electrical test programs designed to maintain these insulation qualities originally provided by the unit. Some company test programs used AC voltages, others used DC voltages, while others used a combination of both. It seems reasonable to believe that the insulation level of these devices can be readily determined. Not so easy is the determination of the physical or mechanical soundness of the insulated portions of the device. Adoption of glass fiber-reinforced plastic materials in construction of insulated portions of the devices introduced some mechanical problems for which no answers were immediately available. Much discussion centered around the useful life of fiberglass. Would the continued flexing produce fatigue? Did the aging process end at some point in time or continue forever? These and other important questions indicated a real need for a system of testing the insulated components of aerial devices to determine at what age it becomes unsafe to continue use of the devices. To be successful, any mechanical testing system must be nondestructive, easy to use, positive in results, and economical to procure. The purpose of this presentation is to describe such a system.

1425 A. A. Pollock and G. Lane. "Prediction of the Strength of Adhesive Bonds and Diagnosis of Poor Adhesion by Means of Acoustic Emission." Information Sheet. Nondestructive Testing Centre, Atomic Energy Research Establishment, Harwell, Oxfordshire, England. Aug 19, 1968.

With the increasing use of bonded structures in the aerospace industry, the nondestructive testing of adhesive bonds is of growing importance. In the last few years, methods have been developed to test the cohesive strength of bonds (i.e. the strength of the cured resin layer), but no way has been found to measure the adhesive strength (i.e. the strength of the adhesive-metal interface). Poor adhesion may be caused by inadequate surface preparation or by contamination in the resin or on the bonding surfaces. Any of these factors can lead to serious reductions in bond strength. Recent experiments indicate that acoustic emissions (stress wave emissions) may be used both to recognise poor adhesion and to predict bond strength. These experiments were carried out at the Nondestructive Testing Centre at the Atomic Energy Research Establishment, Harwell, using adhesive bond specimens prepared by CIBA and instrumentation developed at Imperial College, London.

1426 **A. A. Pollock and J. C. Radon. "Acoustic Emissions in the Fracture Toughness Test of a Mild Steel." Report No. FG 26. Imperial College, Mechanical Engineering Department, London, England. Apr 1970; IIW X-595-70. International Institute of Welding. 1970.**

Acoustic emissions generated during the fracture toughness test of mild steel specimens were investigated over a range of temperatures. At higher temperatures (−70 °C and above) and with increasing amounts of plastic deformation, the emission rate was dependent on the stress intensity factor K. However at low temperatures (below −130 °C), this dependence was not found. The total number of emissions decreased considerably with lower test temperatures; it was also strongly influenced by a change of loading rates applied in comparative tests. In the second part, the acoustic emission rate is discussed in terms of crack growth. It is suggested that a brittle condition of a metal might be indicated by acoustic emissions preceeding catastrophic crack propagation. Further work on emission activity in a range of materials is proposed.

1427 **A. A. Pollock and B. Smith. "Stress-Wave-Emission Monitoring of a Military Bridge."** *Non-Destructive Testing*, **NDETA, 5(6):348-353. Dec 1972.**

Stress-wave-emission monitoring was carried out during the static proof loading trials of a prototype, transportable tank bridge. Analysis of emission counts was performed on-line during loading periods, hold periods and repeat tests, and post test analysis yielded further information on amplitude distributions and source locations. The results were correlated with other engineering measurements and direct observations of local plasticity. Laboratory tests and an acoustic survey were conducted in support of the bridge tests.

1428 **R. B. Pond, Sr. "Acoustic Emission Study of Twinning in Indium Crystals and Lead-Tin Alloys." pp 482-502 in "Proceedings of the Interdisciplinary Workshop for Quantitative Flaw Definition." D. O. Thompson, Program Manager. AFML-TR-74-238, SC595.2ITR. Science Center, Rockwell International, Thousand Oaks, California. Nov 1974. [NTIS: AD/A-003-672]. Workshop held at the Science Center, Rockwell International, Thousand Oaks, California. June 17-20, 1974.**

1429 **M. Popovic. "Raziskava akusticne emisije zvarov (Research of Acoustic Emission of Welds)." Diplomsko delo st. 2044. Fakulteta za strojnistvo, Univerza v Ljubljani, Ljubljana, Yugoslavia. 1974. (in Slovenian).**

1430 **A. W. Porter, M. L. El-Osta, and D. J. Kusec. "Prediction of Failure of Finger Joints Using Acoustic Emissions."** *Forest Products Journal,* **FPJOA, 22(9):74-82. Sep 1972.**

Using a relatively new nondestructive testing technique called acoustic emissions, estimates were made of the ultimate bending strength of 2- by 6-inch Douglas-fir finger joints. The accuracy with which failure could be predicted varied from 1.8 to 25.0 percent, depending upon: (1) load at which prediction was made, and (2) the nature of a finger joint. For normal commercial finger-joint stock, it would appear that obtaining load- and acoustic-emission data to a load level just beyond the proportional limit should permit estimates of failure load accurate to ±10 percent. For reasons of economy, the technique was developed on finger joints, but the results indicate that it deserves to be tried on a practical scale with full-size glulam beams as test material.

1431 A. M. Portevin. "Correspondence: The 'Cry' of Twinning." *Metal Progress,* **MEPOA, 72(2):120–121. Aug 1957.**

It has been known for a long time that when a bar of tin is bent a characteristic sound is heard called the *tin cry*. This behavior has served to determine rapidly whether soft brazing wires or bars of solder have a tin or lead base. Lead, which is of cubic crystallinity, does not exhibit this phenomenon which is common for hexagonal cadmium and zinc, as well as for tetragonal tin. Investigations by both E. N. Andrade and Bruce Chalmers have shown that the phenomenon is caused by the formation of mechanical twins which inevitably accompany the deformation of metallic materials which crystallize in other than the cubic system, if the crystal lattice dimensions are larger than a certain minimum. This deformation twinning produces a large number of twins in noncubic metals, and they resemble the polysynthetic twins known to mineralogists. In cubic metals, these twins form after an anneal subsequent to deformation. Examples are alpha brass and nickel-containing austenites. One can agree with Professor Andrade who reasons that the production of sound on mechanical twinning is a general phenomenon and that even when no audible *cry* is perceived, the sound exists nevertheless, but at a frequency far too high to be picked up by the human ear. It may likewise be theorized that the phenomenon of a *cry* accompanying the formation of martensite from austenite in steels is similar to the effect just described. Forster and Scheil studied the matter by taking a motion picture (with sound effects), which certainly supports the concept of martensite formation as a result of mechanical twinning. Thus, it is seen that a generic term *cry of twinning* may be given to a group of metallurgical phenomena.

1432 A. Portevin and F. LeChatelier. "Heat Treatment of Aluminum–Copper Alloys." *Transactions of the American Society for Steel Treating,* **TASTA, 5:457–478. May 1924.**

The authors observed repeated momentary decreases in the load of a hard tensile machine during deformation of aluminum-copper-magnesium alloys. Each oscillation of the load appeared to correspond to the release of a series of slips (Luders lines) which propagated as a series of waves along the length of the test specimen. Each wave train was accompanied by a *little dry noise,* clearly audible at a distance of several feet.

1433 **A. Portevin and F. LeChatelier. "Sur un Phenomene Observe lors de l'Essai de Traction d'Alliages en Cours de Transformation (On the Phenomenon Observed during Tension Tests on Alloys during Transformation)."** *Comptes Rendus Hebdomadaires des Seances de l'Academie des Sciences,* **COREA, 176:507-510. Feb 19, 1923. (in French).**

The authors report a small sharp noise (*petit bruit sec*) accompanying stress jumps in aluminum alloys undergoing large deformation. The stress jumps were accompanied by the appearance of striations or Luders lines on the surface of the specimen, the noise occurring each time the lines appeared.

1434 **G. J. Posakony. "Generation of a Standards Document for an Emerging Nondestructive Evaluation Technology." pp 260-268 in** *Nondestructive Testing Standards: A Review, ASTM STP 624.* **H. Berger, Editor. American Society for Testing and Materials, Philadelphia, Pennsylvania. 1977. proceedings of a symposium held at the National Bureau of Standards, Gaithersburg, Maryland. May 19-21, 1976.**

Nondestructive testing standards, recommended practices, specifications, procedures, definitions, and classifications are described in technical documents used by both producers and consumers. These documents provide the common language for the testing and evaluation of raw materials, products, structures, and components. Preparing technical documents which are acceptable to involved parties requires a major effort on the part of professional organizations and governmental agencies. Meeting the industrial need of maintaining pace with accepted test methods is a formidable task. Developing documents for new and evolving test methods is a challenge in technology, language, acceptance, patience, and perseverance. This article reviews the time and effort expended to develop acceptable recommended practices within the volunteer consensus mode followed by the American Society for Testing and Materials. [reprinted by permission of the American Society for Testing and Materials, Copyright].

1435 **G. Possa and F. Tonolini. "Appliczioni dell'emissione acustica al controllo della saldatura (Application of Acoustic Emission to Control Welding Operations)."** *Rivista Italiana della Saldatura,* **RISAA, 27(5):275-280. Sep-Oct 1975. (in Italian).**

The main characteristics of welding quality control techniques based on acoustic emission monitoring are briefly outlined. Applications are divided into two groups, namely those utilizing acoustic emission during the welding process and those utilizing acoustic emission during the mechanical test after welding. Typical instrumentation needed for acoustic emission detection and analysis is also presented. A brief survey of specific applications is given. [Abstract 22-0854. *Metals Abstracts,* Vol. 9, 1976].

1436 **G. Possa and F. Tonolini. "Esperienze di applicazione dell'emissione acustica alla rivelazione di difetti in recipienti a pressione durante la prova idrostatia (Experience in the Application of Acoustic Emission to Detect Defects in Pressure Vessels during Hydrostatic Testing)."** *Metallurgia Italiana,* **MITLA, 68(2):85–91. Feb 1976. (in Italian).**

Discussed is the development over the past 15 years of instrumentation and techniques for monitoring acoustic emission in pressure vessels during hydrostatic testing. Details of instrumentation, sensor array, and signal analysis used in source location are discussed.

1437 **D. V. Power. "Acoustic Emissions Following Hydraulic Fracturing in a Gas Well." pp 291–308 in** *Proceedings First Conference on Acoustic Emission/ Microseismic Activity in Geologic Structures and Materials.* **H. R. Hardy, Jr. and F. W. Leighton, Editors. Trans Tech Publications, Clausthal, West Germany. 1977. conference held at Pennsylvania State University, University Park, Pennsylvania. June 9–11, 1975.**

A 20- to 30-foot thick gas-bearing sandstone seam at a depth of 3000 feet was hydraulically fractured with 26,000 gallons of gelled water utilizing a sand proppant. A hydrophone and geophones were emplaced in nonconcentric steel tubing in the well bore of the fracturing well to detect seismic and acoustic signals throughout and following the fracturing process. Background noise from the pumping operations, and fluids passing through small orifices obscured signals during the fracturing process. However, after the fracturing operation, a shutdown of equipment permitted detection of acoustic signals by the hydrophone above the reduced background. These signals contained significant energy above the 1 kilohertz range and were readily detectable as far as 400 feet away from the injection perforations in the fractured zone. An analysis of signal frequency characteristics has been hampered by the frequency limitations of available data reduction equipment. Relationships between signal amplitudes and the number of events to distance from the fracturing perforations and time following pump equipment shut down are presented.

1438 **J. Powers and A. Van Valkenburg. "Pressure-Induced Transformations in Polymers."** *Bulletin of the American Physical Society,* **BAPSA, 8:243. 1963. [abstract only].**

Visual observation of a variety of polymers as they are compressed between diamond anvils up to 70 kbars has revealed changes in optical appearance on compression. The microscopic field of some polymers (Teflon, Kel-F, and polyethylene terephthalate) at some point during compression changes abruptly from one with structure showing shifting interference colors associated with increasing photoelastic strain to one showing reticulated, low birefringent structure. The transition point is usually accompanied by a snapping noise. Other polymers (natural rubber and the polyoletins) develop grooves that spread throughout the field as pressure is increased and finally result in a reticulated, low birefringent field. The transitions appeared to be irreversible, at least on the time scale studied. The X-ray-diffraction patterns of the compressed area after removal from the diamond anvils were compared with patterns of uncompressed material. Crystalline rings, which were not present before compression were observed in compressed natural rubber, while polyethylene displayed a new ring (Bragg angle 9.7) associated with a psuedotriclinic unit-cell modification. These and other results are discussed, a description of the apparatus given, and a partial interpretation of the data presented.

1439 **M. E. Prado. "Predicting Crack Surface Area from Acoustic Emission." UCRL-76168. Lawrence Livermore Laboratory, Livermore, California. Nov 11, 1974; paper presented at the Society for Experimental Stress Analysis Spring Meeting, Chicago, Illinois. May 11-17, 1975.**

The crack-growth characteristics as a function of acoustic emission count were studied for HY-TEN steel. An expression relating crack growth and stress intensity factor to AE count was developed using dislocation energy and the crack energy release rate. The parameter $\Delta K/E)^2\ \dot{a}/(1 - R)^2$ relates the crack surface area per fatigue cycle to AE count per cycle. The AE was plotted with previously published data and was found to exhibit the same AE transition. In plotting AE count per cycle for aluminum data obtained from published data as a function of crack surface area per cycle, it was found that aluminum did not show the same AE transition indicated by published data. Although AE was detected during fatigue cracking, there was no AE during loading of the specimen to failure. The AE during this phase was restricted prior to failure. Tensile tests indicated that all emission was confined to the elastic portion of this test with no further emission occurring past the yield strength of the material.

1440 **F. Press and W. F. Brace. "Earthquake Prediction."** *Science,* **SCIEA, 152 (3729):1575-1584. June 17, 1966.**

1441 **C. C. Price and J. R. Karvinen. "EBR-II Noise-Signature Analysis during the First Half of Fiscal Year 1971." ANL/EBR-036, Argonne National Laboratory, Argonne, Illinois and Idaho Falls, Idaho. Feb 1971. [NTIS].**

The noise-signature analysis done to date (February 1971) on EBR-II and its auxiliary systems is presented. This effort began with a search for the source of the *10-Hz* noise present in the ion-chamber signals. The search provided a signature of primary-sodium pump No. 1 which could be compared with the signature after an apparent binding of the pump. The comparison indicated more eccentric rotation of the pump after the apparent binding. The cause was apparently failure of the lower labyrinth seal. On November 4, 1970, an unusual noise was discovered emanating from the intermediate heat exchanger (IHX). A signature obtained on the secondary-sodium outlet pipe of the IHX on August 21, 1970, had not shown any unusual noise present. Further signature-analysis work, described with preliminary results, helped identify the cause of the noise.

14 **W. Primak and T. T. Anderson. "Metamictization of Lithium Niobate by Thermal Neutrons."** *Nuclear Technology,* **NUTYB, 28(2):235–248. Feb 1976.**

Lithium niobate plates intended for application as sensors for acoustical monitoring of nuclear reactors were subjected to thermal-neutron irradiation near room temperature. By 8×10^{19} n/cm^2, a crystal had become highly disordered, as shown by the back-reflection Laue picture, by loss of piezoelectric response, and by loss of optical birefringence. This transformation appears to be a metamictization. Changes in optical obsorption, tritium loss, postirradiation lithium migration, surface crazing, and effects of postirradiation storage at room temperature and of heating to 140 °C are also observed.

1443 **W. Primak, T. T. Anderson, and S. L. Halverson. "Ionizing Radiation Effects in Lithium Niobate."** *Nuclear Technology,* **NUTYB, 10:76–84. Jan 1971.**

Techniques for accerlerated ionizing radiation tests, utilizing electrons from a Van de Graaff generator, are described and applied to lithium niobate. A broad optical absorption band of small optical density (<0.1 per mm) developed from the middle of the visible to the ultraviolet cutoff. Small dilalations ($\sim 10^{-5}$) occurred; in some specimens positive, in others negative. Small changes were seen in the electro-acoustical properties: −7% in electromechanical coupling coefficient, −6% in electrical permittivity, and a slight decrease in mode purity. These changes annealed partially at room temperature over a period of several weeks. Further annealing occurred on heating for 15-minute intervals at 100 and 140 °C; annealing was completed after a heating at 180 °C. These effects may be associated with storage of charge in a piezoelectric material. A crystal plate was operated effectively as a piezoelectric acoustical sensor during electron bombardment even at a dose rate 10^5 R per second but showed a small decrease in signal-to-noise ratio. The duration of these irradiations shows that no significant radiation damage caused by ionization will accumulate in a century of immersion in coolant sodium in a fast breeder reactor. The lithium niobate is extremely stable toward radiation, behaving as the close-packed oxides (e.g., sapphire), rather than as a salt.

1444 **D. W. Prine. "Inspection of Nuclear Power Component Welds by In-Process Acoustic-Emission Monitoring."** *NDT International,* **NDITD, 9(6):281–284. Dec 1976.**

The author discusses the results of using in-process acoustic-emission monitoring on nuclear power plant piping welds. The technique was applied to laboratory as well as production welds. The acoustic emission results were compared with standard NDT results. This work was sponsored by the U.S. Nuclear Regulatory Commission.

1445 **D. W. Prine. "Inspection of Nuclear Power Plant Piping Welds by In-Process Acoustic Emission Monitoring/Inspection des soudures des tuyaux dans les centrales nucleaires par ecoutement des emissions acoustiques." GATX**

Corporation, GARD, Inc., Niles, Illinois; Paper 2A/8 in *Proceedings of the Eighth World Conference on Nondestructive Testing.* volume containing Sections 3D, 2A, 1B. International Committee for Nondestructive Testing, Secretariat du Cofrend, Institut de Soudure, 32 Boulevard de la Chapelle, 75880 Paris Cedex 18, France. 1976. conference held in Cannes, France. Sep 6–10, 1976. (paper in English). [NTIS: INIS-mf-3585].

The results of using in-process acoustic emission monitoring on nuclear-power plant piping welds are discussed. The technique was applied to good and intentionally flawed test welds as well as production welds, and the acoustic emission results are compared to standard NDT methods, and selected metallographic cross-sections.

1446 D. W. Prine. "Inspection of Nuclear Reactor Welding by Acoustic Emission." NUREG-0035-2. GATX Corporation, GARD, Inc., Niles, Illinois. Sep 1976. [NTIS: PB-258-128].

The capability of acoustic emission to detect cracking in nuclear quality piping welds is documented in the final report for Phase I of this Project (NUREG-0035-1) (see Reference 1450). Experimental acoustic emission data on simulated nuclear pressure vessel welds which were developed since the previous reporting periods are presented in this current report. Some analysis of these data is included, however, the analysis is incomplete due to the current unavailability of confirming NDE and metallographic results. The first simulated nuclear pressure vessel weld test was conducted at Babcock and Wilcox, Barberton, Ohio. The test weld was a multipass semi-automatic submerged arc weld (SAW). The welded plate was typical nuclear pressure vessel steel approximately 5 feet in length by 5 feet wide. Welding procedures, weld geometry, and fit-up were typical of a nuclear pressure vessel weld. Temperature approximately 1 foot from the weld bead was maintained at 350 to 400 °F during fabrication of the weld. Flaws were intentionally introduced into the weld during welding. Preliminary ultrasonic and radiographic data are available now on this weld. The second simulated nuclear pressure vessel weld test was conducted at Combustion Engineering, Chattanooga, Tennessee. Plate material was A-533 steel and again typical nuclear welding procedures and fit-up were followed. The weld was fabricated with semi-automatic submerged arc. Approximately 65 passes were used to complete the 41-inch long 6.75-inch thick weld. The temperature of the weld as in the B&W test was maintained at approximately 400 °F through the test. Intentional flaws were introduced during welding. No confirming NDE data are available. Eventually this plate will receive extensive NDE (radiography and UT) and after completion of the NDE a thorough metallographic examination will be made. Acoustic emission was monitored during both PV tests using the GARD developed flaw detection and noise discrimination techniques previously applied to nuclear piping welds. In addition to the flaw detection circuitry, a GARD modified Dunegan/Endevco two-channel time of arrival flaw-location system was utilized. In addition to detection of flaws, an NDE method should provide some means of characterizing flaws in terms of size and type. In this report, previous as well as current, AE data are considered in terms of the capability of AE to discriminate between various flaw types and to provide information on flaw size. The results of this analysis are compared to current ASME Code requirements to determine those areas where additional laboratory testing is required.

1447 **D. W. Prine. "NDT of Welds by Acoustic Emission." Technical Report DE 73-X. Dunegan/Endevco, San Juan Capistrano, California. 1973; paper presented at the Society for Experimental Stress Analysis 3rd International Congress on Experimental Mechanics, Los Angeles, California. May 13–18, 1973; paper presented at the American Society for Metals Materials Testing and Quality Control Conference, Detroit, Michigan. Oct 21–24, 1974.**

The manufacture and assembly of tank cars employ considerable welding. Currently, these welds are inspected with film radiography. Film radiography is effective in detecting the presence of weld defects. However, it is expensive, does not lend itself to real time on line inspection, and requires skilled personnel for both application and interpretation. In this paper, the feasibility of applying acoustic emission monitoring techniques to the real-time inspection of tank-car welds is discussed. A series of welds was made in the welding laboratory at the GATX Transportation Division. The acoustic emission results from these welds are compared to the radiographic data for the same welds and excellent correlation is shown to exist. In addition, background noise on an actual tank undergoing welding was monitored and found to be low enough in rate and amplitude to allow successful acoustic monitoring of the welds during manufacture. The experimental results definitely show the feasibility of applying acoustic emission to tank-car weld inspection.

1448 **D. W. Prine. "NDT of Welds with Acoustic Emission." pp 171–199 in** *Proceedings of the U. S.-Japan Joint Symposium on Acoustic Emission*. **English volume. Japan Industrial Planning Association, Kawamura-Building 6F, 3-21-6 Akasaka, Minato-ku, Tokyo, Japan 107. 1972. symposium held in Tokyo, Japan. July 4–6, 1972.**

The manufacture and assembly of tank cars employ considerable welding. Currently, these welds are inspected with film radiography. Film radiography is effective in detecting the presence of weld defects, however, it is expensive, it does not lend itself to real time on line inspection, and it requires skilled personnel for both application and interpretation. In this paper, the feasibility of applying acoustic emission monitoring techniques to the real-time inspection of tank car welds is discussed. A series of welds were made in the welding laboratory at the GATX Transportation Division. The acoustic emission results from these welds are compared to the radiographic data for the same welds and excellent correlation is shown to exist. In addition, background noise on an actual tank undergoing welding was monitored and found to be low enough in rate and amplitude to allow successful acoustic monitoring of the welds during manufacture. The experimental results definitely show the feasibility of applying acoustic emission to tank-car weld inspection.

1449 **D. W. Prine and R. N. Clark. "Inspection of Nuclear Reactor Welding by Acoustic Emission." NUREG-75/101. General American Transportation Corporation, General American Research Division, Niles, Illinois. May 1975. [NTIS: PB-245-725].**

This is an interim report (on the work performed from November 1, 1974 through April 30, 1975) on the first phase of a three-year research program sponsored by the Nuclear Regulatory Commission (NRC) under Contract E(11-1)-2504. The goal of this phase is to show feasibility and develop equipment design parameters for real-time production monitoring of nuclear power piping welding by acoustic emission. The report covers introductory material (based upon previous acoustic work by General American Research Division) and the project work to date. The work includes the incorporation of GARD equipment into a field-portable acoustic emission weld monitor, performance of one *calibration* field test, and one production field test. The calibration field test consisted of 96 weld passes, some of which were intentionally flawed, and all of which were monitored and recorded for laboratory study. The production field tests consisted of 17 multipass welds on actual production pieces. The welding in both tests was performed at Southwest Fabricating and Welding Corporation, an N-Stamp pipe fabricating facility. The welding monitored was standard nuclear practice using code-approved materials for the calibration work. The production welding was on actual code-certified pipe. Calibration data on 20 miltipass welds (96 weld passes) showed 7 classes of weld types, 1 with excellent to good acoustic emission detection properties, 2 with good properties, 1 with good properties, and 4 that need further study. No weld type has proven to be completely without *some* acoustic emission monitorability. The production tests have not been completely analyzed; initial analyses show results in line with the calibration data.

1450 **D. W. Prine and R. N. Clark. "Inspection of Nuclear Reactor Welding by Acoustic Emission." NUREG–0035–1. GATX Corporation, GARD, Inc., Niles, Illinois. Feb 1976. [NTIS: NP-20616 or PB–250–030].**

The report covers work accomplished in the first year (November 1974 to November 1975) of a three-year research program sponsored by the United States Nuclear Regulatory Commission, Reactor Safety Research Division, Metallurgy and Materials Research Branch under Contract E(11-1)-2504. The goal of this phase is to show feasibility and develop equipment design parameters for real time, in process, NDE of nuclear power piping welds by means of acoustic emission. Included is some introductory material that was previously published in the first interim report, May 1975. In addition, extensive documentation of one calibration and two production tests is included. The test results definitely show that in-process NDE of nuclear power piping welds by means of acoustic emission is feasible and practical. Results show that acoustic emission is effective at detecting cracking-type flaws during the welding process and that simple digital logic in conjunction with frequency discrimination can eliminate the many interferring background noises encountered in a production environment.

1451 **V. Yu. Proskurin and V. S. Kortov. "Acoustic Emission in the Stretching Type lKh18N9T Steel."** *Izvestiya Vysshikh Uchebnykh Zavedenii, Chernaya Metallurgiya*, **IVUMA, (6):108–109. 1972. (in Russian); Trans. BISI–10698. British Iron and Steel Industry Translation Service, The Iron and Steel Institute, 1 Carlton House Terrace, London, SW1Y 5DB, England. Sep 1972. (in English).**

Samples of commercial plate were notched and deformed at 20 and 200 °C at rates in the range 0.06 to 1.9 mm per minute. The working sample length was 37 mm. The level of acoustic emission was shown to pass through a maximum at intermediate elongations and immediately before fracture. Temperature had little effect on noise levels, but increased them slightly. [Abstract 32-0165. *Metals Abstracts,* Vol. 6, 1973.]

Q

1452 **M. Quivy. "Moyens d'Ecoute Utilisables en Acoustique Sous-Marine (Means of Listening Utilized in Underwater Acoustics)." pp 36-55 in** *Journees d'Etudes sur l'Emission Acoustique.* **P. F. Gobin, R. Goutte, and C. Guillaud, Organizers. Institut National des Sciences Appliquees de Lyon, Avenue Albert Einstein, 69621 Villeurbanne, Lyon, France. Mar 1975. proceedings of a conference held in Lyon, France. Mar 17, 1975. (paper in French).**

R

1453 **D. Rader. "A Stress Wave Study of Brittle Fracture." Ph.D. Thesis. Brown University, Division of Engineering, Providence, Rhode Island. June 1966; University Microfilms 67–2280.**

1454 **D. Rader. "On the Dynamics of Crack Growth in Glass."** *Experimental Mechanics,* **EXMCA, 7(4):160–167. Apr 1967.**

An experimental investigation of crack growth has been performed using plates of glass subjected to impulsive loading by means of small explosive charges. For comparison, some experiments were also performed with two *glass-like* polymers. The formation of the resultant crack patterns in glass was observed by the use of high-speed photography during the period following the detonation. Information has been obtained about the physical features of crack growth with regard to crack velocity, bifurcation, and the interaction between growing cracks and propagating stress waves. The existence of a terminal velocity of crack propagation is considered and, in particular, the degree to which crack bifurcation provides an explanation for this phenomenon is discussed.

1455 **J. C. Radon and A. A. Pollock. "Acoustic Emission and Energy Transfer during Crack Propagation." Report No. FRR-31. Imperial College, Mechanical Engineering Department, London, England. July 1970;** *Engineering Fracture Mechanics,* **EFMEA, 4(2):295–310. June 1972.**

Acoustic emissions in mild steels and Al alloys may be conveniently investigated during fracture toughness tests. This paper describes the development and application of a highly sensitive acoustic instrumentation with a wide dynamic range (100 db). Equipment suitable for sorting, counting, and analysing emissions covering frequencies from 25 to 250 kHz was used in the tests on mild steel at low temperature (−135 °C) and on Al alloy at 21 °C. The correlation of emission amplitudes with the energy released is presented. Also, dynamic surface energy which has a major influence on crack growth is discussed in detail; an interesting relationship between the energy released during the movement of a crack and the length of the crack offers distinct advantages over the usual explanations of crack propagation in toughness tests. It is shown that each material gives off a distinctive emission; thus in failure investigations a material could be recognized by its characteristic *emission signature*; thus emission records may indicate the origin of catastrophic fractures. [reprinted by permission of Pergamon Press Ltd from *Engineering Fracture Mechanics*].

1456 **J. C. Radon and A. A. Pollock. "Development of Fast Fracture in a Low Alloy Steel." pp 15–30 in** *Fracture Toughness and Slow-Stable Cracking, ASTM STP 559.* **American Society for Testing and Materials, Philadelphia,**

Pennsylvania. 1974. proceedings of the 1973 National Symposium on Fracture Mechanics, Part–1, sponsored by ASTM Committee E–24 on Fracture Testing of Metals, held at University of Maryland, College Park, Maryland. Aug 27–29, 1973. [copy available as Technical Report DE 74–9 from Dunegan/Endevco, San Juan Capistrano, California].

Fracture mechanics, acoustic emission, and fractography were used to study the process of crack growth in 2-inch thick double cantilever beam specimens of a low-alloy steel over a wide temperature range. The K_{Ic} values ranged from 33 ksi $\sqrt{\text{inch}}$ at −200 °C to 240 ksi $\sqrt{\text{inch}}$ at +18 °C. The acoustic emission rate rose steadily with load up to fast fracture. Changes in the emission amplitude distribution were observed shortly before fracture at one of the warmer temperatures. Emission during load holds was measured and emission waveforms were photographed. Fractography showed a small region of ductile tearing prior to fast fracture at warmer temperatures. A model of the interaction of ductile tearing and cleavage fracture is proposed. Ductile tearing is seen as a process taking time and strongly dependent on temperature. Cleavage is seen as a rapid process averted by ductile flow and tearing. Emissions are believed to be produced by plastic zone growth and by cleavage. The experimental facts can be consistently interpreted through this small set of assumptions. [reprinted by permission of the American Society for Testing and Materials, Copyright].

1457 **J. B. Radziminski, F. V. Lawrence, T. W. Wells, R. Mah, and W. H. Munse. "Low Cycle Fatigue of Butt Weldments of HY–100(T) and HY–130(T) Steel." Structural Research Series No. 361. Final Report. University of Illinois, Department of Civil Engineering, Urbana, Illinois. July 1970. [NTIS: AD–872–644 L].**

An evaluation of the axial fatigue behavior of plain plates and full penetration butt-welded joints in HY-130(T) steel is presented. The weldments were fabricated using GMA and SMA welding processes. Fatigue tests were conducted with sound weldments and weldments containing internal defects including slag, porosity, and lack of fusion. Radiographic and ultrasonic inspection techniques were used to study the initiation and propagation of fatigue cracks originating at internal weld flaws. Acoustic emission measurements were taken for smooth and notched HY-130(T) specimens tested in static tension and in fatigue. The results of preliminary tests of plain plates and butt weldments of HY-100(T) (HY-110) steel are presented.

1458 **M. A. Raharinaivo. "Utilisation de l'Emission Acoustique pour l'Etude de la Fissuration et de la Rupture des Aciers a Haute Resistance Employes en Genie Civil (Utilisation of Acoustic Emission to Study Cracking and Rupture of High Strength Steels Employed in Civil Engineering)." pp 399–412 in** *Journees d'Etudes sur l'Emission Acoustique.* **P. F. Gobin, R. Goutte, and C. Guillaud, Organizers. Institut National des Sciences Appliquees de Lyon, Avenue Albert Einstein, 69621 Villeurbanne, Lyon, France. Mar 1975. proceedings of a conference held in Lyon, France. Mar 17, 1975. (paper in French).**

It is shown that acoustic emission can be used to study the behavior of steels subjected to stress corrosion cracking tests or hydrogen embrittlement tests.

1459 **R. L. Randall. "Application of Noise Analysis Techniques to Detection of Incipient Malfunctions." pp 71-86 in** *Incipient Failure Diagnosis for Assuring Safety and Availability of Nuclear Power Plants.* **E. P. Epler and D. P. Roux, Coordinators. U.S. Atomic Energy Commission, Division of Technical Information, Washington, D.C. Jan 1968. [NTIS: CONF-671011]. proceedings of conference sponsored by Oak Ridge National Laboratory, held in Gatlinburg, Tennessee. Oct 30-Nov 1, 1967.**

Techniques are described which utilize noise or fluctuation signals from electronic transducers sensitive to vibrations, temperature, power, conductivity, pressure, sound level, flow rate, etc., to detect and study abnormal conditions in reactor systems or test models. Applications which are summarized range from feasibility tests, to development of automatic on-line, incipient malfunction detection systems. Specific phenomena encountered in these applications include: boiling; cavitation; gas entrainment through a reactor core; movement of bubbles through pipes, hydraulic transients; mechanical vibrations; and breakdown of electrical insulators.

1460 **D. K. Rathbun, A. G. Beattie, and L. A. Hiles. "Filament Wound Materials Evaluation with Acoustic Emission." SCL-DC-70-260. Sandia Laboratories, Livermore, California. Apr 1971. [NTIS]; pp 48-58 in** *Proceedings of the Eighth Symposium on Nondestructive Evaluation in Aerospace, Weapons Systems, and Nuclear Applications.* **Western Periodicals Company, North Hollywood, California. 1971. symposium held in San Antonio, Texas. Apr 21-23, 1971.**

The paper describes experiments which establish that (a) broken strands generate detectable acoustic emission in tensile specimens, (b) acoustic emission pulses originate from areas undergoing structural damage in spherical test shapes, and (c) the region which ultimately fails emits large quantities of acoustic waves prior to failure. It has not been possible to establish a statistical relationship between acoustic emission events and broken or damaged strands at this time. Further work with better controlled samples will be required to discover this relation, if one exists.

1461 **F. O. Rathburn, Jr., W. B. Giles, D. E. Wood, and C. Miller. "Spectral Analysis of Pump Cavitation Noise."** *Experimental Mechanics,* **EXMCA, 7(9):19A-29A. Sep 1967; paper presented at the Society for Experimental Stress Analysis 2nd International Congress on Experimental Mechanics. Washington, D.C. Sep 28-Oct 1, 1965.**

Pressure-transducer and accelerometer data accumulated during several water tests of a liquid-rocket-engine pump were subjected to spectral analysis. The analyzer used allows signal amplitude to be displayed as a continuous function of frequency for each instant of time. Various filter band widths, integration times, frequency spans, and display formats have been used to present the data for visual inspection. Results of attempts to discern the signature of various phases of the cavitation phenomenon occurring within the pump are reported.

1462 **J. G. Rawlings. "ASME Acoustic Emission Code Status." pp 103–111 in** ***Acoustic Emission.*** **Edited by R. W. Nichols. Applied Science Publishers Ltd, Ripple Road, Barking, Essex, England. 1976. proceedings of a colloquium on acoustic emission held during the 1975 Annual Assembly of the International Institute of Welding, Tel Aviv, Israel. July 7–13, 1975.**

1463 **J. G. Rawlings. "Methodology for Use of Acoustic Emission Examination during Application of Pressure."** ***Transactions of the ASME, Journal of Pressure Vessel Technology,*** **JPVTA, 97(2):126–131. May 1975.**

Article includes a brief discussion on the formation of the Ad Hoc Working Group on Acoustic Emission by the ASME Subcommittee on Nondestructive Examination to develop the methodology for the use of acoustic emission during hydrostatic test and as an in-service inspection tool. Included is the text of the *Proposed Standard for Acoustic Emission during Application of Pressure* which was approved by the ASME Policy Board, Codes and Standards on Jan 30, 1975.

1464 **W. N. Redstreake. "Metals 'Sound Off' on Fatigue."** ***Iron Age,*** **IRAGA, 192(12):97–99. Sep 19, 1963.**

Article briefly discusses the work of C. A. Tatro, R. B. Engle, and R. G. Liptai conducted at Tulane (New Orleans, Louisiana) and Michigan State (East Lansing) Universities. Investigation of microscopic deformation and slip phenomena and metal fatigue by monitoring acoustic emission of Al or steel specimens under stress. Special test devices, such as a thermal loader, water loader, and a floatation tank shear loader, apply loads quietly enough to allow recording of high-frequency sound waves.

1465 **C. R. Reese. "Sonic Evaluation of Thermal Shock Resistance." M.S. Thesis. Clemson University, Clemson, South Carolina. 1969.**

A sonic method of studying thermal-shock resistance of refractory materials is introduced. Test equipment which allowed the recording on paper of sound patterns emitted during controlled thermal abuse of a sample was constructed. Commercial refractory bricks of known in-service thermal-shock resistance were evaluated. The sonic test is shown to be reproducible. Under certain thermal environments, the test is capable of ranking the samples according to their ability to withstand thermal shock.

This ranking is comparable to that determined by the manufacturer's in-service evaluation. The effect of cooling rate and temperature differential upon the test method was not determined.

1466 J. A. Regalbuto. "Nondestructive Testing Techniques for Diffusion Bonded Titanium Structures." AMMRC CTR 73-53. General Dynamics Corporation, Convair Aerospace Division, Fort Worth, Texas. Dec 1973. [NTIS: AD-775-466].

The report describes the establishment of nondestructive testing techniques for diffusion-bonded structures fabricated from Ti - 6 Al - 4V sheet and forgings. Ultrasonic shear waves, radio frequency and signal-averaged pulse-echo techniques, X-ray, fluorescent penetrant, and blue-etch-anodize were used to inspect a series of diffusion-bonded defect specimens. Analysis of NDT, mechanical property, and fractographic data indicated that shear-wave inspection of the sheet material could be correlated with the observed decrease in fatigue life with defect concentration. Shear strength of the sheet specimens was not affected by the presence of the internal defects to any measurable degree. Grain structure noise limited inspectability of the small (0.006-inch diameter) defects in the diffusion-bonded forging.

1467 J. A. Regalbuto, D. E. Gordon, B. O. McCauley, J. S. Kunselman, R. H. McDaniel, J. R. Jones, J. W. Hagemeyer, and A. D. Crowe. "Nondestructive Testing of Diffusion Bonded Titanium Alloys for Engine and Airframe Components." AFML-TR-74-215. General Dynamics Corporation, Fort Worth Division, Fort Worth, Texas. May 1975. [NTIS: AD/A-014-361].

The development and evaluation of ultrasonic instrumentation for the nondestructive testing of diffusion bonded titanium alloy components are described. Results of a survey to define potential diffusion-bonding deficiencies are given. A signal-averaging pulse-echo system developed to increase defect signal-to-grain noise ratio is described. The AFML Computer-Automated Ultrasonic Inspection System was used to evaluate displaced and angularized bond planes. The fabrication of 600 diffusion-bonded specimens containing intentional deficiencies such as voids, inclusions, lack of deformation, and grain size variations is described. Diffusion-bonded internal defect specimens were evaluated by blue-etch-anodize, fluorescent penetrant, signal-averaged pulse-echo, and delta-scan techniques. Ultrasonic attenuation measurements versus grain size are reported. Static tensile and axial fatigue data are reported and correlated with nondestructive evaluation and fractographic results. Acoustic emission behavior of parent material specimens is compared with that of internal defect specimens. Results are documented of the nondestructive evaluation of a large difussion-bonded titanium structure utilizing blue-etch-anodize, fluorescent penetrant, signal-averaged pulse-echo, delta-scan, and variable tilt-multiple scan pulse-echo techniques. Defect data resulting from sectioning the large structure are documented.

1468 K. J. Reiman [*sic*]. "Nondestructive Testing: High-Temperature Ultrasonic Transducers." pp 57–58 in "Instrumentation and Control Activities at Argonne. Annual Report for Period Ending June 30, 1971." ANL-7905. Argonne National Laboratory, Argonne, Illinois. Dec 1971. [NTIS].

Described is the development of a sensor of wide applicability to the LMFBR Program. The sensor is lithium niobate constructed by brazing the supports to the crystal, thus extending the operating frequency range to several megahertz. Air-brazed transducer assemblies have been successfully operated at temperatures up to 600 °C. Stainless steel-jacketed transducers of this type are suitable for operation immersed in liquid sodium or NaK. This type of transducer has been evaluated for monitoring acoustic emission on the CCTL sodium test loop at 1100 °F and a flow rate of 536 gpm, and monitoring stainless steel tensile specimens undergoing fatigue cycling at 800 °F.

1469 W. W. Reinhardt. "Acoustic Emission Handles Variety of Monitoring Jobs." *Metal Progress*, **MEPOA, 102(2):42. Aug 1972.**

Presented is a brief discussion of applications that range from detection of defects in pressure vessel welds in proof testing to spotting cracks in ceramic materials.

1470 E. R. Reinhart. "A Summary Report on the EEI RP-79 Program." *Transactions of the American Nuclear Society,* **TANSA, 17:403–404. Nov 1973.**

Given is a brief summary of the Edison Electric Institute and Tennessee Valley Authority joint-sponsored project, RP-79 Applied Research Program. The program was divided into the following seven projects: I. Review of Recent Utility Experience, II. Development of a Nondestructive Test Method Evaluation Facility, III. Acoustic Spectrometer, IV. Acoustic Emission, V. Conventional Ultrasonics, VI. Acoustic Holography, and VII. Information Handling. Under Project IV a prototype acoustic emission system was developed for evaluation under simulated reactor operating conditions. The EBOR (Experimental Beryllium Oxide Reactor) vessel was used to study operation at elevated temperatures and in the presence of simulated reactor coolant flow noise. The results of this work have been published in eight biannual progress reports by Southwest Research Institute, all of which are available from NTIS. All of the acoustic emission work are listed in this bibliography under the author who reported the work. The progress reports may be found listed in the Author Index under the editors, R. D. Wylie for progress reports, Nos. 1 through 7, and E. R. Reinhart for No. 8.

1471 E. R. Reinhart and W. D. Jolly. "The EEI-TVA Acoustic Emission Program Results." pp 450–479 in *Topical Meeting on Water-Reactor Safety, Salt Lake City, Utah, March 26–28, 1973.* **Compiled by G. A. Freund. Aerojet Nuclear Company, Idaho Falls, Idaho. 1973. [NTIS: CONF-730304 (entire proceedings)].**

A prototype acoustic emission system for evaluation under simulated reactor operating conditions is described. Experimental results of operation of the prototype system on the Experimental Beryllium Oxide Reactor (EBOR) vessel, modified to simulate a light water reactor, at elevated temperature and in the presence of simulated reactor coolant flow noise, indicate that on-line reactor surveillance is feasible. A successive approximation triangulation technique and a coincident detection location technique were evaluated. The coincident detection location technique was used to locate simulated emission sources at a signal-to-noise amplitude ratio of −12 dbV. Fatigue cracks grown from the inner and outer surface of the EBOR pressure vessel exhibited similar acoustic emission response. These results all tend to show that this technique, when properly utilized, could be a powerful supplement to existing in-service inspection methods.

1472 **E. R. Reinhart and R. D. Wylie. "Nondestructive Testing Techniques Applied to In-Service Inspection." pp 199–217 in** *Nuclear Metallurgy, Volume 16: The Technology of Pressure-Retaining Steel Components.* **M. S. Wechsler, Editor. The Metallurgical Society of the American Institute of Mining, Metallurgical, and Petroleum Engineers, New York. 1970. proceedings of a symposium held in Vail Village, Colorado. Sep 21-23, 1970.**

The paper is a compilation of work performed by various laboratories under the Edison Electric Institute (EEI) and Tennessee Valley Authority Research Program RP 79, *In-Service Inspection Program for Nuclear Reactor Vessels,* and in general represents the work completed from September 1, 1969 to March 1, 1970. The program is building upon basic nondestructive testing techniques developed by the Atomic Energy Commission (AEC) and others, in order to adapt this technology to the specific needs of monitoring the integrity of pressure boundaries of operating nuclear steam-generating units. The various work phases are being carried out at a number of organizations and the overall program is being managed by Southwest Research Institute (SwRI), San Antonio, Texas.

1473 **J. J. Reis. "Instrumentation Applications of Videotape Recording."** *EID–Electronic Instrumentation,* **EIDIA, 7(11):9–14. Nov 1971.**

Article describes the use of a videotape recorder in acoustic emission work. The videotape recorder employs a rotating head with a helical scan which provides a bandwidth up to 4.2 MHz and stop-action replay.

1474 **J. J. Reis. "Unusual Applications of Videotape Recorders."** *Research/Development*, **REDEA, 23(2):24–26. Feb 1972.**

Acoustic emission and failure mode analysis of composites, shock wave and sonic emissions in wind tunnels, and motion of specimens in a scanning electron microscope are among studies benefiting from the unique abilities of videotape recording. This is according to studies made at Northrop Corporate Laboratories in Hawthorne, California.

1475 **C. J. Renken, Jr. (to the U.S. Atomic Energy Commission). "Method and Means for Measuring Acoustic Emissions." U.S. Patent Application 506-819. Filed Sep 17, 1974. [NTIS: PAT-APPL-506-819]; (to the U.S. Energy Research and Development Administration). U.S. Patent 3,930,405. Jan 6, 1976.**

The detection of acoustic emissions emanating from an object is achieved with a capacitive transducer coupled to the object. The capacitive transducer is charged and then allowed to discharge with the rate of discharge being monitored. Oscillations in the rate of discharge about the normally exponential discharge curve for the capacitive transducer indicate the presence of acoustic emissions.

1476 **J. M. Restaino. "A Tensile Testing Experiment Correlating Load with Acoustic Emission." Report ITC-C2-28-75-102. Maintenance Effectiveness Graduate Engineering Program, U.S. Army Material Command Intern Training Center, Red River Army Depot, Texarkana, Texas. Nov 1975. [NTIS: AD/A-024-813].**

By correlating a change of load with a change in acoustic emission counts, a possibility exists that an alternate method of incipient failure detection can be predicted. This will allow the engineer to schedule preventive maintenance during low production time or machine idle time. This presentation is made in the form of a literature review and a first-hand experimental investigation. Fifteen specimens were procured and tested. Load was recorded by the Instron Tensile Test Machine and the acoustic emissions were recorded on tape. A linear regression analysis was then performed on the data collected. The conclusions drawn from the research indicate that the correlation of load versus acoustic emissions indeed exists. Also, a good indication of incipient failure prediction is the engineering yield point.

1477 **T. W. Rettig and M. J. Felsen (to Lockheed Aircraft Corporation). "Acoustic Emission Method for Detection and Monitoring of Corrosion." U.S. Patent 3,946,600. Mar 30, 1976.**

Described is a method for detecting and monitoring corrosion by sensing the spontaneous propagation of elastic waves produced in materials as the result of corrosion phenomena. Certain corrosion reactions generate elastic waves (sounds) which may be detected by sufficiently sensitive instrumentation to provide an identifiable acoustic signature. Correlation of these signatures with known standards provides a nondestructive inspection technique for monitoring invisible or hidden corrosion. In addition to acoustic emissions generated by the low-level oxidation-reduction reactions characteristic of corrosion processes, the invention includes the application of a thermal input to the test specimen to enhance the acoustic signals produced by the corrosion phenomena. By measuring the rate of acoustic pulse emission, the cumulative count of acoustic pulses generated, and the pulse amplitudes of the acoustic emissions, correlation may be made with empirically obtained data to permit failure prediction. While primarily directed to the monitoring of corrosion phenomena, the invention may also be applied to other chemical reactions characterized by acoustic emissions, such as etching, painting, electroplating, adhesive bonding, and chemical milling.

1478 T. W. Rettig and M. J. Felsen. "Acoustic Emission Method for Monitoring Corrosion Reactions." Lockheed-California Company, Burbank, California. 1974; Paper No. 65 presented at Corrosion/74, Chicago, Illinois. Mar 4–8, 1974; *Corrosion,* CORRA, 32(4):121–126. Apr 1976.

The application of acoustic emission techniques for monitoring corrosion phenomena, a unique new NDI approach, is described. Investigations in several areas successfully correlating signal generation and corrosion phenomena are presented. The data showing a correlation between selected signal monitoring and corrosion processes indicate the feasibility of this method for a broad range of laboratory mechanism investigations as well as field-service NDI.

1479 W. G. Reuter, P. P. Crimmins, and C. E. Hartbower. "SWAT Measurements of Stress-Relief Cracking in 2¼ Cr - 1 Mo Weldments." Aerojet-General Corporation, Sacramento, California. Dec 31, 1968.

The Aerojet-General Corporation's Stress Wave Analysis Technique (SWAT) was used during the testing of highly restrained weldments of 2¼ Cr - 1 Mo steel to determine its application as a nondestructive inspection technique for detecting cracking during post-weld stress relieving operations. The detection of stress-wave emissions, created by discrete increments of crack growth, is a unique means of sensing and locating origins of crack growth. A stress wave is the elastic wave which travels through the structure and originates from the energy release associated with fracture and failure porcesses. Sensors which detect this wave as it arrives can be used to detect the flaw extension and determine its origin through the use of triangulation techniques. In this program, SWAT was applied during the stress relieving of modified Lehigh-weld-restraint specimens of 2¼ Cr - 1 Mo steel, which were fabricated by the Standard Oil Company of California and tested at the Materials Research Center, Aerojet-General Corporation, Sacramento, California. Subsequently, SWAT was also applied during the stress relieving of a test weldment which simulated the nozzle section of a reactor pressure vessel. The latter test was performed at the Standard Oil Company facility in Richmond, California. This memorandum report discusses the results of both tests.

1480 **W. G. Reuter, A. T. Green, C. E. Hartbower, and P. P. Crimmins. "Monitoring of Crack Growth in Ti-6 Al-4 V Alloy by the Stress Wave Analysis Technique." NASA CR-101888. Aerojet–General Corporation, Sacramento, California. Dec 1968. [NTIS: N69-36435].**

The results of investigations to establish stress-wave analysis technique (SWAT) system parameters for the monitoring of thin-walled Ti-6 Al-4 V pressure vessels such as the Apollo SPS tank are presented. Sensor attachment techniques which overcome previous problems encountered when bonding cement was used to attach the sensor to the test article are described. Stress-wave emission characteristics which can be employed as precursors of failure in air, distilled water inhibited with sodium dichromate (500 ppm) and methyl alcohol environments were determined, and a SWAT system possessing sufficient sensitivity to detect and monitor stress-wave emissions associated with subcritical crack growth is discussed. Critical stress-intensity data and relationships between applied stress intensity and failure time for the various material thicknesses (0.030 and 0.060 inches) and environments investigated are presented. Application of the results obtained to the proof testing of critical tankage is discussed and recommendations for follow-on studies are presented.

1481 **W. G. Reuter and C. E. Hartbower. "Stress Corrosion Cracking in an Air-Melted and Vacuum-Arc-Remelted Grade-250 Maraging Steel."** *Engineering Fracture Mechanics*, **EFMEA, 3(4):493-510. Dec 1971; paper presented at the 4th National Symposium on Fracture Mechanics, Carnegie–Mellon University, Pittsburgh, Pennsylvania. Aug 24-26, 1970.**

The stress-wave analysis technique (SWAT) was used to monitor subcritical cracking in two heats of grade-250 18 percent nickel maraging steel exposed to circulating water at 70 and 165 °F. The SWAT system consisted of sensors, amplifiers, filters (to eliminate extraneous noise), and an electronic counter and printer. The material investigated was from an air-melted (induction-stirred, ladle-vacuum degassed) 70-ton heat which was split to provide a second vacuum-arc-remelted heat. The rolling practices employed in producing the air-melt and vacuum-arc-remelt sheet products were the same. The stress-corrosion tests were conducted using fatigue precracked, single-edge notched, tension specimens in conjunction with linear-elastic fracture-mechanics to determine effect of melting practice and long exposure times. In addition to the detailed information on the stress-corrosion crack-growth process made possible by SWAT, there were two noteworthy observations; viz. (1) the vacuum-arc-remelt material was more susceptible to stress-corrosion cracking than the air-melt material, and (2) the mechanism of subcritical cracking was different at the two test temperatures. In 165 °F water, the crack growth process was discontinuous indicating a diffusion-controlled mechanism; whereas, in 70 °F water, there were no crack jumps until the onset of instability. [reprinted by permission of Pergamon Press Ltd from *Engineering Fracture Mechanics*].

1482 **M. C. Reymond. "Emission Acoustique sur Dalle en Beton Arme (Acoustic Emission from a Slab of Reinforced Concrete)." pp 387-398 in** *Journees d'Etudes sur l'Emission Acoustique.* **P. F. Gobin, R. Goutte, and C. Guillaud,**

Organizers. Institut National des Sciences Appliquees de Lyon, Avenue Albert Einstein, 69621 Villeurbanne, Lyon, France. Mar 1975. proceedings of a conference held in Lyon, France. Mar 17, 1975. (paper in French).

It is established that a correlation exists between deformations and micronoises in a concrete slab subjected to flexural stresses. The monitoring of support walls or containment shells of nuclear plants using acoustic emission techniques could be envisaged.

1483 **M. C. Reymond. "Emission Acoustique sur les Roches (Acoustic Emission from Rocks)." pp 361-386 in** *Journees d'Etudes sur l'Emission Acoustique.* **P. F. Gobin, R. Goutte, and C. Guillaud, Organizers. Institut National des Sciences Appliquees de Lyon, Avenue Albert Einstein, 69621 Villeurbanne, Lyon, France. Mar 1975. proceedings of a conference held in Lyon, France. Mar 17, 1975. (paper in French).**

The possible analogy between the behavior of a test bar from a stratified layer of a pit subjected to uniaxial stresses and that of the whole mountain mass was investigated using acoustic emission. Such an analogy exists and could be used for pit monitoring.

1484 **M. C. Reymond. "Etude de la fissuration des eprouvettes sous contraintes en laboratoire et des massifs rocheux en carrieres par une methode d'emission acoustique (Mechanism of Brittle Fracture of Rock during Uniaxial Compression in Laboratory in a Quarry by an Acoustic Emission Technique)."** *Rock Mechanics, Felsmechanik, Mecanique des Roches,* **RMFMA, 7(1):1-16. Mar 1975. (abstract in French, English, and German; article in French).**

Different following stages of brittle fractures of rock were established during experiments in the laboratory by using audible noises. The latter stage is interesting and an analogy was established with premonitory crackles preceding the rupture of rock in quarries. A method for the investigation of the evolution of rocks in quarries was established by technical seismic refraction. Two stages are observed: A sensitive crackle, then die oscillations damp down which displays a stabilisation of rock, except when a falling takes place a few minutes after a crash. This theory is applied in rocks mechanism by means of a detection crackles device of which fundamental mechanics are determined by technical acoustic emission.

1485 **M. C. Reymond and N. Chretien. "Analyse en laboratoire et en carriere du comportement des roches sous sollicitations mecaniques par etude de l'emission acoustique (Laboratory and Field Investigation of the Behavior of Rock under Mechanical Stress by Means of Acoustic Emission)."** *Industrie Minerale-Mine* **(St. Etienne, France), INMNC, Supplement (1):33-40. Aug-Sep 1972. (in French).**

The detection of acoustic emission applied to the laboratory study of granitic rocks under compressive loading permits a display of the different periods of the material's evolution until fracture. The recording of information induced by the change of rock masses after blasting shows that a landslide following the explosion is predictable. The application of this new technique to the supervising of geological sites is considered.

1486 **M. C. Reymond and N. Chretien. "Etude de l'emission acoustique des roches soumises a des sollicitations mecaniques (Study of Acoustic Emission from Rocks under Mechanical Stress)."** *Revue d'Acoustique,* **RACQA, 6(23):333–339. 1973. (in French).**

1487 **W. N. Reynolds. "Some New Developments in Nondestructive Testing."** *Journal of Physics E: Scientific Instruments,* **JPSIA, 6(1):8–13. Jan 1973.**

In recent years, there has been a steady growth of industrial interest in nondestructive testing. The scope of the subject is widening, new methods are being developed, and some earlier ideas have been brought to the hardware stage. One branch, which could be called mechanical-acoustic testing, is achieving new precision through such techniques as ultrasonic holography; and the development of fracture mechanics has made it possible to exploit acoustic emission and internal damping capacity measurements to the full. This is also a result of greater reliability in electronics and transducers, which can now be more accurately characterized. In radiography, several developments of possible interest to those with problems of post-irradiation examination have recently taken place. Successful radiographs using 8 MeV X-rays from a linear accelerator have been obtained with 5-second exposures. Some new fine focus sources suitable for projection radiography are becoming available. New developments in neutron radiography permit the achievement of virtually the same quality as by X-radiography. Other important developments have been made in the use of eddy currents, and for the examination of surfaces and insulating materials, there now exists some expertise in the use of microwaves. Details are given of some of the more significant advances in this wide field.

1488 **D. L. Robinson. "Acoustic Analysis Locates Defects in Buried Pipe."** *Metals Engineering Quarterly,* **MENQA, 15(1):17–21. Feb 1975; paper presented at American Society for Metals Second Annual Materials/Design Forum: "Prevention of Structural Failure: The Role of Quantitative NDE." Port St. Lucie, Florida. Apr 9–11, 1974.**

Nondestructive testing by acoustic analysis has been applied to piping systems, as well as large containment and pressure vessels. Tests have been conducted on both exposed and buried piping systems with good results, both leaks and discontinuities having been detected and located. Techniques involved in detection and location of leaks utilize signal attenuation. Location of discontinuities is accomplished through measurement of signal-arrival times. Structural geometry generally allows straight-line triangulation techniques to be used for source location. Applied to piping systems, acoustic analyses provide an evaluation of the response of the system to imposed dynamic-stress conditions. The system shows considerable promise as a tool for evaluation of either exposed or buried piping systems. [Copyright 1975 by the American Society for Metals].

1489 E. Y. Robinson. "A Basic Model for Acoustic Emission from Fiber-Reinforced Material." JPL Technical Memorandum 33-564. Jet Propulsion Laboratory, Pasadena, California. Sep 1, 1972. [NTIS: NASA CR-128344 or N72-33562].

Acoustic emission from fiber-reinforced composites can often be conveniently interpreted by use of normalized coordinates in graphical data display. Many aspects of the shape of the acoustic emission pattern are invariant with the signal amplification ratio, and the use of normalized coordinates allows simultaneous comparison of AE pattern shapes from different experiments. In this paper, the first-order model of AE from fiber composites, based on filament breaking rates, is cast into a normalized form useful for correlating experimental data. The general features of the normalized model are shown and compare favorably with available data.

1490 E. Y. Robinson. "Acoustic Emission Studies of Large Advanced Composite Rocket Motor Cases." Section 12-D, pp 1-6 in ***28th Annual Technical Conference, Reinforced Plastics/Composites Institute.*** **Society of the Plastics Industry, Inc., Washington, D.C. Feb 1973.**

Acoustic emission patterns were measured during pressure testing of advanced composite rocket motor cases made of boron-epoxy and graphite-epoxy. Both accelerometers and high frequency AE transducers were used, and both frequency spectrum and amplitude distribution were studied. The AE patterns suggest that precursor emission might be used in certain cases to anticipate failure. The technique of hold-cycle AE monitoring was also evaluated and could become a valuable decision gate for test continuation-termination. Data presented show similarity of accelerometers and AE transducer responses despite the different frequency response, and suggest that structural AE phenomena are broadband.

1491 G. C. Robinson, C. R. Reese, and E. A. La Roche, Jr. "Determination of Thermally Induced Acoustic Emissions of Ceramics." ***American Ceramic Society Bulletin,*** **ACSBA, 53(6):482-485, 488. June 1974.**

Equipment and procedures are described for determining the acoustic emission produced by heating and cooling ceramics. The equipment can detect noises arising from grain movement, crystalline inversions, microcracking, and cracking. The equipment is useful in evaluating thermal shock resistance of refractories, designing firing schedules and study of the firing process.

1492 G. C. Robinson, J. E. Smith, R. W. Derby, and G. D. Whitman. "Destructive Testing of Flawed 6-In.-Thick Pressure Vessels." ASME Publication 74-Mat-5. American Society of Mechanical Engineers, New York. 1974; paper presented at the American Society of Mechanical Engineers Pressure Vessels and Piping Conference with Nuclear Engineering and Materials Division, Miami Beach, Florida. June 24-28, 1974.

A series of vessels are being destructively tested under the AEC-sponsored, ORNL-managed Heavy-Section Steel Technology (HSST) Program to evaluate the failure characteristics of thick-walled vessels. The design, material, fabrication, and inspection requirements of these vessels are representative of those of current light-water reactor vessels. The test facility, capable of controlled temperatures from −50 to +350 °F (−45.6 to 176.7 °C) and pressurization output to 50,000 psi (344.8 MN/m^2), is being utilized to produce failure in frangible, transitional, or tough modes. Part circular and semielliptical flaws are machined at various inside and outside locations in base materials and welds and are fatigue-sharpened by a hydraulic pulsing technique to the required depth. Up to 100 channels of strain output are periodically recorded at frequent intervals up to the failure point during each test. Strain-gaging techniques have been developed to permit recording of strains to 8 percent with the gage exposed to 35,000 psi (241.3 MN/m^2) and 200 °F (87.7 °C) water environment. Both acoustic emission and ultrasonic techniques are being used to monitor material behavior in the vicinity of the flaw. As of July 1973, five vessels have been tested with the remainder of the tests scheduled for completion during 1974. Data from the completed tests are reviewed and discussed.

1493 G. S. Robinson. "Methods of Detecting the Formation and Propagation of Microcracks in Concrete." pp 131-145 in ***The Structure of Concrete and Its Behavior under Load: Proceedings of an International Conference, London, September 1965.*** **A. E. Brooks and K. Newman, Editors. Cement and Concrete Association. 52 Grosvenor Gardens, London, England. 1968.**

Audible cracking sounds and numerous surface cracks are almost invariably detected in a concrete specimen before it attains its ultimate strength in uniaxial compression. Many investigators have observed and studied these cracks which occur also when the concrete is tested in biaxial and triaxial compression. To date, studies have shown that these minute cracks have an important influence upon the behavior of concrete and are largely responsible for the failure process. Unfortunately, because they are so small, the cracks, usually referred to as microcracks, cannot easily be detected. As a result, there is little information available on the formation and propagation of microcracks or of their influence upon the strength and elastic properties of concrete. As part of the work of the Concrete Materials Research Group at Imperial College, London, the author initiated a testing program to study microcracks. This paper describes the testing techniques developed to detect and locate microcracks and summarizes the results of preliminary tests made on various mortars and concretes. To study the formation and propagation of microcracks and their influence upon the properties of concrete, three testing techniques were used on various mortar and concrete specimens tested to failure in uniaxial compression. Microcrack propagation was studied with acoustic, pulse-velocity, and X-ray tests on similar specimens from the same mix. In this way, it was possible to correlate the data from the three tests with the strength and elastic properties of the concrete. The effect of varying the proportion of coarse aggregate upon the formation of microcracks was also considered. Sample results are shown in the paper and, on the basis of the entire series of tests, a failure mechanism of concrete is suggested.

1494 G. S. Robinson. "The Failure Mechanism of Concrete with Particular Reference to the Biaxial Compressive Strength." Ph.D. Thesis. London University, London, England. 1964.

1495 R. Roche. "Techniques de Controle non Destructif au Cours de la Fabrication des Enceintes sous Pression en Acier (Nondestructive Testing Procedures during Fabrication of Steel Pressure Vessels)." pp 433–439 in "Non-Destructive Testing for Reactor Core Components and Pressure Vessels." IAEA-145. International Atomic Energy Agency, Vienna, Austria. 1972. report of a panel meeting held in Vienna, Austria. Nov 29–Dec 3, 1971. (in French). [NTIS: CONF-711114].

According to current practice, radiographic examination is performed on all welded joints. This inspection is no longer regarded as sufficient and ultrasonic examination is used in addition in order to detect cracks which are not detected by radiography. It is believed that ultrasonic inspection is the best method when reproducible recordings can be obtained. The advantages and disadvantages of ultrasonic and radiographic inspection are examined. In France, ultrasonic examination of austenitic welds was developed for the first French fast neutron nuclear plant (Phenix). Longitudinal waves, focalized transducers, and automatic testing apparatus were used. This method was successful and specified defects were easily detected. Acoustic emission is a promising crack detection method, but research and development are still necessary, particularly in order to overcome the ambient noise perturbation. In France, research and development have been in progress for two years and a commercially available device has been developed.

1496 J. Rodgers and S. Moore. "The Use of Acoustic Emission for Detection of Active Corrosion and Degraded Adhesive Bonding in Aircraft Structures." Sacramento Air Logistics Center, McClellan AFB, California. 1975.

Procedures have been developed by engineers at Sacramento Air Logistics Center to detect active corrosion in aluminum structure and moisture-degraded adhesive bonding in composite honeycomb using acoustic emission monitoring. Simple heating methods employing a hot-air gun or heat lamp are used to increase emissions from active corrosion sources, and to create the stresses necessary to break moisture-degraded adhesive bonds. Corrosion detection in aluminum skin and honeycomb core has proven successful. Detection of moisture-degraded bonding between aluminum skin and phenolic core on the F-111 vertical stabilizer leading edge has been substantially improved and simplified. Acoustic emission is replacing X-ray and ultrasonic inspection procedures in these applications, with direct benefits realized in a 75% reduction of inspection time and costs, fewer work-flow interruptions, and better defect area definition.

1497 **F. R. Rollins, Jr., "Acoustic Emission from Boron–Aluminum Composites during Tensile Fracture and Fatigue." Technical Report No. 6. Midwest Research Institute, Kansas City, Missouri. Oct 1971. [NTIS: AD-731-710].**

Severe deformation of boron-aluminum composite material produces acoustic emission events that are useful in evaluating the degree of damage and in predicting imminent failure. When boron filaments are fractured, the resultant acoustic emissions are energetic and can usually be detected even in noisy environments such as those associated with fatigue testing. During the fatigue of composites with unidirectionally aligned filaments, it is relatively common to obtain shear cracks which become sources of fretting noise. Such fretting noise is not always an indication of continuing damage and it can obscure acoustic emission events of more interest unless some form of isolation or discrimination is used. One technique for accomplishing this task is discussed.

1498 **V. V. Romanov, S. G. Trubnikova, and V. A. Konstantinov. "Effect of Amplitude-Response Variations on the Nonlinear Properties of Acoustic-Emission Systems."** *Soviet Journal of Nondestructive Testing,* **SJNTA, 12(5):493–496. Sep–Oct 1976; translated from** *Defektoskopiya,* **DEFKA, (5):40–44. Sep–Oct 1976. (in Russian).**

The nonlinear properties of the postdetection channel in acoustic emission systems are considered during the passage of signals having probability distributions that are described by an exponential law and by a Rayleigh law.

1499 **H. E. Romine. "Critical Crack Extension Force and Examination of Weld and Base Metal in Failed First Stage Engine Chamber, American Car and Foundry Corporation, Milton, Pa. Aerojet Serial No. 506338. SP Task Assignment No. 71402 (P.O. No. 790)." NRL Technical Memo No. 74, NPG Technical Memo No. T-11/59. U.S. Naval Proving Ground, Warhead and Terminal Ballistics Laboratory, Dahlgren, Virginia. Apr 1959. [available through interlibrary loan from Director, Naval Research Laboratory, Code-2627, Washington, D.C. 20375. reference report as code 264, 616].**

By using a contact microphone on the G_c specimen, it was possible to listen through an audio system to the sounds made as the fracture enlarged slowly from the central notch to the critical length at which fast fracture occurred and to note the corresponding loads on the specimen. Two transverse tests gave good sound emission results. Sound studies of these G_c slow cracks are just beginning and the above values are presented merely to show that a crack will probably creep at a stress considerably below that at which fast fracture occurs. The loading rate was about 10,000 pounds per minute. The extent to which a crack will creep under a constant subcritical load needs investigation. Hydrotest failures have occurred after holding times at test pressure.

1500 **H. E. Romine. "Determination of the Driving Force for Crack Initiation from Acoustic Records of G_c Tests on High-Strength Metals for Rocket Motor Casings." NWL Report No. 1779. U.S. Naval Weapons Laboratory, Dahlgren, Virginia. Oct 4, 1961. [NTIS: AD-263-911].**

Fracture instability tests with a double-inking procedure were employed with the center-notched G_c test specimen to establish conditions at crack initiation. With this information, an acoustic method was used to determine the driving force G^*_{Nc} required to initiate cracking from a machined notch. The audio instrumentation can be added to the G_c test without affecting standard procedure. Tests with reduced length of starting slot indicated that G^*_{Nc} for a given type of notch is a relatively constant material property similar to the critical $G_c{}^*$ value for fast fracture. The G^*_{Nc} value can be used in fracture toughness equations for the motor case to predict a slot size at which cracking should just initiate in a single cycle of stressing the material to yield strength. The general equivalence of K^*_{Nc} to K_{Ic} was an indication of adequate notch sharpness.

1501 **H. E. Romine. "Tape Recording of Microsound Accompanying the Propagation of Slow Fracture in the G_c Test and Discussion of Fracture Sound in Relation to Hydrotest of Rocket Casings. S. P. Task Assignment No. 71402 (P.O. No. 970)." NRL Technical Memo No. 76, NPG Technical Memo No. T-13/59. U. S. Naval Proving Ground, Warhead and Terminal Ballistics Laboratory, Dahlgren, Virginia. Apr 1959. [available through interlibrary loan from Director, Naval Research Laboratory, Code-2627, Washington, D.C. 20375. reference report as Code 264-624].**

The report contains a brief statement of background for using a contact microphone to monitor rocket motor engine chambers during hydrotest to detect developing cracks prior to rupture. The author describes the fracture sound as a sudden local release of stress that occurs when an increment of crack is formed by layers of atoms separating beyond their spheres of mutual atomic attraction in a stress field acting across the path of fracture. A description of the instrumentation is given. Also included with the report is a reel of magnetic recording tape on which are recorded the emission from fracture in steel and titanium tensile specimens.

1502 **H. E. Romine. "The Acoustics of Straining and Fracture in the Notched Sheet Specimen." pp 10-13 in *Symposium on the Testing and Evaluation of Materials for Solid Propellant Rocket Motor Casings.* Report MAB-156-M. Division of Engineering and Industrial Research, National Academy of Sciences, National Research Council, Washington, D.C. Nov 15, 1959. proceedings of a symposium sponsored by the Materials Advisory Board, held in Washington, D.C. Feb 2-3, 1959. [NTIS: AD-230-340].**

The paper discusses the use of a contact microphone and tape recorder for recording sound waves emitted during fracture test on high-strength steel and Ti alloy rocket casing materials. Possibilities of use of sound wave characteristics to locate and identify flaws is also discussed.

1503 D. M. Romrell. "Acoustic Emission Monitoring of the FFTF Reactor Fuel Pin End Closure Welds." HEDL-SA-381S. Hanford Engineering Development Laboratory, Richland, Washington. Feb 22, 1972; *Materials Evaluation*, MAEVA, 30(12):254–258. Dec 1972.

Acoustic emission is defined as the elastic waves produced in solid materials by an energy release as the material is deformed or fractured. These elastic waves travel through the solid material at the speed of sound and can be detected at the surface with sensitive acoustic emission sensors. This provides a basis for a technique of monitoring metal structures during welding processes to detect the formation of anomalies such as micro-cracks. On many structures, acoustic emission monitoring can provide both quantitative and flaw-location data. Acoustic emission monitoring augments other nondestructive testing methods and is sometimes applicable when other tests cannot be conducted. This is, in part, due to the high sensitivity of acoustic emission monitoring. Acoustic emission monitoring is only sensitive to active flaw-growth, however, and will not detect a flaw in equilibrium. This paper describes the application of acoustic emission monitoring to end closure welds on the Fast Flux Test Facility reactor fuel pins. It illustrates the correlation that has been made between destructive metallurgical microscopy of welds and the emissions that were recorded during the welding processes.

1504 D. M. Romrell. "Acoustic Emission Weld Monitoring." HEDL-SA-611. Hanford Engineering Development Laboratory, Richland, Washington. 1973. [NTIS].

Acoustic emission monitoring techniques are utilized to detect the elastic pressure waves that are produced as energy is released in solid materials during deforming or fracturing processes. The detected pressure waves are converted to electronic signals that are processed to provide an indication of the rate and extent of material degradation. By utilizing sophisticated logic circuitry and a small digital computer, additional processing of the output data can be performed to greatly enhance AE methodologies. This includes deriving the emission source location data on a statistical basis and providing a real-time video display of the emission source locations. The source location spectrum is of particular interest when monitoring welds that are being fabricated in larger structures such as large diameter piping or plate sections. In these applications, after suspected crack areas are located, other NDT methods can be utilized and the necessary repairs can be made. Because some emissions occur during most welding processes, that are probably due to the welding arc or stress releases during metal solidification and cooling process, the emission source location spectrum is valuable for indicating areas of unusually high emission activity. AE monitoring can be sensitive to micro-crack formation at levels that cannot be detected by other NDT methods. This often makes data verification difficult. Also, the sensitivity level adjustment can be critical during weld monitoring and should be determined for each welding process. This presentation describes and illustrates the results achieved by utilizing mini-computers in AE weld monitoring systems.

1505 D. M. Romrell. "Acoustic Emission Weld Monitoring of Nuclear Components." HEDL-SA-354S. Hanford Engineering Development Laboratory, Richland, Washington. Jan 25, 1972; *Welding Journal*, WEJUA, 52(2):81s–87s. Feb 1973.

Acoustic emission is defined as the elastic waves produced in solid materials by an energy release as the material is deformed or fractured. These elastic waves travel through the solid material at the speed of sound and can be detected at the surface with sensitive acoustic emission sensors. This provides a basis for a technique of monitoring metal structures during welding processes to detect the formation of anomalies such as microcracks. On many structures, acoustic emission monitoring can provide both quantitative and flaw-location data. Acoustic emission monitoring augments other nondestructive testing methods and is sometimes applicable when other tests cannot be applied. This is, in part, due to the high sensitivity of acoustic emission monitoring. Acoustic emission monitoring is only sensitive to active flaw-growth, however, and will not detect a flaw in equilibrium. This paper describes the application of acoustic emission monitoring to nuclear reactor fuel pin end closure welds and other weldments of the reactor piping.

1506 D. M. Romrell and L. R. Bunnell. "Acoustic Emission Monitors Crack Growth in Ceramic." BNWL-SA-3064. Battelle-Northwest, Richland, Washington. 1970; paper presented at the American Society for Nondestructive Testing National Spring Conference, Los Angeles, California. Mar 9–12, 1970, [see Reference 1508 for publication].

Discussion of the application of acoustic emission monitoring techniques to detect cracking in ceramics. The acoustic emission system is described, and specimen material is discussed, together with some experiments conducted to (1) detect the onset of cracking before a catastrophic failure occurs; and (2) monitor the growth of existing cracks. The experiments, although of a preliminary nature, indicate the feasibility of many possible applications using acoustic emission techniques to monitor ceramic materials.

1507 D. M. Romrell and L. R. Bunnell. "Acoustic Emissions Let You Listen for Defects." *Ceramic Age*, CEAGA, 86(11):17, 18, 20. Nov 1970.

At Battelle-Northwest, during thermal shock testing of ThO_2-Y_2O_3 tubes in sodium, acoustic emissions were readily detected during both early crack growth and catastrophic failure. Cracked tubes were easily distinguishable from perfect tubes when both were monitored during thermal cycling. In other experiments, Al_2O_3 surface damaged by conventional diamond sawing emitted strongly during immersion in 90 °C water; similar samples annealed at 400 °C for one hour did not produce emissions. The sensitivity of the method suggests several uses including: (1) in-service or simulated service monitoring of ceramics or ceramic-coated materials to detect failures caused by cracking or unbonding of the ceramic; (2) thermal shock test monitoring of many ceramics, from technical ceramics to refractories, to detect cracks as they form, permitting easy correlation with temperature cycling; and (3) crack formation propagation in brittle materials as an analytical tool. The real-time detection could greatly aid in understanding the fundamental processes involved in failure of ceramics.

1508 D. M. Romrell and L. R. Bunnell. "Monitoring of Crack Growth in Ceramic by Acoustic Emission." ***Materials Evaluation*****, MAEVA, 28(12):267–270, 276. Dec 1970.**

Acoustic emission is defined as the elastic wave produced in a solid material by energy released as the material deforms or fractures. These waves are detectable at the surface of the material using high sensitivity ultrasonic sensors, thus providing a basis for monitoring flaw growth in solids. Furthermore, incipient failure of the solid material is detectable using the acoustic emission generated. Previously, integrity surveillance using acoustic emission technology was primarily related to metals; this paper describes an extension of the technique to ceramics. Application of this wave phenomenon to ceramic materials has shown that (1) acoustic emission waves are readily detected both during early stages of crack growth and catastrophic failure, and (2) waves are correlatable with crack growth resulting from thermal shock. The work has shown the acoustic emission technique to be a useful tool for studying the effects of thermal shock in ceramic materials.

1509 J. L. Rose. "Panel 1 Report on: Flaws and Defects in Structural Materials." pp 411–430 in "Proceedings of the Interdisciplinary Workshop on Nondestructive Testing–Materials Characterization." D. O. Thompson, Program Manager. AFML-TR-73-69. North American Rockwell Science Center, Thousand Oaks, California. Apr 1973. [NTIS: AD-760-174]. Workshop held at North American Rockwell Science Center, Thousand Oaks, California. Nov 13–15, 1972.

1510 J. L. Rose and P. A. Meyer. "Ultrasonic Procedures for the Determination of Bond Strength." AFOSR-TR-74-0779. Drexel University, Department of Mechanical Engineering and Mechanics, Philadelphia, Pennsylvania. June 1974. [NTIS: AD-780-128].

Theoretical wave-propagation models that treat ultrasonic wave interaction with an adhesively bonded system have been developed. These models allow the selection of appropriate ultrasonic transducers for bond inspection analysis. Such problems as a variation in bondline thickness, the presence of density gradients in the adhesive bond, and improper surface preparation are treated in detail. Preliminary results indicate that the potential for success appears high for obtaining a correlation between a bond performance parameter and some specific ultrasonic test parameter.

1511 **A. Rotem and J. Baruch. "Determining the Load-Time History of Fibre Composite Materials by Acoustic Emission." MED Report No. 44. Israel Institute of Technology, Haifa, Israel. Mar 1974. [NTIS: AD-780-122];** *Journal of Materials Science,* **JMTSA, 9(11):1789-1796. Nov 1974.**

Acoustic emission was monitored during the axial loading of unidirectional fiber-composite tensile specimens. The material consisted of strong, brittle fibers (E glass) embedded in a viscoelastic matrix (epoxy). It was found that when the load was held constant the acoustic emission output continued, but at a decreasing rate with time at load. As the load level was increased, the acoustic emission output at load continued for a longer period. It is suggested that the acoustic emission under constant load is a result of fiber fracture which continues after loading ceases because of the viscoelastic nature of the matrix which allows stress redistribution with time. The experimental results from acoustic emission are compared with computer calculations for fiber fracture based on theoretical considerations. Good agreement is noted between the theoretical and experimental results. [reprinted by permission of Chapman & Hall Ltd].

1512 **R. L. Rothman. "Acoustic Emissions in Rock Stressed to Failure." pp 109-133 in** *Proceedings First Conference on Acoustic Emission/Microseismic Activity in Geologic Structures and Materials.* **H. R. Hardy, Jr. and F. W. Leighton, Editors. Trans Tech Publications, Clausthal, West Germany. 1977. conference held at Pennsylvania State University, University Park, Pennsylvania. June 9-11, 1975.**

The paper presents results of recent laboratory experiments. Cylindrical specimens of Crab Orchard (Tennessee) sandstone were incrementally stressed to failure under uniaxial load. At each loading stage the outputs of piezoelectric transducers mounted on the specimens' surface were recorded. The acoustic emissions were analyzed, using techniques similar to those employed in earthquake seismology, to determine the hypocenters of the events and their radiation patterns. Seventy-two of the events which occurred just prior to failure were randomly selected for analysis. It was found that these events originated near the central axis of the cylinder and had P-wave first motions indicative of tension cracks oriented parallel to the direction of applied compressive stress.

1513 **R. Rothwell and M. Arrington. "Acoustic Emission and Micromechanical Debond Testing."** *Nature Physical Science*, **NPSCA, 233(43):163-164. Oct 25, 1971.**

The mechanical properties of composites are determined to a great extent by the nature of the fibre-matrix interface. A high interfacial bond strength usually results in a brittle high-strength composite, but a lower interfacial bond strength may reduce the ultimate strength of the composite although greatly enhancing its fracture toughness. Only the tensile bond test specimen will be considered in this communication. A compressive axial load on the specimen results in a radial expansion in accordance with Poisson's ratio. Poisson's ratio of the polymer is greater than that of the glass, so it expands transversely more than the glass and interfacial stress is created normal to the interface to preserve continuity. At some critical stress(s), debonding will occur. Using an acoustic emission detection system in conjunction with the test piece, it was demonstrated that the onset of debonding can be accurately and reproducibly detected and recorded.

1514 **D. Rouby, P. Fleischmann, J. C. Baboux, and F. Lakestani. "Spectral Analysis of Acoustic Emission during Plastic Deformation in Aluminium/Densite spectrale energetique de l'emission acoustique lors de la deformation plastique dans l'aluminium." Institut National des Sciences Appliquees de Lyon, 69621 Villeurbanne, Lyon, France; Paper 3K/6 in** ***Proceedings of the Eighth World Conference on Nondestructive Testing.*** **volume containing Sections 3K1, 3H1, 3G. International Committee for Nondestructive Testing, Secretariat du Confrend, Institut de Soudure, 32 Boulevard de la Chapelle, 75880 Paris Cedex 18, France. 1976. conference held in Cannes, France. Sep 6–10, 1976. (paper in French).**

Discussed is the study of the influence of plastic strain and grain size on spectral characteristics of acoustic emission related to dislocation motion.

1515 **J. Ruge and V. Jurgens. "Einsatz der Schallemission zur Beurteilung von Schweissverbindungen an niedriglegierten Baustahlen (Reponse of Acoustic Emission in the Evaluation of Weld Joints in Low Alloy Structural Steels)." pp 310–321 in** ***Schallemission: Anwendung bei der Untersuchung, Prufung und Uberwachung metallischer Werkstoffe.*** **Deutsche Gesellschaft fur Metallkunde eV, 637 Oberursel/Taunus bei Frankfurt/Main, Adenauerallee 21, Germany. 1974. proceedings of symposium held in Munich, Germany. Apr 18–19, 1974. (paper in German).**

1516 **J. Ruge and V. Jurgens. "Einsatz der Schallemissionsanalyse zur Beurteilung der Qualitat von Schmelzschweissverbindungen an niedriglegierten, hochfesten Baustahlen (Use of Acoustic Emission Analysis to Evaluate the Quality of Fusion-Welded Joints in Low-Alloy High-Strength Structural Steels)."** ***Schweisstechnik*** **(Berlin), SCTCA, 25(9):410–411. Sep 1975. (in German).**

The principles of acoustic-emission analysis are outlined and its use as a method of nondestructive testing is described. Acoustic emission gives more information regarding crack initiation and propagation than can be obtained with customary NDT methods, and in particular allows the dynamic processes taking place during cracking to be followed. Examples are given of the use of the method to follow delayed H-induced cracking in a weld, and to follow cracking during the welding of steel by fully utilized if interfering frequencies are filtered out. Future applications are indicated. [Abstract 22-0555. *Metals Abstracts,* Vol. 9, 1976].

1517 J. Ruge and V. Jurgens. "The Analysis by Acoustic Emission of Delayed Cracking." IIW-IX-744-71, IIW-X-624-71. International Institute of Welding, Miami, Florida. 1971.

From numerous cases of loss, the phenomenon of delayed cracking is known. They are brittle fractures occurring all of a sudden in the heat-affected zone or in the weld some time after the welding process has been completed, without any failures having previously been noticeable. The causes of this phenomenon are manifold and so far they have not yet completely been cleared up. However, it seems to be sure that the damages are attributable to hydrogen. The object of the present investigation is to study the period between the completion of the welding process and the beginning of crack formation in order to render it possible to draw conclusions with respect to the reactions taking place in the test specimen, with a view to more fully understand the mechanism of the delayed cracks. For studying delayed fractures in the laboratory, a cruciform test specimen was used and was biaxially loaded with two hydraulic cylinders in a special loading fixture.

1518 J. Ruge and V. Jurgens. "Untersuchungen uber den verzogerten Bruch beim Schweissen niedriglegierter Stahle (Research on Delayed Cracking in Welding Low Alloy Steel)." ***Schweisstechnik*** **(Vienna), SWTEA, 26:225-229. Oct 1972. (in German).**

1519 W. D. Rummel and F. B. Cooke. "Acoustic Emission as a Nondestructive Testing Tool." Martin Marietta Corporation, Denver, Colorado. 1970; paper presented at the American Society for Nondestructive Testing National Fall Conference, Cleveland, Ohio. Oct 19-22, 1970. [paper never submitted for publication].

Acoustic emission is one of the most promising new nondestructive testing tools for improved discontinuity detection and for cost savings in quality assurance of pressure vessels. A preliminary program to evaluate acoustic emission as a tool to be used in conjunction with *proof pressure testing* to fracture mechanics criteria is described. Experimental results of detecting slow crack growth in flawed titanium and aluminum specimens are presented. Acoustic emission is concluded to be a material property which may be used in conjunction with fracture mechanics technology to provide maximum inspectability of high performance pressure vessels. The results of a literature survey on the technique are abstracted and indicate considerable effort in reducing acoustic emission technology to practice.

1520 W. D. Rummel, P. H. Todd, Jr., S. A. Frecska, and R. A. Rathke. "The Detection of Fatigue Cracks by Nondestructive Testing Methods." NASA CR-2369, MCR-73-145. Martin Marietta Aerospace, Denver, Colorado. Feb 1974. [NTIS: N74-17285].

The X-radiographic, penetrant, ultrasonic, eddy current, holographic, and acoustic emission techniques were optimized and applied to the evaluation of 2219-T87 aluminum alloy test specimens. One hundred eighteen specimens containing a total of 328 fatigue cracks were evaluated. The cracks ranged in length from 0.500-inch (1.27 cm) to 0.007-inch (0.018 cm), and in depth from 0.178 inches (0.451 cm) and 0.001 inches (0.003 cm). Specimen thicknesses were nominally 0.060 inches (0.152 cm) and 0.210 inches (0.532 cm), and surface finishes were nominally 32 and 125 rms and 64 and 200 rms, respectively. Specimens were evaluated in the *as-milled* surface condition, in the chemically milled surface conditon and, after proof loading, in a randomized inspection sequence. Results of the nondestructive test evaluations were compared with actual crack size obtained by measurement of the fractured specimens. Inspection data were then analyzed to provide a statistical basis for determining the threshold crack-detection sensitivity (the largest crack size that would be missed) for each of the inspection techniques at a 95% probability and 95% confidence level.

1521 R. J. Runck. "Nondestructive Testing of Beryllium." MCIC-74-22. Metals and Ceramics Information Center, Battelle Columbus Laboratories, Columbus, Ohio. July 1974. [NTIS: AD-784-942].

Because beryllium is extremely expensive to produce in comparison with other metals, it is most necessary that quality assessment be performed on this material by nondestructive methods prior to costly machining and other processing operations. Beryllium, however, has unique properties, that make several of the conventional nondestructive tests especially difficult. Among these are its high modulus of elasticity, high velocity of sound transmission, and high transparency to X-rays. The high velocity of sound in beryllium reduces the separation of reflected signals and therefore special procedures are required to interpret ultrasonic tests. The high transparency to X-rays necessitates the use of special long-wave (low voltage) X-rays to produce useful contrasts in exposures of the metal. The purpose of this report is not only to identify the nondestructive tests that are useful in evaluating beryllium, but also to describe the special modifications in some of these tests that are necessary for their application to beryllium. While the primary purpose of the report is to provide a convenient collection of information on nondestructive testing techniques being applied to beryllium, some general descriptions also have been included in the Appendix to provide a convenient description of some of the terms used. Radiography with X-rays is the *workhorse* as a method of nondestructively testing beryllium. It is especially capable of revealing inclusions and it can detect voids in moderately thick sections. Ultrasonic and eddy-current tests can be used to supplement liquid-penetrant tests for identification of discontinuities near the surface. Ultrasonic tests also are useful for inspecting for voids when the section thicknesses are too great for effective radiography. Acoustic emission is a relatively new nondestructive testing technique that has proven to be useful for evaluating beryllium under load, and more extensive use of this technique is likely in the future. Holographic interferometry has been used to detect strain, and thereby identify debonds, in beryllium structures. However, the method is not especially effective with materials having high modulus of elasticity, such as beryllium, because of the high stresses necessary to produce appreciable strain.

1522 **H. Rusch. "Physikalische Fragen der Betonprufung (Physical Problems in the Testing of Concrete)."** *Zement-Kalk-Gips,* **ZMKGA, 12(1):1-9. Jan 1959. (in German); C&CA Library Translation Cj 86. Translated by Cement and Concrete Association, 52 Grosvenor Gardens, London, England. Mar 1960.**

Some preliminary results of basic physical research on the deformation of concrete under load are reported. Investigations into the time behavior of the stress-strain diagram of concrete show that the strength under sustained load is 75 per cent of the usual strength obtained under short-time application of load. From measurements of the change of volume of concrete, of the absorption of ultrasonic waves, and of the noise produced in the course of loading, it appears that in the usual short-time test considerable disruption of the internal structure of the concrete occurs already at about 75 per cent of the failure load. The *creep* of concrete can be split up into the actual permanent strain, or *delayed plasticity,* and the delayed elastic strain. Tests carried out under well-defined conditions of atmospheric humidity on small cylindrical specimens made of *pure* cement show that the delayed plastic strain and the delayed elastic strain increase with increasing water content of the specimen. For a load duration of up to 2 hours, the delayed elastic strain is greater than the delayed plastic strain.

S

1523 V. E. Sackschewsky and L. A. Lott. "Acoustic Emission Source Location on a Nonspherical Surface of Revolution." RFP-2090. Dow Chemical U.S.A., Rocky Flats Division, Golden, Colorado. Nov 19, 1973. [NTIS].

Described is a procedure to calculate the source location of an acoustic emission signal generated on a nonspherical surface, based on the arrival times of the signal at five transducers. The process assumes a smooth, closed surface of revolution, symmetric with respect to the equator and such that the radial distance, measured perpendicularly from the polar axis, is a strictly increasing function of the polar angle ϕ for $0° \leqslant \phi \leqslant 90°$. Listings and sample runs of the computer programs developed are included. The accuracy of the procedure was tested using simulated timing data for a sphere. The average distance between the true and calculated source location was 0.05% of the radius. The cycle time required to process one set of timing data is estimated to be 10 to 30 milliseconds, depending on which version of the program is used.

1524 M. A. Saifi and S. J. Vahaviolos (to Western Electric Company). "Method and Apparatus for Evaluating Welds Using Stress-Wave Emission Techniques." U.S. Patent 4,007,631. Feb 15, 1977.

Spot welds are evaluated using stress-wave emission techniques by measuring the stress-wave energy emitted from the weld area during n time intervals of the weld cycle, where $n \geqslant 4$. Each of the time intervals corresponds to a different aspect of the weld cycle such as, for example, the initiation of heating, the separate solid-to-liquid phase transformation in each of the articles being welded, material combination within the weld nugget, cooling stresses occurring during resolidification, and post-weld cracking. The energy values obtained for each of the time intervals can be compared with predetermined acceptable energy value ranges for corresponding ones of the intervals or compared with predetermined ratio values between two or more of the obtained interval values to determine the quality of a weld.

1525 C. St. John and W. W. Gerberich. "The Effect of Loading Mode on Hydrogen Embrittlement." ***Metallurgical Transactions,*** **MTGTB, 4:589–594. Feb 1973.**

Hydrogen embrittlement is shown to occur easily in notched-round bars under opening Mode I (tension), but not under antiplane shear Mode III (torsion). The stress tensor invariants under Mode I, II, and III loadings and how these affect interstitial diffusion are discussed. It is suggested that long-range diffusion of hydrogen, down orthogonal trajectories to the vicinity of the crack tip, which can occur under Mode I but not Mode III, is a key part of any hydrogen embrittlement mechanism. This premise was evaluated with AISI 4340 steel heat-treated to ultrahigh strength levels. It was found that an initial Mode I stress intensity level of 17,000 psi-inch$^{1/2}$ produced failure in several minutes. Mode III stress intensity levels three times this produced no crack initiation in 300 minutes. Further analysis of the time-dependent hydrogen concentrating effect utilize a stress-wave emission technique. This produced plausible critical hydrogen concentrations even though the present elastic analysis is a first order approximation of the stress field. [Copyright 1973 by the American Society for Metals and the Metallurgical Society of AIME].

1526 **J. Samman. "Acoustic Emission: A Tool to Study Some Metallurgical Behaviours of Welding Structures/Utilisation de l'emission acoustique pour etudier certains comportements metallurgiques des assemblages soudes." Electricite de France, Saint-Denis, France; Paper 3K/16 in** *Proceedings of the Eighth World Conference on Nondestructive Testing.* **volume containing Sections 3K2, 3H2, 5A. International Committee for Nondestructive Testing, Secretariat du Cofrend, Institut de Soudure, 32 Boulevard de la Chapelle, 75880 Paris Cedex 18, France. 1976. conference held in Cannes, France. Sep 6-10, 1976. (paper in French). [NTIS: FRNC-CONF-172].**

Metallurgical behavior such as delayed cracking, lamellar tearing, and stress releaving by heat treatment were studied by recording acoustic emission signatures during comparative tests.

1527 **N. G. Sankar. "Unload Emission Behavior of Materials and Its Relation to the Bauschinger Effect." Ph.D. Thesis. University of Michigan, Ann Arbor, Michigan. 1969; University Microfilms 70-14631.**

Materials exhibit considerable acoustic emission when loaded in tension, and a correspondingly smaller amount of acoustic emission when a tensile load is removed. The purpose of this investigation was to study the unload emission behavior of certain materials with various heat-treatments and to determine if a correlation exists between the Bauschinger effect and the acoustic emission during unloading. The materials tested were 99.99% aluminum, 6061 and 2024 aluminum alloys, copper-7.9% aluminum alloy, 70-30 brass and 99.95% magnesium. A piezoelectric transducer was used to detect the acoustic emission. The transducer was placed on the face of the test section of the flat specimen. The unload emission from all the materials tested was of the burst type. When the applied stress was increased, the unload emission increased, reached a peak value, and then decreased. Among the fcc materials tested, 99.99% aluminum showed the least unload emission and copper-7.9% aluminum alloy, the largest unload emission for the same plastic strain. A 99.95% magnesium showed even higher unload emission than a copper-7.9% aluminum alloy. Among the fcc materials tested, only copper-7.9% aluminum and 70-30 brass showed significant acoustic emission on repeated loading. Bauschinger tests were conducted on materials having the same chemical compositon as those used in the acoustic emission study. Copper-7.9% aluminum alloy showed a large Bauschinger effect and 99.99% aluminum showed the least effect. The following conclusions are drawn from the test results. Unload emission is the result of reverse plastic deformation. Materials with high-stacking fault energies, such as 99.99% aluminum, give low acoustic emission during unloading. This is due to an increasing tendency for cross slipping in such materials. On the other hand, planar glide materials, such as copper-7.9% aluminum alloy, with low-stacking fault energy, show considerable acoustic emission during unloading. The unload emission in large grain size samples is larger, if other conditions remain the same. Materials that show large acoustic emission during unloading show acoustic emission during repeated loading, even before reaching the maximum stress level of the first loading. It is likely that the dislocations are confined to primary slip planes and, hence, can move back and forth during unloading and repeated loading. There is a good correlation between the Bauschinger effect and the unload emission. Materials that show a large Bauschinger effect show a large unload emission.

1528 **N. G. Sankar, J. R. Frederick, and D. K. Felbeck. "Acoustic Emission from Metals during Unloading and Its Relation to the Bauschinger Effect."** *Metallurgical Transactions*, **MTGTB, 1:2979–2980. Oct 1970.**

The work reported is associated with an effort to find an explanation for the unload emission. Tests were carried out under conditions such that only dislocation motion or twinning would be expected. The results of the work have shown that the unload emission from several annealed metals correlates with the magnitude of the Bauschinger effect in these metals. Hence, it appears that a plausible explanation for the presence of unload emission is that it is the result of the abrupt spring-back of piled-up dislocations when the load on a specimen is removed.

1529 **K. Sano, T. Imanaka, T. Fuakoshi and K. Fujimoto. "The Kinetics of Martensitic Transformation in Steels Studied with Acoustic Emission." Paper 4–1, pp 251–270 in** *The Third Acoustic Emission Symposium.* **Japan Industrial Planning Association, Dobashi Building, 3–17–8, Akasaka, Minato-ku, Tokyo, Japan. 1976. proceedings of a symposium held in Tokyo, Japan. Sep 16–18, 1976.**

The martensitic transformation in Fe-Ni-C and Fe-Ni-Mn-C alloys with different constitution has been studied with acoustic emission. Both amplitude and number of acoustic emission pulses vary widely with the change in the kinetics or morphology (or both) of martensite. The amplitude distribution of acoustic emission which has been determined using a specially designed amplitude distribution analyzer obeys the Ishimoto-Iida statistical relation regardless of the kinetics of transformation and the morphology of martensite. The parameters in this relation are shown to be correlated to the metallographic observations.

1530 **K. Sano, T. Imanaka, T. Funakoshi, and K. Fujimoto. "Acoustic Emission during Deformation of Tempered Martensitic Steels." Session 5, pp 23–45 in** *The Second Acoustic Emission Symposium.* **Japan Industrial Planning Association, Dobashi Building 5F, 3–17–8 Akasaka, Minato-ku, Tokyo, 107 Japan. 1974. proceedings of a symposium held in Tokyo, Japan. Sep 2–4, 1974.**

Acoustic emission during deformation was studied using vacuum melted 0.3% C, 0.3% C-1.5% Si, and 0.3% C-1% V steels which were quenched and tempered at various temperatures. A peak in AE was observed at about 0.05% strain in steels tempered at low temperatures; this peak was associated with the precipitation of epsilon carbides. Other features of AE were found to depend sensitively on the microstructures observed through an electron microscope. Change in internal friction with plastic deformation was investigated to clarify the dislocation mechanism of AE. A term of *Acoustic Yielding* was introduced to describe an initial stage of deformation. The effect of strain rate on the activity of AE was also discussed. It was concluded that the rate of energy dissipation as AE for tempered martensitic steel was proportional to the strain rate.

1531 **W. H. Sansom. "Non-Destructive Testing. Part VIII: Comment and Conclusion."** ***Welder,*** **WLDRA, 140(196):21-23. 1974.**

1532 **I. Satoh, S. Sasaki, S. Kirihara, T. Yoshioka, and Y. Fukuda. "Some Techniques for Signal Discrimination in Acoustic Emission Testing." Paper 3-2, pp 194-211 in** ***The Third Acoustic Emission Symposium.*** **Japan Industrial Planning Association, Dobashi Building, 3-17-8, Akasaka, Minato-ku, Tokyo, Japan. 1976. proceedings of a symposium held in Tokyo, Japan. Sep 16-18, 1976.**

In practical applications of acoustic emission to integrity assessment of structures or materials, signal descrimination is the most important problem since various sources of acoustic noise which interrupt the detection of AE lie in the object and the environment. For instance, noises such as those which arise at the specimen supports in a tensile test obstruct the obtaining of the true acoustic emission characteristics of the material. In monitoring for weld cracking, it is necessary to discriminate the acoustic emission signals which are caused by weld cracking from those which are caused by slag cracking. In the case of a fatigue test, it is helpful if the detected signals are restricted to those that originate within a limited region of the object under observation. For the purpose of on-stream inspection of large structures, simple, low-cost monitoring systems are needed instead of complex, expensive multichannel systems. This paper describes the following techniques, which are designed to meet the above requirements, and some examples of their applications: (1) Spatial Filtering technique including its applications to strength test; and (2) Frequency Filtering technique which is useful in selecting acoustic emission signals in weld crack monitoring test. Two techniques were developed as the means of spatial filtering, one is named the *Analog Spatial Gating technique* and the other the *Zone Location technique.* Using the Analog Spatial Gating technique, it is possible to restrict the received signals to those which take place within a limited region of the object under observation. In tensile and fatigue test, the Analog Spatial Gating technique proved effective in distinguishing the signals from noises such as those which arise at the specimen supports. The Zone Location technique is used in a multichannel acoustic emission monitoring system for the purpose of restricting the acoustic emission signals to those which arise within each appointed zone among the number of object zones. This technique is of use for continuous monitoring of large structures such as pressure vessels and pipelines. The Frequency Filtering technique is used to distinguish the acoustic emission signals caused by delayed cracking and slag cracking in TIG welding.

1533 **I. Satoh, S. Sasaki, and I. Masaoka. "Observation of Acoustic Emission in Tensile Test Using Source Location Technique." Session 3, pp 30–52 in** *The Second Acoustic Emission Symposium.* **Japan Industrial Planning Association, Dobashi Building 5F, 3-17-8 Akasaka, Minato-ku, Tokyo, 107 Japan. 1974. proceedings of a symposium held in Tokyo, Japan. Sep 2–4, 1974.**

The experimental studies of acoustic emission in the tensile tests described, were performed using an acoustic emission source location technique, since this technique was considered useful for distinguishing the signals from noises as well as for observing the local yields or the stress concentrations of the test materials. A number of thin plates and a wide welded plate of mild steel were prepared as the test materials. In the case of testing the thin-plate specimens, acoustic emission detection equipment provided with two channel receivers and a one-dimensional source location system was employed. In order to observe the locations of acoustic emission events with respect to the progress of Luders band in the materials, this equipment was provided with a storage type CRT monitor on which acoustic emission signals can be displayed in relation to elapsed experiment time as the *dynamic acoustic emission pattern.* For the wide-plate tensile test, the specimen had a weld zone which included some lamellar tearing. The acoustic emission detection equipment was provided with four-channel receivers and a two-dimensional source location system. The objective of the wide tensile test was to observe the behavior of acoustic emission in relation to the process of slow-crack growth. This paper describes the acoustic emission source-location technique in tensile tests and the experimental results obtained by the technique.

1534 **J. C. Savage and H. S. Hasegawa. "A Two-Dimensional Model Study of the Directivity Function."** *Bulletin of the Seismological Society of America,* **BSSAA, 55(1):27–45. Feb 1965.**

The directivity method of Ben-Menahem has been applied to the study of Rayleigh waves generated by an internally induced tensile fracture in a glass plate. Independent studies show that velocity of fracture propagation in glass initially increases over a short distance before a terminal rupture velocity is attained. The directivity function has been calculated for such a fracture in terms of four parameters: the length a and duration T_1 of the accelerated phase of fracture, the terminal rupture velocity v, and the total length of the fracture b. Theoretical directivity functions have been chosen to fit observed directivity functions in five cases. Although the theoretical functions do not furnish a detailed fit to the observed directivity functions, they do reproduce most of the essential features. Independent measurements of b and v indicate that those parameters may be determined by the directivity method to within about 15 per cent.

Values of a and T_1 obtained by the directivity method cannot be checked directly but appear to be reasonable. It is suggested that the discrepancy between the lengths obtained from the phase and from the modulus of the directivity function for the Chilean earthquake of May 1960, may be associated with an increase of velocity of rupture over perhaps the first 100 km of the total fault length.

1535 J. C. Savage and L. Mansinha. "Radiation from a Tensile Fracture." *Journal of Geophysical Research,* **JGREA, 68(23):6345–6358. Dec 1, 1963.**

The radiation pattern for elastic waves from a tensile fracture has been observed in a two-dimensional model. The fracture is induced by a local concentration of thermal stresses. The observed radiation pattern for the first motions is compared with that predicted by two different theoretical calculations, one by Ang and Williams and the other by Knopoff and Gilbert. The experiment does not reproduce all of the conditions of either theory, but arguments are advanced which suggest that the theoretical predictions are applicable to the observations. Although there is generally a qualitative agreement between experiment and theory, the quantitative agreement is poor. The principal criticism of the experiment is that the first motions are emergent, and the experimentally measured amplitudes may not characterize the true first motion. On the other hand, linear elasticity is assumed in both theories. Inasmuch as fracture involves nonlinear processes, it may be that the theories cannot properly be applied to radiation from a fracture. [reprinted by permission of *Journal of Geophysical Research,* copyright by American Geophysical Union].

1536 W. F. Savage, E. F. Nippes, and J. M. Sawhill, Jr. "Hydrogen Induced Cracking during Implant Testing of Alloy Steels." *Welding Journal,* **WEJUA, 55(12):400–s to 407–s. Dec 1976.**

The initiation and propagation of hydrogen induced cracking in HY-80 and HY-130 steels and E10018 and E14018 weld metals were investigated with metallographic and acoustic emission techniques. The implant specimens were welded with an AX-140 electrode and moisture enriched shielding gas to introduce hydrogen. Both unmixed zones and partially melted zones were observed near the weld interface in the HY-130 steel, and these regions were approximately equal in size to those observed in the HY-80 steel. An unmixed zone was also observed at the weld interface between the E14018 weld deposit and an AX-140 gas metal arc weld bead. A partially melted zone, similar to those observed in HY-80 and HY-130, was observed in the fusion boundary region between two E14018 weld beads. The average hardness of the coarse grained region in the true heat-affected zone of HY-80 and HY-130 was found to be nearly the same. Microhardness variations in the true heat-affected zone were associated with microsegregation in the form of banding. A sharp transition in microhardness, observed at the weld metal interface in both HY-80 and HY-130 weldments, is believed to contribute to the cracking sensitivity of this region. In HY-80, crack initiated at elongated sulfide inclusions located on grain boundaries and propagated preferentially through alloy-rich bands. In HY-130, cracks initiated at the weld metal interface and propagated along the weld interface, along solute-rich grain boundaries in the partially melted zone, and along martensite laths in the true heat-affected zone of HY-130. The lower density of elongated inclusions in the true heat-affected zone of HY-130 is believed to reduce the incidence of cracking in this region. In E14018 weld metal, cracks usually initiated at the interface between the AX-140 implant test weld and the E14018 weld metal specimen. Crack propagation occurred almost exclusively along grain boundaries, particularly along the liquated grain boundaries in the partially melted zone. Acoustic emission data indicated that crack initiation in weld metal implant specimens occurred immediately after loading, and crack growth occurred during the entire test by the relatively slow intergranular fracture process. By contrast, no cracking was observed in wrought base metal specimens until shortly before failure. Crack propagation in HY-130 occurred by the relatively fast quasi-cleavage and microvoid coalescence fracture modes. The hydrogen-assisted cracking model proposed by Beachem can be employed to explain the changes in both fracture mode and crack growth rate that were observed in the material studied in this investigation.

1537 **R. F. Saxe. "Survey of Boiling Detection Methods in Reactors." pp 41–55 in** ***Incipient Failure Diagnosis for Assuring Safety and Availability of Nuclear Power Plants.*** **E. P. Epler and D. P. Roux, Coordinators. U.S. Atomic Energy Commission, Division of Technical Information, Washington, D.C. Jan 1968. [NTIS: CONF–671011]. proceedings of conference sponsored by Oak Ridge National Laboratory, held in Gatlinburg, Tennessee. Oct 30–Nov 1, 1967.**

To date, the detection of boiling in nuclear reactors has received relatively little attention, but it presents a problem which has increasing importance as a result of the building of reactors like HFIR, and the projected building of large liquid metal cooled fast reactors for power production. This problem is, of course, not applicable to reactors in which boiling occurs as a designed operating condition, but only to reactors operating at as high a temperature as possible for high efficiency, but in which boiling would give rise to potentially hazardous conditions. It follows, therefore, that there is considerable incentive to develop a reliable and sensitive boiling detector. Certain general observations regarding the possible methods for detection present themselves. In the most general terms, detection methods are based either on the receipt and recognition of signals from the phenomenon to be detected or are based on the interaction between the phenomenon and a *probe,* using the term *probe* in its most general form. As far as is known at present, the only signals emitted by boiling are acoustic signals. The types of probe which may be used, in general, fall into those with a characteristic length small compared with that of the phenomenon to be detected and into those with a characteristic length large compared with that of the phenomenon to be measured. Efficient detection requires that the former of the two cases be used, since the effects of the phenomenon on the probe are much greater for these conditions. When the characteristic length of the probe is large compared with that of the phenomenon, the effect caused by the phenomenon is in general small and efficiency is low.

1538 **R. F. Saxe. "The Acoustic Emission from Boiling Bubbles." ORO–3911–3. Progress Report, Feb–Nov 1970. North Carolina State University, Raleigh, North Carolina. Dec 1970. [NTIS].**

Activities are reported on programs to obtain information on parameters associated with generation of acoustic energy by boiling bubbles and to a further understanding of boiling bubble dynamics. Results of investigations on bubble-sound emission, and bubble-acoustic characteristics are included.

1539 **R. F. Saxe. "The Detection of Boiling in Nuclear Reactors."** *IEEE Transactions on Nuclear Science*, **IETNA, NS-18 (Part-1): 337-339. Feb 1971; paper presented at the Institute of Electrical and Electronics Engineers 2nd Nuclear Power Systems Symposium. New York. Nov 4-6, 1970.**

The subject of the detection of boiling in nuclear reactors has become important with the impending construction and operation of Liquid Metal Fast Breeder Reactors for power production. Several methods for the possible detection of boiling by equipment located outside the reactor have been proposed and investigated and have, in general, been reduced to two methods. Experiments were performed to assess the relative sensitivities of these two methods. It is shown that a large difference in sensitivity exists between the two methods.

1540 **R. F. Saxe and R. K. Cothren. "The Characteristics of the Acoustical Pulses Emitted by Boiling Bubbles in Water."** *Journal of the Acoustical Society of America*, **JASMA, 48(5, Part 2):1257-1265. 1970.**

An investigation of the acoustic emissions from boiling bubbles in water has been made using two methods: frequency analysis and pulse-height analysis. Pulse-height analysis, using coincidence techniques, shows that the pulse-height spectrum of boiling bubbles in water has a distribution which is independent of the amount of boiling, indicating that the distribution is governed by liquid, vapor, and possible wire-surface characteristics. The addition of a wetting agent to the water has an appreciable effect on the pulse-height distribution. Frequency analysis is shown to produce no information about pulse characteristics. [reprinted by permission of the American Institute of Physics].

1541 **R. F. Saxe and L. W. Lau. "Cavitation Noise in Nuclear Reactors."** *Nuclear Engineering and Design*, **NEDEA, 8:229-240. 1968.**

An experimental investigation was made of the acoustical noise emitted by flow-induced cavitation in reactor fuel element coolant channels. Spectra of the acoustical noise emitted are shown and are correlated with optical measurements taken by means of short-exposure photographs. It is shown that the amount of acoustical noise emitted by cavitation increases rapidly with increase of flow rate and is approximately proportional to the number of cavitation bubbles present. In addition, it is deduced that the spectrum of the acoustical noise emitted by cavitation is approximately flat up to frequencies in excess of those measured and that peaks in the recorded spectra were due to microphone characteristics.

1542 **R. F. Saxe, W. H. Sides, Jr., and R. G. Foster, Jr. "Detection of Boiling in Nuclear Reactors."** *Journal of Nuclear Energy*, **JNCEA, 25:139–153. Apr 1971.**

The relative sensitivities of two possible methods for the detection of boiling in reactors have been measured, under near ideal conditions, and it has been shown that the acoustical method is more sensitive that the neutron-flux noise method by a factor of at least 10^3 or 10^4. An upper limit for the amount of boiling void used in this experiment was measured as 0.4 ml and the neutron-flux noise measurements were calibrated in terms of reactivity fluctuation so that these results could be translated to other reactors.

1543 **H. J. Saxton, B. C. Odegard, A. R. Willis, L. W. Dahlke, J. C. Keilman, and W. E. Jorgenson. "Behavior of Flawed Pressure Vessels." SAND 75-8768. Sandia Laboratories, Livermore, California. Aug 1976. [specified external distrubtuion only]; pp 489–505 in** *Proceedings of the Third International Conference on Pressure Vessel Technology. Part I: Analysis, Design and Inspection.* **American Society of Mechanical Engineers, New York. 1977. conference held in Tokyo, Japan. Apr 19–22, 1977.**

A study has been conducted to determine the extent that well-characterized flaws degrade the performance of spherical pressure vessels and to evaluate the ability of advanced nondestructive test techniques to detect these flaws. This interdisciplinary study provides a major step towards establishing quantitative relationships between nondestructive acceptance criteria and the actual performance of flawed pressure vessels. When applied during proof testing, holographic techniques were found to detect inner wall defects spanning 10% or more of the wall thickness. Burst performance data and growth data obtained by radiographing under load forms the basis of analytical studies directed at evaluating the performance of tough materials in the presence of flaws. Both elastic-plastic finite element and J-integral techniques have been applied.

1544 **O. Schadel. "Zur Ermittlung der Schadigungsgrenze und Einfluss der Harzrissbildung auf das Beanspruchungsverhalten von Glasfaser-Kunststoffen (On the Determination of the Limit of Damage and Effect of the Formation of Resinous Coal on the Stress Behavior of Glass Fiber Plastic)."** *Kunststoffe,* **KUNSA, 59(12):892–898. 1969. (in German).**

This article is concerned with research on the limit of damage and influence of the formation of resinous coal on the stress behavior of glass-fiber plastic.

1545 **G. Schaffer. "Nondestructive Testing."** *American Machinist,* **AMMAA, 120(11):SR-1 to SR-24. Nov 1976.**

1546 E. Scheil. "Uber die Umwandlung des Austenits in Martensit in gehartetem Stahl (Transformation of Austenite into Martensite in Hardened Steel)." *Zeitschrift fur Anorganische und Allgemeine Chemie,* ZAACA, 183:98–120. 1929. (in German).

The steels studied included a Mn range from 0.24 to 10%, a 4% Cr steel and a 15% W steel. Conclusion: The *austenite-martensite* transformation is caused by thrust strains. If the strains are diminished the beginning of the transformation falls and the amount of transformation is diminished. In its extent and microscopic form, the austenite-martensite transformation resembles the twin formation produced by mechanical deformation of crystals. The influence of annealing upon the beginning of the transformation was studied. [abstract from *Chemical Abstracts,* 24:45, 1930. Copyright by American Chemical Society]

1547 M. V. Scherb, G. P. Kazokas, J. A. Zelik, J. R. Mastandrea, and D. C. MacKallor. "Experimental Evaluation of Three Leak Detection and Location Concepts for Space Stations." NASA CR-112066, MDC-G2911. McDonnell-Douglas Astronautics Company, Huntington Beach, California. June 1972. [NTIS: N72-26795].

Three leak (or precursor damage modes) detection and location concepts for space station overboard leakage were evaluated experimentally. The techniques are: (1) static and dynamic seal leak detector sensing of moisture or all gases in space cabin atmosphere, (2) active utlrasonic Lamb-wave detection of flaws or cracks in cabin wall, and (3) impact gage detection of stress waves induced in cabin pressure wall by meteoroid or orbital impact. The experimental results obtained in the program demonstrated that all three leak detection and location concepts are feasible. With further development, the methods can be integrated into an effective damage control system for advanced manned earth-orbital systems.

1548 H. Schiffelgen. "Messung akustischer Signale bei der Verformung von Metallen (Measurement of Acoustic Signals during Deformation of Metals)." Ph.D. Thesis. Technische Hochschule Munchen, Munich, Germany. 1963. (in German).

1549 R. J. Schliekelmann. "Nondestructive Testing of Adhesive Bonded Metal-to-Metal Joints - Part 2." *Non-Destructive Testing,* NDETA, 5(3):144–153. June 1972.

This is the second part of the author's discussion of methods of evaluation of the adhesion of joints. He discusses acoustic emission, radiography, and sonic and ultrasonic techniques. The Fokker Bond Tester is discussed in some detail.

1550 J. A. Schliessmann and R. R. Corle. "Evaluation of New NDT Techniques for Detection of Weld Defects." Lockheed Propulsion Company, Redlands, California. Mar 30, 1972; paper presented at the 47th Poseiden–Minuteman–Pershing Propulsion Nondestructive Test Committee Meeting, Aerojet Solid Propulsion Company, Sacramento, California. Apr 4–5, 1972.

The objective of this investigation was to evaluate three modern NDT methods which might be used to detect flaws in solid rocket motor cases, viz., laser holography, acoustic emission, and X-ray enhancement. The subsequent results were compared with standard radiographic and magnetic particle inspection test results. Preliminary tests conducted on uniaxial 4335V steel specimens indicated that acoustic emission would be optimum for detecting tight motor-case defects, and could detect flaws as small as 0.02 inches deep and 0.3 inches long. Other tests involving a D6aC steel case show that acoustic emission can be used in conjunction with proof testing to detect flaws before catastrophic fracture occurs.

1551 E. Schmid and M. A. Valouch. "Uber sprunghafte Translation von Zinkkristallen (The Discontinuous Translation of Zinc Crystals)." ***Zeitschrift fur Physik,*** **ZEPYA, 75:531–538. 1932. (in German).**

1552 P. M. Schmidt and R. S. Snow. "Development of Optimum TIG Welding Procedure for Production of Weldments in 12% Nickel Maraging Steel." NASA CR–72981. General Dynamics, Electric Boat Division, Groton, Connecticut. July 1971.

An optimum gas-tungsten-arc welding procedure has been developed for 0.75-inch thick 12% Ni-5 Cr-3 Mo maraging steel and used to fabricate five 36-inch diameter test vessels made of this material. Data are reported on the results of extensive studies of base metal, weld metal and weld heat-affected zone mechanical properties, and fracture toughness. These studies included heat-treatment response studies of both base plate and weldments. Tensile, conventional Charpy, precracked Charpy, single-edge notch, surface-flawed, and metallographic specimens were tested and the 5 fabricated vessels were cyclically tested and burst. The results of these tests demonstrated that vacuum-melted 12% Ni-5 Cr-3 Mo maraging steel, welded using the developed gas-tungsten-arc welding procedure, is capable of meeting a leak-before-burst failure design criteria for 260-inch diameter motor cases at a yield strength of 180,000 psi. Two of the vessels were monitored for acoustic emission during hydrotest. The test procedure and results are discussed in detail in Appendix B of the report. (see also H. L. Dunegan. "Acoustic Emission Testing of 12-Nickel Maraging Steel Pressure Vessels," Reference 406).

1553 Kh. G. Schmitt-Thomas and H. M. Tensi. "Schallemissionsmessung bei metallkundlichen Untersuchungen und Schallemissionsanalyse in der

Sicherheitstechnik (The Measurement of Acoustic Emission with Metallographic Investigations and Acoustic Emission Analysis in the Safety Technique)." *Metall,* **MTLLA, 27(10):965-971. Oct 1973. (in German).**

The variation of emission with phase changes in metals and alloys and during application of mechanical stresses is discussed and diagrams are given to show the observed effects. Two of the diagrams show curves for AlMg3 and AlZnMg3 indicating the effects of mechanical stress and pulse frequency on the observed strain. Examples of use of the techniques are given, and include examination of primary components of nuclear energy plants, and of aircraft parts. The taking, transposition, and evaluation of data are discussed. [Abstract 22-0438. *Metals Abstracts,* Vol. 7, 1974].

1554 **Kh. G. Schmitt-Thomas, H. M. Tensi, and H. Zeitler. "Moglichkeiten der Schallemissionsmessungen bei werkstoffkundlichen Untersuchungen (Feasibility of Acoustic Emission Measurements in Known Materials Research)." pp 1-23 in** *Schallemission: Anwendung bei der Untersuchung, Prufung und Uberwachung metallischer Werkstoffe.* **Deutsche Gesellschaft fur Metallkunde eV, 637 Oberursel/Taunus bei Frankfurt/Main, Adenauerallee 21, Germany. 1974. proceedings of symposium held in Munich, Germany. Apr 18-19, 1974. (paper in German).**

1555 **Kh. G. Schmitt-Thomas, H. M. Tensi, and H. Zeitler. "Schallemissionsmessungen an einer Legierung vom Typ Al Cu Mg2 mit unterschiedlichen Aushartungszustanden (Acoustic Emission Measurements on an Al Cu Mg2-Type Alloy in Different Hardened Conditions)."** *Aluminium,* **ALUMA, 51(8):520-524. Aug 1975. (in German). [English translation of article available as English Supplement of journal from publisher, Aluminium-Verlag GmbH].**

The alloy chosen for the investigation was technical Al Cu Mg2 and condition T42. The material was solution treated for 2 hours at 495 °C, quenched, and aged up to 14 days at 180 °C. Hardness determinations were made on specimens aged 180 °C for various periods and the values are shown by a curve. Other diagrams show load/time stress and acoustic emission-time/stress curves on solution-treated material aged at 170 °C for different periods of time. The emission curves show two maxima. The reasons for these maxima are discussed, the first one observed being considered to occur in the range of the proof stress and to be attributable to inhomogeneous microscopic deformations. The second of the maxima is thought to be due to microcracks associated with intermetallic phases. [Abstract 32-0573. *Metals Abstracts,* Vol. 9, 1976].

1556 **Kh. G. Schmitt-Thomas, H. M. Tensi, and H. Zeitler. "Schallemissionsmessungen an einer unterschiedlich ausgeharteten Al Cu Mg2-Legierung (Acoustic Emission Measurements in an Al Cu Mg2 Alloy after various Hardening Treatments)."**

pp 117–120 in *6. Internationale Leichtmetalltagung Leoben/Wien 1975.* **Aluminium–Verlag, Dusseldorf, Germany. 1975. (paper in German with English and French abstracts).**

Tests were carried out on sound emission during plastic deformation of Al Cu Mg2 alloy specimens after various age-hardening treatments with a view of using sound emission characteristics to draw conclusions regarding the age-hardened condition of the material. Hardness measurements and electron microscope examination were used as comparative test criteria. [Abstract 32-0928. *Metals Abstracts,* Vol. 9, 1976].

1557 **G. L. Schmitz. "Nondestructive Testing of Adhesive Bonds."** *Design News,* **DIGNA, 21(19):182–185. Sep 1966.**

Article is abstracted from a paper of the same title and coauthored with L. M. Frank and presented at the 3rd Annual Aerospace Conference on Nondestructive Testing, held at Georgia Institute of Technology, Atlanta, Georgia. May 2-3, 1966.

1558 **G. L. Schmitz and L. M. Frank. "Nondestructive Testing for Evaluation of Strength of Bonded Material." NASA CR–67983. General American Transportation Corporation, Research Division, Niles, Illinois. Sep 1965. [NTIS: N66–11761].**

The theoretical and experimental investigation of various techniques for nondestructive testing for bond strength are described. Existing nondestructive test techniques are mainly effective for detecting complete lack of bond in adhesively bonded structures. Techniques investigated included: acoustic emission, bond-line electrical parameters, strain sensitive coatings, and ultrasonic attenuation as a function of bond stress. The ultrasonic emission of a bond line under stress was found to be a reliable indication of bond strength. A definite need was found for a widely applicable bond-stress method.

1559 **G. L. Schmitz and L. M. Frank. "Nondestructive Testing of Adhesive Bonds." General American Research Division, General American Transportation Corporation, Niles, Illinois. 1966; paper presented at the 3rd Annual Aerospace Conference on NDT, sponsored by Magnaflux Corporation, held at Georgia Institute of Technology, Atlanta, Georgia. May 2–3, 1966.**

Several important points in bonding theories help in the understanding of poor bonds. The theoretical strength of actual bond between the adhesive and the adjoining surface is stronger than that actually realizable and the bond is actually stronger than the weaker of the cohesive strengths of either the cured adhesive material or the bonded surface. The acoustic emission test has been found successful in detecting *adhesive* failures. This technique consists of listening to the sound of the bond when the bond line is put under tensile stress.

1560 G. L. Schmitz and L. M. Frank. "Ultrasonic Emission Method Enables Testing of Adhesive Bonds." NASA Tech Brief 66-10341. NTIS, Springfield, Virginia. Aug 1966.

The Tech Brief outlines the work of the authors covered in report NASA CR-67983 (see Reference 1558). Request to Technology Utilization Officer, Marshall Space Flight Center, Code A&PS-TU, Marshall Space Flight Center, Alabama 35812 for further information on this Tech Brief will result in *Technical Support Package Brief 66-10341: Ultrasonic Emission Method Enables Testing of Adhesive Bonds.* This is report NASA CR-67983.

1561 G. L. Schmitz and L. M. Frank. "Ultrasonic Emission Testing of Adhesive Bonds." pp 16 in *Testing Methods and Techniques: Quality Control and Nondestructive Testing.* **NASA SP-5952 (02). NASA Technology Utilization Office, Washington, D.C. 1973. [NTIS].**

This is a summary of NASA Tech Brief 66-10341. See References 1560 and 1558.

1562 B. H. Schofield. "A Study of the Applicability of Acoustic Emission to Pressure Vessel Testing." AFML-TR-66-92. Lessells and Associates, Inc., Waltham, Massachusetts. Apr 1966. [NTIS: AD-809-369].

The report describes a series of experiments designed to assess the practical capability of acoustic emission to indicate stress or deformation in a structure, as well as to offer a means of assessing structural integrity. The tests were conducted on split-ring samples and specimens of cylindrical form internally pressurized. The integrity of the cylindrical vessels was altered by inserting machined notches of various shapes at predetermined locations on the vessel surface. Specimens were fabricated from 4130 steel and 6061 aluminum. The studies were extended to tests in which the surface of the specimens was coated with a brittle film. This was done to provide an artificial source of high acoustic energy which conceivably could offer an alternate method comparable in reliability with the metal induced emission and one which would be much less susceptible to the environmental disturbances.

1563 B. H. Schofield. "Acoustic Emission Development and Current State of the Art." Teledyne Materials Research, Waltham, Massachusetts; *IEEE Transactions on Sonics and Ultrasonics*, **IESUA, SU-17:56. Jan 1970. [abstract only].**

The detection of acoustic emission is accomplished through the use of piezoelectric transducers and associated electronic components capable of system gains of 10^5 to 10^6 at threshold noise levels pressing the state-of-the-art. Acoustic emission is a coined phrase relating to the generation and propagation of discrete elastic pulses or waves induced in materials subjected to stress and deformation. For most materials, the pulse energy is far below the audible threshold; hence, this necessitates the sensitive transducers and high gains. A historical review of the phenomenon is presented along with a survey of the early research experiments and theoretical interpretations. More recent experiments and current developments in techniques and instrumentation directed to the practical utilization of the phenomenon as a nondestructive tool are noted. This latter effort is presently of particular interest in the area of continuous surveillance of nuclear vessel integrity. The current state-of-the-art in research and application is summarized. [reprinted from *IEEE Transactions on Sonics and Ultrasonics*].

1564 **B. H. Schofield. "Acoustic Emission from Metals: Its Detection, Characteristics and Source." pp 63-78 in** ***Proceedings of the Symposium on Physics and Nondestructive Testing.*** **Southwest Research Institute, San Antonio, Texas. 1963. proceedings of the 4th Symposium on Physics and Nondestructive Testing, held in San Antonio, Texas. Oct 1-3, 1963.**

The existence of acoustic emission, that is, the generation of elastic pulses in metals during their deformation, has become more widely known during the past few years. This has been due, primarily, to the increased activity and interest stimulated by the ubiquitous need for additional tools and techniques which may provide the elusive pieces to the composite puzzle of fundamental materials behavior. Fatigue of metals has been acknowledged for half a century; creep and brittle fracture have been major design limitations for almost as long a period. Obviously, remarkable advances have been made in these areas over the years through both *art* and theory. Nevertheless there is no doubt, as evidenced by the unprecedented amount of basic research effort in industry and government, that there exists a considerable void in knowledge of fundamental behavior of materials, and that in order to reduce, if not remove, this gap, all aspects of material phenomena must be thoroughly investigated. Acoustic emission is significant therefore, not only since it exists as a physical characteristic of many materials, but probably more important as it offers a possibility of obtaining new and useful knowledge, thereby expanding an understanding of the basic behavior of materials. The details of the historical background of sound observed from metals has been adequately done in other papers. The emission of sound per se has, as has been noted previously, been known for many years; documented at least as early as 1928. These audible sounds were, however, *special* sources, as stated by Liptai and Tatro, and are not of concern in the subject investigations. Generally the energy level of acoustic emission is far lower than that produced by the audible mechanisms of twinning, grain boundary reorientation, and metallurgical transformations. Amplification of the emission pulses in the order of 10^6 to 10^7 is required to obtain energy levels satisfactory for analysis or audible reproduction. This, of course, suggests the operation of mechanisms microscopic relative to those associated with the audible processes; a mechanism such as that related to slip deformation.

In previous work by the author, direct experimental evidence was presented to demonstrate that the origin of the observed emission was within the test section of the specimen. Notwithstanding the many phenomenological characteristics which show the existence of such emission from metals, a priori, considering the extreme sensitivity of the measurements, the basic premise remains so without positive experimental evidence. Inasmuch as the energy level of the sound pulses is extremely low, it is apparent that the special techniques necessary to both induce and detect the emission can be of significance with regard to the emission characteristics observed and particularly the consclusions established thereby. This paper will, therefore, present the experimental methods employed in inducing and detecting the emission as well as exhibit certain of the generic sound characteristics. The emphasis will, however, be concerned with the results of most recent studies and their significance relative to the source of emission. The majority of the experimental work to be described involved static tensile testing directed toward establishment of the fundamental source or sources of the emission. The results of initial studies to verify the existence of the phenomenon and investigate the frequency and amplitude characteristics are presented in a previous report. A separate program of more recent vintage, exploring the existence and behavior of emission under cyclic stress on polycrystalline aluminum, will also be discussed briefly. Finally, some of the attempts that have been made to employ the phenomenon as a tool will be noted.

1565 **B. H. Schofield. "Acoustic Emission Predicts Static and Fatigue Failure."** *Space/ Aeronautics,* **SPAEA, 43:63, 64, 66. Apr 1965.**

Progress report on the use of sound emissions as applied to the nondestructive evaluation of metals and structures. Although test techniques are still being developed under USAF and NASA sponsorship, it is suggested that an immediate use of such techniques could be in the testing of a missile casing. It is pointed out that the acoustic technique does not provide quantitative data, but does sense plastic deformation and provide qualitative information in the plastic deformation range. Current tests on seamless steel tanks are discussed. It is concluded that the technique provides a sensitive indication of fatigue crack growth in its early stages.

1566 **B. H. Schofield. "Acoustic Emission Research and Its Application to Materials Research and to Industry." pp 254–269 in** *Schallemission: Anwendung bei der Untersuchung, Prufung und Uberwachung metallischer Werkstoffe.* **Deutsche Gesellschaft fur Metallkunde eV, 637 Oberursel/Taunus bei Frankfurt/Main, Adenauerallee 21, Germany. 1974. proceedings of symposium held in Munich, Germany. Apr 18–19, 1974.**

The author discusses selected topics of acoustic emission research noting the potential applications of the phenomenon as a research tool in materials behavior studies, particularly in the kinetics of twinning and corrosion. Practical application of the method for evaluating the integrity of pressure vessels is discussed including examination of certain limitations of the method. Specific results of an experimental pressure vessel study are presented to demonstrate the potential value of the acoustic emission technique as a unique nondestructive test tool.

1567 **B. H. Schofield. "Acoustic Emission under Applied Stress." Contract AF-33(616)-5640. Progress Report No. 2. Lessells and Associates, Inc., Boston, Massachusetts. Oct 15, 1958.**

The progress report covers the period of July 15, 1958 to Oct 15, 1958. See Reference 1575 for final report on this contract. Also see Reference 89 for Progress Report No. 1.

1568 **B. H. Schofield. "Acoustic Emission under Applied Stress." Contract AF-33(616)-5640. Progress Report No. 3. Lessells and Associates, Inc., Boston, Massachusetts. Jan 15, 1959.**

The progress report covers the period of Oct 15, 1958 to Jan 15, 1959. See Reference 1575 for final report on this contract.

1569 **B. H. Schofield. "Acoustic Emission under Applied Stress." Contract AF-33(616)-5640. Progress Report No. 6. Lessells and Associates, Inc., Boston, Massachusetts. Oct 15, 1959.**

The progress report covers the period of July 15, 1959 to Oct 15, 1959. See Reference 1575 for final report on this contract.

1570 **B. H. Schofield. "Acoustic Emission under Applied Stress." Contract AF-33(616)5640. Progress Report No. 7. Lessells and Associates, Inc., Boston, Massachusetts. Jan 15, 1960. [NTIS: PB-158-743, AD-232-172].**

The progress report covers the period of Oct 15, 1959 to Jan 15, 1960. See Reference 1575 for final report on this contract.

1571 **B. H. Schofield. "Acoustic Emission under Applied Stress." Contract AF-33(616)-5640. Progress Report No. 8. Lessells and Associates, Inc., Boston, Massachusetts. Apr 15, 1960. [NTIS: PB-158-744, AD-235-448].**

The progress report covers the period of Jan 15, 1960 to Apr 15, 1960. See Reference 1575 for final report on this contract.

1572 **B. H. Schofield. "Acoustic Emission under Applied Stress." Contract AF-33(616)-5640. Progress Report No. 9. Lessells and Associates, Inc., Boston, Massachusetts. July 15, 1960.**

The progress report covers the period of April 15, 1960 to July 15, 1960. See Reference 1575 for final report on this contract.

1573 **B. H. Schofield. "Acoustic Emission under Applied Stress." Contract AF-33(616)-5640. Progress Report No. 10. Lessells and Associates, Inc., Boston, Massachusetts. Oct 15, 1960. [NTIS: PB-158-745, AD-248-500].**

The progress report covers the period of July 15, 1960 to Oct 15, 1960. See Reference 1575 for final report on this contract.

1574 **B. H. Schofield. "Acoustic Emission under Applied Stress." Contract AF-33(616)-5640. Progress Report No. 11. Lessells and Associates, Inc., Boston, Massachusetts. Mar 15, 1961. [NTIS: AD-255-144].**

The progress report covers the period of Oct 15, 1960 to Mar 15, 1961. See Reference 1575 for final report on this contract. A series of tests was conducted on aluminum single crystals of the so-called 0.5 orientation. The data obtained to date were studied relative to the physical behavior of the 0.5 specimens and also with regard to that previously obtained on (100) specimens. Various characteristics are presented of the emission and the physical parameters. Again, the data show a definite correlation between the emission phenomenon and deformation behavior. It is clear from the information obtained up through the present, that the emission is a consequence of, and is produced by, a particular phenomenological characteristic of the deformation process. Several observations were undertaken to determine the particular aspect of the deformation process acting as the source of this emission energy.

1575 **B. H. Schofield. "Acoustic Emission under Applied Stress." ARL-150. Lessells and Associates, Inc., Bostion, Massachusetts. Dec 1961. [NTIS: AD-274-484].**

This is the final report on Contract AF-33(616)-5640 which covers the period from Mar 31, 1958 to Aug 31, 1961. The report continues the work conducted under Contract AF-33(616)-2800 and reported in WADC Technical Report 58-194 (see Reference 1587). The report discusses the studies of the acoustic emission phenomenon of metals. The general nature of the emission of materials is presented, but primary emphasis is placed on the behavior of the phenomenon in various crystallographic orientations of both zinc and aluminum single crystals; the intent of the investigations being the determination of the source of the emitted sound. The results are examined and interpreted in the light of fundamental deformation mechanisms, i.e., dislocations, and the particular manner in which they may induce such emission. All experiments were conducted under tensile loading of the specimens. Descriptions of the equipment, testing and analysis techniques, and findings with interpretation are presented in detail.

1576 **B. H. Schofield. "Acoustic Emission under Applied Stress." ASD-TDR-63-509, Part I. Lessells and Associates, Inc., Waltham, Massachusetts. Apr 1963. [NTIS: AD-419-177].**

The report concerns specific investigations undertaken to assess the role of the surface oxide layer in the generation of acoustic pulses in metals, i.e., acoustic emission. Experiments were conducted on single crystals of aluminum of different crystallographic orientation. The specimens were pulled in tension both in atmosphere and exposed to etchant solutions. Results demonstrated that the oxide layer is not itself an emission source and that the emission characteristics are significantly modified as a result of the influence of the oxide on the deformation behavior of the specimens. Specific observations suggest the operative mechanism of emission generation and the importance of the role played by the surface condition in the deformation of metals.

1577 **B. H. Schofield. "Acoustic Emission under Applied Stress." ASD-TDR-63-509, Part II. Lessells and Associates, Inc., Waltham, Massachusetts. May 1964. [NTIS: AD-600-233].**

The report is an extension of the investigations described in Part I (see Reference 1576). These studies were made to assess the role of the surface and surface films in the generation of acoustic pulses, i.e., acoustic emission of metals. Experiments were conducted on single crystals of pure gold of varying crystallographic orientation. Specific crystallographic orientations were chosen to evaluate the effect of screw and edge dislocation mobility separately. Specimens were deformed in tension in ambient atmosphere. Results indicate that the surface and its condition play a minor role in the generation of the acoustic pulses; the phenomenon being related to the volumetric condition. Specific observations suggest the development of structural faults during deformation, reminiscent of elemental twins. Emission characteristics peculiar to gold are discussed for the several experiments.

1578 **B. H. Schofield. "Comments on Acoustic Emission and Its Status in Pressure Vessel Evaluation." Technical Report No. E-1264. Teledyne Materials Research Company, Waltham, Massachusetts. Nov 5, 1970; Task 1, Project IV, pp 41-80 in "In-Service Inspection Program for Nuclear Reactor Vessels." R. D. Wylie. SwRI Project 17-2440, EEI Project RP-79. Biannual Progress Report No. 5. Southwest Research Institute, San Antonio, Texas. May 28, 1971. [NTIS: PB-200-518].**

The report is a comprehensive history of acoustic emission.

1579 **B. H. Schofield. "Investigation of Applicability of Acoustic Emission." AFML-TR-65-106. Lessells and Associates, Inc., Waltham, Massachusetts. May 1965. [NTIS: AD-468-359].**

The report describes initial investigations directed toward the application of acoustic emission to nondestructive evaluation of structural integrity. Experiments were conducted on internally pressurized cylinders. Variation in the cylinder integrity was accomplished by inserting machined notches in certain specimens. Other studies were to evaluate the possibility of employing the acoustic methods and brittle coatings on similar structures, the acoustic pulse in this case arising from the cracking of the coating. The latter technique is anticipated as a tool to determine the occurrence and location of high stress regions in complex structures. Both methods showed sufficient promise in these preliminary studies to warrant further exploration. The concluding phase of the fundamental research is also reported.

1580 **B. H. Schofield. "Kaiser Effect." p 151 in *Encyclopaedic Dictionary of Physics, Supplementary Volume 1.* J. Thewles, Editor-in-Chief. Pergamon Press, Oxford England. 1966.**

The *Kaiser Effect* is the irreversible characteristic of the acoustic emission phenomena in metals named after its discoverer, *J. Kaiser,* in 1950. The extremely low energy acoustic pulses produced during initial deformation of a metal are not observable in subsequent restressing up to the original limit of stress. The effect corresponds to the elimination of the plastic region of the stress-strain curve due to work hardening of the metal. Exceeding the previous stress limit or reversal of the stress as in the Bauchinger [*sic*] effect reintroduces the acoustic emission response. The Kaiser effect shows promise as a sensitive research tool in the study of physical deformation and work-hardening mechanisms. [reprinted by permission of Pergamon Press Ltd].

1581 **B. H. Schofield. "Noise from Plastically Deformed Materials." pp 196–197 in *Encyclopaedic Dictionary of Physics, Supplementary Volume 1.* J. Thewles, Editor-in-Chief. Pergamon Press, Oxford, England. 1966.**

1582 **B. H. Schofield. "Research on the Sources and Characteristics of Acoustic Emission." pp 11–19 in *Acoustic Emission, ASTM STP 505.* American Society for Testing and Materials, Philadelphia, Pennsylvania. 1972. proceedings of symposium held in Bal Harbour, Florida, Dec 7–8, 1971.**

A review of some of the less familiar acoustic emission characteristics and anomalies is presented with the objective of suggesting new or alternate approaches to better understand the phenomenon. Topical discussions include slip- and twinning-induced emission, acoustic responses during metal etching, residual stresses, and influence of grain size and fatigue. Specific shortcomings in acoustic emission research at present are noted, as well as the need for additional research to keep up with the ever increasing and unusual practical applications.

1583 **B. H. Schofield. "Review of State-of-the-Art Methods and Equipment Employed in the Utilization of Acoustic Emission." Task 1, Project IV. pp 42–44 in "In-Service Inspection Program for Nuclear Reactor Vessels." R. D.**

Wylie. SwRI Project 17-2440, EEI Project RP-79. Biannual Progress Report No. 2. Southwest Research Institute, San Antonio, Texas. Dec 2, 1969. [NTIS: PB-188-347].

Visits were made to several facilities engaged in the use of or development of acoustic emission monitoring systems. During these visits, the systems were observed and discussions were held with the investigators responsible for their use and development. The data obtained on these visits have been thoroughly evaluated and analyzed with respect to the requirements for an in-line reactor monitoring system and the current state of acoustic emission technology. The results of the analysis have been published in a final report, thus bringing this task to a close. The conclusion of the study is that an engineered prototype system be developed for reactor pressure vessel monitoring as there is no single system now available which can meet all of the requirements for this application.

1584 **B. H. Schofield. "Tin Cry." p 355 in** ***Encyclopaedic Dictionary of Physics, Supplementary Volume 1.*** **J. Thewles, Editor-in-Chief. Pergamon Press, Oxford, England. 1966.**

Tin Cry (cry of tin) is a descriptive phrase referring to the crackling and snapping sounds heard when tin is plastically deformed. The origin of the phrase is unknown, however, knowledge of the phenomenon undoubtedly dates back over 4000 years to the earliest known use of tin. The source of these acoustic pulses is the nucleation of mechanical twins induced during deformation. Such twins can form in a time as short as a few microseconds and a part of the energy associated with this sudden transformation occurs as a click or loud report (also referred to as acoustic emission). It follows that metals which twin readily exhibit the *cry* phenomenon to some degree. Metals in this class belong to hexagonal and tetragonal crystallographic structures, e.g., zinc, cadmium, magnesium, and tin. At cryogenic temperatures (4K), the face-centered cubic metals, copper and gold-silver alloys which are not ordinarily considered to deform by mechanical twinning, have produced such twins accompanied by the *tin cry* response. Generally this acoustic effect is accompanied by a stepped or separated stress-strain curve. The generic term *twin cry* would appear more appropriate than the parochial *tin cry.* [reprinted by permission of Pergamon Press Ltd].

1585 **B. H. Schofield. "Utilization of Acoustic Emission for In-Service Inspection." Paper C40/72, pp 76-82 in** ***Periodic Inspection of Pressure Vessels.*** **Institution of Mechanical Engineers, London, England. 1972. proceedings of conference arranged by the Applied Mechanics Group of the Institution of Mechanical Engineers, held in London, England. May 9-11, 1972.**

The utilization of acoustic emission for reactor vessel in-service inspection shows considerable promise. System development is currently active among several commercial sources. Notable is the support given by the utility industry in the U. S. to the development of acoustic emission, holography, and the refinement of ultrasonic nondestructive testing methods through the Edison Electric Institute and its Steering

Committee. This paper describes the emission characteristics significant to pressure vessel surveillance and their utilization in precluding catastrophic failure, and the detection and locating of defects in the structure. The typical components of an emission surveillance system are discussed, followed by a review of the current state-of-the-art.

1586 **B. H. Schofield. "Why Acoustic Emission–Why Not?" pp 3-10 in** *Monitoring Structural Integrity by Acoustic Emission, ASTM STP 571.* **Edited by J. C. Spanner and J. W. McElroy. American Society for Testing and Materials, Philadelphia, Pennsylvania. 1975. proceedings of a symposium held in Ft. Lauderdale, Florida. Jan 17-18, 1974.**

The relative apathy of the industrial community to take advantage of the significant benefits of acoustic emission is discussed against the background of the current state of the technology. Examples of immediate applications are noted. It is suggested that developing trends necessitate timely initiation of industrial utilization and that such efforts and the experience gained therein are a prerequisite to the realization of the technical benefits of acoustic emission and the establishment of proper and adequate guidelines. [reprinted by permission of the American Society for Testing and Materials, Copyright].

1587 **B. H. Schofield, R. A. Bareiss and A. A. Kyrala. "Acoustic Emission under Applied Stress." WADC Technical Report 58-194. Lessells and Associates, Inc., Boston, Massachusetts. Apr 30, 1958. [NTIS: AD-155-674].**

The initial experiments on commercial grades of carbon steel, 24ST-4 aluminum, and cast tin verified the existence of the acoustic emission effect, but did not verify the correlation of the frequency and amplitude spectrums with applied stress in the specimen, as reported by Kaiser in Germany. The frequency and amplitude spectrums did show shaped distributions, as opposed to random behavior; however, the distributions were not consistent for a given material. The frequency spectrums indicated a relatively narrow range of predominant frequencies and a consistency, from specimen to specimen in a group, of the frequency values. Preliminary investigations on pure aluminum specimens of varying grain density did not conclusively verify dependency of the acoustic generation on grain boundaries, as predicted in previous work. More work is to be done on this phase. Experiments on pure aluminum single crystals showed the existence of the acoustic phenomenon. These tests indicated that the phenomenon consists of two primary effects, the high amplitude bursts of intermittent occurrence and the relatively higher frequency signals of a continuous nature. Present results, based primarily on calculated values, indicate a possible connection between the acoustic signals and the generation and movement of dislocations within the material. The work in progress is directed toward verification of this hypothesis. It is speculated that the high frequency continuum is the motion of the multitude of free dislocations and the high amplitude bursts correspond to the breaking away of dislocation blocks.

1588 C. H. Scholz. "Experimental Study of the Fracturing Process in Brittle Rock." *Journal of Geophysical Research,* JGREA, 73(4):1447–1454. Feb 15, 1968.

1589 C. H. Scholz. "Frequency–Magnitude Relationship of Microfracturing Events during the Triaxial Compression of Rock." *Transactions of the American Geophysical Union,* TAGUA, 48(1):205. Mar 1967.

1590 C. H. Scholz. "Microfracturing and the Inelastic Deformation of Rock in Compression." *Journal of Geophysical Research,* JGREA, 73(4):1417–1432. Feb 15, 1968.

1591 C. H. Scholz. "The Frequency–Magnitude Relation of Microfracturing in Rock and Its Relation to Earthquakes." *Bulletin of the Seismological Society of America,* BSSAA, 58(1):399–415. 1968.

1592 J. J. Schuldies. "Surface Evaluation of Dense Silicon Nitride Using Acoustic Emission to Monitor the Response of Applied Thermal Stresses." pp 416–423 in *Proceedings of the Ninth Symposium on Nondestructive Evaluation.* Southwest Research Institute, San Antonio, Texas. July 1973. symposium held in San Antonio, Texas. Apr 25–27, 1973.

The surface integrity of dense silicon-nitride forms may be determined by utilizing acoustic emission techniques. Surface degradation of test specimens can be achieved by preferentially introducing surface and/or near surface cracks with a CO_2 laser. The presence of these artificially introduced flaws may be substantiated by monitoring the acoustic emission response of the material as the laser thermal energy is transmitted to the specimen. Subsequent flexural testing in four-point bending, and monitoring the AE response to stress, then defines the extent of surface damage as depicted by the reduction in strength and the magnitude of recorded AE. Slow crack growth has been observed at room temperature during these tests, indicating that a time dependence with applied stress exists prior to failure. Reducing the power density of the laser beam produces a means for thermally shocking the near surface of the material. Since flaw size influences the magnitude of detected acoustic emission, the response of the material to thermal stresses results in an evaluation of surface integrity. These parameters should result in obtaining a nondestructive evaluation procedure which can be applied to structural ceramic components to define maximum allowable service tensile stress.

1593 J. J. Schuldies. "Surface Evaluation Using Acoustic Emission." pp 189–200 in *Fracture Mechanics of Ceramics.* R. C. Bradt, D. P. H. Hasselman, and F. F. Lange, Editors. Volume 1: Concepts, Flaws, and Fractography.

Plenum Press, New York. 1974. proceedings of a symposium held at Pennsylvania State University, University Park, Pennsylvania. July 11-13, 1973.

Artificially induced surface flaws in dense silicon nitride may be activated by applying a thermal stress with a CO_2 laser to the near surface of the material. The extension of these flaws may then be recorded by acoustic emission insturmentation and this response then serves as an indication of surface integrity. Surface degradation of test specimens was achieved by thermal shocking with the laser to produce surface cracks. The laser was then used to thermally stress the near surface and activate the small cracks. Flexural tests were used to determine the extent of damage by noting the reduction in strength associated with various levels of AE recorded during laser thermal shocking. Utilizing these parameters on an unknown surface should result in obtaining meaningful data in terms of the presence of surface defects and their influence on material strength.

1594 **J. J. Schuldies. "The Acoustic Emission Response of Mechanically Stressed Ceramics." Ford Motor Company, Turbine Research Development, Dearborn, Michigan. Oct 19, 1972; paper presented at the American Society for Nondestructive Testing National Fall Conference, Cleveland, Ohio. Oct 16-19, 1972;** ***Materials Evaluation*****, MAEVA, 31(10):209-213. Oct 1973.**

Acoustic emission techniques can be applied to detect the onset of catastrophic failure in ceramic materials. The AE response to stress for specimens tested in four-point bending has exhibited a precursor to final failure. This implies a transition from microcracking to unstable macrocracking resulting in catastrophic failure that occurs within a finite period of time. Utilizing this time dependency, results in obtaining characteristic AE versus stress curves for a given material are useful in determining design limits for structural components.

1595 **D. R. Schuyler, II and T. H. Feiertag. "Acoustic Emission Testing Program: Progress Report on Tensile Testing through October 1970." LA-4724-MS. Los Alamos Scientific Laboratory, Los Alamos, New Mexico. Mar 1972. [NTIS].**

Development of an acoustic emission testing program was initiated to broaden the nondestructive testing capability of the Laboratory. Acoustic emission is created in materials as they undergo stress relief mechanisms and may be detected by high-gain electronic amplification equipment. A program of conducting tensile tests on various materials while monitoring with acoustic emission equipment was undertaken. This furnished background information which would be useful for predicting and evaluating acoustic emission behavior of structures more complicated than tensile samples.

1596 H. J. Schwalbe. "Zwischenbericht fur die Deutsche Forschungsgemeinschaft (Bonn, Bad-Godesberg) zu den Forschungsvorhaben Ah12/6 ***Untersuchungen zur Risszahigkeit Faserverstarkter Kunststoffe*** **und Ah 12/10** ***Anwendung neuer Zertorungsfreier Prufverfahren zur Fruherkennung von Schadigungen in Verbundwerkstoffen bei wechselnder Belastung*** **(Interim Report for the German Research Agency (Bonn, Bad-Godesberg) on the Projects: Ah 12/6: Study of the Crack Resistance of Fiber-Reinforced Plastics, and Ah 12/10: Use of New Nondestructive Tests for Early Detection of Damage in Composite Materials Subjected to Alternating Loads, Period Covered: 1 May 1973 through 30 April 1974)." Battelle-Institut eV, Frankfurt am Main, Germany. (in German); UCRL-Trans-10929. translated for Lawrence Livermore Laboratory, Livermore, California. Sep 1975.**

Four types of failure in GFK and KFK laminates were investigated. Those are distinguished by macroscopic course of rupture, stress-strain behavior, and associated microrupture processes. Unstable crack growth, as is known for homogeneous isotropic materials, is observed with the investigated laminates only in the case of damage Type 4 (crack front and notch in the laminate plane). In this case, it also seems to be possible to determine a K_{IC} value. Specimens with other notch orientations are characterized by stable crack growth. In these cases, the energy which must be supplied to bring about complete failure after initial fiber rupture seems to be a measure of the *crack resistance* of the composite. These assertions are valid both for CT specimens subjected to static load and for notched three-point bend specimens subjected to static and sudden loads. The GFK specimens of different laminate structure show the same rupture behavior under cyclic loading (pulsating compression loading in the three-point bend test) as they do in the static test. A criterion for evaluating unstable crack propagation is available in the form of acoustic emission analysis. The cumulative total of acoustic emission pulses shows a constant value at ultimate failure. For the same loaded cross section, this value is independent of the loading rate, upper-lower load, number of cycles, and position of the notch relative to fiber direction when the notch is perpendicular to the layers of the laminate. As far as can be ascertained from the present results, this value, although not a material constant, is a criterion by which danger to the specimens can be evaluated. For composite materials, acoustic emission analysis has proved to be a test method which provides early indication of any material change. In particular, it is capable of distinguishing between fiber rupture and rupture between fibers (damage such as resin crack formation, debonding, and delamination).

1597 E. B. Schwenk and G. D. Shearer. "Measuring Projection Weld Strength by Acoustic Emission." ***Non-Destructive Testing,*** **NDETA, 6(1):29-33. Feb 1973.**

Data are presented that show the feasibility of nondestructively measuring resistance projection weld strength in Zircaloy-2 by measuring the acoustic emission count during the weld. The data were obtained by mounting one or two piezoelectric sensors on one electrode and conditioning and recording the resultant acoustic emission on one or two high-frequency digital counters. The acoustic emission appears to be generated primarily by the consolidation of the metal joint and the number of emissions appears to be a function of the weld zone size.

1598 E. B. Schwenk and G. D. Shearer. "Measuring Resistance Weld Strength by Acoustic Emission." Battelle-Northwest, Richland, Washington. Dec 1969.

The strength of a spot weld in 6061-T6 aluminum is monitored by counting the number of emission pulses occurring during the weld period. An acoustic emission transducer is mounted on one of the weld electrodes and is connected to a digital counter. Calibration of the counter is accomplished by making a series of spot-welded or projection-welded lap shear specimens using different weld currents to make different size welds. These specimens are then loaded to failure and referenced to the number of acoustic emission pulses measured during welding.

1599 E. B. Schwenk, G. D. Shearer, P. H. Hutton, and R. F. Klein. "Monitoring Electroslag Welding by Acoustic Techniques." Paper 6-2, pp 428-441 in *The Third Acoustic Emission Symposium.* Japan Industrial Planning Association, Dobashi Building, 3-17-8, Akasaka, Minato-ku, Tokyo, Japan. proceedings of a symposium held in Tokyo, Japan. Sep 16-18, 1976.

The electroslag (ES) welding process is attractive because of low joint cost and low distortion. Present process control however, is operator dependent. The primary intent of the present work was to determine if indicated acoustic emission is sensitive to changes in ES weld parameters, such as fluxing and wire-feed rate. A portion of the work also involved detection and location of defects as they formed during welding. Investigation of these effects could provide a basis for enhancing the consistency of electroslag welds and future direction for development of ES weld-process monitoring and control.

1600 I. G. Scott. "Stress Wave Emission (SWE)." ARL/MET-72. Australian Department of Supply, Aeronautical Research Laboratories, Melbourne, Australia. Sep 1970. [NTIS: N71-27857].

The nature of recorded stress wave emission was examined in terms of frequency, pulse amplitude, and count rate. Frequency and pulse amplitude were found to be strongly affected by instrumentation methods, whilst wave shape and count rate were dependent on the mode of, and proximity to, specimen failure. A partly successful attempt was made to identify, in general terms, the sources of SWE, by examining experimental results from a number of workers.

1601[1] I. G. Scott and J. J. Carter. "Acoustic Emission." Paper No. 5 in *Nondestructive Testing for Quality Control.* Australian Institute for Nondestructive Testing, Melbourne, Australia. 1974. proceedings of the 4th National Conference, held in Melbourne, Australia. Aug 19-21, 1974.

A general discussion on acoustic emission and its detection is presented. Applications are reviewed, citing the advantages and limitations. The application of acoustic emission to quality control and its relation to other NDT methods are discussed.

1602 **I. G. Scott and J. J. Carter. "Quality Control with Acoustic Emission."** *Testing, Instruments and Controls,* **TICOA, 12(4):15–17. Apr 1975.**

Acoustic emission is now an established NDT method. Some of its applications are considered, especially those which have direct relevance to quality control. The advantages of acoustic emission over traditional NDT methods and its current limitations are also discussed.

1603 **L. Seaman. "One-Dimensional Stress Wave Propagation in Soils." DASA-1757. Stanford Research Institute, Menlo Park, California. Feb 1966. [NTIS: AD-632-106].**

Soil behavior during stress wave propagation was studied on a sand and two clays by making one-dimensional wave propagation tests on 5-metre long columns of the soils. Attempts were made to predict this behavior by determining soil properties in dynamic compression tests on small samples and by using these properties in a variety of mathematical models for soils. In all the wave propagation tests, stress and acceleration records were similar, showing that the three soils differ in degree, not in kind. Peak stress and particle velocity attenuated to 20-40% of the peak value in the length of the 5-metre column. Two theoretical soil models were analyzed: one to investigate the effect of combined time-dependent and time-independent dissipation; and one to study the effects of nonlinear stress-strain relations and geostatic stress. Comparison of the theoretical predictions from the first of these and two previously studied models (using properties obtained from compression tests on soil samples) with the wave propagation results shows: (1) for clays the arrival time of the wave at the column base was within 10% of that calculated from the tangent modulus and for sand it was within 25%, and (2) attenuation of peak stress and particle velocity was predicted within ±50% at the base of the column (5-metre length).

1604 **D. Sebenik. "Raziskave akusticne emisije kompozitov (Research of Acoustic Emission from Composites)." Diplomsko delo st. 2303. Fekulteta za strojnistvo, Univerza v Ljubljani, Ljubljana, Yugoslavia. 1975. (in Slovenian).**

This work deals with acoustic emission from composites used in dental medical practice.

1605 **R. T. Sedgwick. "Acoustic Emission from Single Crystals of LiF and KCl."** *Journal of Applied Physics,* **JAPIA, 39(3):1728–1740. Feb 15, 1968.**

Single crystals of LiF and KCl were subjected to incremental compressive stress along a [100] direction. A continuous record of stress and strain was kept, together with data from ultrasonic damping and acoustic emission measurements. The experimental techniques and the electronic equipment used are discussed in detail. It was found that specimens of LiF exhibited acoustic emission peaks in the macroscopic elastic range having their maximums at a strain of approximately $\epsilon_0 = 10^{-3}$ and skewed toward larger values of ϵ. The analysis of this acoustic emission distribution forms the basis for a microscopic deformation model for LiF. It is hypothesized that this distribution whould be identical to the distribution of dislocation-source inverse loop-lengths. An expression for the probability of observing N acoustic emission bursts while the crystal is being strained from ϵ to $(\epsilon + d\epsilon)$ is written in terms of inverse loop-length and used to calculate average values of ϵ, ρ_M, and (A_1/A_2), the dislocation strain, mobile dislocation density, and ratio of successive ultrasonic wave amplitudes, respectively. Since no such peak was observed for KCl, the differences in the microscopic deformation mechanisms for that material are emphasized.

1606 R. T. Sedgwick. "An Investigation of Acoustic Emission from Coated and Uncoated Ionic Crystals." Ph.D. Thesis. Michigan State University, East Lansing, Michigan. 1965; University Microfilms 66-6169.

Single crystals of LiF and KCl, some of which were coated with either NaCl or CaF_2, were subjected to incremental compressive stress along a [100] direction. A continuous record of stress and strain was kept together with data from ultrasonic damping and acoustic emission measurements. The experimental techniques and the electronic equipment used are discussed in detail. It was found that the presence of a thin-surface film (500 Å) raised the entire stress-strain curve for KCl and increased the slope of the linear poriton of the elastic range of the curve for LiF. This phenomenon, called the Roscoe effect, is attributed to dislocation pile-ups created when leading dislocations emanating from active sources are prevented by the film from leaving the crystal through the surface. The release of these pile-ups was evidenced by the large number of acoustic emission bursts following the removal of a film from its strained substrate crystal. Both coated and uncoated specimens of LiF exhibited acoustic emission peaks in the macroscopic elastic range having their maximums at a strain of approximately $\epsilon_0 = 10^{-3}$ and skewed toward larger values of ϵ. The analysis of this acoustic emission distribution forms the basis for a microscopic deformation model for LiF. It is hypothesized that this distribution should be identical to the distribution of dislocation source inverse loop-length. An expression for the probability of observing N acoustic emission bursts while the crystal is being strain from ϵ to $(\epsilon + d\epsilon)$ is written in terms of inverse loop-length and used to calculate average values of ϵ, ρ_M, and (A_1/A_2), i.e., the dislocation strain, mobile dislocation density, and ratio of successive ultrasonic wave amplitudes, respectively. The effect of the thin films on the acoustic emission peak is discussed qualitatively and interpreted in terms of the model. Since no such peak was observed for KCl, the differences in the microscopic deformation mechanisms for that material are emphasized. Finally, suggestions are made for further work concerned with the effects of thin films on the properties of substrate crystals and related acoustic emission phenomena.

1607 **M. G. Seitz. "Moglichkeiten der Werkstoffkontrolle durch Schallemissions-messungen beim Zugversuch (Feasibility of Material Control Through Acoustic Emission Measurement from Tensile Testing)." pp 45–58 in** *Schallemission: Anwendung bei der Untersuchung, Prufung und Uberwachung metallischer Werkstoffe.* **Deutsche Gesellschaft fur Metallkunde eV, 637 Oberursel/Taunus bei Frankfurt/Main, Adenauerallee 21, Germany. 1974. proceedings of symposium held in Munich, Germany. Apr 18–19, 1974. (paper in German).**

1608 **B. J. Selig, H. A. Runde, and R. P. Kosky. "N–Plant Monitoring Improves Reliability."** *Electrical World,* **ELWOA, 181(10):44–48. May 15, 1974.**

New monitoring techniques and requirements have been developed in recent years to improve the safety and reliability of nuclear power plants. These are in addition to the functions normally monitored in the control room, and they include, but are not limited to, these areas: In–service inspection, internals-vibration monitoring, loose-parts monitoring, and measuring primary-coolant boron concentration in pressurized water reactors. In-service inspection of nuclear power plants is defined in the ASME Boiler and Pressure Vessel Code, Section XI, "Rules for Inservice Inspection of Nuclear Reactor Coolant Systems." This section of the code stipulates that all nuclear plants must have 100% nondestructive examination of welds in the primary-coolant circuit (Class 1) before plant operation, and subsequent periodic re-examination of selected areas as defined by applicable tables in the code. Addenda require that Class systems be examined in a manner similar to that already required for Class 1, and action has been initiated to extend the examination to Class 3 systems. *Ultrasonic Testing,* is the principal technique used for volumetric examinations. To ensure the detection of defects regardless of their orientation, Section XI requires inspection by beams propagating perpendicular to the surface, and in directions 45 and 60 degrees from the perpendicular, both parallel to and perpendicular to the weld. The objective of performing these extensive nondestructive tests periodically is to assure the integrity of the overall system. By examining pressure-boundary welds and their adjacent base metal once before startup, and by re-examining representative areas after periods of power operations, any changes in existing reflectors, or the occurrence of new reflectors, can be determined. Section XI requires plant design to provide access either for mechanical-equipment or manual-personnel inspections. Due to radiation limitations, remote mechanical equipment is required to position the ultrasonic test heads to inspect the longitudinal and circumferential welds, the nozzle welds and inner radii, and the flange-to-shell welds of the reactor pressure vessel. An addendum to Section XI provides an optional method of evaluating reflectors found during in-service examinations by using fracture-mechanics techniques. A committee has been formed under Section V of the code to establish the rules for using acoustic emission as a nondestructive technique to perform inservice inspections. Piezoelectric transducers are placed on the primary-coolant boundary (reactor pressure vessel) to detect acoustic emission, and the data are processed through computers to determine the location and approximate growth rate of indications in the system, on a real-time basis. Acoustic-emission testing has been successful on hydrostatic tests, and work is under way to verify its use during reactor operations. This technique is expected to augment or replace utlrasonic testing in in-service inspections.

1609 **G. W. Sevall, Jr. "Nondestructive Testing of Construction Materials and Operations." CERL-TR-M-67. Department of the Army, Construction Engineering Research Laboratory, Champaign, Illinois. Dec 1973. [NTIS: AD-774-847].**

The report reviews nondestructive testing methods applicable to the construction industry. Established methods and methods still in development are included. Recently developed methods were evaluated from initial reports and prototype applications. A number of the new methods show promise in reducing inspection time or presenting information previously unobtainable. Existing and developing methods are also classified according to the construction materials and operations they test. The report is designed to help not only those interested in the field of nondestructive testing, but also those concerned with specific applications.

1610 **N. R. Sewall. "Acoustic Emission Coincidence Detector for Monitoring High Residual Stress Areas in Symmetrical Pressure Vessels." UCRL-74515. Lawrence Livermore Laboratory, Livermore, California. Mar 7, 1973. [NTIS]; pp 247-250 in** ***Proceedings of the Nineteenth International ISA Aerospace Instrumentation Symposium.*** **B. Washburn, Editor. Instrument Society of America, Pittsburgh, Pennsylvania. 1973. symposium held in Las Vegas, Nevada. May 21, 1973.**

In pressure vessels where high residual stress areas are suspected, acoustic emission can be a valuable nondestructive testing tool in the process of proofing the vessel. Naturally, the suspected areas of failure would be in the inlet-tube connection weld or braze, or a waist weld in the part itself. Welding destroys the homogeneity of the base metal itself; hence, one can suspect high residual stresses left in the base metal after welding has taken place. In proof testing, by judicious placement of acoustic emission transducers along with the described flaw detection coincidence circuitry, a vessel can be checked or somewhat certified against high-residual stress areas causing microcracks, by the acoustic emission addition to the proof test. Discussed is the problem and the proposed solution including the acoustic flaw window, acoustic velocity in the material, reflection or reradiation of bursts, the placement of transducers, and the final circuit. The results have been helpful in the certification of the vessels worked on.

1611 **R. S. Sharpe. "Research, Development and Applications at the British Nondestructive Testing Centre." pp 37-42 in** ***Proceedings of the 7th Symposium on Nondestructive Evaluation of Components and Materials in Aerospace, Weapons Systems and Nuclear Applications.*** **Western Periodicals Company, North Hollywood, California. July 1969. symposium held in San Antonio, Texas. Apr 23-25, 1969.**

A Centre for research and development to assist industry in solving materials inspection problems has been set up by the British Ministry of Technology at the Atomic Energy Research Establishment, Harwell, England. The Centre is expected to demonstrate industrial interest and commercial viability by attracting and carrying out sponsored work in all areas of nondestructive testing, to complement what is already available in Britain from the organised structure of the NDT industry and the industrial Research Assoications. In addition to sponsored work, the Centre initiates research and development in areas where there is a commercial or a *national interest* justification, and the program of work in this category is regularly reviewed by an Advisory Committee of specialists. The Centre has also set up an advisory service for NDT which collects and disseminates information on all aspects of the subject. The paper describes how the Centre has grown from a development section originally devoted to the requirements of the British nuclear power program and discusses how the current programs of work have evolved naturally from this earlier rather specialized expertise.

1612 **R. S. Sharpe. "Research in Non-Destructive Testing."** ***British Journal of Non-Destructive Testing,*** **BJNTA, 9(1):10–15. Mar 1967.**

1613 **R. S. Sharpe. "The Application of NDT Techniques to the Problem of Stress Corrosion Cracking." Paper No. 10, pp 10–1 to 10–11 in "Engineering Practice to Avoid Stress Corrosion Cracking." AGARD Conference Proceedings No. 53. North Atlantic Treaty Organization, Advisory Group for Aerospace Research and Development, Paris, France. Feb 1970. [NTIS: AD-701-789]; paper presented at the Structures and Materials Panel 29th Meeting, held in Istanbul, Turkey. Sep 30–Oct 1, 1969.**

The paper examines the areas where improved nondestructive testing methods could assist in determining the susceptibility of components to stress corrosion supporting laboratory investigations into the mechanism of stress corrosion and monitoring the progression of cracking during the service life of components to replace the present practice of extrapolating from laboratory simulation tests. Emphasis is directed towards methods of residual stress measurement which seem closest to a practical test to replace the destructive stress relaxation methods currently used.

1614 **B. J. Shaw. "On the Use of Piezoelectric Films for Ultra Sensitive Strain Gauges, Metal Failure Warning Devices and for Acoustic Emission." pp 237–239 in** ***1973 Ultrasonics Symposium Proceedings.*** **Catalog No. 73 CHO 807-8 SU. Institute of Electrical and Electronics Engineers, Inc., New York. 1973. symposium sponsored by IEEE Group in Sonics and Ultrasonics, held in Monterey, California. Nov 5–7, 1973.**

With the establishment of techniques for making reliable piezoelectric films and contacts to the films, three devices have been evolved: (1) An ultra sensitive strain gauge with time, spacial, and strain resolution far superior to currently available gauges; (2) A metal failure warning device for monitoring in-service machinery; and (3) A superior acoustic emission monitor for research and industrial applications. Applications of the devices and limitations are discussed.

1615 M. Shimizu, T. Imanaka, and K. Fujimoto. "Acoustic Emission Characteristics during Stress Relaxation in Low Carbon Steels." Paper 5-4, pp 376-392 in *The Third Acoustic Emission Symposium.* Japan Industrial Planning Association, Dobashi Building, 3-17-8, Akasaka, Minato-ku, Tokyo, Japan. 1976. proceedings of a symposium held in Tokyo, Japan. Sep 16-18, 1976.

Recent studies on acoustic emission have suggested that the generation of acoustic emission is related to the dynamic motion of dislocations during elementary processes of deformation. It has been shown that the generation of acoustic pulses is associated with the increase of internal friction due to the multiplication of mobile dislocations at the early stage of plastic deformation in copper single crystals. It is also recognized that strain aging in low carbon steel could be interpreted in terms of segregation of interstitial impurities around dislocations. Acoustic emission generated during plastic deformation has not yet been quantitatively interpreted in terms of dislocation parameters, such as pinning length, density, and velocity of dislocations. In low carbon steels, it is expected that dislocation density dependence of rms acoustic emission can easily be determined as well as that of other mechanical properties because of monotonic decrease of dislocation density during stress relaxation. Strain amplitude dependences of internal friction suggest that acoustic emission detected through air gap varies during aging. Acoustic emission observed in the Granato-Lucke region contains the third harmonic which is caused by nonlinear dislocation motion. In the present study, rms acoustic emission during plastic deformation and stress relaxation has been measured and an attempt has been made to correlate dislocation reaction with the emission observed during stress relaxation. Both time dependents of internal friction and of third harmonic acoustic emission have also been determined in the strain amplitude region exceeding a critical breakaway strain. Most of the experimental observations are concerned with the pinning down of dislocations by interstitials during aging.

1616 M. Shimizu, K. Sano, T. Imanaka, and K. Fujimoto. "Acoustic Emission during Plastic Deformation in Aluminum, Copper and 3% Silicon-Iron." Session 5, pp 1-22 in *The Second Acoustic Emission Symposium.* Japan Industrial Planning Association, Dobashi Building 5F, 3-17-8 Akasaka, Minato-ku, Tokyo, 107 Japan. 1974. proceedings of a symposium held in Tokyo, Japan. Sep 2-4, 1974.

In the paper, an attempt is made to assign specific dislocation reactions to the emission observed in the loading, strain rate change, relaxation and unloading regimes of uniaxial tensile or compressive tests on copper single crystals, aluminum single crystals, and 3% Si-Fe single crystals with the simultaneous measurement of ultrasonic attenuation in some cases. In addition, special attention has been given to the effect of strain rate on rms voltage of acoustic emission.

1617 O. A. Shinaishin, M. S. Darlow, and S. J. Acquaviva. "Acoustic Emission Detection of Fatigue Crack Initiation and Propagation in Notched and Unnotched Titanium Specimens." *Materials Evaluation,* MAEVA, 34(6):137-143. June 1976; paper presented at the American Society for Nondestructive

Testing National Fall Conference, Atlanta, Georgia. Oct 13–16, 1975. [see also report AMMRC–CTR–75–1, by M. S. Darlow *et al.* with same title, Reference 352].

Acoustic emission from test specimens was used in this investigation to warn of initiation or presence of cracks. The titanium specimens used in these tests were of high strength and were brittle, causing the process of acoustic emission, from its initiation to specimen failure, to be short. Static and dynamic (fatigue) tests were conducted. The fatigue specimens were run in unnotched and notched configurations in reverse bending and in tensile fatigue. Crack initiation or the nucleation of small cracks was clearly and consistently defined by the commencement of detectable acoustic emissions throughout the entire tests conducted. In general, emission from notched specimens started considerably earlier in the test than that from unnotched specimens. One explanation for this phenomenon is presented. Cracks formed in notched specimens from both reverse bending and tensile fatigue tests were metallurgically examined before failure and after sectioning in order to define the relationship between acoustic emission behavior and fatigue crack morphology. Fatigue crack arrest can be attained and should be valuable in many fatigue test runs.

1618 T. Shiraiwa, Y. Sakamoto, K. Hirakawa, H. Yamaguchi, and T. Sakai. "Detection of Crack Initiation by Means of Acoustic Emission Techniques." Session 6, pp 1–14 in *The Second Acoustic Emission Symposium.* Japan Industrial Planning Association, Dobashi Building 5F, 3–17–8 Akasaka, Minato-ku, Tokyo, 107 Japan. 1974. proceedings of a symposium held in Tokyo, Japan. Sep 2–4, 1974.

The detection of crack initiation is important for the study of mechanical behaviors of materials. In the present study, the acoustic emission technique is applied for two types of fracture behaviors of steel, fatigue cracking, and hydrogen embrittlement, and its applicability for the detection of crack initiation is investigated.

1619 A. K. Shoemaker. "The Embrittling Effect of Small Elastic Stress Waves on Crack Toughness of a Structural Steel." ***Journal of Materials,*** **JMLSA, 2(3):597–624. Sep 1967.**

The reduction in static low-temperature crack toughness, as measured by fracture mechanics K_{Ic} values, was studied in an A-201B structural grade steel for the embrittlement caused by small-amplitude elastic stress waves superimposed on the static stress state of an edge-notched 0.75-inch thick specimen. The stress waves were generated by five different methods of impacting the edge of the specimen on the side opposite the crack tip when a static load existed on the specimen. The data showed an average reduction of 26 percent in the static K_{Ic} over a temperature range of −220 to −190 °F for impacting the specimen with a 1.312-inch diameter steel ball, a 12-inch steel bar, or a ball-peen hammer. A transient tensile strain of 40 microinches per inch normal to the crack was shown to cause crack extension. This transient tensile strain was measured before and during the fracture test by high-resistance strain gages fixed to the specimen ahead of the crack tip. A four-point bending specimen was also fractured by this method of embrittlement to show that this phenomenon is not just peculiar to one specimen geometry. The embrittlement can be attributed to two effects: (1) a reduction in crack toughness value necessary for crack extension due to the rate at which the transient strain increased, and (2) the superposition on the static K_I value of the stress intensity factor, $K_{I(transient)}$, caused by the stress wave amplitude. A discussion is given of the possible damage of waves at the crack tip even though crack extension does not occur. Data for 0.0005-inch radius and fatigue cracked notches showed consistent crack toughness values for the temperature range investigated. [reprinted by permission of the American Society for Testing and Materials, Copyright].

1620 **P. S. Shoemaker. "Acoustic Emission, An Experimental Method." M.S. Thesis. Michigan State University, East Lansing, Michigan. 1961.**

The purpose of the investigation is to examine an experimental method that would enable one to study acoustic emission and to show some experimental results. These acoustic emissions are at such low energy levels that amplification is necessary to detect them. Great care must be taken in the method of detection so that any experimental process that generates mechanical or electrical vibrations will not interfere or be confused with the actual acoustic emission signals. The transducer employed for sound pulse detection is an ammonium-dihydrogen phosphate piezoelectric crystal. The loading mechanism used to subject a specimen to stress must be completely silent. This silent load is produced by cooling contraction of a preheated aluminum rod. The results presented are in the form of histograms with acoustic emissions per minute versus specimen strain. There is ample verification that emission signals are detected and that the experimental method is satisfactory. Analysis of the results indicates that there is an obvious difference in acoustic emission between different engineering materials; and some evidence that there could be a difference in the same engineering materials. There are obvious patterns present in the histograms that indicate possible correlation between acoustic emission and engineering design properties. Certainly, there is evidence present that indicates a relation between acoustic emission and metallurgical and crystal properties in metals.

1621 **M. Kh. Shorshorov, O. V. Gusev, A. G. Penkin, S. L. Dubashev, and M. M. Rybal'cenko. "Investigation of Plastic Deformation Processes and Destruction in Composite Materials Using the Method of Acoustic Emission." in *Second International Symposium on Composite Metallic Materials.* 1975. (in Russian). proceedings of a symposium held in Bratislava-Smolenice, Czechoslovakia. Oct 29-31, 1975.**

Description is given of equipment for recording acoustic emission parameters. The kinetic laws governing the processes of plastic-deformation and destruction in fibre composites Al-steel and Ni-W were studied by this method. The experimental tests of flat specimens were carried out on the Instron-type apparatus at constant deformation rate of 7.10^{-5}/s with simultaneous recording of acoustic emission parameters.

Evaluated were the experimental results of acoustic emission intensity–pulse number in time unit–and histograms of acoustic emission pulse number distribution according to their amplitudes in terms of specimen deformation degree. [Abstract 22-0921. *Metals Abstracts,* Vol. 9, 1976].

1622 **J. B. Shortall and C. P. Yarwood. "An Annotated Bibliography on Acoustic Emission from Composite Materials." P. G. Report No. 29. University of Liverpool, Department of Metallurgy and Materials Science, Liverpool, England. Oct 1972.**

The bibliography contains 67 references which are divided into seven sections: (1) Books of general interest, (2) Acoustic emission from composite materials, (3) Instrumentation-transducers and electronic instrumentation, (4) Review of stress wave analysis, (5) Experimental technique, (6) Papers of general interest, and (7) Conference reports.

1623 **L. Z. Shuck and T. W. Keech. "Monitoring Acoustic Emission from Propagating Fractures in Petroleum Reservoir Rocks." pp 309–338 in** ***Proceedings First Conference on Acoustic Emission/Microseismic Activity in Geologic Structures and Materials.*** **H. R. Hardy, Jr. and F. W. Leighton, Editors. Trans Tech Publications, Clausthal, West Germany. 1977. conference held at Pennsylvania State University, University Park, Pennsylvania. June 9–11, 1975.**

The results presented are from a continuing petroleum reservoir stimulation research program at the Morgantown Energy Research Center. This research is related to hydraulic fracturing of oil and gas reservoir rocks. The paper describes a field experiment which was conducted in a petroleum reservoir in Bradford, Pennsylvania. Twelve hydrophone type sensors were placed at the oil-bearing rock elevation of 2000 feet from the surface in adjacent wells surrounding the fractured well. Complete directional acoustic velocity distribution and acoustic background noise levels were evaluated. The resulting measurements made during the hydraulic fracture operations are discussed.

1624 **E. Siegel. "Acoustic Emission Study of Fatigue Crack Growth in Unirradiated Zircaloy." Public Service Electric and Gas Company, Energy Laboratory, Maplewood, New Jersey. (no date shown).**

This short note summarizes the recent demonstration that acoustic emission is feasible as a technique for studying fatigue crack growth in through-cracked, single-edge notched (SEN) specimens of unirradiated zircaloy. As such, it is possible that this technique can be extended to SEN specimens of other hcp nuclear metals and alloys, such as hafnium, and other bcc nuclear metals, such as niobium and tantalum, in addition to stainless and pressure vessel steels. Its application to non-SEN specimens (smooth round bar, square bar, etc.) and to actual inservice studies of fatigue in nuclear reactor applications will hinge considerably on detailed metallurgical studies in the laboratory, such as this one, and a maturity of interpretation, which will only come about through long experience.

1625 E. Siegel. **"Burst Acoustic Emission during the Bauschinger Effect in FCC and HCP Metals and Alloys." pp 204–208 in** *1973 Ultrasonics Symposium Proceedings.* **Catalog No. 73 CHO 807–8 SU. Institute of Electrical and Electronics Engineers, Inc., New York. 1973. symposium sponsored by IEEE Group on Sonics and Ultrasonics, held in Monterey, California. Nov 5–7, 1973; paper presented at the American Society for Metals Materials Testing and Quality Control Conference, Detroit, Michigan. Oct 21–24, 1974;** *Acta Metallurgica,* **AMETA, 25(4):383–394. Apr 1977.**

Unload burst-type acoustic emission during the tension cycle of the stress-strain hysteresis loop during cyclic deformation was studied and related to grain size, Bauschinger strain, and stacking fault energy. The Bauschinger effect in polycrystalline 99.99% aluminum, 6061 and 2024 aluminum alloys, copper-7.9% aluminum alloy, 70-30 brass, and 99.95% magnesium was studied at ultrasonic frequencies using acoustic emission. These relations lead to a natural identification of acoustic emission during cyclic deformation with the low frequency strain burst phenomenon, explained by Neumann's edge dislocation dipole dissociation-trapping avalanche theory. Comments are made on the applicability of this theory to an explanation of burst acoustic emission during cyclic deformation and the Bauschinger effect.

1626 E. Siegel. **"Correlation of Microyielding Strain Bursts and Acoustic Emission Pulses in BCC Alloys Exhibiting Yield Drops and Serrated Yielding."** *Scripta Metallurgica,* **SCRMB, 8(6):617–620. June 1974.**

The author indicates that the recent study of microplastic strain bursts during cyclic deformation of a carbon steel by Abel and Muir is directly indentifiable as a low (zero) frequency component of acoustic emission as made by Fisher and Lally in bcc iron alloys in the macro-plastic region. This correspondence is an extension of the correlation of Neumann's cyclic deformation strain burst phenomenon in fcc and hcp single crystals at low temperatures to the acoustic emission burst phenomenon during macro-plastic cyclic deformation in fcc and hcp polycrystalline metals and alloys of Sankar and Siegel. The study corresponds to the analogy pointed out by Siegel and and by Day between the Tinder-Trzil non-cyclic, torsional deformation millimicroplastic strain burst phenomenon in Zn single crystals and the Siegel study of acoustic emission during macroplastic creep in fcc and hcp polycrystalline metals and alloys. [reprinted by permission of Pergamon Press Ltd].

1627 **E. J. Siegel. "Kilocycle Acoustic Emission during Creep in Lead, Aluminum, and Cadmium, Part I. Experimental."** ***Physica Status Solidi (a)*, PSSAB, 5(3):601–606. June 1971.**

Acoustic emission in the 100 to 200 kHz frequency band during simple, axial tensile loading of polycrystalline lead, aluminum, and cadmium has been observed. A general correspondence and mirroring of the usual strain-time creep curve in the total cumulative counts-time curve at room temperature were found, both at constant load and during loading increments. Stress was not a constant in these experiments. For Part II, see Reference 1628.

1628 **E. J. Siegel. "Kilocycle Acoustic Emission during Creep in Lead, Aluminum, and Cadmium, Part II. Theoretical."** ***Physica Status Solidi (a)*, PSSAB, 5(3):607–615. 1971.**

For Part I, see Reference 1627. The general mechanism independent features of the Sedgwick theory of acoustic emission and the Gilman approach to microplasticity are used to analyze acoustic emission creep results. A more detailed theory will be presented at a later time.

1629 **E. Siegel. "Relation of the Neumann Strain Burst Phenomenon to Acoustic Emission during Cyclic Deformation and Fatigue of FCC and HCP Metals."** ***Scripta Metallurgica*, SCRMB, 8(6):587–600. June 1974.**

The paper discusses the fundamental relation between the strain burst phenomenon during cyclic deformation of fcc and hcp single crystals (Al, Ag, Au, Pb, Zn, Cd, Mg, Cu) performed recently by Neumann and the phenomenon of acoustic emission as observed during cyclic deformation and fatigue of many fcc and hcp metals and alloys. Neumann observed that during the rising bias-stress cyclic deformation of fcc and hcp single crystals, periodic series of strain bursts of short duration were emitted with ever increasing amplitude, separated by long periods of relatively little strain increase. The author discusses the similarities between the low-frequency strain burst phenomenon during cyclic deformation and the higher frequency acoustic-emission phenomenon during cyclic unloading, especially in regard to their similar dependence upon grain size, prestrain, stacking fault energy, theoretical interpretation, and number of bursts per unit time which indicate that these are closely related, if not identical, phenomena.

1630 **E. Siegel. "Strain Bursts and Acoustic Emission as Fundamental Plastic Resonance Phenomena: Theory of the Phonon Maser." Paper 4–5, pp 533–558 in *The Third Acoustic Emission Symposium*. Japan Industrial Planning Association, Dobashi Building, 3-17-8, Akasaka, Minato-ku, Tokyo, Japan. 1976. proceedings of a symposium held in Tokyo, Japan. Sep 16–18, 1976.**

In the paper, it is argued that acoustic emission and the Neumann strain-burst phenomenon are both manifestations of fundamental plastic resonance phenomena of a solid, and these phenomena are treated as a phonon maser source. In addition, examined will be a Shockley partial dissociating total dislocation and its associated stacking fault ribbon as a possible harmonic oscillator in the mechanical spectrum which can store and discharge, or emit, acoustic phonon radiation in discrete bursts after some energy storage time. This is in exact analogy with the classical cavity theory of the infrared maser or optical laser. Further it is expounded that such an inclusion of the stacking fault in acoustic emission or strain burst theory *ab initio* agrees with Siegel's recent exhibition of the similar dependence of acoustic emission pulse amplitudes and strain burst magnitudes on stacking fault energy in fcc and hcp metals and alloys.

1631 **M. G. Silk, N. R. Williams, and K. F. Bainton. "The Potential Role of NDT Techniques in the Monitoring of Fixed Offshore Structures."** ***British Journal of Non-Destructive Testing,*** **BJNTA, 17(3):83-87. May 1975.**

For the NDT of drilling and production platforms, it is suggested that monitoring systems may have to replace divers and possible methods are considered. Methods discussed are witness devices, television viewing, sonar, ultrasonic Lamb waves, low frequency ultrasonics for concrete, acoustic emission, and vibration analysis. Some of these techniques could readily be installed on existing structures; others would be more limited and would have to be designed into new platforms.

1632 **M. G. Silk, N. R. Williams, C. H. Jones, B. Watkins, and G. E. Cassie. "The Continuous Monitoring of Fixed Offshore Platforms for Structural Failure: A Review and Assessment of Techniques." AERE-R 8055. United Kingdom Atomic Energy Authority, Nondestructive Testing Centre, Materials Physics Division, Atomic Energy Research Establishment, Harwell, Oxfordshire, England. Sep 1975. [NTIS].**

1633 **J. A. Simmons and R. B. Clough. "Theoretical Aspects of Acoustic Emission Spectral Measurements/Les aspects theoriques des measures des spectres d'emission acoustique." National Bureau of Standards, Washington, D.C.; Paper 3K/1 in** ***Proceedings of the Eighth World Conference on Nondestructive Testing.*** **volume containing Sections 3K1, 3H1, 3G. International Committee for Nondestructive Testing, Secretariat du Cofrend, Institute de Soudure, 32 Boulevard de la Chapelle, 75880 Paris Cedex 18, France. 1976. conference held in Cannes, France. Sep 6-10, 1976. (paper in English). [NTIS: INIS-mf-3450].**
Improvements in acoustic emission systems may be possible through a consideration of the theory of acoustic emission.

1634 **A. B. Simpkins. "Silent Sounds of Industry."** *Design News,* **DIGNA, 21(6):50–53. Mar 16, 1966.**

Article discusses the Hewlett-Packard Company, Delcon Division's ultrasonic translator detector.

1635 **G. D. Sims. "Stress Wave Emission." pp 5.1 to 5.8 in "Newsletter on the Non–Destructive Testing of Composite Materials and Adhesive Bonded Joints." Issue Number 3. Royal Aircraft Establishment, Farnborough, Hants, England. June 1974.**

Interest at the National Physical Laboratory in the use of emission techniques as a method of detecting, characterising, and continuously monitoring the occurrence of damage in composite materials originated with work by Clarke and Sims on electrical emission in glass-fibre reinforced resins. Electrical emission techniques required the specimen and test apparatus to be carefully screened against spurious external electrical effects and the examination of stress wave or acoustic emission was preferred in later studies. Recent areas of interest have included: (a) Equipment development, a portable and simplified SWE system; (b) Characterisation of failure processes by waveform analysis; (c) Determination of the effect of transducer response and specimen attenuation, etc., on the detection and analysis of SWE; and (d) Detection and assessment of microdamage during short and long term testing.

1636 **A. C. E. Sinclair, C. L. Formby, and D. C. Connors. "Acoustic Emission from Defective C/Mn Steel Pressure Vessel."** *International Journal of Pressure Vessels and Piping,* **PRVPA, 3(3):153–174. July 1975.**

Location analysis has been used to examine acoustic emission from a partial thickness slit in a test vessel under pressurisation. Pulse arrival times at three sensors were used to locate emission sources. The requirement for near-coincidence in arrival times of all sensors reduced spurious background events to low levels. In spite of high background count rates at each individual sensor, the slit was clearly delineated by its emissions. Emission activity moved from the centre to the ends of the slit as the stress on the slit increased. Significantly, the emission count appeared to be directly proportional to the volume of plastically yielded material associated with the defect based on a simple model for the increase in plastic zone size with pressure. Application of acoustic emission techniques to structural validation is discussed in terms of the measured amplitude spectrum of the detected acoustic emissions. A relationship is derived between the detected emission count, the threshold amplitude for detection, the sensor spacing, and the created plastic volume for the steel studied.

1637 **A. C. E. Sinclair, C. L. Formby, and D. C. Connors. "An Assessment of Acoustic Emission Techniques on Medium–Size Pressure Vessels Tested to Failure." Paper G3/9 in** *Structural Mechanics in Reactor Technology.* **Compiled by T. A. Jaeger. Volume 3, Part G–H. North–Holland Publishing**

Company, Oxford, England. 1975. proceedings of the 3rd International Conference on Structural Mechanics in Reactor Technology, held in London, England. Sep 1-5, 1975. [distributed in the U.S. and Canada by American Elsevier Publishing Company, Inc., New York].

Location analysis has been used to examine acoustic emission from a partial thickness slit in a test vessel during pressurization. Preliminary investigations showed that sharp emission pulses at a source would propagate as sharp pulses to a sensor several metres distant, and that intrinsic material attenuation was negligible at the 165-kHz operating frequency. Pulse arrival times at three sensors were used to locate emission sources and the requirement for near-coincidence in arrival times at all sensors reduced background noise to low levels. The slit was clearly delineated by its emissions. Emission activity moved from the centre to the ends of the slit as the stress on the slit increased, suggesting that emissions were generated during the initial yielding of the steel. This indication was strengthened by the fact that the emission count appeared to be directly proportional to the volume of plastically yielded material assoicated with the defect, based on a simple model for the increase in plastic zone size with pressure. Application of acoustic emission techniques to structural validation is discussed in terms of the measured amplitude spectrum of the detected acoustic emissions. A relationship is derived between the detected emission count, the threshold amplitude for detection, the sensor spacing, and the created plastic volume for the steel studied.

1638 **A. C. E. Sinclair, C. L. Formby, and D. C. Connors. "Some Examples of Laboratory Application and Assessment of Acoustic Emission in the United Kingdom. Part II: Fatigue Crack Assessment from Proof Testing and Continuous Monitoring." pp 51-64 in *Acoustic Emission*. Edited by R. W. Nichols. Applied Science Publishers Ltd, Ripple Road, Barking, Essex, England. 1976. proceedings of a colloquium on acoustic emission held during the 1975 Annual Assembly of the International Institute of Welding, Tel Aviv, Israel. July 7-13, 1975. [paper presented by B. J. L. Darlaston].**

This part of the paper deals with the detection and assessment of cracks in structural components and considers two possible styles of acoustic emission monitoring, namely during a proof test of the inservice component and during normal stress-cycling of the component with attendant fatigue. (See Reference 813 for Part I.)

1639 **A. C. E. Sinclair, A. Tobias, and D. C. Connors. "Operation of an Acoustic Emission Defect Location System in Industrial Conditions." Paper No. 11 in *Reactor Inspection Technology*. British Nuclear Energy Society, London, England. 1975. proceedings of the Symposium on Reactor Inspection Technology, sponsored by the British Nuclear Energy Society, Western Branch, held in Bristol, England. Feb 26, 1975.**

An acoustic emission analysis system (ACEMAN) designed for acoustic source location was used to monitor the proof test of a large industrial pressure vessel. The extensive surveillance region included a representative sample of welds and nozzle array complexity. In spite of a large noise background (since the test was conducted according to standard industrial procedures), sophisticated signal discriminators and multisensor timing correlation permitted location of several distinct areas of acoustic activity. Differing emission patterns with pressure pointed to a variety of source mechanisms. Factors important to the acoustic emission assessment of structural integrity are discussed.

1640 **J. J. Singh. "Acoustic Emission from Aluminum Alloys Subjected to Low Level Periodic Stress."** *IEEE Transactions on Sonics and Ultrasonics*, **IESUA, SU-19:392. July 1972. [abstract only; see Reference 1641 for complete report].**

Examined was the acoustic emissions attendant on a propagating crack in high strength structural aluminum alloys, using a PZT accelerometer cemented on the test specimen with RTV. The acoustic emission frequency ranges from audible to several megacycles. By choosing narrowly banded, high frequency (100 to 200 kHz) and high amplitude signals for counting most of the background noise as well as the test specimen murmur due to microcracks, etc., were eliminated. Measurements on double-notched (7075-T6) plate aluminum samples indicate that the summation acoustic emission is related to: (Stress Intensity Factor)^X, where the value of X depends upon the discriminator level setting in the counting system. The simultaneous use of a narrow frequency band and the amplitude discrimination permits this technique to be used for structures subjected to low frequency cyclic stress. [reprinted from *IEEE Transactions on Sonics and Ultrasonics*].

1641 **J. J. Singh. "Detection of Crack Growth by Acoustic Emission Technique." National Aeronautics and Space Administration–Langley Research Center, Hampton, Virginia. 1971; paper presented at the Institute of Electrical and Electronics Engineers Ultrasonic Symposium, Miami Beach, Florida. Dec 6–8, 1971.**

According to recent studies from pressure vessels and tensile specimens subjected to increasing stress, acoustic emission occurs only when the crack size increases. Re-examination of the acoustic emission phenomenon in cases of flawed and unflawed plate tensile specimens subjected to steadily increasing static load, as well as constant level cyclic loading, was done. The paper discusses some results obtained for 7075–T6 and 2024–T3 Al plate specimens.

1642 **J. J. Singh and W. T. Davis. "Acoustic Emission: A New Technique for Monitoring Structural Integrity." Langley Working Paper LWP–998. National Aeronautics and Space Administration–Langley Research Center, Hampton, Virginia. Oct 6, 1971.**

A new technique of nondestructive monitoring of stressed structures, based on the acoustic transducer action of a propagating crack, has been developed. It is applicable to metallic as well as nonmetallic structures. Present investigation has shown that the acoustic emissions from stressed specimen with flaws are strongly related to the stress intensity factors at the flaw. It is concluded that this dependence of acoustic emission on the flaw size can be used to calculate the crack size for assumed crack shape.

1643 **J. J. Singh and W. T. Davis. "Evaluation of Acoustic Emission Technique for Crack Growth Measurement in Aeronautical Structures." ISA Reprint. Instrument Society of America, Pittsburgh, Pennsylvania. 1974; paper presented at the 20th International Instrumentation Symposium, Albuquerque, New Mexico. May 21–23, 1974.**

The applicability of acoustic emission technique for measuring fatigue crack growth in aluminum alloy specimen has been investigated. Extensive data presented on center notched, thin plate, specimen have shown that ($\Sigma N \propto K^m$) type of acoustic emission law is applicable in subcritical crack growth region and can be used to verify the occurrence of crack length changes. Although acoustic emission signals are always detected whenever the crack dimensions change, acoustic emission technique should be supplemented by other NDE methods to obtain quantitative data on crack growth.

1644 **J. J. Singh and W. T. Davis. "Fatigue Crack Growth Measurement in Aluminum using Acoustic Emission Technique." National Aeronautics and Space Administration-Langley Research Center, Hampton, Virginia; paper presented at the 10th Acoustic Emission Working Group Meeting, Pasadena, California. Dec 6–8, 1972.**

The paper deals with the study of the acoustic emission rate as a function crack propagation rate in thin 2024-T3 Al plate specimens. Also reported are subsequent studies of acoustic emission rate as a function of crack length extension rate.

1645 **J. J. Singh, W. T. Davis, and J. H. Crews, Jr. "The Application of Acoustic Emission Technique to Fatigue Crack Measurement." Technical Note NASA TN D-7695. National Aeronautics and Space Administration, Langley Research Center, Hampton, Virginia. Oct 1974. [NTIS].**

The applicability of acoustic emission technique to measure fatigue cracks in aluminum alloy specimens was investigated. There were several variables, such as the metallurgical and the physical treatment of the specimen, that can affect the level of acoustic activity of a fatigue specimen. It is therefore recommended that the acoustic emission technique be supplemented by other nondestructive evaluation methods to obtain quantitative data on crack growth.

1646 **D. E. Siskind, H. R. Hardy, Jr., and S. S. Alexander. "A Study of Pressurized Underground Storage Cavities: Theoretical and Laboratory Investigations."** *International Journal of Rock Mechanics and Mining Sciences,* **IJRMA, 10:133–150. Mar 1973.**

The increasing use of oil and gas reservoir traps, salt cavities, and other man-made underground cavities for the storage of natural gas and other fluids necessitates a better understanding of the loading and failure characteristics of such storage areas. In the present study, cylindrical models containg minature reservoir cavities were fabricated for the experimental simulation of storage reservoir behavior. These were loaded under typical *in situ* stress fields and subjected to internal pressurization. Tangential and radial stresses on the cavity walls were inferred from theoretical solutions and the known boundary conditions and were compared with those predicted from measured levels of acoustic emission. The reservoir failures observed were tensile in form and occurred as predicted at or near the north or south poles of the cavities. The internal pressures necessary to cause reservoir failure, however, were about 2.8 to 3.0 times those predicted on the expectation of failure occurring when the tangential tensile stresses exceed the tensile strength of the rock surrounding the cavity. It was observed that the onset of acoustic emission began early in the pressurization cycle, at about one third of the pressure required for the eventual ultimate failure of the model. The acoustic activity and high failure pressures observed indicated that some kind of stress-relieving inelastic phenomenon probably commenced at a relatively low pressure, with the ultimate failure, easily recognized by the sudden drop in reservoir pressure, occurring considerably later. These results suggest the possible use of acoustic emission as a useful means of monitoring reservoir stability in both models and full-scale field structures. [reprinted by permission of Pergamon Press Ltd].

1647 **L. O. Slind and W. R. Sturrock. "The Acoustic Emission Characteristics of Ti - 6 Al - 4 V during Cyclic Loading." Boeing Company, Seattle, Washington. Feb 23, 1968. [NTIS: AD-836-482].**

The investigation studied the acoustic emission characteristics of Ti-6 Al-4V alloy sheet under varying conditions of cyclic stress, temperature, specimen configuration, and surface treatment. Specimens were loaded in flexure while transducers detected acoustic emission from deformation or fracture processes occurring within the specimens. The band width used was from 50 kHz to 1 MHz. Analysis of the acoustic emission, recorded for each test and plotted versus percent fatigue life, produced the following: (1) The acoustic emission rate and amplitude began at a high level and then decreased in the first ten percent of the fatigue life. Similarly, the acoustic emission rate and amplitude increased during the last ten percent of the fatigue life; (2) An increase in acoustic emission was noticed in some tests, in the range of 20 to 50 percent fatigue life. This could be related to crack initiation; (3) The magnitude of acoustic emission in the initial stages of fatigue indicated the magnitude of the fatigue life; and (4) Varying the specimen configuration or surface condition, cyclic stress, or test temperature had no noticeable effect on acoustic rate or amplitude.

1648 **B. Smith. "Stress Wave Emission Monitoring Applied to Proof Testing of a Military Bridge." Report No. 74510. Military Vehicles and Engineering Establishment (Christchurch), Barrack Road, Christchurch, Dorset, England. June 1974.**

The application of stress wave emission monitoring to the proof testing of a military bridge is described. The results are discussed in relation to supporting laboratory tests on specimens. Possible applications of the technique to service use of military bridges are discussed and a program for future work is briefly described.

1649 **D. Smith. "Reactor Pressure Vessel Integrity: Heavy Section Steel Technology Vessel Tests."** *Nuclear Engineering International,* **NEINB, 19(220): 751–757. Sep 1974.**

Fingers of suspicion have been pointed by opponents of nuclear energy in many countries at doubts about the adequacy of the large pressure vessels required to maintain high pressures over periods of 40 years or more in light water reactor plants. Recent work in the practical experimental testing of pressure vessels at Oak Ridge National Laboratory (Tennessee), a report by the ACRS in the U.S.A. on the probability of reactor vessel failure, and an important international conference on inservice testing of reactor pressure vessels in London have all now contributed to greatly reinforced confidence.

1650 **K. A. Smith. "Mechanical Signature Analysis of Rotating Machinery." General Electric Company, Corporate Research and Development, Schenectady, New York. 1972; paper presented at the American Society for Nondestructive Testing National Fall Conference, Cleveland, Ohio. Oct 16–19, 1972.**

The report describes how acoustic and vibration signals generated by operating machinery can be used to diagnose its internal condition without using internal sensors which might perturb the system of interest. It is shown how characteristic *signatures* can be electronically extracted from the overall noise and vibration signals. By interpreting these signatures in the light of engineering knowledge of the machinery under study, it is often possible to identify incipient failure modes long before final failure. This technology opens up many new possibilities to avoid unscheduled maintenance, improve product quality, and reduce testing time. These techniques have been implemented in a number of application areas ranging from jet engines to automobiles.

1651 **L. M. Smith. "Elevated Temperature Burst Test." Report 1148M-R. Aerojet-General Corporation, Sacramento, California. Nov 18, 1964.**

1652 **R. W. Smith and G. E. Adamson (to U.S. Atomic Energy Commission). "Akustischer Wandler und Verfahren zu seiner Herstellung (Acoustic Transducer and Method of Production)." German Patent 2,436,328. Feb 20, 1975. (in German).**

Patent concerns the design and fabrication of a piezoelectric transducer for use at high temperatures in a corrosive environment such as a nuclear reactor.

1653 S. Smith and T. M. Morton. "Acoustic Emission Detection Techniques for High-Cycle-Fatigue Testing." *Experimental Mechanics*, **EXMCA, 13(5):193-198. May 1973; Paper No. 2059A presented at the Society for Experimental Stress Analysis Fall Meeting, Seattle, Washington. Oct 17-20, 1972.**

A technique for the study of acoustic emission during high-cycle fatigue crack propagation has been developed. Difficulties resulting from complex emission behavior and the low signal-to-noise ratio encountered in this type of test are rectified by three methods: minimization of mechanical noise; separation of noise causes according to position in the loading cycle; and rejection of noise from sources other than the crack by a short-time correlation method. The approaches are discussed and data from a fatigue test are presented demonstrating acoustic emission behavior for crack-growth rates from 3×10^{-6} to 70×10^{-6} inches per cycle.

1654 B. V. Sokolov and V. V. Shemyakin. "Determination of the X, Y Coordinates of the Pulse Ultrasonic Signal Source." IAE-2573. I. V. Kurchatov Institute of Atomic Energy, Moscow, U.S.S.R. 1975. (in Russian and English).

A number of problems has been investigated, entering the total problem of predicting the emergency condition of a large-scale vessel, which are solved using acoustical emission. The place of each problem considered has been determined. The choice of the minimum numbers of signal detecting transducers in the array, required for the unambiguous calculation of the source site, has been substantiated. The relations have been obtained connecting the X,Y coordinates of the pulse signal source with the experimentally measured differences in the activation times of the transducers of the array. The criterion for the selection of the actual combination of signal groups when waves from several sources arrive to the transducers, has been shown. The form of the experimental information to be fed to a computer for further handling has been determined.

1655 B. V. Sokolov and V. V. Shemyakin. "On the Processing of the Surface Topography of the Acoustic Emission Field." IAE-2382. I. V. Kurchatov Institute of Atomic Energy, Mascow, U.S.S.R. 1974. (in Russian and English).

A method for the investigation of surface distribution of energy release for detecting the location of the source of acoustic emission signals has been described. The possibility for creating a *reference* background field has been discussed.

1656 V. N Sosedov and V. E. Vainberg. "Inspection of Pressure Vessels by the Acoustic Emission Method." *Soviet Journal of Nondestructive Testing,* **SJNTA, 10(2):245-247. Mar-Apr 1974; translated from** *Defektoskopiya,* **DEFKA, (2):142-144. Mar-Apr 1974. (in Russian).**

1657 **N. K. Sowards. "Acoustic Analysis of Stressed-Metal Components."** *Nuclear Safety,* **NUSAA, 8(5):503–504. Sep–Oct 1967.**

The objective was to perform developmental tests of acoustic-analysis techniques for application to structural-integrity evaluation of reactor pressure vessels and other stressed-metal components of nuclear power plants. Experiments pertaining to the detection and analysis of acoustic emissions from flaws in reactor pressure vessels and other stressed-metal components of nuclear power plants were started by Phillips Petroleum Company in 1966. Results of work to date have established a good foundation for developing a safety system that will provide data, either continuously or intermittently, on the structural integrity of such components. The technique was used in monitoring the PM-2A reactor vessel during the series of pressurization tests that culminated in vessel rupture. Analysis of the detected acoustic signals revealed a direct correlation between the degree of material stress and the intensity of acoustic energy emanating from the material at the point where rupture ultimately occurred. After the PM-2A vessel test program, experiments with tensile specimens of pressure-vessel steels were initiated to determine the following: (1) the effect on acoustic-signal characteristics of the degree of metal embrittlement, (2) signal reproducibility in the elastic and plastic regions of material deformation, and (3) threshold of acoustic-detection sensitivity as a function of flaw size, stress, material characteristics, and material environmental history. Data obtained through mid-April 1967 on SA-212 steel revealed a factor-of-16 difference between the amount of acoustic energy emitted from annealed and from embrittled specimens. Acoustic analysis will be applied to the SM-1A reactor primary system during a pressure-vessel annealing program presently scheduled for June 1967. The signal-detection and data-recording system has been prepared for shipment to the SM-1A site at Fort Greeley, Alaska, in May. This program provides an opportunity to evaluate the usefulness of the system for reactor-vessel-integrity analysis in an operating nuclear power facility. Future plans are to apply acoustic analysis to the SM-1A reactor vessel during in-place annealing of the vessel; evaluate the effectiveness of acoustic-analysis techniques when applied during plant operation; and evaluate the application of a mathematical model to the prediction of flaw size and growth rate.

1658 **H. Spanheimer and A. J. De Herrera. "Prediction of Composite Pressure Vessel Performance by Application of the Kaiser Effect in Acoustic Emission." Hercules Incorporated, Magma, Utah and Thiokol Company, Brigham City, Utah. 1975.**

Cases of solid-propellant rocket motors usually are fabricated from continuous fibers embedded in an epoxy matrix. Quality acceptance testing consists of conducting a hydrotest at the maximum expected operating pressure or greater, for the operating time of the motor. Such testing does not define the ultimate strength of the motor case, nor even whether the case could withstand a second pressurization at the same conditions. Further, as the safety factor is reduced for high performance cases such testing could cause static fatigue of the composite. Evaluation of acoustic emissions during testing could provide assurance of chamber integrity, predict ultimate performance, and improve reliability from testing conducted at lower pressure levels. Acoustic emission evaluation is being used to predict the performance of metal pressure vessels but the technique is much less advanced. A large amount of work has been conducted on small laboratory composite specimens and these results have been applied to evaluation of large cases in the Trident I (C4) Program. In laboratory specimens, the construction is simple and the location of flaws is well-defined. Actual motor cases have complex construction and the flaw(s) location is unknown. The methods of data acquisition are similar to those used for metal cases. The results have shown that cases have an identifiable acoustic emission output identified as the signature, that acoustic emission in good quality cases are minimal during periods of constant pressure, and that in successive pressurizations the acoustic emissions are minimal until the previously achieved maximum pressure is achieved. This last effect is known as the Kaiser effect. These effects have been used to approximately identify the flawed region, to identify acoustic emission with events during testing, and to qualitatively define case quality. The evaluation of data from more than 60 large cases and the correlation of the data to the above effects to define case performance will be presented.

1659 J. C. Spanner. "A Summary of Recent Acoustic Emission Applications." BNWL-SA-2262S. Battelle-Northwest, Richland, Washington. 1970.

1660 J. C. Spanner. *Acoustic Emission: Techniques and Applications.* **Intex Publishing Company, Evanston, Illinois. 1974. [available from American Society for Nondestructive Testing, Columbus, Ohio].**

The book provides a selective summary of reported activities on the development and application of acoustic emission techniques and equipment, based on a review of the most significant contributions to the technical literature through mid-1973. The book is intended to serve as a reference for those actively engaged in acoustic emission technology and as a guide to some of the capabilities and limitations of this method. In addition, the topical coverage and extensive referencing are designed to provide an introduction for those less familiar with acoustic emission and to enable the book to be used as a text for short courses on this subject. The opening chapter contains a brief historical review and qualitative discussion of the acoustic emission phenomenon. Next the reader is introduced to some of the terminology and definitions that have found unique usage within this special field. In the third chapter, a quantitative description of the total acoustic emission process is developed using systems analysis concepts employed in electrical engineering. Important contributions to the literature in the areas of materials research investigations, structural integrity applications, and weld monitoring are reviewed in three major chapters. Theoretical studies are emphasized in another chapter, which outlines some of the analytical and empirical models that have been proposed by various authors. Geological applications of acoustic emission are covered to illustrate the similarity of observed response from geologic and engineering materials. A separate chapter is devoted to the different types of instrumentation systems, sensors, and calibration methods in industrial and development uses. A bibliography of 295 entries is included.

1661 J. C. Spanner. "Selective Review on the Utilization of Acoustic Emission Techniques for Materials Research and Structural Integrity Analysis (A Literature Survey)." M.S. Thesis. Washington State University, Pullman, Washington. 1970.

Thesis includes a brief historical review and qualitative discussion of the acoustic emission phenomenon. Insofar as possible, an analytical description of the acoustic emission process is developed using a systems analysis approach. Significant recent contributions to the literature in the areas of materials research investigations and structural integrity applications are reviewed. A summary of analogous work from the geological field is included to illustrate the similarity of observed emission response from geologic and engineering materials. A discussion of the latest available research reports is included to indicate the status of current work. Conclusions are drawn regarding present and projected capabilities of this method, and an extensive bibliography was compiled to provide a useful index to the literature.

1662 C. R. Sparks and J. C. Wachel. "Quantitative Signature Analysis for On-Stream Diagnosis of Machine Response." Southwest Research Institute, San Antonio, Texas. 1972; paper presented at the American Society for Nondestructive Testing National Fall Conference, Cleveland, Ohio. Oct 16–19, 1972.

The art of signature analysis for machine monitoring diagnostics and control is being significantly augmented by advancing sophistication in electronic instrumentation and by advancing analytical technology for describing machine vibratory response. Historically, machine signature analysis has relied strongly on empirical data obtained from implanted defects and destruction tests. More recently, correlation techniques have been evolved using extensive multi-point monitoring of various machine performance paramenters. Digital data-acquisition systems using minicomputers have extended this approach through their ability to handle large amounts of data and to mathematically analyze and compare data to past history or present conditions. Various signal enhancement and correlation techniques such as power spectral density, cross- and auto-correlation and cross-power spectral analysis provide extended capabilities for extracting vibration signature data which have previously been hidden in high background noise. With recent developments in analytically describing machine vibration response, another significant step has been made by relating single point measurements to incipient failure in high speed rotary equipment. By pre-programing vibrational mode shapes and resonant frequencies, vibration levels at a few points can be correlated to shaft stress, bearing instabilities, blade vibrations, wheel runout, and seal failure. Through such techniques, protective criteria can often be evolved without extensive empirical data, failure implants, and destructive tests.

1663 W. Spath. "Fliesserscheinungen an belasteten Metallen (Creep Phenomena in Metals under Load)." ***Metall,*** **MTLLA, 10(19–20):910–915. Oct 1956. (in German).**

The article is concerned with elementary processes in metals under load, initiation and effect of ultrasonic waves, heating of plastically deformed materials, and phenomena during creep.

1664 W. Spath. "Zur Entstehung von Gleitlinien auf plastisch verformten Metalloberflachen (Formation of Slip Lines on Plastically Deformed Metal Surfaces)." *Metalloberflache,* **MOFEA, Ausgabe A, 7(12):177–183. Dec 1953. (in German).**

The article presents a literature review on slip lines and bands. The mechanism of slip is shown to generate ultrasonic waves. Also discussed are the effect of temperature on slip band width, and the correlation between slip-band width and ultrasonic frequencies of different metals.

1665 J. H. Speake and G. J. Curtis. "Characterization of the Fracture Processes in CFRP Using Spectral Analysis of the Acoustic Emissions Arising from the Application of Stress." Paper No. 29 presented at the International Conference on Carbon Fibers, their Place in Modern Technology, London, England. Feb 18–20, 1974. [reprint available from the Plastics Institution, 11 Hobart Place, London SW1, England or from the authors at the Nondestructive Testing Centre, AERE Harwell, Oxfordshire OX11 ORA, England].

A frequency analysis has been made of the stress waves arising from fracture of 60% loaded unidirectional CFRP and tensile fracture of a number of different types of fiber. Tensile and torsion tests were conducted on CFRP to produce two distinct types of failure. The frequencies recorded from CFRP were shown to be dependent on both the material type and test-piece geometry which resonated in a number of different modes as a result of the stress-wave emission. Discrete frequencies from torsional failure which were higher than any fundamental specimen mode were also recorded, and the source of these is not at present known. Tensile tests on Type I and Type II carbon fiber, glass roving, and polyacrylonitrile fibers showed that the end piece, in which the fibers were secured and the transducer positioned, was stimulated by stress-wave emission, but that the mode type and amplitude of vibration were dependent on the material from which the emissions were derived.

1666 G. R. Speich and R. M. Fisher. "Acoustic Emission during Martensite Formation." pp 140–151 in ***Acoustic Emission, ASTM STP 505.*** **American Society for Testing and Materials, Philadelphia, Pennsylvania. 1972. proceedings of symposium held in Bal Harbour, Florida. Dec 7–8, 1971.**

Martensite formation in an Fe-28 Ni-0.11 C alloy has been studied using acoustic emission, electrical resistivity, and quantitative metallography techniques. Comparison of these three techniques indicated that about 15 martensite plates are involved in each acoustic emission and that the volume of martensite formed per acoustic emission decreases with increasing volume fraction. The first observation is consistent with the autocatalytic effect and the second with the geometric partitioning effect which are both present in martensitic transformations. Because acoustic emission gives an almost plate-by-plate record of the martensitic transformation, more detailed studies of the transformation kinetics can be made with this technique than by other techniques, such as dilatometry or electrical resistivity.

1667 **G. R. Speich and A. J. Schwoeble. "Acoustic Emission during Phase Transformation in Steel." pp 40–58 in** *Monitoring Structural Integrity by Acoustic Emission, ASTM STP 571.* **Edited by J. C. Spanner and J. W. McElroy. American Society for Testing and Materials, Philadelphia, Pennsylvania. 1975. proceedings of a symposium held in Ft. Lauderdale, Florida. Jan 17–18, 1974.**

Acoustic emission was monitored during phase transformations that occur during cooling in a wide variety of steels. Acoustic emission was generated during the formation of martensite but not during the formation of ferrite, bainite, or pearlite. This observation is consistent with the rapid, diffusionless, shear-like nature of martensite formation and the slow, diffusion-controlled growth of ferrite, bainite, or pearlite. The martensite start temperatures, and the temperature range of martensite formation determined by acoustic emission were in good agreement with those determined by metallographic or dilatometric methods. The intensity of acoustic emission generated during martensite formation decreased markedly as the carbon content of the steel decreased, becoming nearly undectable in a maraging steel. This decrease in intensity correlates with a morphological change from large plate-shaped martensite units to smaller lath-shaped martensite units as the carbon content of the steel is decreased. [reprinted by permission of the American Society for Testing and Materials, Copyright].

1668 **B. W. Speller. "Acoustic Emission Energy to Pulse Height Converter." Y-1926. Union Carbide Corporation, Y-12 Plant, Oak Ridge, Tennessee. Dec 18, 1973. [NTIS].**

An energy-to-pulse-height converter was developed for use in acoustic emission testing. This instrument accepts an amplified signal from a piezoelectric crystal and generates a pulse whose amplitude is proportional to either the energy or width of the input signal. Either linear or logarithmic responses can be selected. Theory, circuit design, and performance specifications have been determined, based on calculations and measurements.

1669 **B. W. Spencer, T. T. Anderson, R. E. Holtz, and A. P. Gavin. "Acoustic Surveillance in LMFBR Safety Experiments."** *Transactions of the American Nuclear Society,* **TANSA, 21:377–378. June 1975.**

As part of the LMFBR safety-instrumentation program at Argonne National Laboratory, work has been under way on development of a high-temperature acoustic pressure transducer for in-sodium applications. Such an acoustic sensor may find application in an LMFBR upper plenum for detecting sodium boiling resulting from subassembly malfunctions. Acoustic surveillance using such sensors is potentially an economic detection technique to prevent local malfunctions from leading to substantial fuel failure. Development of sensors successfully meeting the service requirements has proceeded at ANL to the point at which testing in various sodium systems is under way. The R-series in-reactor safety experiments have recently provided an opportunity to test the detection capability under simulated LMFBR accident conditions.

1670 J. E. Srawley and W. F. Brown, Jr. "Fracture Toughness Testing Methods." pp 133–198 in *Fracture Toughness Testing and Its Applications, ASTM STP 381.* American Society for Testing and Materials, Philadelphia, Pennsylvania. 1965. proceedings of a symposium held in Chicago, Illinois. June 21–26, 1964.

A comprehensive survey is presented of current methods of fracture toughness testing that are based on linear elastic fracture mechanics. General principles are discussed in relation to the basic two-dimensional crack-stress field model, and in relation to real, three-dimensional specimens. The designs and necessary dimensions of specimens for mixed mode and opening mode (plane-strain) crack toughness measurements are considered in detail. Methods of test instrumentation and procedure, including an acoustic method, are described. Expressions for calculation of crack toughness values are given for the common types of specimens.

1671 J. E. Srawley and J. B. Esgar. "Investigation of Hydrotest Failure of Thiokol Chemical Corporation 260–Inch–Diameter SL–1 Motor Case." NASA TM X–1194. Lewis Research Center, Cleveland, Ohio. Jan 1966.

The Thiokol Chemical Corporation 260-inch diameter SL-1 motor case failed during hydrotest at a pressure of 542 pounds per square inch, which was about 56 percent of proof pressure. The motor case was constructed of 250 grade maraging steel plate joined mostly by submerged arc automatic welding. An investigation of the failure was conducted by a committee composed of members from Industry and Government. Investigation showed that the failure originated from a defect that was not detected by nondestructive testing techniques prior to aging. Nondestructive testing was not conducted between aging and hydrotest. The defect causing the failure was in the heat-affected zone of a longitudinal submerged arc weld on the cylindrical section of the motor case. The area where the defect was loacted had been repaired by a manual gas-tungsten-arc (TIG) weld. The defect, which was submerged within the vessel wall and oriented longitudinally, was approximately 1.4 inches long and had a width of about 0.10 inches. Four other undetected defects of significant size were discovered after the motor case failure. All of these defects were also located beneath manual TIG weld repairs. One of these defects, which was in the same longitudinal weld as the defect causing the motor case failure, was involved in the fracture as a secondary origin. The other defects were discovered during reinspection of all welds by nondestructive testing techniques after hydroburst. Although the detailed mechanism of the formation of these defects has not been established, it is reasonably certain that their formation was a consequence of the TIG weld repairs to the original welds. The fracture toughness of the submerged arc welds was inadequate to tolerate cracklike defects as large as those that occurred in the motor case. It had been believed that much smaller defects could be detected with high reliability by the nondestructive inspection procedures that were used. The fracture toughness would have been sufficient if no defects had been present larger than what had been considered to be the detectable size. As part of the failure investigation, a limited test program was conducted to determine the sensitivity of radiographic and ultrasonic procedures for detecting tight fatigue cracks in flat plate specimens. It was found that these nondestructive testing methods were much less sensitive and reliable than expected. It is concluded that much more research is required on nondestructive testing techniques for detection of small defects in thick pressure vessel walls before materials and weld procedures having the toughness found in the Thiokol motor case can be successfully utilized.

1672 **A. Stadeler. "Fortschritte der Metallographie."** *Stahl und Eisen,* **STEIA, 37(21):502-504. May 24, 1917. (in German).**

Review of article by J. Czochralski appearing in *Metall und Erz* (1916). (see Reference 344.

1673 **R. W. Staehle and G. E. Kerns. "Acoustic Emissions and Slow Crack Growth in High Strength Steel." pp 182-193 in "Specialists Meeting on Stress Corrosion Testing Methods." AGARD Conference Proceedings No. 98. North Atlantic Treaty Organization, Advisory Group for Aerospace Research and Development, Paris, France. Jan 1972. [NTIS: AD-740-724]; paper presented at the Structures and Materials Panel 33rd Meeting, held in Brussels, Belgium. Oct 5-6, 1971.**

Acoustic emission techniques were used to study stress corrosion cracking in high strength material. Pre-fatigue cracked single-cantilever specimens were loaded and exposed to gaseous and aqueous environments. Acoustic emission signal-shape and total resonance counts were recorded for specimens of different strength levels tested in different environments. Also, the phenomenon of wave reflection was reduced in order to examine the frequency content and energy of the emission. The results show that a higher strength level produces more acoustic activity, regardless of environment. Also, the generated stress wave is of a high frequency and low energy nature.

1674 **L. Stagni and A. Congiu Castellano. "Experiments on Acoustic Emission Generated during Rapid Heating of Tin Single Crystals."** *Journal of Materials Science,* **JMTSA, 12(7):1431-1437. July 1977.**

In most acoustic emission experiments, acoustic emission is excited by either of two methods: (1) direct mechanical deformation; and (2) temperature change. The latter method has mainly been used to study phase transformations of a martensitic kind. Such experiments are performed with heating (or cooling) rates of ~1 °C per minute or less. These rates permit the transformation process to start and to be maintained; at the same time they are low enough so that a study of the transformation versus temperature is not meaningless. Papa *et al.*, have shown that temperature changes can be used to generate acoustic emission in a different way, i.e., by heating with rates as high as 1 °C per second. It has been found that such a method produces detectable acoustic emission activity in tin and zinc single crystals. These materials do not undergo martensitic phase transformations. In any case, no acoustic emission is generated by heating (or cooling) them with the same rates employeed for materials that do show martensitic transformations. It should be noted that acoustic emission has also been detected during welding and its cooling period. So far, however, no attempt has been made, except by Papa *et al.*, to use acoustic emission produced by fast temperature changes as an independent method for materials investigation. The aim of this work is to point out some applications of this experimental method, which presents itself as simple and versatile. The experiments are carried out on Sn, which showed emission features more suitable for this purpose. First, results concerning the measurement of characteristic temperatures of the acoustic emission generating process are reported, then the response to an impulsive load is examined. [reprinted by permission of Chapman & Hall Ltd].

1675 **K. E. Stahlkopf and G. J. Dau. "Acoustic Emission: A Critical Assessment."** ***Nuclear Safety,*** **NUSAA, 17(1):33–42. Jan–Feb 1976.**

The purpose of this article is to highlight the limitations of acoustic emission for use in hydrostatic test monitoring and continuous monitoring of pressure vessels as determined from a review of prior work. With the present state-of-the-art for monitoring structural integrity, acoustic emission can probably reveal abnormal incidents and their locations, but it cannot describe the incident or its seriousness. Specific conclusions about the maturity of acoustic-emission monitoring are drawn, and suggestions are given for future research and development efforts.

1676 **K. E. Stahlkopf and A. T. Green. "Nondestructive Pressure Vessel Testing by Acoustic Emission."** ***EPRI Journal,*** **EPRJD, 1(1):20–23. Feb 1976.**

1677 **K. E. Stahlkopf, P. H. Hutton, and E. L. Zebroski. "Detection of Stress Corrosion Cracking in Stainless Steel Piping by Acoustic Emission." Electric Power Research Institute, Palo Alto, California and Battelle–Northwest, Richland, Washington. 1976; paper presented at the 3rd Conference on Inspection of Pressurized Components, London, England. Sep 20–22, 1976.**

The objective of this investigation was to determine if intergranular stress corrosion cracking (IGSCC) of the type experienced with stainless steel LWR piping could be detected by acoustic emission. A favorable initial result was obtained under laboratory conditions, but further work is required before such techniques may be considered for field application. Test arrangement consisted of four pipe sections prepared by welding together several heats of 48-inch long sections of 4-inch schedule 80, Type-304 stainless steel pipe. The tests were performed in a refreshed loop of high purity water at 289 °C (550 °F) with an externally applied bending moment. The AE data were collected using a simplified prototype monitor instrument developed for the U.S. Federal Highway Administration. The instrument is capable of long-term continuous monitoring and rapid data retrieval through the use of a unique digital data storage technique. Prior to the pipe testing, specialized ultrasonic examination showed no indication of pipe cracking. Results showed distinct AE response throughout the progress of IGSCC from initiation to pipe-wall penetration. In the course of the 23-day test period, a sustained increase in ambient AE occurred on the fourth day apparently indicating initiation or acceleration of the IGSCC. On the 14th day, intermittant high AE periods began to appear indicating more extensive crack increments. Ambient AE increased by a factor of about 10 six hours before pipe penetration. Post-test metallographic examination confirmed the failure was caused by IGSCC.

1678 R. M. Stateham and J. S. Vanderpool. Microseismic and Displacement Investigations in an Unstable Slope." Report of Investigations 7470. U.S. Bureau of Mines, Washington, D.C. Jan 1971.

Microseismic noise rates and slope movements were measured in an unstable area along the north wall of the Tripp-Veteran pit, Kennecott Copper Corporation, Ruth, Nevada. Low-noise, wide-band microseismic geophones were located at various points in the wall, and the change in distance across open fractures, resulting from instability, was determined by repeated measurements between fixed points set on each side of the fractures. The microseismic noise rate increased and decreased with corresponding increases and decreases in the rate at which the fractures opened. The microseismic noise rates and changes in width of the fractures show excellent agreement, and the two types of measurements complement each other in monitoring the progress of pit-wall failure.

1679 C. R. Stauffis, G. A. Malone, and R. A. Duscha. "Nondestructive Evaluation of Regeneratively Cooled Thrust Chambers for Rocket Engines." Bell Aerospace Company, Buffalo, New York, and National Aeronautics and Space Administration–Lewis Research Center, Cleveland, Ohio. 1973; pp 65–79 in *Proceedings of the 1973 Symposium on Electrodeposited Metals for Selected Applications.* MCIC Report 74–17. Metals and Ceramics Information Center, Battelle–Columbus Laboratories, Columbus, Ohio. 1974. symposium held in Columbus, Ohio. Nov 14–15, 1973.

Electroformed Ni–TD (thoria dispersion strengthened) Ni bonds were tested over the total bonded surface area–full bond, weak bonds, and nonbonds. The testing methods evaluated were ultrasonics, acoustic emission, and holographic interferometry. The sequence in which the three test methods are applied is most important. The correlations of the three methods with the three areas being tested are discussed. [Abstract 22-0801. *Metals Abstracts,* Vol. 7, 1974].

1680 **R. B. Steel (to General Electric Company). "Method of Non-Destructively Testing Aluminum-to-Copper Welds." U.S. Patent 3,937,073. Feb 10, 1976.**

Discloses a method of nondestructively testing an aluminum-to-copper butt weld in a bar comprising an aluminum section and copper section joined together by said weld. Forces are applied to the bar in such a manner that the weld is subjected to total stresses that are at least 80 percent, and preferably at least 90 percent, shear stresses. An acoustic emission signal is produced by the stresses at the weld, and this signal is sensed by a suitable transducer to provide an indication of weld quality.

1681 **R. B. Steel and C. C. Popadick. "Nondestructive Testing of Aluminum-Copper Welds using Acoustic Emission Techniques." General Electric Company, Philadelphia, Pennsylvania. 1974;** ***Materials Evaluation,*** **MAEVA, 34(3):64–69, 72. Mar 1976; paper presented at the American Society for Nondestructive Testing National Spring Conference, Los Angeles, California. Mar 11–14, 1974.**

An experimental program was conducted to evaluate acoustic emission techniques in predicting the weld quality of aluminum to copper flash-butt welds. The quality of such welds for long-term service in electrical switchgear depends on the thickness and nonuniformity of the intermetallic precipitate layer. The hypothesis advanced for such a study was that a high acoustic emission activity recorded during a nondestructive stress test would indicate that the intermetallic transition zone is sufficiently thick to cause embrittlement in the weld area. This embrittlement can presently be detected by destructive testing of the aluminum-copper welds. Results of the study show that acoustic emission techniques are excellent in nondestructively predicting the quality of such flash-butt welds.

1682 **R. K. Steele. "Stress Wave Analysis Technique." Aerojet Solid Propulsion Company, Sacramento, California. Mar 1970.**

The stress wave analysis technique (SWAT) system was developed to fill a void in nondestructive testing technology for qualifying the integrity of structures under dynamic loading conditions. The particular need for the system at the Aerojet-General Corporation initially was to monitor the structural integrity of rocket cases during hydrotest pressurization. As initially developed, the system was used to detect stress waves emitted when filament-wound rocket cases were hydrotested. This was accomplished by attaching multiple microphones on the external surface of the rocket case, *listening* to the noises, and recording the sound waves caused by the emissions as the vessel was internally pressurized. It was found that a correlation existed between the wave forms recorded for each rocket case and the approach of chamber failure at elevated pressures. Analysis of the data obtained in testing the filament wound cases was extremely difficult because in the audible frequency range of the microphone sensor, audible background noises occurring during the hydrotest also appeared on the tape recording. Therefore, tests were conducted on a modified SWAT system in which the microphone pick-ups were initially replaced by piezoelectric crystals, and later by accelerometers. Since the frequency range of the sensors was much higher than the audible range, much of the objectionable background noise was eliminated. Electronic filters were also added to the system to exclude other undesirable background noise which could interfere with signal detection and analysis. Subsequently, studies were made relative to the propagation of stress wave emissions through various pressure vessel materials. The technology of locating the source of the emission by determining the travel time of the propagating wave to the three nearest sensors and triangulating to the origin was also developed. Further technology development provided the capability of forecasting critical crack growth and catastrophic failure through application of material characterization curves for the emissions in the specific material used in the vessel. Later, electronic equipment and circuitry, including a digital computer, was added, which permitted real-time determination of fracture source location and the onset of failure thereby permitting termination of loading prior to failure. These features are included in today's system and have been installed in a mobile instrumentation van. This mobile facility permits utilization of the system at remote field locations or at customer sites.

1683 R. K. Steele. "SWAT Application to the Testing of Pressure Vessels." Aerojet–General Corporation, Sacramento, California. 1969; Paper No. 216 presented at the American Society of Mechanical Engineers 24th Petroleum Mechanical Engineering Conference, Tulsa, Oklahoma. Sep 21–25, 1969.

1684 R. K. Steele. "SWAT Testing of Hydroformer R–2 Reactor Pressure Vessel for Humble Oil and Refining Company." Aerojet–General Corporation, Sacramento, California. 1969.

The program was performed in two phases: Phase I: Material Characterization Studies Performed at Aerojet's Materials Research Laboratory in Sacramento, California; and Phase II: Hydrotest of Hydroformer Reactor R-2 Performed at the Humble Refinery in Baytown, Texas. Results of both phases are described in the report.

1685 R. K. Steele, H. P. Briar, C. Morais, and J. Stanfield. "Design of a System to Detect and Locate the Incipient Development of Flaws in Materials." Naval Ship Research and Development Center Contract No. N00600–69–C–0308. Aerojet–General Corporation, Sacramento, California. May 1969.

The work performed under this contract consisted of the design of a SWAT System to meet the requirements for Naval Ship Research and Development Center applications. The overall system consists of three subsystems identified as (1) Master SWAT System, (2) Material Characterization System, and (3) Tape Loop System. These systems are described as to operation and component requirements. In addition, complete drawings and schematics delineating each subsystem are included.

1686 R. K. Steele, A. T. Green, and C. S. Lockman. "Acoustic Monitoring of Hydrotests." pp 361–370 in *Weld Imperfections.* A. R. Pfluger and R. E. Lewis, Editors. Addison-Wesley Publishing Company, Reading, Massachusetts. 1968. proceedings of a symposium held in Palo Alto, California. Sep 19–21, 1966.

Discussion of the following aspects of the stress-wave analysis technique (SWAT): (1) the types of stress waves that are initiated by flaws and their characteristics; (2) the system used for SWAT and its application to the hydrotest; (3) typical stress-wave spectra from flaws and the manner in which they are influenced by flaw shape; and (4) methods of flaw location during the hydrotest. The essential SWAT components are shown schematically. Sensitive accelerometers are bonded to the chamber at various locations. Stress waves emanating from propagating flaws are recorded on a multichannel-hf-response tape recorder. These data are later played back to an oscillograph with a time expansion equivalent to recording at 16,000 inches per second. This process captures the stress wave on the oscillogram, and this information is then used for locating the flaw by triangulation.

1687 R. K. Steele, A. T. Green, and C. S. Lockman (to Aeroject–General Corporation). "Method and Apparatus for Nondestructive Testing of Pressure Vessels." U. S. Patent 3,545,262. Dec 8, 1970.

The disclosure concerns a method and apparatus for nondestructively testing pressure vessels to detect the presence and location of any flaws in the walls. In practicing the method, a plurality of stress-wave sensors, such as accelerometers, for example, are attached to the external surface of a pressure vessel in spaced relationship with respect to each other so as to be arranged over a substantial portion of the pressure vessel. Pressure fluid is then introduced into the pressure vessel to continually increase the internal pressure. If a flaw is present in a wall of the pressure vessel, the extension of the flaw under load by the continually increasing internal pressure propagates a stress wave which is detected by the plurality of stress-wave sensors. The flaw-propagated stress wave travels through the wall of the pressure vessel at the velocity of sound in the material of which the wall of the pressure vessel is made. Signals are emitted by the stress-wave sensors as the stress wave arrives at various sensor locations, and the location of the flaw in the wall of the pressure vessle is determined by a triangulation technique which relies upon the time differences as to the arrival of the stress wave at various sensor locations. For each particular pressure vessel whose wall structure is made of a given material, a critical stress-wave signature is inscribed in a translucent template which may be overlaid on an oscilloscope to permit an operator to visually observe when the pressure vessel undergoing testing has been internally pressurized to the danger point, whereupon the operator manually actuates a mechanism relieving the internal pressure within the pressure vessel. Alternately, the critical stress-wave signature of the pressure vessel is inscribed on an opaque template attached to an oscilloscope such that coincidence between the signal level received by the oscilloscope and the critical stress-wave signature etched in the template will be detected by a photocell which automatically actuates the mechanism for relieving the internal pressure of the pressure vessel undergoing testing.

1688 R. K. Steele, A. T. Green, and C. S. Lockman. "Use of Acoustic Techniques for Verification of Structural Integrity of Polaris Filament-Wound Chambers." Aerojet-General Corporation, Sacramento, California. 1963; paper presented at the Society of Plastic Engineers 20th Annual Technical Conference, Atlantic City, New Jersey. Jan 27-30, 1964. [see Reference 634 for publication].

The advent of the Polaris Model A3 filament-wound, solid-rocket motor case has required the development of unique methods for verification of structural integrity of the chamber. Conventional techniques of experimental stress analysis normally used in pressurization testing of steel and titanium rocket cases could not be utilized to their fullest advantage because of local resin crazing and the random nature of deflection perturbations occurring during tests of filament-wound structures. To overcome these limitations, the Structures Laboratory at the Aerojet-General plant in Sacramento has developed an acoustic technique for verification of the structural integrity of Polaris filament-wound structures.

1689 R. K. Steele, A. T. Green, and C. F. Morais. "Results of SWAT Application to Hydrotest of R-3 Reactor." Aerojet-General Corporation, Scaramento, California. Dec 10, 1968.

Results are summarized in a letter of the application of the stress wave analysis technique (SWAT) to the hydrotest of Humble's R-3 Reactor at Baytown, Texas. The Stress Wave Analysis Technique (SWAT) is a system for detecting the propagation of flaws in pressure vessels in real time. It is a form of structural seismology wherein waves initiated by propagating flaws are detected by transducers attached to the pressure vessel. In addition, the arrival times of these waves are measured such that conventional triangulation techniques may be used for locating the flaws.

1690 R. K. Steele, C. S. Lockman, and A. T. Green. "Acoustic Analysis of Filament Wound Chambers." Aerojet-General Corporation, Sacramento, California. 1964. [abstract only].

The paper discusses the development of an acoustic technique for verification of structural integrity of filament wound chambers. A brief review is presented of methods used to obtain the frequency, acceleration, and pressure spectra utilized for analysis. The main discussion will be directed toward presentation of results and analysis of their significance. In particular, typical frequency acceleration, pressure spectra will be shown for the experimental chambers tested during the Special Projects Research Program on Acoustics. The construction features of these chambers are presented in tabular form to illustrate the differences in the chambers selected to represent the typical experimental acoustic family of chambers. Missileprints, a new technique of analysis by Dr. Kersta of Voiceprint Laboratories will be used to show acoustic definition of interlaminar shear and filament failure. Statistical linear repression analysis used for establishing the relationship between data points is discussed, as well as resulting burst prediction curves based on proof as well as burst data. Recent developments and difficulties of applying burst prediction curves generated on the experimental program to the P-Design Chambers will be analyzed relative to reasons for change and suggested data reduction techniques required for better correlation with burst pressure of P-Design chambers. The paper will conclude with the presentation of a technique which would permit the presentation of the significant data in real time at the conclusion of the hydrotest in accordance with the improved data reduction methods previously discussed.

1691 **H. D. Steffens and H. A. Crostack. "Einflussgrossen bei der Analyse von Schallemissionen (Factors Influencing Acoustic Emission Analysis)."** *Zeitschrift fur Werkstofftechnik/Journal of Materials Technology,* **ZWKTA, 4(8):442–447. 1973. (in German).**

The interpretation of AE measurements requires careful consideration of the prevailing conditions, to ensure reproducible results. By the use of simulated signals in a small steel plate, an analysis of the effect of these various conditions, on the frequency and amplitude distribution of these signals, is possible. In this way the accuracy of location of emission sources can be improved. [reprinted by permission of Verlag Chemie GmbH].

1692 **H. D. Steffens and H. A. Crostack. "Ermittlung des Einflusses von Fehlern in Widerstandspunktschweissverbindungen durch Schallemissionsanalyse (Determination of the Effect of Defects in Resistance Spot Welded Joints by Means of Acoustic Emission Techniques)."** *Schweissen und Schneiden,* **SCSCA, 25(8):284–285. Aug 1973. (in German).**

The crack propagation behavior of various defects present in resistance spot-welded sheet specimens of a low-alloy, high-strength steel, X38 CrMoV 51, was investigated by means of acoustic emission analysis. The influence exerted by the flaws on the individual stages of cracking was studied for three states of heat-treatment, viz., solution annealed, tempered, and hardened. Compared with radiographic examination, acoustic emission enables the progress of cracks to be monitored well below the point of fracture which begins in the region of the joining plane. It is shown that flaws at the spotweld center do not affect crack initiation. Fracture-mechanics aspects concerning the results of the investigation are discussed. [Abstract 22-0650. *Metals Abstracts,* Vol. 7, 1974].

1693 **H. D. Steffens and H. A. Crostack. "Untersuchungen zur Gutekontrolle des Punktschweissprozesses durch Schallemissionsanalyse (Investigation into the Quality Control of Spot Welding Processes by Acoustic Emission Analysis)." pp 292–309 in** *Schallemission: Anwendung bei der Unterschung, Prufung und Uberwachung metallischer Werkstoffe.* **Deutsche Gesellschaft fur Metallkunde eV, 637 Oberursel/Taunus bei Frankfurt/Main, Adenaueralle 21, Germany. 1974. proceedings of symposium held in Munich, Germany. Apr 18–19, 1974. (paper in German).**

1694 **H. D. Steffens, H. A. Crostack, and K. Seifert. "Moglichkeiten zur Beurteilung der Rissanfalligkeit unterschiedlicher Schweissgutgefuge durch Schallemissions-analyse (Methods for Evaluating the Sensitivity of Weld Deposits with Different Structures Using Acoustic Emission Analysis)."** *Schweissen und Schneiden,* **SCSCA, 27(7):257–261. July 1975. (in German).**

Tungsten inert gas welds in 18 Ni – 8 Co – 5 Mo – 0.4 Ti maraging steel were investigated after different heat treatments by fracture mechanics methods. Crack opening and acoustic emission measurements were compared with the results of light and electron microscopy. The results, which prove the usefulness of acoustic measurements in following crack propagation, show why the high toughness values of the matrix can not be maintained by the structurally changing welds in the maraging steel. [Abstract 044010. *Engineering Index,* Vol. 74, 1975].

1695 **R. W. Steffens, J. C. Crowe, G. L. Borsheim, and J. C. Wormeli. "Nondestructive Testing of Encapsulated Radioactive Cesium and Strontium."** *Nuclear Technology*, **NUTYB, 11(2):213–221. June 1971.**

A nondestructive testing program is currently being developed at Battelle-Northwest, under the sponsorship of the Atlantic Richfield Hanford Company, to provide integrity assurance for waste capsules containing solidified radioactive cesium and strontium. Initiation of this project early in the design program for the waste capsules has permitted advanced nondestructive testing techniques to be included in the testing program. An ultrasonic inspection system has been developed and successfully tested in the laboratory using simulated specimens. Acoustic emission techniques, examined for potential use as a weld quality monitor, are promising but development is not yet complete.

1696 **R. W. B Stephens and A. A. Pollock. "Waveforms and Frequency Spectra of Acoustic Emission."** *Journal of the Acoustical Society of America,* **JASMA, 50(3, Part 2):904–910. Sep 1971.**

Illustrations are given of information carried in the waveforms and frequency spectra of acoustic emissions. The effects of multiple reflections and resonances are discussed. A model of the acoustic emission source wave is developed, and arguments are given why this wave should be a pulselike function, rather than an oscillatory function of stress. Further use of this model may allow more quantitative treatment of emission amplitudes, energies, and spectra. Experimental results show the use of two instruments for the evaluation of emission spectra. The feasibility of using frequency analysis to obtain information about source events is demonstrated. [reprinted by permission of the American Institute of Physics].

1697 **R. Stern. "Experimental Techniques in Acoustic Emission Detection Systems." pp 1–3 in** ***1972 Ultrasonics Symposium Proceedings.*** **Catalog No. 72 CHO 708–8SU. Institute of Electrical and Electronics Engineers, Inc., New York. 1972. symposium sponsored by Institute of Electrical and Electronics Engineers Group on Sonics and Ultrasonics, held in Boston, Massachusetts. Oct 4–7, 1972.**

A review of experimental techniques for detecting acoustic emission will be presented. Emphasis will be placed on current methods. The inherent advantages and disadvantages of the various systems and components will be discussed.

1698 **B. Stiefeld. "NDT and the Ubiquitous Digital Computer."** ***ASTM Standardization News,*** **STDNA, 3(3):14–20. Mar 1975.**

To best illustrate how the NDT field applies computers, this paper covers three general NDT areas describing applications where the digital computer adds a significant new dimension to the field.

1699 **W. W. Stinchcomb, K. L. Reifsnider, L. A. Marcus, and R. S. Williams. "Effects of Cyclic Frequency on the Mechanical Properties of Composite Materials." VPI–E–73–25, AFOSR–TR–73–1907. Virginia Polytechnic Institute and State University, Department of Engineering Science and Mechanics, Blacksburg, Virginia. June 1973. [NTIS: AD–768–614].**

Strain-controlled cyclic stress-strain and fatigue tests were conducted on specimens of boron-epoxy and boron-aluminum at cyclic frequencies between 0.5 and 45 Hz. The specimens were cross-ply laminates with a [0°, ±45°, 0°]s fiber orientation and had a 0.25-inch hole in the center. The results of data obtained from both types of tests showed the response of the material to be frequency and cycle dependent. The marked difference in behavior of the material at 15 and 30 Hz suggests the presence of a transition frequency (or frequency range) at which the properties of the material are greatly influenced by frequency controlled mechanisms. At a given frequency, changes in properties occur with increasing cycles and correlate with acoustic emissions monitored during the test.

1700 **G. R. Stoeckinger. "Pulsed DC High Frequency GTA Welding of Aluminum Plate."** ***Welding Journal,*** **WEJUA, 52(12):558s–567s. Dec 1973.**

The pulsed high frequency GTA welding process was examined to determine applicability to the welding of 2219-T87 and 2014-T6 aluminum plate alloys. A shunt regulated high frequency power source and an externally connected oscillator high frequency pulsation unit were utilized in this investigation. These systems were applied to the welding of 0.500-inch thick 2014-T651 and 2219-T87 aluminum plate. The process was found to improve weld penetration and reduce heat input requirements when used in the 4 to 8 kHz pulsation range. With the external oscillator unit, the room temperature static fracture toughness, (K_{Ic}), F_{tu}, F_{ty} and reduction-in-area weld mechanical properties were improved over steady state GTA welds on the 0.500-inch 2219-T87 aluminum plate. A unique correlation was found to exist between these properties and detectable acoustic energy in the workpiece. The shunt regulator system exhibited an improvement only in reduction-in-area weld mechanical properties over steady state values.

1701 **V. E. Stolin, N. I. Brazhnikov, A. M. Dukhanin, and V. G. Kasoev. "Main Trends and Future Prospects for the Development of Ultrasonic Noise Diagnostics in Nondestructive Testing: Review of Patented Material and Publications from the USA, Japan, France, Italy, and Canada."** *Soviet Journal of Nondestructive Testing,* **SJNTA, 7(4):433–436. July–Aug 1971; translated from** *Defektoskopiya,* **DEFKA, (4):82–86. July–Aug 1971. (in Russian).**

The article gives results of reviewing and analysis of patented material, and publications on methods and instruments for ultrasonic noise diagnostics in nondestructive testing. The main trends in the development of the systems, and the design of these instruments, as well as the special features of their operation in relation to the field of application are determined.

1702 **D. E. W. Stone and P. F. Dingwall. "Acoustic Emission Parameters and Their Interpretation." RAE-TR-76121. Royal Aircraft Establishment, Farnborough, Hants, England. Sep 1976;** *NDT International,* **NDITD, 10(2):51–62. Apr 1977.**

The need for improved methods of analysing acoustic emission signals and relating observed data to the original failure events is identified. The potential and limitations of current methods of frequency and amplitude analysis are discussed and some suggestions made. Mechanisms of transfer of energy from the source event via acoustic waves in the structure to electrical energy in the instrumentation are investigated; the theoretical relationship between various electrical and acoustic parameters are described, and the electrical and acoustic energies are related to the loss in stored strain energy by the failure event. A need for improved understanding of transducer response is demonstrated and a number of practical problems are discussed.

1703 **D. E. W. Stone and P. F. Dingwall. "The Kaiser Effect in Stress Wave Emission Testing of Carbon Fibre Composites."** *Nature Physical Science,* **NPSCA, 241:68–69. Jan 15, 1973.**

Reported is a brief discussion on tensile specimens fabricated from Type II, treated, carbon fiber in an epoxy resin matrix (unidirectional, 60% volume fraction) in which was observed a most pronounced Kaiser effect. The specimen used was the standard RAE tensile specimen.

1704 **D. E. W. Stone and P. F. Dingwall. "Transducer Characteristics in Stress Wave Emission." Royal Aircraft Establishment, Farnborough, Hants, England; paper presented at the Institute of Physics Conference on Acoustic Emission, Imperial College, London, England. Mar 14, 1972. [proceedings were never published].**

Considerable effort has been expended in processing and interpreting stress wave emission data; but such data depend entirely on the transducer which, while it should ideally give a faithful analogue of the mechanical disturbance, can often color or modify the signal considerably. The transducers used are usually either piezoelectric or capacitive. The former (and possibly the latter) are force-responding devices. Also, piezoelectric crystals may be resonant or damped, but although higher sensitivity is claimed for the former no evidence exists for this. They respond to longitudinal waves normal to the transducer face, but also detect waves parallel to it through the Poisson's ratio effect. The claimed detection of shear waves has not been observed. Experiments have been conducted with the impulse response of various transducers and the effects noted of mechanical loading and electrical coupling networks. The propagation of waves along long bars has shown that the response of a transducer to waves moving normal to its face is confirmed by surface strain measurements, while the response to parallel waves is anomalous. Although the former situation is often used for calibration, it is the latter that is usually encountered in practical measurement. Directional effects were observed in a differential transducer and the sensitivities of certain transducers were different of different stimuli.

1705 **D. H. Stone and A. T. Green. "Acoustic Emission Monitoring of Shop and Field Welds of Rail." Session 9, pp 1–26 in *The Second Acoustic Emission Symposium.* Japan Industrial Planning Association, Dobashi Building 5F, 3-17-8 Akasaka, Minato-ku, Tokyo, 107 Japan. 1974. proceedings of a symposium held in Tokyo, Japan. Sep 2–4, 1974.**

A short program was conducted utilizing state-of-the-art acoustic emission instrumentation to monitor both thermit cast welding and electric-flash butt welding of rail. The primary objective of the program was to investigate the applicability of acoustic emission techniques for establishing the quality of the weld. Continuous welded rail strings were monitored during electric-flash butt welding, weld cool-down, and weld anneal. The specimen welds were fabricated to represent a variety of good and poor conditions. Piped rail sections, for example, were joined to good rail and electrode burns were purposely made, as were good welds. The acoustic emission data, both location and number, have been correlated with mechanical properties established by physical tests of the welds. The acoustic emission data gathered during the weld anneal cycle show a tentative relationship with the strength of the weld. A total of seven welds were monitored. Thermit cast welding, as used in the field repair of rail, was also monitored with acoustic emission instruments. Variations in the weld procedure were used to produce good and poor quality welds during this portion of the program. The acoustic emission data were obtained both during the pour and subsequent weld cool-down. A total of three welds were monitored. The instruments, techniques, and test results present a clear description of the ease of performing acoustic emission monitoring. Due to the limited number of thermit welds monitored, the data are inconclusive. The data obtained during the electric-flash butt welding show a good-to-excellent possibility of usefulness as a quality control monitor.

1706 A. J. Stosick. "Data Processing for Locating Acoustic Disturbances." Progress Report, Appendix A. Aerojet-General Corporation, Sacramento, California. July 9, 1968.

The SWAT (Stress Wave Analysis Technique) method is a system intended to locate the origin of an acoustic type of disturbance in solid materials by measuring the clock time of arrival of the disturbance at the members of an array of sensors, and, from these times and the known sensor locations, to deduce the location of the source of the disturbance. This report is for the purpose of describing data processing methods which can be used for applying the SWAT method to problems in structural testing and to problems of anti-intrusion detection of interest both for commercial and military situations. Except for the direct algebraic method, all schemes have been successfully tested on a small digital computer. Some programs had to be segmented because of the limited memory size (4K-12 bit) of the available computer. The methods used are adaptable either for the case of physical testing of pressure vessels of simple geometry or to the anti-intrusion problem which has become important in both civil and military situations. This report will not concern itself with questions concerning the selection of the sensors or the associated preamplifiers, and only to a small degree will the question of the rise-time of the signals be considered. The latter must be considered because there is a selective attenuation of the high frequency components of the pulse, especially in nonmetallic media such as rock or soil, so that the rise-time of the disturbance pulse becomes longer for the more distant sensors than for the close-in sensors. This difference in rise-times is equivalent to an error in the time of arrival of the pulse at the sensors farther from the source. This matter of selective attenuation of high frequencies may require that all channels be filtered to reduce the high frequency content to the same degree, followed by overdriven amplifiers to convert all signals to fast rising pulses of an equivalent kind.

In any complete data processing scheme, several different steps must be undertaken. A choice is available as to whether they will be sequential or concurrent steps, and a choice of the order can be made if sequential processing is selected. The choice will be determined by the complexity permitted in the computer system, by the maximum time allowed for the data processing, and by experience gained with testing several alternate approaches. The first essential step is a calculation system to deduce the location of the source from the known locations of the sensors. The expression *clock-time* is used to denote the reading of a clock common to the sensor array. It does not denote the transit time of the pulse from source to sensor, since not known is the time at which the disturbance occurs at the source; as known only is when it reaches each sensor. This calculation of the location of the source point will be referred to as *triangulation* regardless of the details of the computational scheme employed. The actual schemes make no use of triangulation in the classic sense. The other major step in the data processing is a cross-correlation calculation to verify that the signals which arrive at the sensors and are used for triangulation are indeed from the same disturbance event. Two events closely spaced in time and at different locations could be mistaken for a single event, and any complete system should have a means for screening out the true events from close pairs and from noise. The major part of the report will concern the triangulation problem, and only incidental discussion of the cross-correlation problem will be given.

1707 **M. Stupica. "Akusticna emisija fiber glassa." Diplomsko delo st. 2034. Fakulteta za strojnistvo, Univerza v Ljubljani, Ljubljana, Yugoslavia. 1974. (in Slovenian).**

The work describes some fundamental properties of acoustic emission from fiber glass.

1708 **F. E. Sugg, C. C. Kammerer, F. J. Moskal, and E. S. Scherba. "On-Board Checkout of Structural Integrity of Cryogenic Tanks." SD71-264. Final Report. Space Division, North American Rockwell Corporation, Downey, California. Feb 15, 1972.**

The document reports a technology study (Study 4) conducted to evaluate potential on-board nondestructive evaluation concepts that can reveal significant information about the integrity of cryogenic tank structures for possible use on existing and future space vehicles. It presents 13 potential concepts, weighs the merits of each, then selects two with the highest potential for subsequent development and a more thorough evaluation. The two concepts selected were an active ultrasonic multiplexing system with bonded-in-place shear wave transducers, and an acoustic emission passive system also using bonded-in-place sensors. The ultrasonic system developed and tested uses a specially designed sensor that can detect flaws and monitor growth from a fixed position. A multiplexer sequentially pulses an array of these sensors then analyzes the signal return at a rate of 1000 times per second. A warning light is illuminated when any transducer detects flaw growth to identify the location of the flaw. The acoustic emission system developed and tested is a passive system which detects flaw growth from bonded on sensors listening for acoustic signals emanating from stress risers in flawed areas. The system senses the signals, amplifies, and conditions them, then passes them to an analog unit. This unit processes the signals by establishing the arrival sequence and delay time between the arrivals. The first arrival signal is also analyzed for signal envelope to add more significance before it and the previously described data are converted to a binary form and passed to an on-line digital computer. The computer performs an extremely fast statistical analysis of all data input upon command. The output data are printed on a standard teletypewriter. Extensive cryogenic environmental testing data for both system concepts are reported.

1709 **F. E. Sugg and F. J. Moskal. "Real Time Statistical Analysis of Acoustic Emission Signals for Flaw Monitoring Systems." NASA Tech Brief B73-10212. NTIS, Springfield, Virginia. June 1973.**

(Request to Technology Utilization Officer, Marshall Space Flight Center, Code A&PS-TU, Marshall Space Flight Center, Alabama 35812 for further information on this Tech Brief will result in a copy of disclosure T.U. No. SID-94571, NT Control No. MFS-24402.) Described is an acoustic emission flaw monitoring system for determining structural flaw location and growth rate. The structure under observation is divided into small areas. Each area is monitored separately by selectively positioned transducers. The system is divided into two data handling segments which are analog and digital. The analog segment continuously monitors input signals from each transducer and obtains the approximate location of the acoustic event. The digital segment utilizes delay information from the analog unit to provide an exact location of the acoustic event. The innovation described in this disclosure is considered moderately novel in its use of both digital and analog modes to obtain a real-time statistical analysis of acoustic signals produced by a flaw monitoring system. The statistical capabilities of the computer allows a high degree of random noise rejection and can thus easily identify a growing flaw. Acoustic flaw detection systems can be used to monitor critical portions of most any structure. The information contained in this disclosure should be of interest to those industries which require real-time analysis of possible flaw locations on structural systems or subsystems.

1710 **F. E. Sugg, F. J. Moskal, C. C. Kammerer, and D. R. Dixon. "Nondestructive Evaluation (NDE) System for Structural Integrity Monitoring." SD73-SA-0110. Rockwell International, Space Division, Downey, California. Aug 20, 1973.**

The document presents the evaluation of two advanced nondestructive testing methods to monitor and survey structures to reveal information about the integrity of the structure. The two methods, Acoustic Emission Concept and Ultrasonic Multiplex Concept, were developed into test systems that have the potential of evaluating critical structural areas of aerospace vehicles while in extended outerspace; on the ground between flights; during proof loads; and during fatigue testing. Prime consideration was given to the remote inspection of primary structures in inaccessible areas. Preliminary system development testing is also discussed. The evaluation testing of the systems was performed on large aluminum test panels, 3 by 8 feet. Cracks were machined into the test panels, then propagated by application of fatigue cycling and proof loading the test panels. The nondestructive testing systems were evaluated for their ability to detect and locate the induced flaws (cracks) during the controlled flaw growth operations. The crack growth was monitored by crack opening displacement (COD) gauges throughout the testing. Post test analysis of flaw activity compared the NDT equipment response with actual flaw growth. Negative and positive test results are presented and discussed. The potential applications and limitations are presented.

1711 **S. Suguri, T. Fujimura, S. Miyazono, K. Shibata, T. Ohba, R. Kawamura, M. Matsumoto, H. Ohno, H. Nakasa, T. Machida, K. Ishida, M. Onoe, K. Yamaguchi, H. Ichikawa, T. Shimada, and Y. Shimoma. "Application of**

Acoustic Emission Techniques to Monitoring of Fatigue Processes of Reactor Pressure Vessels." CRIEPI-74003. Central Research Institute of Electric Power Industry, Tokyo, Japan. June 1974. (in Japanese). [work performed in cooperation with the Japan Atomic Energy Research Institute, Ibaragi, Japan and the University of Tokyo, Tokyo, Japan].

The objectives of this program are: (1) to develop the flaw-location technique by the acoustic emission signals, (2) to monitor the process of fatigue failure in nozzles of a model pressure vessel, and (3) to investigate usefulness of the acoustic emission technique for the inservice inspection and continuous monitoring of the reactor pressure vessel. Semisphere pressure vessels and one-third scale model of the pressure vessel of JPDR are used for the AE measurement in the fatigue tests. It was found that the flaw-location techniques in use were applicable to monitoring the fatigue-flaw growth, and that the AE characteristics, such as amplitude distribution, count rate, and summation of emissions, varied corresponding to the fatigue progress.

1712 **T. Suzuki, I. Satoh, and R. Iwaya. "Acoustic Emission Observed in Tensile Test of Mild Steel." Paper No. 9 in** ***Proceedings of the U.S.-Japan Joint Symposium on Acoustic Emission*****. Japanese volume. Japan Industrial Planning Association, Kawamura-Building 6F, 3-21-6 Akasaka, Minato-ku, Tokyo, Japan 107. 1972. symposium held in Tokyo, Japan. July 4-6, 1972. (paper in Japanese, abstract available in English).**

Noise elimination and discrimination of signal from noise are the most important techniques in the acoustic emission study. With this in mind, acoustic emissions observed in the tensile test of mild carbon steels SS41 and SM41 (JIS G3010 and G3106, both are commonly used in Japan) are reported. The AE measuring system employed is featured as follows. Both sensor and preamplifier are the balanced type, so that their electrical noise rejection capability is fairly good. It was shown by comparison of the balanced system and an unbalanced one, that the former has better noise rejection capability by 20 db than the latter, even in the worst case. In tensile tests, a high level of mechanical noise was observed from the chucking portions of the specimen. However, these noises were almost completely eliminated by loading with two cemented-through rods, i.e., two tie-plates were cemented with an epoxy adhesive to the chucking portions of the specimen, then preloaded. A typical pattern of acoustic emission and stress-strain curve showed that the rise of continuous emission coincided with the yield point of the specimen, and the buildup of the emission was clearly observed as strain increased. Similar emission patterns were obtained from both SM41 and SS41. Burst-type emission was observed to occur by itself, or to be superposed on continuous emission, and their occurrences were random. These random emissions may have resulted from the unannealed specimens. The same thing may also be said of the irregularities of continuous emissions. Generally speaking, only burst-type emissions were observed before reaching the yield point of the specimen. Then at the peak of stress, the level of continuous emission with random bursts increased considerably. Beyond the yield point, both types of emission decreased their levels gradually.

1713 W. A. Svekric. "Electronic Focusing for Electron Beam Welding." OSU Project RF 2199. Ohio State University, Columbus, Ohio. June 14, 1968.

It is now possible to determine within close limits the current required to focus an electron beam at the surface of a work-piece. It is further possible to determine quantitatively the amount of unfocus either above or below the surface of the specimen. This technique may be applied to a target block or directly to the work with equal case. An inexpensive phono-cartridge is positioned with its needle in contact with the specimen. Since the cartridge is designed to detect lateral deflections, the side of the needle is used against the underside of the part. The output signal is brought out of the work chamber at an electrical feedthrough and amplified. Presently, the signal is being displayed on an oscilloscope, but it could be used to operate warning lights or a buzzer to warn of an out-of-focus condition. Due to the extreme sensitivity of the cartridge, shock absorbers must be used to eliminate vibrations from the roughing pump and other outside sources. The screws on the target block are used to adjust the position of the block relative to the cartridge; obviously the cartridge could have been made adjustable and the target fixed. The correct adjustment is determined by maximizing signal output before closing the chamber.

1714 P. R. Swann and J. D. Embury. "Microstructural Aspects of Stress-Corrosion Failure." Chapter 8, pp 327–355 in *High-Strength Materials.* V. F. Zackay, Editor. John Wiley and Sons, Inc., New York. 1965. proceedings of the 2nd International Symposium on Materials, held at the University of California, Berkeley, California. June 15–18, 1964.

The authors include a brief discussion on the work of Fisher and Embury on the detection of stress corrosion failure by acoustic emission which was subsequently published by R. M. Fisher and J. S. Lally (see Reference 472).

1715 G. D. Swanson and J. R. Hancock. "Effects of Interfaces on the Off-Axis and Transverse Tensile Properties of Boron-Reinforced Aluminum Alloys." Technical Report No. 2 to Office of Naval Research, Contract No. N00014-70-C-0212, NR 031-743. Midwest Research Institute, Kansas City, Missouri. Apr 1971. [NTIS: AD-722-020].

Tension specimens of boron-reinforced 7075 aluminum and 6061 aluminum were tested at angles of 0, 30, and 90 degrees to the unidirectionally aligned filaments. Metallurgical structures in the matrix and at interfaces were correlated with the stress-strain curves, with acoustic emission from boron-filament breakage during the tests, and with fractographic studies. The effects of aging heat treatments and subzero cooling on tensile properties were related primarily to the changes in metallurgical structures produced in the 7075 aluminum matrix. The filament-matrix interfacial bond strengths exceeded the strength of the matrix alloys. Mildly degraded boron filaments had no apparent effect on the ultimate tensile strength parallel to the filaments, but drastically altered the fracture mode in off-axis tensile specimens of 7075 aluminum-boron. In the 0-degree samples, fracture appeared to initiate in the core of boron filments or at the core-boron interface. Diametral splitting of boron filaments was concluded to occur as a result of the stress state developed in the filaments during their manufacture, and limited the ultimate strengths of off-axis and transverse, tension specimens. Recommendations for future research are made.

1716 **W. E. Swindlehurst. "Acoustic Emission-1: Introduction."** *Non-Destructive Testing,* **NDETA, 6(3):152–158. June 1973. [This is the first in a series of five articles on acoustic emission. See References 1404, 212, 336, and 712 for Parts 2, 3, 4, and 5, respectively].**

The author discusses acoustic emission, its causes, and detection. Covered are types of emissions and sources. Also the author emphasizes the complications introduced by the pathway of the emission and its conversion into a signal. The various practical systems used to extract useful information from such complex data are discussed.

1717 **W. E. Swindlehurst and T. R. Wilshaw. "Acoustic Emission from Microcracks during Sliding Contact."** *Journal of Materials Science,* **JMTSA, 11(6):1183–1186. June 1976.**

A letter. The relationship between the size and density of microcracks and the acoustic emission was investigated during sliding contact between a tungsten carbide spherical indenter and a soda-lime-silica-glass block. A significant reduction in the emission strength with increasing microcrack density was observed. This effect is explained in terms of indentation fracture mechanics. [Abstract 22-0079. *Metals Abstracts,* Vol. 10, 1977].

1718 **W. E. Swindlehurst and T. R. Wilshaw. "Acoustic Emission in Brittle Solids."** *Journal of Materials Science,* **JMTSA, 11(9):1653–1660. Sep 1976.**

A signal-source correlation study of the stress waves emitted during unstable microscopic Hertzian fracture in glass is described. A theoretical analysis of the variation in excess strain energy with applied load is made and the results compared with experimental data covering a wide range of crack sizes. [reprinted by permission of Chapman & Hall Ltd].

T

1719 **O. Takano and K. Ono. "Acoustic Emission during Electro and Electroless Plating." UCLA-ENG-7473. University of California, School of Engineering and Applied Science, Los Angeles, California. Sep 1974. [NTIS: AD-787-087]; Session 6, pp 31-49 in *The Second Acoustic Emission Symposium.* Japan Industrial Planning Association, Dobashi Building 5F, 3-17-8 Akasaka, Minato-ku, Tokyo, 107 Japan. 1974. proceedings of a symposium held in Tokyo, Japan. Sep 2-4, 1974.**

The paper reports acoustic emission characteristics during electroplating of chromium and copper, as well as those during electroless nickel plating. Acoustic emission pulses are produced by the generation of cracks in the deposit layer and can be utilized in monitoring plating quality. It is known that cracks develop in a chromium plating layer, when it exceeds a certain thickness, typically 0.5 μm. The density and distribution of the cracks vary greatly depending on the condition of the plating bath used and are important parameters governing the quality of the deposited layer. Extensive development efforts have been reported to obtain a finer and more uniform distribution of cracks in a deposit. To date, however, a nondestructive detection method of crack formation during plating has not been known. Acoustic emission is well-suited for this application. The results of acoustic emission experiments, varying the type of plating and bath conditions and examining the microstructures of deposit layers, are described. Chromium plating utilized a chromic-sulfuric acid bath at the current density of 0.45 A/cm^2. The bath temperature was varied between 50 and 65 °C. Conventional acid baths were used for copper electroplating and for electroless nickel plating. Conventional acoustic emission instrumentation was employed using a resonance transducer and a threshold crossing counter with the threshold voltage of 62 μV referred at the transducer output. During chromium plating at 59 °C, acoustic emission count rate was low for the first-two minutes (corresponding to the deposit thickness of 0.5 μm), but it increased after this incubation period to a constant count rate at 1.5×10^4 counts/cm^2-minutes. When the temperature was raised, observed count rates decreased and vice versa. The acoustic emission results correlated with microscopic observations, which detected numerous cracks in chromium deposit at 51 °C, but none at 65 °C. During copper and nickel plating, count rates were higher than that of hydrogen bubble formation only, but were a hundred times lower than that of chromium plating at 59 °C. As expected, no crack was detected in copper or nickel layer. These and other experimental findings are reported in full and discussed in terms of the current microstructural theories of plating layers. Also, it is shown that the acoustic emission technique is a valuable tool in quality control of plating processes.

1720 **M. Takehana and I. Kimpara. "Experimental Investigations of Failure Process of Fiberglass Reinforced Plastic Composites by the Acoustic Emission Method." Paper No. 4 in *Proceedings of the U.S.-Japan Joint Symposium on Acoustic Emission.* Japanese volume. Japan Industrial Planning Association, Kawamura-Building 6F, 3-21-6 Akasaka, Minato-ku, Tokyo, Japan 107. 1972. symposium held in Tokyo, Japan. July 4-6, 1972. (paper in Japanese, abstract available in English).**

It is well-known that composite materials emit peculiar and distinguishable noise during deformation, because of two-phase materials. As this noise is believed to be closely related to formation and extension of cracks or crazings in matrix (breaking of reinforcements embedded in matrix and interfacial debonding), it will yield useful information to investigate the failure process of composite materials. In this paper, the application of the acoustic emission method to investigate the failure process of fiberglass reinforced plastic (FRP) composites is discussed. FRP test specimens used are unsaturated polyester resin reinforced with several types of combinations of woven rovings, unidirectional woven rovings, and chopped strand mats. The acoustic emission of various FRP test specimens under tensile loading is detected with two methods utilizing microphone and accelerometer as sensors. The overall acoustic emission properties of FRP are correlated to the stress-strain behaviors in terms of the acoustic emission count rates or total counts. The difference in the failure process of FRP due to the different types of glass reinforcements is discussed in relation to the acoustic emission properties. The generating mechanism of the acoustic emission in FRP is considered experimentally in conjunction with the failure mechanism of the material. The possibility of the acoustic emission method as one of the acoustic non-destructive testing methods of FRP is also considered from the acoustic emission properties at lower stress level.

1721 **M. Takehana and I. Kimpara. "Internal Fracture and Acoustic Emission of Fiberglass Reinforced Plastics." pp 156–167 in** *Mechanical Behavior of Materials.* **Volume V: Composites, Testing and Evaluation. Society of Materials Science, Japan, Kyoto, Japan. 1972. proceedings of the International Conference on Mechanical Behavior of Materials, held in Kyoto, Japan. Aug 15–20, 1971.**

In the paper, the application of the acoustic emission method to investigate the fracture mechanism of fiberglass reinforced plastics (FRP) is discussed. The acoustic emission of FRP test specimens under tensile loading is detected by microphone and analyzed by pulse counter, cathode ray oscilloscope, 1/3 octave band pass filter and high speed level recorder. The difference in fracture process of FRP due to different types of reinforcements is discussed in relation to the acoustic emission properties. The acoustic emission is related to the stress-strain behavior. The generating mechanism of the acoustic emission is considered experimentally in conjunction with the fracture mechanism.

1722 **H. Tanaka and R. Horiuchi. "Acoustic Emission Accompanied by Twin Formation in Titanium and Titanium Alloy."** *Bulletin of the Institute of Space and Aeronautical Science, University of Tokyo,* **TDUKA, 11(2A): 427–435. 1975. (in Japanese).**

Acoustic emission was measured during tensile test of titanium and Ti – 6 Al – 4 V alloy rolled sheets. The purpose of this paper is to investigate the relationship between acoustic emission and the deformation twinning which is considered to play an important role in the mechanical anisotropy in titanium. The results obtained are as follows: (1) In the case of titanium of small grain size, only low-amplitude acoustic emission pulses were observed. From this result, it is considered that twinning does not occur in this condition and the material deforms only by slip; (2) In the case of titanium of large grain size, a number of large bursts of pulses was observed, which is thought to correspond to the twin formation. Moreover, from acoustic emission characteristics, an estimate was made of the contribution of twins to the mechanical anisotropy of titanium; (3) In Ti - 6 Al - 4 V alloy, acoustic emission pulses were observed in correspondence with twin formation; and in (4) It was found that the amount of twin formation increased with increase in the Schmid factor calculated on the assumption of a so-called quasi-single crystal model, but the twinning mode was not confirmed.

1723 **H. Tanaka and R. Horiuchi. "Acoustic Emission Due to Deformation Twinning in Titanium and Ti-6Al-4V Alloy."** *Scripta Metallurgica,* **SCRMB, 9(7):777–780. July 1975.**

It is recognized that the release of large elastic energy occurs rapidly when deformation twins are formed, since dislocations move quickly at twin formation. Therefore, by continuous monitoring of acoustic emission, twin formation may be detected sensitively as so-called burst-type signals, without interrupting the deformation. An alpha-titanium rolled sheet shows the mechanical anisotropy due to the existence of rolled texture, which depends on the grain size of the material. Slip and twinning are thought to be factors of the mechanical anisotropy in titanium alloys. The purpose of this note is to distinguish these two deformation mechanisms by use of acoustic emission techniques. [reprinted by permission of Pergamon Press Ltd].

1724 **K. N. Tandon and K. Tangri. "Acoustic Emissions during Deformation of Polycrystalline Silicon-Iron."** *Materials Science and Engineering,* **MSCEA, 20(1):47–54. July 1975.**

Acoustic emission technique has been used to study the dislocation processes occurring at various strain levels during deformation of polycrystalline Fe-3% Si samples. In the microyielding and macroyielding regions the detectable AE pulses are due to dislocation generation from grain boundary sources for which etch-pit evidence is also presented. During the work-hardening region, AE is subdued owing to the lack of high energy dislocation processes. By increasing the grain size of samples, AE activity increases in the microyielding region; however, the trend is reversed during macroyielding. The results are analysed in terms of dislocation source activation and mobile dislocation density necessary in the sample during yielding. In the work-hardening region, rate of AE events increases with decrease in grain size.

1725 **K. N Tandon and K. Tangri. "An Acoustic Emission Study on Deformation Behaviour of Strain-Aged Polycrystalline Samples of Fe-3% Si."** *Materials Science and Engineering,* **MSCEA, 29(1):37–40. June 1977.**

Acoustic emissions were monitored during deformation of polycrystalline Fe-3% Si samples which were strained by 5% and aged at 353 K for various lengths of time. The aim of this study was to resolve whether the deformation mechanism in such samples is by dislocation unpinning from impurities, or by the creation of dislocations at points of stress concentration, or both. After comparing AE from well-annealed samples (where the creation of dislocations takes place at a grain boundary only) and from strain-aged samples, it is interpreted that for aging times 0 to ~200 minutes both the unpinning of weakly pinned dislocations and the creation of dislocations at stress concentrations contribute to deformation. However, with longer aging times, the latter seems to contribute more and more until beyond ~200 minutes, when all the deformation could occur by such a process. The AE was analyzed in terms of the average amplitude, peak amplitude, and frequency spectrum.

1726 **K. Tangri, J. P. Toronchuk, and D. J. Lloyd. "Acoustic Emission from Polycrystalline Copper and Iron–Silicon." University of Manitoba, Winnipeg, Manitoba, Canada. 1972;** *Transactions of the Indian Institute of Metals,* **TIIMA, 25:33–40. Dec 1972.**

The acoustic emission resulting from the deformation of polycrystalline copper and iron-silicon is discussed in terms of possible dislocation mechanism giving rise to the emissions.

1727 **N. Taniguchi, E. Isono, and T. Udagawa. "AE Technique for Exploring Crack Initiation and Propagation." Nippon Steel Corporation, Products Research and Development Laboratories, Sagamihaha, Kanagawa, Japan. June 1973; paper presented at the 11th Acoustic Emission Working Group Meeting, Richland, Washington. July 18–20, 1973.**

The experiments reported here were conducted as a part of a series of new developments and applications of AE technique. In the first report of this paper, AE was applied to be a monitor of crack initiation and propagation. Material used here was AISI 4340 steel of 160 kg/mm^2 level, and hydrogen charged to obtain delayed precracking under a cantilever-type bending machine. The purpose of introducing the precrack by hydrogen embrittlement was to study the feasibility of substituting for fatigue cracking, and to study correlation between AE and the cracked areas. Results were such that there is a difference in the pattern of crack initiation and propagation at the notch tip with corresponding different AE count signals; and that there is a linear correlation between AE counts and cracked areas as a whole. Instantaneous spectrum analysis was attempted, too. The second application of AE was made to notched bend test at low temperatures (COD test) with high strength steel of 60 kg/mm^2 grade. Here, the specific purpose of AE application was to pinpoint in the load-COD curve the initiation of slow crack or ductile tearing which is formed at the notch root. In this case, the result indicated it was possible to point out the initiation of slow cracks. However, further study should be done in order to eliminate mechanical noise from the cracks.

1728 **O. Taraba. "Le testing des defauts des coussinets glissants par une analyse spectrale de l'emission du champ d'ultrason genere par la perturbation du coussinet (The Testing of the Defects of Sliding Bearing by a Process of Spectral Analysis of an Ultrasonic Field Generated by the Perturbation of a Bearing)." Faculte d'Electrotechnique, Prague, Czechoslovakia. 1973. [in French and English]; Paper E-09 presented at the 7th International Conference on Nondestructive Testing, Warsaw, Poland. June 4-8, 1973.**

The diagnosis without dismantling a machine is conducted by means of an ultrasonic process in a band of 300 kHz. The instrument used for this observation is the Diagnost 01 FEL. The subject of discussion related to the ultrasonic emission is based on: (1) the friction of matting (sliding) surfaces in a bearing, and (2) the influence of cavitation in a bearing. Several typical cases of perturbation characteristics have been recorded and a general description of the inspection of a bearing without a need for dismantling is explained.

1729 **C. A. Tatro. "A Welder's Introduction to Acoustic Emission Technology." UCRL-76505-Rev. 1. Lawrence Livermore Laboratory, Livermore, California. Mar 18, 1975. [NTIS]; pp 1-9 in *Acoustic Emission.* Edited by R. W. Nichols. Applied Science Publishers Ltd, Ripple Road, Barking, Essex, England. 1976. proceedings of a colloquium on acoustic emission held during the 1975 Annual Assembly of the International Institute of Welding, Tel Aviv, Israel. July 7-13, 1975.**

Reviewed are the history, background, fundamentals, and techniques employed in the use of acoustic emission to detect and study defects during welding and brazing operations. The author also addresses the problem of extraneous noise which must be distinguished from relevant signals in order to differentiate between good and poor welds and brazes.

1730 **C. A. Tatro. "Acoustic Emission from Crystalline Materials Subjected to External Loads." Progress Report. Division of Engineering Research, College of Engineering, Michigan State University, East Lansing, Michigan. Apr 1, 1960.**

The report describes in a qualitative way some of the behavior of acoustic emission from polycrystalline specimens in tension and creep tests, and for surfaces subjected to static friction forces. Some instrumentation and apparatus requirements are discussed. Since acoustic emission signals are so weak, special means for loading specimens are required. Plans for proceeding to testing of single crystals are projected.

1731 **C. A. Tatro. "Acoustic Emission Related to Nondestructive Testing." UCRL-73441 (Rev. 1). Lawrence Radiation Laboratory, Livermore, California. Sep 22, 1971. [NTIS]; Paper 720175 presented at the**

Society of Automotive Engineers National Automative Engineering Congress, Detroit, Michigan. Jan 10–14, 1972.

Acoustic emission technology has developed to the state where broad applications to materials testing and evaluation are evident. The unique characteristics of acoustic emission as they bear on applicability to nondestructive testing are emphasized. Among the advantages anticipated for the acoustic emission method, two are especially noteworthy: (1) Moving inspection upstream in the production process so that bad production can be eliminated before large fabrication and assembly costs are accumulated, and (2) Economical 100% NDT inspection of finished components or products.

1732 **C. A. Tatro. "Acoustic Emission Sensors and Instrumentation." UCRL-51320. Lawrence Livermore Laboratory, Livermore, California. Jan 3, 1973. [NTIS].**

This is a formalization of a two-hour lecture that has frequently been given as an orientation in acoustic emission technology. It describes acoustic emission sources as to range of amplitudes, range of frequencies, and other aspects of the source signal pertinent to the design and selection of a satisfactory sensor. Alternatives in sensor selection are discussed, and the rationale for selection of the PZT-5 piezoelectric ceramic element is presented. Following the selection of the sensor and discussion of its limitations, the rationale for the classical method of instrumentation to process and display acoustic emission results is presented. This method includes correlation of the rate of occurrence of acoustic emissions with pertinent macroscopic engineering parameters, such as stress, strain, pressure, temperature, and time.

1733 **C. A. Tatro. "Design Criteria for Acoustic Emission Experimentation." UCRL-73230. Lawrence Radiation Laboratory, Livermore, California. June 15, 1971. [NTIS]; pp 84–99 in** *Acoustic Emission, ASTM STP 505.* **American Society for Testing and Materials, Philadelphia, Pennsylvania. 1972. proceedings of symposium held in Bal Harbour, Florida. Dec 7–8, 1971.**

Acoustic emission experiments that can be performed in the laboratory using a single transducer and involving specimens or simple components are discussed. The emphasis is on experimental design rather than on single experiments or results produced. The experiments described are restricted to those where application of mechanical load is used to condition the material, and where commercially available equipment is adequate for instrumentation. The magnitudes and characteristics of the acoustic emission response are described. The compromises in signal detection and data acquisition methods follow from this description. Those aspects of experimental design which are most likely to be foreign to the new user of acoustic emission techniques are pointed out, as are ways to assure that a proper design has been achieved. Results of an experiment which reveals the unique power of acoustic emission technology to reveal subtle changes in material are given as an example.

1734 C. A. Tatro. "Evaluation of the Integrity of a Structure using an Acoustic-Emission - Instrumented Proof Test." UCID-16084. Lawrence Livermore Laboratory, Livermore California. July 18, 1972. [NTIS].

An evaluation of a single-material structure that must survive a dynamic load superimposed on a large static load is presented. The use of acoustic emission detection equipment to detect flaws and flaw growth during proof testing is described and the correlation of proof-test results and structural behavior discussed. It is concluded that acoustic emission detection techniques can give higher assurance of structural integrity under specified conditions than can classical nondestructive testing methods alone.

1735 C. A. Tatro. "Experimental Considerations for Acoustic Emission Testing." UCRL-72758. Lawrence Radiation Laboratory, Livermore, California. Oct 10, 1970; ***Materials Research and Standards,*** **MTRSA, 11:(3)17-20,44. Mar 1971.**

Acoustic emission technology has advanced through the laboratory curiosity and research phases in the last few years. It should now be considered, along with other available techniques, as a normal materials testing tool. The paper is addressed to those who have access to a well-equipped laboratory but who have not previously attempted to use acoustic emission as a tool. It describes an experimental design for such a laboratory and discusses those aspects of the experiment that would most likely lead to difficulty for a person attempting such a test for the first time. Design precautions, pitfalls and how to avoid them, special considerations for equipment and instrumentation, and general advantages and disadvantages of acoustic emission test methods are discussed. The unusual aspects of the acoustic emission test have been emphasized in the discussion, with greatest emphasis being placed on those matters which most often cause acoustic emission tests to fail or to result in the generation of fallacious data.

1736 C. A. Tatro. "Sonic Techniques in the Detection of Crystal Slip in Metals." Progress Report. "Engineering Research," Volume 1. Engineering Experiment Station, College of Engineering, Michigan State University, East Lansing, Michigan. 1957. [author's reference ER-57].

A study has been made to determine whether ultrasonic pulses are emitted by metal samples when they are loaded in tension. Some correlation of ultrasonic pulses in a frequency range about 350 kHz has been achieved. The activity (total number of pulses emitted per unit of load) seemed to be a complicated function of the state of stress or strain of the sample. Results reported here should be considered both speculative and qualitative. A resumé of another work on the same subject, which was carried out at lower frequencies, is given.

1737 C. A. Tatro. "Sonic Techniques in the Detection of Crystal Slip in Metals." Progress Report. Division of Engineering Research, College of Engineering, Michigan State University, East Lansing, Michigan. Jan 1, 1959. [author's reference TA-59].

Continuing efforts to evolve a sonic technique for detection of crystal slip in metals have uncovered serious obstacles. Among these are: the problem of high-frequency mechanical vibrations in modern testing machines, electromechanical matching problems, and the many problems involved in gleaning intelligence from low-level electrical signals of the order of a few microvolts. The report deals with several steps in the sophistication of the technique originally used a year ago, and outlines further steps that must be taken.

1738 C. A. Tatro and B. R. Bass. "Final Report on Acoustic Emission Research at Tulane University under UCRL P.O. 2412203. Part 2." UCRL-13231, Part 2. Lawrence Radiation Laboratory, Livermore, California. June 1966.

The objective of this investigation is to determine the acoustic attenuation properties of three liquids for frequencies from 25 kHz to 75 kHz, when the liquid media are confined in a container of limited volume. It is desired to correlate these results with established material properties in order to predict the relative acoustical transmission merits of other liquids over the prescribed range of frequencies. A piezoelectric crystal is utilized to introduce a pulsed stress wave into one end of a long buffer rod, the other end of which terminates in a tank containing the fluid medium under study. A second buffer rod, with piezoelectric crystal attached, is employed to receive the transmitted acoustic wave. The amplitude of the transmitted strain pulse is measured for two different lengths of travel between the source and the receiver rods. The ratio of these two amplitudes, measured for two prescribed gap lengths, is determined as a function of frequency for each liquid. An acoustic attenuation coefficient is introduced through a solution to the wave equation. Spherical wave theory is used to express the coefficient as a function of the ratio of transmitted strain-pulse amplitudes and distance of travel. Values of the acoustic attenuation coefficient are then plotted from the measured amplitude ratios for frequencies from 25 kHz to 75 kHz. Part II is in thesis format. It is the M.S. Thesis of Bennett R. Bass, "Absorption of Low Ultrasonic Sound in Three Liquids by a Pulse Technique."

1739 C. A. Tatro, R. P. Dickinson, and T. F. Freeman. "Central Computer Analysis of Minicomputer Acquired Data for Acoustic Emission Flaw Detection in Solid Bodies." UCRL-76135. Lawrence Livermore Laboratory, Livermore, California. Oct 21, 1974. [NTIS].

A method of on-line safety assurance monitoring for the LLL 345 MPa autoclave has been developed. It utilizes a multisensor acoustic emission based flaw-detection system with minicomputer aided-data acquisition capability. While this system is capable of revealing repeatedly incrementing flaws, it is incapable of calculating flaw location accurately in the virtually solid end caps of this device. To achieve flaw location, a three-dimensional flaw location theory was developed. The mini-computer acquired data were transferred to a CDC 7600 computer, where flaw-location calculations were performed and graphical displays of results were made. A viable method of achieving shut down of this device to prevent an approaching catastrophic failure is demonstrated.

1740 **C. A. Tatro and R. G. Liptai. "Acoustic Emission from Crystalline Substances." pp 145–158 in *Proceedings of the Symposium on Physics and Nondestructive Testing.* Southwest Research Institute, San Antonio, Texas. 1962. proceedings of the 3rd Symposium on Physics and Nondestructive Testing, held in San Antonio, Texas. Oct 2–4, 1962.**

Acoustic emission is the term used to denote spontaneous emission of sound pulses from crystalline substances subjected to external stress. Two distinct components of the phenomenon can be recognized: (1) a low-level high frequency component observed as *growth of noise,* and (2) a discontinuous burst type component observed as a ring-down pulse. The observations are made with apparatus described in detail in the report. Several acoustic emission spectra of the burst type are to be discussed, along with implications the results infer for basic studies of crystalline materials. The precise source of the acoustic emission cannot be given as yet, although evidences are presented which imply that it is a surface phenomenon and is connected in some way with slip-line formation or appearance of cracks. The sound level of the acoustic emissions is low, and extreme care must be taken in designing a loading mechanism that does not produce sufficient noise to mask the acoustic emission noises. Conventional testing machines have been found generally unsatisfactory for these studies. Some special load producing mechanisms are discussed. The detection system is far from ideal in that it produces an indication only that an acoustic emission event has occurred. It is not a high fidelity system in the sense that details of the output electrical pulse have a one-to-one correspondence with details of the primary mechanical pulse. The *ring down* nature of the output pulse suggests that the primary mechanical pulse is a fast impulse phenomenon. A complete description of the detection system is given, along with some suggestions for improvement. Some anomalous behaviors of acoustic emissions during several tests are given. Correlation of some of these with the Fitzgerald effect appears possible. Aside from the problem of deducing the mechanism which causes acoustic emission, several possibilities for application to testing problems are given. One preliminary investigation dealing with the dry friction process is briefly discussed.

1741 **C. A. Tatro, R. G. Liptai, and D. W. Moon. "Acoustic Emission from Formation and Advancement of Cracks." UCRL-74607. Lawrence Livermore Laboratory, Livermore, California. Mar 5, 1973. [NTIS]; paper presented at the International Confernece on Stress Corrosion Cracking and Hydrogen Embrittlement of Iron Base Alloys. Unieux-Firminy, France. June 10–16, 1973.**

Stable crack advancement, microcracking, dislocation breakaway from pinning sites, martensitic transformations, and many other such phenomena in material deformation and failure are accompanied by an energy release, known as *acoustic emission,* in the form of impulsive elastic stress waves. The paper describes the detection and use of acoustic emission in materials research, emphasizing its unique advantages as a research tool. For instance, acoustic emission can follow the development of material failure in real time; it can sort destructive processes into categories by amplitude and frequency discrimination; it can stop a destructive process at an optimum time to perform a metallurgical inspection of material damage by sectioning techniques; and it can be the basis for a practical, economic method of selecting materials for a specific application. The paper also discusses the disadvantages and limitations of acoustic emission technology in research. Acoustic emission technology has not been used widely in hydrogen embrittlement and stress corrosion studies. A summary of the available literature is included here, along with summaries of work pertinent to a better understanding of the usefulness of acoustic emission for these studies. Examples of acoustic emission data from stress corrosion studies in high strength steels, not yet reported elsewhere, are included and are compared with results from the literature. Those deformation and failure mechanisms known or suspected of operating during hydrogen embrittlement or stress corrosion cracking of iron-based alloys are discussed as acoustic emission sources. Those phenomena that do not serve as acoustic emission sources are identified. The various important sources are classified in terms of the amplitude and frequency content of the emitted stress wave, though this classification cannot be done on a quantitative basis.

1742 **C. A. Tatro and P. S. Shoemaker. "Acoustic Emission from Polycrystalline Steel and Aluminum."** *Bulletin of the American Physical Society,* **BAPSA, 7(3):216. 1962. [abstract only].**

Acoustic emission is the term used to describe spontaneous bursts of sound issuing from crystalline substances when they are stressed. It is essentially an irreversible phenomenon. Acoustic emission has two principal components; a relatively continuous high-frequency component appearing as growth of noise on an oscilloscope screen, and a component emitted in relatively large-amplitude bursts. The large burst acoustic emission is discussed in this paper. The phenomenon is presumed to be associated with dislocation connected processes—probably with break-away of dislocation clusters induced by applied stress. Extreme care must be taken in producing loads so that noise from the loading mechanism does not mask the extremely weak acoustic emissions. A way of accomplishing the loading is demonstrated. The character of the acoustic-emission spectrum from two polycrystalline solids is discussed. The strong dependence of surface treatment on acoustic emission from steel is discussed. Some remarks concerning possible driven and self-excited vibrations from steel due to the acoustic-emission source are made. Plans for extending this work to single crystals with the view to specifying the source more precisely is described.

1743 **H. M. Tensi. "Acoustic Emission Measurements during Crystallization and Melting of Metals and Binary Alloys." Session 5, pp 46–57 in** *The Second*

Acoustic Emission Symposium. **Japan Industrial Planning Association, Dobashi Building 5F, 3-17-8 Akasaka, Minato-ku, Tokyo, 107 Japan. 1974. proceedings of a symposium held in Tokyo, Japan. Sep 2-4, 1974.**

Previously, many experiments have been done to use the experiences about acoustic emission known till now, in order to control quality during welding processes. As yet, the different reasons for sound emission during welding were scarcely explored. The different sources for sound emission are the plastic deformation and the final crack forming in the not-melted zone, produced by the high temperature gradient (thermal stresses); the possible phase-transformation in the solid state of the weld; and the not-fused zones; and further the melting and solidification-process of the weld seam. The described experiments intend to enlighten the fundamental phenomena of sound emission during phase transformation solid ⇌ liquid. The first experiments related to these phenomena were made by H. Borchers and J. Kaiser.

1744 **H. M. Tensi. "Piezoelektrische Impulsmessungen zur Untersuchung von Vorgangen in Metallen bei Phasenanderungen und bei mechanischer Beanspruchung (Piezoelectric Impulse Measurements for the Investigation of Processes in Metals during Phase Changes and Mechanical Stressing)." Ph.D. Thesis. Technische Hochschule Munchen, Munich, Germany. 1961. (in German).**

1745 **O. Terada. "Acoustic Emission and Prediction of Material Failure."** *Mitsui Technical Review,* **(81):2-15. 1973. [text in Japanese, abstract in Japanese and English; available from Mitsui Shipbuilding and Engineering Company Ltd., 6-4 Tsukjii 5-chome, Chuo-ku, Tokyo, Japan].**

In designing various structures of late, a high reliability in strength has been required. To cope with such a condition, a new testing method that can predict the failure of the materials, such as fatigue fracture, etc., is necessary. Acoustic Emission Technique (AET) explained herein is a new approach to the prediction of the material failure which satisfies the above requirements. It seems that the technique, though still in its infancy, can be widely used and applied to various purposes. The AE is a sensitive phenomenon to yielding and cracking of many materials and so will enable detection of a fracture at an early stage. In this paper, several basic natures of the AE and its measurement system are explained on the basis of the author's experiments. And then, some relations between the fracture mechanics and the AE are studied. Fracture phenomena of small scale yielding are presented as experimented by other investigators. Phenomena of large scale yielding and ductile fracture are further considered from the standpoint of the author's experimental results. Last of all, an application of the AET to the spherical tank under hydrostatic test is presented.

1746 A. S. Tetelman. "Acoustic Emission and Fracture Mechanics Testing of Metals and Composites." UCLA-ENG-7249. University of California, School of Engineering and Applied Science, Los Angeles, California. July 1972. [NTIS: AD-746-261]; pp 1-46 in ***Proceedings of the U.S.-Japan Joint Symposium on Acoustic Emission.*** **English volume. Japan Industrial Planning Association, Kawamura-Building 6F, 3-21-6 Akasaka, Minato-ku, Tokyo, Japan 107. 1972. symposium held in Tokyo, Japan. July 4-6, 1972; paper presented at the Society for Experimental Stress Analysis 3rd International Congress on Experimental Mechanics, Los Angeles, California. May 13-18, 1973;** ***Pressure Engineering, Journal of Japan High Pressure Institute (Kotsu Gasu),*** **KOGAA, 11(2):92-103,125. 1973. (in Japanese).**

Acoustic emission testing has found many uses in recent years and new applications are constantly being uncovered. One of the most fruitful areas for application lies in the characterization of microscopic processes of yielding and fracture, and macroscopic processes of slow crack growth and onset of fast crack propagation. This paper has described these processes and emphasized that the total counts and count rate depend on the energy released per event and the density of events per unit of deformation. The models presented herein are first-order approximations which await confirmation.

1747 A. S. Tetelman. "Acoustic Emission Testing and Microfracture Processes." UCLA-ENG-0370. University of California, School of Engineering and Applied Science, Los Angeles, California. Oct 1970. [NTIS: AD-715-019]; ***Materials Research and Standards,*** **MTRSA, 11(3):13-16, Mar 1971.**

Acoustic emission testing offers a means of studying microcrack initiation, microcrack propagation, and final fracture. These events are recorded on line as a function of applied stress, strain, time, and environment. Typical graphs of acoustic emission rates are illustrated for low-C steel, 4340 steel, and Al-Al_3Ni eutectoid composites. [Abstract 22-0735. *Metals Abstracts,* Vol. 4, 1971].

1748 A. S. Tetelman. "The Application of Acoustic Emission to Stress Corrosion Cracking." ***Corrosion,*** **CORRA, 28(12):472. Dec 1972. [abstract only]; paper presented at the Corrosion Research Conference, Anaheim, California. Mar 19-21, 1973.**

Acoustic emission testing has found many uses in recent years and new applications are constantly being uncovered. One of the most fruitful areas for application lies in the characterization of microscopic processes of yielding and fracture, and macroscopic processes of slow crack growth and onset of fast crack propagation. Of particular interest is the use of acoustic emission testing to characterize environmentally induced cracking such as stress corrosion cracking and hydrogen embrittlement. An analysis of the acoustic emission signals can be used to determine the time between bursts of discontinuous crack growth more accurately than by conventional clip-gauge methods. This analysis helps elucidate the kinetics of the mocroscopic processes responsible for slow crack growth. Susceptibility to stress corrosion cracking may be noted quite early (within minutes) during a static test on precarcked specimens. This allows an investigator to avoid long-time (48 hours) testing in determining the severity of a particular environment or the susceptibility of a particular environment or the susceptibility of a particular materials. Since the acoustic emission rate is a direct function of the stress intensity factor, K, it is possible to use a critical count rate to predict (and thereby prevent) the failure of structures in reactive environments or after cathodic charging. Examples of these applications of acoustic emission testing will be given in the presentation.

1749 **A. S. Tetelman. "The Use of Acoustic Emission Testing to Monitor Hydrogen Embrittlement in Steels." pp 78–97 in** ***Effects of Chemical Environment on Fracture Processes.*** **C. J. Osborn and R. C. Gifkins, Editors. Department of Metallurgy, University of Melbourne, Parkville, Victoria, Australia. 1974. proceedings of the Third Tewksbury Symposium on Fracture, held at the University of Melbourne. June 4–6, 1974. [proceedings available from D. S. Mansell, Department of Mining and Metallurgy, University of Melbourne, Parkville, Victoria 3052, Australia].**

Acoustic emission is a relatively new method of nondestructive inspection that can be used to monitor microcrack formation and crack growth. The principles of acoustic emission are discussed. Recent models of hydrogen embrittlement are then presented and related to fracture mechanics parameters. The two concepts are combined and the acoustic emission accompanying hydrogen cracking is discussed. Particular emphasis is given to cathodically charged specimens and components and to welded specimens in which the hydrogen is introduced during the welding process. The use of acoustic emission to monitor plating baths is described. Finally, some discussion of the use of acoustic emission as a tool for studying hydrogen cracking is presented.

1750 **A. S. Tetelman and R. Chow. "Acoustic Emission Testing and Microcracking Processes." pp 30–40 in** ***Acoustic Emission, ASTM STP 505.*** **American Society for Testing and Materials, Philadelphia, Pennsylvania. 1972. proceedings of symposium held in Bal Harbour, Florida. Dec 7–8, 1971.**

The various stages of microcrack initiation, microcrack growth, and final fracture are reviewed and the relation between their processes and acoustic emission is discussed. A theoretical model relating acoustic emission to the microcrack density of low carbon steel is presented, based upon experimental results of the microcrack density in low carbon iron. The model predicts that the total number of acoustic emissions increases rapidly with applied stress above a threshold value.

1751 **A. S. Tetelman, H. L. Dunegan, R. A. Wullaert, and D. R. Ireland. "Acoustic Emission Evaluations." Appendix D, pp D-1 to D-30 in "Examination and Evaluation of Capsule A for the Connecticut Yankee Reactor Pressure-Vessel**

Surveillance Program" D. R. Ireland and V. G. Scotti. Final Report on work performed under the Nuclear Utilities Technical Assistance Plan (NUTAP). Battelle Columbus Laboratories, Columbus, Ohio. Oct 30, 1970.

A considerable amount of supplementary data was obtained in this program. The results of these evaluations are presented separately in the appendices. This work included fracture-mechanics evaluations of weld-metal and base-metal WOL specimens, acoustic emission studies of the WOL and tensile specimens, and instrumented impact test evaluation of the Charpy specimens. Battelle Columbus Laboratories (BCL) in conjunction with Failure Analysis Associates and Dunegan Research Corporation recognized the need to investigate the acoustic emission characteristics of irradiated reactor-grade steels. This information is required to determine whether it is possible to use acoustic emission to detect incipient failure in pressure vessels. Previous and current test programs have attempted to apply acoustic emission techniques directly to pressure vessels before determining whether growing flaws in these materials emit characteristic signals strong enough to be detected and adequately analyzed with respect to the relative severity of the flaw. It has been well-established that the energy associated with acoustic emission signals is somewhat a function of the yield strength of the materials involved. The mechanical properties of the steel in a reactor pressure vessel will be altered by the service exposure to fast neutrons. There is no guarantee at this time that the reactor-grade steels will emit sufficient acoustic signals for use in flaw detection and that the acoustic emission information obtained from unirradiated material will be adequate for analysis of the similar data obtained from irradiated material. Acoustic emission tests were performed on the irradiated WOL and tensile specimens that were contained in Capsule A.

1752 **A. S. Tetelman and A. G. Evans. "Failure Prediction in Brittle Materials Using Fracture Mechanics and Acoustic Emission." pp 895–924 in** *Fracture Mechanics of Ceramics.* **R. C. Bradt, D. P. H. Hasselman, and F. F. Lange, Editors. Volume 2: Microstructure, Materials, and Applications. Plenum Press, New York. 1974. proceedings of a symposium held at Pennsylvania State University, University Park, Pennsylvania. July 11–13, 1973; UCLA-ENG-7365. University of California, School of Engineering and Applied Science, Los Angeles, California. 1973.**

Acoustic emission testing has found many uses in recent years and new applications are constantly being uncovered. The interesting areas for application lie in failure prediction and the characterization of the microscopic processes of yielding and fracture, and the macroscopic processes of slow-crack growth and onset of fast-crack propagation. The paper describes these applications and emphasizes that the total counts and count rate depend on the energy released per event and the density of events per unit of deformation. Where models are presented, they should be regarded as first order approximations which await confirmation.

1753 **D. O. Thompson. "Advanced NDE Techniques." pp 114–145 in** *Prevention of Structural Failure: The Role of Quantitative Nondestructive Evaluation.*

T. D. Cooper, P. F. Packman, and B. G. W. Yee, Editors. Materials/Metalworking Technology Series, No. 5. American Society for Metals, Metals Park, Ohio. 1975.

The capabilities and limitations of some new nondestructive evaluation techniques are reviewed, including ultrasonic spectroscopy, acoustic emission, and the measurement of the bond strength of adhesively bonded parts and composite materials. The improvements in ultrasonic spectroscopy include: diffusion-bonded controlled flaws, ultrasonic refraction coefficients for various geometries of flaws leading to flaw size and shape determinations, and digitized signal processing for improved resolution. Discrepancies in the acoustic emission count rate to electronic gain ratio show that for many situations an insufficient number of parameters are measured. Recording the emissions on a wide-band videotape recorder and pickup give wider spectroanalysis capabilities. The energy of emission bursts by-pass some problems found in counting. [Abstract 22-1152. *Metals Abstracts,* Vol. 8, 1975].

1754 **D. O. Thompson. "Requirements for and Advances in Nondestructive Evaluation." pp 642–652 in *1974 Ultrasonics Symposium Proceedings.* Catalog No. 74 CHO 896-1SU. Institute of Electrical and Electronics Engineers, Inc., New York. 1974. symposium held in Milwaukee, Wisconsin. Nov 11–14, 1974.**

A limited review is given of current problems in nondestructive evaluation and approaches that are being taken in some of these problem areas. It is pointed out that a serious need exists to *bridge the gap* between NDE and the materials and engineering research communities. Because of various changes in structural design philosophy that have been introduced in recent years, new NDE techniques must be much more quantitative in nature than previously required. Several new approaches which exemplify these requirements in ultrasonics, acoustic emission, and adhesive bond strength measurements are discussed.

1755 **J. L. Thompson (to Esso Research and Engineering Company). "Acoustic Leak Location and Detection System." U.S. Patent 3,838,593. Oct 1, 1974.**

A system is described for locating and detecting acoustic noise sources, such as liquid or gas leaks in the presence of conflicting ambient (i.e., background) noise by implementing one point correlation. In a preferred embodiment, the device employed comprises a pair of spaced sensors mounted on a movable connecting arm, each of which provides an output signal representative of a noise source being sent to an electronic circuit including correlator. Therein the signal is processed and the correlator output is monitored on a meter. The time average of the instantaneous product of the detected signals will be low except when nearly indentical signals are processed. When the sensors are equidistant from the noise source, the correlator output signal is at a maximum, thus providing an indication that the noise source is located in a central plane normal to the axis of the mounting arm. By scanning the suspect area in two or more planes, the noise source may be precisely located from a distance.

1756 **J. L. Thompson. "Summary Report on Acoustic Emission Weld Monitoring in the Welding of Rail." Y–40647. Battelle–Northwest, Richland, Washington. Sep 1969.**

The purpose of this program was to determine the feasibility of applying acoustic emission monitoring as a quality control technique in the welding of ribbon rail sections. Three types of welds were examined: (1) electric flash butt welding, (2) oxyactylene butt welding, and (3) thermit cast welding. Acoustic emission monitoring is based on the fact that discontinuities within the weld generate stress waves during cooling. The acoustic emission generated by this release of energy is evidently due to thermal stress gradients generated in the weld or around inclusions which, if large enough, can cause cracking which relieves much of the stress. The acoustic emissions are detected as the weld cools by an ultrasonic transducer placed near the weld zone. Appropriate signal processing circuitry enables detection of the rate of release of acoustic energy and, with two or more transducers, can pinpoint the location of the defect. All three welding techniques were investigated with prototype equipment during trips to the Norfolk and Western Railroad ribbon rail welding plants at Bellevue, Ohio and Roanoke, Virginia. The data obtained indicate that acoustic emission weld monitoring as a quality control method is feasible in all three cases, and especially so in the case of the thermit welding process. This statement must be qualified to some extent in the case of electric flash butt welding however. Severe electrical interference at the Bellevue plant prevented obtaining physical evidence in the form of chart recordings of emission activity. The conclusion of feasibility in this case is based on the visual observation of emission bursts on a monitor oscilloscope by the investigator. The benefits of employing acoustic emission rail weld monitoring would include increased safety on the road, as each weld in trackage would be tested nondestructively for integrity. Such checking by conventional radiographic means is normally not done due to cost and time required. Another important advantage would be reduction of welder training time by making the welding of rail, especially oxyacetylene welding, a process controlled by objective standards rather than an acquired art and subject to the operator's skill and attention.

1757 **N. Thompson and D. J. Millard. "Twin Formation in Cadmium."** ***Philosophical Magazine,*** **PHMAA, 43:422–440. Apr 1952.**

Experiments on the formation of twins in single crystal wires of cadmium under tension demonstrate the essentially discontinuous nature of the process of twin growth, and produce evidence in favor of a critical resolved shear stress law for twinning. Twinning can contribute to creep at room temperature but not at 90 degrees absolute. A dislocation mechanism for the growth of a twin is discussed in some detail.

1758 **R. F. Tinder. "Flow Mechanisms and Acoustical Emission during Initial Yielding in Crystals." AFOSR 70–0801 TR. Final Report. Washington State University, Pullman, Washington. Apr 13, 1970. [NTIS: AD-705–686].**

A broad-band acoustic emission monitoring system with a sensitivity of 8×10^{-15} watts/cm^2 was used in an attempt to detect acoustic emission from highly-perfect tubular zinc monocrystals which were deformed in the torsional mode at room temperature. Millimicroplastic bursts ranging from 10×10^{-9} to 760×10^{-9} in strain magnitude were recorded by the loading apparatus during the initial stages of yielding, but no acoustic emission accompanying the bursts could be detected. This seems to suggest that the burst behavior occurring in the initial stages of yielding of zinc is controlled by low-energy release mechanisms, such as dislocation node diffusion. [The report briefly summarizes the work of C. K. Day in his M. S. Thesis (see Reference 360)].

1759 **A. Tobias. "ACEMAN–A PDP-11 Disk Operating System for Acoustic Emission Analysis." RD/B/N3188. Central Electricity Generating Board, Berkeley Nuclear Laboratories, Berkeley, England. Nov 1974.**

A versatile software system for acoustic emission analysis from plate structures with an array of sensors has been developed. The main features include: (1) A display of the spatial resolution of the system for any transducer configuration, (2) Real-time display of emissions on a developed diagram of the test area, (3) Facility to analyze data over pressure or cycling ranges, (4) Display of emission source as event count, sum of event amplitude, or sum of event amplitude square, and (5) Amplitude spectrum analysis of emissions from a located source. Limitations in the present system are discussed and remedies in progress are outlined.

1760 **A. Tobias. "ACEMAN (II): A PDP-11 Software Package for Acoustic Emission Analysis." RD/B/N-3553. Central Electricity Generating Board, Berkeley Nuclear Laboratories, Berkeley, England. Jan 1976.**

A powerful, but easy-to-use, software package (ACEMAN) for acoustic emission analysis has been developed at Berkeley Nuclear Laboratories. The system is based on a PDP-11 minicomputer with 24 K of memory, an RK05 DISK Drive and a Tektronix 4010 Graphics terminal. The operation of the system is described in detail in terms of the functions performed in response to the various command mnemonics. The ACEMAN software package offers many useful facilities not found on other acoustic emission monitoring systems. Its main features, many of which are unique, are summarised. The ACEMAN system automatically handles arrays of up to 12 sensors in real-time operation during which data are acquired, analysed, stored on the computer disk for future analysis, and displayed on the terminal if required.

1761 **A. Tobias. "Acoustic Emission Source Location in Two Dimensions by an Array of Three or Four Sensors." RD/B/N3218. Central Electricity**

Generating Board, Berkeley Nuclear Laboratories, Berkeley, England. Jan 1975. [see Reference 1762 for publication].

Various triangulation techniques have been used for locating acoustic emission sources. Some analyses systems utilize iterative procedures, while others require particular sensor configurations. A general method has been developed for calculating the location, in two dimensions, of an acoustic emission source from the differences in arrival time of the emission waves measured by an array of three sensors on a structure. The method yields an exact analytic solution for any sensor configuration. Examples of the resolution capability of selected sensor arrays are given. The most precise location for a given timing capability is shown to be achieved when the source lies within the area bounded by the sensors. Examination of the solutions for different sensor configurations shows the existence of ambiguous locations; i.e., the same differences in arrival time of an emission wave at three sensors can arise from two distinct source locations. The areas of ambiguity are shown to lie around the three sensors. It is also shown how a fourth sensor may be employed to resolve these ambiguous solutions and validate located events.

1762 **A. Tobias. "Acoustic-Emission Source Location in Two Dimensions by an Array of Three Sensors."** *Non-Destructive Testing,* **NDETA, 9(1):9-12. Feb 1976. [see also Reference 1761].**

Sensors on the surface of a material under stress can detect acoustic emission from a defect within the material. The difference in time of detection of an emission from the defect at different sensors gives a way of finding where it is. This paper shows a general method of calculating the location of defects in two dimensions from the arrival times at the sensors. The ACEMAN system, which uses this method can derive the resolution properties of a sensor array in about 7.5 minutes.

1763 **A. Tobias. "On-Line PDP-11 Acoustic Emission Location Analysis for a System of 3 Transducers." RD/B/N2854. Central Electricity Generating Board, Berkeley Nuclear Laboratories, Berkeley, England. Dec 1973. [NTIS].**

An acoustic emission location system based on a PDP-11 computer is described in terms of the operation of the present software. Limitations of the system are discussed and possible improvements are outlined. Despite these limitations, it is shown that clearly defined source locations can be obtained from the analysis of acoustic emissions. A data input rate of ~1400 signals per second can be handled by the system without the introduction of dead time. It is therefore anticipated that the conditions of high noise, which are characteristic of reactors in service, will be accommodated by the system.

1764 E. G. Tomachevsky. "Appliance of Stress Wave Emission in Phenix Reactor." pp 48–54 in "EWGAE–ISPRA: Proceedings of the Third Meeting of the European Working Group on Acoustic Emission." A. C. Lucia, Editor. Report EUR-5513e. Commission of the European Communities, Directorate General Scientific and Technical Information and Information Management, Luxembourg, Belgium. 1976. meeting held at the Joint Nuclear Research Center, Ispra, Italy. Sep 25–26, 1974.

1765 E. G. Tomachevsky. "Surveillance par emission d'ondes de contrainte de la propagation d'une fissure a l'aide d'un localisateur Leanord transforme (Acoustic Emission Testing of Crack Propagation Using a Modified Leanord Localizer)." *Bulletin d'Informations Scientifiques et Techniques, Commissariat a l'Energie Atomique,* BUIAA, (211):81–84. Feb 1976. (in French); "Crack Propagation Surveillance by Stress Wave Emission Using a Converted Leanord Localizer." BNWL-TR-192. translated for Battelle-Northwest, Richland, Washington. July 29, 1976. [NTIS].

The acoustic emission testing of cracks is perturbed by mechanical parasites and by electrical parasites. Since these parasites are perfectly localized, they can be eliminated by means of a localizer, if special adjustments are made and the apparatus is slightly modified.

1766 E. G. Tomachevsky. "SWE Testing on Reactor Phenix." Note Technique EMT 73-270. CEA Centre d'Etudes Nucleaires de Saclay, 91-Gif-sur-Yvette, France. Aug 13, 1973; paper presented at the 2nd European Working Group on Acoustic Emission Meeting, Frankfurt-am-Main, Germany. Sep 25–26, 1973. [NTIS: CEA-CONF-2579].

Stress wave emission tests were carried out on the Phenix main vessel in order to investigate the possibilities of the method.

1767 E. G. Tomachevsky, A. Drouet, and P. J. Despresles. "Recherche de la Diminution de Resistance en Traction du Beton par Detection d'Emission d'ondes de Contrainte (Research in the Decrease in Tensile Strength of Concrete by Stress Wave Emission)." pp 336–360 in *Journees d'Etudes sur l'Emission Acoustique.* P. F. Gobin, R. Goutte, and C. Guillaud, Organizers. Institut National des Sciences Appliquees de Lyon, Avenue Albert Einstein, 69621 Villeurbanne, Lyon, France. Mar 1975. proceedings of a conference held in Lyon, France. Mar 17, 1975. (paper in French).

The recording of the cumulated stress wave emission counting from a concrete specimen under increasing load shows angular points which appear at well-defined fractions of the breaking strength and mark the limits of 4 distinct phases: compression stage for: 0<sigma<0.2R; assumed microcracking phase for: 0.2R<sigma<0.75R; assumed macrocracking phase for: 0.75R<sigma<0.96R; prefailure phase for: 0.96R<sigma<R. The beginning of the 3rd phase is indicated by an easily identifiable elbow bend which announces impending failure, whereas the specimen shows no apparent external sign of damage. A sample previously loaded under monoaxial compression still keeps an appreciable breaking resistance in traction if the compression has remained below the appearance of this characteristic elbow. Finally the Kaiser effect distinguished by an SWE appearing only above the maximum load previously attained can supply information on earlier loads.

1768 J. P. Toronchuk. "On the Nature of Acoustic Emission and Its Application to the Study of Deformation Mechanisms in Metals." Ph.D. Thesis. University of Manitoba, Department of Mechanical Engineering, Winnipeg, Manitoba, Canada. May 1974. [available as Canadian Thesis on Microfiche No. 20458 from Canadian Thesis Division, Cataloguing Branch, National Library of Canada, Ottawa, Canada K1A ON4].

Stress waves which are generated in solids during deformation are termed acoustic emissions. Detection of acoustic emissions in this study is by piezoelectric crystal transducers and accelerometers. The effect of various test parameters on acoustic emission was determined. These parameters are specimen length, strain rate, and grain size. Single crystals of zinc were tested to study acoustic emission during twinning and aluminum single crystals were tested to study the effect of specimen orientation on acoustic emission. At a constant strain rate, the effect of a change in volume of the specimen should give a proportional change in total emissions over a given strain interval. However the total emissions increased by more than a factor of two for the same elongation when the specimen length was doubled. It was observed that the amplitude of emissions increased with increasing gage length which contributed to this effect since the counts were made above a fixed trigger level. Most of the energy in the signal was at the resonant frequency of the transducer used with smaller amounts at other frequencies which depended on specimen length. An increase in strain rate should give a proportional increase in emission rate and a constant number of emissions per unit strain if the dislocation mechanisms which cause slip are not affected by the change in strain rate. However the counts per unit strain were observed to increase. This is mainly due to an increase in amplitude of acoustic emissions. During twinning, a strain step formed before the load drop occurred. The amplitude of the acoustic emission events was more than an order of magnitude greater during the strain step than that of the events during the load drop. This indicates that the type of mechanism causing the strain step is different than that which occurs during the load drop. From the above, it may be concluded that nucleation of the twin occurs during the strain step formation and then it extends across the width of the crystal during the load drop. The rise time of the acoustic emissions during the strain step formation indicates this to be a very rapid event. These two factors are in agreement with the homogeneous nucleation model for twinning. In aluminum single crystals of soft orientation the acoustic emission amplitude is constant during Stage 1 hardening and increases during Stage 2 and reaches the highest value in Stage 3. This is ascribed to different dislocation mechanisms occurring in each stage. Single crystals of hard orientations gave emission rate curves of approximately the same form and amplitude as those obtained from polycrystalline specimens.

1769 W. C. Townsend, V. B. Watwood, D. M. Romrell, and C. W. Hunter. "Defect Study: Pressure Vessel CSE Reactor Vessel Simulator." BNWL-1554. Battelle-Northwest, Richland, Washington. June 1971. [NTIS].

A large laminar-type defect was detected by ultrasonic tests in the 150-cubic foot pressure vessel used for coolant blowdown experiments, following a thrust support failure. Extensive testing and analysis of core samples indicated that the defect was an inclusion containing calcium and aluminum oxide and that the base metal surrounding the flaw had adequate strength and ductility. Analytical studies of potential crack propagation led to the conslusion that further growth into the sound metal was unlikely and the decision was made to continue the testing program. Careful acoustic emission monitoring followed by ultrasonic examination of the defect region was carried on following each of about 50 blowdowns made after detection of the flaw. No further growth of the defect region was noted.

1770 R. L. Trantow. "Feasibility of Evaluating the Integrity of FTR Duct-to-End Fixture Welds using Acoustic Emission Monitoring." HEDL-TME 75-88 Hanford Engineering Development Laboratory, Richland, Washington. Nov 1975. [NTIS].

A method of applying real-time acoustic emission source location to evaluate the integrity of duct-to-end fixture welds was developed and evaluated using the HEDL R8-4 AE monitor coupled with a PDP-8e minicomputer. Operational software was developed to control the system's data acquisition, storage, and display functions. Performance was evaluated on the basis of comparisons between AE source location data and subsequent destructive examination of 6 duct-to-transition ring qualification welds. Ten separate discontinuities seen in the metallography performed at 35 locations along these welds, correlated with the averaged acoustic emission activity revealed by the AE source location maps. This relationship is not presently considered to be usable as a system calibration method however, because of the limited range of discontinuity lengths (from 0.003 to 0.004 inches long) that were correlated. The presence of 6 other discontinuities, found in regions showing low levels of emission activity, indicated that significant stress risers can go undetected under the conditions investigated during this study.

1771 T. L. Triffet. "Acoustic Emission from Crystalline Materials." NSF G-14650, Final Report. Division of Engineering Research, Michigan State University, East Lansing, Michigan. Apr 1, 1969.

The primary objective of this research was to relate acoustic emission to some of the fundamental phenomena in imperfect crystals. Other objectives of the study were: (1) To develop a transducer to detect displacements associated with acoustical emission; (2) To study the magnitude and duration of the stress pulse produced in a gross specimen by a dislocation motion; (3) To observe acoustic emission from single crystals of lithium fluoride loaded in direct shear; and (4) To propose a mechanism for acoustic emission in terms of interaction between dislocations and obstacles in their motion.

1772 **T. Triffet, R. B. Engle, and R. G. Liptai. "Acoustic Emission from Crystalline Materials." Division of Engineering Research, College of Engineering, Michigan State University, East Lansing, Michigan. 1963.**

See Reference 1771 for final report of work performed under National Science Foundation grant NSF-G-14650.

1773 **Trodyne Corporation. "Acoustic Emission Crack Monitor." British Patent 1,344,315. Jan 23, 1974.**

1774 **M. Trombik and W. Zuberek. "Microseismic Research in Polish Coal Mines." pp 169–194 in** ***Proceedings First Conference on Acoustic Emission/Microseismic Activity in Geologic Structures and Materials.*** **H. R. Hardy, Jr. and F. W. Leighton, Editors. Trans Tech Publications, Clausthal, West Germany. 1977. conference held at Pennsylvania State University, University Park, Pennsylvania. June 9–11, 1975.**

Due to the great depth of mining and the specific geological conditions in some of Polish coal mines, a relatively high degree of rock-burst hazard has been observed. Considerable effort has been made to reduce this hazard by applying and developing suitable methods of prevention, prediction, and fighting of the rock-burst hazard. In particular, the microseismic method has been introduced into several mines. Several types of microseismic equipment (stationary, portable, and automatic) have been developed and tested under underground and laboratory conditions. Essentially this equipment is based on microseismic noise rate and cumulated relative energy determinations. During some years of observations directly in the mines, several coal bursts have been recorded and the results of microseismic research are encouraging although several difficulties in data analysis have been found. This paper describes recent underground studies and the direction of future research in the microseismic area.

1775 **B. Tucker. "New Approach to Reactor Pipe Problems: Listening to Cracks Crack."** ***Nucleonics Week,*** **NUWEA, 16(31):7–8. July 31, 1975.**

1776 **T. R. Tucker and C. T. Fujii. "Acoustic Emissions and Stress–Corrosion Cracking in High–Strength Alloys." NRL Memorandum Report 2879. Naval Research Laboratory, Washington, D.C. Aug 1974. [NTIS: AD–785–009].**

The usefulness of acoustic emission or stress wave emission (SWE) to studies of stress-corrosion cracking of high-strength alloys was explored. Single edge-notched, precracked cantilever specimens were used to study the stress-corrosion-crack growth and SWE characteristics of a high-strength stainless steel and a titanium alloy. The SWE data correlate reasonably well to crack-growth measurements by conventional beam-deflection techniques for high-strength stainless steel, but are too insensitive for reliable detection of crack extension in the titanium alloy. The use of SWE data to define the energetics of discrete cracking events is currently beyond the capabilities of existing equipment and analytics.

1777 **M. Ya. Tutans and Yu. S. Urzhumtsev. "Seismoacoustic Prediction of the Fracture Processes in Glass–Reinforced Plastics."** *Polymer Mechanics,* **PLYMA, (3):377–383. May–June, 1971; translated from** *Mekhanika Polimerov,* **MKPLA, (3):421–429. May–June 1971. (in Russian).**

A technique and equipment for investigating the seismoacoustic emission associated with the fracture of a loaded glass-reinforced plastic are described. It is shown that the emission is present from the beginning of loading and increases regularly up to failure. Glass strands consisting of 100 plain elementary fibers of varying diameter and resin-impregnated strands have been investigated in tension and specimens of glass-reinforced plastic and unreinforced epoxy resin in tension and bending. Experimental data on short-time, static, and repeat loading are presented for the glass-reinforced plastic. It is shown how the integral pulse distribution and the pulse amplitude frequency function depend on the applied load level.

U

1778 H. Ucisik and K. Ono. "Acoustic Emission Characteristics of High Strength Aluminum Alloys during Fracture Toughness Test." Session 4, pp 64–65 in *The Second Acoustic Emission Symposium.* Japan Industrial Planning Association, Dobashi Building 5F, 3–17–8 Akasaka, Minato-ku, Tokyo, 107 Japan. 1974. proceedings of a symposium held in Tokyo, Japan. Sep 2–4, 1974. [abstract only, paper not published in proceedings].

The paper reports acoustic emission studies of five high strength aluminum alloys, 7001-T75, 7049-T6, 7049-T73, 7075-T651, and 7075-T761. Fatigue precracked compact tension specimens were tested according to ASTM Standard E 399-72. Stress-intensity factor and crack-opening displacement were measured during the tests in addition to the following acoustic emission parameters; threshold crossing counts, and their dependence on the threshold level, rms voltages, and power density function spectra. The AE signals were detected using a commercial wide-band transducer (FC500, AET Corporation), and a PZT-5 element (¼-inch diameter compression) with 3 MHz fundamental resonance. The former responded with nearly uniform sensitivity to longitudinal waves in the measurement range of 50 to 2000 kHz. The latter appeared to have a flat response to longitudinal waves, but exhibited a radial resonance peak near 400 kHz. The radial resonance peak can be excited by shear-mode acoustic emission signals and provides a convenient means of detecting their presence. Amplified and filtered AE signals were recorded on a wide band instrumentation tape recorder (Bell & Howell 3700A) together with applied load. Both AM (to 2 MHz) and FM (to 500 kHz) recording modes were utilized. A power-density spectrum was obtained by first determining autocorrelation function, using a hard-wired analyzer and by performing its Fourier transform in a minicomputer. When the stress intensity K_I exceeded that used for precracking, total counts started to increase rapidly. In 7001-T75 alloy, this value was 7 ksi$\sqrt{\text{inch}}$. The rate of increase was reduced significantly after the initial stage in the range of 10 to 19 ksi$\sqrt{\text{inch}}$. For $K_I < 19$ ksi$\sqrt{\text{inch}}$, total counts were proportional to (K_I^m-constant), where m of 2 to 4 provided a good fit. The data from the FC500 transducer required much smaller values of m in the same range of K_I. At higher values of K_I up to $K_{Ic} = 26.5$ ksi$\sqrt{\text{inch}}$, total counts increased rapidly with K_I with m values of 20 or more. For this material, the frequency spectra of K_I below 19 ksi$\sqrt{\text{inch}}$ indicated the predominance of 50 to 100 kHz components corresponding to the longitudinal wave resonance. At higher K_I values, the shear-wave mode centered at 400 kHz became significant. These differences are due to the expansion of plastic zone at low K_I values and to slow-crack growth at higher K_I values. In another material, such as 7049-T6, total counts versus K_I can almost be characterized by a single m value (4 to 6), except near the precracked load. The shear-wave mode was predominant from early stages of fracture toughness testing. The frequency response of signal with the FC500 transducer was flat, but that due to a PZT-5 element transducer always showed strong radial-mode resonances. This is a much stronger alloy and slow-crack growth is expected to start at a lower K_I than that for 7001-T75 alloy. Experimental results from tests on other alloys are also described and discussed in terms of various mechanisms of acoustic emission. The paper describes these and other AE parameters determined in fracture toughness tests of aluminum alloys and correlated to microscopic fracture processes. In particular, it demonstrates the importance of vibration mode and frequency analysis in acoustic emission research.

1779 N. Uesugi, K. Tatsuno, K. Uchida, and T. Miyazawa. "Preliminary Investigations of On-Line AE Monitoring System for Nuclear Power Plants." Paper 2-7, pp 167-179 in *The Third Acoustic Emission Symposium.* Japan Industrial Planning Association, Dobashi Building, 3-17-8, Akasaka, Minato-ku, Tokyo, Japan. 1976. proceedings of a symposium held in Tokyo, Japan. Sep 16-18, 1976.

In nuclear power plants, applications of acoustic emission techniques have been expected to be one of the most promising methods for on-line integrity monitoring systems. The AE techniques have been considered to be excellent in both incipient indications of faults of structural materials and their locations. Many research efforts have been expended for realizing an on-line AE monitoring system, based on the fundamental research on materials. However several problems impede these developments, owing to nuclear power plant conditions such as high temperature, high irradiation dose, and electrical and mechanical noises. In principle, the assumed system for this purpose is similar to a well-developed inspection system which is composed of several sensors, signal transmission cables and preamplifiers corresponding to each sensor, main amplifiers, signal analysis circuits, and signal read-out and display circuits. Of these components, some cannot be employed without great improvements for nuclear power plants. Thus each component should be developed in relation to nuclear power plant conditions for realizing an on-line integrity monitoring system employing AE techniques. In developing a reliable and suitable on-line AE monitoring system for nuclear power plants, the following preliminary study has been set up: (1) high temperature acoustic sensors development, (2) evaluation of acoustic couplants at high temperature, (3) on-line calibration method development for the AE detection system, and (4) evaluation of mechanical background noises at BWR.

V

1780 S. J. Vahaviolos (to Western Electric Company, Inc.). "Method and Apparatus for the Real–Time Evaluation of Welds by Emitted Stress Waves." U.S. Patent 3,965,726. June 29, 1976.

Spot welds are evaluated using stress-wave emission techniques by detecting and measuring the stress waves emitted from a weld area during a first solid-to-liquid phase transformation period and a second liquid-to-solid phase transformation period of the weld. The stress waves emitted during the first transformation period provide an indication of the weld nugget size, while the stress waves emitted during the second transformation period provide an indication of the amount of post-weld cracking in the weld area. By subtracting the stress-wave energy measured during the second transformation period from the stress-wave energy measured during the first transformation period, an indication of the strength of the weld is obtained.

1781 S. J. Vahaviolos (to Western Electric Company, Inc.). "Method and Apparatus for the Real–Time Monitoring of a Continuous Weld Using Stress–Wave Emission Techniques." U.S. Patent 3,986,391. Oct 19, 1976.

Continuous welds are monitored by detecting and processing stress-wave signals emitted from the moving weld area to provide an indication of both the weld energy supplied to the weld area by the welding apparatus, and the development of microcracks in the cooling weld area. According to the present invention, the detected stress-wave signals emitted from the weld area are amplified and filtered to provide excellent signal-to-noise ratios and then applied to either one, or both, of (a) an rms voltmeter, and (b) a spectrum analyzer to provide an indication of the weld energy supplied to the weld area and the development of cracks in the cooling weld area.

1782 S. J. Vahaviolos (to Western Electric Company, Inc.). "Method and Apparatus for the Real–Time, Non–Destructive Evaluation of Adhesion Bonds Using Stress–Wave Emission Techniques." U.S. Patent 4,004,456. Jan 25, 1977.

The strength of an adhesion bond can be determined using stress-wave emission techniques. One method measures the stress-wave emitted from the bonding area during the compression phase of the formation of the bond and compares this measurement with a predetermined substantially linear relationship between the emitted stress-wave energy and the strength of the adhesion bond. For more accuracy, a second method measures the stress-wave energy during both the compression phase and a pressure relief interval and determines the difference between the two measurements, and also compares the difference value with the substantially linear relationship. An alternative to the second method subtracts a predetermined threshold value from the pressure-relief interval measurement, reduces the compression phase measurement by the excess determined by said subtraction step, and compares the reduced compression phase measurement with the predetermined substantially linear relationship to determine the strength of the bond.

1783 **S. J. Vahaviolos (to Western Electric Company, Inc.). "Methods and Apparatus for Detecting the Presence of Cracks in a Workpiece by the Use of Stress Waves Emitted Therefrom." U.S. Patent 3,924,456. Dec 9, 1975.**

Microcracks in ceramic substrates are detected by positioning a piezoelectric transducer proximate a supporting base for the substrate. The stress waves which are emitted by the cracks which form in the substrate when it is subjected to a load are converted into electrical pulses by the transducer. Because ceramic is a brittle material rather than ductile these pulses are of low amplitude, short duration, and fast rise time. Special electronic circuitry is disclosed to detect these pulses and to distinguish them from the background noise.

1784 **S. J. Vahaviolos. "Real Time Detection of Microcracks in Brittle Materials Using Stress Wave Emission (SWE)." pp 34–42 in** *Proceedings of the 24th Electron Components Conference.* **Institute of Electrical and Electronics Engineers, Inc., New York. 1974. conference held in Washington, D.C. May 13–15, 1974;** *IEEE Transactions on Parts, Hybrids and Packaging,* **IEPHA, PHP-10(3):152–159. Sep 1974.**

A series of experiments in SWE were conducted to determine the characteristics of the emissions, the instrumentation requirements, and to better understand the phenomenon for its application to brittle materials [glass and ceramic (Al_2O_3) thin film substrates]. It was found that the detected SWE released have a frequency spectrum much like acoustic white noise over the frequency range of 0.1 kHz to approximately 100 MHz. The detection of SWE in ceramic substrates was accomplished using a broadband 700 kHz piezoelectric transducer.

1785 **V. E. Vainberg. "Acoustic Emission in the Deformation of Steels at Different Strain Rates and with Change of Sign."** *Soviet Journal of Nondestructive Testing,* **SJNTA, 11(5):654–656. Sep–Oct 1975; translated from** *Defektoskopiya,* **DEFKA, (5):133–135. Sep–Oct 1975. (in Russian).**

A note. The acoustic emission of C-steel, duralumin, and brass samples subjected to tensile or compressive strain was analysed with a view to using such emission in the defectoscopy of metal parts. Acoustic emission was also studied for the case of alternating deformation (tension–compression, etc.). On first loading in every case, the intensity of acoustic emission was at a high level and depended substantially on the sample material. On loading for the second time in the same direction after reducing the applied stress to zero, the Kaiser effect (sharp changes in emission) developed. On loading in the opposite direction the intensity of acoustic emission equaled that obtained in the first loading. [Abstract 22-0596. *Metals Abstracts,* Vol. 9, 1976].

1786 V. E. Vainberg. "Temperature Dependence of Acoustic Emission." *Soviet Journal of Nondestructive Testing,* **SJNTA, 12(5):564–566. Sep–Oct 1976; translated from** *Defektoskopiya,* **DEFKA, (5):124–126. Sep–Oct 1976. (in Russian).**

The acoustic-emission method of testing steels and other alloys for defects is outlined. One characteristic of this method is the sensitivity of the output signals to temperature; this is a valuable property since the materials under examination may work at either cryogenic or high temperature. The problem has been previously studied in the cryogenic range; and now the temperature dependence of acoustic emission is studied at high (e.g., 300 °C) temperature. Examples are given for Cr steel 20Kh, and also aluminum alloy D16T. The acoustic-emission method is sensitive to a change from the ductile-to the brittle-fracture mechanism. [Abstract 22-0643. *Metals Abstracts,* Vol. 10, 1977].

1787 V. E. Vainberg. "The Acoustic Emission Study during Programmed Loading." *Strength of Materials,* **SMTLB, 6(9):1157–1158. Sep 1974; translated from** *Problemy Prochnosti,* **PPCNB, (9):117–118. Sep 1974. (in Russian).**

In the recent years, a considerable attention has been devoted to the acoustic emission studies during mechanical testing of materials. This method is primarily used in nondestructive testing of welds and also stressed samples and parts. However, since the nature of the acoustic emission is closely connected with the deformation process and the structure of the material studied, it can also be used in the metallophysical studies. In the present work, the results of the Kaiser effect studies of loading and unloading of the samples made from different materials are presented. The effect of irreversibility of acoustic emission (Kaiser effect) was first described by J. Kaiser in 1953. It is known that acoustic emission is an ultrasonic emission from the material during deformation. This emission is usually generated when the region of plastic deformation is reached. If after the generation of acoustic emission the load applied to the deformed body is reduced to zero, acoustic emission is ceased. Acoustic emission can again be generated by exceeding the previous stress level experienced by the specimen. Kaiser observed this effect on the samples of steel, copper, aluminum, and wood. The accuracy of determining the previous stress level according to the generation of acoustic emission was several percent. Subsequently other investigators also observed the irreversible effect of acoustic emission. The studies described here show that the Kaiser effect exhibits some characteristics, previously not observed, and also that acoustic emission dependence on the previous stress differs sometimes completely from that as reported by Kaiser.

1788 **V. E. Vainberg and A. Sh. Kantor. "Temperature Investigations of Acoustic Emission."** *Soviet Journal of Nondestructive Testing,* **SJNTA, 11(6):783-784. Nov-Dec 1975; translated from** *Defektoskopiya,* **DEFKA, (1):129-131. Nov-Dec 1975. (in Russian).**

Recently, the acoustic emission method has been used for the inspection of high-pressure vessels, atomic reactors, piping, etc. Usually such equipment operates at higher temperatures. Therefore, there is much interest in temperature investigations of acoustic emission, which until now have not received sufficient attention. In this article, the relationship of the acoustic emission parameters to temperature are studied using the combination of the AE-5 (AF-20) apparatus with the UME-10TM universal test machine equipped with a Type UN2S-1100 heater, developed in the All-Union Scientific-Research Institute of Nondestructive Testing. The design of the unit permits heating of the samples and measurement, control, recording, and signalling of the extreme values of temperature.

1789 **V. E. Vainberg, A. Sh. Kantor, and R. G. Lupashku. "Using the Kinetic Concept of Fracture for Calculating the Acoustic Emission Intensity."** *Soviet Journal of Nondestructive Testing,* **SJNTA, 12(3):299-304. May-June 1976; translated from** *Defektoskopiya,* **DEFKA, (3):89-96. May -June 1976. (in Russian).**

A phenomenological model of acoustic emission is proposed and an expression for the emission intensity is derived.

1790 **V. E. Vainberg, R. G. Lupashku, A. Sh. Kantor, and A. M. Kushnir. "The Noise Intensity in Cracking."** *Strength of Materials,* **SMTLB, 7(9):1139-1141. Sep 1975; translated from** *Problemy Prochnosti,* **PPCNB, (9):92-94. Sep 1975. (in Russian).**

Noise emission during the cracking of metal structures is discussed theoretically. A phenomenological model is proposed for the development of acoustic emission during crack initiation and extension. Methods based on the existing kinetic concept of the strength of materials provides a simple relationship between the intensity of sound emission, the length of the crack, and the physicomechanical properties of the metal. Practical tests with Al alloys confirm the validity of the theory. [Abstract 31-0543. *Metals Abstracts,* Vol. 9, 1976].

1791 **V. E. Vainberg, V. N. Sosedov, and A. M. Kushnir. "Acoustic-Emission Study of Crack Growth."** *Soviet Journal of Nondestructive Testing,* **SJNTA, 11(3): 374-376. May-June 1975; translated from** *Defektoskopiya,* **DEFKA, (3): 127-129. May-June 1975. (in Russian).**

The possibility of using acoustic emission as a means of determining the parameters of fatigue cracks in steel and other metal components, and in particular the instant at which such cracks become dangerous, are discussed, and an instrument based on this principle is described. Typical cases of acoustic emission from cracks of various shapes and sizes are presented, together with an analysis of the corresponding signals. By incorporating filters and differential amplifiers in the instrument, acoustic emission may be employed to determine the shape, size, and velocity of fatigue cracks more accurately than by traditional methods. [Abstract 22-0159. *Metals Abstracts,* Vol. 9, 1976].

1792 K. B. Vakar. "An Evaluation of Parameters of the Acoustic Emission Signal Detection Circuit in Noise Background." IAE-2383. I. V. Kurchatov Institute of Atomic Energy, Moscow, U.S.S.R. 1974. (in Russian and English).

The principle of estimating the parameters of the acoustic emission signal detection-tract circuit in noise background has been considered as applied to the problem of the acoustic emission monitoring the structural integrity of the power reactor vessel. The incidental nature of signals and noises is taken into account.

1793 K. B. Vakar. "On Estimation of Detection Range Acoustic Emission Signals of the Noise Background." I. V. Kurchatov Institute of Atomic Energy, Moscow, U.S.S.R. 1975.

The way of analytical estimation of operational range of the four-points detection system and space-time processing of information is considered. Simple approximations of initial functional dependences are used in numerical models. The practical realization of the monitoring system of big surfaces of metal for casings of high-pressure vessels poses the problem of estimation of possible sizes of the area controlled, which, in its turn, defines the complexity of the system and its cost. Outlined are ways to analytically estimate the detection range of acoustic emission signals on a noise background with the use of four-points receiving groups of acoustic sensor and the processing circuit described in a previous work.

1794 K. B. Vakar. "Principles of Construction of Acoustic-Emission Monitoring Systems." I. V. Kurchatov Institute of Atomic Energy, Moscow, U.S.S.R. 1975.

A possibility and ways of realization of acoustic-emission methods for monitoring the metal pressure vessel and equipment of the reactor first loop of an atomic power plant in operating conditions are considered. The sequence of solution of the problem and ways of optimization for the information processing system are determined under conditions of high-level background noises being available. Block diagrams of four versions of acoustic-emission monitoring systems are presented. Given are difficulties, possible ways of realization of the methods, and the field of investigations which have to precede creation of automatic test sets.

1795 **H. H. Vanderveldt. "Monitoring Stress Wave Emissions." Report No. 3809. Department of the Navy, Naval Ship Research and Development Center, Bethesda, Maryland. June 1972. [NTIS: AD-745-729].**

Basic principles of the acoustic emission technique used at the Center are discussed, and a review of some applications of the method are presented. Potential capabilities and drawbacks to naval applications are summarized; finally, a description of the design and construction of the present monitoring capability at the Naval Ship Research and Development Center is presented.

1796 **H. H. Vanderveldt and Q. Tran. "Acoustic Emissions from Synthetic Rope."** *Naval Engineers Journal*, **NVEJA, 83(6):65-68. Dec 1971.**

The applicability of the stress wave emission phenomena from synthetic ropes to indicate imminent failure is demonstrated. An increase of at least an order of magnitude in the slope of the curve which shows the number of stress wave emissions versus the applied load is a good indicator of impending catastrophic failure. However, no significant differences in the stress wave emission characteristics was observed for the three types of materials considered.

1797 **S. L. Van Doren, R. B. Pond, Sr., and R. E. Green, Jr. "Acoustic Characteristics of Twinning in Indium." AFOSR-TR-75-0607. Johns Hopkins University, Department of Mechanics and Materials Science, Baltimore, Maryland. Nov 1974. [NTIS: AD/A-009-163]; paper presented at the American Society for Metals Materials Testing and Quality Control Conference, Detroit, Michigan. Oct 21-24, 1974;** *Journal of Applied Physics,* **JAPIA, 47(10):4343-4348. Oct 1976.**

With the increased use of acoustic emission as a nondestructive testing tool, it has become extremely important to document and to understand the emissions characteristic of various modes of deformation and fracture. Results are reported of acoustic emission measurements made during mechanical deformation of indium single crystals. Although the metal indium is of little structural importance, it develops well-delineated twins with associated acoustic emissions, and therefore affords an excellent medium with which to develop techniques and to generate interpretations which can be applied to structurally important materials such as chromium, iron, magnesium, molybdenum, tin, titanium, tungsten, and zinc. By recording the acoustic emission signals on a video tape recorder, it was possible to play them back as often as desired, to observe them in slow and stop motion, and to play any desired portion of the signal through a spectrum analyzer for frequency analysis. The experimental results show that a correspondence exists between the duration of the acoustic signal and the volume of twinned material as determined by cinematographic observation of the generation and growth of twin bands. Repetition of some wave-form features was demonstrated not to be the result of multiple reflections of the emitted elastic waves from bounding surfaces, but to be related to actual physical events associated with the twinning process. It was also demonstrated that the direction of the stress, tensile or compressive, induced in the body of the crystal due to twinning can be determined from the polarity of the elastic wave emitted by the twin as detected by the transducer.

1798 **J. B. Van Erp, D. R. Macfarlane, and H. K. Fauske. "Protection against Local Core Accidents in Liquid Metal Fast Breeder Reactors."** ***Nuclear Engineering and Design,*** **NEDEA, 15:441–457. 1971.**

An analysis is presented concerning the performance characteristics of sensors considered for use in protection systems of liquid metal fast breeder reactors. Alternate solutions to protection system having one or more reactor-trip channels per fuel subassembly are proposed. Primary emphasis is placed on pressure pulse sensors, installed in the upper (and possibly also lower) coolant plenum, capable of sensing the collapse of sodium vapor bubbles. It is shown that, even for the limit-case of a complete coolant blockage of one or more fuel subassemblies, a reactor protection system based on pressure pulse sensors should be capable of effecting reactor-shutdown prior to generation of any substantial amount of molten fuel. Some preliminary experimental data regarding vapor bubble collapse are presented, showing reasonable agreement with analytical predictions. Recommendations are made regarding some potentially promising areas for further research and development with respect to protection systems for liquid metal fast breeder reactors.

1799 **A. Vary. "Nondestructive Evaluation Technique Guide." NASA SP-3079. National Aeronautics and Space Administration, Lewis Research Center, Cleveland, Ohio. 1973. [NTIS].**

A total of 70 individual nondestructive evaluation (NDE) techniques are described, each in a standardized single-page format for quick reference. Information is presented in a manner that permits ease of comparison of the merits and limitations of each technique with respect to various NDE problems. An NDE technique classification system is presented. It is based on the system that was adopted by the National Materials Advisory Board (NMAB). The classification system follows the NMAB system closely with the exception of additional categories that have been added to cover more advanced techniques presently in use. The rationale of the technique description format is explained. The format provides for a concise description of each technique, the physical principles involved, objectives of interrogation, example applications, limitations of each technique, a schematic illustration, and key reference material. Cross-index tabulations are also provided so that particular NDE problems can be referred to appropriate techniques.

1800 A. Vary and S. J. Klima. "A Potential Means of Using Acoustic Emission for Crack Detection under Cyclic-Load Conditions." NASA TM X-68228. National Aeronautics and Space Administration, Lewis Research Center, Cleveland, Ohio. 1973. [NTIS: N73-21848]; pp 258-266 in *Proceedings of the Ninth Symposium on Nondestructive Evaluation.* Southwest Research Institute, San Antonio, Texas. July 1973. symposium held in San Antonio, Texas. Apr 25-27, 1973.

A preliminary investigation was conducted to assess the feasibility of monitoring acoustic emission signals from fatigue cracks during cyclic bend tests. Plate specimens of 6Al-4V titanium, 2219-T87 aluminum, and 18-nickel maraging steel were tested with and without crack-starter notches. It was found that significant acoustic emission signals could be detected in the frequency range from 100 to 400 kHz. Cracks emanating from starter notches were monitored by the ultrasonic pulse-echo technique and periodically measured by micro-optical examination. Methods used to reduce the effects of extraneous noises (i.e., machine noises, fretting) are described. A frequency spectrum analyzer was used to characterize the emissions and to evaluate methods used to acquire the signals (i.e., transducer location, bandwidth selection). The investigation indicated that it was possible to extract meaningful acoustic emission signals in a cyclic-bend machine environment.

1801 C. L. Veach. "Development of Coincidence Acoustic Emission Monitoring Unit." ERR-FW-1175. General Dynamics, Convair Aerospace Division, Fort Worth, Texas. 1971. [GDPI: report contains General Dynamics private or proprietary information and is not available on the open literature].

1802 C. L. Veach. "Development of MK II (Rise Time) Acoustic Emission Monitoring Unit." ERR-FW-1021. General Dynamics, Convair Aerospace Division, Fort Worth, Texas. Aug 31, 1970. [GDPI: report contains General Dynamics private or proprietary information and is not available on the open literature].

1803 C. L. Veach and Y. Nakamura. "Development of Acoustic Emission Monitoring Techniques." pp 313-315 in "1970 Applied Research Program." Research Summary ARR-15. General Dynamics, Convair Aerospace Division, Fort Worth, Texas. [unpublished].

1804 M. Veaux. "Surveillance d'un Roulement a Rouleaux en Service (Surveillance of a Roller Bearing in Service)." pp 81-91 in *Journees d'Etudes sur l'Emission Acoustique.* P. F. Gobin, R. Goutte, and C. Guillaud, Organizers. Institut National des Sciences Appliquees de Lyon, Avenue Albert Einstein, 69621 Villeurbanne, Lyon, France. Mar 1975. proceedings of a conference held in Lyon, France. Mar 17, 1975. (paper in French).

1805 P. Veber. "Raziskave akusticne emisije pri rezalnem procesu aluminijevih zlitin." Diplomsko delo st. 2254. Fakulteta za strojnistvo, Univerza v Ljubljani, Ljubljana, Yugoslavia. 1975. (in Slovenian).

1806 J. B. Vetrano. "Procure Acoustic Emission System." Task 2, Project IV. pp 45–53 in "In-Service Inspection Program for Nuclear Reactor Vessels." R. D. Wylie. SwRI Project 17-2440, EEI Project RP-79. Biannual Progress Report No. 2. Southwest Research Institute, San Antonio, Texas. Dec 2, 1969. [NTIS: PB-188-347].

In keeping with the recommendations resulting from the completion of Task 1, Project IV, specifications were written for an engineered prototype of an acoustic emission system. The specifications were based on an analysis of the best operating features of the existing acoustic emission systems insofar as these features were revealed by their owners. Bids on the system were solicited from industry, but no potential suppliers elected to bid to supply the entire system or any part thereof.

1807 J. B. Vetrano. "Procure Acoustic Emission System." Task 2, Project IV, pp 64–71 in "In-Service Inspection Program for Nuclear Reactor Vessels." R. D. Wylie. SwRI Project 17-2440, EEI Project RP-79. Biannual Progress Report No. 3. Southwest Research Institute, San Antonio, Texas. July 30, 1970. [NTIS: PB-193-433].

At an EEI RP-79 Steering Committee meeting on September 18, 1969, an official action was taken to have Battelle-Northwest proceed with the construction of the prototype system. Final design was immediately initiated, and the entire system was completed, programed, and checked out by February 2, 1970.

1808 J. B. Vetrano and W. D. Jolly. "In-Service Acoustic Emission Monitoring of Reactor Pressure Vessels." *Materials Evaluation,* MAEVA, 30(1):9–12. Jan 1972; paper presented at the American Society for Nondestructive Testing Spring Conference, Los Angeles, California. Mar 8–11, 1971.

In general, periodic nondestructive evaluation of nuclear reactor pressure vessels is recognized as a necessity. Such inspections will be performed while the reactor is shut down for refueling. Secondly, it would be desirable to continuously monitor the nuclear reactor pressure vessel for flaw growth during operation. This report summarizes a research program directed toward the development of an acoustic emission system for continuous surveillance of reactors. The program results to date indicate that it would be feasible to detect and locate growing subcritical flaws in operating reactor pressure vessels.

1809 **J. B. Vetrano and W. D. Jolly. "Inservice Acoustic Emission Monitoring of Reactor Pressure Vessels."** *Nuclear Engineering and Design,* **NEDEA, 17(1): 177–178. 1971. [abstract of Paper No. 28 presented at the 5th Annual Information Meeting of the Heavy Section Steel Technology Program, held at Oak Ridge National Laboratory, Oak Ridge, Tennessee. Mar 25–26, 1971].**

Under the sponsorship of the EEI RP-79 Research Program, Battelle-Northwest is currently developing engineering models of the critical components needed for an acoustic emission system capable of inservice monitoring of nuclear reactor pressure vessels. The basic technology of this nondestructive evaluation method is based on detection of the ultrasonic wave front emanating from the tip of a growing flaw. This application imposes several unique, severe, and complicating boundary conditions on the acoustic emission system. In an operating nuclear reactor, there are constant hydraulic flow noises whose amplitude at the lower end of the frequency spectrum is many times greater than that of typical acoustic emission signals. Furthermore, the acoustic emission signals to be processed are expected to occur infrequently. In this paper, various approaches to the solution of this complex signal-to-noise problem are discussed. Preliminary results of in-plant noise monitoring are also presented. Another critical area being investigated is the type of acoustic emission sensor that will be coupled to the reactor wall. This sensor must be able to resist the hostile environment of the reactor and still function over extended periods of time. Two approaches to this problem are to use a higher temperature transducing technique or to locate the sensor out of the hostile environment and sonically couple it to the vessel with a waveguide. Both of these approaches are being evaluated, and the highlights to this study are presented. The third area discussed in this paper is the determining of the source location of the acoustic emission signals. Basically, this problem is one of combining the data of transducer spacing (relative to an imaginary coordinate or grid system), sonic velocity in the pressure vessel wall (accounting for propagation mode), and relative time of arrival of signals in the sensor to make a seismic-like conputation of the origin of the source relative to the imaginary grid system. Many variations of transducer array patterns were evaluated and the one having the optimum benefits for this particular application is described.

1810 **J. B. Vetrano, W. D. Jolly, and P. H. Hutton. "Continuous Monitoring of Nuclear Reactor Pressure Vessels by Acoustic Emission Techniques." Paper C58/72, pp 221–226 in** *Periodic Inspection of Pressure Vessels.* **Institution of Mechanical Engineers, London, England. 1972. proceedings of conference arranged by the Applied Mechanics Group of the Institution of Mechanical Engineers, held in London, England. May 9–11, 1972.**

The integrity of the primary circuit nuclear central station power plants that are now operating in the U.S. is of utmost concern. As part of an overall effort to resolve this concern, an association of investor-owned utilities in the U.S. is supporting research at Battelle-Northwest Laboratory in Richland, Washington. The purpose of this research is to develop a system that will continuously monitor nuclear reactor pressure vessels during operation to detect any incipient flaw growth. The system that has been developed is based on the acoustic emission phenomenon and is capable of detecting subcritical cracks in pressure vessel steels in the presence of temperatures, fluxes, and hydraulic flows that are characteristic of those in operating nuclear power plants.

1811 **J. B. Vetrano and P. A. Witt, Jr. "Acoustic Emission: A Powerful New Tool for the Petroleum Industry." Battelle-Northwest, Richland, Washington. 1971; paper presented at the American Institute of Chemical Engineers 68th National Meeting, Houston, Texas. Feb 28–Mar 4, 1971.**

1812 **A. Vetters, N. R. McDonald, and A. R. Wilson. "Application of NDT in Pressure Vessel Research." Paper No. 18 in *Nondestructive Testing for Quality Control.* Australian Institute for Nondestructive Testing, Melbourne, Australia. 1974. proceedings of the 4th National Conference, held in Melbourne, Australia. Aug 19–21, 1974.**

Details are given of NDT used in research on a full-scale pressure vessel withdrawn from service due to severe weld cracking. Visual and ultrasonic methods were mainly used to determine initial cracks in full-size nozzles. The use and effectiveness of acoustic emission monitoring to supplement ultrasonic testing and resistance strain gauging in determining crack growth under cyclic conditions, are discussed. It is believed that these methods provide a powerful combination of testing methods both for research and for practical application. [Abstract 22-0213. *Metals Abstracts,* Vol. 9, 1976].

1813 **G. Volta. "Incipient Failure Detection by Stress Wave Emission Techniques." Commission of the European Communities, Euratom Study Contract No. 043-71-10 ECIE (F). Status Report. Euratom CCR, Ispra, Italy. 1972.**

This report has two aims: (1) to present an inventory of the effort being made today in Europe for the development and application of acoustic emission technology, and (2) to indicate the limitations and merits of this technique from the point of view of its practical application to a nuclear power plant. As this technique is still in its infancy, the conclusions of the report are confined to discussing the applicability more than the applications. It is worthwhile emphasizing that this technique shows interesting perspectives not only for the surveillance of nuclear components.

1814 **R. von Klot. "Ortung von Ultraschallemissionen in Grossen Bauteilen (Localization of Ultrasonic Emissions in Large Structural Components)." Battelle-Institut eV, Frankfurt am Main, Germany. 1975. (in German); paper**

presented at the 4th Common Meeting and Colloquium of the Deutsche Arbeitsgemeinschaft fur Akustik, held in Braunschweig, Germany. Apr 7, 1975. [NTIS/INIS: AED-CONF-1975-054-001].

Sound emission analysis (SEA) is a new nondestructive method of testing for the detection and localization of faults (e.g., cracks, slag inclusions, or corrosion damage) in large constructional units. When under strain, these faults emit sound pulses (sound emission) of a frequency spectrum that goes far into the ultrasonic region. If a number of probes is attached to a constructional unit under analysis, the signals emitted by one source will reach the probes at different times. The location of the fault can then be calculated from the measurable differences in the delay of transmission.

W

1815 **J. R. Wadin and H. L. Dunegan. "The Use of Spatial Filtering and Distribution Analysis of Acoustic Emission Signals to Isolate and Characterize Subcritical Crack Growth." Paper 3-1, pp 183-193 in** ***The Third Acoustic Emission Symposium.*** **Japan Industrial Planning Association, Dobashi Building, 3-17-8, Akasaka, Minato-ku, Tokyo, Japan. 1976. proceedings of a symposium held in Tokyo, Japan. Sep 16-18, 1976.**

Results are shown, from a fatigue specimen containing several stress concentrations in the form of saw cuts, that recent developments in acoustic emission instrumentation provide the ability to locate the region of crack initiation and growth at one of the stress concentrations. It is further shown that acoustic emission signals from this particular region can be selected for amplitude distribution analysis while ignoring signals from other soruces. The slope of the log-sum amplitude distribution function *b* can then be output and recorded in real time for failure prediction purposes. Results are also shown that distribution analysis of acoustic emission pulse width data can be used to separate signals due to crack growth from those due to electromagnetic interference and fretting type background noise induced in the specimen grip region. Other potential applications for distribution analysis of rise time, energy, pulse width, and amplitude are discussed with special emphasis on the ability of this type of data presentation to separate and record signals due to a particular mechanism while ignoring signals due to other mechanisms.

1816 **E. V. Waite. "Acoust-S: A Digital Program for Acoustic Triangulation of Spherical Vessels." IN-1369. Idaho Nuclear Corporation, Idaho Falls, Idaho. Mar 1970. [NTIS].**

The ACOUST-S program was written as an extension of the original triangulation program ACOUST (Report IDO-17280, see Reference 1817) for the purpose of locating structural flaws on spherical pressure vessels. The program operates on time-of-arrival differences between unique acoustic transducers to locate points of structural flaw on spherical vessels. Flaw location is achieved by hyperbolic triangulation techniques. The program is designed to permit the location of sensors on inlet-outlet piping connected to the test vessel. Any portion of a spherical shape may be tested for acoustic emission points. The ACOUST-S is written in FORTRAN IV for use on the IBM-360 computer. The program requires 32,632 bytes of storage for execution. Standard input-output assignments are used in the program. An average triangulation problem will require about 0.04 minutes execution time on the IBM 360/75 computer when using the FORTRAN-IV H level 2 compiler.

1817 **E. V. Waite and K. V. Moore. "ACOUST: A Digital Program for Acoustic Triangulation of Nuclear Vessels." IDO-17280. Phillips Petroleum Company, Idaho Falls, Idaho. Nov 1968. [NTIS].**

A structural fault, when deformed, will generate acoustic emissions. Any given acoustic emission can be detected at different times by acoustic sensors located at different positions on the test structure. The program, ACOUST, was written to locate the structural fault through the implementation of hyperbolic triangulation on the time-of-arrival differences between a fault and three acoustic sensors. The code is designed to operate on a cylindrical vessel with (or without) capped end sections. The program is written to permit any type of end section geometry, and the program is designed to permit acoustic sensor locations on pipes leading to or from the test vessel. One transducer (per set of three) may be located on a vessel end section. The ACOUST is a FORTRAN IV digital computer program written for usage on the IBM-7044. The program occupies 10,343 octal memory locations. Standard input-output assignments are used in the program. The computer usage time required for each solution varies from five seconds to one minute, depending on the complexity of the particular problem. At the present time, only the 7044 version of the program is available. An IBM 360-75 version of the code will be available by early 1969.

1818 **E. V. Waite and D. L. Parry. "Field Evaluation of Heavy-Walled Pressure Vessels Using Acoustic Emission Analysis."** *Materials Evaluation*, **MAEVA, 29(6):117-124. June 1971; paper presented at the American Society for Nondestructive Testing National Fall Conference, Cleveland, Ohio. Oct 19-22, 1970.**

A portable acoustic analysis system has been developed by Idaho Nuclear Corporation which detects, locates, and analyzes microacoustic energy emissions originating from flaws or cracks within a stressed structure. The system was developed for the U.S. Atomic Energy Commission for potential use as a nondestructive test technique for the safety assessment of pressure retention envelopes of nuclear power reactors. To aid in the development and evaluation of the acoustic system, acoustic emission data were obtained and on-site analyses made of the integrity of several large industrial chemical reactor pressure vessels in conjunction with the hydrostatic acceptance testing of the vessels. The system was shown to be capable of detecting and locating acoustic emission sites (flaws) as small as 0.1 inches (2.54 mm) in size with acoustic signal transmission distance of up to 60 feet (18.28 m) from the emission site to the detector.

1819 **R. M. Wallace, C. G. Annis, and M. C. Van Wanderham. "Experimental Methodology for Accurate Predictions of Total Low Cycle Fatigue Life." Pratt & Whitney Aircraft, Florida Research and Development Center, West Palm Beach, Florida. 1976.**

An experimental methodology for accurate predictions of total low cycle fatigue (LCF) life in gas turbine rotor components is presented which combines the disciplines of acoustic emission, fracture mechanics, and strain controlled LCF testing to provice a unified approach to component life prediction. The feasibility of the proposed methodology has been demonstrated experimentally.

1820 **S. A. Wallace. "Acoustic Emission: An Overview." Y-DA-6376. Oak Ridge National Laboratory, Y-12 Palnt, Oak Rige, Tennessee. July 22, 1975. [NTIS].**

The mechanisms and applications of acoustic emission testing are reviewed. A bibliography of selected literature covering the period 1967 to 1975 is presented.

1821 **T. Watanabe. "Flaw Inspection System by AE Technique and Its Application to Large Pressure Vessels." Nippon Steel Corporation, Plant Engineering and Technology Center, Edamitsu, Yawata-higashi-ku, Kitakyushu-city, Japan. June 25, 1973; paper presented at the 11th Acoustic Emission Working Group Meeting, Richland, Washington. July 18–20, 1973.**

A flaw inspection system has been developed by making use of the AE technique. The report deals with its technological composition and an example of its application to large pressure vessels. The system is a 28-channel AE location system. Main amplifiers, time-difference measuring devices, mini-computer, CRT display unit, and other measuring instruments are mounted in a large trailer. Each PZT sensor is directly connected to its small preamplifier, and for the coaxial cables between the preamplifiers and the main amplifiers, wind-up drums are provided below the trailer body. The sensors are placed in multirhombic arrangement on the surface of the test object for locating AE sources. After selecting a rhombus in which an AE source exists, the coordinates of the emission source on the Cartesian coordinate system in the selected rhombus are calculated by the mini-computer and the result is immediately displayed as a dot in the multirhombus sensor arrangement diagram on the screen of the CRT display unit. As an example of the several tests conducted, the result of an AE source location made with a hydraulic test up to 60 kg/cm^2 on the pressure vessel, which is about a 3-metre diameter X 22 metres long X 54 millimetres thick, is discussed. The positions at which AE source location was effected during the test with regard to the AE waves exceeding a given signal strength level were 21 points. After completion of the hydraulic test, they were examined by ultrasonic testing, 15 of 21 points, and it was found that many definite small flaws existed at 14 of the 15 points. The accuracy of this location test was about 100 mm.

1822 **T. Watanabe, S. Hashirizaki, and H. Arita. "Acoustic Emission Inspection during Water-Pressure Testing of Pressure Vessels."** ***NDT International,*** **NDITD, 9(5):227–232. Oct 1976.**

The article describes the advantages of acoustic-emission inspection over other methods of NDT. It reports results of two water-pressure fracture tests using actual pressure vessels with various defects. In these tests, acoustic emissions from the vessels were measured and located by a 32-channel acoustic-emission inspection system, the NAIS 32. From the results of these tests, it may be said that a good correlation exists between acoustic-emission behaviour and changes in various defects under stress. At present, however, it seems impossible to predict plastic fracture although current results hold some promise for the future.

1823 T. Watanabe, S. Hashirizaki, and H. Arita. "Effectiveness of AE Inspection Method at Water Pressure Test of Pressure Vessels." Paper 2-2, pp 85-100 in *The Third Acoustic Emission Symposium.* Japan Industrial Planning Association, Dobashi Building, 3-17-8, Akasaka, Minato-ku, Tokyo, Japan. 1976. proceedings of a symposium held in Tokyo, Japan. Sep 16-18, 1976.

In recent years, AE measuring technology has found wide application as an effective means of studying plastic deformation or fracture behavior of various materials. Efforts are now being made toward the application of this technology for industrial use; for example, monitoring of welding operations, flaw detection in large structures at the time of proof test, and application to safety surveillance systems of equipment in operation. In some areas, practical application of this technology has been already started. The most significant example is a new inspection method to be applied to pressure test of large pressure vessels. This method is designed to judge the presence, locations, and severity of such defects, which may propagate under stressing, in pressure vessels by measuring acoustic emission signals when pressure tests are performed, thereby to inspect and confirm the safety of such vessels. The AE method has come to find application in various fields to meet users' requirements for reliable information which can make up for the shortcomings of conventional inspection methods. As an understanding of the advantages of the AE method is increased, movement of the standardization of this method in the American Society of Mechanical Engineers (USA) or High Pressure Institute (Japan) will progress. However, the data obtained from the AE inspection of small specimens conducted in laboratories do not agree with those of AE inspection of actual large vessels in many respects. It is not adequate to evaluate the practicality of the AE method from these data only. As pressure tests of large pressure vessels are usually conducted as a final quality assurance test, major defects detected by conventional methods before the pressure test is undertaken are removed during frabrication processes. Accordingly, there are only a few cases where major defects leading to failure of the vessel are detected at the time of pressure test. For this reason, sufficient quantities of data to demonstrate the effectiveness of the AE method have not been collected through its application to actual structures. Such being the case, there are many people who have some doubt about the interpretation of the data obtained by the AE method. To establish the effectiveness of the AE method and to determine the applicable range by solving these problems, it will be necessary to accumulate more information on the relation between the data obtained by the AE method and fracture behaviors by conducting many fracture tests of large pressure vessels. As such tests involve many difficulties in the aspects of cost and safety, it is difficult to conduct such tests on a large scale. This may be the reason why many reports of such tests have not been published to date. Since 1969, Nippon Steel has continued fundamental research on acoustic emission and application of AE technology to the study of characteristics of materials or to welding operations at its Fundamental Research Laboratories and Products R and D Laboratories, obtaining various results in the respective fields. Studies on the practical application of AE technology, development of inspection method of large structures employing AE technology, and development of associated AE inspection techniques have been undertaken at the Plant Engineering and Technology Center since 1972, as a result of which the 28- and 32-channel AE inspection systems incorporating the minicomputer have been developed. The AE data have been accumulated by inspecting 24 large welded structures (as of November 1975) using these systems. The introduction of the AE inspection system (Nippon Steel AE inspection system) being employed by this company and examples of inspection of large pressure vessels by this system were reported previously. In addition to the inspection of actual structures, several fracture tests of large pressure vessels of the size comparable to actual structures were conducted with a view to obtaining the clear correlation between the AE information and the degree of severity of defects. In this report, two examples of such tests are introduced to demonstrate the effectiveness of the AE inspection method.

1824 **T. Watanabe, S. Hashirizaki, and H. Arita. "Flaw Inspection System of Large Pressure Vessels by Acoustic Emission Technique."** *Journal of N.D.I. (Hi-Hakai Kensa,* **Japan), HIHKA, 22(12):757–766. Dec 1973. (in Japanese).**

A new flaw inspection system was developed making use of the acoustic emission technique which has of late been attracting particular attention as a new nondestructive inspection technique. The principle, technological composition, hardware, and software of AE testing are discussed and an example given of applications to large pressure vessels. In more than ten cases in which pipes and pressure vessels were tested by this system, the presence of a flaw was confirmed by ultrasonic flaw detection with a success rate of more than 90%. The present system, when combined with conventional nondestructive inspection methods, can establish a highly sensitive and efficient new inspection system. Further, experimental work should be continued to improve the reliability of the AE technique and to make it applicable to equipment other than pipes and pressure vessels.

1825 **T. Watanabe, S. Hashirizaki, and H. Arita. "Flaw Inspection Test of Large Spherical Tanks by Acoustic Emission Technique." Session 3, pp 53–66 in** *The Second Acoustic Emission Symposium.* **Japan Industrial Planning Association, Dobashi Building 5F, 3-17-8 Akasaka, Minato-ku, Tokyo, 107 Japan. 1974. proceedings of a symposium held in Tokyo, Japan. Sep 2–4, 1974.**

Since March 1973, flaw inspection has been conducted of the entire surface of large pressure vessels employing the multichannel AE inspection system developed by the company. In previous reports, the inspection system developed by the company was introduced and the test results of a cylindrical pressure vessel, which is about 3 m in diameter, 22 m long, 54 mm thick, were presented as an example of application of this system to flaw inspection. Since then, improvements to the system were made to extend its applicability be conducting the inspection by acoustic emission method on large pressure vessels during hydraulic tests after manufacture. Recently, the application has been successful in the acoustic emission method to flaw inspection of the entire surface of 18 to 20 m diameter large spherical tanks. The inspection is outlined.

1826 **T. Watanabe, S. Hashirizaki, and H. Arita. "Inspection of Large Pressure Vessels by the Acoustic Emission Technique."** *Welding and Metal Fabrication,* **WLMFA, 43(7):525–527. Sep 1975.**

With the recent rapid progress of studies on acoustic emission phenomena, the AE inspection technique has been considerably improved. It is now finding wider practical application as a means of inspection, diagnosis, safety surveillance, tests, and research of various structures and materials in conjunction with various conventional methods. The AE inspection technique is nearing the stage of industrial application to the inspection of pressure vessels. The object of this technique is to provide a method of evaluating the safety of pressure vessels in which acoustic emission (stress wave emissions) generated from unstable flaws, such as cracks, are detected during pressure tests; the generating points of such emissions are located, and the severity of flaws is estimated based on the AE generation pattern with respect to location and time of generation. In the US and various European countries, AE test systems for large pressure vessels have been introduced and examples of the application of such systems have been reported. In Japan, the multichannel AE inspection system was developed by the Plant Engineering and Technology Centre of Nippon Steel. Using this system, the inspection of the entire surface of various large pressure vessels has been conducted since March 1972. In 1974, the Corporation succeeded in applying this system to the inspection of the entire surface of a cylindrical pressure vessel of 4-m diameter and 31-m long, and to a spherical tank of 20-m diameter. This article reports the current stage of development of the AE inspection technique by reference to the two multichannel AE inspection systems NAIS-28 and NAIS-32, now being used in the works, and by outlining the results of tests on a spherical tank conducted in September 1973 as an example of AE tests using such systems.

1827 **T. Watanabe, S. Hashirizaki, and H. Arita. "Non-Destructive Inspection of Spherical Tanks by Acoustic Emission Technique."** *Pressure Engineering, Journal of Japan High Pressure Institute (Kotsu Gasu),* **KOGAA, 13(2):57–64. 1975. (in Japanese).**

1828 **J. L. Weaver and G. C. Knollman. "Evaluation of an Acoustic Emission Monitor for On-Line Quality Control in Spotwelding of Electronic Components." LMSC-D 358302. Lockheed Palo Alto Research Laboratory, Lockheed Missiles & Space Company, Palo Alto, California. Feb 1974.**

Feasibility of characterizing the strength of small resistance (fusion) welds by detection of acoustic emission is demonstrated. A commercially available acoustic emission monitor from Trodyne Corporation was used to analyze the acoustic emission with respect to amplitude and time of occurrence. Test results obtained with this system show that, in spotwelding of nickel wires, weak welds are indicated by the large number of high-amplitude bursts of acoustic emission which are generated by expulsion of molten metal from the weld. Weak welds also occur if an insufficient number of acoustic emission pulses are produced as the weld is formed. The acoustic emission monitor is found to be applicable to on-line control of weld quality in spotwelding of electronic components, and may be utilized on a time-share basis.

1829 **D. Webber and S. J. Maddox. "Problems in the Detection and Monitoring of Fatigue Cracks in Al–Zn–Mg Alloy Fillet Welds." Chapter 4, pp 16–21 in *The Detection and Measurement of Cracks.* The Welding Institute, Abington Hall, Abington, Cambridge, England. 1976.**

Consideration by means of fracture mechanics of the fatigue behavior of Al-Zn-Mg alloy transverse fillet welds, particularly failure from the weld toe, shows that cracks are small (<1 mm) for a large part of their lives. In order to apply fracture mechanics to the prediction of the service life of structures containing cracks, it is necessary to be able to detect small cracks. A study of available crack-detection methods shows that acoustic emission offers the greatest potential because of its sensitivity to important defects. [Abstract 22-1394. *Metals Abstracts,* Vol. 9, 1976].

1830 **A. E. Wehrmeister. "Acoustic Emission Monitoring of Multi–Pass Submerged–Arc Welding." *Materials Evaluation,* MAEVA, 35(6):45–47. June 1977; paper presented at the 15th Acoustic Emission Working Group Meeting, Ft. Lauderdale, Florida. Jan 20–23, 1976.**

The acoustic emission method has been used to detect the formation of flaws during submerged arc welding. This paper describes a technique that detects cracks and entrapped slag as the weld metal solidifies. Emissions from slag are also detected on subsequent weld passes over the defective region. Multipass welds in 6-inch thick carbon steel plates have been monitored using acoustic emission sensors that travel with the welding electrode. The detection system provides flaw location capability for immediate weld repair, reducing post-weld repair expenses. Weld monitoring parameters, experimental techniques, and data analysis are discussed. Typical experimental data are presented.

1831 **A. E. Wehrmeister, G. C. Efthymiou, and A. E. Holt. "Flaw Evaluation in Thick Section Sub–Arc Welds." Babcock & Wilcox Company, Lynchburg Research Center, Lynchburg, Virginia. 1974; paper presented at the Instrument Society of America 20th International Instrumentation Symposium, Albuquerque, New Mexico. May 21–23, 1974.**

Extensive ultrasonic and radiographic examinations of 6 to 8 inches thick submerged arc welds have been performed throughout the years to detect the presence of flaws such as lack of fusion, slag entrapments, cracks, and porosity. Recent technological advances in the fields of acoustic emission, ultrasonic spectrum analysis, and acoustic holography have presented new dimensions to the examination and evaluation of thick section weldments. Three phases in weld examination are normally pursued; the detection of the flaw as it is being produced during on-line welding, the detection of the flaw after it is in a finished weldment, and the evaluation of the flaw to determine its type, size, and location within the weldment. The acoustic emission method is shown to be useful for on-line detection of cracking and slag entrapment. Standard ultrasonic and radiographic detection of flaws in a completed weldment are not discussed, since they have been presented in other references. The evaluation of flaw type, size, and location is presented in terms of ultrasonic frequency spectrum analysis and acoustic holographic examination. These methods are directed at the improved analysis of flawed areas for the determination of repair requirements. Development efforts are presently under way to use these new techniques for flaw evaluation. This paper presents the initial results of those developments and illustrates the instrumentation used and the techniques developed to data to detect and evaluate the discontinuity in question.

1832 **B. Weichbrodt. "Mechanical Signature Analysis: A New Tool for Product Assurance and Early Fault Detection." in** ***Annals of Reliability and Maintainability.*** **Volume 5, Achieving System Effectiveness. American Institute of Aeronautics and Astronautics, New York. 1966; paper presented at the American Institute of Aeronautics and Astronautics, Society of Automotive Engineers, and American Society of Mechanical Engineers 5th Reliability and Maintainability Conference, New York. July 18–20, 1966; pp 314–339 in** ***Incipient Failure Diagnosis for Assuring Safety and Availability of Nuclear Power Plants.*** **E. P. Epler and D. P. Roux, Coordinators. U.S. Atomic Energy Commission, Division of Technical Information, Washington, D.C. Jan 1968. [NTIS: CONF-671011]. proceedings of conference sponsored by Oak Ridge National Laboratory, held in Gatlinburg, Tennessee. Oct 30–Nov 1, 1967.**

Sound and vibration signals have been useful for many years in judging the internal condition of machinery and structures. An automobile mechanic listens to an engine for internal defects. A machinist checks a grinding wheel for cracks by rapping it and listening for a clear tone or a dull thud. At General Electric's Research and Development Center, the interpretation of sound and vibration signals is being developed from an art into a scientific technique. The technique, Mechanical Signature Analysis, uses external measurement of sound and vibration signals to diagnose internal condition or malfunctions and to detect incipient failure. Mechanical Signature Analysis has great potential as a unique tool for increasing the reliability and maintainability of a wide range of components, assemblies, and systems.

1833 **W. Weidemann. "Research on Continuous Control of Crack Propagation in Pressure Vessels. Part I: Basic Research on Sound Waves Emitted during Crack Propagation." pp 48–49 in "Progress Report on Research Projects Sponsored by the Federal Ministry of Scientific Research and Applicable to the Reactor Safety Aspect." Compiled by A. Tietze. Research Report Monograph Series**

IRS–F–2. Institut fur Reaktorsicherheit der Technishen Uberwachungs–Vereine eV, Cologne, West Germany. July–Dec 1968. (in German); ORNL–tr–2244. translated for Oak Ridge National Laboratroy, Oak Ridge, Tennessee. June 1969. [NTIS].

A warning system for an early indication of the failure of structural components by repture is considered not only ideal but also feasible. The following is necessary for the development of a warning system: (1) Creation of principles for the quantitative analysis of the recorded acoustic waves; this is based on a knowledge of the elementary processes taking place in reactor construction steels during crack propagation as a function of the operating stresses; (2) The useful signals generated by these elementary processes must be clearly separated from the general background noise of an industrial nuclear plant; and (3) Coupling of the acoustic analysis with mechanical rupture studies. In this manner, one can gain quantitative information on crack propagation in real structural components. To achieve the above long-term objective (development of a warning system), close collaboration with organizations dealing with the same problems, particularly in the U.S., is planned; at the present, the available literature is being analyzed; and urgent questions concerning the mechanism of rupture are being discussed.

1834 W. Weidemann. "Untersuchungen auf dem Gebiet der Schallemissionsanalyse (Investigations in the Field of Acoustic Emission Analysis)." Battelle–Institut eV, Frankfurt am Main, Germany. 1970. (in German); paper presented at the Fachkonferenz des Instituts fur Reaktorsicherheit (Conference of the Institute for Reactor Safety). Koln, Germany. Mar 6, 1970.

1835 D. B. Weir. "Acoustic Emission and Thermal Scanning for Structural Sandwich." RHR–75–016. Rohr Industries, Inc., Chula Vista, California. 1975; paper presented at the American Society for Nondestructive Testing Spring Conference, Los Angeles, California. Mar 10–12, 1975.

A description is given of how acoustic emission technology can be used in conjunction with a moving thermal load to make a nondestructive evaluation of sandwich strength. Current NDT techniques and pertinent sandwich characteristics are discussed. Correspondence of acoustic emission data with ultrasonic inspection and flatwise tension strength test data are given and a follow-on program described.

1836 D. B. Weir. "Acoustic–Emission NDT Technique for Structural Sandwich." Rhr–73–138. Rohr Industries, Inc., Chula Vista, California. 1973; paper presented at the Society for Experimental Stress Analysis Third International Congress on Experimental Mechanics. Los Angeles, California. May 13–18, 1973.

The paper describes an NDT method for predicting the strength of honeycomb sandwich structure. The method makes use of a flatwise tension load applied gradually to the structure during which time the structure is monitored for acoustic emission. Total acoustic emission is plotted against a time axis and the resulting plot, defined as the acoustic signature, is used to predict sandwich strength. Prediction is made by comparing signatures of test articles with predetermined standards. Correlation of sandwich strength and acoustic emission data is shown.

1837 **D. B. Weir. "Loading Methods for Acoustic Emission Testing of Structural Sandwich." RHR-74-508. Rohr Industries, Inc., Chula Vista, California. 1974; paper presented at the 13th Acoustic Emission Working Group Meeting, Laguna Hills, California. Aug 19–21, 1974.**

Loading methods which can be used in conjunction with acoustic emissions monitoring for testing structural honeycomb sandwich are discussed. Correspondence between acoustic emissions and sandwich strength is described. Prospects for correlation between acoustic emission, stress or deformation, and mechanical and thermal loading methods are discussed.

1838 **V. Weiss, G. Krause, C. Chave, Jr., and G. Marchetto. "Crack Initiation in Metallic Materials." SURI Report No. MET-E-1289-1566P2-3. Quarterly Report June 1 through Nov 30, 1965. Syracuse University Research Institute, Department of Chemical Engineering and Metallurgy, Syracuse, New York. Jan 1966. [NTIS: N66-32184].**

Progress is reported on the development of an acoustic detection method for determining crack initiation and propagation in metallic materials. Since the previous pickup arrangement seemed insufficient, a new compensation method for acoustic detection was used which has a good signal-to-noise ratio and directly gives an acoustic signal versus load diagram on the screen of a storage oscilloscope. This recording system is described in detail and examples of typical first fracture-sound records are discussed. The measuring setup was tested in connection with a vacuum environment chamber which is used to standardize different specimens before changing the environment.

1839 **V. Weiss, G. Krause, and G. Marchetto. "Crack Initiation in Metallic Materials." SURI Report No. MET-E-1289-3766P4. Quarterly Progress Report No. 4. Syracuse University Research Institute, Department of Chemical Engineering and Metallurgy, Syracuse, New York. Mar 1966.**

The effect of different environments on the re-initiation of crack growth under tensile load was investigated for 7075-T6 aluminum. Characteristic changes in the acoustic emission versus load spectrum caused by the environments were found for vacuum, air, argon, water, and electrolytic NaCl solutions.

1840 V. Weiss, G. Nash, G. Krause, G. Marchetto, and J. Sessler. "Crack Initiation and Crack Propagation in Structural Metal Alloys." SURI Report No. MET-E-1289-0666F. Final Report. Syracuse University Research Institute, Department of Chemical Engineering and Metallurgy, Syracuse, New York. May 1967.

Studies were conducted to determine the effect of stress concentration factor, K_t, and test environment on crack initiation, crack reinitiation, and crack growth in structural metal alloys. The test program included stress-controlled fatigue tests on D6aC steel, notch tensile tests on D6aC and 9 Ni-4 Co steels and monotonic tensile tests on 7075-T6 aluminum alloy. Acoustical emission techniques were employed to detect the stress required for crack reinitiation in precracked 7075-T6 specimens. The fatigue test results indicated that the test environment has a pronounced effect on fatigue crack initiation time but little effect on crack growth rate. Little effect of environment on notch strength was observed in the tests on D6aC and on 9 Ni-4 Co. However, the greatest effect was observed at high K_t factors. The studies on 7075-T6, using acoustical emission techniques, indicate that these techniques can be developed into a powerful tool to study crack characteristics and environmental effects. However, additional effort is required to obtain quantitative information by means of acoustic emission methods.

1841 V. Weiss, G. Nash, and K. Schroder. "Crack Initiation in Metallic Materials." SURI Report No. MET-E-1289-63065P1. Quarterly Report No. 1. Syracuse University Research Institute, Department of Chemical Engineering and Metallurgy, Syracuse, New York. June 1965. [NTIS: N66-20851].

The values of the crack initiation parameter, N_o, obtained in these studies were found to agree well with the relation between the applied stress, stress concentration factor, and N_o value postulated in the previous final report. Also the value of the parameter n +, which is related to the crack propagation characteristics of the material, were in excellent agreement with the previous values obtained. Studies of crack initiation in 7075 aluminum under monotonic tensile loading have commenced using an acoustical method and electrical resistance measurements for the detection of the pop-in load. The acoustical method thus far seems to be the most promising; the major difficulty experienced with the resistance measurements has been a too high electrical noise level. Photographs are being taken while the specimen is under load in an attempt to correlate the observed acoustic signals with crack initiation and slow growth.

1842 D. Wells. "An Acoustic Apparatus to Record Emissions from Concrete under Strain." ***Nuclear Engineering and Design*, NEDEA, 12(1):80-88. May 1970.**

The paper presents the design of an apparatus to detect the sound emitted by small concrete specimens under strain. The acoustic equipment assembled to detect and record these minute sounds is described together with the steps taken to minimize the effects of background noises. An account is given of the machinery developed to apply load at a noise level low enough not to be detected by the acoustic equipment. Some typical results are shown. [reprinted by permission of North-Holland Publishing Company from *Nuclear Engineering and Design*].

1843 V. N. Whittaker. "A Review of Non-Destructive Measurement of Flaw Size." *Non-Destructive Testing,* **NDETA, 5(2):92–100. Apr 1972. [This is the fourth in a series of six articles on fracture mechanics.]**

Critical defect sizes for various engineering situations are established before asking whether modern NDT techniques can find these flaws. Radiography is shown to be a poor technique for detecting cracks, but readily detects the much less dangerous defects: porosity and rounded inclusions. It cannot estimate the depth of cracks. Ultrasonics is much dependent on operator skill, but reliably detects surface cracks 5 mm in length and can make reasonable estimates of depths greater than 1.75 mm. Ultrasonics also detects internal cracks if they are more than 2.5 mm in diameter but accurate estimates of size are still difficult. Acoustic emission is claimed to detect smaller flaws than ultrasonics. In view of these limitations, it would seem prudent not to use material of such low toughness that the critical defect size is less than 3 mm.

1844 N. Wilde. "Acoustic Emission: Model." pp 26–32 in "Instrument Development Branch Annual Report, 1967." R. G. Morrison, Editor. IDO–17269. Phillips Petroleum Company, Idaho Falls, Idaho. Aug 1968. [NTIS].

Transducer development, including eddy current techniques, ultrasonic techniques, acoustic emission, neutron and reactor fuel temperature measuring techniques, flow detection, and a capacitance sensing height gage for nonconductive liquids, is reported. Systems engineering development, including computer interfaces, control systems, and specialized components, is described. Instrumentation for the ex-reactor ATR in-pile tube vibration simulation apparatus, seismic switch calibration system, vibration instrumentation during the ATR startup, and Zircaloy-2, zirconium oxide emissivity are discussed.

1845 D. Wildermuth. "Pressure Testing of AFRM–017 Service Propulsion System Fuel Tank Utilizing Aerojet–General Corporation's Stress-Wave Analysis Technique (SWAT)." Report NAS 9–6766–003. Aerojet–General Corporation, Aerovessels Division, Downey, California. May 16, 1967.

The report was prepared by Aerojet-General Corporation under Contract NAS 9-6766, administered by the National Aeronautics and Space Administration's Manned Spacecraft Center. It describes the hydrostatic test of an Apollo service propulsion system fuel tank (S/N 023), conducted at Aerojet's Downey Plant on 14 February 1967. The tank was removed from airframe-017 after its mating fuel tank had ruptured at a pressure of 240 psig during a system pressure test. Both tanks had previously been proof pressure tested with inhibited water to 325 psig. Methyl alcohol had been used as the pressurization medium for the system tests, and stress corrosion, resulting from exposure to alcohol, was suspected to be the cause of failure. After removal from airframe-017, the fuel tank was subjected to a fluorescent penetrant inspection, but indication of surface defects was found. The purpose of this test was to determine if defects existed in the tank that were too small to be detected by the penetrant inspection, and, if there were, to demonstrate the ability of Aerojet's Stress-Wave Analysis Technique (SWAT) to detect crack propagation as the tank was pressurized. An additional objective was to arrest the test prior to a suspected failure. The first test was to be a proof pressure test to 320 psig, but failure occurred prematurely in the forward hemisphere at 290 psig. Twenty-six acceleration transients were recorded by the SWAT instrumentation. Post-test data analysis disclosed that only three stress waves were emitted by propagating defects. One acceleration transient was attributed to mechanical noise and the remaining 22 resulted from cracking in the adhesive used to bond accelerometers to the tank. The first of the three stress waves occurred at 153 psig and originated in the aft hemisphere. The second and third waves occurred at 287.3 and 287.6 psig, and both originated in the forward hemisphere. The origin of the acceleration transient recorded at the instant of failure coincides with the origin of the last two stress waves. A post-test inspection of selected areas in the tank was performed using high-resolution penetrant inspection materials marketed under the trade name, Spotcheck Hy-Rez. Indications of surface defects were found adjacent to the fractured surface and throughout the aft hemisphere. No defects were indicated in the other areas checked. Although the number of stress waves detected from the SWAT data was less than anticipated, the feasibility of the concept was verified.

1846 **L. E. Willertz and D. O. Hunter. "An Acoustical Emission–Damping Fatigue Study of a 17–4 PH Steel."** ***Journal of Testing and Evaluation,*** **JTEVA, 2(3):190–195. May 1974.**

The acoustic emissions emitted from a 17-4 PH stainless steel were monitored, while the specimen was undergoing fatigue testing in the torsional mode. Below the cyclical stress sensitivity limit (CSSL) of the material, no abnormal emissions were detected, while at and above CSSL emissions were detected which could be attributed to fatigue damage and resulting in failure of the specimen. This test was the first in a series which will give more knowledge about fatigue damage in various materials. [reprinted by permission of the American Society for Testing and Materials, Copyright].

1847 **J. A. Williams. "A Simple Model for Analysis of Acoustic Emission Data." Laboratory Note No. RD/L/N 203/71. Central Electricity Research Laboratories, Leatherhead, Surrey, England. Oct 11, 1971.**

A brief review is made of analyses for correlating acoustic emission data with fracture behavior. The model of Gerberich and Hartbower (Reference 548) is extended to include known-crack growth laws.

1848 **J. H. Williams, Jr. and G. P. Emerson. "Constitutive Acoustic-Emission Elastic-Stress Behavior of Magnesium Alloy." NASA CR-2839. Massachusetts Institute of Technology, Cambridge, Massachusetts. June 1977.**

Repeated loading and unloading of a magnesium alloy below the macroscopic yield stress result in *continuous* acoustic emissions which are generally repeatable for a given specimen and which are reproducible between different specimens having the same load history. An acoustic emission Bauschinger strain model is proposed to describe the unloading emission behavior. For the limited range of stress examined, loading and unloading stress delays of the order of 50 MN per m^2 are observed, and they appear to be dependent upon the direction of loading, the stress rate, and the stress history. The stress delay is hypothesized to be the manifestation of an effective friction stress. The existence of acoustic emission elastic stress constitutive relations is concluded, and provides support for a previously proposed concept for the monitoring of elastic stresses by acoustic emission.

1849 **P. G. Williams. "Acoustic Emission Testing: 1950–February 1972, A Comprehensive Bibliography." CE-BIB-222. Central Electricity Generating Board, Information Services, London, England. Apr 1972. [NTIS: N72-26549].**

A bibliography, containing 119 references on acoustic emission, is presented. These data provide comprehensive coverage of acoustic emission as a nondestructive testing technique for flaws in metals and nonmetals, locating defects in pressure vessels in service, inspecting welds, proof testing of structures, stress corrosion detection, fracture mechanics analysis, and pahse transformation investigations.

1850 **P. G. Williams. "Acoustic Emission Testing: A Selective Bibliography, 1972–January 1975." CE-BIB-222.1. Central Electricity Generating Board, Information Services, London, England. Feb 1975.**

The work is an up-date of the author's previous bibliography CE-BIB-222 (see Reference 1849) and contains 81 abstracted references, 62 of which are unrestricted and are available to the public. A list of 19 restricted references are available only to authorized staff within the electricity supply industry. Included in the bibliography is an author index.

1851 **R. S. Williams and K. L. Reifsnider. "Characterization of Acoustic Emission during Fatigue Loading of Composite Specimens." AFOSR-TR-74-1200. Virginia Polytechnic Institute and State University, College of Engineering, Blacksburg, Virginia. May 1974. [NTIS: AD-783-227]. [see Reference 1852 for journal publication].**

Strain and load-controlled fatigue tests were run on boron-aluminum and boron-epoxy angle-plied specimens. Acoustic emission data were recorded using a gating technique that eliminated most extraneous noise. Various material parameters were monitored during the test. A good correlation between acoustic emission and damage extent and propagation was obtained. This was evidenced by an apparent first-order linear relationship between dynamic compliance and totalized acoustic emission. A basis for an energy-based failure model was formualted and discussed.

1852 **R. S. Williams and K. L. Reifsnider. "Investigation of Acoustic Emission during Fatigue Loading of Composite Specimens."** *Journal of Composite Materials,* **JCOMB, 8(4):340–355. Oct 1974. [see Reference 1851 for report to the Air Force Office of Scientific Research].**

Strain and load-controlled fatigue tests were run on boron-aluminum and boron-epoxy angle-plied specimens. Acoustic emission data were recorded using a gating technique that eliminated most extraneous noise. Various material parameters were monitored during the test. A good correlation between acoustic emission and damage extent and propagation was obtained. This was evidenced by an apparent first-order linear relationship between dynamic compliance and totalized acoustic emission. A basis for an energy-based failure model is formulated and discussed. [reprinted by permission of Technomic Publishing Company, Inc., 265 Post Road West, Westport, CT 06880].

1853 **T. R. Wilshaw and R. Rothwell. "Instrumented Scratch Test for Measuring the Fracture Behaviour of Strong Solids."** *Nature Physical Science*, **NPSCA, 229(5):155–157. Feb 1, 1971.**

A technique is described in which the intensity of scratching is measured using the acoustic emissions generated by microcracks and simultaneously measuring the normal force on the indentor. A diagram of the instrument is given and its mechanism and mode of use are described. Possible uses for this scratch test are indicated. [Abstract 22-0804. *Metals Abstracts,* Vol. 4, 1971].

1854 **T. R. Wilshaw and W. E. Swindlehurst. "The Application of Acoustic Emission to the Measurement of Surface Mechanical Properties of Brittle Solids." University of Sussex, Brighton, Sussex, England; paper presented at the Institute of Physics Conference on Acoustic Emission, Imperial College, London, England. Mar 14, 1972. [proceedings were never published].**

The mechanical strength of strong brittle solids is governed by cracks and other stress concentrations to be found in the near surface material. Conventional methods for measuring the mechanical properties of these materials usually involve the fracture of many pieces with subsequent statistical analysis. An alternative method which has not been widely used, involves testing a small area of surface with a localized stress field induced by, for example, the contact loading of a sphere on a plane, the Hertzian test. This test has many practical advantages since it is easy to perform and requires only a small quantity of material. More recently the test has been developed a stage further by sliding the sphere on the surface to facilitate the *search* for the surface defects. This test has only been made possible by the detection and recording capabilities offered by the acoustic emission technique. The source of the emissions is an individual microcrack which is propagated under controlled conditions and which may be observed directly afterwards under the optical microscope. This provides a means of evaluating the performance of the equipment and exploring the potential of the technique while at the same time using it as a tool to help with a particular problem. Described are the apparatus and some results which show the correlation between sources (microcracks) and the emission records. The performance of the apparatus will be discussed together with the analysis of the data as related to the specific problem of measuring surface mechanical properties of glass.

1855 **G. H. Wilson and C. T. Herakovich. "Application of a Microprocessor for Acquisition of Load, Strain and Acoustic–Emission Data."** *Experimental Mechanics,* **EXMCA, 16(3):111–115. Mar 1976; paper presented at the Society for Experimental Stress Analysis Spring Meeting, Chicago, Illinois. May 11–16, 1975.**

The paper describes an automatic system for acquisition, logging, and analysis of stress, strain, and acoustic-emission data. The complete system is a combination of several standard items of equipment and a newly developed intelligent data logger which has the capability to send data directly to a large-scale computer for reduction and plotting. The data logging and all interfacing with peripheral equipment are accomplished using digital electronics. The heart of the system is a four-bit microprocessor and digital cassette-tape unit. The system, which is controlled by instructions from a cathode-ray-tube terminal, has the capability to transmit data from the cassette unit to a large computer via a time-sharing option. Typical results are presented for acoustic-emission count rate and total-count curves superimposed on stress-strain diagrams for boron-epoxy advanced composites.

1856 **G. H. Wilson, III and R. S. Williams. "Real–Time Data Acquisition and Reduction of Critical Dynamic Fatigue Parameters."** *Proceedings of the IEEE,* **IEEPA, 63(10):1445–1450. Oct 1975.**

Recent interest in characterizing fatigue mechanisms and damage propagation in the field of materials science has led to more closely controlled and sophisticated tests which have resulted in greater amounts of data to be recorded and processed. The application of real-time data acquisition and reduction to deal with these data in conjunction with an ongoing research project dealing with the effect of frequency on the fatigue of composite materials are discussed. Utilizing a Tektronix WP 1100, which consists of a digital processing oscilloscope and a programable calculator connected via an intelligent interface, load, strain, and acoustic emission signals are monitored continuously. The method of operation of this system and the ensuing software is detailed. This type of system is compared with other systems, and its effect on the research project is described.

1857 **P. M. Wingfield. "Structural Validation (4th Progress Report) Weld Quality Monitoring." AML/24/71. Admiralty Materials Laboratory, Holton Health, Poole, Dorset, England. Dec 1971. [NTIS: AD-745-822].**

Stress-wave emission techniques have considerable potential for monitoring the quality of welds, either as they are made, or within a short time of their completion. Two low-cost prototype weld quality monitors have been designed, and several have been made for laboratory evaluation in parallel with a broad research program into the application of stress wave emission techniques to materials and structures. This report describes these two instruments, their performance and limitations, and the principles involved, and makes brief reference to the application of stress-wave emission procedural tests such as the CTS test. The devices, and the wider implications of weld quality monitoring in practice, are being studied at AML and at The Welding Institute under contract.

1858 **P. M. Wingfield. "Weld Quality Monitoring by Stress Wave Emission."** *Welding and Metal Fabrication*, **WLMFA, 40:351-355. Oct 1972.**

Stress-wave emision techniques have considerable potential for monitoring the quality of welds, either as they are made, or within a short time of their completion. Two low-cost prototype weld quality monitors have been designed, and several have been made for laboratory evaluation in parallel with a broad research program into the application of stress-wave emission techniques to materials and structures. Described are the two instruments, their performance and limitations, and the principles involved, and a brief reference to the application of stress-wave emission procedural tests such as the CTS test. The devices, and the wider implications of weld quality monitoring in practice, are being studied at AML and at The Welding Institute under an extramural research contract.

1859 **P. M. Wingfield and R. H. Warren (to National Research Development Corporation). "Stress Wave Emission Defect Location System." British Patent 1,343,694. Jan 16, 1974.**

1860 **W. H. Winn, F. B. Sawyer, and C. M. Richards. "Preliminary Tests in the Use of Stress-Wave Emission Analysis for the Detection of Cracking during Welding." NCRE/R-568. Naval Construction Research Establishment, Dunferm-**

line, Scotland. Feb 1971. [NTIS: AD-730-391]; IIW IX-740-71 and IIW X-623-71. International Institute of Welding. 1971.

Stress-wave emission analysis techniques were used to detect cracking which occurs immediately after welding in CTS tests, butt-weld cracking tests, and in the root pass of full-size butt welds. There is a correlation between the length of crack and the total area of the bursts of energy as recorded on an ultraviolet trace.

1861 **G. C. Winston, J. J. Lombardo, and R. B. Tatge. "A New Diagnostic Technique for Hydraulic Systems."** *Control Engineering,* **CENGA, 18(5):38-42. May 1971.**

Listening is an old diagnostic technique and in its simplest form it requires only an experienced mechanic. This article describes a powerful new *listening* method that has been successfully applied to high-performance hydraulic drive systems by National Aeronautics and Space Administration. The technique, called vibration signature analysis, automates the skilled mechanic, so to speak, by giving him the benefit of engineering evaluation and modern instrumentation.

1862 **P. A. Witt, Jr. "Applications of Acoustic Emission Technology."** *Chemical Engineering Progress, Technical Manual: Loss Prevention,* **5:106-108. 1971;** *Chemical Engineering Progress,* **CEPRA, 68(1):56-57. Jan 1972.**

Brief discussion is made of the application of acoustic emission as a nondestructive testing technique for on-stream inspection of pressure vessels and piping systems in the chemical and petrochemical industries. Included is a discussion of various filtering systems to overcome high background noise levels.

1863 **B. Woodward. "Acoustic Emission for Initial and In-Service Inspection."** *Testing, Instruments and Controls,* **TICOA, 12(8):13-21. Aug 1975.**

Acoustic emission is a dynamic phenomenon which may be used as the basis of a real-time nondestructive testing technique for detecting cracking, porosity, or inclusions in weld joints. During the past ten years, acoustic emission has been applied to both in-process and post-weld monitoring, and also to subsequent in-service inspection. A brief review of these applications is given, together with an introduction to acoustic emission aspects of the first stage of an Australian Welding Research Association co-operative research project in which a steam receiver vessel withdrawn from service due to severe weld cracking is being used to check acceptance standards for nozzle welds. In a post-test analysis emission, sources were located for various pressure steps up to 35.8 MPa or 79% of the expected yield stress of the vessel shell. Randomly distributed sources indicated that no significant defect extension occurred.

1864 B. Woodward. "Identification of Acoustic Emission Source Mechanisms by Energy Spectrum Analysis." *Ultrasonics,* **ULTRA, 14(6):249–255. Nov 1976.**

The purpose of energy spectrum analysis is to calculate random signal parameters which, when considered statistically, may show characteristic differences between similar signals and hence indicate different source mechanisms generating the signals. A computer program has been written to compute mean and median frequency, mean amplitude spectral density, mean energy spectral density and amplitude spectrum variance. Plots of amplitude spectra, normalized energy spectra, and energy difference spectra (the difference between pairs of spectra) may also be selectively generated by the program. The results obtained were from real data, from specially prepared reactor-grade zirconium test pieces.

1865 B. Woodward and N. R. McDonald. "Flaw Identification Using Acoustic Emission." Paper G3/7 in *Structural Mechanics in Reactor Technology.* **Compiled by T. A. Jaeger. Volume 3, Part G–H. North-Holland Publishing Company, Oxford, England. 1975. proceedings of the 3rd International Conference on Structural Mechanics in Reactor Technology, held in London, England. Sep 1–5, 1975. [distributed in the U.S. and Canada by American Elsevier Publishing Company, Inc., New York].**

Acoustic emission *signatures* contain information about the fine structure of metallurgical *source events* and their interpretation may provide a means of assessing the severity of internal flaws as well as surface flaws. The ultimate aim of this research on signature analysis is to develop a real-time nondestructive testing technique having the capability of flaw *recognition,* as well as flaw location in nuclear reactor components and structures under stress. Thus the requisite, unlike that in most acoustic emission work to date, is for a technique which affords discrimination between acoustic emission from different types of flaws propagating simultaneously. The approach described here requires detailed analysis of the emission signatures in terms of a specific statistical parameter, energy spectral density. In the acoustic emission monitoring and recording system, the observed energy spectrum for the waveform of a single emission *burst* is the result of a chain of modification of the frequency content of the original source event imposed by the frequency responses of the test-piece, sensor, and electronic equipment. The principle of the analysis procedure is that if all the intermediate responses in this chain remain constant then any differences in observed spectra should indicate different types of source event. Thus the method applies for groups of bursts which occur during small changes of strain, so that the test piece geometry, and hence its spectral characteristics, remain unchanged. In order to realize the full inspection potential of acoustic emission monitoring, data obtained from zirconium and steel test pieces have been correlated with metallurgical condition and mechanical behavior, since the nature of emission signatures is strongly affected by the physical characteristics and internal structure of the material. During experiments, signals were tape recorded and the surface of each test piece was recorded on movie film or videotape so that acoustic and visual information could be correlated. Large numbers of tape-recorded bursts have been analysed with a real-time spectrum analyser, and statistical parameters (such as mean energy density, mean frequency, and variance) derived from the spectra were calculated by an IBM 360/50 computer and selectively displayed on a line plotter. In the case of zirconium, observed differences in these parameters were an indication that the emission signals were generated by three different metallurgical mechanisms. A movie film of test piece surface deformation revealed the occurrence of twin initiation, twin broadening, and slip. Fracture events either in the zirconium matrix or in second phase particles are also possible although not observed directly. A direct correlation was confirmed between twin initations and emission signals. Work is proceeding on establishing a different correlation between emission signals and stress corrosion cracks.

1866 **B. Woodward, N. R. McDonald, and M. J. Hincksman. "Application of Acoustic Emission Monitoring to Pressure Tests of a Steam Receiver Vessel with Flawed Nozzle Welds." AAEC/E371. Australian Atomic Energy Commission, Research Establishment, Lucas Heights, Australia. Jan 1976. [NTIS].**

As part of the first stage of an Australian Welding Research Association co-operative research project, acoustic emission monitoring has been applied to a steam receiver vessel withdrawn from service owing to severe weld cracking. This technique is used to check acceptance standards for defects in nozzle welds and to apply modern methods of assessing the integrity of a pressurised plant. Acoustic emission monitoring has been used, together with strain gauge measurements and ultrasonic scanning, to detect the occurrence of any significant defect growth during cyclic pressurisation of the vessel. During this first stage, no significant defect growth has been produced by 1000 cycles of pressure up to 24.1 MPa (3500 psi), subsequent pressurisation up to 35.8 MPa (5200 psi), or 97 per cent of the expected yield stress of the vessel shell. The small amount of acoustic emission detected was consistent with this result.

1867 **B. Woodward and R. W. B. Stephens. "Some Aspects of Boiling Noise Detection in Sodium Reactors by Means of a Mechanical Waveguide."** ***Ultrasonics*****, ULTRA, 9(1):21-25. Jan 1971.**

Some of the problems associated with the detection of boiling in a liquid-metal reactor are presented. A possible method of detecting the boiling lies in the use of a stainless steel waveguide as an acoustic transformer which vibrates in response to boiling noise signals. The design of a laboratory apparatus for simulatory experiments is discussed and results reported which were obtained using waveguides in the form of solid rods, tubes, and wires in a variety of configurations, as well as for several materials.

1868 **D. C. Worlton. "A Review of Selected Nondestructive Testing Activities at Battelle Northwest Laboratory." BNWL-SA-1395. Battelle-Northwest,**

Richland, Washington. May 13, 1968. [NTIS]; paper presented at the Inter-American Conference on Materials Technology, San Antonio, Texas. May 20–24, 1968.

A review of the nondestructive testing facilities and equipment at Battelle-Northwest Laboratory is presented. Acoustic emission, eddy current, and critical angle ultrasonic methods of nondestructive testing are described.

1869 **D. C. Worlton. "Recent Nondestructive Testing Advances at Battelle–Northwest Laboratory." BNWL–SA–1772, Part 1. Battelle–Northwest, Richland, Washington. May 1, 1968. [NTIS]; paper presented at the Colloquium on Nondestructive Testing and Control in the Field of Nuclear Metallurgy and Technology. Saclay, France. June 24–26, 1968.**

Acoustic emissions internally released in materials undergoing deformation or fracture may provide early warning of incipient failure of critical nuclear components, such as reactor pressure vessels or piping, during routine operations. A practical monitoring system based on this principle requires detection of acoustic emission signals against normal background noises, location of emission point of origin, characterization of failure or flaw growth mechanisms, and prediction of time of failure. Experimental data using this method to study crack propagation in large diameter steel pipe are presented. Further development work on the monitoring equipment and study of the relationship between emission signal characteristics and the irreversible metallurgical events which give rise to these signals are recommended.

1870 **D. C. Worlton. "Systems Analysis Study of Acoustic Emission Nondestructive Testing System." Task 1, Project IV. pp 29–32 in "In–Service Inspection Program for Nuclear Reactor Vessels." R. D. Wylie. SwRI Project 17–2440, EEI Project RP–79. Biannual Progress Report No. 1. Southwest Research Institute, San Antonio, Texas. June 2, 1969. [NTIS: PB–184–359].**

During the quarter, a 5-month contract was awarded on December 9, 1968, to Teledyne Materials Research Corporation for performance of Task I of Project IV, EEI-TVA Program RP-79. The objective of this Task is to perform an analysis of existing Acoustic Emission Technology, and recommend an Instrumentation Configuration for use in subsequent phases of this Project. Teledyne is scheduled to have a draft of the final report completed by mid-March. Tests have been initiated for the acquisition of base acoustic emission data by use of existing equipment applied to fracture studies being performed by the Heavy Section Steel Program and other related programs. A 1.25-inch thick steel plate in which flaws were induced by stress corrosion was monitored at Southwest Research Institute.

1871 **C. Wright. "Propagation of Ultrasound in Linear Structures." Internal Report. Cambridge Consultants Limited, Bar Hill, Cambridge, England. Aug 1972.**

Using an accelerometer mounted in the same plane as a detector and subjected to electrical pulses, two types of waves–compression and shear–are produced in linear structures. However, when an Electromechanisms exciter is mounted in a plane normal to a detector, only a compression wave is detected. Both types of waves are detected by all types of transducers used, except for shear wave transducers, which only detect the shear wave component. The two-component waveform detected is a good simulation of an acoustic emission signal, but in most cases there is a large risetime to the peak of the shear wave component (which is generally twice to four times the amplitude of the compression wave component). This can be as high as 0.4 milliseconds, and means that the time delay between an event and its detection some distance away (measured by the *firing* of a CCL detector module) varies with input amplitude. This is because increasing the size of the whole wave effectively lowers the threshold and causes the trigger to *fire* earlier. Under limited conditions, the use of a damping medium (Blu-Tack adhesive) around the surfaces of a specimen cut the response of a Dunegan detection system substantially. From the experiment carried out, it is not possible to infer how much or which parts of a wave are affected by surface damping. For waves travelling through the specimens used, general signal loss in mild steel appears to be 0.4 db/m for the compression wave, and 3.0 db/m for the shear wave (variations are due to an amplitude effect caused by the way a *ringdown* pulse is formed). Corresponding values for cast iron were found to be about 2.0 db/m and 3.5 db/m; however for both specimens there were several factors such as wave spreading which affected wave propagation, and these values are not an accurate measurement of true attenuation. For waves which do not suffer a great deal of attenuation, reflections from material boundaries may be detectable although in some cases the ringdown time to the exciting pulse may obscure reflections. The shape of signals detected is dependent firstly on the types of detecting accelerometer used, as each type has its own operating frequency. Secondly the preamplifiers and signal processors used also have frequency characteristics. Comparison of results from different specimens is rather limited in this report; however the only set of results comparing similar specimens of different sizes show that increasing the size decreases the number of counts detected using a Dunegan totaliser, especially at higher amplitudes. The calibration measurements show that quantitative data obtained during acoustic emission testing are dependent on the type of material tested, the specimen geometry, and the conditions of the test. Therefore comparisons of emission data from different specimens can only be made on the basis of appropriate calibration measurements.

1872 **C. P. Wright. "Development of an Acoustic Emission Inspection Technique for Hydraulic Components with Moveable Parts." ISA Reprint 76238. Instrument Society of America, Pittsburgh, Pennsylvania. 1976; paper presented at the 22nd International Instrumentation Symposium, San Diego, California. May 25–27, 1976. [copy of reprint also available as Technical Report DE 76–4 from Dunegan/Endevco, San Juan Capistrano, California].**

Hydraulic components subject to fatigue cracking can be difficult to inspect using periodic proof testing, if noise sources of *apparent* acoustic emission are present. This paper defines a systematic approach for noise identification and suppression used in the development of a successful inspection technique on an actual hydraulic component. It was found that the primary noise levels were due to moveable parts in contact with *pressurant* medium (e.g., O-rings, Teflon seals, and felt contamination seals). This effort was involved with the identification and suppression of noise levels, while retaining flaw-related acoustic emissions. Techniques were developed for this noise suppression using variable system gain and frequency range. A test method for quieting movable seals is presented. The result was a technique appropriately insensitive to these noise levels, yet with significant gain margin for the detection of fatigue-crack related emissions. The technique is applicable to in-place inspection and requires minimal judgment on the part of inspection personnel.

1873 **R. E. Wright. "Acoustic Emission of Aluminum Titanate."** *Journal of the American Ceramic Society*, **JACTA, 55(1):54. Jan 1972.**

The acoustic emission on cooling of stablized aluminum titanate is reported as a function of temperature. The results indicate that microcracking in aluminum titanate begins within the temperature range predicated on the basis of the model of Buessem *et al.* Acoustic emission appears to be a useful technique for monitoring microcracking in ceramic materials. [Abstract 29381. *Physics Abstracts*, 75(940), 1972].

1874 **R. D. Wylie. "A General Review of the Status of the EEI–TVA Research Program on In-Service Inspection." Southwest Research Institute, San Antonio, Texas. 1970; Paper No. 31 presented at the Heavy Section Steel Technology Program 4th Annual Information Meeting, Oak Ridge, Tennessee. Mar 31–Apr 1, 1970. [CONF–700347–Paper 31 available from Technical Information Center, P.O. Box 62, Oak Ridge, Tennessee 37830].**

The Edison Electric Institute–Tennessee Valley Authority research program has been previously described. The objective of the program is to evaluate potential techniques which may be applicable to in-service inspection of reactor pressure vessels. It is presently planned to evaluate these techniques both in the laboratory and in the field by using the Experimental Beryllium Oxide Reactor facility at the National Reactor Test Site in Idaho. This facility has been decommissioned and will be provided to the project by the Atomic Energy Commission. The program originally included work on acoustic emission procedures, the acoustic spectrometer, and improved ultrasonic systems. After the first phase of study, work has been discontinued on the acoustic spectrometer. An acoustic emission system has been constructed and is ready for laboratory evaluations. Work is proceeding on studying the influence of defect detection and discrimination by conventional ultrasonics. Studies are also in progress on the influence of the reactor environment on the reproducibility of ultrasonic signals as influenced by crud deposits or neutron irradiation. A new project has been initiated on acoustic holography as a tool for in service inspection of reactor components.

1875 **R. D. Wylie. "Developments in Nondestructive Testing Techniques in Reactor Pressure Vessels." Southwest Research Institute, San Antonio, Texas; Paper No. 1 presented at the Heavy Section Steel Technology Program 6th Annual Information Meeting, Oak Ridge National Laboratory, Oak Ridge, Tennessee. Apr 25-26, 1972. [NTIS: CONF-720468-1].**

The EEI-TVA research program on nondestructive testing techniques for reactor pressure vessels is to be completed in early 1972. This paper describes the results of this program up to the initiation of the field tests on the acoustic emission test facility in Idaho. The EBOR vessel was used.

1876 **R. D. Wylie. "In-Service Inspection of Steel Nuclear Reactor Pressure Vessels: Survey of Present Experience."** ***Atomic Energy Review*****, AERVA, 8(1):127-172. Mar 1970.**

The development of nuclear power has brought about an interest in measures for assuring the continued structural reliability of the pressure components of the nuclear reactor system. Discussed is the present experience in the application of non-destructive testing techniques to the in-service inspection of steel reactor pressure vessels, primarily for water-cooled reactor service. Consideration is given to the limitations of the various techniques and some areas of required development are presented. Examples are given of some early reactor inspections in the United States, Italy, and Japan and some of the equipment which has been used in making these inspections is described. Some new approaches are indicated and, in fact, must be developed if economical reliability assurance is to be obtained.

Y

1877 **K. Yamaguchi, H. Ichikawa, T. Shimada, and H. Atoh. "An Acoustic Emission Location System and Its Application to Hydrostatic Test of Pressure Vessel." Paper 2-1, pp 67-84 in** ***The Third Acoustic Emission Symposium.*** **Japan Industrial Planning Association, Dobashi Building, 3-17-8, Akasaka, Minato-ku, Tokyo, Japan. 1976. proceedings of a symposium held in Tokyo, Japan. Sep 16-18, 1976.**

Since 1972, a multichannel acoustic emission location system has been under development. A nine-channel AE location system was built in April 1973. After some improvement, the nine-channel system was applied to fatigue tests of model reactor vessels in July and September 1973, and satisfactory results were obtained. Soon after these experiments, a new system was developed which could equip a large number of input channels and could be applied to large structures at high efficiency. The AE location system has unitized time difference detectors and adopts rapid locating process by correction coefficient table. The time-difference detector system has such features as reduction of measurement loss, flexibility of channels arrangement, and convenience for increasing channels. The location method by correction coefficient table satisfies high locating speed and high location accuracy at the same time. The AE location system was successfully used in hydrostatic fracture tests of pressure vessels conducted jointly by the University of Tokyo and Nippon Kokan Kabushiki Kaisha under the sponsorship of the High Pressure Institute of Japan in November, 1975.

1878 **E. Yamamoto, N. Uetake, R. Motegi, and S. Takehashi. "Relation between the Ultrasonic Velocity Changes and AE Phenomena of the Low Carbon Steel during Application of Stress." Paper 5-5, pp 393-410 in** ***The Third Acoustic Emission Symposium.*** **Japan Industrial Planning Association, Dobashi Building, 3-17-8, Akasaka, Minato-ku, Tokyo, Japan. 1976. proceedings of a symposium held in Tokyo, Japan. Sep 16-18, 1976.**

Regarding the occurrence of AE due to the stress-strain characteristics of low carbon steel, it is generally known that less AE occurs in the elastic zone, and much AE occurs in the Luders deformation or in the plastic zone. That this phenomenon continues until a fracture occurs is widely known. The authors have concentrated their researches mainly on the stress-ultrasonic resonance frequency and velocity change in the elastic zone. However, this time, the authors have studied the relation between the above and AE. When any stress is applied to a metal specimen, the reasonance frequency of the ultrasonic wave changes due to changes in strain, density, properties of the material, etc. This change was measured accurately as a change of ultrasonic continuous wave-type resonance frequency. Also, at the same time, AE was measured and the relation between the two was obtained. The results showed the occurrence of much AE in the nonlinear section of the stress-resonance frequency characteristics. The authors expect to continue their experiments on this subject in order to fully study the relations.

1879 M. E. Yancey. "Acoustic Emission: Instrumentation for Acoustic Emission Studies." pp 32–35 in "Instrument Development Branch Annual Report, 1967." R. G. Morrison, Editor. IDO-17269. Phillips Petroleum Company. Idaho Falls, Idaho. Aug 1968. [NTIS].

1880 S. P. Ying. "Acoustic Emission Experiments for Safety of Nuclear Reactor Vessels." Southwest Research Institute, San Antonio, Texas. 1976. [NTIS: CONF-760431-1].

1881 S. P. Ying. "Acoustic Emission Monitoring of the 6-Inch Thick Flawed Tensile Specimen." *Nuclear Engineering and Design,* NEDEA, 17(1):176. 1971. [abstract of Paper No. 23 presented at the 5th Annual Information Meeting of the Heavy Section Steel Technology Program, held at Oak Ridge National Laboratory, Oak Ridge, Tennessee. Mar 25–26, 1971. (see Reference 1882 for entire paper)].

Acoustic emissions due to crack growth were monitored by Southwest Research on four of the 6-inch flawed tensile specimens. Emissions in the 200-kHz range were monitored with an on-line data process system and recorded for permanent record on a wideband magnetic tape recorder. The experimental results of accumulative counts (N) fit the equation $N = AK^n$ where K is the stress intensity factor and A and n are constants. Also, the number of counts and the amplitude of each acoustic emission pulse are presented as a function of load.

1882 S. P. Ying. "Acoustic Emission Monitoring of the 6–Inch Thick Flawed Tensile Specimens." Southwest Research Institute, San Antonio, Texas. 1971. [NTIS: CONF-710342-21]; Paper No. 23 presented at the Heavy Section Steel Technology Program 5th Annual Information Meeting. Oak Ridge, Tennessee. Mar 25–26, 1971.

Acoustic emissions due to crack growth were monitored by Southwest Research Institute on four of the 6-inch flawed tensile specimens. Emissions in the 200 kHz range were monitored with an on-line data process system and recorded for permanent record on a wideband magnetic tape recorder. The experimental result of accumulative counts, N, fits the equation $N = AK^n$, where K is the stress intensity factor, and A and n are constants. Also, the number of counts and amplitude of each acoustic emission pulse are presented as a function of load.

1883 S. P. Ying. "Acoustic Emission Monitoring of One–Inch–Thick and Six–Inch–Thick Flawed Tensile Specimens." Southwest Research Institute, San Antonio, Texas. 1972; Paper No. 15 presented at the

Heavy Section Steel Technology Program Sixth Annual Information Meeting, Oak Ridge, Tennessee. Apr 25-26, 1972. [CONF-720468-Paper 15 available from Technical Information Center, P. O. Box 62, Oak Ridge, Tennessee 37830].

In order to apply acoustic emission technology to safety inspections of nuclear reactor vessels, basic acoustic emission data have been obtained from heavy-section flawed steel specimens during tensile tests. There is a characteristic of the emission rate which is typical of linear elastic materials, but is distinguishably different from nonlinear elastic materials. It was also found that the rate of change of emission with respect to the strain depends on the test temperatures. The temperature-dependent effect can be explained by a theoretical model based on the theory of dislocations.

1884 **S. P. Ying. "Characteristics and Mechanisms of Acoustic Emission from Solids under Applied Stress."** *CRC Critical Reviews in Solid State Sciences,* **CCSCB, 4(1):85-123. Dec 1973.**

Theoretical and experimental work on acoustic emission from solids under applied stress is reviewed with emphasis on the characteristics and mechanisms of the emission. The experimental data on acoustic emission from single crystals, alloys, and ceramic materials are discussed along with the corresponding stress-strain relationships. An effort is made to summarize the rates of acoustic emission and the peaks of the emission rates of the above solids for a comparison. The data for the emission obtained from flawed spceimens and metal fatigue are also reviewed. From recent developments in the concepts and models used to describe acoustic emission, a theoretical model of acoustic emission based on the experimental evidences and the predications of dislocation dynamics is presented. This theory correlates the number of emissions with the probability of operation of a dislocation at a given strain. Using this theory, the Kaiser effect, which is an irreversible behavior of acoustic emission, can be explained. Based on this theory, the temperature dependence and the radiation effect of acoustic emission are shown to be consistent with the temperature dependence of yield stress and the radiation hardening effect, respectively. The mechanism of crack propagation and the size of a crack associated with acoustic emission are then discussed. A brief introduction of the experimental techniques and the applications are also included.

1885 **S. P. Ying. "Controllo con l'Acoustic Emission del comportamento sotto carico di provini di trazione precriccati da 1 a 6 pollici di spessore."** *Metallurgia Italiana,* **MITLA, 64(7):203. 1972. (in Italian). [abstract only].**

1886 **S. P. Ying. "Effect of Crack Size on Acoustic Emission Related to Moving Dislocations."** *Journal of Applied Physics,* **JAPIA, 46(7): 2882-2888. July 1975.**

Acoustic emission from a crack vicinity in a uniformly stressed solid is correlated with the crack size based on a continuous theory of dislocations. It is assumed that the total number of acoustic emissions is directly proportional to the total number of nonuniformly distributed moving dislocations in the plastic zone near the crack. The resultant equation of the relationship between acoustic emission and crack size can be expanded into a power series of applied stress. The first-order approximation is consistent with that of a macroscopic fracture mechanics consideration on the emission related to the plastic zone size. The theoretical calculation is verified with experimental results obtained from a flawed carbon steel vessel by a multichannel acoustic emission monitoring system and agrees reasonably well with the experimental data. Further calculations by utilizing the theory and by analyzing available experimental data of acoustic emissions from center-surface-flaw specimens and wedge-opening load specimens of various crack sizes also show agreement between the theory and the experimental data.

1887 **S. P. Ying. "Review of Acoustic Background Noise of a Boiling Water Reactor." Task 12, Project IV, pp 197–214 in "In-Service Inspection Program for Nuclear Reactor Vessels." R. D. Wylie. SwRI Project 17-2440, EEI Project RP-79. Biannual Progress Report No. 7, Southwest Research Institute, San Antonio, Texas. Aug 25, 1972. [NTIS: PB-212-144].**

For the purpose of developing an acoustic emission monitoring system for nuclear power reactors during the period of operation, the measurements of acoustic background noise of a boiling water reactor (BWR) have been reviewed. Both the noise spectra of the BWR and a pressurized water reactor (PWR) were normalized for comparison. In the 500 kHz range, the noise of the PWR is about 0 db for the selected normalization, while the noise of the BWR varies from 0 to 200 db for various locations.

1888 **S. P. Ying and S. C. Grigory. "Acoustic Emission from Flawed Heavy Section Steel Plates." Southwest Research Institute, San Antonio, Texas. 1973; *Materials Evaluation,* MAEVA, 33(2):30–36. Feb 1975.**

In order to apply acoustic emission technology to safety inspections of pressure vessels and piping, basic acoustic emission data have been obtained from heavy-section flawed steel specimens during tensile tests. When the emission rate is plotted as a function of applied stress or the corresponding value of strain, characteristically, the acoustic emission rate has a peak prior to the elastic limit. It was also found that the rate of change of emission with respect to strain depends on the test temperatures. The temperature dependence of acoustic emission is investigated by a theoretical model based on the theory of dislocations.

1889 **S. P. Ying and S. C. Grigory. "Tests of 6–Inch Thick Tensile Specimens. Fifth Technical Summary Report: Acoustic Emission Monitoring of One–Inch–Thick and Six–Inch–Thick Tensile Specimens." HSSTP-TR-24. Southwest Research Institute, San Antonio, Texas. Nov 1972. [NTIS].**

Results of tensile tests by acoustic emission on 1- and 6-inch thick flawed specimens of A533B steel are reported. Results showed that the acoustic emission increased before the yield point of the material and reached a plateau near the yield point. The emission increased again prior to failure. The rate of change of emission with respect to the strain prior to yielding depends on the test temperature. The rate of emission from the base metal was greater the higher the test temperature. During the few steps at the beginning of the tensile tests, the acoustic emission stopped when the tensile machine was stopped to record strain data; however, the emissions continued after the tensile machine was stopped when the stresses were near and beyond the yield point. This was probably due to an increase in strain at constant load, a phenomenon observed during these tests. The emission counts obtained from a six-inch thick specimen were greater than that from a 1/6-scale model. A heat-treated specimen had lower acoustic emission in comparison with a hydrogen baked-out specimen tested at the same temperature.

1890 **S. P. Ying, D. R. Hamlin, and D. Tanneberger. "A Multichannel Acoustic Emission Monitoring System with Simultaneous Multiple Event Data Analyses."** *Journal of the Acoustical Society of America,* **JASMA, 55(2):350–356. Feb 1974.**

A multichannel acoustic emission monitoring system has been developed recently. The system possesses high sensitivity for the sound pressure level of the emission and is concerned with the frequency selection to overcome various background noise disturbances. The signal and data processing unit can process multiple acoustic emission events simultaneously for analyses. The cathode-ray tube screen of a computer is used for a real-time graphic display of the number of events at individual locations of the emission sources.

1891 **S. P. Ying, D. R. Hamlin, and A. R. Whiting. "Acoustic Emission from Welds, Ceramic Materials, and Radiation Effect of the Emission from Nuclear Reactor Steel." pp 217–223 in** *Proceedings of the Ninth Symposium on Nondestructive Evaluation.* **Southwest Research Institute, San Antonio, Texas. July 1973. symposium held in San Antonio, Texas. Apr 25–27, 1973.**

Crack growth in welds was detected by acoustic emission technique. The performance of the technique was evaluated at two frequency ranges (160 and 450 kHz). The high-frequency channel was less sensitive to the welding noise in comparison to the low-frequency channel and detected the number of counts for each burst due to cracking being much greater than that of other noises. During a cooling period, the variation of the emission rate of a growing crack was distinguishable from minor deformation due to thermal stress. Characterization of acoustic emissions from plastic and glass has also been observed at constant stresses and during variation of applied load. There were peaks of the emission rates which occurred at the stress levels far below the fracture stress of the specimens. The rate of change of emission with respect to time at a constant stress decreased slowly. The Kaiser effect also characterized the emission which was dominated by plastic deformation. Further experiments on acoustic emission from irradiated nuclear reactor steel were conducted at temperatures near −20 °F. The data obtained from wedge-opening load (WOL) specimens are compared with that with a different radiation dosage. When accumulation of acoustic emission is plotted versus the stress intensity factors, the radiation-hardening effect causes the slopes of the curves to decrease. The radiation effect of acoustic emission is discussed.

1892 **S. P. Ying, J. E. Knight, and C. C. Scott. "Background Noise for Acoustic Emission in a Boiling Water and a Pressurized Water Nuclear-Power Reactor."** *Journal of the Acoustical Society of America,* **JASMA, 53(6): 1627–1631. June 1973.**

The acoustic noise spectra from 0 to 700 kHz of a boiling water reactor (BWR) have been normalized and compared with the available noise data of a pressurized water reactor (PWR). At frequencies near 500 kHz, which is the interesting frequency range for acoustic emission technology, the hydraulic turbulent and cavitation noises of the PWR were about zero db for the selected normalization and the noises of the BWR varied from 0 to 15 db for the various locations that were monitored. Since the intensity levels of acoustic emission signals are on the order of 0 to 20 db, the PWR background noise should not completely mask them; however, the BWR noise would obliterate some weak acoustic signals in some locations, but allow them to be observed in other locations.

1893 **S. P. Ying, J. A. Meyer, and A. R. Whiting. "Acoustic Emission Monitoring of a Cyclically Pressurized Vessel." ASME Publication 72–PVP–22. American Society of Mechanical Engineers, New York. 1972; paper presented at the American Society of Mechanical Engineers Petroleum Mechanical Engineering and Pressure Vessels and Piping Conference, New Orleans, Louisiana. Sep 17–21, 1972.**

Acoustic emission data obtained from a cyclically pressurized vessel are correlated with data obtained by laboratory tensile tests. Acoustic emission techniques and their ability to determine the location and nature of flaws were evaluated by comparison with magnetic particle and ultrasonic results. A general acoustic emission technique is also described.

Z

1894 V. F. Zackay, W. W. Gerberich, and E. R. Parker. "Structural Modes of Fracture." Chapter 6, pp 395–440 in *Fracture.* H. Liebowitz, Editor. Volume I. Academic Press, New York. 1968.

The structural modes of brittle and ductile fracture are discussed in terms of both static and cyclic stresses and in terms of the type of environment. Current concepts of brittle fracture are presented within the framework of defect, microstructural and macrostructural theories. The initiation and subsequent growth of microcracks in ductile materials is examined in terms of the elements of microstructure. It is concluded, primarily from the results of fractographic studies, that crack growth can occur both by continuous tearing and by void formation ahead of the advancing crack tip. Fractographic studies of fatigue fractures suggest that fatigue-crack growth can be described by a critical crack-tip-displacement model similar to that developed for ductile fracture. The crack growth of environmentally embrittled materials is followed by the relatively new technique of monitoring the stress waves emanating from the growing tip. Utilizing the results of this technique, a continuum model for predicting the time to failure in a hydrogen-embrittled specimen is deduced from first principles. Attempts are made throughout the chapter to relate the continuum and microstructural aspects of fracture. The experimental techniques of fractography and stress wave emission (SWE) are shown to be especially helpful.

1895 W. J. Zarella, Jr. "Stress Wave Emission from Non– and Irradiated Hydrogen Embrittled Reactor Steels." M.S. Thesis. University of Idaho, Moscow, Idaho. May 1969.

The purpose of the investigation was to (1) design, test, and evaluated stress wave emission equipment; and (2) use the stress wave equipment as a research tool to measure the following: crack incubation times, crack size as a function of cumulative bursts, and signal variances between non- and irradiated hydrogen embrittled A-542 steel. Equipment capable of monitoring crack growth in the notch of a buttonhead specimen was designed and evaluated. Specimens were cathodicly charged with approximately 3.50 ppm of hydrogen and loaded with predetermined stresses using a Satec Stress Rupture Tester. All Testing was accomplished remotely in a specially designed enclosure. The equipment was used to test 25 unirradiated A-542 specimens of which 7 were heat-tinted to obtain crack areas and one was metallurgically sectioned. Seventeen irradiated A-542 specimens, from 5×10^{18} to 1×10^{20} nvt, were also tested. Data of stress wave emission, relative energy per burst, burst rate, and cumulative emission were obtained. The relationship of heat tinting of the crack to stress wave emissions was attained. Delayed failure curves of the non- and irradiated steel are shown.

1896 M. Zeman. "Emisja akustyczna jako nowa metoda wykrywania inicjacji pekniec (Acoustic Emission as a New Method of Detecting Crack Formation)." *Biuletyn Institutu Spawalnictwa,* BISWA, (53):38–40. 1973. (in Polish).

1897 **M. B. Zisfein and W. B. Tarpley. "Critical Thoughts on Structural Mechanics and NDE." AFOSR-70-2870TR. The Franklin Institute Research Laboratories, Philadelphia, Pennsylvania. Sep 9, 1970. [NTIS: AD-720-899]; Keynote address presented at the Symposium on Advanced Experimental Techniques in the Mechanics of Materials, San Antonio, Texas. Sep 9-11, 1970.**

A critical review of nondestructive evaluation is given in general terms and specific examples of NDE techniques are given. These employ radiography (X-ray and neutron), ultrasonics, penetrants, eddy currents, thermography, laser interferometry, and holography. Other techniques not discussed in detail include magnetics, Mossbauer spectra, and spin resonance. The state-of-the-art is presently technique-oriented, rather than problem oriented due, in part, to ineffective communication links between designer, NDE researcher, and user. Problem areas are identified which include better data handling and interpretive techniques for the user as well as detectability and resolution criteria; and certain materials problems (e.g., limited ductility materials); and an especially urgent need is expressed for control indices for surface contamination by NDE. Methods are suggested for improving the communications link and the idea of *analogue* acceptability or evaluation is stressed in favor of *binary* go-no-go type decisions. In this way the NDE techniques would enjoy more user acceptance, as well as utility.

1898 **V. F. Zlepko, V. A. Luk'yanenko, and A. V. Fedoseenko. "Investigating the Possibility of Monitoring Discontinuity of Steam Pipe Metal by the Acoustic Emission Method under Creep Conditions."** *Teploenergetika/Thermal Engineering,* **THENA, 22(10):27-29. Oct 1975; translated from** *Teploenergetika,* **TPLOA, 22(10):19-21. Oct 1975. (in Russian).**

The possibility of using an acoustic-emission method as a means of periodically monitoring steam pipes in power stations to detect any discontinuities is discussed. Tests carried out with standard Cr-Mo-V steel steam pipes show that this procedure is rapid, reliable, and sensitive. Examples of the kind of signal to be expected from defects in 12Kh1MF Cr-Mo-V pipes are presented and analysed. Substantial changes appear in the noise level when rupture is imminent. Warning of impending failure may be obtained well in advance of actual danger. [Abstract 22-0722. *Metals Abstracts,* Vol. 9, 1976].

1899 **B. M. Zlobinsky and V. A. Murav'ev. "Factors Affecting Sound Emission of Cast Irons."** *Izvestiya Vysshikh Uchebnykh Zavedenii, Chernaya Metallurgiya,* **IVUMA, (3):154-157. 1974. (in Russian).**

Investigated were the material properties affecting the sound emission of cast-irons. The sound-pressure level of an impact-produced pulse is influenced by the elastic modulus but not other mechanical properties. Various cast-iron compositions were studied for their noise damping ability.

1900 **B. M. Zlobinsky and V. A. Murav'ev. "Sound Emission of Cast Irons."** *Metal Science and Heat Treatment,* **MSHTA, pp 950–952. June 1972; translated from** *Metallovedenie i Termicheskaya Obrabotka Metallov,* **MTOMA, 13(11):49–51. Nov 1971. (in Russian).**

One means of reducing noise at the source is to manufacture parts and units from materials with low sound radiation. Sound radiation can be characterized by the sound pressure level resulting from impact or the damping rate of sound, i.e., the change in the sound pressure level per unit time. To study the effect of alloy compositions on sound radiation, the vibrations were investigated of a body of simple shape that are excitable by impact. Investigated also was the sound radiation of plates induced by the impact of a ball in the center. As the result of the impact, the center of gravity of the plate shifted. The plate turned as the result of the angular velocity of rotation around the center of gravity, in which case a sonic pulse occurs. Part of the energy of impact is transformed into energy of elastic deformations that accumulate in the plate. Free elastic vibrations occur after the impact, which are responsible for the noise of free vibrations. The sound power on impact depends on the sound pulse and the duration of the noise of free vibrations. The formation of the sound pulse depends greatly on the elastic properties and mass of the colliding bodies and also the velocity of collision. The variation of sound radiation of cast iron with the chemical composition was investigated.

1901 **B. M. Zlobinsky and V. A. Murav'ev. "Sound Emission of Metals."** *Metal Science and Heat Treatment,* **MSHTA, 16(1–2):58–59. Jan–Feb 1974; translated from** *Metallovedenie i Termicheskaya Obrabotka Metallov,* **MTOMA, (1):56–57. Jan 1974. (in Russian).**

The sound emission of a wide range of metals and alloys (Al, Ni, V, Ti, Zr, Nb, Ta, etc.) was studied and correlated with other physical-chemical properties; in the case of alloys the correlation was with the mutual solubility of the components, and the form of the phase diagrams. Changing the hardness of a metal by heat treatment made little difference to its sound emission on receiving a sharp impact, provided that the elastic modulus remained constant. In alloy systems, the sound emission was fairly constant when a continuous series of solid solutions was formed, but exhibited anomalies associated with intermetallic compounds. [Abstract 32-0596. *Metals Abstracts,* Vol. 7, 1974].

1902 **J. R. Zurbrick. "Nondestructive Testing: Acoustic Emission." p 42 in** *Yearbook of Science and Technology–1970.* **McGraw-Hill Book Company, New York. 1970.**

1903 **"A Second Pressure Tube Leak."** *Nuclear News,* **NUNWA, 18(8):148. June 1975.**

The entire article appears as follows:

"A second pressure tube leak turned up last month at Ontario Hydro's Pickering station, this time at Unit 4. The utility said it considers the leaks similar to those that shut down Unit 3 from last August until March 31. Using devices developed during the earlier leak problem, the company scanned one reactor face with acoustic emission equipment and found two suspect tubes, then detected a defect in one of them with ultrasonic scans. The two tubes were then removed and sent to the Atomic Energy of Canada Ltd.'s Chalk River Laboratories for metallurgic testing. Plant manager Kevin Keane said it was too early to tell how long the unit might remain shut down, but a spokesman thought less time should be needed because of the test devices and experience gained as a result of the earlier leaks. The Unit 3 problems were traced to 17 tubes leaking as a result of improper installation methods (*Nuclear News,* May 1975, p. 54)."

1904 **"A Sound Way to Detect Faults."** *Industrial Research,* **IDRSA, 16:34. Jan 1974.**

Briefly described is the work of R. M. Koerner and A. E. Lord, Jr., at Drexel University, Philadelphia, Pennsylvania, on monitoring soil in earth dams by acoustic emision techniques.

1905 **"Acoustic and Ultrasonic Tests." pp 226–228 in** *Fundamental Nuclear Energy Research – 1969: A Supplemental Report to the Annual Report to Congress for 1969 of the U.S. Atomic Energy Commission,* **Government Printing Office, Washington, D.C. Jan 1970.**

Brief description of what acoustic emission is and its application to in-service monitoring to detect onset of fatigue cracking. Also included is a summary of the investigation work by Idaho Nuclear Corporation, Idaho Falls, on irradiated and unirradiated ASTM-A542 steel specimens to detect hydrogen-caused cracks.

1906 **"Acoustic Emission and Microstructural Damage."** *Metals and Materials,* **MEMTA, pp 48–49. Mar 1977.**

Brief report on one of a series of Round Table meetings organized by the Metallurgical Engineering Committee of The Metals Society and held at the Society's headquarters in London, England, January 19, 1977. Speakers included: H. Wadley, C. Scruby, G. Green and P. McIntyre, J. Holt and I. Mogford, J. F. Knott, and P. M. Bartle.

1907 *Acoustic Emission, ASTM STP 505.* **American Society for Testing and Materials, Philadelphia, Pennsylvania. 1972.**

The ASTM Special Technical publication contains the proceedings of the Symposium on Acoustic Emission at the ASTM Committee Meeting of December 7–8, 1971, in Bal Harbour, Florida. The papers cover a wide range of topics, from general reviews of the field to specific reports on research findings, technological advances, and applications in different industries. Their collective references, cited in the papers, include much of the principal published material on acoustic emission technology. The first section is primarily introductory. It is intended for those who seek a general understanding of acoustic emission. Included are the historical background and discussions of the relevant aspects of dislocation dynamics, microcracking processes, rock mechanics, materials studies, and structural evaluation studies. The remaining sections comprise a state-of-the-art review of specific research and application areas. Much of the material is previously unpublished and at the forefront of present technology. Described are applications in various industries and the results of research intended to advance the capabilities of acoustic emission. Experimental considerations and instrumentation are emphasized in several of the papers, ranging from direct application of commerical equipment to research efforts pushing the limits of instrument capabilities. (for individual papers see References 25, 83, 217, 328, 417, 491, 505, 691, 721, 743, 829, 1086, 1087, 1143, 1250, 1327, 1582, 1666, 1733, 1750, and 1918).

1908 "Acoustic Emission for Non–Destructive Testing." Report WSG:1:73. The Welding Institute, Abington, Cambridge, England. 1973.

Given is a bibliography of 48 references obtained from The Welding Institute's computer index system and which covers the period 1967 to 1972.

1909 "Acoustic Emission from a Flexible Composite." *NTIAC Newsletter,* **4(4):3–5. Oct 1976.**

Article summarizes the work reported by P. H. Hutton in Department of Transportation, U.S. Coast Guard, Office of Research and Development Report No. CG-D-37-76. (see Reference 845).

1910 "Acoustic Emission from Titanium." *NTIAC Newsletter,* **3(11):1–4, May 1976.**

Article summarizes report AMMRC–CTR–75–1 by M. S. Darlow *et al.* (see Reference 352).

1911 "Acoustic Emission Instruments in Stress/Strain Testing." *Quality,* **QUASA, 16(6):35. June 1977.**

Article announces the introduction of acoustic emission instrumentation by Instron Corporation, Canton, Massachusetts.

1912 "Acoustic Emission May Predict Dam Collapse." *Research/Development,* **REDEA, 24:10. Dec 1973.**

A brief note reporting a $50,000 grant by the Environmental Protection Agency to R. M. Koerner and A. E. Lord, Jr., at Drexel University, Philadelphia, Pennsylvania, to continue development and field tests to monitor earth dams with acoustic emission.

1913 "Acoustic Emission Monitoring of Structures during Controlled Stimulation." ASTM Standard E569–76. American Society for Testing and Materials, Philadelphia, Pennsylvania. Sep 1977; pp 624–627 in *1977 Annual Book of ASTM Standards, Part 11: Metallography; Nondestructive Testing.* **American Society for Testing and Materials, Philadelphia, Pennsylvania. 1977.**

The recommended practice provides guidelines for acoustic emission examination or monitoring of structures, such as pressure vessels, piping systems, or other structures that can be stressed by mechanical or thermal means. The basic functions of an AE monitoring system are to detect, locate, and classify emission sources. Other methods of nondestructive testing (NDT) may be used to further evaluate the significance of acoustic emission sources. Acoustic emission examination of a structure usually requires application of a mechanical or thermal stimulus. Such stimulation produces changes in the stresses in the structure. During stimulation of a structure, AE from discontinuities (such as cracks and inclusions) and from other areas of stress concentration, or from other acoustic sources (such as leaks, loose parts, and structural motion) can be detected by an instrumentation system, using sensors which, when stimulated by stress waves, transform mechanical excitations into electrical signals. The sensors are acoustically coupled to the surface of the structure, usually by means of a couplant material that improves the transmission of stress waves to the sensor. The detected AE signals can be electronically conditioned and processed, to facilitate recording or visual display of the amplitude and location of the emission sources. Increasing the internal pressure of a vessel is a common stimulus. Application of localized forces to the exterior boundary of piping and similar structural components is an example of another possible stimulus. In addition to immediate evaluation of the emissions detected during the application of the stimulus, a permanent record of the number and location of emitting sources and the relative amount of AE detected from each source provides a basis for comparison with sources detected during the test and during subsequent stimulation. [reprinted by permission of the American Society for Testing and Materials, Copyright].

1914 "Acoustic Emission: New NDT Prospect." ***Materials Engineering,*** **MAENB, 69(3):58. Mar 1969.**

Brief note outlines the ongoing work in acoustic emission of welds at Battelle-Northwest and comments on a discussion with J. R. Frederick at the University of Michigan, Ann Arbor.

1915 "Acoustic Emission: New Testing Method Requires No Scanning." ***Instruments and Control Systems,*** **INCSA, 48(1):21-22. Jan 1975.**

The article is the result of an interview with J. Becht and J. Eisenblatter of Battelle-Institut, Frankfurt-Main, Germany by P. J. Brennan, Editor-in-Chief of this journal. Presented is a brief outline of the basic principles of acoustic emission. Listed are several advantages of acoustic emission testing over other nondestructive test methods.

1916 "Acoustic Emission Support Studies for Hydrotest Monitoring." Task 11, Project IV. Jersey Nuclear Company, Richland, Washington. pp 135-159 in "In-Service Inspection Program for Nuclear Reactor Vessels." R. D. Wylie. SwRI Project 17-2440, EEI Project RP-79. Biannual Progress Report No. 6, Volume 1. Southwest Research Institute, San Antonio, Texas. Jan 7, 1972. [NTIS: PB-207-591].

The following task was established to investigate the feasibility of using acoustic emission monitoring during the initial shop hydrotests of a nuclear reactor pressure vessel. Although this overall project has primarily been directed toward developing an on-line system, an initial acoustic mapping of a vessel was considered helpful as a baseline test, much the same as an initial ultrasonic mapping of a vessel has been advocated. This report therefore documents results of performing acoustic emission analysis of a Pressurized Water Reactor pressure vessel, and a Boiling Water Reactor pressure vessel, as these vèssels were subjected to shop hydrostatic proof testing. In contrast to other pressure vessels successfully examined by acoustic emission technique, these vessels employed temporary nozzle plugs to seal the nozzle openings. It was expected that the plugs would be a source of interferring noise to the acoustic emission tests. The work reported here was primarily done to elucidate plug behavior during vessel pressurization, and to determine the extent to which they interferred with acquisition and interpretation of meaningful acoustic emission data. The test reports are issued under a single cover to facilitate comparison of results. The acoustic emission analysis of the Arkansas Nuclear-One Pressure Vessel was funded by the EEI/TVA RP-79 Research Program.

1917 **"Acoustic Emission Used in Plating Processes."** *NTIAC Newsletter,* **2(7):4, 6. Feb 1975.**

Article summarizes report UCLA-ENG-7473 by O. Takano and K. Ono. (see Reference 1719).

1918 **"Acoustic Emission Working Group Subcommittee Report: Recommended Acoustic Terminology." pp 335-337 in** *Acoustic Emission, ASTM STP 505.* **American Society for Testing and Materials, Philadelphia, Pennsylvania. 1972.**

The report if the AEWG Terminology Committee defines eight acoustic emission terms.

1919 **"Acoustic Integrity Analysis of EBOR Reactor Vessel." Jersey Nuclear Company, Richland, Washington. Mar 2, 1972.**

In November 1971, the EBOR nuclear reactor pressure vessel was hydrostatically tested prior to being placed in service as the In-Service Inspection Evaluation Facility for the Edision Electric Institute, RP-79 research program. An acoustic integrity analysis was conducted in conjunction with the hydrotest using the mobile acoustic analysis test laboratory. The objectives of the acoustic analysis test were: First, to prevent failure of the vessel during the hydrotest if critical defects were present in the structure; and second, to detect and locate defects that were insignificant to the structural integrity under the imposed stress condition, but yet of sufficient magnitude as to be cross-correlatable with other NDT techniques. This report outlines the test procedures used, describes the data analysis procedures, and presents the results and conclusions of the EBOR acoustic integrity analysis.

1920 **"Acoustical Die Monitor Will Warn of Faults."** *Electronic Design,* **ELODA, 24(18):21-22. Sep 1, 1976.**

Brief news item concerns an acoustical die monitor built by Westinghouse Electric, Control Products Division, Beaver, Pennsylvania. The device will detect and interpret the vibrations creating these sounds from punch press operations. (see Reference 1267).

1921 **"An Acoustic-Depressurization System for Prevention of Failure during Hydrotest of Large Solid Rocket Chambers." Proposal No. SRR-64515 to Air Force Flight Test Center, Edwards AFB, Edwards, California. Aerojet-General Corporation, Solid Rocket Plant, Sacramento, California. Jan 1964.**

Chamber size and cost, associated with the development of large solid rockets have encouraged increased emphasis on developing nondestructive testing techniques for preventing catastrophic failure of the chambers during hydrostatic proof test. The cost of a high-nickel-content steel, 260-inch diameter chamber exceeds $2,000,000; and catastrophic failure during hydrotest would seriously jeopardize the entire effort on the 623A Program large solid rocket at Aerojet. The most probable reason for such failures is the propagation of a flaw that goes undetected despite the careful inspection techniques employed. The probability of overlooking a critical flaw is aggravated by the sheer magnitude of the surface to be inspected on a 260-inch diameter chamber. The purpose of this proposal is to present a program for establishing the feasibility of using an acoustic technique to detect the initiation of growth of a critical flaw, locate the flaw, and trigger a system that would arrest the further growth of the located flaw that could cause the destruction of a 260-inch diameter rocket chamber during its proof hydrotest. Sensitive accelerometers would be bonded directly to the 260-inch diameter chamber exterior. These accelerometers would detect the *signature* of a critical flaw propagating in the 260-inch diameter chamber and initiate a rapid depressurization of the chamber that would arrest further propagation of the flaw. The flaw could be located through a triangulation technique that would utilize the time-space relationship of the recordings of the flaw-propagating signature by the several accelerometers mounted to the case.

1922 **"Behavior of Metals Studied at –423 F."** *Materials Engineering,* **MAENB, 69(3): 59. Mar 1969.**

A unique testing facility at the National Aeronautics and Space Administration Lewis Research Center in Cleveland, Ohio, features a cryostat-equipped 1,200,000-pound capacity universal testing machine from Tinius Olsen which will be used in tension testing to study crack propagation in metals in liquid hydrogen (–423 °F) and other cryogenic environments. The Center is conducting studies to characterize materials quantitatively in terms of their tolerance for cracks to determine which materials can best meet certain space vehicle applications. Specimens are tested with cracks of controlled size and shape to gain a better understanding of a material's capabilities in service. Because of the potentially hazardous nature of the liquid hydrogen, the load frame and control console are located in widely separate rooms, and the entire test area is isolated from other facilities. At Lewis, as well as at Battelle and the University of Michigan, engineers are using acoustic emission devices to listen to a piece of metal as it breaks. Sound gathered through a phonograph-type pick-up mounted on the test specimen often signals when a crack starts before it can actually be seen with the naked eye. To provide the noise-free environment necessary for this type of testing, the machine is equipped with two loading systems. [reprinted by permission of Reinhold Publishing Company].

1923 **"Bibliography: Acoustic Emission Testing Applied to Welded Joints."** *Welding Research International,* **WASIA, 5(2):78–85. 1975.**

The bibliography of publications on acoustic emission testing applied to welded joints includes 95 references, mainly covering the period 1969 to 1974. Seventeen of the references relate specifically to nuclear industry, covering topics such as the periodic inspection of steel pressure vessels for reactors, monitoring of reactor fuel pin-end closure welds, and evaluation of radioactive waste capsule and welds.

1924 **"Comparison of NDT Methods."** *NTIAC Newsletter,* **2(1):6–8. Aug 1974.**

Article summarized report NASA CR–2369 by W. D. Rummel *et al.* (see Reference 1520).

1925 **"Defects in Materials Make Themselves Heard."** *Iron Age Metalworking International,* **IAMIA, 13(8):27–28. Aug 1974.**

Article consists of a series of questions and answers with P. F. Packman of Vanderbilt University concerning what is acoustic emission and its present status as a nondestructive testing method.

1926 **"Detecting Flaws by Ear."** *Business Week,* **BUWEA, (2433):80B and 80F. May 24, 1976.**

1927 **"Development of a Nondestructive Testing Technique to Determine Flaw Criticality." Proposal No. SR691711 to the Advanced Research Projects Agency and the U.S. Air Force, Wright-Patterson AFB, Ohio. Aerojet General Corporation, Solid Rocket Division, Sacramento, California. Nov 6, 1969.**

1928 **"Development of a Nondestructive Testing Technique to Determine Flaw Criticality." Proposal No. SR701416 to the U.S. Air Force. Aerojet Solid Propulsion Company, Sacramento, California. Apr 1970.**

1929 **"Device Hears Corrosion."** *Quality,* **QUASA, 15(5):8. May 1976.**

The entire article appears as follows:

"Aircraft are being inspected with a process that *hears* corrosion long before it is visible. This nondestructive testing method is being used through a licensing agreement between Acoustic Emission Technology Corporation of Sacramento, California and Lockheed-California Company. Several years ago Lockheed scientists discovered that certain corrosion reactions are accompanied by high-frequency, low-amplitude sound generation. Acoustic Emission Technology now markets equipment of the practical application of this method of monitoring corrosion." [reprinted by permission of Hitchcock Publishing Company].

1930 **"Do Welds Have Nerves?"** *Metal Construction and British Welding Journal,* **MCBWA, 1(10):492–493. Oct 1969.**

The article is an extraction of material published by W. D. Jolly in Battelle-Northwest Reports BNWL-SA-2727 (see Reference 942), BNWL-SA-2212 (see Reference 941), and BNWL-SA-1967 (see Reference 925).

1931 **"Early Warning Test Equipment for Engineering Structures."** *EM&D Journal of Engineering Materials Components and Design,* **EMJEB, 17(8):30. Aug 1973.**

This is a brief article describing the Admiralty Materials Laboratory Defect Location System.

1932 **"Experimental Study of Pop-in Behavior of Surface Flaw Type Cracks." Proposal RT-69115. Aerojet-General Corporation, Sacramento, California. Mar 1969.**

The purpose of the program is to study the flaw-growth behavior in 6 Al-4 V titanium alloy resulting from stress profiles and environments similar to those which occur during proof testing of Apollo spacecraft pressure vessels. Through the results obtained, it will be possible to establish the proof testing conditions causing crack extension and their effect on the subsequent service life of a typical Apollo pressure vessel. In order to accomplish the objectives, it will be necessary to accurately determine the occurrence of flaw growth, relate its occurrence to the loading and environmental variables under investigation, and thereby determine if flaw growth during proof testing occurs due to environmental effects (stress corrosion) or pop-in crack growth under plane-strain conditions which subsequently arrests when reaching a plane-stress state. Testing will be performed using Aerojet's Stress Wave Analysis Technique (SWAT) instrumentation system and a high sensitivity compliance gage. The test data will be related to typical pressure vessel operating conditions and anticipated service life.

1933 **"Final Report NDT–Acoustics Inservice Inspection of the NSSS of Kernkraftwerk–RWE–Bayernwerk GmbH (KRB) Gundremmingen, West Germany." XN–234. Exxon Nuclear Company, Richland, Washington. June 1974.**

NDT-ACOUSTICS was utilized to evaluate the physical integrity of the nuclear steam supply system (NSSS) of the Kernkraftwerk-RWE-Bayernwerk GmbH. This evaluation was conducted in conjunction with a scheduled inservice hydrostatic test of the NSSS on May 18 and 19, 1974. Three *insignificant* (Grade 1 and Grade 1+) acoustic sources (defects) were detected, located, and analyzed during the test. The NDT-ACOUSTICS evaluation showed the tested NSSS components to be of high structural integrity under the stress conditions imposed by the hydrostatic test. Ultrasonic evaluation of the Grade 1+ sources following the test provided positive confirmation of the accuracy and reliability of the NDT-ACOUSTICS inservice inspection.

1934 **"Finding Foreign Particles as Small as a Grain of Salt."** *Quality Assurance,* **QUASA, 8:36–37. Feb 1969.**

Brief article describes the work of M. W. Schulz at General Electric's Research and Development Center to acoustically detect loose parts and foreign particles in small electronic assemblies by vibrational signature analysis.

1935 **"Gas Contaminants Trigger Crack Detector for Helicopter Blades."** *Product Engineering,* **PRENA, 41:54, 56. Nov 9, 1970.**

Article includes a brief description of Trodyne Corporation's Structural Integrity Monitoring System (SIMS) for in-flight detection of cracks in helicopter blades and fixed-wing aircraft according to an interview with K. R. Notvest of Trodyne.

1936 **"Heard Any Bad Knees Lately?"** *Physics Today,* **PHTOA, 29(6):20. June 1976.**

Brief note on the acoustic emission work being done for *in vitro* studies on cow's knee joints by L. Zavodney and M. L. Chu at the University of Akron, Akron, Ohio. Their work consists of diagnosing the progress of joint diseases such as arthritis by monitoring the change in acoustic signature during flexure of the joint.

1937 **"Investigation of Correlation of Stress–Wave Emission Characteristics with Fracture in Selected Aluminum Alloys. Proposal No. RT681003 to George C. Marshall Space Flight Center. Aerojet–General Corporation, Sacramento Plant, Sacramento, California. Mar 1968.**

1938 **"Language of Materials Translated at the Bal Harbour Symposium."** *Materials Research and Standards*, **MTRSA, 12(3):48. Mar 1972.**

The article is a synopsis of the American Society for Testing and Materials Symposium on Acoustic Emission held in Bal Harbour, Florida, Dec 7–8, 1971. The symposium was held in conjuction with the 8th Acoustic Emission Working Group meeting.

1939 **"Listening in on Teeth Chattering."** *Industrial Research,* **IDRSA, 15(1):31. Jan 1973.**

In this brief note, H. R. Hardy, Jr., at Pennsylvania State University, University Park, Pennsylvania, speculates that in the future acoustic emission may detect a decaying tooth.

1940 **"Listening in on those Welding Flaws."** *Product Engineering,* **PRENA, 45(5):5. May 1974.**

The entire article appears as follows:

"A more reliable way of detecting flaws in fusion welding may come out of a research project under way at Battelle-Northwest. The work is being supported by ten Japanese companies, seeking to advance acoustic emission methods of nondestructive testing. At present acoustic techniques can detect flaws. The goal is to measure the significance of these flaws. Four types of weld flaws—cracking, porosity, lack of fusion, and inclusion—are under investigation. Acoustic information is being recorded, stored on magnetic tape, and analyzed to isolate the unique characteristics of each flaw type."

1941 **"Listening to Flaws in Pressure Vessels."** *New Scientist*, **NWSCA, 47:22. July 2, 1970.**

Briefly described is a promising method of monitoring the growth of cracks and flaws in metal pressure vessels by listening to the stress wave emission (SWE) of *noise* associated with flaw growth in materials. When perfected, this technique will be ideal for nuclear power stations where a method of continuously monitoring a reactor and detecting a crack at its inception is extremely valuable. If a flaw is discovered, the pressure in the suspect area can be relieved until a convenient time for shutdown and repair.

1942 **"Listening to the Sounds of Structural Failure." (Developments to Watch).** *Product Engineering,* **PRENA, 44:5. Apr 1973.**

Brief note on the status of acoustic emission as reported by C. A. Tatro at Lawrence Livermore Laboratory, Livermore, California.

1943 "Loose-Part Detection Program for the Primary System of Light-Water-Cooled Reactors." Regulatory Guide 1.133. U.S. Nuclear Regulatory Commission, Office of Standards Development, Washington, D.C. Sep 1977. [available from U.S. Nuclear Regulatory Commission, Washington, D.C. 20555, Attn: Director, Division of Document Control].

1944 "Measure Shear Strength without Damaging Weld." *Welding Design & Fabrication,* **WDEFA, 43:35-36. Feb 1970.**

Briefly described is the work of G. D. Shearer and E. B. Schwenk, Battelle-Northwest, to measure the shear strength of a spot weld without breaking the weld by counting acoustic emission pulses given off by the metal during welding.

1945 "Measurement: Talking Semiconductors Tell their Structure." *Industrial Research,* **IDRSA, 15(13):23, 25. Dec 1973.**

Brief note on the work being done in the Electrical Transport Division, Sandia Laboratories, Albuquerque, New Mexico on the use of acoustic emission to monitor and evaluate glass semiconductors during the switching process.

1946 "Metals Whisper Their Secrets." *Iron Age,* **IRAGA, 200(19):86-87. Nov 9, 1967.**

The article is a review of P. H. Hutton's paper, "Acoustic Emission in Metals as a NDT Tool" which was presented at the 27th National Fall Conference of ASNT in Cleveland, Ohio, Oct. 18, 1967. (see Reference 833 for entire paper).

1947 "Microseismic Devices that Listen for Mine Cave-ins, Tooth Cavities?" *Research/Development,* **REDEA, 24:11, 12. Jan 1973.**

This is a brief note reporting comments by H. R. Hardy, Jr., Pennsylvania State University, University Park, Pennsylvania, on work in microseismic studies of rock burst in coal mines and underground gas reservoirs. The author also predicts that some day acoustic emission techniques will reveal a decaying tooth.

1948 *Monitoring Structural Integrity by Acoustic Emission, ASTM STP 571.* **Edited by J. C. Spanner and J. W. McElroy. American Society for Testing and Materials, Philadelphia, Pennsylvania. 1975.**

The continued increase in the number of reported applications of acoustic emission to monitor structural integrity influenced ASTM to authorize this special technical publication to publish the papers presented during an ASTM Symposium on Monitoring Structural Integrity by Acoustic Emission. This symposium was held in Fort Lauderdale, Flordia, in January 1974, under the sponsorship of the ASTM E-7 Committee on Nondestructive Testing, and was a sequel to an introductory ASTM Symposium on Acoustic Emission which was held in December 1971, documented in ASTM STP 505. The purpose of the 1974 symposium, and of this STP, is to present a collection of papers selected to provide a representative coverage of recent activities in applying acoustic emision to monitor the integrity of engineering structures. It is significant that many of the speakers at this symposium are among the leading U. S. experts in this new and rapidly expanding area of technology. The first few papers provide background information on the acoustic emission method and its applications, and discuss specific characteristics of the signals that are emitted by structural materials. The next series of papers describe the techniques that were used, and the results that were obtained, when commercial and developmental acoustic emission instrumentation systems were employed to monitor the integrity of a wide variety of engineering structures and components. The last paper is a bibliography containing 412 references on acoustic emission that were published during the years 1970–1972. This publication is intended to provide a permanent record on the technological status of monitoring structural integrity by acoustic emission as it existed in early 1974. It is expected to be of value to those who are actively engaged in this field, as well as to those with structural integrity monitoring applications requiring the unique capabilities offered by this relatively new nondestructive testing method. (for individual papers see References 389, 395, 572, 591, 693, 826, 984, 1160, 1216, 1364, 1586, and 1667).

1949 **"Navy Sells Know How."** *Ultrasonics,* **ULTRA, 11(6):245, 247. Nov 1973.**

The brief article is a news release reporting on a 20-channel acoustic emission defect detection and location system that was developed by the Admiralty Materials Laboratory of the British Royal Navy. The system will be marketed through the National Research Development Corporation, London, England, which deals with commerical use of systems developed by the AML.

1950 **"Navy to Sell Know How."** *Non–Destructive Testing,* **NDETA, 6(5): 225, 229. Oct 1973.**

This is a news release similar to Reference 1949.

1951 **"NDT-Acoustic Test of Hydrogeneration Sphere." Jersey Nuclear Company, Richland, Washington. 1972.**

The NDT-Acoustics method was applied to the customers hydrogenerator sphere for the purpose of determining the physical integrity of the structure and to evaluate the significance of known discontinuities in the vessel welds. Analysis showed the vessel to be of good integrity under the imposed stress conditions. Several discontinuity regions were detected and located; these regions were confirmed by cross-checking with customer ultrasonic test records and recent dye penetrate, and visual inspections. None of these regions were considered to be significant to the vessel integrity under the stress conditions imposed by the test.

1952 **"NDT-Acoustics Demonstration Tests–Bilbao, Spain." XN-108. Exxon Nuclear Company, Richland, Washington, Feb 1973.**

On February 19, 1973, a special test was conducted in Bilbao, Spain on a hydro-desulfurizer pressure vessel designed by Foster Wheeler Iberia and built by Babcock and Wilcox Espanola for REPESA. The test was intended as a demonstration of the new volumetric nondestructive testing technology known as NDT-ACOUSTICS. The test was conducted by Exxon Nuclear Company with the permission of REPESA and the combined assistance and efforts of Foster Wheeler Iberia and Babcock and Wilcox Espanola. Forty-four observers representing twenty-eight European companies were present at the test. Twelve minor vessel acoustic sources (discontinuities) were detected, located, and analyzed during the test, using the ACOUST-II nondestructive testing system. These discontinuities were confirmed by conventional nondestructive test techniques immediately following the test. The vessel was analyzed to be of high integrity under the imposed test conditions.

1953 **"NDT-Acoustics Test of Hortonspheroid Vessel." XN-109. Exxon Nuclear Company, Richland, Washington. Feb 1973. [report is confidential to Mobil Oil Company, Beaumont, Texas].**

In August 1972, NDT-ACOUSTICS was applied to Hortonspheroid No. 2149 located at the Mobil Oil Company's Beaumont, Texas Refinery. The acoustic analysis non-destructive test technology was utilized in conjunction with a fill and pressurization operation which was directed at requalification of the storage vessel. The vessel emitted acoustic signals at a slow rate throughout the test; the signals were found to be emitted by minor discontinuities within the vessel. These minor discontinuities were detected, located, and analyzed during the test as being insignificant to vessel structural integrity. The vessel was determined to be of good integrity under the conditions of the imposed test.

1954 "NDT of Beryllium." *NTIAC Newsletter,* 2(3):3–4. Oct 1974.

Article summarizes report MCIC-74-22 by R. J. Runck. (see Reference 1521).

1955 "NDT of Diffusion Bonded Titanium Alloys." *NTIAC Newsletter,* 3(4):1–3. Oct 1975.

Article summarizes report AFML-TR-74-215 by J. A. Regalbuto *et al.* (see Reference 1467).

1956 "NDT on Adhesive Joints May Open Up New Jobs." *Steel,* STELA, 160:59–62. Mar 27, 1967.

The General American Transportation Corporation Research Division in Niles, Illinois reports that acoustic emission *listens* to noise emitted by a bonded joint placed under stress. Strength is signaled by a sudden peak that is correlated to a specific percentage of the bond's ultimate strength.

1957 "New NDT Methods Probe: the Look, Sound and Smell of Quality." *Steel,* STELA, 158: S-1 to S-7. Feb 14, 1966.

1958 "New to NDT: Dynamic Testing." *Welding Design and Fabrication,* WDEFA, 48(11):75, 77. Nov 1975.

1959 "No Nosie from Liberty's Flaw." *NDT International,* NDITD, 9(5):257. Oct 1976.

Brief news item reports monitoring of the Liberty Bell for acoustic emission to determine if the Bell's large crack propagated during moving of the Bell. Acoustic emission monitoring was performed by Dunegan/Endevco engineers.

1960 *Nondestructive Evaluation.* Publication NMAB-252. Prepared by the *Ad Hoc* Committee on Nondestructive Evaluation, National Materials Advisory Board, Division of Engineering – National Research Council. National Academy of Sciences – National Academy of Engineering, Washington, D.C. Jan 1969. [NTIS: AD-692-491].

Nondestructive evaluation (NDE) must be incorporated into every phase of the design-production-service cycle, including the screening and qualification of materials for new design criteria, in order to design materials to their limits to satisfy the ever-increasing demands of sophisticated military systems. Even in the presence of adequate technology, the key to NDE implementation and maximum hardware reliability is the proper application of suitable specifications. To achieve the necessary reliability at reduced costs, a general management-control specification on the application of NDE to the design-production-service cycles should be prepared and included among the pertinent specifications in DOD contracts. The majority of the technical problem areas facing DOD are well characterized and could be best handled by the proper use of NDE during the development and manufacture of materials and products. Solving these problems requires selecting test parameters, designing tooling, defining acceptance levels and using more than one technique for evaluation. Automatic data processing equipment must be used to attain quantitative capabilities through rapid feedback of data for process control, utilization of all data, comparison with prior data for early detection of trends and to evaluate operational changes. Review of special phenomena did not reveal any new energy form which could have impact on NDE techniques. Among areas recommended for further study were failure mechanism, data processing, lasers, holography, acoustic propagation, and acoustic emission. A greater number of NDE professionals and technicians are needed and could be produced through increased university and other technical training. The establishment of national NDE research and application centers was recommended as was the expansion of the present DOD NDT Information Analysis Center.

1961 "Nondestructive Inspection of Pressure Vessels and Pipelines. pp 380–395 in *Metals Handbook, 8th Edition, Volume 11: Nondestructive Inspection and Quality Control.* **American Society for Metals, Metals Park, Ohio. 1976.**

1962 "Nondestructive Inspection of Weldments." pp 340–355 in *Metals Handbook, 8th Edition, Volume 11: Nondestructive Inspection and Quality Control.* **American Society for Metals, Metals Park, Ohio. 1976.**

1963 "Nondestructive Testing Research and Development in Inflight Aircraft Monitoring via Acoustic Emission." Battelle-Northwest, Richland, Washington. 1968.

See Reference 361 for detailed report on in-flight detection of fatigue of aircraft structures.

1964 "Nonmetallics Outlook: Ceramics Warn Listeners before Failing." *Materials Engineering*, **MAENB, 71(8):17. Aug 1970.**

"By monitoring the sound of ceramics, scientists at Battelle-Northwest have successfully detected crack formation and growth in the materials. Elastic sound waves produced from the energy released during deformation or fracture are detected during initial cracking and before complete failure occurs. According to researchers, 'Since the emissions are produced simultaneously with crack events, the experimental conditions necessary to originate and propagate cracks can be correlated with cracking on a real-time basis.' Furthermore, 'Other experiments with surface-damaged aluminum oxide rods indicate that extremely small emissions can be monitored. These emissions are associated with events that are difficult to detect by other means.' The sensitivity of the acoustic method may be useful in several ways, including: (1) in-service or simulated service monitoring of ceramics or ceramic-coated materials to detect cracking failure or ceramic unbounding; (2) detecting cracks under thermal shock conditions; and (3) as an analytical tool to detect the fundamental processes involved in crack formation and propagation." [reprinted by permission of Reinhold Publishing Company].

1965 **"Now You can Hear the Flaws."** *Quality Management & Engineering*, **QMEGA, 11:13. Nov 1972.**

"Flaw propagation in structures under stress is pinpointed by a system which *hears* a defect enlarging and displays information on its location and propagation in real time, while the flaw is propagating, and in real time for corrective action. The computerized system is particularly useful in monitoring structures in which failures cannot be tolerated, including nuclear or petrochemical pressure vessels and piping and local areas of large structures. The system employs sensors mounted in an array at various points on the structure. These sensors detect the energy release which occurs when a crack or other flaw enlarges. The energy pulses are converted to electrical signals which are processed and transmitted to the monitoring system for triangulation and display of data." [reprinted by permission of *Quality Management & Engineering*, Hitchcock Publishing Company].

1966 **"NPRA Maintenance Q&A-3: Plant Maintenance Men Look at Inspection, Vibration, Instruments."** *Oil and Gas Journal,* **OIGJA, 72(24):74, 77–81. June 17, 1974.**

Article is the reproduced transcript of a question-and-answer panel during the 1974 NPRA Refinery and Petrochemical Plant Maintenance Conference in Houston, Texas on Jan 20–Feb 1, 1974. Brief comments concerning acoustic emission analysis were included.

1967 **"Planar Location of Flaws by Use of Two Model 902 Flaw Locators." Technical Report DC 72-2. Dunegan Corporation, Livermore, California. 1972.**

The Model-902 Flaw Locator accepts inputs from two acoustic emission preamplifiers and determines the difference in time of signal arrival between the two. The time is displayed on a digital display in terms of 0 to 1000 scale reflecting the distance between transducers. If the emission source is located on the straight line between the two transducers, the digital readout provides a direct measure of the source location. If the emission source is not located on the line, then the digital readout corresponds to the point of intersection of the line with a hyperbola passing through the source, and having foci at the two transducer locations. In this case, it is necessary to have a third transducer to provide an independent time measurement. This second independent time measurement will define a second hyperbola, and the source can be located by determination of the point of intersection of the two hyperbolas.

1968 "Preliminary Tests to Evaluate Attenuation of Stress-Wave Emissions in a Pipeline." AGC Report No. 2-4831. Aerojet-General Corporation, Sacramento, California. Dec 1967.

The program was prompted by the interest of the Natural Gas Pipeline Company of America in exploring the feasibility of applying the SWAT (stress-wave analysis technique) system to the hydrotest of pipelines. Covered herein are: (1) tests made with tensile specimens of X65C steel alloy to characterize the stress-wave emissions peculiar to this material from which piplines are currently fabricated, and (2) tests conducted to determine attentuation of stress-wave emission (SWE) versus distance, on sections of a 24-inch diameter pipeline being constructed by the National Gas Pipeline Company of America. The tests of tensile specimen showed (1) that stress-wave emissions in the parent material, in the weld affected zone, or in the weld deposit were of amplitudes approximately 0.005 g, (2) that the rate of occurrence of stress wave emissions increases with approaching failure, and (3) that, as in many other materials, a large number of stress waves were observed during initial loading. The field tests were made on sections of bare pipeline, pipeline after being wrapped, and pipeline wrapped and buried in the trench. The wave frequencies investigated were above 10 kHz and the single sensors employed were Endevco accelerometers Model 2215C. The tests indicated that stress-wave transmission was best in bare pipeline, that the attenuation increased in the wrapped pipeline, but that signal transmissibility in the wrapped pipeline improved after being covered with soil. The sensors used in these tests with the maximum gain available in our present equipment will detect frequencies above 10 kHz from the crack propagation emissions, characteristic of the X65C material, at distances of 460 feet in bare pipe. This is more than sufficient to use the present system during hydrotest of individual sections of pipe which are usually 40 feet long. The present system with single sensors and amplification gains on hand would detect crack propagation in pipeline wrapped and buried up to a distance of 250 feet. This would be economically inadequate for use in field hydrotests of pipeline sections of say 8 miles long since it would require a large number of sensors. The use, however, of higher sensitivities, amplifications, and multiple sensor packages designed to enhance the signal-to-noise ratio will allow a substantial increase in the spacing between sensor stations. This can provide the basis for an economically acceptable SWAT system for field hydrotest of pipeline sections.

1969 *Proceedings of the U.S.–Japan Joint Symposium on Acoustic Emission.* **English volume. Japan Industrial Planning Association, Kawamura Building 6F, 3-21-6 Akasaka, Minato-ku, Tokyo, Japan. 1972. (in English).**

The proceedings consist of two volumes, one in English and one in Japanese. This is the English volume containing six papers which are listed separately (see References 574, 608, 622, 860, 1448, and 1746).

1970 *Proceedings of the U.S.–Japan Joint Symposium on Acoustic Emission.* **Japanese volume. Japan Industrial Planning Association, Kawamura Building 6F, 3-21-6 Akasaka, Minato-ku, Tokyo, Japan. 1972. (in Japanese).**

The proceedings consist of two volumes, one in English and one in Japanese. This is the Japanese volume containing eleven papers which are listed separately (see References 34, 513, 517, 775, 898, 965, 1198, 1203, 1253, 1712, and 1720).

1971 **"Proposed Standard for Acoustic Emission Examination during Application of Pressure." Book No. E00096. American Society of Mechanical Engineers, New York. 1975.**

1972 **"Proposed Stress-Wave Tests on the LEM Ascent Stage Feasibility Tank." Special Report 1-4081-01-5.4-002. Aerojet-General Corporation, Downey, California. Dec 19, 1966.**

The Stress Wave Analysis Technique (SWAT) has been introduced by Aerojet-General Corporation as a nondestructive test on LEM ascent stage propellant tanks. The technique employs instrumentation that will sense and record energy emitted when flaws grow under proof loads. This report describes a proposed series of SWAT tests on the LEM ascent-stage feasibility tank that will serve the following purposes: (a) To provide additional data for use in analysis of oscillograms obtained during SWAT testing of production hardware; and (b) To provide data to be used as a basis for Material Review Board (MRB) decisions on ascent tanks containing scratches.

1973 **"R&D in Welding."** *Welding Design & Fabrication,* **WDEFA, 49(1):39-42. Jan 1976.**

Research and development activities being conducted in the field of welding by The Welding Institute in England and the Welding Research Council in the U.S., that being performed by private industry such as Babcock & Wilcox and by universities such as Lehigh University are discussed. Acoustic emission testing and crack opening displacement testing used by The Welding Institute are described.

1974 **"Reactor Pressure Vessel Integrity: In-Service Examination."** *Nuclear Engineering International,* **NEINB, 19(220):757–761. Sep 1974.**

Article is a report on the Institution of Mechanical Engineers Conference on Periodic Inspection of Pressurized Components, held in London, England, on June 4-6, 1974. For papers on acoustic emission presented at conference, see References 69, 102, 164, 439, 812, 570, and 1052.

1975 **"Sandia Switch to Acoustic Emission."** *Non-Destructive Testing,* **NDETA, 7(3):121. June 1974.**

The news release reports the work at Sandia Laboratories, Albuquerque, New Mexico, on using acoustic emission to study the switching characteristics of glass semiconductors. For details see Reference 122.

1976 **"Sounding Out a Way to Discover Faults in Material Acoustically."** *Engineer,* **ENGIA, 241(6232):33. Aug 21, 1975.**

Stored elastic energy releases stress waves of a frequency within audio range, when a controlled load induces local deformation. The *cry* of Sn is quoted as an example. Gains of ~80 to 100 db are used from an electronic noise level of ~20 μV. Usually amplification is needed. Apparatus is mentioned which can pin-point a source of stress wave emitted from twins, slip bands, phase change, and movement of dislocations. The method can be used when defects are too small for conventional nondestructive testing.

1977 **"Stress-Wave Analysis of LEM Ascent Tanks during Proof and Burst Tests." Special Report 1–4081–01–5.4–001. Aerojet–General Corporation, Downey, California. Dec 13, 1966.**

Since SWAT can detect and locate much smaller propagating flaws than other nondestructive test methods used on LEM ascent-stage propellant tanks, Aerojet began on its own initiative to utilize the method during proof-pressure testing of these articles. No firm acceptance criteria have been agreed upon with Grumman. However, Aerojet proposes that any tank that contains a propagating crack detectable by SWAT be considered a rejectable article. The feasibility of this criteria is based on the assumption that a simple spherical tank, containing a single girth-seam weld and constructed from machined forgings, should exhibit no flaw growth at proof loads that develop maximum stress levels less than two-third the ultimate strength of the tank material. The SWAT tests made so far at the Aerojet Downey Plant on 6 LEM ascent tanks and 12 Minuteman chambers (which are structurally similar to the LEM units and made from the same material) have confirmed the validity of this assumption. No propagating defects were detected in any of these 18 units. This report contains information on the SWAT results obtained from the first four of the six LEM ascent tanks tested. Data on the last two tanks have been reviewed, but the written analysis was not completed in time to include a detailed description of the results.

"Stress Wave Analysis Technique Demonstration Program for the Fast Flux Test Facility (Phase I)." Proposal No. NPS-681105. Aerojet-General Corporation, Nuclear Division, Sacramento, California. Jan 1968. [NTIS: NPS-681105].

A program for structural-failure warning, nondestructive testing, leak and boiling detection, and component malfunction-signature analysis at the Fast Flux Test Facility is described.

1979 **"Testing and Inspection."** *Metal Progress,* **MEPOA, 105(3):96, 98, 100, and 103. Mar 1974.**

1980 **"Testing News: Spot Weld Strength Now Measured Nondestructively."** *Materials Engineering.* **MAENB, 71(3):58. Mar 1970.**

Briefly described is a method for measuring the shear strength of a spot weld without breaking the weld, developed by Gordon D. Shearer and Earl B. Schwenk of Battelle-Northwest, Richland, Washington. The new method measures weld shear strength by counting the acoustic emission pulses given off by the metal during welding.

1981 **"Testing Triggers Big Savings: Acoustic Emission–Savings in Rails, Pipes, Ships and Vessels."** *Welding Design & Fabrication,* **WDEFA, 48(2):32–35. Feb 1975.**

The brief article reports on the work that Acoustic Emission Technology Corporation did with the Association of American Railroads. Acoustic emission was used to monitor thermit and flash butt-welded rail joints. A correlation was found between annealing noises and weld strength in flash butt-welds, but no specific conclusions were found for thermit welds. Also brief mention is made of the use of acoustic emission for testing reactors, boilers, and other pressure vessels by Westinghouse Canada, for testing LNG tanks in ships, for testing pipelines by Philadelphia Electric Company, and for monitoring spot welding of appliances by Whirlpool Corporation.

1982 **"The Amychometer: A New Tool for Materials-Testing."** *Design Engineering,* **DEMCB. Mar 1971. pp 108–109. [copy of article available from Manager of**

Reader Service, Morgan–Grampian Company Ltd., Morgan–Grampian House, Calderwood Street, London SE18 68H, England].

Article describes the Amychometer, an acoustic emission device for detecting microcracks, which was developed by T. R. Wilshaw and the Applied Materials Science Group at the University of Sussex, England. This instrument detects microcracking in the surface of a hard brittle solid when a small spherical scribe is dragged along the surface with sufficient pressure to scratch the surface. Intensity of scratching and the resultant crescent-shaped cracks formed in the wake of the sliding indenter are measured by detecting and measuring the acoustic emission generated by the microcracks while simultaneously measuring the normal force of the indenter. The results are related to many of the properties of solids which affect fracture behavior, e.g., fracture toughness, surface crack-size distribution which is dependent on surface finish, and surface residual stress. Work to date has been on glass microscope slides using a 0.7 mm diameter tungsten carbide ball as an indenter. Sliding velocity is maintained at a constant 2 mm per minute.

1983 *The New Encyclopaedia Britannica.* **Volume 3, p 851 and Volume IX, p 1020. 15th Edition. Encyclopaedia Britannica, Inc., Chicago, Illinois. 1974.**

In Volume 3 under properties of tin, one of the characteristics of white or beta tin resulting from the transformation of gray or alpha tin at 13.2 °C is: "As a consequence of internal friction, white tin emits a characteristic sound *(tin cry)* when bent." In Volume IX under *Tin* it is stated: "When bent, tin makes an eerie, crackling *cry* as its crystals crush each other."

1984 *The Second Acoustic Emission Symposium.* **Japan Industrial Planning Association, Dobashi Building, 3–17–8, Akasaka, Minato–ku, Tokyo, Japan. 1974.**

This volume is the proceedings of the second symposium held in Tokyo, Japan, September 2-4, 1974. The symposium was cosponsored by the High Pressure Institute of Japan (HPI) and the Japanese Society for Non-Destructive Inspection (NDI). Thirty-one papers are included in the proceeding and are listed separately. One paper by J. Eisenblatter *et al.*, is contained in a separate volume (see Reference 441).

1985 *The Third Acoustic Emission Symposium.* **Japan Industrial Planning Association, Dobashi Building, 3–17–8, Akasaka, Minato–ku, Tokyo, Japan. 1976.**

This volume is the proceedings of the third symposium held in Tokyo, Japan, September 16-18, 1976. The symposium was cosponsored by the High Pressure Institute of Japan (HPI), the Japanese Society for Non-Destructive Inspection (NDI), and the Pan Pacific Committee for Non-Destructive Testing (PPCNDT). Thirty-one papers are included in the proceedings and are listed separately.

1986 "The Ultrasonic Detection Handbook." Application Journal DC-101. Hewlett-Packard Company/Delcon Division. 333 Logue Avenue, Mountain View, California. 1966.

Booklet describes the theory and applications of the Hewlett-Packard/Delcon Ultrasonic Translator.

1987 "Tin" p 235 in ***Encyclopaedia Britannica.*** **Volume 22. Encyclopaedia Britannica, Inc., Chicago, Illinois. 1953.**

"Where bent, the cast metal emits a creaking sound called the *cry of tin.* This is thought to be caused by the grinding of crystals of the metal against one another when the bar is bent." –taken from C. L. Mantell, *Tin: Its Mining, Production, Technology, and Applications* (see Reference 1129).

1988 "Ultrasonic Detector Aids Precision Machining." ***Machinery,*** **MACYA, 74(8): 104–105. Apr 1968.**

Briefly described is the Hewlett-Packard, Delcon Division's Ultrasonic Translator Detector Model 4917A and its uses in machining operations.

1989 "Users' Forum: Acoustic Emission." ***NTIAC Newsletter,*** **3(10):5–6. Apr 1976.**

Article gives a review of the book *Monitoring Structural Integrity by Acoustic Emission, ASTM STP 571* published by the American Society for Testing and Materials. (see Reference 1948).

1990 "Users' Forum: Acoustic Emission from Corroding Aluminum Foil." ***NTIAC Newsletter,*** **3(9):4. Mar 1976.**

Article presents a brief report on work being performed by L. Wilson, I. G. Scott, and G. G. Martin at the Aeronautical Research Laboratories in Melbourne, Australia.

1991 "Users' Forum: Acoustic Emission Study." ***NTIAC Newsletter,*** **2(9):4–5. Apr 1975.**

Article highlights the work reported by A. F. Grenis and F. J. Rudy in report AMMRC TR 73-57. (see Reference 646).

1992 "Users' Forum: Pressure Vessels Monitored." *NTIAC Newsletter,* **2(10):3. May 1975.**

Article presents a brief description of the work Dunegan/Endevco, San Juan Capistrano, California is doing on a Nuclear Regulatory Commission sponsored test program on heavy section steel pressure vessels at Oak Ridge National Laboratory, Oak Ridge, Tennessee.

1993 "Vibrations Warn of Potential Bearing Failure." *Product Engineering,* **PRENA, 43(11):16. Nov 1972.**

A defective bearing produces acoustic and vibrational warnings, that, if detected in time, can avert catastrophic failure of machines. A considerable amount of work has been going on to gain insight into the nature of these warnings and ways of detecting them. F. H. Schildreth at Inficon, Inc., (East Syracuse, N. Y.) has worked on a contract to develop techniques for anticipating machine failure through vibration analysis, especially on the analysis of ball bearings. Pits or spalls on the surface of the bearing or the race will eventually cause failure. With pitting, there will be small vibrations and these can be detected with a sensor on the housing of the bearing. Schildreth has demonstrated that it is possible to attach an accelerometer to the housing. As the balls pass by, the accelerometer can be made to produce a waveform that, when analyzed, will indicate anomolies that signal damage. Recently, Computer Signals Processors, Inc., (Burlington, Mass.) announced that development of a digital processing system for analyzing the frequency spectrum of bearings. The frequency spectrum of a good bearing will differ from that of one having cage, ball, inner or outer race defects, or an imbalance. By comparing the spectrum of a normal bearing with that of a test bearing, the condition of the test bearing can be established. Once spectral information is obtained using a Fast Fourier Transform method of the Zoom technique, it is plotted so it can be stored for later reference. The prediction system is software-controlled.

1994 "Weld Defects Sound Off." *Iron Age,* **IRAGA, 203:85–87. Mar 27, 1969.**

Survey article outlines the work of W. D. Jolly at Battelle-Northwest on acoustic emission testing of in-process welding. Detailed reports will be found in References 925, 926, 941, and 942.

1995 "What Future for Ultrasonics." *Ultrasonics,* **ULTRA, 11:14–17. Jan 1973.**

H. L. Carson of the Central Electricity Research Laboratories reports his views on the future of acoustic emission as a nondestructive testing method as follows: "The method has already been established as a very powerful extractor of information about the dynamic behaviour of flaws and is being applied to aerospace and nuclear-power components. It shares with ultrasonic testing the problem of data interpretation in terms of defect severity, but with the difference that its basic data is more representative of likely behaviour in service as it is obtained under dynamic instead of static conditions. Once again, the computer approach seems the obvious one and there can be little doubt that ten years will see this method established as one of the most useful in nondestructive testing."

1996 "27 Geophones Protect Miners from Rock Bursts." ***Electronic Design,*** **ELODA, 24(18):21. Sep 1, 1976.**

Brief news item appeared as follows:

"A network of 27 geophones monitored by an IBM System 7 computer has been installed a mile below the earth's surface in the Sunshine Mine in Idaho to detect, pinpoint, and measure the intensity of rock noises which normally precede a rock burst. The typical noise pattern before a burst is a series of snapping and popping noises which build up to a certain point and suddenly stop. The silence is usually broken by a rock burst. The computer's job is to count noises and to warn of build-ups so that the specific area can be evacuated. If enough warning is given, the stress might be relieved by drilling holes to disperse the stress."

LIST OF JOURNALS

The list is arranged in alphabetical order by title. Each title is followed by the CODEN. Numbers in brackets indicate the number of articles published in each journal.

A

Acta Metallurgica, **AMETA [2]**
Acustica, **ACUSA [3]**
Aerospace Technology, **AETEB [1]**
Akademiya Nauk Ukrayins'koyi SSR, Metallofizika, **MFIZA [1]**
Akusticheskii Zhurnal, **AKZHA (see** *Soviet Physics–Acoustics* **for English translation) [2]**
Aluminium, **ALUMA [1]**
American Ceramic Society Bulletin, **ACSBA [4]**
American Machinist, **AMMAA [1]**
American Society of Civil Engineers, Journal of the Soil Mechanics and Foundations Division, **JSFEA [1]**
Annual Review of Earth and Planetary Sciences, **AREPC [1]**
Annual Review of Materials Science, **ARMSC [1]**
Applied Electrical Phenomena, **AECPB (English translation of** *Elektronnaya Obrabotka Materialov*) **[1]**
Archiv fur das Eisenhuttenwesen, **AREIA [1]**
Archiv fur Technisches Messen und Industrielle Messtechnik, **ATMIA (now** *Archiv fur Technisches Messen und Messtechnik Praxis,* **ATMPC) [1]**
Archiwum Akustyki, **ARAKB [1]**
Army Research and Development Newsmagazine **[1]**
ASM Transactions Quarterly, **ASMQA [1]**
ASTM Standardization News, **STDNA [1]**
Atomic Energy Review, **AERVA [1]**
Atomnaya Energiya, **AENGA (see** *Soviet Atomic Energy* **for English translation) [1]**
Atomwirtschaft, Atomtechnik, **AWAKA [1]**
Australian Welding Journal, **AUWJA [1]**
Automatic Welding, **AUWEA (English translation of** *Avtomaticheskaya Svarka*) **[2]**
Aviation Week & Space Technology, **AWSTA [1]**
Avtomaticheskaya Svarka, **AVSVA (see** *Automatic Welding* **for English translation) [2]**

B

Biuletyn Institutu Spawalnictwa, **BISWA [1]**
Blech, Rohre, Profile, **BLCHA [1]**
British Chemical Engineering and Process Technology, **BCPTA [1]**
British Journal of Non-Destructive Testing, **BJNTA [8]**
Bulletin of the Academy of Sciences of the USSR, Geophysics Series, **BASRA (English translation of** *Izyestiya Akademii Nauk SSSR, Seriya Geofizicheskaya*) **[3]**
Bulletin of the American Physical Society, **BAPSA [4]**

Bulletin of the Earthquake Research Institute, University of Tokyo, **TDJKA [5]**
Bulletin of the Geological Society of America, **BUGMA [1]**
Bulletin d'Informations Scientifiques et Techniques, Commissariat a l'Energie Atomique, **BUIAA [3]**
Bulletin of the Institute of Space and Aeronautical Science, University of Tokyo, **TDUKA [2]**
Bulletin of Japan Institute of Metals **(see** *Nippon Kinzoku Gakkai Kaiho***)**
Bulletin of the Seismological Society of America, **BSSAA [4]**
Business Week, **BUWEA [1]**

C

California Geology, **CGEOA [1]**
Canadian Consulting Engineer, **CCEGB [1]**
Canadian Journal of Chemical Engineering, **CJCEA [1]**
Canadian Journal of Physics, **CJPHA [3]**
Canadian Mining and Metallurgical Bulletin, **CMMBA [1]**
Canadian Research and Development, **CRDVA [1]**
Carbon, **CRBNA [1]**
Centre de Documentation Siderurgique, Circulaire d'Informations Techniques, **CDSCA [1]**
Ceramic Age, **CEAGA [1]**
Chartered Mechanical Engineer, **CHMGA [2]**
Chemical Engineering Progress, **CEPRA [1]**
Collection of Czechoslovak Chemical Communications, **CCCCA [2]**
Composites, **CPSOA [1]**
Comptes Rendus Hebdomadaires des Seances de l'Academie des Sciences, **COREA [3]**
Comptes Rendus Hebdomadaires des Seances de l'Academie des Sciences, Serie C: Sciences Chimiques, **CHDCA [2]**
Control Engineering, **CENGA [1]**
Corrosion, **CORRA [6]**
CRC Critical Reviews in Solid State Sciences, **CCSCB [1]**
Crystal Lattice Defects, **CLADA [1]**

D

Defektoskopiya, **DEFKA (see** *Soviet Journal of Nondestructive Testing* **for English translation) [24]**
Denki Gakkai Ronbunshi, C, **DGRCA (see** *Electrical Engineering in Japan* **for English translation) [1]**
Design Engineering, **DEMCB [1]**
Design News, **DIGNA [2]**
Doklady Akademii Nauk Belorusskoi SSR, **DBLRA [3]**
Doklady Bolgarskoi Akademii Nauk, **DBANA (also called** *Comptes Rendus de l'Academie Bulgare des Sciences***) [1]**

E

EID–Electronic Instrumentation, **EIDIA (now** *Electronic Instrument Digest***) [1]**
Electrical Engineering in Japan, **EENJA (English translation of** *Denki Gakai Ronbunshi, C***) [1]**
Electrical World, **ELWOA [1]**
Electronic Design, **ELODA [3]**
Elektrokhimiya, **ELKKA (see** *Soviet Electrochemistry* **for English translation) [1]**
Elektronnaya Obrabotka Materialov, **EOBMA (see** *Applied Electrical Phenomena* **for English translation) [1]**
EM&D Journal of Engineering Materials Components and Design, **EMJEB [1]**
Energia Nuclear **(Madrid), ENNCA [1]**
Energia Nucleare **(Milan), ENNLA [3]**
Engineer, **ENGIA [3]**
Engineering, **ENGNA [2]**

Engineering Fracture Mechanics, **EFMEA [15]**
Engineering Geology, **EGGOA [1]**
EPRI Journal, **EPRJD [1]**
Experimental Mechanics, **EXMCA [14]**

F

Felsmechanik und Ingenieurgeologie, **FEINA [1]**
Fizika Metallov i Metallovedenie, **FMMTA (see** *Physics of Metals and Metallography* **for English translation) [1]**
Fizika Tverdogo Tela, **FTVTA (see** *Soviet Physics–Solid State* **for English translation) [9]**
Flight International, **FLIND [1]**
Forest Products Journal, **FPJOA [1]**
Forschung auf dem Gebiete des Ingenieurwesens, **FGIAA [1]**
FWP Journal, **FWPJA [1]**

G

Genshiryoku Kogyo (Nuclear Engineering), **GKOGA [1]**
Growth of Crystals, **GRCRA (English translation of** *Rost Kristallov*) **[1]**

H

Highway Research Record, **HIRRA [1]**
Hi-Hakai Kensa, **Japan (see** *Journal of N.D.I.*)

I

IEEE Transactions on Magnetics, **IEMGA [1]**
IEEE Transactions on Nuclear Science, **IETNA [7]**
IEEE Transactions on Parts, Hybrids and Packaging, **IEPHA [1]**
IEEE Transactions on Sonics and Ultrasonics, **IESUA [8]**
Industrial Laboratory, **INDLA (English translation of** *Zavodskaya Laboratoriya*) **[6]**
Industrial Research, **IDRSA [4]**
Industrie Minerale-Mine, **INMNC [1]**
Ingenieur Digest **[1]**
Instruments and Control Systems, **INCSA [2]**
Interface **[2]**
International Journal of Fracture Mechanics, **IJFMA [3]**
International Journal of Nondestructive Testing, **IJNTA [5]**
International Journal of Pressure Vessels and Piping, **PRVPA [3]**
International Journal of Rock Mechanics and Mining Sciences, **IJRMA [3]**
International Journal of Solids and Structures, **IJSOA [1]**
Iron Age, **IRAGA [5]**
Iron Age Metalworking International, **IAMIA [2]**
ISA Transactions, **ISATA [2]**
Izvestiya Akademii Nauk SSSR, Metally, **IZNMA (see** *Russian Metallurgy* **for English translation) [1]**
Izvestiya Akademii Nauk SSSR, Seriya Geofizicheskaya, **IAGFA (see** *Bulletin of the Academy of Sciences of the USSR, Geophysics Series* **for English translation) [3]**
Izvestiya Vysshikh Uchebnykh Zavedenii, Chernaya Metallurgiya, **IVUMA [2]**
Izvestiya Vysshikh Uchebnykh Zavedenii, Tsvetnaya Metallurgiya, **IVUTA [1]**

J

Jemna Mechanika a Optika, **JMKOA [1]**
JETP Letters, **JTPLA (English translation of** *Zhurnal Eksperimental'noi i Teoreticheskoi Fiziki, Pis'ma v Redaktsiyu*) **[1]**
Journal of the Acoustical Society of America, **JASMA [22]**
Journal of the Acoustical Society of Japan **(see** *Nippon Onkyo Gakkaishi*)
Journal of Adhesion, **JADNA [1]**

Journal of the American Ceramic Society, **JACTA [4]**
Journal of the American Concrete Institute, **JACIA [1]**
Journal of Applied Mechanics, Transactions ASME, **JAMCA [1]**
Journal of Applied Physics, **JAPIA [8]**
Journal of Applied Physics, Japan (see *Oyo Butsuri*)
Journal of the Australian Institute of Metals, **JAMTA [1]**
Journal of the British Nuclear Energy Society, **JBNSA [2]**
Journal of Composite Materials, **JCOMB [4]**
Journal of the Electrochemical Society, **JESOA [1]**
Journal of Engineering for Power, Transactions ASME, **JEPOA [1]**
Journal of Geophysical Research, **JGREA [4]**
Journal of the Institute of Metals, **JIMEA [1]**
Journal of the Iron and Steel Institute, **JISIA [1]**
Journal of the Iron and Steel Institute of Japan (see *Tetsu-To-Hagane*)
Journal of Japan Institute of Metals (see *Nippon Kinzoku Gakkaishi*)
Journal of the Japan Society of Mechanical Engineers (see *Nippon Kikai Gakkai Shi*)
Journal of Materials, **JMLSA [2]**
Journal of Materials Science, **JMTSA [10]**
Journal of Metals, **JOMTA [1]**
Journal of N.D.I. (Hi-Hakai Kensa, **Japan***)*, **HIHKA [19]**
Journal of Nuclear Energy, **JNCEA [1]**
Journal of Nuclear Materials, **JNUMA [1]**
Journal of the Physical Society of Japan, **JUPSA [1]**
Journal of Physics D: Applied Physics, **JPAPB [3]**
Journal of Physics E: Scientific Instruments, **JPSIA [1]**
Journal of Polymer Science, Part A-2: Polymer Physics, **JPLPA [1]**
Journal of the Society of Environmental Engineers, **JSEEB [1]**
Journal of the Society of Materials Science, Japan (Zairyo), **ZARYA [1]**
Journal of Steel Structure C **(Japan)**, **JSSCD [1]**
Journal of Testing and Evaluation, **JTEVA [9]**

K

Kawasaki Seitetsu Giho (Kawasaki Steel Technical Report), **KWSGB [1]**
Keikinzoku Yosetsu, **KEYOB [1]**
Keramische Zeitschrift, **KERZA [1]**
Kerntechnik, **KTAPA [1]**
Kotsu Gasu (see *Pressure Engineering, Journal of Japan High Pressure Institute*)
Kunststoffe, **KUNSA [1]**

L

Lastechniek **(Hague)**, **LASTA [1]**
Letters in Applied and Engineering Sciences, **LAESB [1]**
Loss Prevention **[1]**

M

Machine Design, **MADEA [3]**
Machinery, **MACYA [1]**
Marine Technology Society Journal, **MTSJB [1]**
Maschinenschaden, **MSCNA [2]**
Materialprufung, **MTPRA [2]**
Materials Engineering, **MAENB [5]**
Materials Evaluation, **MAEVA [48]**
Materials Research and Standards, **MTRSA [13]**
Materials Science and Engineering, **MSCEA [9]**
Mecanique Electricite, **MQELA [1]**
Mecanique-Materiaux-Electricite, **MMXEA [1]**
Mekhanika Polimerov, **MKPLA (see** *Polymer Mechanics* **for English translation) [1]**
Memoires Scientifiques de la Revue de Metallurgie, **MRMTA [4]**
Mesures, Regulation, Automatisme, **MRAUA [1]**

Metal Construction and British Welding Journal, **MCBWA [2]**
Metal Progress, **MEPOA [7]**
Metal Science and Heat Treatment, **MSHTA (English translation of** *Metallovedenie i Termicheskaya Obrabotka Metallov*) **[4]**
Metal Stamping, **MLSMA [1]**
Metall, **MTLLA [3]**
Metall und Erz, **MERZA [1]**
Metalloberflache, **MOFEA [1]**
Metallovedenie i Termicheskaya Obrabatka Metallov, **MTOMA (see** *Metal Science and Heat Treatment* **for English translation) [4]**
Metallurgia Italiana, **MITLA [3]**
Metallurgical Transactions, **MTGTB [3]**
Metallurgical Transactions A, **MTTAB [2]**
Metallurgist and Materials Technologist, **MMTCA [1]**
Metals and Materials, **MEMTA [3]**
Metals Australia, **MAUSB [1]**
Metals Engineering Quarterly, **MENQA [3]**
Metaux **(Corrosion, Industrie), MTUXA [1]**
Mitsui Technical Review **[1]**
Modern Plastics, **MOPLA [1]**

N

Nature, **NATUA [3]**
Nature Physical Science, **NPSCA [5]**
Naturwissenschaften, **NATWA [1]**
Naval Engineers Journal, **NVEJA [1]**
Naval Research Reviews, **NARRA [1]**
NDT International, **NDITD (formerly** *Non-Destructive Testing*) **[10]**
New Scientist, **NWSCA [1]**
Nippon Kikai Gakkai Shi (Journal of the Japan Society of Mechanical Engineers), **NKGKA [1]**
Nippon Kinzoku Gakkai Kaiho (Bulletin of Japan Institute of Metals), **NKZKA [1]**
Nippon Kinzoku Gakkaishi (Journal of Japan Institute of Metals), **NIKGA [3]**
Nippon Onkyo Gakkaishi (Journal of the Acoustical Society of Japan), **NIOGA [1]**
Non-Destructive Testing, **NDETA (now** *NDT International*) **[23]**
NTIAC Newsletter **[10]**
Nuclear Engineering, **NUENA (now** *Nuclear Engineering International*) **[1]**
Nuclear Engineering **(Tokyo) (see** *Genshiryoku Kogyo*)
Nuclear Engineering and Design, **NEDEA [5]**
Nuclear Engineering International, **NEINB (formerly** *Nuclear Engineering*) **[3]**
Nuclear Instruments and Methods, **NUIMA [1]**
Nuclear News, **NUNWA [1]**
Nuclear Safety, **NUSAA [4]**
Nuclear Technology, **NUTYB [4]**
Nucleonics Week, **NUWEA [1]**

O

Oil and Gas Journal, **OIGJA [3]**
Oyo Butsuri (Journal of Applied Physics, Japan), **OYBSA [1]**

P

Paper-Maker and British Paper Trade Journal, **PMBPA [1]**
Philosophical Magazine, **PHMAA [1]**
Physica Status Solidi (a) Applied Research, **PSSAB [3]**
Physica Status Solidi (b) Basic Research, **PSSBB [1]**
Physical Review, **PHRVA [3]**
Physics Today, **PHTOA [1]**
Physics of Metals and Metallography, **PHMMA (English translation of** *Fizika Metallov i Metallovedenie*) **[1]**
Plastics and Rubber Weekly, **PLRWB [1]**
Pollution Engineering, **PLENB [1]**
Polymer Mechanics, **PLYMA (English translation of** *Mekhanika Polimerov*) **[1]**

Pressure Engineering, Journal of Japan High Pressure Institute (Kotsu Gasu), **KOGAA [6]**
Problemy Prochnosti, **PPCNB (see** *Strength of Materials* **for English translation) [5]**
Proceedings of the British Ceramic Society, **PBRCA [1]**
Proceedings of the IEEE, **IEEPA [1]**
Product Engineering, **PRENA [6]**

Q

Qualite, Revue Pratique de Controle Industriel, **QRPCA [4]**
Quality, **QUASA [1]**
Quality Assurance, **QUASA [1]**
Quality Management & Engineering, **QMEGA [1]**

R

R/D, Research/Development, **REDEA [4]**
Research News **(Canadian Department of Forestry) [1]**
Review of Scientific Instruments, **RSINA [4]**
Reviews on Coatings and Corrosion, **RVCCB [1]**
Revista de Soldadura, **RSLDB [1]**
Revue d'Acoustique, **RACQA [1]**
RILEM (Reunion Internationale des Laboratories d'Essais et de Recherches sur les Constructions) Bulletin, **RILBA [1]**
Rivista Italiana della Saldatura, **RISAA [2]**
Rock Mechanics, Felsmechanik, Mecanique des Roches, **RMFMA [2]**
Rost Kristallov, **RKANA (see** *Growth of Crystals* **for English translation) [1]**
Russian Metullurgy, **RMLYA (English translation of** *Izvestiya Akademii Nauk SSSR, Metally*) **[1]**

S

SAE Journal of Automotive Engineering, **SJAUA [1]**
SAMPE Quarterly, **SAMQA [2]**
Schweissen und Schneiden, **SCSCA [3]**
Schweisstechnik **(Berlin), SCTCA [1]**
Schweisstechnik **(Vienna), SWTEA [3]**
Science, **SCIEA [4]**
Sciences et Techniques de l'Armement, **STQAB [1]**
Scripta Metallurgica, **SCRMB [12]**
Solid-State Electronics, **SSELA [1]**
Soudage et Techniques Connexes, **SOTCA [3]**
Soviet Atomic Energy, **SATEA (English translation of** *Atomnaya Energiya*) **[1]**
Soviet Electrochemistry, **SOECA (English translation of** *Elektrokhimiya*) **[1]**
Soviet Journal of Nondestructive Testing, **SJNTA (English translation of** *Defektoskopiya*) **[24]**
Soviet Physics–Acoustics, **SOPAA (English translation of** *Akusticheskii Zhurnal*) **[2]**
Soviet Physics–Solid State, **SPSSA (English translation of** *Fizika Tverdogo Tela*) **[9]**
Space Aeronautics, **SPAEA [1]**
Stahl und Eisen, **STEIA [2]**
Steel, **STELA [2]**
Strain, **STRNB [1]**
Strength of Materials, **SMTLB (English translation of** *Problemy Prochnosti*) **[5]**
Strojirenstvi **(Machine Industries), STRJA [1]**
Strojniski Vestnik/Mechanical Journal, **STJVA [2]**
Svarochnoe Proizvodstvo, **SVAPA (see** *Welding Production* **for English translation) [5]**

T

Technica, **TCHNA [1]**
Technology Reports of the Osaka University, Japan, **TROUA [1]**
Teploenergetika, **TPLOA (see** *Thermal Engineering* **for English translation) [1]**
Testing, Instruments and Controls, **TICOA [3]**

Tetsu–To–Hagane (Journal of the Iron and Steel Institute of Japan), **TEHAA** [2]
The Welding Institute Research Bulletin [3]
Thermal Engineering, **THENA (English translation of** *Teploenergetika*) [1]
Traitement Thermique/Heat Treatment, **TRTHA** [1]
Traitements de Surface, **TRSUA** [1]
Transactions of the AIME, **TAIMA** [2]
Transactions of the American Geophysical Union, **TAGUA** [1]
Transactions of the American Nuclear Society, **TANSA** [13]
Transactions of the American Society for Steel Treating, **TASTA** [1]
Transactions of the ASME, Journal of Pressure Vessel Technology, **JPVTA** [2]
Transactions of the Indian Institute of Metals, **TIIMA** [1]
Transactions of the Society of Mining Engineers of AIME, **TAMMA** [1]
Trudy Uralskogo Politekhnicheskogo Instituta, **TUPIA** [2]

U

Ukrainskii Fizicheskii Zhurnal, **UFIZA** [1]
Ultrasonics, **ULTRA** [10]
Umschau, **UMSCA** [1]
U.S. Naval Journal of Underwater Acoustics [1]

W

Water Power and Dam Construction, **WAPOA** [1]
Welder, **WLDRA** [1]
Welding and Metal Fabrication, **WLMFA** [2]
Welding Design & Fabrication, **WDEFA** [2]
Welding in the World/le Soudage dans le Monde, **WDWRA** [3]
Welding Journal, **WEJUA** [12]
Welding Production, **WEPRA (English translation of** *Svarochnoe Proizvodstvo*) [5]
Welding Research International, **WASIA** [1]

Z

Zairyo, **ZARYA (see** *Journal of the Society of Materials Science, Japan*)
Zavodskaya Laboratoriya, **ZVDLA (see** *Industrial Laboratory* **for English translation)** [6]
Zeitschrift fur Anorganische und Allgemeine Chemie, **ZAACA** [1]
Zeitschrift fur Metallkunde, **ZEMTA** [5]
Zeitschrift fur Physik, **ZEPYA** [3]
Zeitschrift fur Werkstofftechnik/Journal of Materials Technology, **ZWKTA** [3]
Zement–Kalk–Gips, **ZMKGA** [1]
Zhurnal Eksperimental'noi i Teoreticheskoi Fiziki, Pis'ma v Redaktsiyu, **ZFPRA (see** *JETP Letters* **for English translation)** [1]
Zvaranie, **ZVARA** [1]

AUTHOR INDEX

Numbers refer to reference numbers appearing at the beginning of each entry. Reference numbers in italics indicate the person was editor or coordinator of the publication.

D

E

F

G

H

L

M

N

O

Q

R

S

SUBJECT INDEX

Numbers refer to reference numbers appearing at the beginning of each bibliographic entry.

B

C

D

E

F

I

O

P

Q

R

S

Y

Z

GPSR Compliance

The European Union's (EU) General Product Safety Regulation (GPSR) is a set of rules that requires consumer products to be safe and our obligations to ensure this.

If you have any concerns about our products, you can contact us on ProductSafety@springernature.com

In case Publisher is established outside the EU, the EU authorized representative is:

Springer Nature Customer Service Center GmbH
Europaplatz 3
69115 Heidelberg, Germany

Batch number: 10683448

Printed by Printforce, the Netherlands